# College Preparatory Mathematics 2 (Geometry)

**Second Edition**


**Editors:**

Brian Hoey
CPM Educational Program
Karen Wootton
CPM Educational Program

**Technical Assistance:**

Thu Pham
The CRESS Center
University of California, Davis
Chris Mikles
CPM Educational Program
Justin Bradley
CPM Educational Program
Glenda Wilkins
San Gorgonio High

**Program Directors:**

Tom Sallee
Department of Mathematics
University of California, Davis
Judy Kysh
Departments of Mathematics and Education
San Francisco State University
Elaine Kasimatis
Department of Mathematics
California State University, Sacramento
Brian Hoey
CPM Educational Program

**Illustrators:**

Eric Ettlin
Menlo-Atherton High
Julien Howe
Merrie Rose Academy

v.4.1

## Credits for the First Edition

**Editors**

Brian Hoey, Christian Brothers High
Karen Wootton, Will C. Wood High

**Consultant**

Joel Teller, College Preparatory School

**Technical Assistance**

Thu Pham, UC Davis
Kristin Sallee, UC Davis
Crystal Mills, Consultant
Kirk Mills, UC Davis

**First Edition Contributors**

| | |
|---|---|
| Doreen Bryant | El Camino High |
| Nancy Clark | Woodland High |
| Virginia Doyle | Elk Grove High |
| James Friedrich | Valley High |
| Jenny Gee | Sacramento High |
| Donald Gernes | Ponderosa High |
| David Goodwin | Florin High |
| Scott Grensted | Grant High |
| Carol Grossnicklaus | Oxnard High |
| Ted Herr | Roseville High |
| Brian Hoey | Christian Brothers High |
| Christopher Scott Holm | Cloverdale High |
| Gail Holt | El Camino High |
| Sylvia Huffman | Del Campo High |
| Tim Jordan | Elk Grove High |
| Jian Lian | Luther Burbank High |
| Yury Lokteff | San Juan High |
| Steve McCullough | Florin High |
| Richard Melamed | El Camino High |
| Crystal Elaine Mills | Sacramento Waldorf |
| Michael Palmer | Center High |
| Jeanne Shimizu-Yost | San Juan High |
| Clark Swanson | Sacramento High |
| Joel Teller | College Prep School |
| Bill Wharton | San Juan High |
| Karen Wootton | River City High |
| Sharon Yamamoto | McClatchy High |



7 8 9 10 11 12 08 07 06 05 05 vesion 4.1

Printed in the United States of America ISBN 1-885145-70-5

Hello. My name is Chris Ott. While I am not an author of the College Preparatory Mathematics (CPM) textbook series, I have tutored dozens of CPM students. Based on this experience, I have written study guides for CPM Algebra 1 and Geometry that are incorporated into each student text. If you want an extended overview of this course (Geometry), read all the PZL problems in the first four units (Units 0-3). They are easy to find, since each PZL problem has a puzzle piece (like the one above) around the problem number in the left margin. These problems will explain the goals of this geometry course, its methods, the structure of the textbook, the role and duties of the teacher, and the responsibilities of the student. It has numerous practical suggestions for students to maximize the prospect for success in this course.

**CPM underwrites online homework help at www.hotmath.com.** Solutions are shown in a tutoirial fashion to help you learn the ideas as well as solve a specific problem. The authors have also written a more extensive unit by unit "Parent's Guide with Review to Math 2 (Geometry)." This document contains annotated solutions for important problems that help you and your child understand the core ideas of each unit. Thee are also additional sample problems and their solutions, suggestions about how to help your child in this course, and several dozen additional practice problems to supplement the textbook. The guide concludes with a brief outline of how the six content threads of the course--algebra, graphing, ratios, geometric properties, spatial visualization, and proof--are developed through the thirteen units in this textbook. Some schools make the guide available through the classroom teacher or school library. You may download each chapter of the Parent Guide free or get an order form at www.cpm.org. Call (916) 681-3611 to order with a credit card. You may mail your request for the "Math 2 Parent Guide with Review" with a check payable to "CPM Educational Program" for $20 plus local tax (CA only) and $3 shipping and handling, to 1233 Noonan Drive, Sacramento, CA 95822-2569 or Allow ten days for delivery.

# College Preparatory Mathematics 2: Second Edition

**(Geometry)**

## Table of Contents

# UNIT 0

# Prelude

## STUDY TEAMS AND COORDINATE GRIDS

# *Unit 0 Objectives*

## *Prelude*: STUDY TEAMS AND COORDINATE GRIDS

Most of the mathematics in this unit is hopefully familiar to you: using the xy-coordinate system, writing and solving algebraic equations, and working with data. What might be new to you is that you will be studying the mathematics in study teams so you will get to know other students in the class as well. Also, pay close attention to the PZL problems. These are special problems to help you understand the goals of the course and why it is designed the way it is. Use this unit to get back into "mathematical shape," so you are ready to go when the new topics begin.

In this unit you will have the opportunity to:

- understand the goals and methods of this course.
- become accustomed to working in a team to solve mathematical problems.
- review the fundamentals of the xy-coordinate system.
- practice combining algebraic expressions, writing equations, and solving simple equations.
- practice basic data analysis, including graphs, and probability.
- organize your notebook for this course.

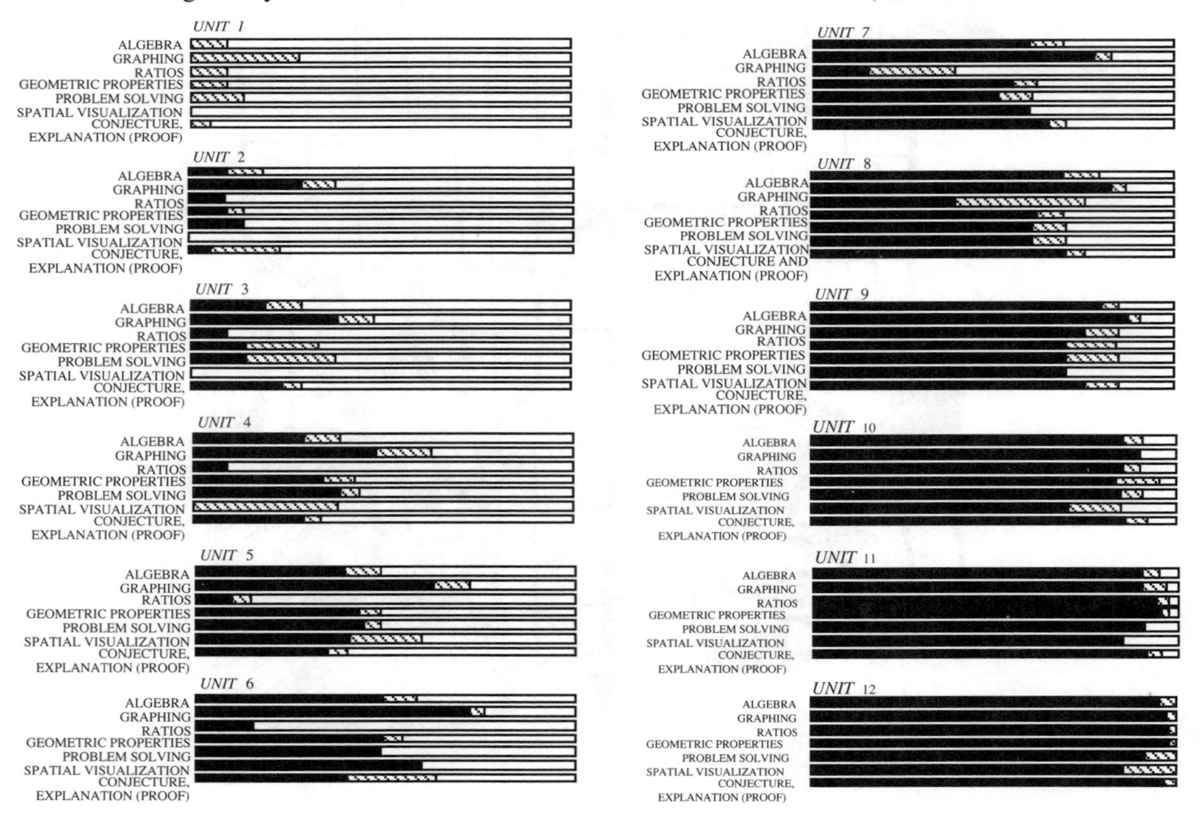

# *Unit 0*

## *Prelude*: STUDY TEAMS AND COORDINATE GRIDS

WELCOME TO YOUR GEOMETRY COURSE!

*Hi! I'm Chris Ott from "The Learning Connection" in Davis, California. As a college student I tutored quite a few students in math and science. I wrote the PZL parts of this program from the point of view of a student, not a teacher. These "puzzle pieces" are an honest attempt to level with you about what you will need to do to be successful. Put all the pieces together and you'll increase your chances of success tremendously. Start by reading the authors' description of the course, then complete the "mission" at the end of the reading. I'll go into more detail later.*

*By the way, if you took our Algebra 1 course, you read my suggestions about how to be successful in that course. Some of you thought I talked too much, so this year the PZL problems will be shorter--just the main points. Be careful to pay attention, then, to the suggestions I make. Students who follow my suggestions usually do well.*

The CPM authors believe that most students can be successful in math classes if the approach blends the best of past practices with some other work based on research about how people learn. **This course will teach you the basic content and skills of geometry,** but do it in ways that will help you understand what you are doing and teach you **strategies for doing math** that will be useful in later math courses and in life.

What can you expect? The course is **built around problems**, just the way you will encounter math in college and the workplace. You will spend more of your class time **doing and discussing** problems in study teams and less time listening to your teacher tell you rules. Rules, of course, are important. They will emerge (and the textbook will highlight them) as you develop an understanding of the mathematics they represent. This book usually gives you several days to develop an idea and several weeks to practice and understand it and eventually use it in challenging application problems.

Your teacher will help you stay on track during classwork by moving from team to team. He or she will also highlight the points of your lessons by giving brief lectures, leading class discussions, and reviewing selected problems. However, **the CPM authors believe that a key part of your learning geometry is thinking carefully about ideas yourself,** so expect your teacher to ask you leading questions that will help you answer many of the questions you ask. By learning, with the guidance of your teacher, to answer the questions you pose, you will become a better, more confident learner in math as well as in other courses. **Expect to study math using several approaches, which will increase your chances of learning geometry and remembering what you learn.**

In this geometry course you will:

- deal with the major concepts taught in every geometry text and apply them in realistic problems;
- develop your problem solving abilities;
- connect geometry to other areas of mathematics and science. In particular, continue to develop your algebra skills, especially writing, solving, and graphing equations;
- extend ratios to trigonometry, area, and volume;
- improve your ability with spatial visualization (3D ideas);
- present logical arguments and justifications of your solutions to problems;
- emphasize understanding mathematics; and,
- explore many problems that require a team effort.

In short, the goal of this course is to **have mathematics make sense.** Making sense of new ideas takes time (sometimes lots of time) and discussion with others. You need to do lots of problems AND to talk about the big ideas of geometry. No one can learn to play a musical instrument by simply watching and listening to others play; they need to ask questions, talk about music and practice to get good at it. Likewise, the **ONLY** way you can become good at geometry is by doing problems and talking about them. **If you do both, you will learn geometry.**

As you go through the year, do your best to meet these expectations each day. You will find that geometry is not only useful, it is fun. Have a successful year!

**MISSION POSSIBLE** (something to do)

Your first written assignment for this course is to find at least three of the main ideas you should know about this course before you do any math problems. Make your first entry in your notebook by listing at least three main ideas in the "Welcome Note" you just read.

PR-1. Much of your class time will be spent working with your fellow students in study teams of two or four. Study teams are the first source of support for learning in this course. Read the guidelines for study teams in the box below, as well as the paragraph that follows.

Record the names of the members of your study team on your paper. Then copy these four guidelines into your notebook.

**GUIDELINES FOR STUDY TEAMS**

1) Each member of the team is responsible for his or her own behavior.
2) Each member of the team must be willing to help any other team member who asks for help.
3) You should only ask the teacher for help when all team members have the same question.
4) Use your team voice.

Often, a problem will direct you to share the work within your study team. This does not mean one person does all the work and then tells the rest of the team members the answers. Decide how to divide the labor, then share each individual's work to help solve the problem. Your team can verify whether your solutions are reasonable and provide an opportunity for all of you to discuss different ways to solve the same problem.

PR-2. KEEPING A NOTEBOOK

You will need to keep an organized notebook for this course. Here is one method of keeping a notebook. Your teacher may alter these guidelines.

- The notebook should be a sturdy three-ring loose-leaf binder with a hard cover.
- It should have dividers in it to separate it into at least six sections:

TEXTBOOK
HOMEWORK/CLASSWORK
TOOL KIT
NOTES
TESTS and QUIZZES
GRAPH PAPER

Because you do not want to lose your notebook, you should put your name inside the front cover. If you lose it on the bus, you will want it returned to you, so you should also put your phone number and address (or the school's) inside the cover. If you cover your book, you may wish to put your name in large clear letters on the outside cover. Also, write your name and the name of your teacher inside the back cover of your textbook.

Finally, read the information in the box on the next page **slowly and carefully.** Having the right tools and a positive attitude will serve you well throughout the year.

## STUDY TOOLS FOR SUCCESS

**MATERIALS:** Your notebook will be your written record of what you have done in the course and will be the chief way to study for tests, so take good care of it. You will also need a scientific calculator and plenty of graph paper throughout the year.

We also recommend that you carry two colored pencils or pens, a #2 pencil and a pen (blue or black ink), and a ruler. All of these tools, along with your calculator, can be kept in a plastic pouch inside your binder.

**ATTITUDE and EFFORT:**

You will increase your chances for success in geometry by using your class time to complete the day's lesson. We expect you will develop your ability to contribute to and benefit from working together in teams. You must be willing to both ask and answer questions within your study team. Furthermore, you will need to work on problems for more than a minute or two--sometimes for several days! In addition, this course will often ask you to explain your thinking--sometimes orally, frequently in writing. In short, YOU must take responsibility for doing your work. Simply stated, you need to do all of your assignments regularly. If you do all of this, you will be much more successful in this course.

## PAPER TOWERS

PR-3. The object of this problem is to work together in your teams to build the highest tower in the class with only the materials provided by your teacher. There are a few simple rules that your team must follow.

a) Only materials provided may be used in building your tower.

b) Towers must be free-standing. They may not lean against any other object or be held up by anyone or any object. "Free standing" means at least five seconds.

c) When you finish building, measure your tower's height and have your teacher verify your results.

d) If you are not satisfied with your tower's height, what might you have done differently in order to build a taller tower? Do it, that is, try again!

## PZL-2 THE KEY TO SUCCESS IN CPM GEOMETRY

*Hello, again. This is Chris. In this and the first three units I'll be talking to you about how to be successful in this course. Remember that it's me talking in any problem numbered with PZL. Read the next section, then complete the "mission" after it.*

One of the skills I never learned in high school was how to read a math textbook. In fact, I didn't think you read a math book; you just used it for the problems. I sure missed out on a lot of valuable information that would have made learning math easier. In this course, **information is the key to success.** Learning how to read a math book will be an extra benefit beyond learning geometry.

**CPM Geometry requires that you read the textbook.** If you understand that, you are on your way to doing well in this class. Start learning how to read a text book NOW. **Information is the key to success in this program.** In this course, you will collect pieces to a puzzle that, when put together, you call geometry. Remember, this is a college-required math course. It will have different expectations than some other courses.

The welcome note in PZL-1 says, "In short our goal is to have **mathematics make sense.**" Just about every responsible adult knows that learning and understanding math will be essential to a young person's success in the 21st century. I know--some of you don't believe this. But listen to what Shannon Springmeyer of Carnegie Middle School wrote to her teacher at the end of CPM Algebra 1. She said, "Above all I have learned how to logically think through a problem and use information to produce a solution." She continued with, "I now have many more skills than I did at the beginning of the year and most important, I understand!" Shannon found ways that made math make sense. She was able to understand the reasons for the teaching method CPM uses and the format of the material. **Shannon took responsibility for her education.** In turn, she was able to learn algebra and get good grades. Shannon figured out the system. Not only will she use these skills in future classes, but she will use them for the rest of her life. If you **expend reasonable effort**, you will get what you want out of geometry.

**MISSION POSSIBLE** (something to do)

Take a few minutes to answer the following questions. Your answers do not have to be long. The questions are for you, not your teacher.

- What feature(s) of math books help you learn math?
- What else would help you learn math?

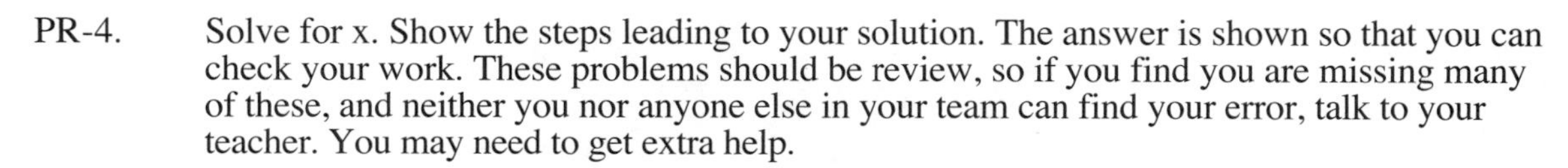

PR-4. Solve for x. Show the steps leading to your solution. The answer is shown so that you can check your work. These problems should be review, so if you find you are missing many of these, and neither you nor anyone else in your team can find your error, talk to your teacher. You may need to get extra help.

a) $5x + 7 = -12$ **[ $-^{19}/_{5} = -3.8$ ]**

b) $3x - 11 = 30$ **[ $13^{2}/_{3} \approx 13.67$ ]**

c) $37 - 6x = 61 - 2x$ **[ $-6$ ]**

d) $13x + 18 = 5 - 7(1 - x)$ **[ $-3^{1}/_{3} \approx 3.33$ ]**

PR-5. Write the following polynomials in shorter form, if possible:

a) $10x + 27 - 8x$

b) $4x^2 + 8x^2 - 5y + 16y$

c) $16x - 8xy + 17x + 3xy$

PR-6. Write a description of what happens to x to produce y.

a)

| x | -3 | -2 | -1 | 0 | 1 | 2 | 3 | x |
|---|---|---|---|---|---|---|---|---|
| y | -5 | -3 | -1 | 1 | 3 | 5 | 7 | ? |

b)

| x | -3 | -2 | -1 | 0 | 1 | 2 | 3 | x |
|---|---|---|---|---|---|---|---|---|
| y | -2.5 | -2 | -1.5 | -1 | -0.5 | 0 | 0.5 | ? |

c)

| x | -3 | -2 | -1 | 0 | 1 | 2 | 3 | x |
|---|---|---|---|---|---|---|---|---|
| y | 14 | 11 | 8 | 5 | 2 | -1 | -4 | ? |

PR-7. Agustin is in line to choose a new locker at school. The locker coordinator has each student reach into a bin and pull out a locker number. There is one locker at the school that all the kids dread! This locker, # 831, is supposed to be haunted, and anyone who has used it has had strange things happen to him or her! When it is Agustin's turn to reach into the bin and select a locker number, he is very nervous. He knows that there are 535 lockers left and locker # 831 is one of them. What is the probability that Agustin reaches in and pulls out the dreaded locker # 831? Should he be worried? Explain.

## PZL-3 BUILDING A SUCCESSFUL MATH COURSE

Hi! Chris again. I thought you ought to know what the authors had in mind when they wrote the this math course. I'll list them first, then comment briefly.

- Math should make sense to you (that is, you should understand what you are doing).
- Problems should be realistic, showing applications of math to the world around you.
- You should learn ideas by doing guided investigations (math labs), then practice the ideas and skills over several weeks and months.
- The course should be built around word problems.
- The teacher should work with you and your team during investigations, helping to answer questions by posing more questions, then help summarize important ideas.

The authors based the CPM courses on the above ideas in the belief that **when you actively seek solutions to problems, you will not only understand how and where to use the math skills, but you will retain these skills as well**. Another advantage of learning math through investigation is **learning how to think logically through a problem**, rather than memorizing steps.

All of this means that **your role as a student will be varied.** You will be required to come up with strategies for solving these problems, but do not despair. The authors also devoted considerable time to teaching you learning strategies that will help you in this course as well as in other courses, college, and your career. At first you may say, "The teacher is not teaching me," but this is only if you expect the teacher to be explaining at the board or overhead, then assigning practice problems. In this course the teacher will lecture and solve examples when appropriate, usually after you have had a chance to explore an idea. Then you should have some real questions based on your initial study to ask your teacher.

Finally, **the book presents examples in several ways.** Many of them come as word problems rather than step by step illustrations of solutions. Later, we will learn how to use this type of example, but it will be important that you keep a good notebook and correct problems as necessary after you have done them. Remember that **one objective of learning is to eliminate the problem of forgetting by helping you develop strong problem solving skills**. This way, when you face slightly different versions of a problem, you will be able to pull out the principle math component and solve it rather than saying, "I don't understand" or "I can't remember."

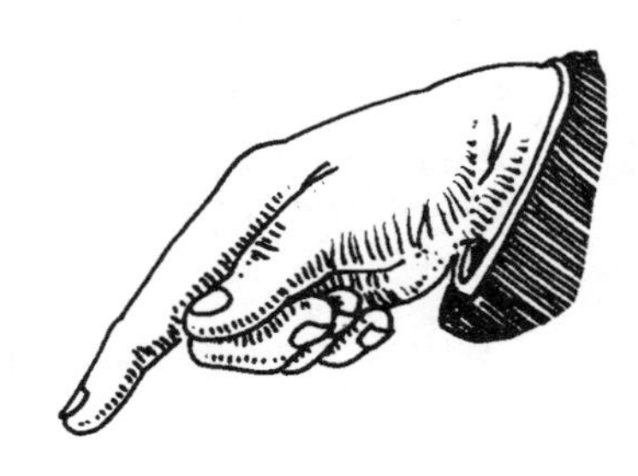

What will all this mean for you as a student? Three things:

- first, you **will have to be actively involved in doing math every day;**
- second, you will have to **read the textbook;** and
- third, you will have to **communicate with the teacher every day.**

I'll say more about all of this in the next unit.

**MISSION POSSIBLE** (something to do)

Use the comments above and possibly PZL-1 to make a list of what you need to do to be successful in this class. Write the list in your notebook or at the start of today's assignment.

## PICK'S THEOREM INVESTIGATION

PR-8 Using square dot paper, draw at least five **different** figures which have **NO** points (or dots) in the interior of the figure. All sides of the figures must be straight line segments.

a) How many dots are on the perimeter (the border) of each shape?

b) Calculate the **area** of each shape and express your result as square units.

PR-9. On your paper, set up a table similar to the one below, and fill in the values you found in the previous problem. Remember that an organized table makes it easier to find patterns in your data.

| Dots in the Interior | Dots on the Perimeter | Area |
| --- | --- | --- |
| | | |

PR-10. When Armando looks at his data, he claims that there is a relationship between the area of the figure and the number of dots on the perimeter. Find the relationship that he sees.

PR-11. This time, draw five different figures with exactly one point (dot) in the interior of the figure. Continue your table from problem PR-9 and fill in this new data.

PR-12. What relationship does Armando see for figures with one interior dot?

PR-13. You have tried zero interior points and one interior point. What situation should you investigate next? DO IT! Continue until you establish a pattern.

PR-14. Write a general rule for finding the area of any figure drawn on dot paper based on the patterns you observed. Finally, turn your rule into a formula for calculating the area (A). Represent the number of interior points with the letter "i" and the number of border points with "b." You should be able to find the area of **any** shape or size figure using your result.

PR-15. Multiply the following.

a) $(x + 3)(2x - 1)$ b) $-3x(4 - 2x)$

PR-16. One plant has a mass of 15 grams and increases its mass by 3 grams each day. Another plant has a mass of 7 grams and increases its mass at the rate of 5 grams each day. On what day will they have the same mass?

PR-17. Five blue sticks of equal length, when laid end to end with one red stick, have a length of 124 cm. Draw a diagram to represent this. If the red stick is 4 cm long, write an equation that will allow you to find the length of one blue stick. Find the length of one blue stick.

PR-18. Match the table of data to the most appropriate graph and <u>briefly</u> <u>explain</u> why it matches the data.

a) Boiling water cooling down.

| Time (min) | 0 | 5 | 10 | 15 | 20 | 25 |
|---|---|---|---|---|---|---|
| Temp (°C) | 100 | 89 | 80 | 72 | 65 | 59 |

b) Cost of a phone call.

| Time (min) | 1 | 2 | 2.5 | 3 | 4 | 5 | 5.3 | 6 |
|---|---|---|---|---|---|---|---|---|
| Cost (cents) | 55 | 75 | 75 | 95 | 115 | 135 | 135 | 155 |

c) Growth of a baby in the womb.

| Age (months) | 1 | 2 | 3 | 4 | 5 | 6 | 7 | 8 | 9 |
|---|---|---|---|---|---|---|---|---|---|
| Length (inches) | 0.75 | 1.5 | 3 | 6.4 | 9.6 | 12 | 13.6 | 15.2 | 16.8 |

Graph 1 Graph 2 Graph 3 Graph 4 Graph 5

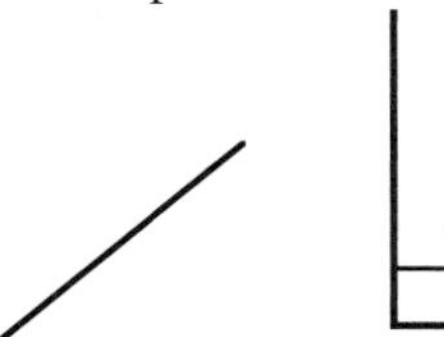

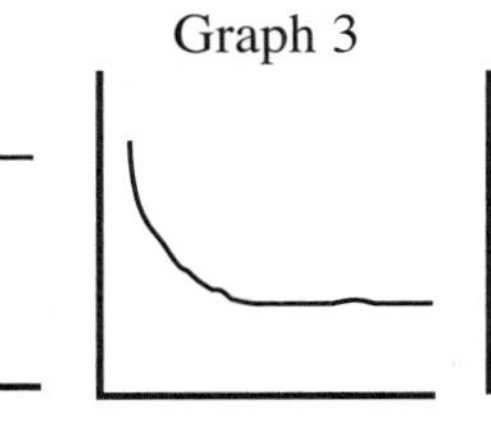

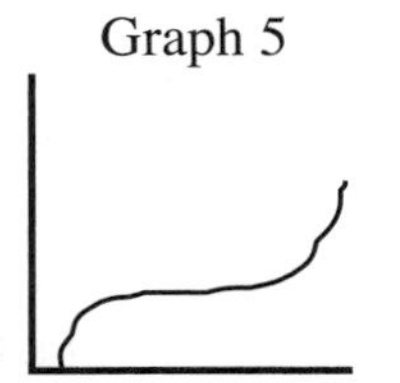

## USING A COORDINATE SYSTEM

PR-19. **DIRECTIONS**: Gather a straightedge (such as a ruler) and a clean sheet of large grid graph paper. Be especially neat and careful in scaling the axes and drawing the lines because this graph will be used for several problems. Work with a partner or your teammates as directed by your teacher and agree what you will do before writing on your paper. Each of you should do the activity on your own sheet of paper. Read the following box, then begin.

> **INEQUALITY SYMBOLS**
>
> $\le$ means "less than or equal to;" for example $3 \le 4$ or $4 \le 4$;
>
> $\ge$ means "greater than or equal to;" for example $7 \ge 5$;
>
> $-3 \le x \le 4$ means x can be any number between -3 and 4, including -3 and 4.

a) Draw a dark, firm line horizontally across the middle of your large grid graph paper. Then draw a line vertically down the middle of your paper. Scale your axes for $-9 \le x \le 9$ and $-9 \le y \le 9$. Label each unit along each axis.

b) Plot (0, 9) and (1, 0) and **connect them** with a straight line segment. Do not extend the segment beyond either endpoint.

c) Plot (0, 8) and (2, 0) and connect them; plot (0, 7) and (3, 0) and connect them. With your partner or team, discuss the pattern (relationship) for how the x-coordinate and y-coordinate change in the three pairs of points in parts (b) and (c). Then agree on how to get the next pair of points. Write your results in one or two sentences. State the next pair of points.

d) Using the pattern, continue plotting pairs of points and drawing line segments, ending with the pair (0, 1) and (9, 0).

e) Describe the visual results of drawing these nine line segments in one or two sentences.

f) Create the same kind of picture in each of the other three regions (quadrants) of the graph. Start with (0, 9) and (-1, 0), (0, -9) and (-1, 0), and (0, -9) and (1, 0) in each respective region.

g) Describe in a sentence or two the visual results of drawing these lines.

PR-20. Do this problem on the same set of axes you used for the previous problem.

a) Mark point (9, 9) with the letter A on the graph. Then draw a horizontal line from A over to the y-axis and a vertical line from A down to the x-axis.

b) Connect the points (0, 9) and (9, 8), then (1, 9) and (9, 7) and so forth until the upper right portion of your figure has a pattern similar to the one in the opposite corner.

c) Describe the visual result in a sentence or two in relation to the results in (d) of the previous problem.

PR-21. Use the same set of axes as you did previously. Complete the other three outer corners of the figure so that they look like the result of the previous problem. Describe the resulting figure in two to three complete sentences. Note that your other corners are (-9, 9), (-9, -9), and (9, -9).

PR-22. Imagine you are on a coordinate grid at the point (3, 2). If you first add 10 to your y-coordinate, then subtract 8 from your x-coordinate, then add 7 to both coordinates, which quadrant are you in? Describe your path to get there.

PR-23. Plot the following points on another sheet of graph paper and connect them in the order given, then connect points A and D: A(5, 3), B(5, -6), C(-4, -6), and D(-4, 3).

a) What is the resulting figure?

b) Find its area.

PR-24. Write the following polynomials in shorter form, if possible.

a) $6x + 7 + 3x$

b) $3(x + 5)$

c) $4x + 8xy - 7$

d) $2x(x - 10)$

e) $(2x + 3)(x - 5)$

f) $3x(x - 4) + 7x + 8$

PR-25. Solve for the variable. Show the steps leading to your solution. The answers are included for you to check. If you are missing many and you cannot find your error, see your teacher!

a) $8x - 22 = -60$ **[ -4.75 ]**

b) $\frac{1}{2}x - 37 = -84$ **[ -94 ]**

c) $\frac{3x}{4} = \frac{6}{7}$ **[ ≈ 1.14 ]**

d) $9a + 15 = 10a - 7$ **[ 22 ]**

## THE SILENT SQUARE

PR-26. Each student in your team will be given three puzzle pieces. When all the pieces are put together correctly, they will form four separate, but identical squares. You must follow these four rules:

1) Absolutely NO talking!

2) You may **give** a piece to any other team member.

3) You may **NOT** **<u>take</u>** a piece from any other player; pieces must be offered to you.

4) You may only touch or point to pieces on **your** own desk or area.

## GUESS AND CHECK

PR-27. To solve the following problem, set up and use a guess and check table. Each member of your team must have all the work and the complete solution on his/her paper. DO NOT solve this problem by writing equations. We will solve plenty of problems by writing equations later in the course.

Today Euclid High School has 1220 students enrolled. The school's population increases by 55 students by the end of each school year.

Socrates High, on the other hand, has only 670 students enrolled today. However, its enrollment increases by 125 students by the end of each school year.

a) Assuming that these growth rates remain the same each year, after how many years will the two school's enrollments be about the same?

b) At the point when they both have about the same enrollment, about how many students will be attending each school?

NOTE: A well organized table is the key to solving this problem.

PR-28. Which is more likely to happen: rolling a four on a normal die OR flipping a fair coin and getting tails? Justify your answer completely.

PR-29. Solve for x. Show all steps leading to your solution. The answers are provided so that you can check. If you are missing many of them and cannot find your error (you can ask your team as well), you may need to see your teacher for extra help.

a) $5x - 2x + x = 15$ **[ $3^3/_4$ or 3.75 ]**

b) $3x - 2 - x = 7 - x$ **[ 3 ]**

c) $3(x - 1) = 2x - 3 + 3x$ **[ 0 ]**

d) $3(2 - x) = 5(2x - 7) + 2$ **[ 3 ]**

e) $\frac{26}{57} = \frac{849}{5x}$ **[ ≈ 372.25 ]**

f) $\frac{4x + 1}{3} = \frac{x - 5}{2}$ **[ -3.4 ]**

PR-30. Solve for x. The solution will be an expression with a and/or b in it.

a) $3x + 2b = a$ **[ $^{(a - 2b)}/_3$ ]**

b) $x + 2a = 3x - a$ **[ $^{3a}/_2$ ]**

PR-31. Make a guess and check table and solve this problem. Show all parts of your solution.

The ratio of girls to boys at Roxborough High School is 6:7. If there are 1,352 students enrolled in the school today, how many girls and how many boys attend Roxborough High?

PR-32. Arrange the following numbers in order from smallest to largest:

$\frac{2}{3}$ $\sqrt{9}$ -4.2 -4 $\frac{16}{3}$ $3(5^2)$ $(3 \cdot 5)^2$

PR-33. Of the graphs below, which do you think best illustrates the relationship between a typical person's age and his/her annual income over the course of their lifetime? Briefly explain why you made your choice.

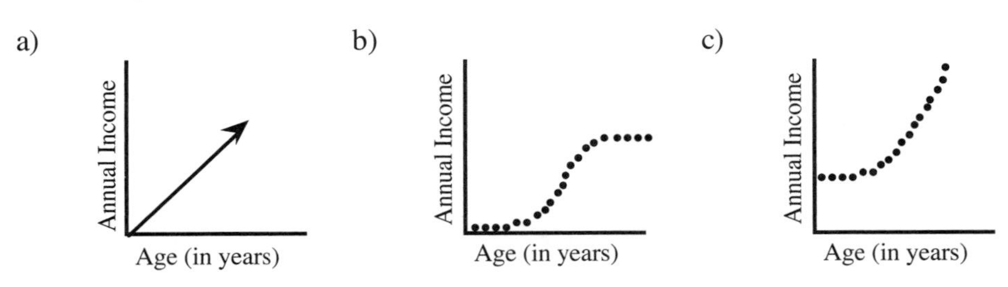


PR-34. Calculate, showing all steps. The answers are provided for you to check. If you miss two or more of these and cannot find your errors, see your teacher!

a) $2 \cdot [3 \cdot (5 + 2) - 1]$ **[ 40 ]**

b) $6 - 2 \cdot (4 + 5) + 6$ **[ -6 ]**

c) $3 \cdot 8 \div 2^2 + 1$ **[ 7 ]**

d) $5 - 2 \cdot 3 + 6 \cdot (3^2 + 1)$ **[ 59 ]**

## COMMUNICATING WITH YOUR TEACHER

Just a quick note about the importance of doing your work in this class. Homework and classwork are the best ways to communicate with your teacher. One purpose behind assigned course work is that **your teachers can use it to evaluate your level of understanding**. If you do not do your work, the teacher can only assume you know nothing. You need to complete all assignments, especially those at the end of a unit to help you pull the ideas together. Otherwise, exams are the only other way that your teacher has to evaluate your knowledge of geometry. We all have bad exam days, so this leaves assigned course work as the best way to regularly demonstrate your level of understanding to your teacher. I will even say that daily assignments are far more important than exams. The moral to this story is DO ASSIGNED WORK EACH DAY.

Now do PR-35 below and you will have completed the first unit in this course!

PR-35. Write a paragraph that describes each of the following:

a) the contributions you can make to team work, (that is, what personal skills and strengths do you possess that will help your team be successful?).

b) the strengths and weaknesses of team learning you observed during this unit.

# UNIT 1

# Riding a Roller Coaster

## PERIMETER, AREA, GRAPHING, AND EQUATIONS

# *Unit 1 Objectives*

***Riding a Roller Coaster:***

**PERIMETER, AREA, GRAPHING, AND EQUATIONS**

In this unit you will encounter geometric concepts you have studied before: perimeter, area, bases and heights, and the Pythagorean Theorem. A major emphasis is helping you understand area formulas and how and why they work. You will also continue to practice fundamentals of solving algebraic equations. The second half of the unit focuses on linear equations and their graphs, using the slope and y-intercept to efficiently graph linear equations. The PZL problems continue to offer suggestions to help you succeed in geometry.

In this unit you will have the opportunity to:

- explore and apply the Pythagorean Theorem.
- develop methods for finding the perimeter and area of polygons, including formulas and subproblems (figure dissection).
- learn some basic properties of and vocabulary for triangles, squares, rectangles, parallelograms, and trapezoids.
- explore relationships between points on a straight line, leading to an understanding of and ability to use slope and y-intercept to graph linear equations.
- extend core ideas of the unit to find the length and midpoint of a line segment.
- continue to review solving and graphing equations, simplifying expressions, and multiplying binomials.

*Read the description below . We will return to it later in the unit and have you answer the question.*

RC-0. The diagram at right represents a roller coaster track. As you travel from left to right, the steepness of the track requires different amounts of "effort" to be exerted on your car.

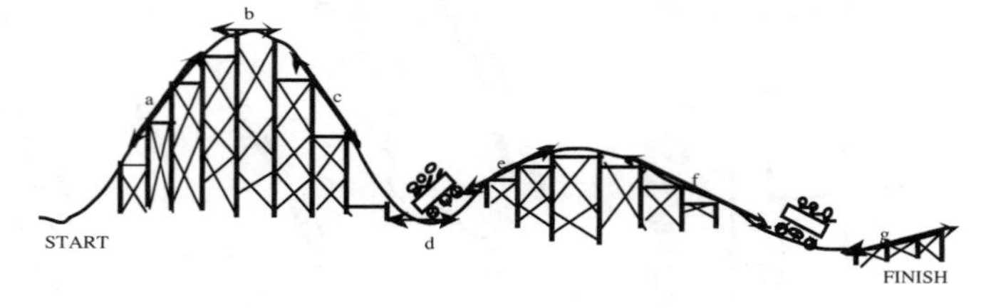


It takes the greatest effort to go up a steep hill. Explain how the notion of effort is similar to the concept of the slope of a straight line

*UNIT 1*

ALGEBRA
GRAPHING
RATIOS
GEOMETRIC PROPERTIES
PROBLEM SOLVING
SPATIAL VISUALIZATION
CONJECTURE, EXPLANATION (PROOF)

# Unit 1

***Riding a Roller Coaster:* PERIMETER, AREA, GRAPHING, AND EQUATIONS**

WHAT TO DO IN THIS COURSE

The nice thing about the beginning of the school year is that everyone gets a fresh start. Just like the student in the drawing, YOU have the primary responsibility for what gets done and how much you will learn. If you want maximum results from this course, keep the following in mind:

- Your primary responsibility is **to think** about mathematics!
- Thinking requires that you **be active** in your learning.
- You will need to **read the book** and **do your homework** every night to practice thinking.

I will not kid you by saying this is easy, but it will get easier as you do it on a regular basis. Other responsibilities were mentioned in the welcome note in PZL-1. Make sure you read this note again. Also, be sure you know all the requirements for this course. If you are still not sure what they are, ask your teacher as soon as possible.

**MISSION POSSIBLE** (something to do)

Re-read the second to last paragraph in the welcome note (PZL-1).

a) Do you agree with the statements about what it takes to learn how to play a musical instrument? Why or why not?

b) Read the boxes below, then begin today's lesson with RC-1.

**REMINDER: Guidelines for Study Teams**

1) Each member of the team is responsible for his or her own behavior.
2) Each member of the team must be willing to help any other team member who asks for help.
3) You should only ask the teacher for help when all team members have the same question.
4) Use your team voice.

Problems that are especially important have a box around the problem number. They help you develop understanding and consolidate ideas. Pay careful attention to these problems, and be sure to revise your work if necessary. Complete, correct solutions should be in your notebook to serve as examples of why a concept or procedure works and as an example of how to do it.

## THE PYTHAGOREAN THEOREM

RC-1. Use the resource page provided by your teacher. **Be neat.** The axes are scaled at intervals of five (that is, every 5th unit: 5, 10, ...) for $-20 \le x \le 18$ and $-15 \le y \le 14$.

a) Fill in the coordinates for points A, B, and C and connect them to form a triangle. What is special about one angle of the triangle?

> A graph paper ruler is useful for measuring lengths in grids, especially when the units are not a standard length. Cut off an edge of the kind of graph paper you are using to make your graph paper ruler. The segment at right would have measure 8 units using the ruler shown.
>
> 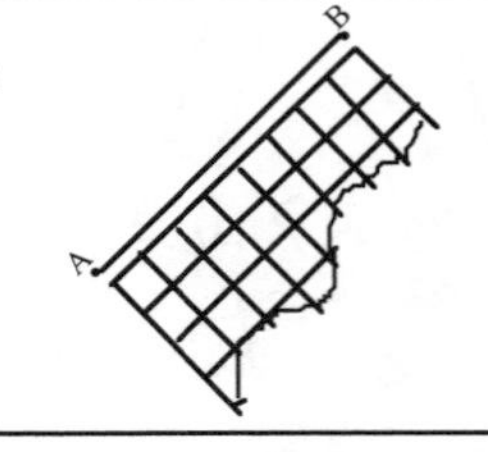
> 

b) Determine the length of each side of ΔABC and mark the length along the inside of each segment. The edge of each square on your graph paper ruler is one unit.

c) Carefully draw a SQUARE on each side of the triangle (build each square so it is outside the triangle and its side lengths are the same as the length of the triangle's side segment). Estimate its area by counting small squares or calculating the area of each square. NOTE: the side of each square is equal to the length of the side of the triangle on which it is drawn.

d) Add the areas of the two smaller squares and compare the sum to the area of the largest square. State your observation in a full sentence.

e) Repeat parts (a) through (d) for ΔDEF: D(-5, -1), E(-11, -1), and F(-11, -9). How do your results compare to those for the first triangle?

f) Repeat parts (a) through (d) for ΔGHJ: G(3, -2), H(15, -2), and J(15, 3). Write one or two sentences describing the relationship between the squares of the two shorter sides of right triangles and the square of the longest side.

g) Read the two boxes on the next page and be sure that your results in parts (a) through (f) above agree with their contents.

In the previous problem you verified for three specific triangles that the sum of the areas of the two smaller squares is equal to the area of the largest square. Notice that in each problem you started by knowing the lengths of the sides of the triangle and found the areas of squares formed on each side. It is usually more convenient to express the relationship between the lengths of the sides of a right triangle and the area model you used algebraically.

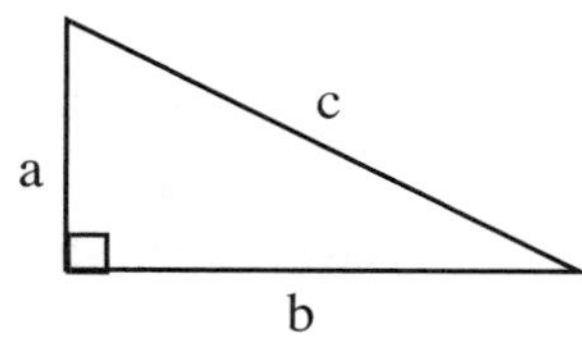


Any triangle that has a right angle is called a **RIGHT TRIANGLE**. The two sides that form the right angle, a and b, are called **LEGS**, and the side opposite (that is, across the triangle from) the right angle, c, is called the **HYPOTENUSE**.

For any right triangle, the sum of the squares of the legs of the triangle is equal to the square of the hypotenuse, that is, $a^2 + b^2 = c^2$. This relationship is known as the **PYTHAGOREAN THEOREM**. In words, the theorem states that:

$$(\text{leg})^2 + (\text{leg})^2 = (\text{hypotenuse})^2.$$

## THE PYTHAGOREAN THEOREM

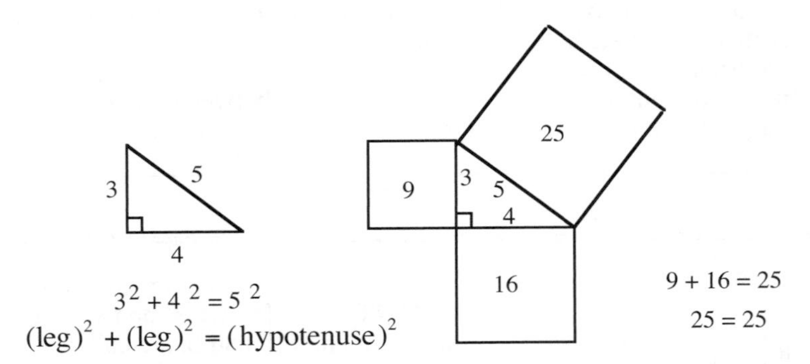


Suppose you did not know the length of the hypotenuse. The example above then becomes an equation:

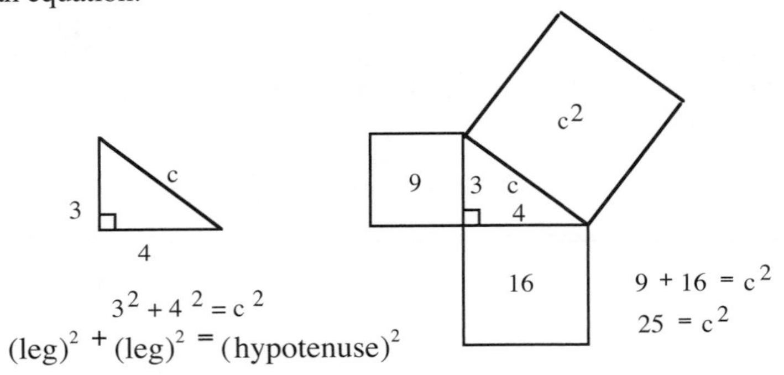


One way to think of the Pythagorean relationship is in terms of **dimensions**. The sides of the triangles are one-dimensional lengths. When each length is squared in the equation, the result actually measures the area of the square on each side of the triangle. These squares are <u>two-dimensional</u>. In order to solve the equation in the above example, you need to return to the one-dimensional length of the hypotenuse itself. This requires <u>reversing</u> the squaring process by using its mathematical inverse, <u>square root</u> (unsquaring). For example:

If $a = 5$, then $a^2 = 5^2 = 25$. (squaring)

If $a^2 = 25$, then $a = \sqrt{25} = \sqrt{5^2} = 5$. (square root)

Until you are comfortable using the Pythagorean Theorem algebraically, use both the geometric model with the area diagram and the algebraic solution as shown in the example above.

**Since many problems will not have whole number solutions, use the square root ($\sqrt{\ }$) key on your calculator to approximate square roots to two decimal places.**

RC-2. Scale a pair of xy-axes on quarter inch graph paper for $-10 \le x \le 10$ and $-10 \le y \le 10$. Plot each set of three points and connect them in the order given. Be sure to connect the first and third points so that you form a triangle. Next find the lengths of the horizontal and vertical segments. Finally, use the Pythagorean Theorem to find the length of the hypotenuse. Each part (a), (b), and (c) is a separate problem, but you may graph them on one set of axes.

a) (2, 4), (10, 4), and (10, 10)

b) (-9, -4), (8, -4), and (8, -9)

c) (-3, 1), (-5, 1), and (-5, 7)

RC-3. For each figure, use the Pythagorean Theorem to write an equation, then solve for the missing length. Draw area pictures on each side of the triangle if you need help solving the equations.

a)

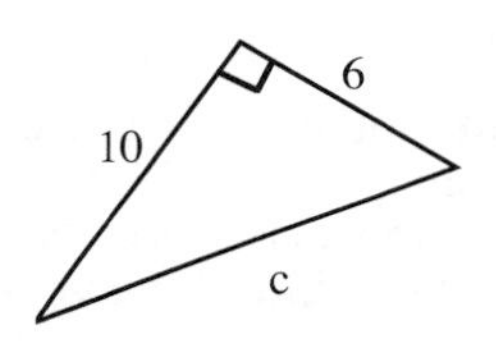


b)

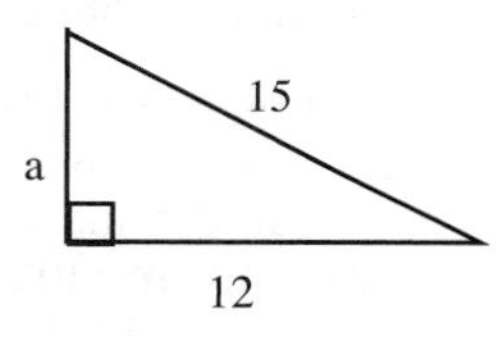


c)

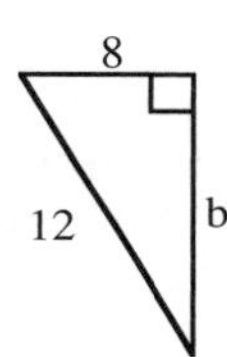


RC-4. Write each expression without parentheses and shorten it as much as possible.

a) $2a + 4(7 + 5a)$

b) $4(3x + 2) - 5(7x + 5)$

c) $x(x + 5)$

d) $2x + x(x + 6)$

RC-5. Solve for x, show all steps leading to your solution, and check your solution.

a) $34x - 18 = 10x - 9$

b) $4a - 5 = 4a + 10$

c) $3(x - 5) + 2(3x + 1) = 45$

d) $-2(x + 4) + 6 = -3$

RC-6. Examine the graph at right. In a sentence or two, state reasons why the graph rises at 11:00 AM and drops at 1:15 PM.

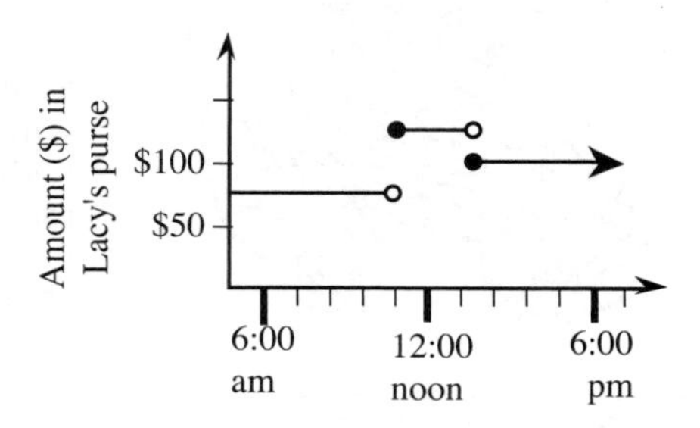

RC-7. If there is a 0.0006% chance of being struck by lightning, and a 0.00000006% chance of winning the lottery, which of the two events has a greater probability of occurring? Why?

## UNDERSTANDING HOW TO USE THIS TEXTBOOK

By now you have noticed that this Geometry course expects you to be an active learner. You work with other students in study teams, the problems guide you to learning the mathematical ideas, you are asked questions about what your work means, and your teacher works with you in several ways during class time. Another key step to get the most out of this course is to understand the textbook. **Understanding the textbook** is quite a bit different from understanding the problems within it. Quickly glance at the welcome note (PZL-1 in Unit 0) and see if you can find the three big ideas it contains about this course.

How did you find them? If you noticed that they are in **bold print**, you have just discovered **one way for the authors to relate big ideas to you**. Flip through a few pages and find a few other examples of using bold print for emphasis. Bold print is one of the most common ways to point out information. **Sentences in bold print contain valuable information**. Always make sure you **add information in bold print to your tool kit**.

There are some other signs to help you find key information in the text. Every unit of this text contains several main ideas. These ideas are the fundamental concepts of geometry. **You will absolutely need to know these ideas to be successful in this course.**

- The mathematical ideas and definitions are in **single** and **double line boxes.** These boxes sometimes contain **examples** of how to use an idea or do a procedure.

- Important problems--usually ones that develop an idea or help pull it together--have a **box around the problem number.**

The information you find in boxes and important problems should also be in your tool kit.

In the front of your book there is a **table of contents** that lists what you will study unit by unit. In the back of each volume there is an **index** that helps you locate where a term or concept is introduced. At the beginning of each unit there is a **list of what you will learn** in that unit. At the end of each unit, everything you should have learned and added to your geometry tool kit is listed in a **"Tool Kit Check-Up."** In addition to your textbook, there is a **"Parent Guide with Review"** that will help your parent(s) understand how to help you with this course. It has additional examples and lots more practice problems. Ask your teacher how to obtain these resources or use the ordering information in the "Note to the Parents" at the beginning of the book.

**MISSION POSSIBLE** (something to do)

Make a list of all the locations in the text that will help you find the important information it contains, then begin problem RC-8.

## INTRODUCING AREA AND PERIMETER

RC-8. **DIRECTIONS**: Read the following definition, then draw an example of a polygon on your homework paper. Have another student verify that your figure meets all the requirements of the definition.

> A **POLYGON** is a two-dimensional closed figure made up of straight line **segments** connected end to end. The segments may not cross. The point where two sides meet is called a **vertex** (plural: vertices). Some familiar polygons are triangles (3 sides), quadrilaterals (4 sides), pentagons (5 sides), and hexagons (6 sides). Some special four-sided polygons include squares, rectangles, parallelograms, and trapezoids.

a) On the resource page provided by your teacher, name as many of the polygons as you can. Set up a list of the polygons on your paper so that you have a place to record your results in parts (b) and (c).

b) Measure the distance around each figure (the **perimeter**) in graph paper units. For some of them, you can simply count the edges of the squares, but for others you will need to cut off an edge of your graph paper to make a graph paper ruler and measure segments. You can also use the Pythagorean Theorem to calculate these lengths.

c) Since **area** is the number of square units inside the boundary of a two-dimensional figure, one way to describe how much of the paper's surface is enclosed by the figure is to count the number of identical squares within a figure's boundary. Use this counting technique to find the area of each figure. Some figures will require that you count parts of squares, but you can still get the exact total. For example, the figure at right is half of three squares, or one and one-half square units. Look for shortcuts.

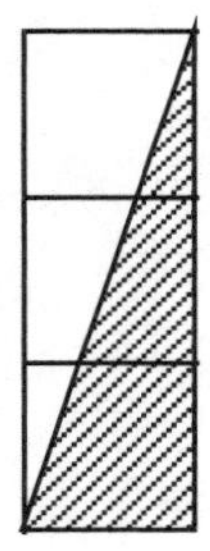

RC-9. Suppose we discover the island shown in the drawing at right and we want to know its area. If we cover the island with a grid we can approximate the area.

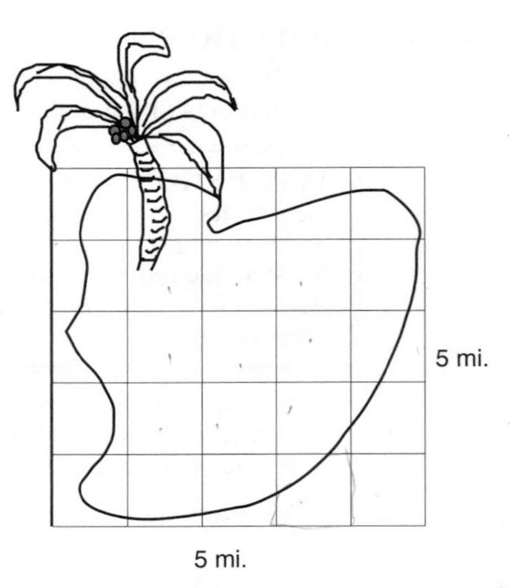


a) What is the area of the grid?

b) The island's area is less than the area of the grid. Why?

c) To approximate the area, count the number of squares entirely inside the island's shoreline.

d) Now count the number of squares partially covering the island. Do not spend too much time deciding whether a square is entirely or partially covering the island. We are only getting an estimate.

e) A useful approximation is to consider a partially covered square as half covered. Add the number of whole squares to half the number of partially covered squares to get the approximate area.

RC-10. To get a better estimate you can use a grid with smaller squares. Each small square is $\frac{1}{2}$ mile on a side, so each square covers $\frac{1}{4}$ of a square mile and it takes 4 squares to cover a square mile.

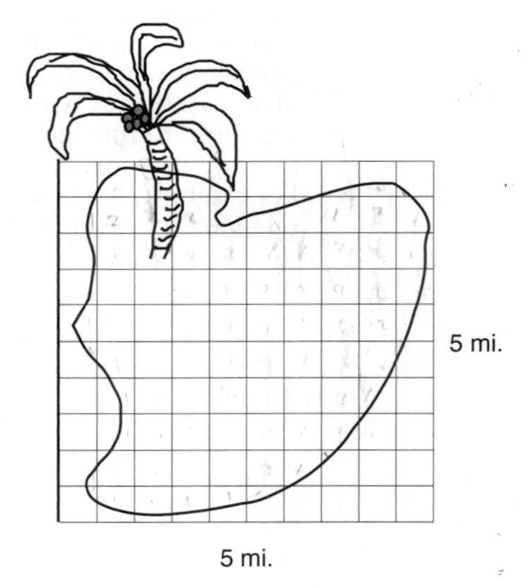


a) Count the number of whole squares inside the shoreline. Next, count the number of squares that partially cover the island and divide by 2.

b) Add the results of part (a) to get the total number of squares, then divide by 4 to get the number of square miles. Explain why you needed to divide by 4.

c) Compare the area of the island in the previous problem to your result in this problem. Write your observations in a sentence or two.

d) Do you think the smaller squares give a better approximation or worse? Explain your answer.

RC-11. Write in simplest form.

a) $7r + 3(6 + 2r) + 9$

b) $5(8y + 4) + 3(2y + 9)$

c) $2x^2 + 7x + 3x + 5$

d) $-6x^2 + 16x - 7$

RC-12. Evaluate each expression if a = -2 and b = 3.

a) $3a^2 - 5b + 8$

b) $\frac{2}{3}b - 5a$

c) $2a^3 - 4b^2 + a$

d) $\frac{a + 2b}{4} + 4a$

RC-13. Copy the table onto your paper, complete it, and write a rule using x that will always result in an appropriate y-value.

| x | 3 | -1 | 0 | 2 | -5 | -2 | 1 |
|---|---|---|---|---|---|---|---|
| y | 0 | | | -1 | | | -2 |

RC-14. While in France last summer I bought a hat for 25.50. A friend bought a similar hat for 5 in the United States. What's going on here? Explain completely.

RC-15. Explain what the graph at right tells you.

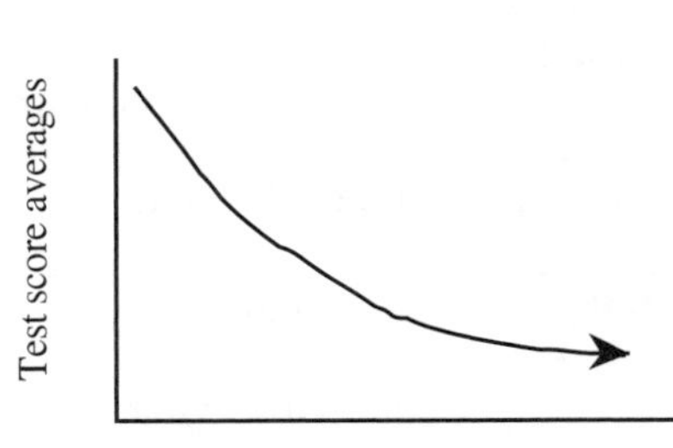


Number of students
in the classroom

RC-16. Palmer S. Friedrich loves to play baseball. His team, the Magical Mathematicians, has a probability of $\frac{1}{3}$ of winning any particular game. In an upcoming tournament, the MMs will be playing at least 9 games. They need to win 5 games to advance to the next tournament. Do you think Palmer's team will advance? Explain why or why not.

## UNDERSTANDING THE ROLE OF THE TEACHER

Earlier I mentioned that your teacher would use several methods to help you learn algebra. The idea is to use several approaches to give you more opportunities to understand geometry. In the past, you may have had math classes where the teacher almost always stood in front of the class and gave you instructions on how to complete different math problems.

In other words, the teacher **told you** exactly what to write, what to say, and what to think. This method of teaching leaves little room for you to think and understand. Being told what to do before understanding the reason(s) behind it makes concepts easy to forget. Past experience by students who learned through reasoning and discovery showed that it helped eliminate the tendency to merely memorize information and then see it quickly slip away.

Your teacher will usually give step by step instructions at key points in the course, frequently when tying together several days of class exploration.

**In most cases, your teacher will ask questions designed to help YOU develop thesteps yourself.** The teacher will not stand up in front of the class and lecture you very often. You will often work with a partner or in a study team, where you can help each other while the teacher moves from team to team. **Your instructor wants you to become a strong thinker, not a tape recorder.** Be patient. Work at asking good questions and sharing ideas with other students. Soon you should find that you are **learning how to learn** in this course. The teacher will make sure your frustration does not become overwhelming. Start thinking about the advantages this approach may offer you.

**MISSION POSSIBLE** (something to do)

a) Summarize what you can expect from your teacher in this class.

b) Compare and contrast the above description (list similarities and differences) to the way class time has been used so far this year.

**RC-17.** Make all measurements to the nearest whole inch or centimeter.

a) Measure a side of the large square in inches and give its area.

b) Measure a side of the small square in centimeters and give its area.

c) How do their numerical answers compare?

d) How do their sizes compare?

e) Explain why your answers to (c) and (d) are different.

RC-18. **DIRECTIONS**: In a previous assignment, RC-8, you measured the areas of five figures using square grid units. If you had used square inches you would have found different numbers for the same five regions. Square A below is approximately one square centimeter. Square B is about one square inch.

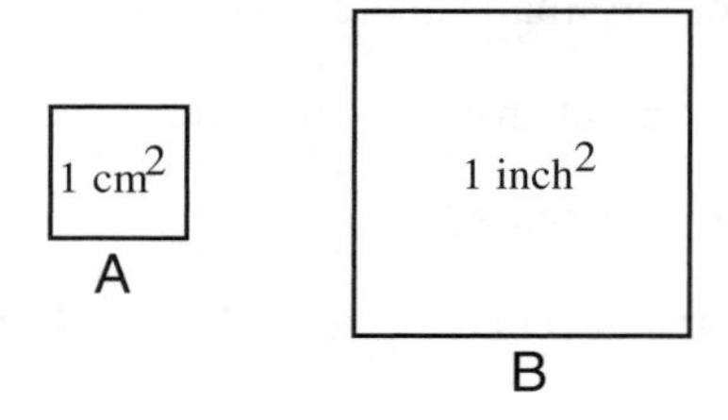


Use the rectangle drawn by your teacher and answer the following questions:

a) Estimate the number of square centimeters it would take to cover the rectangle drawn by your teacher and explain how you determine your answer.

b) Estimate the number of square inches you would need to cover the same rectangle, and explain how you determine your answer.

c) Compare your estimate with the other members of your team and agree on a team estimate to report to the class. You might want to average your individual estimates.

d) Get the actual measurements of the rectangle in both inches and centimeters and calculate the area in both square inches and square centimeters.

e) How close were your estimates? What would you do differently next time to get a more accurate estimate?

RC-19. The last two problems have demonstrated that <u>different</u> <u>regions</u> can have the <u>same</u> <u>number</u> for their areas, and <u>one</u> <u>region</u> can have <u>different</u> <u>numbers</u> that describe its area.

a) Discuss these observations with your partner and/or team and be sure that everyone understands why these differences exist.

b) Write a summary of your discussion that **<u>defines</u>** what area is. "Define" means to state the characteristics that make something what it is in a way that most people understand and that eliminates the chance of confusing it with something else.

RC-20. A grid has a shaded rectangle drawn on it.

a) Find the area of the shaded region.

b) Explain in one or two complete sentences, how you found your answer.

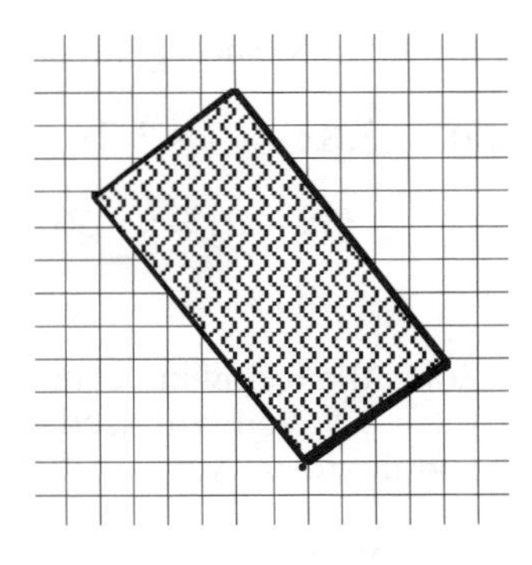

RC-21. Santi draws a rectangle with a length of 5 inches and a width of 3 centimeters.

a) What is the perimeter of Santi's rectangle?

b) What is the area of Santi's rectangle?

RC-22. Plot the set of points on both large grid graph paper **and** quarter-inch graph paper. Connect the points A to B, B to C, and A to C: A(-2, 6), B(5, 6), C(0, 0).

a) Name the figure.

b) Find the perimeter and area of the figure from both graphs.

c) Why is the numerical answer for the area the same while the sizes of the figures different? Explain in one or two complete sentences.

RC-23. A piece of cloth is made in the shape shown. WRITE AN EXPLANATION telling how the area of cloth needed for that piece could be estimated. DO NOT actually find its area!

RC-24. Write the following polynomials in shorter form, if possible.

a) $-4x^2 - 2x + 8 + 7x^2 - x$

b) $5a + 2b - 9 - 3a - 7b$

c) $-13y + 72 - 19y - 38 + 21y$

d) $4(3x - 2x^2) + 6x - 3(5 - x^2)$

RC-25. Solve for x, show your solution steps, and check your solution.

a) $5(8x + 4) + 3(2x + 9) = 53$

b) $3(x + 5) = x + 21$

c) $-(3x + 2) = x - 50$

d) $14 - 2x = 2(7 - x)$

RC-26. Solve the following equations for y (get the y variable by itself). For example: to solve $y + x = 5$, you will need to subtract x from both sides of the equation. This will give you the result $y = -x + 5$. Show the steps leading to your solution.

a) $y - x = 8$

b) $x + y = 4$

c) $2x + y = 3$

RC-27. Sketch a graph with appropriately labeled and scaled axes that represents the numbers of hours of daylight over the course of 12 months (i.e., one year).

RC-28. On a cubical die, there are six numbers, 1 through 6. A four-sided die has the numbers 1 through 4. Assume that both are fair die, so that each number is equally likely to come up on any roll. On which die is a 2 more likely to come up? Explain why.

RC-29. For the next three problems you will need a copy of the resource page "Rectangles, Parallelograms, and Triangles."

a) What is the name of figure A? Give the length of each side and the area. Imagine you are talking to a 4th grader who is just learning about area. Explain to her how to find out the area without actually counting the squares.

b) Draw a line segment between (2, -8) and (9, -2). Shade the region in the rectangle that is below this line and find its area. How is the shaded area related to the area of the rectangle?

c) Draw a rectangle in the box on your resource page. Make one pair of opposite sides horizontal and the other pair vertical. Label the horizontal sides "b" for **base** and the vertical sides "h" for **height.** What would be the formula to find the area of a rectangle?

d) In the rectangle you just drew, connect one pair of opposite corners to make two triangles. Explain why adding the segment shows us that the formula used to find the area of a right triangle is $\frac{1}{2}bh$, where b represents the **base** and h represents the **height** of the triangle.

RC-30. Figure B is called a **parallelogram** (both pairs of opposite sides are parallel).

a) Count the area of the parallelogram.

b) Draw dashed lines (like the one in the figure in the next problem) between (-6, 9) and (-6, 5), then (4, 9) and (4, 5), and then (4, 5) and (1, 5). Write the length of each dashed segment next to it.

c) Count and compare the areas of the two right triangles formed by these dashed lines.

d) Shade the interior (area) of the figure formed when you connect the points (-6, 9), (-6, 5), (4, 5), (4, 9) and back to (-6, 9) in order. Compare the area to your answer in part (a).

e) Explain why this shows that the formula for finding the area of a parallelogram of base b and height h is bh.

RC-31. Write a formula for finding the area of this parallelogram.

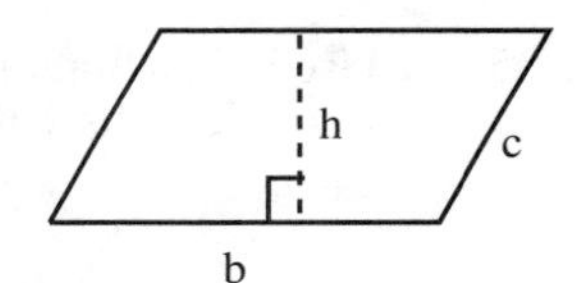


RC-32. Jon's answer to the previous problem was A = bc. If you were the teacher who corrected Jon's work, how would you reply so that he understands why he is mistaken and sees how to do such problems correctly?

RC-33. Use Figure C for the following.

a) Locate a fourth point so you can draw a parallelogram that is twice as large as the triangle. (How many ways are there to do this? Decide on one and draw it.)

b) What are the base and height of the parallelogram? What is its area?

c) How can you use the information in (b) to find the area of the original triangle? What is the area?

d) Write a formula for the area of any triangle with base b and height h. Explain why you know that this procedure will always work.

RC-34. Use your results from today's activities to <u>calculate</u> the exact areas of these polygons (that means no measuring or estimating this time.) All dimensions are in centimeters. Be sure to express your answers in **correct units** (sq. cm).

a)

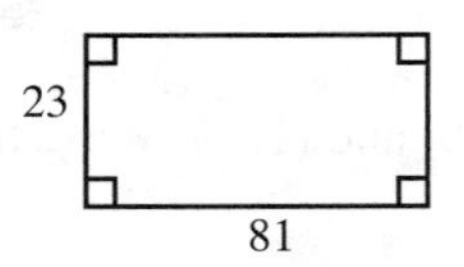


b)

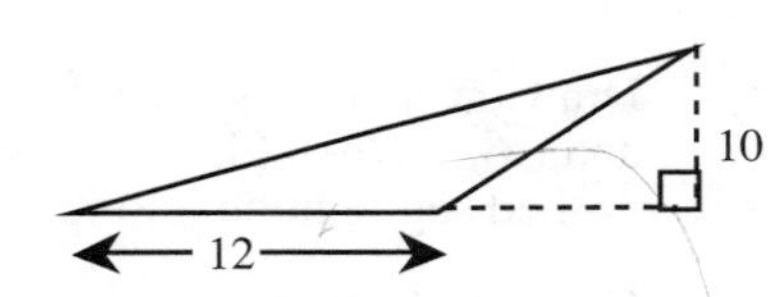


c)

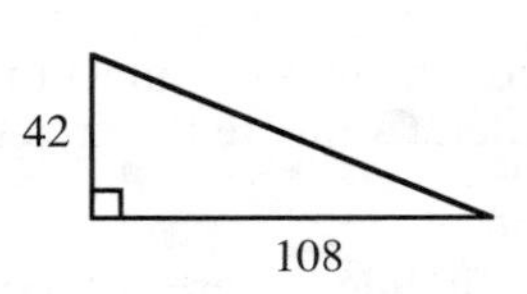


d)

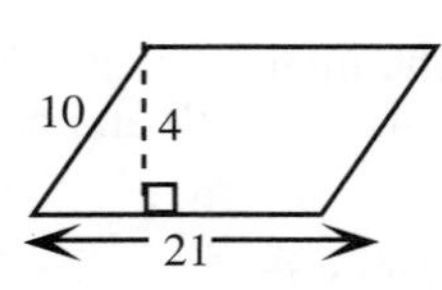


RC-35. Use the Pythagorean Theorem to find the indicated lengths. All dimensions are in feet. Label lengths with the appropriate units.

a)

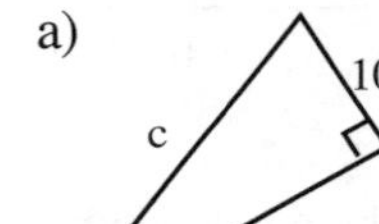


b)

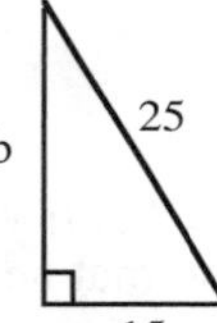


c)

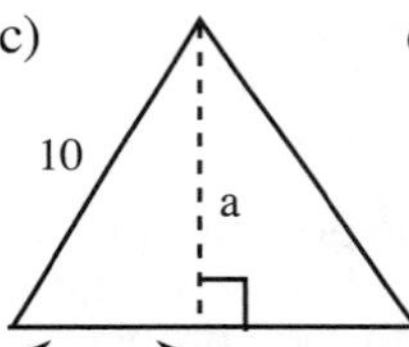


d)

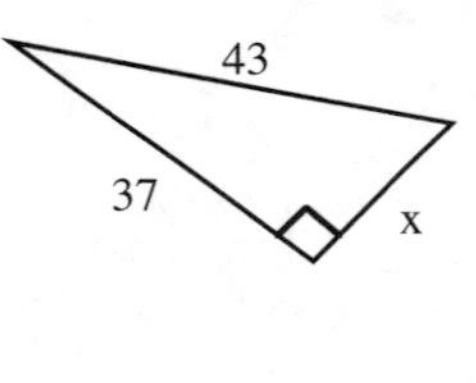


RC-36. Find the length of the diagonal of a square with 5 cm sides.

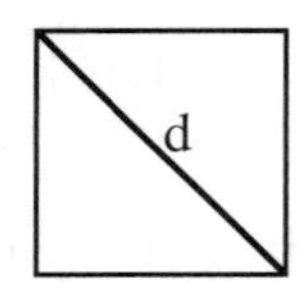


RC-37. Solve for x. Show the steps leading to your solution.

a) $8x - 9 = -2x - 9$

b) $4(x^2 + 6x - 3) = 24 - 8x + 2(2x^2 - 7)$

c) $17x - 329 = 2,409$

d) $\frac{13}{2x - 5} = \frac{4}{x}$

RC-38. Solve for y in terms of x. Keep numbers as integers and fractions (i.e., no decimals).

a) $2x - 3y = 12$

b) $5x + 2y = 7$

RC-39. Copy the table below, complete it, and write a rule using x that will always produce the appropriate y-value.

| x | -3 | -2 | -1 | 0 | 1 | 2 | 3 | 4 |
|---|---|---|---|---|---|---|---|---|
| y | -7 | | | 2 | 5 | | | 14 |

RC-40. Sketch a graph with appropriately labeled and scaled axes that represents the total amount of food a human consumes over an average lifetime.

RC-41. For this problem your teacher will give you two copies of **trapezoids** (a quadrilateral with one pair of parallel sides) on grid paper. Work together with a partner or your team. You will be developing a way, besides counting, to find the area of a trapezoid.

a) The two parallel sides (horizontal in this case) of the trapezoid are called **bases**. Label the top base $b_1$, and the bottom base $b_2$ on both of the trapezoids. Note that some books call them the upper base and lower base.

b) Draw a copy of the first trapezoid immediately to the right of it, BUT draw it so that the first trapezoid is turned upside down and shares a common (equal) side with the first trapezoid. If you draw it correctly, the two bases will switch positions (that is, the top base will become the bottom and the bottom length will become the top base).

c) If your drawing is accurate you now have a parallelogram whose base is $b_1 + b_2$. How long is the parallelogram's base in grid units?

d) Draw in a height of the parallelogram. Label it h. (Refer to RC-31 if you need help here.) Write an algebraic expression (using b's and h's) for the area of the parallelogram and give the numerical result in square units.

e) How can you figure out the area of one trapezoid? Show your result both algebraically and numerically. f) Finally, take the second trapezoid, label the bases as before, and draw a straight line segment from one corner to the other. If drawn correctly, you should have two triangles, each with a labeled base ($b_1$ and $b_2$). Find the area of each triangle, add the two results, and compare the sum to your answer in part (e).

g) Use your results in parts (e) and (f) to write a general rule (formula) for finding the area of any trapezoid using $b_1$, $b_2$, and h.

**RC-42.** Today you received a sample page for your **GEOMETRY TOOL KIT**, essentially a summary of what has been covered so far in this course. YOU will be responsible for the creation of subsequent entries in your tool kit.

We suggest that you summarize each new idea with an appropriate diagram or example as it is presented. **It is also a good idea to label each entry with the textbook page number or problem number where the idea appeared.**

Your tool kit is a handy reference guide you will create during the year. As you identify and organize the important ideas, you will have a useful tool to make geometry more manageable. Whenever you see the tool box icon (above, right), you are expected to add the information to your tool kit. Do so with the information below.

**DIRECTIONS**: It is often useful in mathematics to agree to common notation and ways to express general properties. Read, record in your tool kit, and discuss the following definitions with your team.

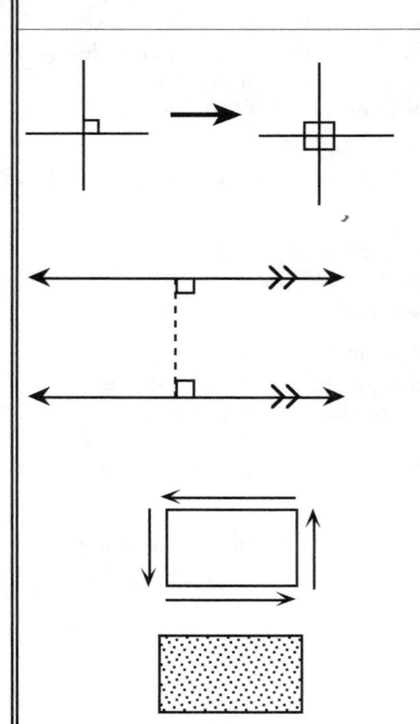

Two lines that meet (intersect) to form a 90° angle are called **PERPENDICULAR** lines. The symbol for perpendicular is $\perp$. The small square indicates that the lines are perpendicular and that the angle there measures 90°.

Note: once you have established one right angle with intersecting lines, you have established four right angles.

Two straight lines on a two-dimensional surface are **PARALLEL** if they do not intersect (cross) no matter how far they are extended. The symbol for parallel is $\parallel$. In figures, pairs of arrows like **>**, **>>**, and **>>>** will mean that the lines are parallel.

Note that all lines extend without limit. This is what the arrowheads indicate.

**PERIMETER** is the distance around a two-dimensional figure.

**AREA** is a number associated with a two-dimensional figure to represent the size of the region covered. The area of any irregularly shaped figure can be divided into several smaller regions. The sum of the areas of these regions is equal to the area of the original figure.

**RC-43.** Read the definitions below and use them to help you complete the Height resource page. Then add them to your tool kit. Directions for the resource page appear in parts (a) and (b) below.

Any side of a two-dimensional figure can be used as a **BASE (b)**.

A **HEIGHT (h)** is the perpendicular distance:

a) in triangles, from a vertex (corner) to a line containing the opposite side.
b) in quadrilaterals, between two parallel sides or the lines containing those sides.

Note that examples of a line containing a side appear in the figure at right below and on the formula page for the third triangle, both parallelograms, and the trapezoid (extensions are dashed).

Some books call the height an **altitude**. You may use either term in this course.

a) **DIRECTIONS:** For figures 1 through 8 on the resource page, draw a height to the side labeled "base." Remember that the height must be perpendicular (at a right angle) to the base. The square corner of a 3 x 5 index card makes an excellent height (or right angle) locator. Just slide one edge of the card along the base as shown below.

**How to Use the Index Card**

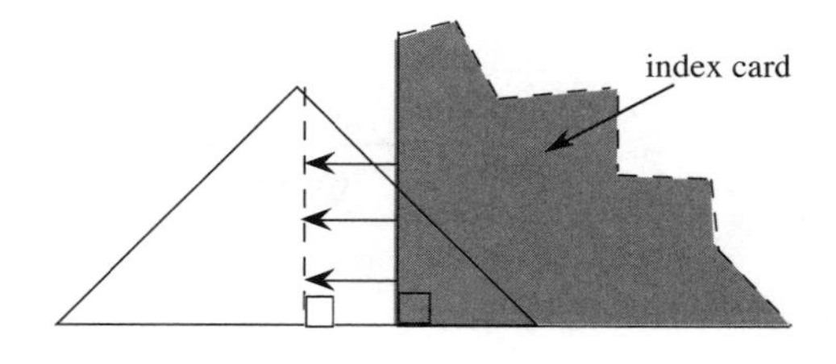


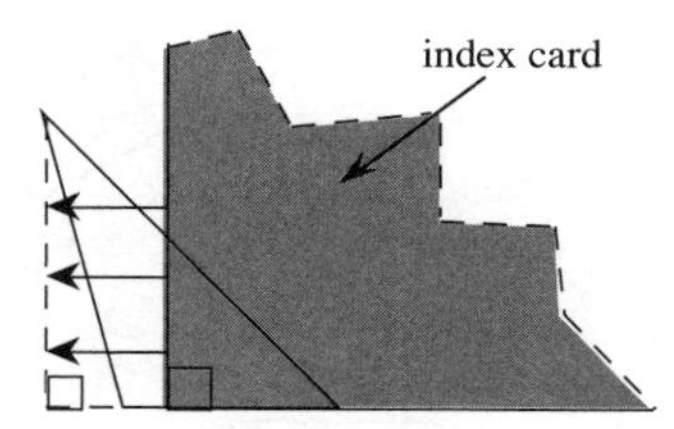


Slide one edge of the card along the base until the vertical (⊥) edge of the card reaches the triangle vertex above the base. In this example, the card is slid from right to left until its vertical edge aligns with the dashed line.

b) For figures 9, 10, and 11 on the resource page, draw a height from each vertex (labeled A, B, and C). You will have three heights in each figure.

RC-44. Use the rectangle to answer the following questions.

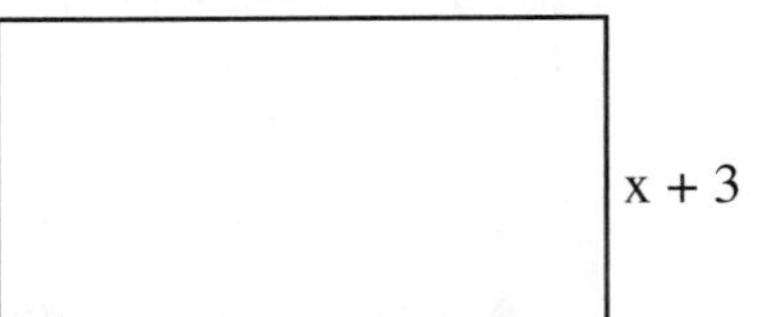


a) What is the perimeter in terms of x? "In terms of x" means that your answer will contain a variable; it will not be a single number.

b) If the perimeter is 78 cm, find the dimensions of the rectangle. Show how you do this. Start by writing an equation using the definition of perimeter.

c) Verify that the area of this rectangle is 360 sq. cm. Explain (prove) how you know this for certain.

RC-45. Is $\overline{XH}$ a height of $\Delta WHS$? Compare the figure to the definition of height to verify your answer. Explain completely.

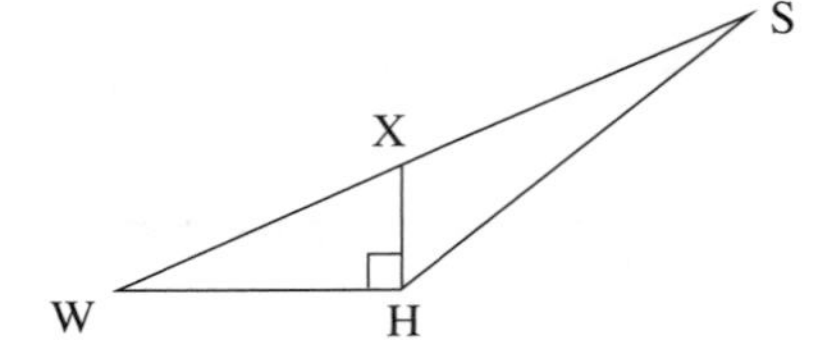


RC-46. Use your results from the last two days' activities to calculate the exact areas of these polygons (that means no measuring or estimating this time.) All dimensions are in centimeters. Be sure to express your answers in **correct units** (sq. cm).

a)

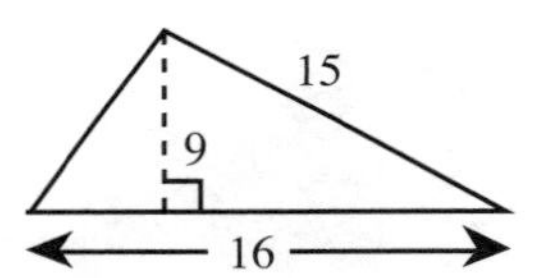


b)

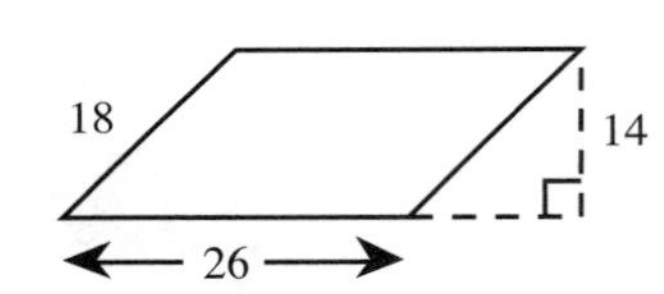


c)

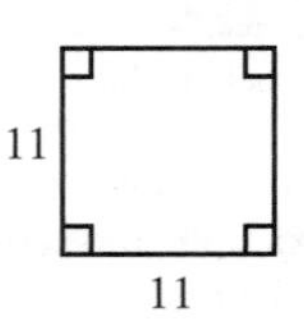


d)

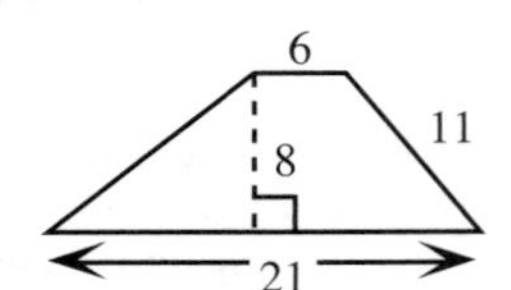

RC-47. Carefully draw a pair of xy-coordinate axes with $-4 \le x \le 9$ and $-3 \le y \le 9$ on quarter inch graph paper.

a) Plot and label the points A(2, 8), B(2, 4), and C(9, 4) and then connect them to form $\Delta ABC$. Add the following definitions and symbols to your tool kit, then complete parts (b) through (e) below the box.

> A straight **LINE SEGMENT** is named by its **endpoints** and is written $\overline{\mathbf{AB}}$ if A and B are its endpoints. The **LENGTH** of a line segment, that is, a number that represents how long the segment is, is usually written without a bar on top, for example, **AB**. In some books, length is represented as |AB|. Examples:
>
> 
> 
> and
> 
> 

b) Find AB and BC (these are lengths). Try to do it by using the coordinates rather than by counting. Write the length of each segment near the middle of its inside edge.

c) Find AC.

d) Repeat the problem for D(6, 2), E(6, -3), and F(3, -3) to find the lengths of $\overline{DE}$ and $\overline{EF}$, then calculate the length of $\overline{DF}$. Be sure to label the points with the letters.

e) Now find the length of the line segment between G(-4, -3) and H(2, 3). Draw in the right triangle and label the right angle J.

RC-48. Evaluate each expression if $p = \frac{1}{2}$ and $q = -1$.

a) $4p^2 - q^3$

b) $12p - 8q - 7$

RC-49. Simplify.

a) $x(x + 2) + 4(x + 2)$

b) $x(x - 5) + 3(x - 5)$

RC-50. Solve for x. Express your answers in terms of a, b, and c.

a) $x + 4b = 3a$

b) $2x - a = b$

c) $ax + 5 = c$

d) $\frac{1}{a}x + b = c$

RC-51. Which triangle has the greatest area? Explain your reasoning completely.

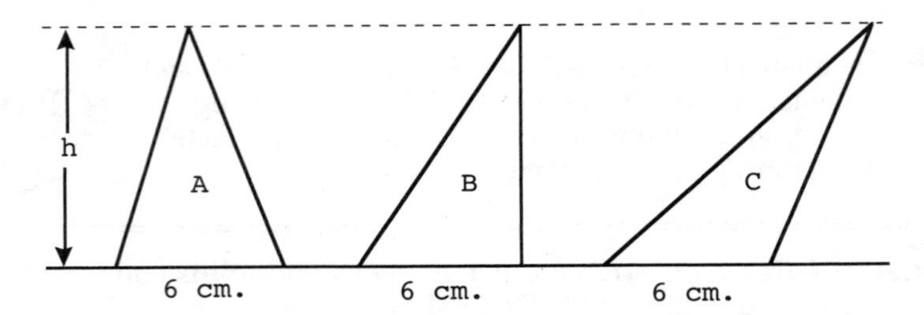


RC-52. If you go 5 miles due north, turn and go 4 miles due east, turn and go 3 miles due south, how far will you be from your starting point? Draw a diagram and label its parts. Note that the question does not ask, "How far have you traveled?"

RC-53. Read the information in the box below, add it to your tool kit, then do the assignment in the last paragraph in the box below.

**DIRECTIONS**: A useful procedure in area problems is to dissect (or break) the figure into parts. You can then work with each part or **SUBPROBLEM.** Suppose you wanted to find the area of the shaded region shown at left below. This **FIGURE DISSECTION** can be done in at least two ways:

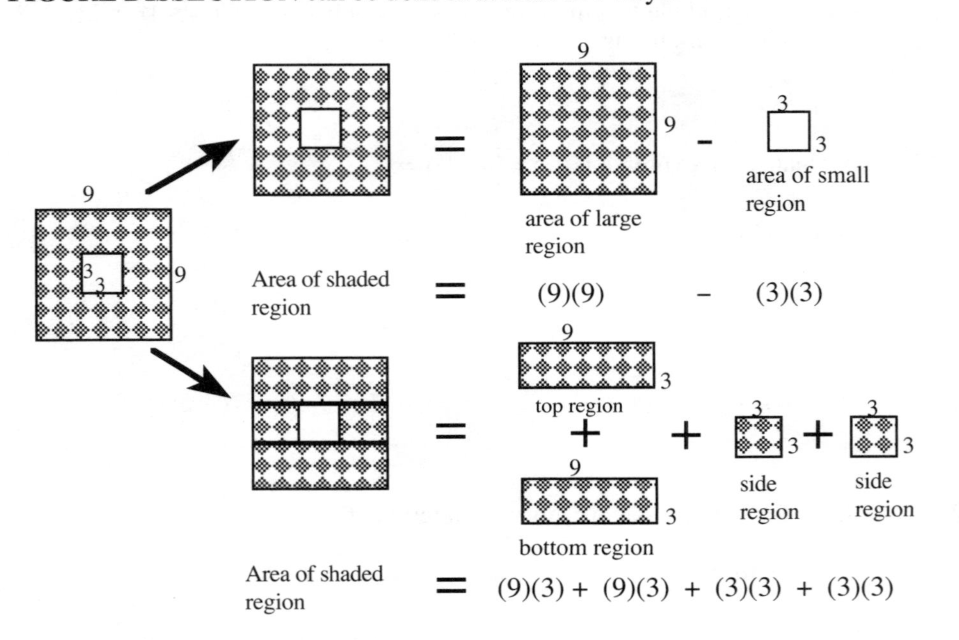


Show that each of the two dissections above will give the same answer by computing the final result. Write one or two complete sentences explaining why this should be true.

RC-54. Find the area of the shaded region. Show how you found the area by using a figure dissection diagram for the subproblems.

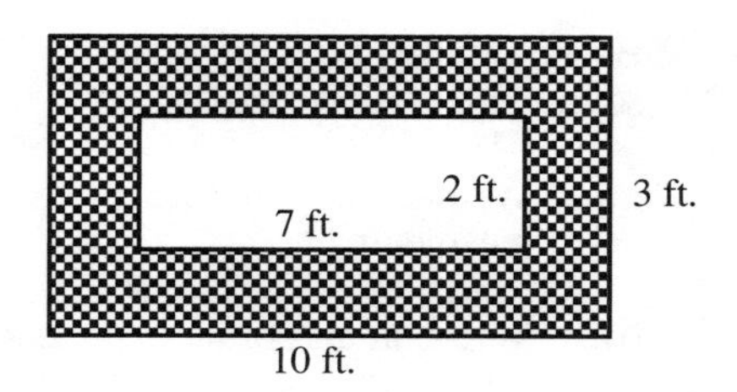


RC-55. Find the perimeter and area of the figure at right. Show your dissection of the figure as part of your solution and show the subproblems you solve.

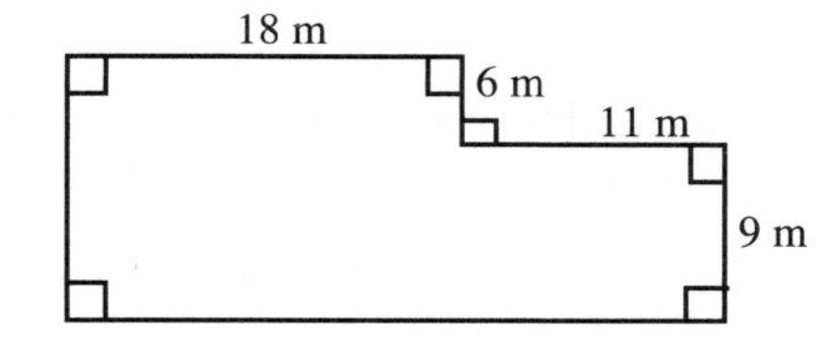


RC-56. A square has an area of 120 square centimeters. Find the length of its sides. Another way to ask the question: "unsquare" the area to find the length of any side of the square.

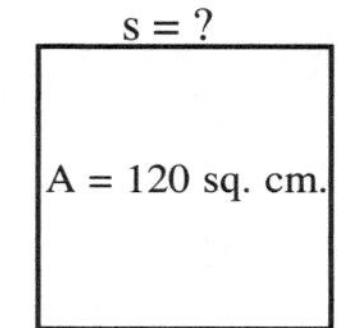


RC-57. Find the area of the shaded region. Sketch the picture on your paper and show the figure dissection and calculations for the subproblems you used in your solution.

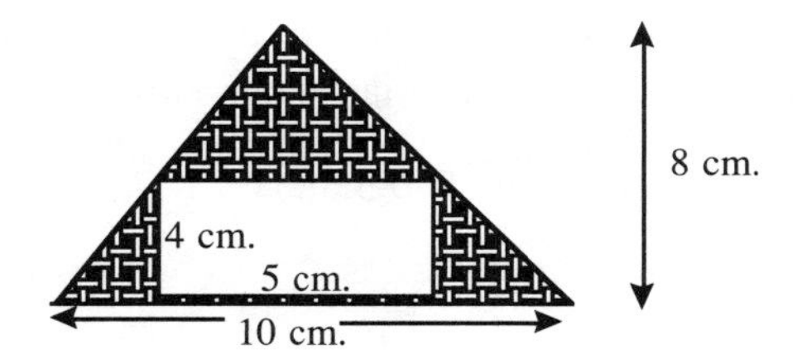


RC-58. In the figure at right, draw an additional height and then solve each problem.

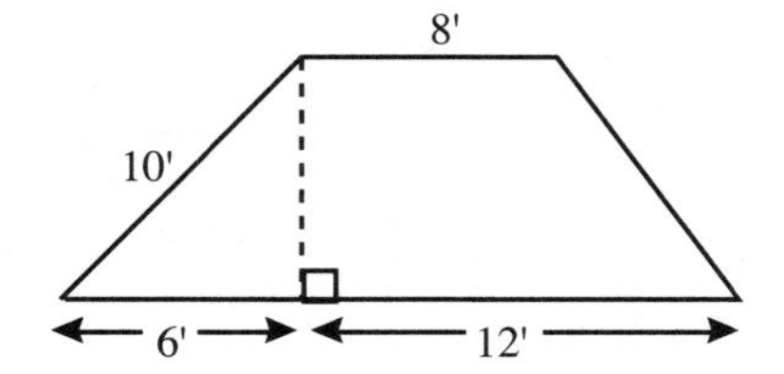


a) Find the perimeter of the trapezoid. Drawing a second height will be helpful.

b) Find the area. What information do you need in order to solve this part? Show a diagram of any figure dissection you use for the subproblems.

RC-59. Based on the given dimensions, label the rest of the figure's segments. Assume all angles are right angles.

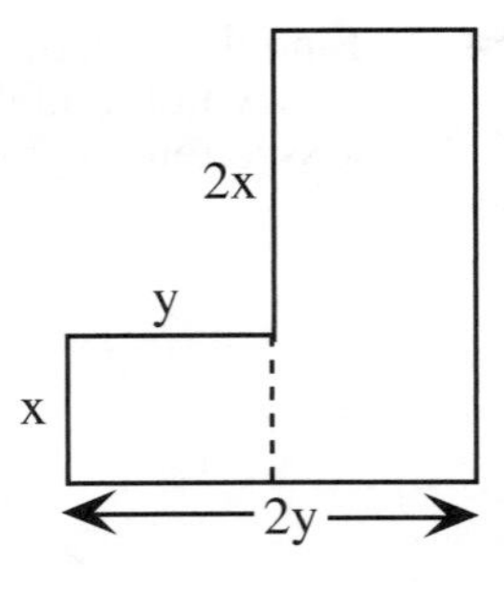


a) Write an algebraic expression to represent the perimeter.

b) Write an algebraic expression to represent the area. Show any figure dissection you use.

RC-60. Label the missing dimensions from what is given. Assume all the corners are right angles.

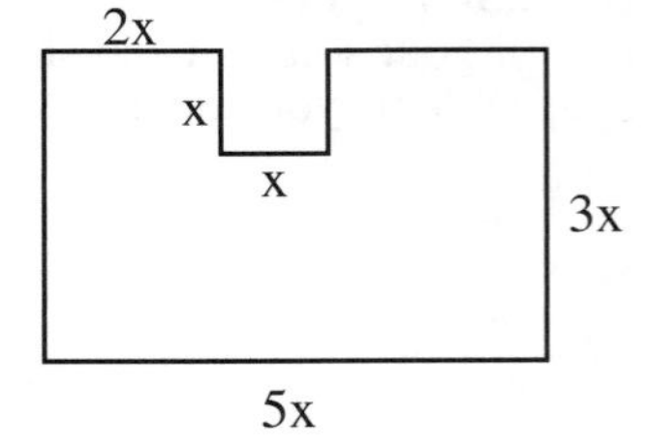


a) Write an algebraic expression for the perimeter.

b) If the perimeter is 72 feet, find x and the actual dimensions. To do this write an equation and solve it.

c) Write an algebraic expression for the area. Show any figure dissection you use to get the subproblems.

d) If the area were 350 square feet, what would the perimeter be?

RC-61. Calculate the area of each figure. All dimensions are in inches. Express your answers in the appropriate square units.

a) parallelogram

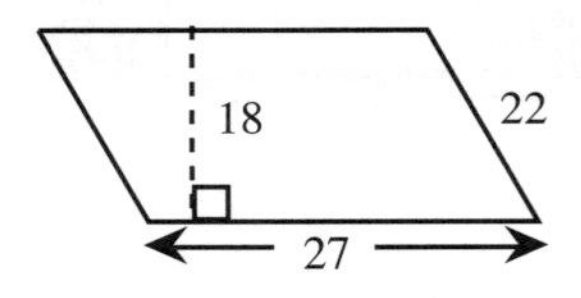


b) trapezoid

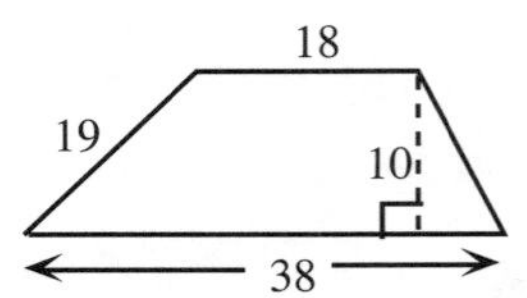


c) triangle

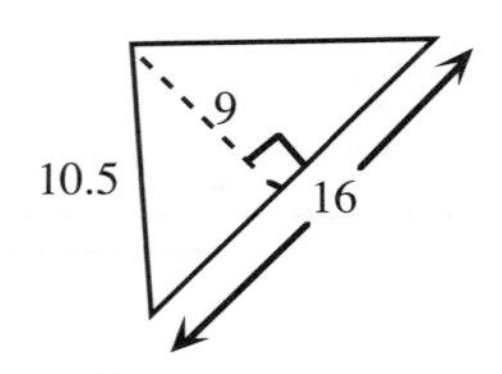


d) rectangle

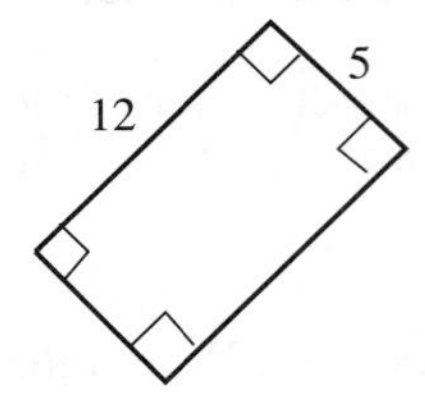

RC-62. Using any type of graph paper, draw a pair of xy-axes and scale them for $-6 \le x \le 6$ and $-10 \le y \le 10$. Complete the table below, substituting the x-values (inputs) provided, to find the y-values (outputs) for $y = x + 2$. Plot the resulting points and connect them.

$y = x + 2$

| x (input) | -4 | -3 | -2 | -1 | 0 | 1 | 2 | 3 | 4 |
|---|---|---|---|---|---|---|---|---|---|
| y (output) | | | | | | | | | |

RC-63. Solve the following equations for y (get the y variable by itself). Show the steps leading to your solution.

a) $y - 3x = -3$ b) $y + 4x = 5$ c) $3y + 2x = 9$

RC-64. According to the graph at right, how much would it cost to speak to an attorney for 2 hours and 25 minutes?

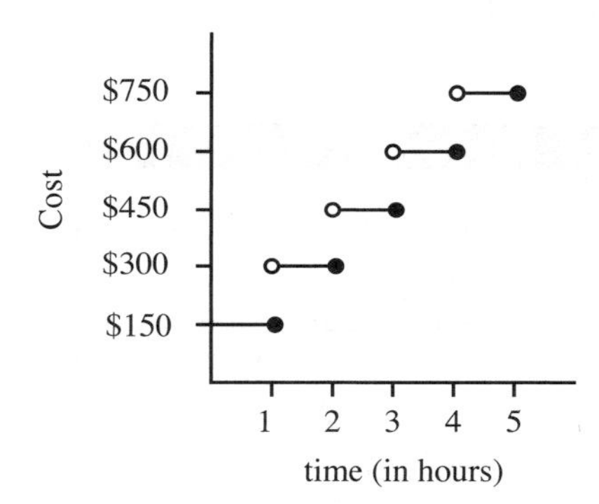


RC-65. What is the probability of drawing each of the following cards from a standard playing deck? A standard deck of cards includes the following: 52 cards of four suits. Two suits are black: clubs and spades; two are red: hearts and diamonds. Each suit has 13 cards: 2-10, ace, jack, queen, and king.

a) A queen b) A heart c) The queen of hearts

HUMAN RESOURCES

In PZL-6 I listed all the printed resources that are available for help. I forgot to mention another kind of resource -- the most important one--YOU and your study team members!

Parts of the CPM textbook are written for you to work with other students, and the authors assume that you will have other students with you so that you can talk about what you are doing, ask questions, and get immediate help from your partner or teammates. They also believe (and have read research that agrees) that talking about the math you are doing will help you learn it--even if you are the so-called "smartest" person in the class.

**>>>Problem continues on the next page.>>>**

And be careful! Don't fall for traps like copying another student's work, visiting with friends during work sessions, or avoiding working with others altogether. Remember what I said at the beginning of this unit: YOU are responsible for and will determine what you learn in this class.

Also be sure to GET THE PHONE NUMBERS OF STUDENTS IN YOUR TEAM so that you have someone to talk to when you are doing work out of class. You may want to form study pairs with a friend who is also taking this course or study teams to work after school, in the evening, and/or on the weekend. The point is that there is plenty of help available if you take advantage of it.

**MISSION POSSIBLE** (something to do)

Discuss with your partner or teammates several ways for everyone to contribute to the daily classwork. Write a brief summary of your conclusions. When you are finished, continue with today's lesson.

## LINEAR EQUATIONS

RC-66. **DIRECTIONS**: There are many possible relationships between x and y. We will explore several families or types of equations and their graphs during the year. For now, we will write all of the equations of lines--called **linear equations**--in **y-form** ($y = mx + b$) so that it is easy to substitute (input) values for x and find each paired y-value (output).

On a sheet of quarter inch graph paper, draw a pair of axes on the <u>upper</u> <u>half</u> of the page and scale them for $-9 \le x \le 9$ and $-9 \le y \le 9$. For each equation, make a table like the model below, use the x-values (inputs) provided and find each paired y-value (output). Then graph the equation by putting the points on the graph and drawing a straight line through all of them. After you graph it, write the equation on the line. Keep the lower half of the graph paper clear for problem RC-68. It is important that you graph these lines carefully; they will be used again tomorrow.

Read the next box for the meaning of "..." and add it to your tool kit.

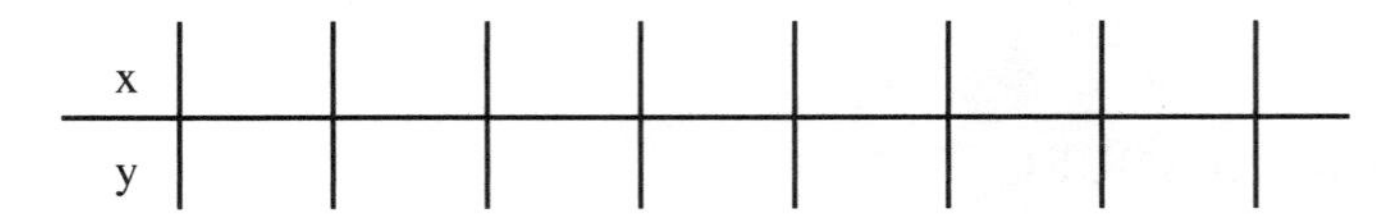

| x | | | | | | | |
|---|---|---|---|---|---|---|---|
| y | | | | | | | |

The symbol "..." is called an **ELLIPSIS** mark and indicates that values have been intentionally omitted to save time and/or space. Missing values follow the same pattern as those values that are listed.

Example: {3, 3.5, 4, ..., 8.5 ,9} means that 4.5, 5, 5.5, 6, 6.5, 7, 7.5, and 8 have been omitted from the <u>listing</u>, but are <u>included</u> in the <u>set</u> of numbers to be used in the problem.

a) $y = x + 2$ for $x = -8, -6, ..., 4, 6.$

b) $y = -2x + 6$ for $x = -1, 0, ..., 6, 7.$

c) $y = \frac{1}{2}x - 2$ for $x = -8, -6, ..., 6, 8.$

d) Name the polygon formed by segments of the three lines. Approximate its area.

RC-67. Make a table with headings as shown at right.

| Equation | Coordinates Where Graph Crosses y-axis |
|---|---|
| | |
| | |
| | |
| | |

a) Fill in the table using the equations from the preceding problem. Carefully examine the information in the table and discuss with your team anything you notice.

b) Use the outcome of your discussion to write a sentence that explains how to determine where the graph of a line crosses the y-axis without graphing it.

c) Add the following term to your tool kit.

> The point where the graph crosses the y-axis is called the **Y - INTERCEPT.**

RC-68. On the bottom half of the graph paper you used for problem RC-66, draw and scale a similar pair of axes. See the directions for problem RC-66 for scaling information.

a) Given the equation $y = 2x + 5$, use your conclusion from the previous problem to find and plot one point of this line, namely, the y-intercept.

b) How many other points are needed to graph the line? Recall your results from the introductory activity today.

c) Choose any x-value other than zero and find its corresponding y-value. Plot this point and draw a line through the two points. This method is called the **two-point method** of graphing linear equations.

d) Which method do you prefer, graphing just two points or a complete table of values?

RC-69. On the same axes you used in the previous problem, neatly graph each linear equation using the two-point method.

a) $y = 3x - 3$  b) $y = -\frac{2}{3}x + 3$  c) $y = -4x + 5$

RC-70. Draw a pair of xy-coordinate axes with $-8 \le y \le 8$ and $-8 \le x \le 8$ on quarter inch graph paper.

a) Find the length of the line segment between A(1, 3) and B(7, 8). Remember the Pythagorean Theorem.

b) Find the distance between C(-7, 7) and D(-3, -3). (Remember that we could represent the length of $\overline{CD}$ as just CD.)

c) Given E(-7, -5) and F(9, -8), find EF.

RC-71. Evaluate the following expression for k = 2.

$\frac{1}{4}k^5 - 3k^3 + k^2 - k$

RC-72. Solve each equation for y.

a) $x + y = 16$

b) $2x - 3y = 6$

## FINDING SLOPES

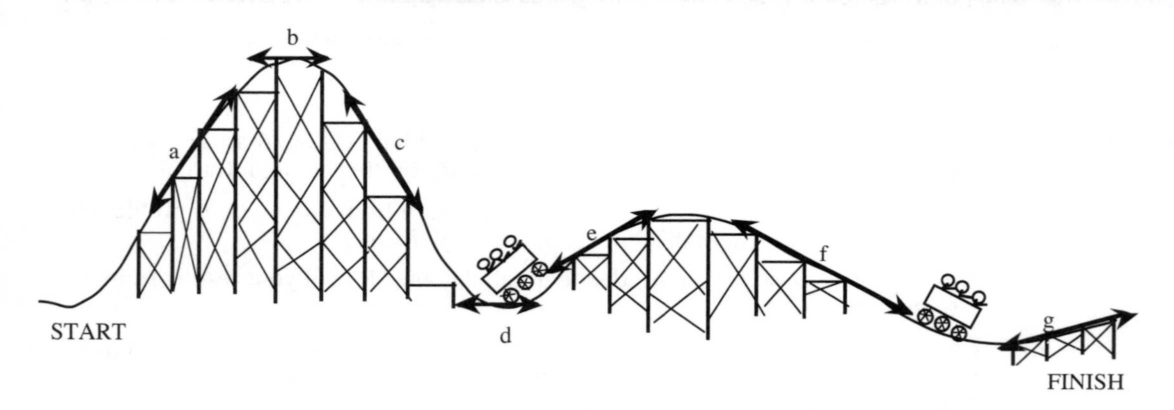


RC-73. **DIRECTIONS**: In this activity you will investigate the mathematical notion of **slope**. The above diagram represents a roller coaster track. As you travel from left to right, the steepness of the track requires different amounts of "effort" to be exerted on your car. It takes the greatest effort to go up a steep hill.

> For the purposes of this discussion, we will use the notion of effort to help you understand the relationship between the slope of a line and the number used to describe it.

a) How would you describe the "effort" it takes to go down a hill? What kind of slope number would you give it?

b) Suppose we assign an effort (slope) number of 5 to location a. Assign reasonable numbers to the remaining letters and arrange the lettered sections of track in order from largest to smallest assigned number.

c) Discuss the results of part (b) with your team. Write your conclusions about the relationship between the size of the slope number and the actual steepness of the line.

RC-74.

**DIRECTIONS**: The notion of effort is fine to help you understand the general idea of slope, but we need a method to find slopes so that everyone will agree on the number we assign to a specific line for its slope.

To find the slope of a line such as $l$, pick any two points on the line. The line segment between these points is the hypotenuse of a right triangle as shown at right. The right triangle is called a **SLOPE TRIANGLE.**

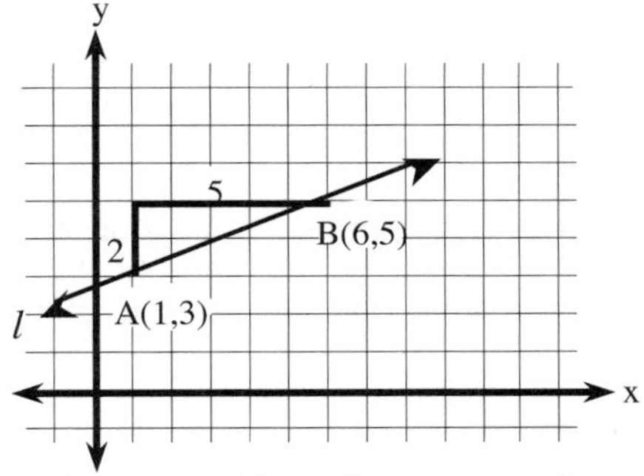


In this example, points A and B and $\overline{AB}$ are used.

a) Copy line $l$ on a sheet of graph paper, then extend it so that you can draw a larger slope triangle using the same point A(1,3) and C(11,7). Draw and label the legs with their lengths.

Add the following definitions to your tool kit.

**SLOPE** is a number that indicates the steepness (or flatness) of a line, as well as its direction (up or down) left to right.

**SLOPE** is determined by the ratio: $\frac{\text{vertical change}}{\text{horizontal change}}$ between <u>any</u> two points on a line.

For lines that go **up** (from left to right), the sign of the slope is **positive.** For lines that go **down** (left to right), the sign of the slope is **negative**.

b) In the above example the slope would be $\frac{2}{5}$ since there is a vertical change of 2 and a horizontal change of 5. Convince yourself by counting the vertical and horizontal changes. Then verify that you get the same slope if you use $\overline{AC}$ as the hypotenuse of your slope triangle.

c) Draw another slope triangle on your paper <u>below</u> $\overleftrightarrow{AB}$. Is the slope the same as in part (a)? Show why or why not.

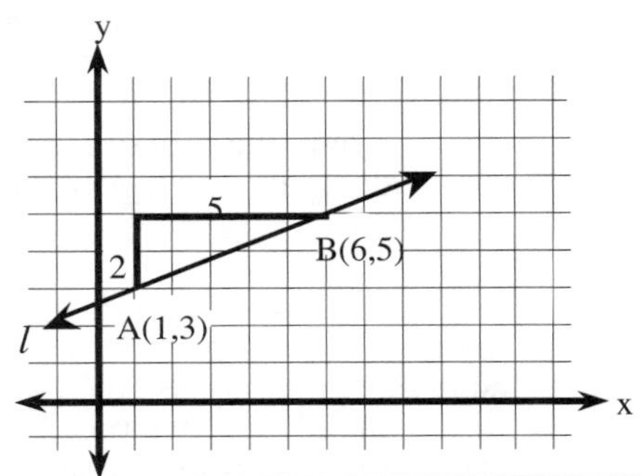

RC-75. Suppose you pick A and B on line $m$ shown at right and draw your slope triangle as indicated. The slope is $-\frac{2}{3}$.

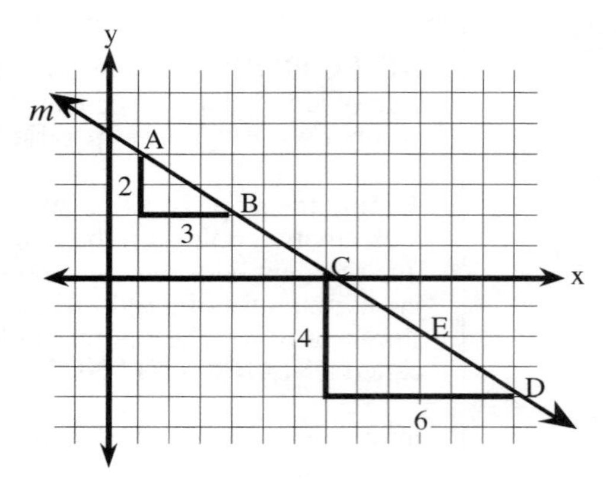


a) If you decide to pick points C and D on line $m$ and draw your slope triangle using them will you get the same slope? Why?

b) Another person in your team wants to use points A and E. Will these give the same slope? Explain.

RC-76. Arvilla thinks that if you pick other points on lines $l$ and $m$ to make a bigger triangle, you will get a larger number for the slope. Use the figure in either of the preceding problems or draw one of your own and use it to explain to her why she is mistaken.

RC-77. Use the figure in problem RC-75.

a) Suppose you start at A and go to B by following the path along the two legs of the right triangle. What direction (+ or -) do you go in the vertical (y) direction? the horizontal (x) direction?

b) Since slope is the ratio of vertical change to horizontal change, that is, $\frac{\text{vertical}}{\text{horizontal}}$, what is the sign when you divide two numbers and one is negative and one is positive?

c) Does your result in part (b) change if you travel from B to A along the legs?

d) Explain why lines that go down left to right always have a negative slope.

e) Use the figure in problem RC-74 and follow the path along the two legs to go from A to B. Explain why this demonstrates that lines that go up left to right always have positive slope.

f) Would the sign of the slope change in problem RC-74 if you followed the path along the two legs from B to A? Explain.

RC-78. Using the Slope resource page provided by your teacher, find the slope of each line by drawing a slope triangle and counting the vertical and horizontal changes. Be sure that you only use the points that are highlighted by dots. As an example, the line a already has two possible slope triangles drawn. Note that you will need the results of this problem again soon. Keep the resource page with your work in your notebook.

RC-79. Find your graphs of $y = x + 2$, $y = -2x + 6$, and $y = \frac{1}{2}x - 2$ from problem RC-66. Complete parts (a) through (d), then add the information in the box below to your tool kit.

a) Draw a slope triangle to find the slope of each line on your graph. Start with the point where the line crosses the y-axis and use the y-axis as the vertical leg of the right triangle. Choose your other point where the line has whole number (integer) coordinates. There are several for each line if your graphs are neat.

b) Make a three column table with the headings, Equations, Slope, and Coefficient of x and fill it in with the equations above. (Example: for 2x, 2 is the **coefficient** of x. Also note that integer slopes like "2" can be written as the ratio $\frac{2}{1}$.

c) Compare the slope you found for each equation to the coefficient of x . Write a sentence stating your observation. d) What is the slope and the y-intercept of each line represented by the following equations? Note: use your results from previous problems! You should not need to graph them.

1) $y = 3x - 4$

2) $y = -4x + 3$

3) $y = -\frac{1}{4}x + 5$

4) $y = x - \frac{3}{5}$

> Any linear equation written as **y = mx + b**, where m and b are any real numbers, is said to be in **SLOPE-INTERCEPT FORM**.
>
> m is the **SLOPE** of the line.
>
> b is the **Y-INTERCEPT**, that is, the point (0, b) where the line intersects (crosses) the y-axis.

RC-80. Write a clear explanation of how to use the slope and y-intercept to graph a linear equation such as $y = \frac{4}{7}x + 2$. Tell where to start and then what to do.

**RC-81.** Test your method from the previous problem to graph $y = \frac{2}{3}x + 1$ on the grid provided on the bottom half of today's resource page. If you are stumped, try the method below.

a) Name the slope and y-intercept for $y = \frac{2}{3}x + 1$.

b) Mark the y-intercept on the y-axis.

c) Since $\frac{2}{3}$ is the ratio of the vertical change to the horizontal change, and since the slope is positive, count up the y-axis 2 units from (0, 1), then go right 3 units. Mark this point. Verify that it is (3, 3).

d) Draw a line through (0, 1) and (3, 3). This is the graph of $y = \frac{2}{3}x + 1$.

RC-82. Use the slope-intercept method of graphing to graph the four linear equations in part (d) of RC-79. Graph them on the bottom of today's resource page. Be sure to read the box below.

> All three methods of graphing linear equations--making a table, two point, and slope-intercept--produce the same results for any line you graph. For linear equations written in the y-form, the two-point method is usually adequate and faster than plotting several points. The slope-intercept method is fastest when the values of m and b are reasonably small.

RC-83. Use the second graph on the bottom of today's resource page for parts (a) and (b) below.

a) Graph each equation by the method of your choice.

1) $y = -2x + 8$ 2) $y = x + 8$

b) Find the base and height, then calculate the area of the triangular figure enclosed by the intersection of the two lines in (a) and their intersections with the x-axis.

RC-84. Solve each equation for y.

a) $3x - 2y = 10$ b) $5y - 6x = 10$

RC-85. **DIRECTIONS**: Suppose we took three points A(1, 1), B(3, 4), and C(-4, -4) and wanted to know if they lie on the same line. You could plot the three points and try to draw a straight line through them and see what happens. This procedure is not always convincing, since it is easy to be slightly inaccurate when sketching graphs, so consider the following question, then complete the other parts below:

a) For any line in general, does the slope of a line change anywhere on the line?

b) Plot the three points on quarter inch graph paper. Draw a slope triangle from A
to B and find the slope. Draw a slope triangle from C to A and find the slope. Are the slopes the same? Carefully extend the line through A and B. Do the same through A and C so that both lines extend across the entire grid.

c) Do all three points lie on the same line? Why or why not?

d) Try point D(-3, -5) instead of point C. Draw a slope triangle from D to A, find the slope and compare it to the slope from A to B. Are the slopes the same?

e) Are points A, B, and D on the same line? Explain why or why not.

RC-86. The figure at right shows two lines, $p$ and $q$. Remember, the general form of any line in the slope-intercept form is $y = mx + b$.

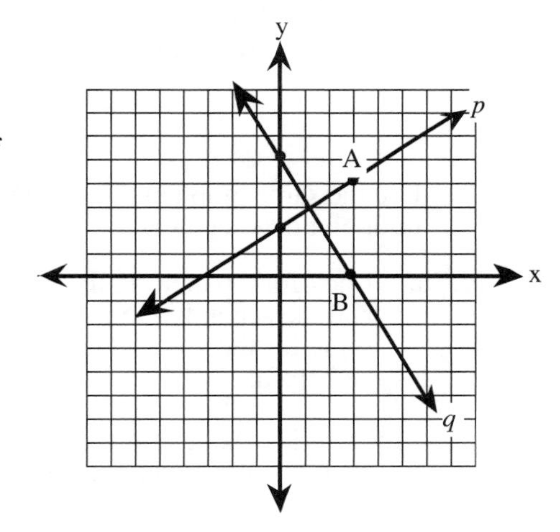


a) Use the graph to find the y-intercept of line $p$, then form a slope triangle from the y-intercept to A and write its slope.

b) Write the equation of line $p$ in slope-intercept form.

c) Repeat parts (a) and (b) for line $q$ using point B.

RC-87. Use the resource page for problem RC-78 and write the equation of each line, a through e, in slope-intercept form.

**RC-88.** Draw a pair of xy-coordinate axes with $-10 \le x \le 10$ and $-10 \le y \le 10$ on quarter inch graph paper.

a) Plot each trio of points and draw a line segment through the three points.

1) (0, 4), (4, 2), and (8, 0).

2) (-2, 2), (0, 5), and (2, 8).

3) (0,- 3), $(2\frac{1}{2}, -1\frac{1}{2})$, and (5, 0).

b) In each case, what appears to be true about the location of the second point?

c) Katie is skeptical about your observation. Convince her (prove) that you are correct.

d) Add the following definition to your tool kit.

> The second point in each set divides the segment into two equal parts (segments). The point is known as a **MIDPOINT**.

RC-89. Use the same graph in the previous problem to help devise a method to calculate the midpoint of a segment.

a) Look for a pattern in the relationship between the first and third x-coordinates and the second x-coordinate in each set of points in the previous problem. Do the same for the y-coordinates in each set of points. Drawing in right triangles might be helpful.

b) Describe the relationship you see. Suggest a method for finding the midpoint of a segment if you know the coordinates of the two endpoints.

c) Plot these pairs of points and use your method from part (b) to find the coordinates of the midpoint of the segment that connects them.

1) (0, 2) and (4, 0).

2) (-6, 2) and (-2, 6).

3) (-2, -3) and (6, -7).

RC-90. Count the vertical and horizontal changes to find the slope of each line segment. Copy them onto your graph paper if you choose to draw slope triangles.

a)

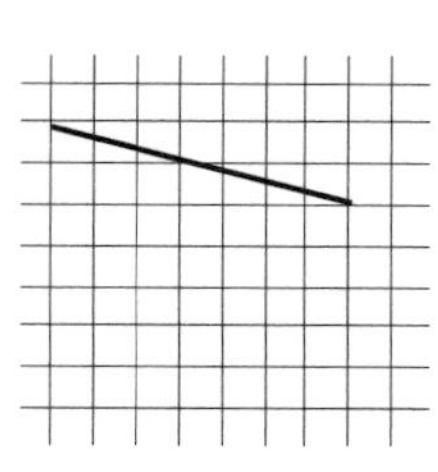

b)

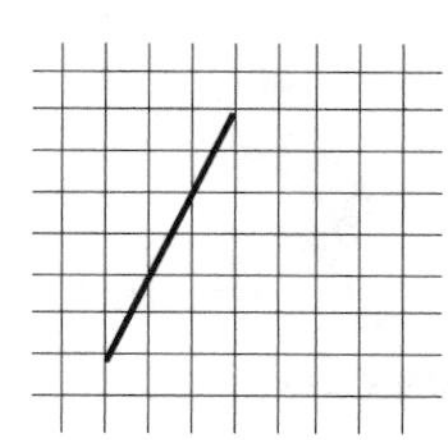

c)

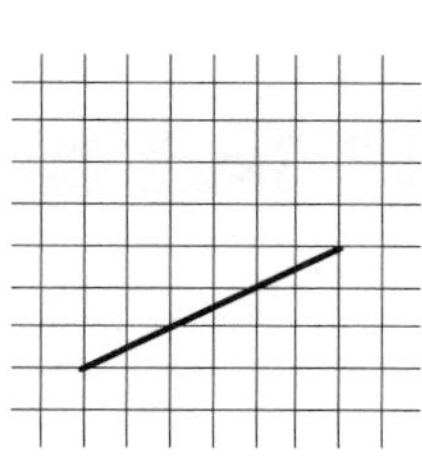

d)

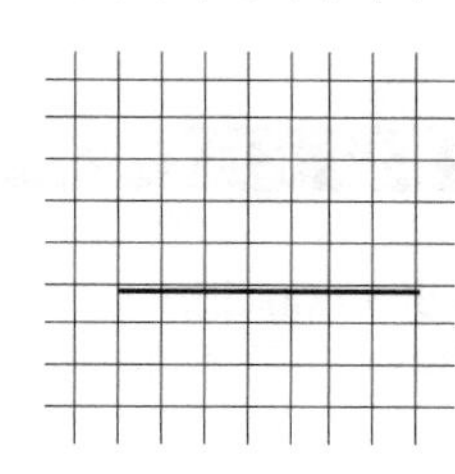

RC-91. Write the equation of the line in slope-intercept form if:

a) $m = \frac{6}{5}$ and b = (0, -3).

b) $m = -\frac{1}{4}$ and b = $(0, 4\frac{1}{2})$.

c) $m = \frac{1}{3}$ and the line passes through the origin (0, 0).

d) m = 0 and b = (0, 2).

RC-92. Find the length of the segment, the coordinates of its midpoint, and the slope of the line containing the segment for the two problems below.

a)

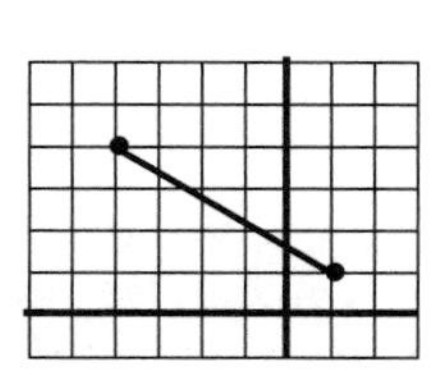

b) A(7, 2) and B(-1, -2).

RC-93. Find the area and perimeter of the figure at right.

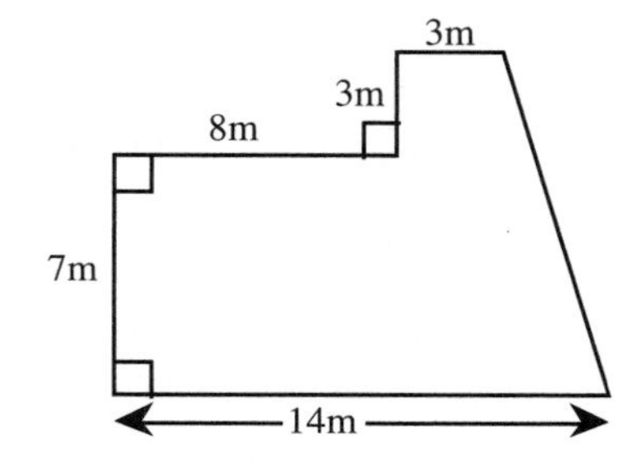

RC-94. Write each equation in slope-intercept form and state the slope and intercept.

a) $3x - 5y = 10$

b) $4y - 6x = 10$

RC-95. Multiply.

a) $3x(5x + 7)$

b) $(x + 2)(x + 3)$

c) $(x + 12)(x - 3)$

d) $(x - 13)(x - 2)$

## UNIT AND TOOL KIT REVIEW

RC-96. State under what conditions you would use the Pythagorean Theorem, that is, what do you need to have, have to know, and want to find?

RC-97. Copy the triangle at right and draw two different heights.

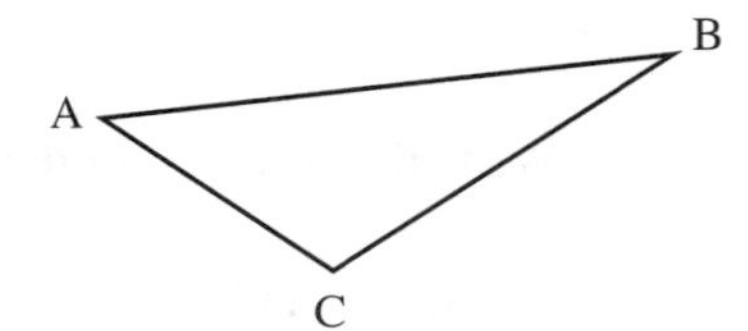


RC-98. Explain in words why a parallelogram and a rectangle BOTH have the same area formula, A = bh. Then explain what is different about how you use the formula with each figure.

RC-99. Locate the points A(-2, 5) and B(4, - 4) on a set of axes scaled for $-7 \le x \le 7$ and $-7 \le y \le 7$.

a) Carefully draw the line segment connecting the points. Write a step by step explanation of how to find the length of this line segment. Be clear enough so that a new student who just transferred into the class will understand what to do. Then find its length.

b) Explain, step by step, how to find the slope of the line that contains the segment. Then find the slope for $\overleftrightarrow{AB}$.

c) Explain how you would find the coordinates of the midpoint of $\overline{AB}$. What are they?

d) If both ends of the line segment were extended, what would be the equation of the line? Explain how to figure out the equation.

e) Finally, find the area and perimeter of the right triangle you drew for part (a).

RC-100. Scale a set of axes $-4 \le x \le 6$ and $-4 \le y \le 10$. Sketch a graph of each of the following equations. Use the slope-intercept method.

a) $y = -3x + 7$

b) $y = \frac{3}{5}x - 2$

RC-101. Use large grid graph paper to draw a trapezoid with vertices (-5, -1), (6, -1), (-3, 5), and (2, 5).

a) Draw a line which connects the midpoints of the two non-parallel sides. Notice that this line cuts the trapezoid into two trapezoids. How does the height of each trapezoid compare to the height of the original?

b) Use scissors to cut out the big trapezoid then cut along the line you drew to get two trapezoids. Now rotate the top trapezoid so it is upside down next to the bottom one, that is, fit two of the non-parallel sides together. You should have a long skinny parallelogram. Tape this figure on your assignment paper. What is its area?

c) Explain how the area of the long skinny parallelogram is related to the area of the original trapezoid.

RC-102. What if the original trapezoid in the preceding problem had height h instead of 6 units and bases $b_1$ and $b_2$? Using h, $b_1$, and $b_2$ explain why the area of the parallelogram is the same as the area of the original trapezoid.

RC-103. Use the given information to find the missing dimensions. All angles that look like right angles are right angles. Find the shaded area for (a) and (b), then write an algebraic expression for the area in (c). Show your figure dissections (subproblems). State your units of measure.

a)

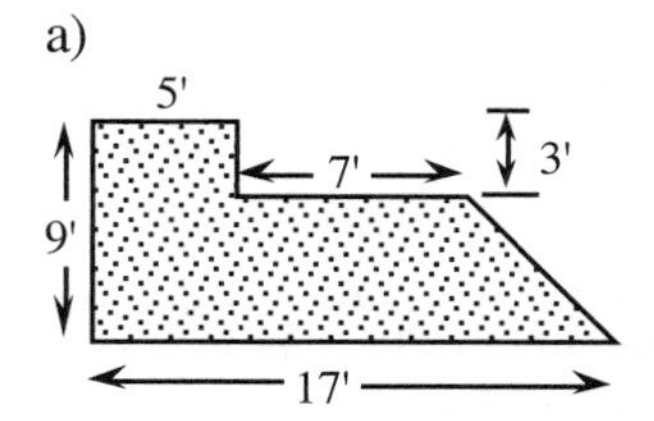


b) Lower regions: squares.

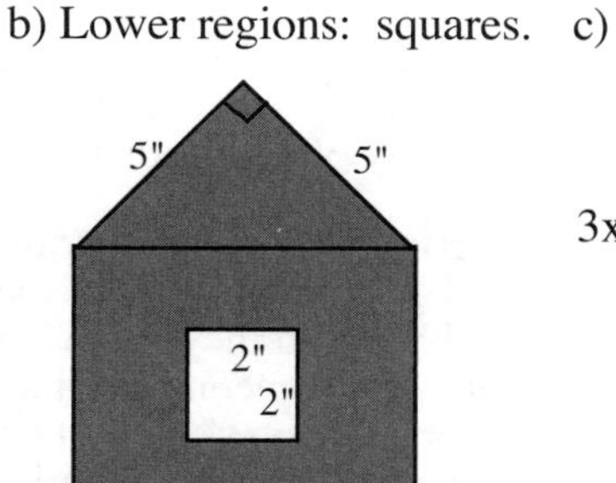


c)

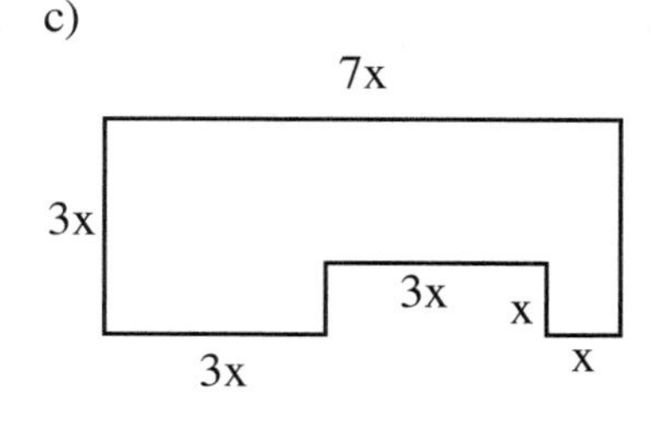

RC-104. Find the area of each figure.

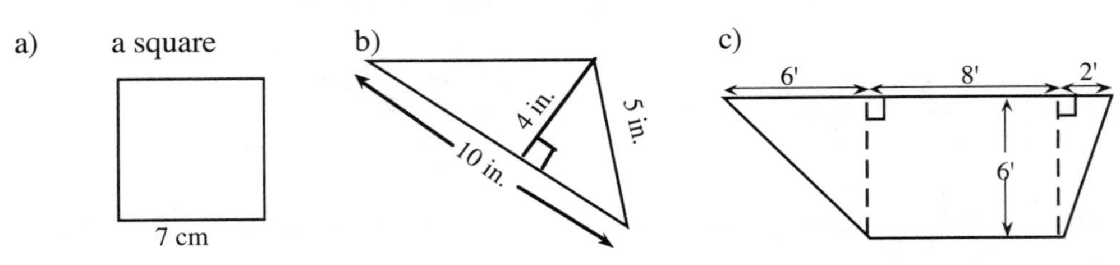


RC-105. Find the length of the missing side of the figure at right.

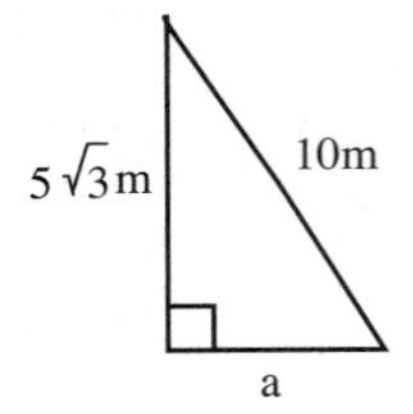


RC-106. Graph the following lines on the same set of axes. Label the lines by writing the equation of the line.

$y = -x + 3$ $\qquad$ $y = x + 3$ $\qquad$ $y = 0$

a) What figure is formed by the three lines?

b) Find the perimeter and area of the figure.

RC-107. Draw an accurate sketch of the face of a clock on graph paper. Connect the center of the clock to each number, 1 through 12, with a line segment. Use what you have learned about slope to write a number on each segment that approximates its slope.

RC-108. Write a team and learning reflection for Unit 1. Use complete sentences and write as honestly as you can. Use the questions below as a guide; you do not have to answer every one of them. Include any other information you like.

a) **Study Team Reflection:** Who was in your study team and how did your team work? Be as detailed as possible. What exactly did you do to help someone? Who helped you and how did they do so? What talents did you find you had? Describe a different talent someone else had. What was disappointing about your team? What are important team skills that help a team work effectively? What team skills do you personally want to work on in your next team?

b) **Learning Reflection:** What part of learning mathematics do you feel good about? What is still difficult for you? What learning are you proudest of? Did you do all your assignments? Did you participate in class discussions? What study skill do you want to improve in the next unit? Describe your plan to do so.

**REMINDERS:** WHERE TO GET INFORMATION and WHAT TO DO IF YOU FALL BEHIND

Before you start the next problem, go back and review PZL-6. Use the list you made to help assemble a COMPLETE tool kit for Unit 1 as directed in the next problem.

By the way, some of you may have had a hard time getting started with this course and are beginning to fall behind. I have good news for you! The information contained in PZL-6 tells you where to look to catch up quickly. But don't wait! You know that math courses build on each unit. Take the time NOW to use the tool kit check-up to guide you to what you should know. Do enough work with these ideas to get a basic idea of them. You can go back and do the rest of the problems to practice them later, if necessary. Finally, I hope one of your goals in the preceding learning reflection was to do your work regularly. The authors designed the homework in this course to be reasonable in terms of both time and degree of difficulty. It must be done daily, though, since it helps you develop the ideas introduced during class.

RC-109. TOOL KIT CHECK-UP

Your tool kit contains reference tools for geometry. Return to your tool kit entries. You may need to revise or add entries.

Be sure that your tool kit contains entries for all of the items listed below. Add any topics that are missing to your tool kit NOW, as well as any other items that will help you in your study of geometry. Terms are listed in the order they appear in the unit.

- right triangle, leg, hypotenuse
- Pythagorean Theorem
- polygon
- perpendicular
- parallel
- perimeter
- area
- base
- height
- line segment and length
- subproblems
- figure dissections
- ellipses
- y-intercept
- slope
- slope-intercept form of a line
- midpoint of a line segment

Algebra Review: solving and graphing equations, simplifying expressions, multiplying monomials and binomials.

# Convincing Your Team

BEGINNING PROOF

# *Unit 2 Objectives*

## *Convincing Your Team:* BEGINNING PROOF

This unit uses games and puzzles to help you practice and develop your ability to present persuasive arguments to justify conjectures you and your team develop. The problems give you an opportunity to discuss solutions, then write clear, step by step explanations that prove that your solution is correct.

The algebra focus extends work with linear equations from Unit 1 to consider systems of equations--usually two equations at a time--and how to solve them graphically and algebraically.

In this unit you will have the opportunity to:

- construct logical arguments to justify your solutions to logic games and simple number theory principles.
- write conjectures based on observations and patterns in games and problems.
- learn how to solve systems of linear equations algebraically using the addition and substitution methods.
- learn how to solve and graph single variable inequalities.
- practice solving word problems by drawing diagrams.
- review and consolidate the algebra topics from the first two units: solving and graphing equations, simplifying expressions, and multiplying monomials and binomials.

*Read the problem below. During this unit you will study what is needed to solve it. Do not attempt to solve it now.*

**BP-0.** Three boys were comparing their house numbers. Each house number contained three digits. It turned out that the sum of the digits of each three-digit number was the same.

1) None of the numbers contained any digits from another number.
2) None of the numbers began with a 4.
3) Bill's number was 252.
4) Ben's number began with a 6.
5) Bart's number ended with a 1.

Use these conditions to find two possible house numbers for Bart's house. List your reasons.

*UNIT 2*

ALGEBRA
GRAPHING
RATIOS
GEOMETRIC PROPERTIES
PROBLEM SOLVING
SPATIAL VISUALIZATION
CONJECTURE, EXPLANATION (PROOF)

# *Unit 2*

## *Convincing Your Team:* BEGINNING PROOF

## LOGICAL ARGUMENTATION, PART 1

HOW TO PREVIEW A UNIT

Your literature or social studies teacher may have suggested that before you read a story or a passage in your history text, you skim it quickly, looking for main ideas. You can do the same thing with your geometry textbook by using the ways the authors have highlighted important information. Take a few minutes--no more than five--to read the list of what you will learn in this unit (see the opposite page), then flip through the unit and skim each single and double line box. Finally, each person in the team should take a topic in bold print from one of the boxes, look it up in the index, and verify that the problem number with the box matches the problem number listed in the index. Once you have done all of this--remember, no more than five minutes, so work quickly--go on with today's lesson.

BP-1.

**INTRODUCTION**: This unit begins the development of the idea of mathematical proof used in this geometry course. The main goal is to encourage you to make educated guesses (that is, based on data, patterns, relationships, etc.), which mathematicians call **CONJECTURES** and scientists call **hypotheses**, and to justify your reasoning. **JUSTIFY** means to use facts, definitions, rules, and/or previously proven conjectures in an organized sequence that convinces your audience that what you claim (or your answer) is valid (true). When you have done this, you have created a **logical argument** called a **proof**. The domino sketch at right will indicate problems in the course that focus on logic and proof.

We will play some games and do some activities which are designed to introduce you to constructing logical arguments to validate your conclusions.

**Read the directions on the next page for the first game.**

## THE DIGIT PLACE GAME

This game is the number version of the logic game called Master Mind.

OBJECTIVE: The object of the game is to figure out a secret number using as few guesses as possible.

RULES: One person thinks of a two-digit number. (We'll use a three-digit number later.) The other players try to figure out what the secret number is by using some very specific clues. No digits are repeated within the secret number and no number begins with 0. A player makes a guess and the person with the secret number responds with two values. The first value (Digit) stands for the number of correct digits in the guess. The second value (Place) stands for the number of correct digits in the guess which are in the correct place. The players continue to make guesses/conjectures and receive responses until they know what the secret number is. YOU SHOULD NOT SAY THE NUMBER, YOU ONLY NEED TO KNOW IT.

PURPOSE: Remember that our purpose is to give logical arguments to support conjectures, so be prepared to explain why you think your conjectures are reasonable.

BP-2. Suppose you are playing the Digit Place game and the first guess is the following:

| Step | Guess | Digits | Place |
|---|---|---|---|
| 1 | 13 | 1 | 0 |

a) Could the answer be 31? Explain to other members of your team why or why not.

b) Could the answer be 15? Explain why or why not.

c) Which would give you more information, a guess of 37 or a guess of 76? Justify your choice. What information does your choice give you?

BP-3.

| Step | Guess | Digits | Place |
|---|---|---|---|
| 1 | 34 | 0 | 0 |
| 2 | 42 | 1 | 0 |
| | | | |

a) Explain why we know that the solution to this problem is in the 20's.

b) Are there any numbers in the 20's that can't be a solution? Explain to the other members of your team which number(s) and why.

c) Why would 51 be a better third guess than 20? Write a complete sentence explaining your reasons.

BP-4. Statement W: **All trees are green.**

ASSUME STATEMENT W IS TRUE.
Justify your response to each of the following questions using complete sentences.

a) Does this statement mean that an oak tree must be green? Explain why or why not.

b) Does this statement mean that anything green must be a tree? Explain why or why not.

c) Are the statements "All trees are green" and "All green things are trees" saying the same thing? Explain why or why not.

BP-5. Nalini and Ravi are playing the Digit Place game. Ravi is thinking of a two-digit number. Nalini's first guess is 91. Ravi responds D = 0 and P = 0. If Nalini is an expert Digit Place game player, what can she conclude? Explain your conclusion in one or two sentences.

BP-6. Find the perimeter and area of each figure. Remember that all answers should be rounded to two decimal places.

a) square

12.5'

b) rectangle

75'

250'

c) parallelogram

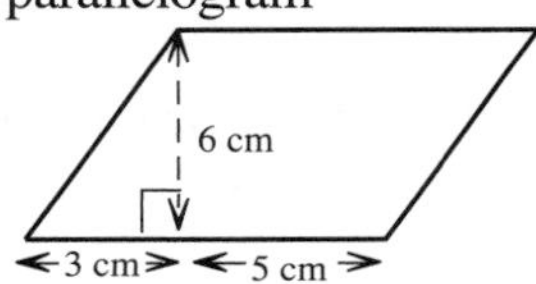


BP-7. For each pair of points, plot the points, draw the slope triangle, and determine the slope.

a) (-3, -4) and (-5, 0)  b) (2, 8) and (6, 5)  c) (0, 3) and (4, 4)

BP-8. Write the equations of the lines with y-intercept (0, 2) and slopes:

a) 7  b) $-\frac{2}{3}$  c) 1.5

BP-9. Given the equation of the line $y = 2x - 3$, test each point to see if it lies on the line.

a) (1, -1)  b) (3, -2)

c) (-2, 1)  d) (0, -3)

BP-10 Draw two intersecting segments that satisfy <u>all</u> <u>three</u> of the following conditions simultaneously (at the same time). You will need the definition of **bisect** (below) for condition (2). Add it to your tool kit.

1) The two segments are equal in length.
2) Neither of them is bisected.
3) They are not perpendicular.

Example: **BISECT** means to divide something into two equal parts.

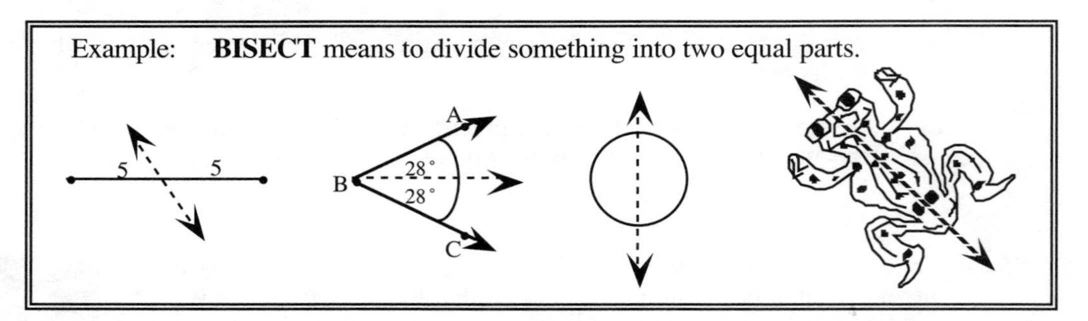


BP-11. The ratios of the sides of a triangle are 2 : 3 : 4. For example, the sides could be 4, 6, and 8. If the perimeter is 45 cm, find the length of the <u>shortest</u> side. The figure shows what is meant by sides in the ratio 2 : 3 : 4.

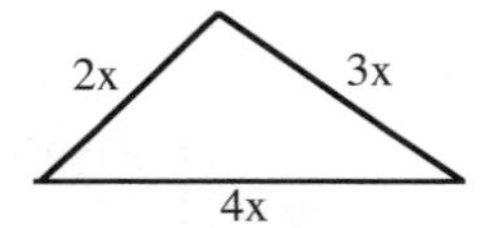


BP-12. Find the total area of each large rectangle in terms of x (remember, this means that x appears in your answer).

a)

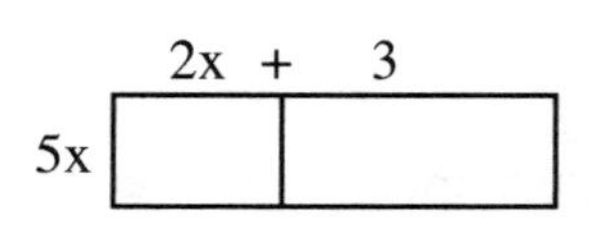


b)

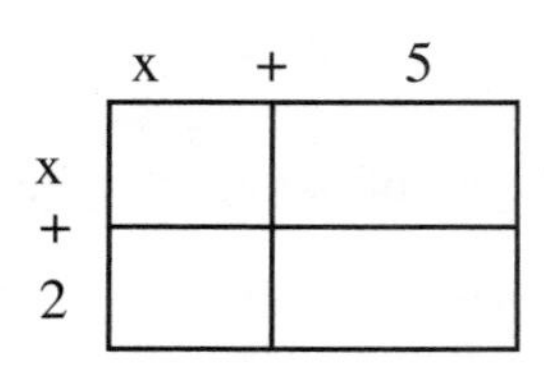


BP-13. Multiply and simplify.

a) $5x(2x + 3)$

b) $(x + 2)(x + 5)$

BP-14. Explain the relationship between your methods and answers in the two previous problems.

## HOW TO ASK YOUR TEACHER QUESTIONS

So far I have pointed out several resources to help you in this class: yourself, your teammates, and keys to information in the textbook. Notice that asking your teacher is suggested only after you have tried to solve a problem yourself. When you do ask your teacher a question, here is how to get the most out of your teacher. **BE PREPARED**.

When you ask the teacher a question, make it a good one. Do not ask, "Can you help me?" **BE SPECIFIC**. A better question will come after you work on a problem with your partner or team and use the resources I described previously. Ask questions like, "Mrs. Cho, I am having a problem with BP-16 part (b). My study team is having the same problem. We have tried re-reading the rules, reviewing other examples of the game, but cannot reach agreement about whether 3 is in the right place or not. We are not sure where to go next. What else could we try?"

Your teacher will probably respond to your question by asking you a leading question. A leading question is one that **makes you think,** but at the same time gives you a hint. **Pay close attention to your teacher's questions. They contain hints and sometimes information** that you can use to move forward.

As you work on today's and future lessons, try to improve the quality of the questions you ask your teacher. The more specific you are about what confuses you, the better questions and hints you will get in return. Now start your work on the next problem.

BP-15. What is wrong with the responses for digit and place in the table below? Explain your observation to other members of your team and record it in one or two sentences.

| Step | Guess | Digits | Place |
|---|---|---|---|
| 1 | 471 | 3 | 2 |

BP-16. Discuss your reasons and conclusions with your partner or team. Write them using one or two complete sentences.

| Step | Guess | Digits | Place |
|---|---|---|---|
| 1 | 321 | 1 | 1 |
| 2 | 309 | 1 | 1 |

a) After these two guesses could you conclude <u>for</u> <u>sure</u> that the 3 is in the hundred's place? Why or why not?

Continue the game:

| | | | |
|---|---|---|---|
| 3 | 456 | 1 | 1 |
| 4 | 342 | 2 | 0 |

b) What can you now conclude about the 3? Why?

c) In the fourth guess which two digits are correct? Why?

d) Since they are not in the proper place where do they belong? Justify your answer.

e) In the second guess, which digit is in the right place? Why?

f) What is the secret number? Be sure that your conclusion is logically supported by your work in parts (a) through (e).

BP-17. Discuss your reasons and conclusions with your partner or team. Write them using one or two complete sentences.

| Step | Guess | Digits | Place |
|---|---|---|---|
| 1 | 470 | 1 | 0 |
| 2 | 910 | 3 | 1 |

a) In the first guess, if 0 were one of the digits where would it have to go? Explain this conclusion in a sentence or two.

b) Can you solve the problem? If so, justify your solution. If not what would your next guess be and why?

BP-18. On graph paper, draw $\Delta ABC$ as it is shown below. Draw $\overline{JK}$ between J(11, 2) and K(20, 2).

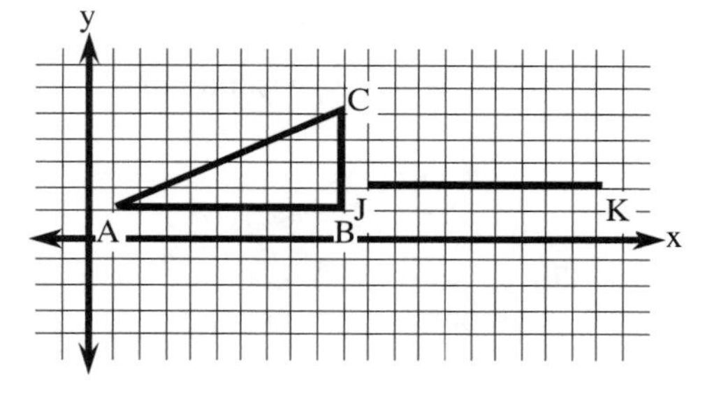


a) Plot point L so that $\Delta ABC$ is the same size and shape as $\Delta JKL$ .

b) Can you find more that one such point (i.e., is there another place you could put L so that $\Delta JKL$ is the same size and shape)?

c) Find the area and perimeter of $\Delta ABC$.

d) Find the area and perimeter of $\Delta JKL$.

e) Compare your results in parts (c) and (d). What do you notice?

BP-19. Match each of the descriptions (a) - (e) below with the most appropriate graph of TIME vs. TEMPERATURE, (1) - (5) below.

a) In February, you enter a cold house and turn up the thermostat to 72° Fahrenheit.

b) In July, you enter a hot house and turn on the air conditioner.

c) You put your hot spaghetti on your plate and then leave it to cool.

d) You put a cup of soup in the microwave and heat it for two minutes.

e) You put ice cubes in your lemonade and then drink it slowly.

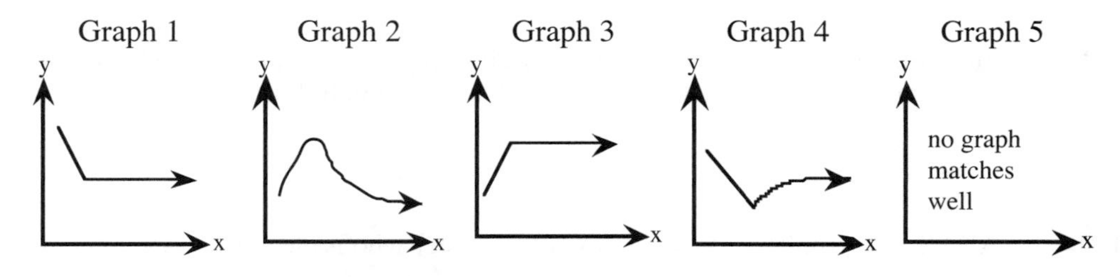


BP-20. Consider the following statements:

All deer like lettuce.
Gail likes lettuce.

a) Do these two statements imply that Gail is a deer? Why or why not?

b) Suppose the second statement said, "Gail is a deer." What, if anything, could you conclude?

BP-21. Draw two segments that satisfy all three of the following conditions simultaneously:

1) The two segments are not equal in length.
2) Only one of them is bisected.
3) They are perpendicular.

BP-22. Find the slope of each segment. Try to do the problem without redrawing the segments on graph paper.

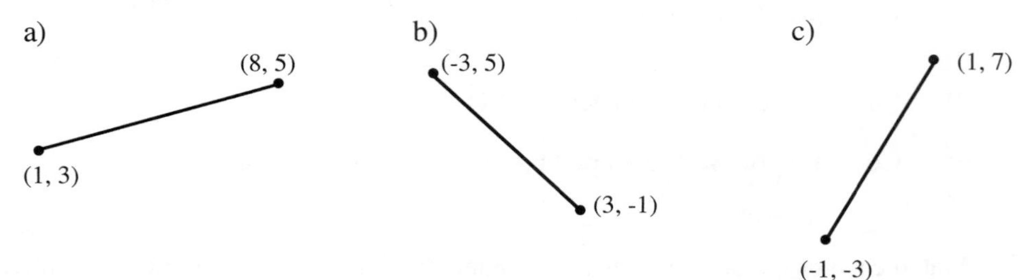


BP-23. Graph A(7, 16), B(4, 8), C(2, 3).

a) Does it appear that these three points lie on a line?

b) Use slope to show that these three points are not on the same line by comparing the slopes of $\overline{AB}$ and $\overline{BC}$. Draw the slope triangles on your graph for each segment.

c) Explain in a short paragraph why checking slopes will verify whether three points that LOOK as though they lie on a straight line actually do or do not.

BP-24. Alfred and Arnold are twins. They have a sister who is eleven years younger than them and an older sister who is four years older. The sum of the ages of all four siblings is 81. Find Alfred's age. Make a guess and check table or written an equation.

## LOGICAL ARGUMENTATION, PART 3

BP-25.

### THE COLOR SQUARE GAME

Read the rules for this game below.

<u>OBJECT</u>: Figure out the arrangement of colored squares on a 3 x 3 grid or a 4 x 4 grid <u>using as few clues as possible</u>.

<u>RULES</u>: In a 3 x 3 Color Square Game, each of the 9 squares are colored: 3 are red, 3 are green, and 3 are blue. However, all squares of the same color must be contiguous (linked along an edge). The diagram below shows you what is meant by contiguous.

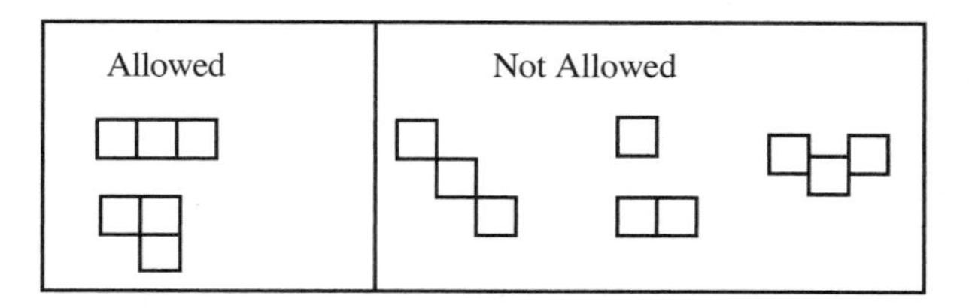


| B | B | G |
|---|---|---|
| B | R | G |
| R | R | G |

In a 4 x 4 Color Square Game there are 4 red squares, 4 green squares, 4 blue squares, and 4 yellow squares. Again, all of the squares of the same color are contiguous.

In order to get information, you ask, "What is in row ___?" or "What is in column ___?" You will then be given the total number of squares of each color in that row/column BUT not necessarily in the order in which they appear in the secret arrangement. For reference, rows are numbered from top to bottom and columns are numbered left to right.

BP-26. After two guesses you know the following information. Complete as much of this color square as you can and explain why you place each square where you do.

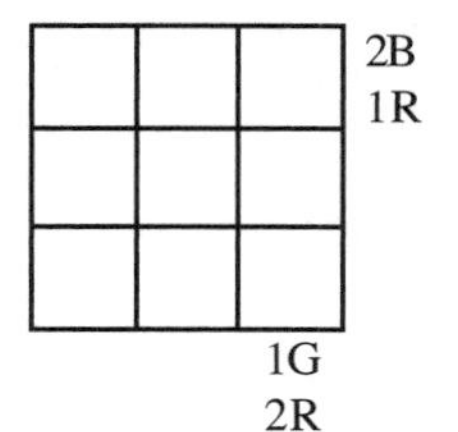

BP-27. Complete as much of this color square as you can from the clues given. Then complete (a) through (d) below.

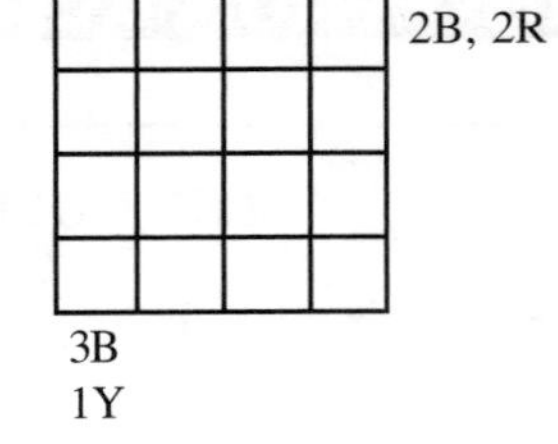

a) What must be in the upper left hand corner? Why?

b) Fill in the far left hand column and the top row.

c) There is one more box you can fill in with the information you have. Find it and write a sentence or two to justify your answer.

d) The third row has one blue, one green, and two yellow squares. Does this new information result in a unique (that is, exactly one) solution? If so, what is it? If not, list at least two possible solutions.

BP-28. Complete as much of this color square as you can from the clues given, then complete (a) through (d) below.

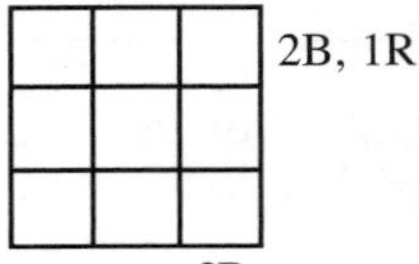

a) What color must go in the upper right hand corner? Why?

b) By the rules of the game, what colors must be in the top row and the far right column? (List them in order from left to right for rows and top to bottom for columns.)

c) Would you be able to put a green in the middle of the far left column? Justify your response using complete sentences.

d) Is there a unique solution to this problem? If so, record your solution. If not, show as many different solutions as you can.

BP-29. Statement G: **All of my students love geometry.**

ASSUME THAT STATEMENT G IS TRUE!

a) If Chantal is my student, what can be concluded?

b) If Ellie is my student, what can be concluded?

c) Complete the following: If you are my student, then ______________________________.

Read the box below and add the definition to your tool kit.

> You have just rewritten statement G as an **IF-THEN** statement. Another name for an IF-THEN statement is a **CONDITIONAL STATEMENT**.

d) If Lisa is <u>not</u> my student, is it possible for her to love geometry?

e) Tina loves geometry. What, if anything, can be concluded from the preceding statements?

f) Glen does not love geometry. What can be concluded?

BP-30. Peter thinks there is a unique solution to this Color Square game. Joaquin disagrees. Who is correct? Justify your response.

| | | | |
|---|---|---|---|
| | | | 3G |
| | | | 2R, 1B |
| | | | |

BP-31. Write the following statement in if-then form:

Rosa needs to score 92% on the final exam to earn an A.

**BP-32.** Read the following definition, then find the two possible slopes of the hypotenuse in the example below. Add the term to your tool kit.

> We have used **slope triangles** on sets of coordinate axes. These right triangles can also be called **GRID TRIANGLES**. These are triangles that have been cut out of graph paper and dropped, face up, on the table. When they are dropped, they can have <u>any</u> **orientation** (position). These grid triangles have <u>two</u> possible slopes for the hypotenuse.
>
> **Example:**
>
> 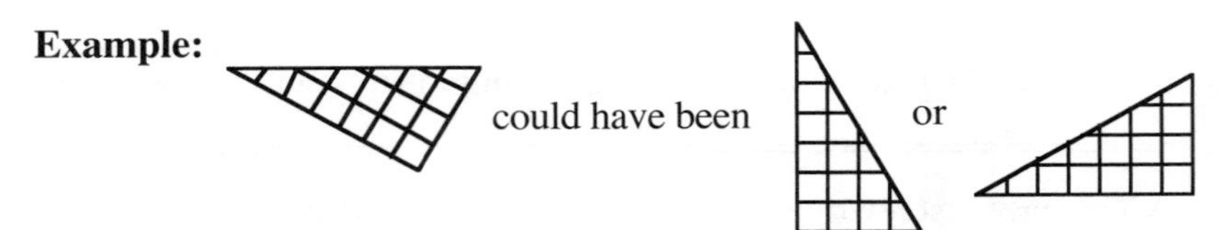
> 
>
> Note that grid triangles may only be rotated, they may never be flipped over. If we allowed them to be flipped, then there would be four possible slopes.

**BP-33.** Find BOTH possible slopes of the hypotenuse for each grid triangle. Remember, you may not turn the triangles over, only turn (rotate) them on the page.

a)

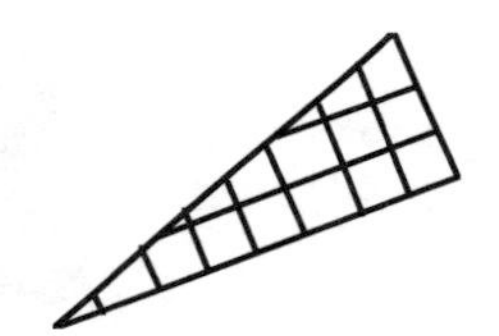

b)

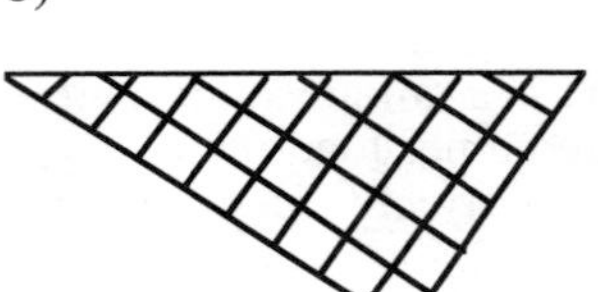

c)

BP-34. Find the length of the hypotenuse of each grid triangle in the previous problem.

BP-35. Write <u>two</u> possible equations of the line that contains the hypotenuse of the grid triangle in part (b) of problem BP-33 if the coordinates of the right angle are (0, 0).

BP-36. Evaluate the polynomial for $t = -1$ (that is, what number does the polynomial represent when $t = -1$?).

$$2t^3 + t^2 - 3t + 4$$

BP-37. Solve for x. Remember to check your answer.

a) $\frac{x}{2} = 8$ b) $\frac{x}{3} = -9$ c) $\frac{x+1}{4} = 16$

## WRITING JUSTIFICATIONS

BP-38. **DIRECTIONS**: In the figure below, the area of square G is 64 square units and the area of square F is 81 square units. Below the figure are two examples of writing **justifications** for finding the areas of other squares. Study these, then write similar justifications for squares H, E, and C.

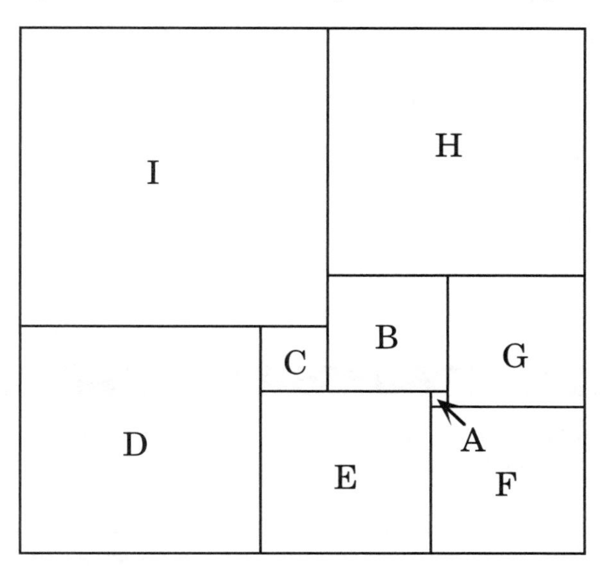


<u>Justification for square A:</u>

If the area of square G is 64 square units, then the length of a side is 8 units.
If the area of square F is 81 square units, then the length of a side is 9 units.
Since the side length of F minus the side length of G = 1,
the length of a side of A is 1.
Therefore, the area of A is 1 square unit.

<u>Justification for square B</u>:

If the length of a side of G is 8 units, and the length of a side of A is 1 unit,
then the length of a side of B is 7 units.
Therefore, the area of B is 49 square units.

a) Write a justification for finding the area of square H. You may use any of the conclusions from the PREVIOUS justifications.

b) Write a justification for finding the area of square E. You may use any of the conclusions from the PREVIOUS justifications.

c) Write a justification for finding the area of square C. You may use any of the conclusions from the PREVIOUS justifications.

d) Find the areas of D and I. Justify your answers.

BP-39. Find the area of the large figure in the previous problem. Is it a square? Justify your answer.

BP-40. Statement Q: **If an animal is a dog, then it has a tail.**

ASSUME THAT STATEMENT Q IS TRUE!

In your teams, discuss whether each of the following statements is true or false. Record your decision and its justification on your homework paper. Be prepared to justify your answers to the class.

a) If an animal has a tail, then it is a dog.

b) If an animal is not a dog, then it does not have a tail.

c) If an animal does not have a tail, then it is not a dog.

d) All dogs have tails.

## ALGEBRAIC INEQUALITIES

BP-41. The next four problems introduce algebraic inequalities. Answer each of the questions and summarize the process for solving and graphing inequalities in your tool kit.

**DIRECTIONS**: The process for solving equations such as $3(2x + 5) = 4x + 23$ uses several fundamental properties of algebra. However, not all relationships with numbers involve equality. Two numbers, represented by a and b, can be related in one of three ways:

$a = b$ $\quad$ $a > b$ $\quad$ $a < b$

Single variable order relations can be represented on a number line. The middle figure shows the solution to $3(2x + 5) > 4x + 23$ and the figure at right shows the solution to $3(2x + 5) < 4x + 23$:

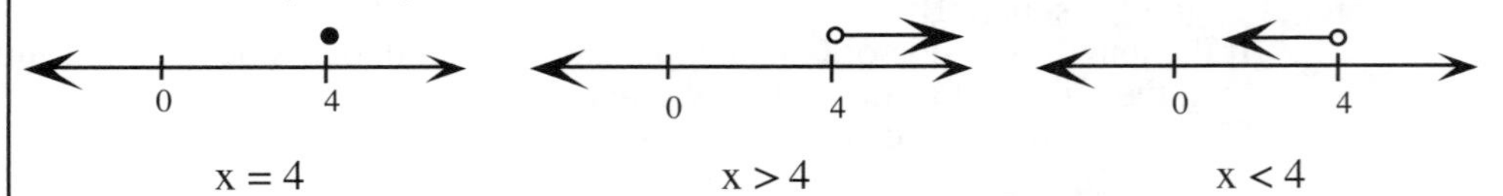


$x = 4$ $\quad$ $x > 4$ $\quad$ $x < 4$

In this example, notice that for inequalities, 4 is the critical point for the statement. In this course we will call it the **DIVIDING POINT**. All of the equations you solved in Unit 1 could have described an inequality if the order relation sign had not been an equal sign (=).

Why do you think the number 4 in the example above is called the dividing point? Why is the dividing point marked by an open dot instead of a solid one in the second and third graphs above?

The arrow above the number line indicates all the points in the solution to an inequality. It is called a **ray**.

BP-42. A simple example of an **ALGEBRAIC INEQUALITY** is $3x - 11 > 7$. Complete each part below to learn how to solve it.

a) Solve $3x - 11 = 7$.

b) Your solution to part (a) is the **dividing point** for this inequality. On a number line graph, indicate this dividing point with an open circle. Why is the point an open circle?

c) On which side, left or right, of your dividing point is the point $x = 10$? Substitute 10 for x in the original inequality. Is the result true or false? Now substitute 8 for x. Next try 15 for x. What do you notice?

d) A graph of an inequality represents all the solutions of the inequality. Draw the ray that shows the complete solution in the direction where points made the inequality true. Then describe the solution in a sentence.

BP-43. On what side of the dividing point is $x = 2$? Substitute 2 for x in the original inequality. Is the result true or false? Try $x = 5$, then $x = -3$.What do you notice?

BP-44. What does your answer to BP-43 tell you about the solution to the inequality?

Note that in general you only need to test one point. Why?

**Inequality symbols:** For $<$ and $>$, the dividing point is NOT included in the solution, so the endpoint of the ray is open. For $\leq$ and $\geq$, the dividing point IS included in the solution, so the endpoint of the ray is a solid point. Always check points on either side of the dividing point by substituting them into the original inequality. Remember, add all of the above information to your tool kit.

BP-45. Solve the following inequalities as if the problems were equations, then test one point on either side of the dividing point in the original inequality. Also sketch a graph of each solution. Note that **sketching a graph** means representing your solution using the number line as illustrated in the previous problems.

a) $4x + 9 < 13$

b) $2(3x - 5) \geq 2x - 7$

c) $5x < 4(3x - 2) - 6$

d) Check with students in your team to see what numbers they used for testing the inequalities. Are some of the numbers easier to use than others?

BP-46. Solve this Digit Place game. Be sure to justify each step in your solution.

| Step | Guess | Digits | Place |
|---|---|---|---|
| 1 | 12 | 1 | 0 |
| 2 | 34 | 0 | 0 |
| 3 | 56 | 1 | 1 |
| 4 | 16 | 1 | 0 |

BP-47. Write the following statement in if-then form:

The varsity soccer coach predicts that winning nine league games will earn the team a berth in the playoffs.

BP-48. Show that X(1, 3), Y(3, 7), and Z(7, 17) are not on the same line. Draw the slope triangles on your graph.

BP-49. Cody has a box of See's Candies. His favorite type is the coconut chews, but there are only five coconut chews in the box of 30 candies. He offers the box to his friend Kasey and says, "Help yourself to one!" Cody crosses his fingers and secretly hopes that Kasey doesn't get a coconut chew!

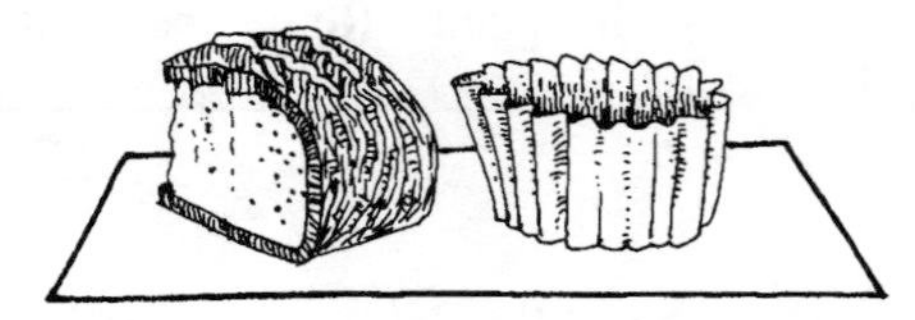

a) What is the probability that Kasey selects a candy at random and it is a coconut chew?

b) What is the probability that he does <u>not</u> get a coconut chew?

c) What is more likely to happen? Should Cody be worried?

BP-50. Over the years the cost of going to the movies has risen. Below is a table showing the average cost for a movie admission in the United States for five different years.

| 1989 | 1991 | 1993 | 1995 | 1997 |
|---|---|---|---|---|
| \$5.70 | \$6.15 | \$6.53 | \$6.95 | \$7.40 |

a) Set up a graph on your paper showing YEAR vs. AVERAGE COST as shown at right. Plot the information above on your graph.

b) What do you predict will be the average admission cost in 2001? Justify your answer.

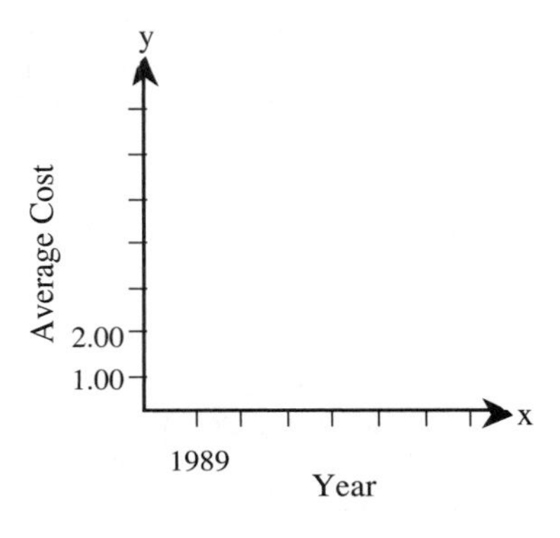


BP-51. Follow the directions in each part to complete the problem.

a) Simplify:

$4xy - 13x + 27y - 21xy + 6x$

b) Multiply:

$(2x + 5)(3x - 8)$

c) Simplify:

$(x + 5)^2$

d) Solve:

$3(2x - 8) + 14 = 2x - 9$

**BP-52.** You will work with a partner to complete the Over the Phone activity. Writing effective explanations and justifications requires clarity in your language. Keep this in mind as you do the activity. You and your partner will each be given a resource page. Each page is different. DO NOT LET YOUR PARTNER OR ANYONE ELSE SEE YOUR PAGE. Follow the directions on your page as closely as possible.

BP-53. Below you have a crossword puzzle and a list of words to fill in. We hope that completing the puzzle itself is easy. However, for each step you must write a reason that justifies why you are able to place a word in a particular space or why a word cannot go there.

a) Which word did you place first? Why?

b) Which word must go in 3-across? Why?

c) Which word must go in 8-down? Why?

d) Which word must go in 8-across? Why?

e) Which word must go in 16-across? Why?

| | | | | | |
|---|---|---|---|---|---|
| | | | | 1 | |
| | | | | 2 | |
| | | 3 | 4 | 5 | 6 |
| | | | | 7 | |
| 8 | 9 | 10 | 11 | 12 | 13 |
| 14 | | | | 15 | |
| 16 | 17 | 18 | 19 | 20 | 21 |
| 22 | | | | | |

MATH
TUNE
TIMBER
MONDAY
MANDATE

BP-54. Hatsumi and Tyrone are playing another game of Digit Place. Hatsumi's first guess is 48. Tyrone, who knows the secret number, responds D = 2 and P = 0. If Hatsumi is an expert Digit Place game player, what can she conclude? Explain your conclusion in one or two sentences.

BP-55. Write the following statement as a conditional statement (that is, if-then form):

All students who have a legal driver's license are at least sixteen years old.

BP-56. Draw a pair of coordinate axes on quarter inch graph paper. The y-intercepts provide a guide for the scale of your axes.

a) Carefully graph $y = -2x + 7$ and $y = 2x - 5$ using the slope-intercept method.

b) Use the graph to find and state the coordinates of the point of intersection of the two lines (that is, where they cross).

c) Take the x- and y-coordinates from the point of intersection you found in (b) and substitute them for x and y in each of the original equations. Simplify the right side of each equation. Are both sides of each equation the same? Should they be the same? What might it mean if the two sides of an equation are not equal?

BP-57. Draw a pair of xy-coordinate axes with $-8 \leq x \leq 8$ and $-8 \leq y \leq 8$ on quarter inch graph paper. Find the coordinates of the midpoint of the segment that connects each pair of points.

a) (-8, 0) and (0, -6)

b) (-6, 5) and (1, -1)

BP-58. On graph paper, plot and connect the four points in the order given and connect A and D. Find the perimeter and area of the resulting figure: A(1, 5), B(1, -2), C(-6, -5), and D(-6, 2).

BP-59. The shortest leg of a right triangle is 10 cm long. The hypotenuse is 2 cm longer than the other leg. Draw and label a diagram, then find the dimensions of the triangle.

BP-60. Solve the following inequalities and sketch a graph of each solution.

a) $3y < 7y + 12$

b) $-5y + 4 \geq y + 13$

c) $3(5 - x) < 2(x - 7)$

BP-61. In each part (a) through (c), multiply, then simplify. You may want to use a rectangle to represent the product.

a) $(2x)(3x + 4)$

b) $(x + 2)(x + 3)$

c) $(4n + 1)(n - 7)$

BP-62. One number is three more than another. The product of the numbers is 54. Find the two numbers.

## SOLVING SYSTEMS OF EQUATIONS BY SUBSTITUTION

BP-63. On large grid graph paper, graph $y = 2x + 1.75$ and $y = -2x + 1.5$ on the same set of axes.

a) Where do the two graphs intersect?

b) Convince your team that in fact the point you wrote in part (a) is really the exact point of intersection.

BP-64. The next three problems review the substitution method for solving systems of linear equations. Read them carefully, answer the questions, then add the method to your tool kit.

**DIRECTIONS**: In problem BP-56, you graphed the equations $\mathbf{y = -2x + 7}$ and $\mathbf{y = 2x - 5}$ and found that they intersect at (3, 1) by examining the graph. How well did this method (examining the graph) work for BP-63? Are **you** convinced you have found the exact point of intersection?

At times it is not convincing to just look at the graph and guess at the coordinates of the point of intersection. As you might suspect, there is a more efficient and accurate method. For problem BP-56, the two graphs intersected at (3, 1). Recall that (3, 1) means $x = 3$ and $y = 1$. When substituted for x and y in the original equations, (3, 1) is the ONLY pair that will result in a true statement for both. (You should have verified this in part (c) of BP-56.) Consider the equations written in this manner:

$$y = -2x + 7$$
$$y = 2x - 5$$

Now we can use algebra to solve for the point of intersection. Since the point of intersection has the same (equal) y-value, the right sides of the equations must equal each other, that is,

$$-2x + 7 = 2x - 5$$

a) Notice what we did. We converted two equations with two unknowns to ONE equation with ONE unknown, namely x. Solve the equation above for x.

b) Once we know $x = 3$, we can replace it with 3 (you did get 3 above, didn't you?) in either original equation and solve for y. Do this and solve for y.

Notice that for this **SYSTEM OF EQUATIONS**, graphing and algebra both give exactly the same result. In general, graphing yields close approximations; algebraic techniques can produce exact results.

c) Use this algebraic technique to find exactly what the coordinates of the point of intersection are in BP-63. How close were you?

d) Which method would you use--graphing or algebra--if you wanted to prove to someone (convince them) that you had the correct point of intersection? Why?

**BP-65.** Solve these linear systems algebraically to find each point of intersection.

a) $y = -x + 8$
$y = x - 2$

b) $y = -3x$
$y = -4x + 2$

c) $y = 3x - 7$
$y = -5x + 1$

d) Why are these problems called **linear** systems of equations?

BP-66.

**DIRECTIONS**: A more general application of the preceding process is known as the **SUBSTITUTION METHOD.** Suppose the equations are not in slope-intercept form.

$$y = \mathbf{-2x + 7} \quad \text{and} \quad 2x - y = 5 \quad \text{(previously } y = 2x - 5)$$

We can still use the **- 2x + 7** from the first equation to replace y in the second one.

a) If we substitute for y, we get: $2x - (\mathbf{-2x + 7}) = 5$. Solve this equation for x.

b) Replace x with 3 (you did get $x = 3$ again, right?) in either original equation and solve for y.

Sometimes equations get more complicated. Suppose you have:

$$y = \mathbf{-2x + 7} \text{ and } 3x + 2y = 9 \text{ (note the second equation is new).}$$

In order to simplify this system from two unknowns to one unknown, carefully substitute **- 2x + 7** for y in the other, new equation. Now the second equation becomes $3x + 2(\mathbf{-2x + 7}) = 9$. Solving this equation yields $x = 5$. Now you can replace x in either equation (e.g., $y = -2(5) + 7$) to get $y = -3$. These two lines cross at (5, -3). You would expect to get a different point of intersection when you use a different second equation.

BP-67. A partial solution for this problem is given below. Complete the blanks, and then finish the rest of the puzzle. Be sure to justify your statements.

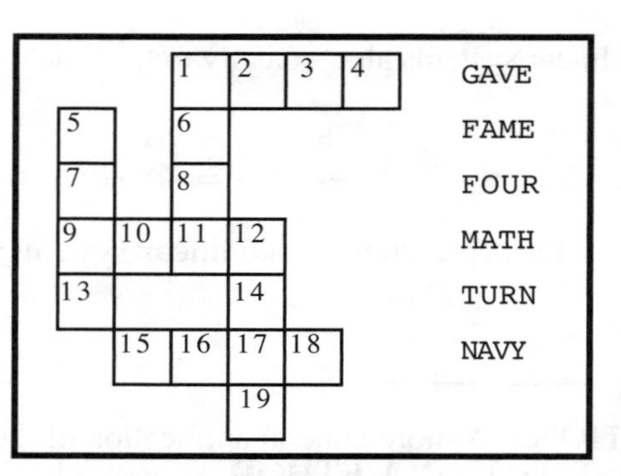

GAVE cannot be 1-across because 1-down would then have to start with the letter G, and there are no other words that begin with G.

For the same reason, ___________, __________, and _________ cannot be 1-across.

Therefore, either __________ or _________ must go in 1-across. __________ cannot be 1-down because ____________________________________________.

Therefore, _________ must be 1-across and _________ must be ___________.

Therefore, 9-across must be _________ because ______________________________.

Finish the solution. Justify where you place each word.

BP-68. Use the substitution method to solve these linear systems algebraically to find each point of intersection.

a) $y = x - 4$
$2x + 3y = -17$

b) $x - y = -4$
$y = 2x - 3$

c) $y = x$
$3x - 10y = 14$

BP-69. For each of these grid triangles, compute the possible slopes of the hypotenuse. Remember that the triangles cannot be flipped, only rotated.

a)

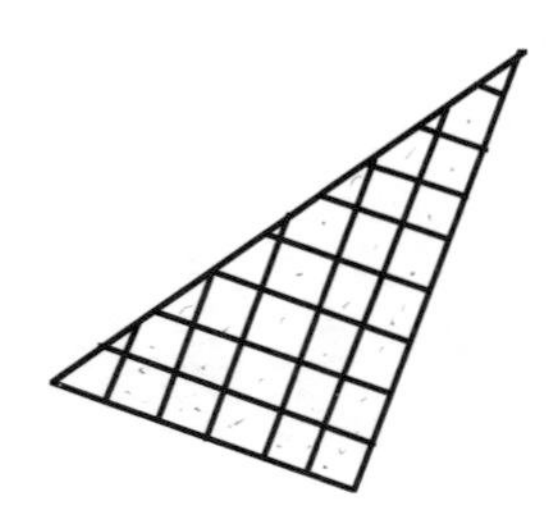

b)

BP-70. Write two possible equations of a line that contains the hypotenuse of the grid triangle in part (b) of the preceding problem if the coordinates of the right angle are (0, 0).

BP-71. A rectangular poster of your favorite movie measures 30" x 42". You plan to frame it yourself (hey, it saves a lot of money!). The frame will be 3" wide and its outer edge will fit the outer edge of the poster. Draw a diagram, show all subproblems, and determine the following:

a) How many feet of frame do you need?

b) How many square inches of the poster will be covered by the frame?

BP-72. Suppose you agreed to do yard work for 10¢ a minute. How much money would you earn in:

a) an hour? b) an 8 hour day? c) a 40 hour week? d) a 50 week year?

BP-73. Using your results from the preceding problem, graph the relationship TIME (in hours) vs. AMOUNT OF MONEY EARNED.

a) What is the slope of the line?

b) What is the y-intercept?

c) Write an equation representing this relationship.

d) Suppose you were paid a $10 bonus to take the job. Write the equation that describes your earnings after x hours.

## WRITING CONCLUSIONS

BP-74. You have probably thought about numbers before. You may even like certain numbers better than others. Do you like even numbers better than odd numbers? Do you ever wonder why you always get an even number when you add two even numbers? Read and add the following definition and notations to your tool kit, then do parts (a) and (b).

> **INTEGERS** are all positive and negative whole numbers, including zero. They can be expressed as: {..., -4, -3, -2, -1, 0, 1, 2, 3, 4,...}
>
> Certain kinds of integers can be represented algebraically:
>
> any integer.............. n
> any even integer......... 2n
> any odd integer............ 2n + 1
> consecutive integers...... n, n + 1, n + 2, ...

a) If n is any integer (positive or negative whole number), then n is either even or odd. In each case below, is 2n even or odd?

- If n = 13, 2n = ? • If n = -3, 2n = ? • If n = 5, 2n = ? • If n = -8, 2n = ?

b) Based on the examples in part (a), write a conjecture about the result--even or odd--of multiplying any integer by 2.

BP-75. Choose a few random integers and square them. Be sure to include negative integers. Then complete the following argument. Note that starting with a statement and developing it to a logical conclusion is often called an "argument" in mathematics.

If x is <u>any</u> <u>positive</u> <u>integer</u>, then $x^2$ is positive.
If x is zero, then $x^2$ is zero.
If x is <u>any</u> <u>negative</u> <u>integer</u>, then $x^2$ is positive.
Therefore, if x is any integer, then $x^2$ is either _________ or _________.

a) Does the argument above work for fractions also?

b) Give two examples of using positive and negative integers and two using fractions.

BP-76. Answer the following questions about consecutive odd integers.

a) If $2n + 1$ is an odd integer, then $2n + 3$ is the next odd integer. Why?

b) Add $2n + 1$ to $2n + 3$. Does this sum represent an odd or even integer? Explain.

c) Write an argument (logical presentation) to convince your team that no matter what value you choose for n in part (b) the result is always an even number. Hint: your result from BC-74 might be useful.

> Note: when you write an **argument**, remember that you are trying to convince your audience that your conjecture is always true. For example, the problem above asked you to show that the sum of two consecutive odd integers is always an even integer. Simply showing several examples using numbers--or even thousands of examples--will not prove it. A proof must show that a conjecture is true for every possible case. We will study proof in detail in Unit 6.

BP-77. On the upper half of a sheet of graph paper, draw a set of coordinate axes with $-6 \le x \le 10$ and $0 \le y \le 15$.

a) Plot the points A(2, 0), B(10, 0), and C(6, 14) and connect them to form a triangle.

b) Calculate the lengths of each side of the triangle.

c) Read the following definitions and add them to your tool kit.

> A **SCALENE** triangle has no sides of equal length.
>
> An **ISOSCELES** triangle has two sides of equal length.
>
> An **EQUILATERAL** triangle has all three sides of equal length. Note that **equi** means equal and **lateral** means side.

d) What type of triangle did you graph? Justify your answer.

BP-78. Consider multiples of 3 (the sequence 3, 6, 9, 12,...). If we want to represent this algebraically, we can do so as follows: Let n be a positive integer. Then 3n is a multiple of 3. How could you represent:

a) The multiples of 5?

b) Multiples of the sides of a triangle that are in the ratio 3 : 4 : 5?

BP-79. When you add two multiples of 3, will you get a multiple of 3? Justify your answer.

BP-80. Solve these systems algebraically.

a) $x + y = 4$
$y = x + 2$

b) $y = x + 3$
$x + 2y = 3$

c) $x = 5 - y$
$x - 2y = -4$

BP-81. Discuss your reasons and conclusions with your partner or team. Write them using one or two complete sentences.

| Step | Guess | Digits | Place |
|---|---|---|---|
| 1 | 425 | 1 | 0 |
| 2 | 567 | 0 | 0 |
| 3 | 489 | 1 | 0 |
| 4 | 130 | 1 | 0 |
| 5 | 913 | 2 | 0 |
| 6 | | | |

a) List the <u>possible</u> solution(s) you can identify at this point.

b) Do you know any digits for sure? If so write a short explanation to convince your team that you are right. If not, what guess would you suggest for step 6? Why would this be helpful?

BP-82. Crossnumber Grid: put the following numbers in the crossnumber grid given.

a) Which four-digit number goes in 4-across? Why?

b) Which three-digit number goes in 3-down? Why?

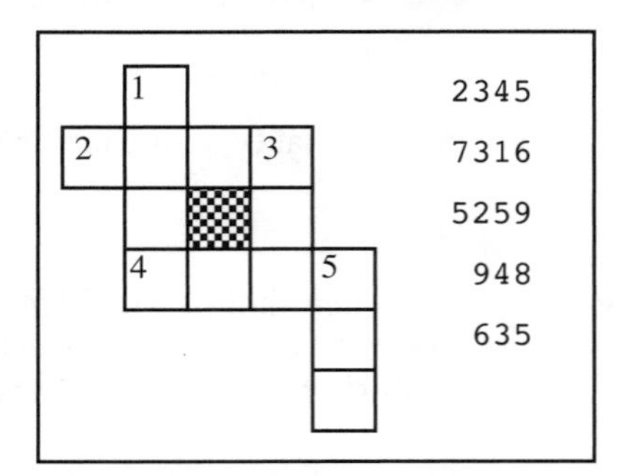

BP-83. Color Square:

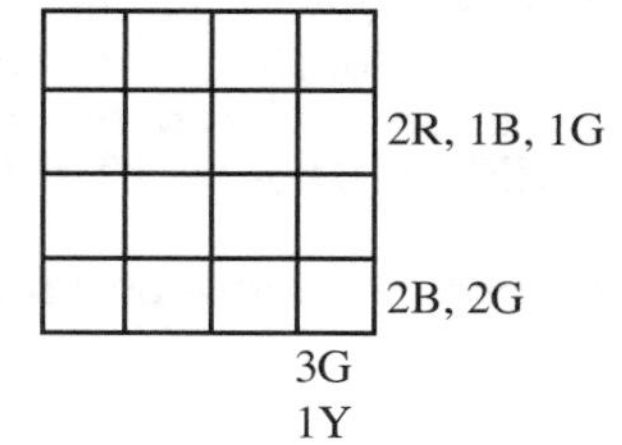

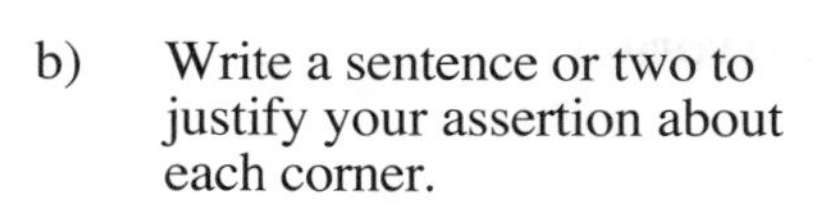

a) Name the colors of the four corners clockwise from the upper left.

b) Write a sentence or two to justify your assertion about each corner.

c) Is it necessary to complete the square in order to know the colors of the corners? Explain.

## SOLVING SYSTEMS OF EQUATIONS BY ADDITION

BP-84. Complete as much of this Color Square game as you can. Discuss possibilities with a partner or your team.

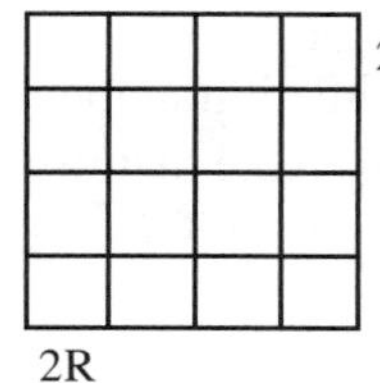

BP-85. Three boys were comparing their house numbers. Each house number contained three digits. It turned out that the <u>sum</u> of the digits of <u>each</u> three-digit number was the same.

1) None of the numbers contained any digits from another number.
2) None of the numbers began with a 4.
3) Bill's number was 252.
4) Ben's number began with a 6.
5) Bart's number ended with a 1.

Use these conditions to find <u>two</u> possible house numbers for Bart's house. List your reasons.

BP-86.

**DIRECTIONS**: Sometimes the substitution method for solving systems of equations is awkward or tedious. Another way to solve systems of equations is to add the two equations to **eliminate** one variable. This is known as the **ADDITION METHOD** or the **ELIMINATION METHOD**.

a) For the equations below, **add them vertically**:

$$\begin{array}{r} 2x + 3y = 8 \\ \underline{5x - 3y = -1} \end{array}$$

b) Solve the resulting equation for x.

c) Substitute the value of x you found in part (b) into either of the two original equations and solve for y.

d) Where do the two lines cross?

Notice that in either method, addition or substitution, the goal is to get from two equations with two unknowns to one equation with one unknown.

BP-87. Use the addition/elimination method to find the point of intersection of the graphs of these equations. Note that the instruction to "solve the system of linear equations" asks you to do the same thing. The solution is usually stated as an ordered pair.

a) $2x + y = 7$
$2x - y = 5$

b) $2x + 3y = 4$
$-2x + 2y = 6$

c) $5x + 2y = 5$
$2x + 2y = 8$
Start by multiplying the second equation by -1 on both sides.

BP-88. A rectangle has one side length of 14' and a diagonal length of 18'. Draw a diagram and find the width and the area of the rectangle.

BP-89. Solve for x and sketch a graph of your solution.

a) $-3(4 - 2x) < x + 2$

b) $5x \geq \frac{1}{3}(9x - 24)$

BP-90. What kind of number (even or odd) would be the sum of an even and odd integer? Write an algebraic argument to support your idea and give several examples, but remember: examples alone will not prove your conjecture.

BP-91. Multiply:

a) $(x + 3)(x + 5)$

b) $(x + 2)^2$

BP-92. As a car is approaching a freeway on-ramp, it is traveling at a constant speed of 35 mph. Once on the on-ramp, the car increases its speed at a constant rate of 3 mph per second.

a) Write an equation relating the speed of the car and the time it has been since it started accelerating.

b) When will the car be traveling at the speed limit of 55 mph?

c) Based on your answer to part (b), do you think the car will be traveling the speed limit by the time it reaches the freeway? Explain.

BP-93. In the rectangle at right, solve for x.

Area = 127 sq. cm

6 cm

x

BP-94. What is the perimeter and area of the rectangle at right?

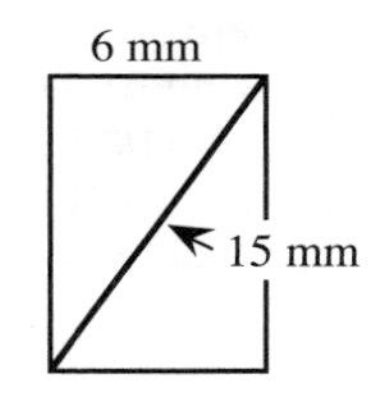

## UNIT AND TOOL KIT REVIEW

**REMINDER:** THE ROLE OF THE TEACHER

Well, you've almost completed another unit of your geometry course. I've found that even after students work with these materials for several weeks, some students are still confused about the teacher's role in the class. Sometimes this confusion--even anger and frustration--is because the student still expects the teacher to tell the class what to do for each problem. This often means the student wants the teacher to use most of the class time telling students how to work examples and homework problems.

Your teacher should do some brief lecturing and demonstrate selected problems. However, we've already established that the authors of the book want you to understand mathematics. Their experience has been that, for most students, telling them what to know is not enough. If you are still uncomfortable with the structure of the class, go back and read PZL-7, just after problem RC-16 in Unit 1. You should also take some time to meet with your teacher--probably outside of class time--to discuss your feelings. I hope at least two things will happen: you will understand more about why the class is run the way it is, and your teacher will learn more about what you need to learn geometry successfully.

Now continue the lesson by doing BP-95.

BP-95. Here is another *Instructions Over the Telephone* problem: Thu wrote the following set of directions for her friend, Steve. (She was not allowed to use the terms rectangle, square, and parallelogram.) "Draw a four-sided figure where the sides are all line segments and which has two pairs of equal sides." Steve drew a parallelogram from these instructions. Thu wanted him to draw a kite. Revise Thu's directions so that the next time Steve follows her directions, he draws a kite.

BP-96. Solve this Digit Place game. Write several sentences to justify your solution.

| Step | Guess | Digits | Place |
|---|---|---|---|
| 1 | 123 | 1 | 0 |
| 2 | 456 | 0 | 0 |
| 3 | 789 | 1 | 1 |
| 4 | 187 | 0 | 0 |
| 5 | 452 | 1 | 0 |

BP-97. Complete as much of the square as you can. Justify the color of each square.

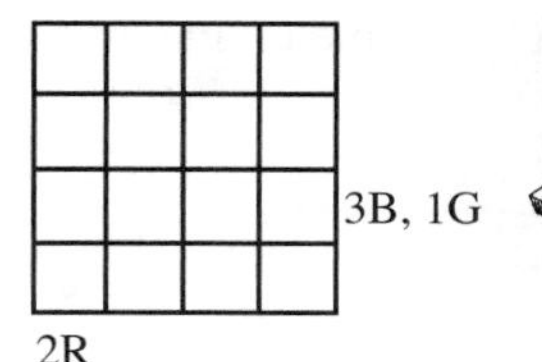

BP-98. Write this statement in conditional (if...then...) form.

All of (insert your teacher's name) students are brilliant.

BP-99. Read the following statement, presume that it is true, and decide whether (a) and (b) are true.

If it rains, then the ground will be wet.

a) If the ground is wet, then it has rained.

b) If the ground is not wet, then it did not rain.

BP-100. Solve each system of linear equations and explain what the solution means if you graph the lines. DO NOT actually graph the lines!

a) $y = 4x - 2$
$y = -2x + 4$

b) $2x - y = 12$
$6x + y = 8$

c) $y = \frac{1}{2}x - 6$
$3x + 4y = 6$

BP-101. On graph paper, plot and connect the four points in the order given. Then connect A and D and find the perimeter and area of the resulting rectangle:

A(3, 0), B(2, -2), C(-2, 0), and D(-1, 2).

BP-102. Solve $2(x + 7) \geq 5(2 - x)$ and draw a number line graph.

BP-103. Find the perimeter and area of each shape below. All lengths are in feet.

a)

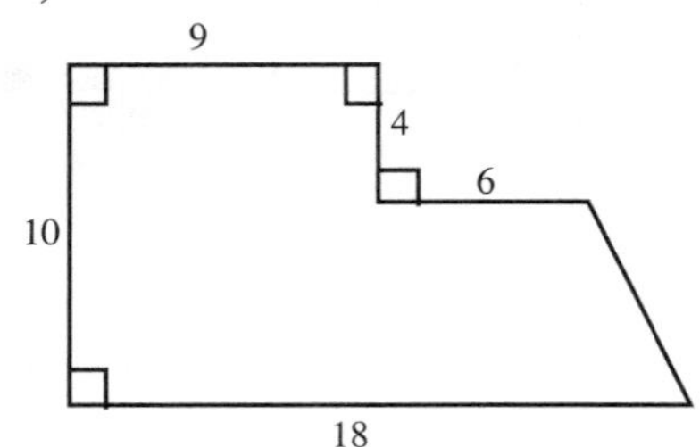


b) The top portion is a trapezoid.

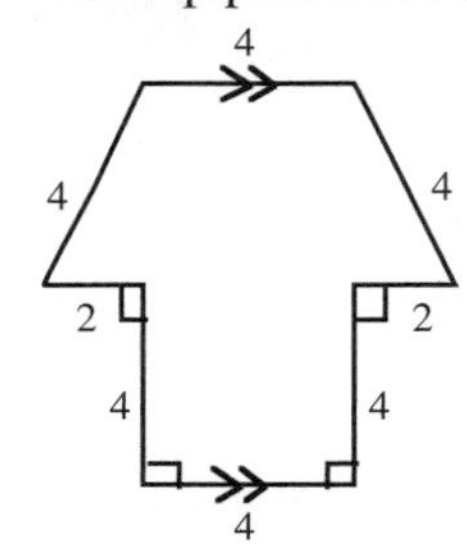


BP-104. The square of a number is 56 more than the number itself. What is the number?

BP-105. Crystal is working on a report for her science class, but she is very frustrated because she believes there is something wrong with the problem. The assignment asks Crystal to predict an outcome of an experiment based on certain probabilities, and to write a clear explanation as to how she arrived at her conclusion. Written on her worksheet is the following:

> When chemical A is placed in a test tube and held over a flame, only two possibilities can happen. Either (1), the chemical ignites and burns away, or (2), the chemical remains a solid, unaffected by the heat. The probability of (1) occurring is $\frac{1}{3}$ and the probability of (2) occurring is $\frac{1}{4}$. Which is the more likely outcome? Why? Justify your answer completely.

Is anything wrong with this problem? Help Crystal out by either explaining to her that the information in the problem is correct and then answer the questions from her worksheet for her, justifying your answer, OR, reassure her that something is wrong with the problem and help her describe the error(s).

BP-106. TOOL KIT CHECK-UP

Your tool kit contains reference tools for geometry. Return to your tool kit entries. You may need to revise or add entries.

How many entries do you have? Examine the list of tool kit entries from this unit listed below and add any topics that are missing.

- conjecture
- justify
- conditional (if-then) statement
- bisect
- grid triangles
- algebraic inequalities
- dividing point
- triangles: scalene, isosceles, and equilateral

Algebra Review: systems of linear equations (substitution and addition/elimination methods), integers.

# UNIT

# Problem Solving and Geometry Fundamentals

# Unit 3 Objectives
## PROBLEM SOLVING AND GEOMETRIC FUNDAMENTALS

This unit introduces many of the fundamental terms and properties of parallel lines and triangles. It does so by using problem solving strategies that are studied in the first part of the unit. You will look for patterns in the guided investigations, make conjectures about the relationship between pairs of angles, and use these conjectures to solve problems. Solutions to these problems will require that you use of the conjectures to justify solutions. You will prove many of the conjectures you make in Unit 6.

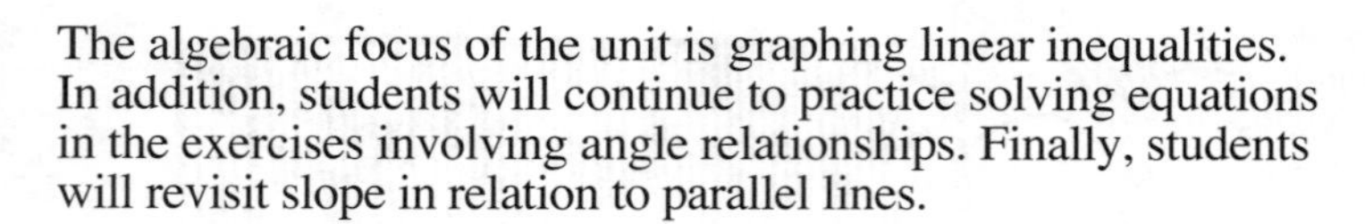

The algebraic focus of the unit is graphing linear inequalities. In addition, students will continue to practice solving equations in the exercises involving angle relationships. Finally, students will revisit slope in relation to parallel lines.

In this unit you will have the opportunity to:

- develop the problem solving skills of looking for patterns, making tables and systematic lists of data, and drawing diagrams.
- build the fundamental vocabulary for angles and angle relationships.
- investigate angle relationships formed by parallel lines and transversals.
- confirm that the sum of the angles of any triangle is 180° and study exterior angles of triangles.
- graph solutions to linear inequalities.
- extend your understanding of slope to parallel lines.

*Read the following problem and make sure everyone in your team understands it. Do not attempt to solve it now. We will come back to it later.*

PS-0. Suppose we want to find the number of dots in the 50th or the 1,281st figure in this sequence. It would be convenient (not to mention useful and powerful) to figure out what the pattern is IN GENERAL, that is, how to find the number of dots in <u>any</u> figure.

fig. 1 fig. 2 fig. 3 fig. 4

Find the number of dots in figure 99. Then show how to get the number of dots in figure n, that is, <u>any</u> figure in the sequence.

*UNIT* 3

ALGEBRA
GRAPHING
RATIOS
GEOMETRIC PROPERTIES
PROBLEM SOLVING
SPATIAL VISUALIZATION
CONJECTURE, EXPLANATION (PROOF)

# *Unit 3*
## PROBLEM SOLVING AND GE0METRY FUNDAMENTALS

## PATTERNS

PS-1. **DIRECTIONS:** Examine the following four examples of patterns, then read the box below them.

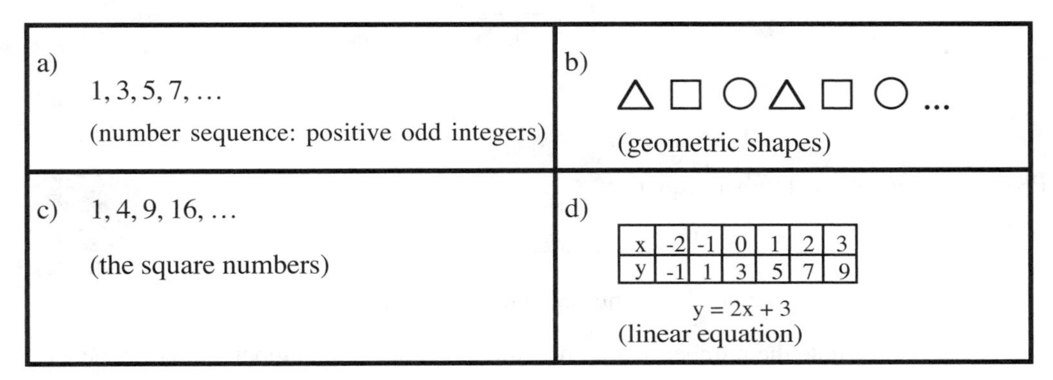


In the last unit, most of the problems unfolded in a logical fashion, that is, once you started your solution, one step led to the next. However, problems often do not have clear or obvious methods which lead to their solutions. One way to handle such problems is to employ **PROBLEM SOLVING STRATEGIES**. Problem solving skills help you to see relationships, break problems into manageable parts, organize information, and write equations. You have already made use of:

LOOKING FOR SUB-PROBLEMS with area and perimeter,
DRAWING A DIAGRAM for verbal problems, and
GUESS AND CHECK for verbal problems.

In this unit you will learn more about:

LOOKING FOR A PATTERN and
MAKING AN ORGANIZED TABLE (LIST) OF DATA.

In addition to focusing on problem solving strategies, this unit will help you to develop your skills of conjecturing, explaining, and justifying what you do. You will also learn some more geometry by using these strategies and skills.

Reminder: figures are not always drawn accurately or exactly. They are meant to be a visual aid in solving the problem, just like the diagrams that you sketch for verbal problems.

**PS-2.** Based on the pattern for the three figures below, <u>sketch</u> the next three figures and <u>explain</u> in a sentence or two how the pattern works in terms of the number of 1 x 1 squares in each figure. Note: This sequence of numbers is known as the **triangular numbers**.

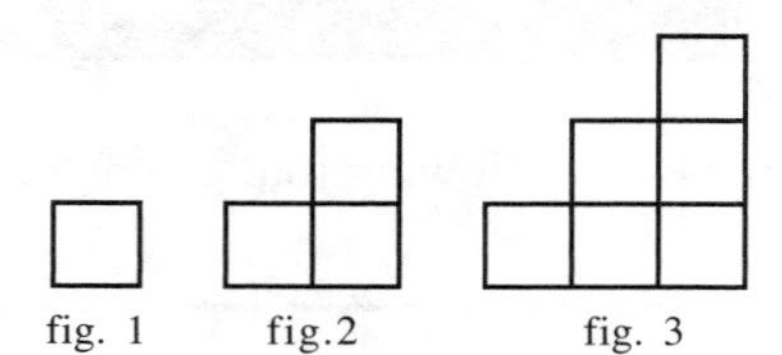

fig. 1 fig.2 fig. 3

PS-3. Given the number sequence:

59, 54, 49, 44, ____, ____, ____

a) Write the next three numbers in this sequence.

b) State the number that comes after each of these numbers in the sequence above:

1) 14 2) 4 3) -16 4) $x$ 5) $64 - 5n$

c) Briefly state the rule for the sequence above in a phrase.

PS-4. Example (c) in problem PS-1 at the beginning of this unit was 1, 4, 9, 16, 25,. ... These numbers are called the **square numbers.**

a) Explain in a sentence or two why you think these are called square numbers.

b) We can represent the number 1 with the square below. Use the same idea, that is, arrangements of squares, to represent the next four square numbers geometrically. A freehand sketch of each figure is fine.

| □ | ? | ? | ? | ? |
|---|---|---|---|---|
| 1 | 4 | 9 | 16 | 25 |

PS-5. The figure below is known as Pascal's Triangle, named after the French mathematician Blaise Pascal.

a) Examine the triangle for patterns and discuss them with your partner or team. Patterns might include number sequences, sums, differences, and/or products. Record at least four patterns and give a brief description of each one.

**1**
**1 1**
**1 2 1**
**1 3 3 1**
**1 4 6 4 1**
**1 5 10 10 5 1**
**1 6 15 20 15 6 1**

b) Find the relationship between each pair of rows. For example, how is the sixth row related to the fifth? Rows are numbered starting at the top of the figure.

c) The second figure shows Christmas stockings drawn around parts of the triangle. Find the same relationship in each stocking. Write a rule for the numbers in the Christmas stockings. Create at least two new Christmas stockings and record them.

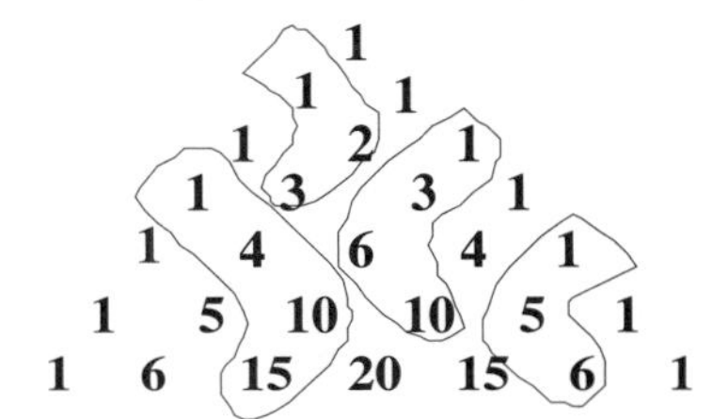


d) The third figure shows two sets of triangles, each forming a Star of David. The triangles in each pair are **congruent** (same size and shape). They also share a numerical equality. See if you can find this relationship for each pair, then write a rule for the Stars of David. Make another star and show that your rule works.

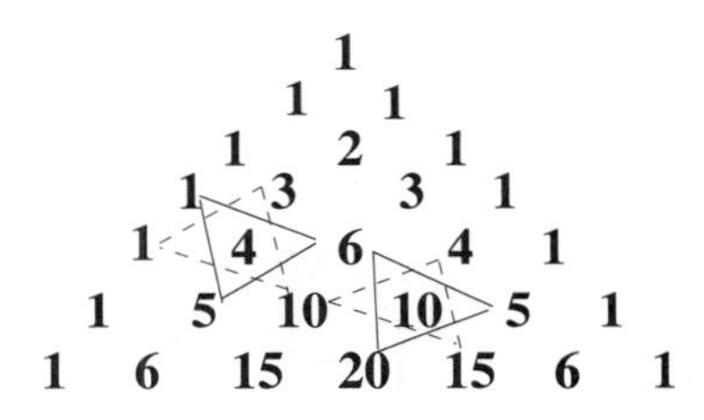

PS-6. Marisa used seven guesses to solve this game.

| Step | Guess | Digits | Place |
|---|---|---|---|
| 1 | 12 | 0 | 0 |
| 2 | 32 | 0 | 0 |
| 3 | 54 | 1 | 0 |
| 4 | 67 | 0 | 0 |
| 5 | 98 | 0 | 0 |
| 6 | 26 | 0 | 0 |
| 7 | 50 | 1 | 1 |

a) What is the secret number for this game?

b) Could it have been solved with fewer guesses? Which guess (or guesses) could you have eliminated? Explain why you would eliminate this guess (these guesses).

PS-7. Franchesca has four blouses, two skirts, and two pairs of shoes reserved for school clothes.

a) If all of these items are color compatible (i.e., they can be worn together), how many different outfits (blouse, skirt, and shoes) can she choose from? Make an organized list to be sure that you do not miss any!

b) If three of her blouses are shades of blue, what is the probability that if one day she blindly reaches into her closet and grabs a blouse, she will be wearing a blue blouse that day?

PS-8. Based on the information in the graph, what would you expect the world population to be in 2005?

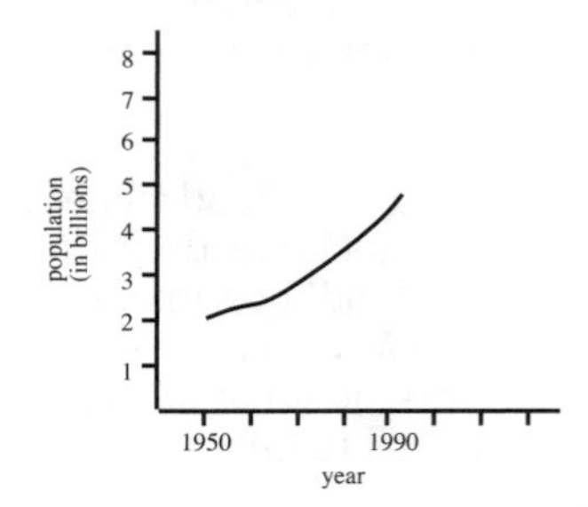


PS-9. Study the figures below, find a pattern, and sketch the next four figures. State the pattern's rule in a sentence or two.

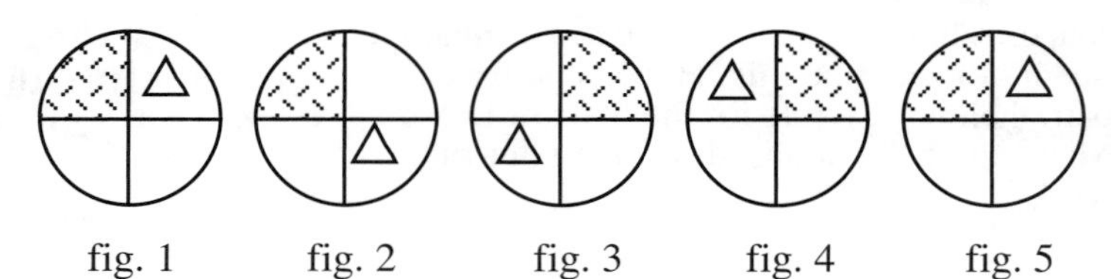


PS-10. Here is another Color Square game. Complete the first row and first column, then answer the questions below.

a) A1 and A2 are red squares. Why?

b) Explain why you <u>cannot</u> place a green in square B2.

c) Complete the Color Square game. Is this the only way the colors can be placed in the square? Explain why or why not.

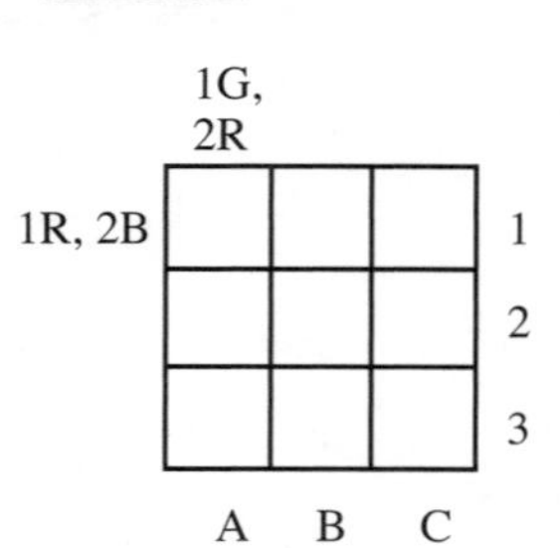

PS-11. Solve the linear systems by any algebraic method you choose.

a) $2x + y = 11$

$3x - y = 4$

b) $y = \frac{1}{2}x + \frac{3}{2}$

$y = -x + \frac{3}{2}$

c) Which of the two systems could also have been solved accurately by graphing? Justify your choice.

PS-12. Write the following statement as a conditional statement, that is, in if-then form.

The slope of the line that runs uphill left to right is positive.

## LOOKING FOR PATTERNS IN TABLES

PS-13. **DIRECTIONS:** Sometimes diagrams can get too complicated to be useful. In this problem, you will use the four diagrams to create an organized table of data then find the pattern in the data to solve the problem. Read the problem, then begin with part (a).

**PROBLEM**: Find the maximum number of points of intersection of seven lines.

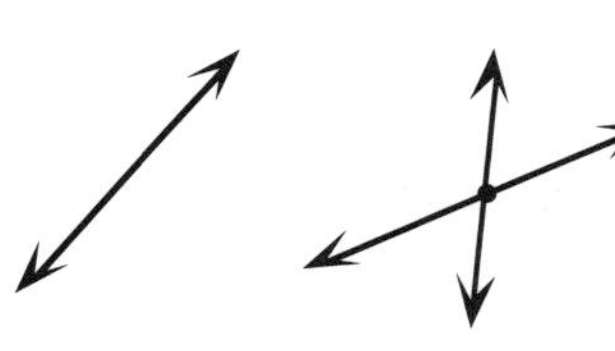

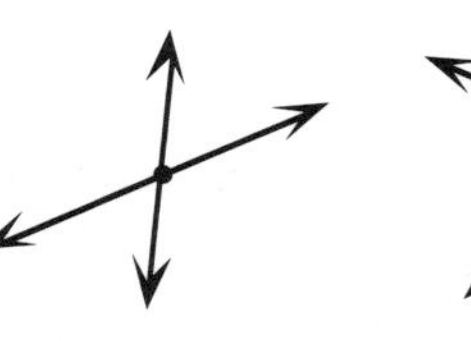

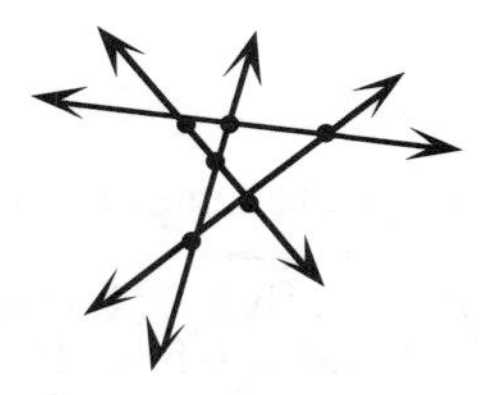

fig. 1 fig. 2 fig. 3 fig. 4

a) Try to draw the fifth figure. You should find 10 points of intersection!

b) Solve the problem by completing the table below for the number of lines 1, 2, 3, ... , 7, and the number of points of intersection. Use the problem solving strategy of looking for a pattern.

| number of lines | 1 | 2 | 3 | 4 | 5 | 6 | 7 |
|---|---|---|---|---|---|---|---|
| number of points | 0 | 1 | 3 | 6 | | | |

c) Write a description of the pattern you found and explain why it is valid.

PS-14. **DIRECTIONS**: Continue the pattern below by drawing the triangles for figures 7 through 11 (i.e., end with legs that are 6 and 6). With a partner or in your team, find the length of each hypotenuse. DO NOT find the decimal approximation of the length.

**KEEP ALL ANSWERS IN RADICAL (SQUARE ROOT) FORM, EVEN IF THEY ARE PERFECT SQUARES!**

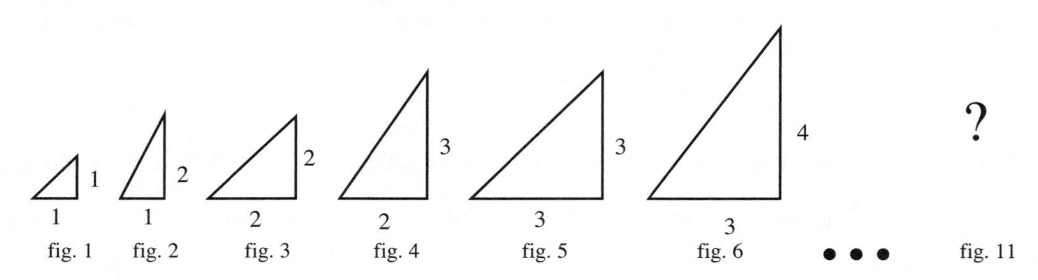


a) Complete the table below through figure 11.

| figure number | 1 | 2 | 3 | 4 | 5 | 6 | 7 | 8 | 9 | 10 | 11 |
|---|---|---|---|---|---|---|---|---|---|---|---|
| hypotenuse length | $\sqrt{2}$ | $\sqrt{5}$ | $\sqrt{8}$ | | | | | | | | |

b) Describe a pattern for the numbers inside the radical signs (called radicands).

c) Copy the radicands for the <u>odd-numbered</u> figures to form a number sequence. Describe the resulting pattern.

d) Repeat part (c) for the even-numbered figures. How are the two results similar? Different?

PS-15. If there are ten points on a line, how many ways can you name the line (e.g. $\overleftrightarrow{AB}$, $\overleftrightarrow{AC}$, $\overleftrightarrow{BC}$...)? STOP! You aren't really going to write out every possibility, are you?!? Use what you have learned about making a table and looking for patterns to solve this problem. Start your table with "1 dot, no (0) name."

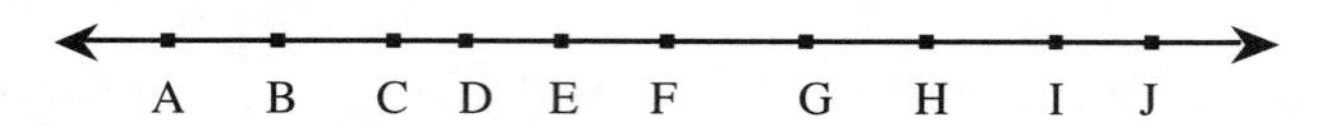

PS-16. Examine the following sequence of rectangles below.

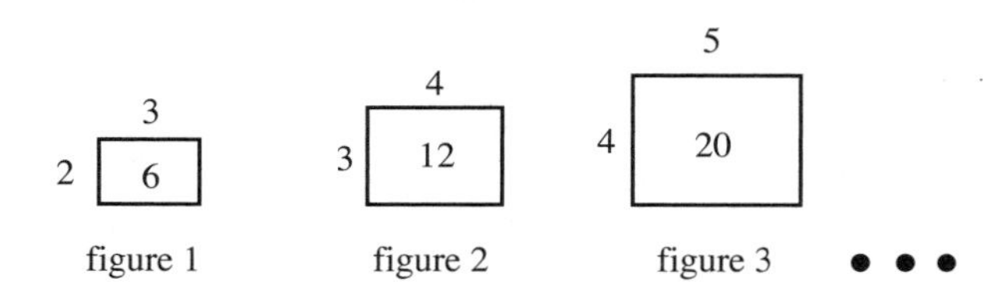


a) Draw figures 4 and 5.

b) Set up a table with headings Figure Number, Area of Rectangle, and Perimeter of Rectangle. Fill in the table through figure 5.

c) What would be the dimensions, area, and perimeter of figure 150?

d) Describe the pattern in each column (or row) of your table.

PS-17. **DIRECTIONS**: Cindy has started to play the Digit Place game. Her guesses and answers are written below. Look at what she has done and write your answers to (a) and (b) clearly and in complete sentences.

| Guess | Digit | Place |
|---|---|---|
| 123 | 0 | 0 |
| 456 | 2 | 0 |
| 765 | 3 | 1 |

a) After Cindy's second guess can she conclude that 4 is not in the number? Why or why not?

b) After Cindy's third guess she concludes that 6 is in the correct place. Explain whether her conclusion is right or not.

PS-18. Fill in the blanks by determining the pattern that connects the boxes. Explain how you were able to complete the figure.

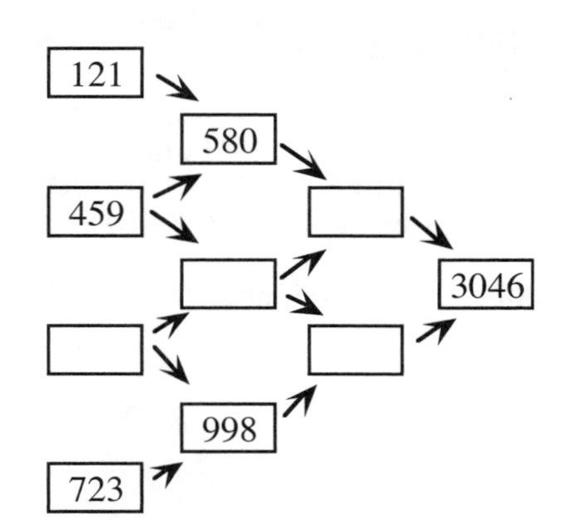

PS-19. Solve the system of equations at right algebraically by the method of your choice. Once you have your solution, demonstrate how you can use your answer and the original equations to prove that the answer is correct.

$y = 2 - x$

$2x + y = 6$

PS-20.
a) Solve for y in terms of x: $3x - 5y = 18$.

b) Solve for y: $y^2 + 24 = 73$.

c) Solve and sketch a graph for: $3(2q - 7) \geq 17q - 9$.

PS-21. Consider the following list of numbers: $2^1$ $2^2$ $2^3$ $2^4$ $2^5$ ...

a) Evaluate each term above and write the results as a new sequence (list) of numbers in ascending order.

b) Find the next three numbers for both sequences.

c) What will be the 100th number in this sequence (list)? Use an exponent to write it as a power of 2.

## FINDING GENERAL TERMS FROM PATTERNS

**NOTEBOOKS:** YOUR COLLECTION OF EXAMPLES

Hey! I just wanted to remind you that you will be much more successful in this course if you keep a good notebook. This is critical, because many of the **examples** in this book are the problems you solve with your partner or study team. So keep track of the problems that you solve. Organize them neatly in a notebook. You will always be able to refer to them if you have them in a safe place. You can no longer throw away problems after you are through with them.
**Make sure they are correct and use them as a resource.** CORRECT PROBLEMS ARE YOUR EXAMPLES. BE SURE TO REVISE AND CORRECT YOUR PROBLEMS AND KEEP THEM IN A NOTEBOOK FOR EASY REFERENCE.

That's all for now. Continue with today's lesson -- and keep a good notebook and tool kit!

**PS-22.** **DIRECTIONS**: The diagrams below show n by n squares divided into 1 by 1 squares (commonly called **unit squares**). Any unit square can have either zero, one, or two outside edges. These examples describe "outside edges:"

- The unit square labeled a is an example of a square with one outside edge;
- The unit square labeled b is an example of a square with two outside edges;
- The unit square labeled c is an example of a square with no outside edges.

Use these squares to help you complete parts (a)-(e) below:

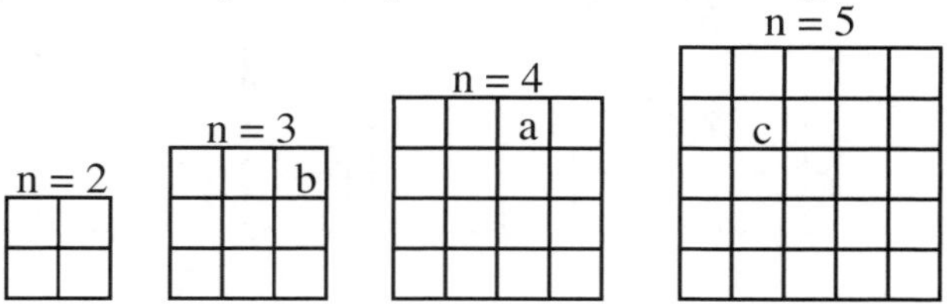

n = 2 n = 3 n = 4 n = 5

a) Use the resource page provided by your teacher or make a table like the one below on your paper and fill in the table. As you complete each section, check with the rest of your team to make sure that everyone agrees. An important part of this activity is to analyze how you get each entry in the table. You want to consider as many different methods as possible for the same entry, so discuss your methods with each other. Be sure to write them down!

| n | # of small squares with 2 outside edges | # of small squares with 1 outside edge | # of small squares with 0 outside edges |
|---|---|---|---|
| 2 | | | |
| 3 | | | |
| 4 | | | |
| 5 | | | |
| | | | |
| 10 | | | |
| n | | | |

b) If you copied the table above, notice that n skipped from 5 to 10. Write out in words how you were still able to find the entries for row 10 despite the jump.

c) Identify one entry you or your team found in which there were at least two different methods to determine the entry. Discuss the two methods and describe each one briefly in writing.

d) After you have filled in each row, check the numbers horizontally. The sum across each row should equal the area of that particular large square. Why? For example, in the row $n = 4$, you should have a total of 16 small squares.

e) The last row, which should be mostly algebraic, can also be checked horizontally. To be sure your algebraic entries are correct, add horizontally. What should the last row equal?

PS-23. Figure 1 has side length 1, therefore its shaded region has an area of one square unit. Create a table to record the figure number matched with the shaded area in each square. Write the area as a fraction (e.g., in fig. 2 the area is $\frac{1}{2}$). Then look for a pattern and predict the shaded areas for parts (a), (b), and (c) below. Hint: 4 can be expressed as $2^2$, 8 as $2^3$, etc.

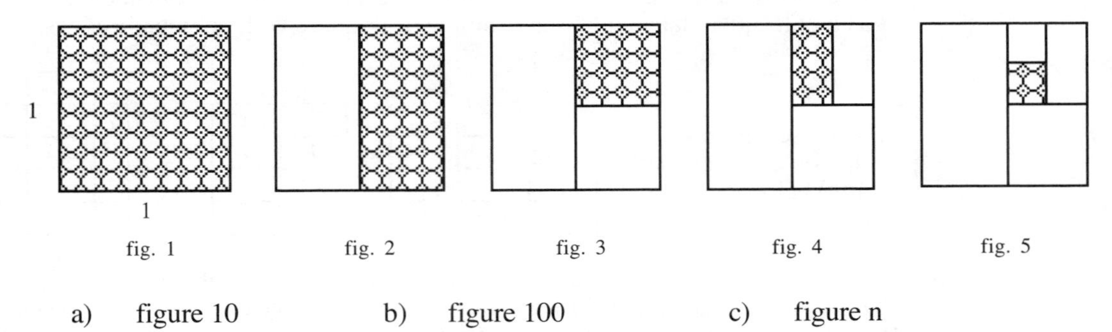


a) figure 10 b) figure 100 c) figure n

PS-24. Complete row 1 and column A and then answer each of these questions:

a) Explain why A1 must be a blue.

b) Suppose you also know that column C has 2R, 1G, 1Y. Can you find a unique solution? Explain why or why not.

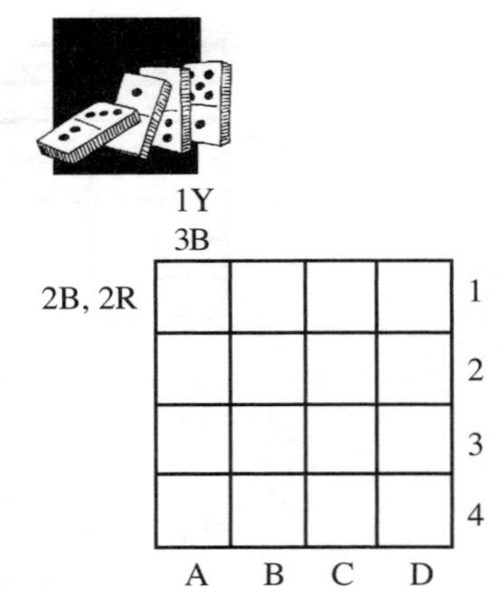

**PS-25.** **DIRECTIONS:**
Examine these figures:

a) Sketch the fourth figure.

b) Copy and complete the table below and use it to predict the number of vertices, triangles, and squares for the seventh figure. Explain how you know your result is correct. (Remember that a **vertex** is a point where the endpoints of two or more segments meet.)

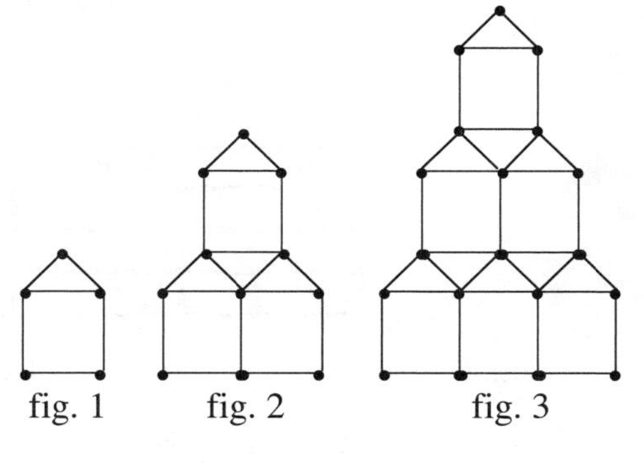


| Figure # | # of vertices | # of triangles | # of squares |
|---|---|---|---|
| 1 | | | |
| 2 | | | |
| 3 | | | |
| 4 | | | |
| 5 | | | |
| 6 | | | |
| 7 | | | |

PS-26. Which of the number patterns in the previous problem look familiar? Describe them.

PS-27. This figure contains two overlapping squares. Find the area of the shaded region.

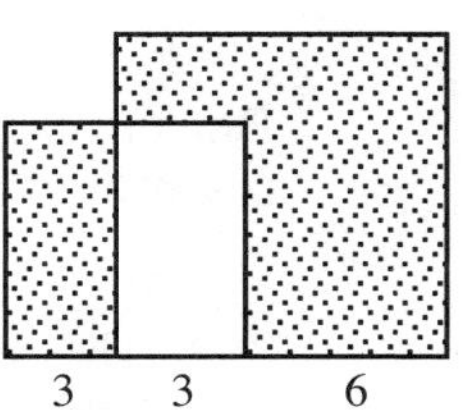

PS-28. Bill played the digit place game. His first three guesses are shown below.

| Guess | Digits | Place |
|---|---|---|
| 531 | 1 | 1 |
| 671 | 0 | 0 |
| 580 | 1 | 1 |

a) After Bill's second guess what numbers can he eliminate? Why?

b) After the third guess Bill believes he has enough information to say that 5 is in the correct place. Is he correct? Explain why or why not.

c) After Bill's third guess explain what Bill can conclude about the number.

PS-29. Solve by any method. Check your solution.

a) $6x - 3y = 18$

$4x + 3y = 12$

b) $x = 3y - 4$

$4y - 2x = 13$

PS-30. Draw two segments that satisfy all of the following conditions simultaneously, that is, one picture that meets all three conditions:

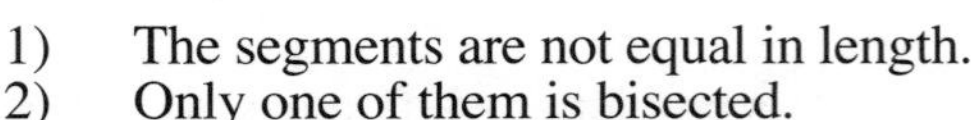

1) The segments are not equal in length.
2) Only one of them is bisected.
3) They are not perpendicular.

PS-31. Draw two segments that satisfy all of the following conditions simultaneously, that is, one picture that meets all three conditions:

1) The segments are not equal in length.
2) They bisect each other.
3) They are not perpendicular.

**PS-32.** Study the figures at right, then sketch the next three figures and make a table that matches the figure number with the number of dots in each figure.

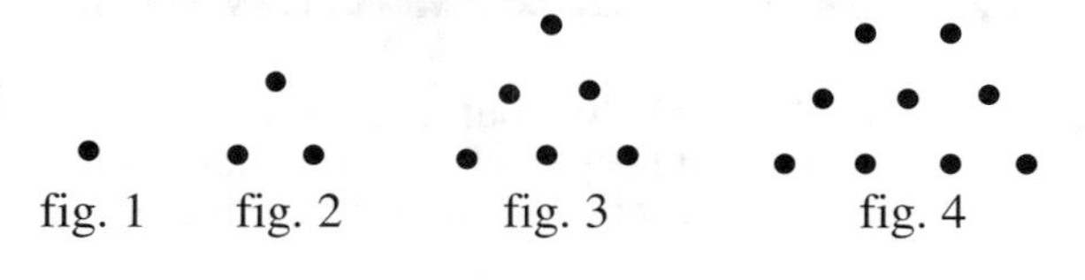


a) The sequence in the dot column should seem familiar since it has appeared as a pattern in several problems so far. Use the figures above to explain why the numbers in this sequence are called the triangular numbers.

To find an algebraic representation that describes the triangular number sequence, we can use the figures from problem PS-2. Recall that you were asked to predict how many squares would be in the next four figures and you found the sequence 1, 3, 6, 10, 15, 21, 28, … . Suppose we want to find the number of dots in the 50th or the 1,281st number in this sequence. Clearly, solving for these terms by drawing diagrams, while it could be done, is not an efficient way to approach such problems! It would be much more convenient (not to mention useful and powerful) to figure out what the pattern is IN GENERAL, that is, how to find the number of dots in any figure. We can use squares to represent the number of dots and arrange them in a convenient way (just like problem PS-2). Below each figure is an expression for the area of the solid line portion of the figure.

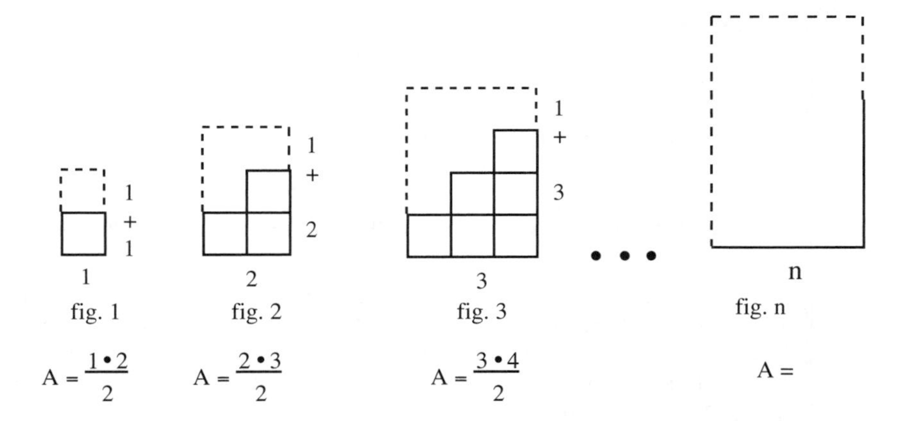


b) In your team, discuss how you could use a similar diagram to find the number of squares in figure 99. Explain what to do and then do it.

c) Now show how to get the number of squares in figure n.

## RAYS AND ANGLES

PS-33. **DIRECTIONS**: Your teacher will give you a resource page to help you keep track of vocabulary in this unit. As you encounter each term, find the picture on the tool kit resource page and write the name of the term and its definition in the box. Start by completing the boxes for terms from the first two units. Read the following definitions and explanations about angle notation, then add everything to your tool kit vocabulary grid.

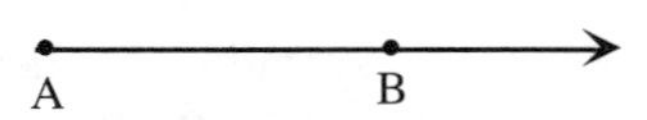


A **RAY,** $\overrightarrow{AB}$, is a part of line $\overleftrightarrow{AB}$ that starts at A and contains all of the points on $\overleftrightarrow{AB}$ that are on the same side of A as B, including A. An **ANGLE** is formed by two rays with a common endpoint. The point where two or more rays (or segments in polygons or three-dimensional figures) intersect is called a **VERTEX** (plural vertices). Note that some angles are formed by segments, which could be extended as rays. For example:

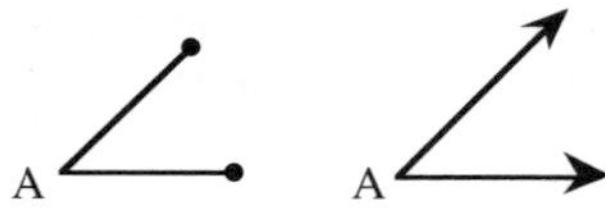


In this course, $\angle A$ will represent angle A, that is, the physical angle itself, while $m\angle A$ will represent the <u>measure</u> of angle A, that is, the number of degrees associated with the angle.

Angles may be named several ways:

$\angle 1$ means (1) $\angle A$ means (A) $\angle ABC$ means (B, A, C)

The third method is used whenever there are more than two rays from a vertex. Since there are three angles in the figure at right (you do see three, don't you?), calling any one of them $\angle B$ would be confusing. Hence, the three angles are named $\angle ABC$ or $\angle CBA$, $\angle CBD$ or $\angle DBC$, and $\angle ABD$ or $\angle DBA$.

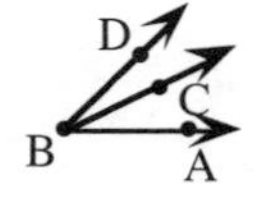


Note that when we want to indicate that two or more segments or angles are equal in a figure, we use an equal number of slash marks as illustrated in the three figures below.

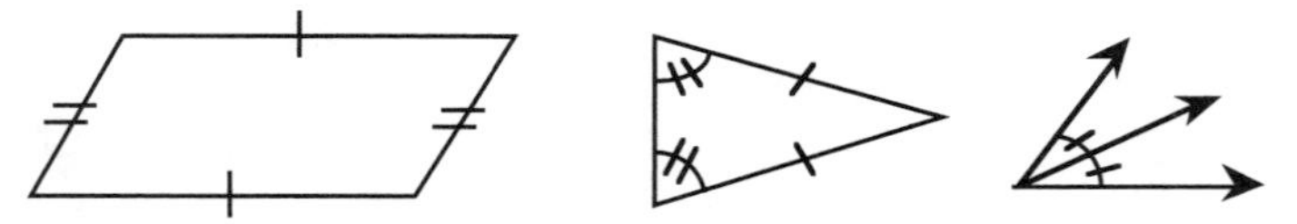

**Remember: add the above definitions to your tool kit vocabulary grid.**

PS-34. **DIRECTIONS**: One way to measure angles is to base their measure on how much of a circle they represent, where one complete revolution (turn) around the circle measures 360°. The geometric tool used to physically measure the number of degrees in an angle is called a **protractor**, and usually measures from 0° to 180°. In this course, we will almost always <u>calculate</u> angle measures from given information. The figure below shows the typical measurement markings on a protractor. Note that the center of the protractor's base (between 0° and 180°) is always placed at the vertex of the angle being measured.

Use the figure to estimate the measure of each angle named below to the nearest degree.

a) m∠AFB    b) m∠AFC    c) m∠AFD

d) m∠EFA    e) m∠CFB    f) m∠DFB

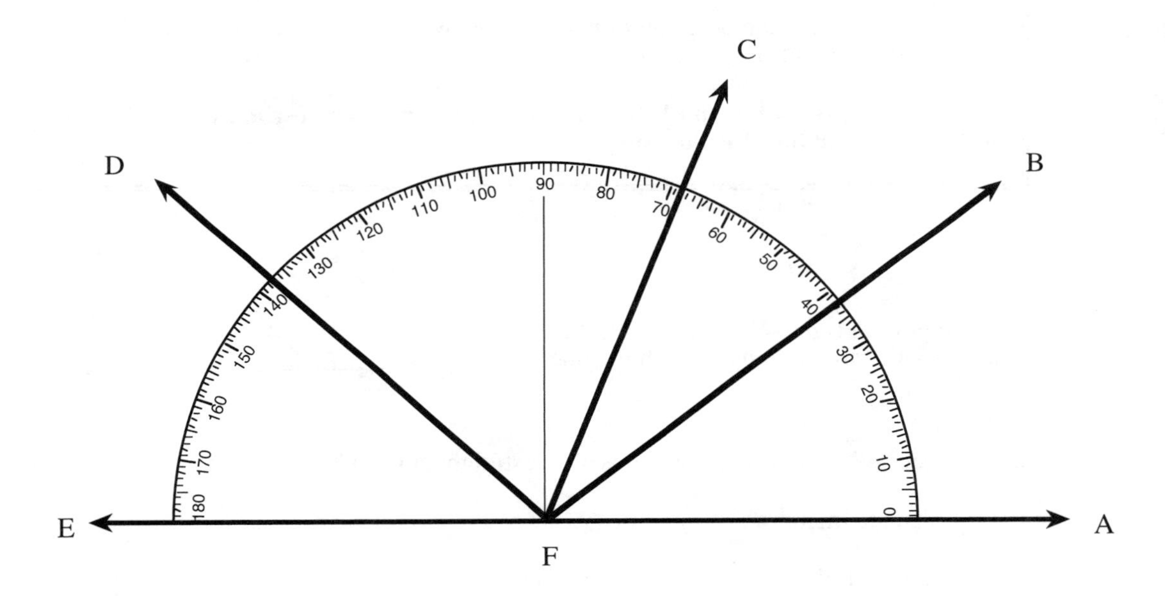

PS-35. Add these definitions to your tool kit vocabulary grid.

It is convenient to classify types of angles. We will use the following terms to refer to angles:

**ACUTE:** Any angle with measure between (but not including) 0° and 90°.

**RIGHT:** Any angle which measures 90°.

**OBTUSE**: Any angle with measure between (but not including) 90° and 180°.

**STRAIGHT**: Straight angles (lines) have a measure of 180°.

Note: While angles can be greater than 180°, we will confine our study to angles between 0° and 180° until Unit 9.

**Draw an example of each of these angles on your paper, estimate its measure, and fill in your tool kit vocabulary grid.**

PS-36. In the figures at right, $\overleftrightarrow{WY}$ is a straight line and $\overrightarrow{XZ}$ is a ray that forms two angles with the line.

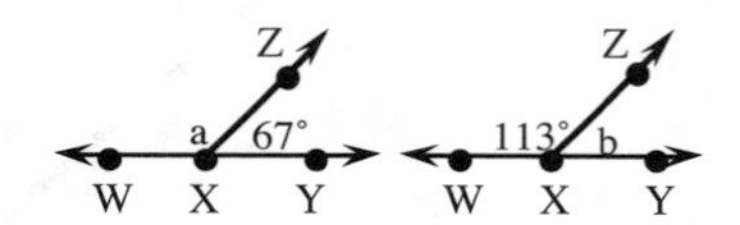


a) Since $\overleftrightarrow{WY}$ is a straight line, it is also a straight angle. Calculate m∠a.

b) Calculate m∠b.

c) Read the following definition and add it to your tool kit vocabulary grid.

Pairs of angles whose sum is 180° are called **SUPPLEMENTARY ANGLES**. Any pair of angles whose sum is 180°, whether or not they are adjacent (i.e., form a straight angle), are said to be supplementary.

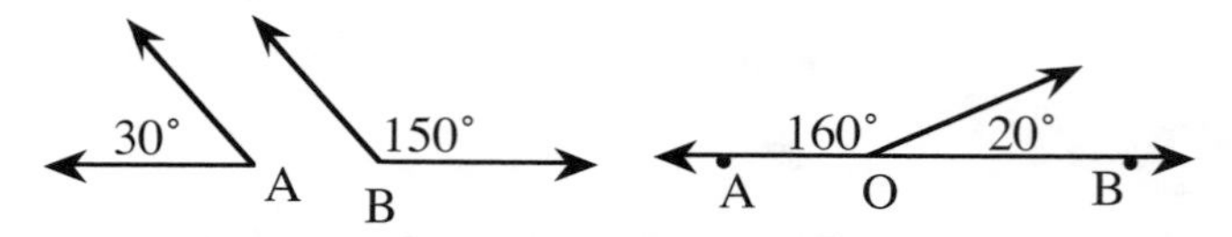

PS-37. Calculate your answers. Parts (a) and (b) are separate problems.

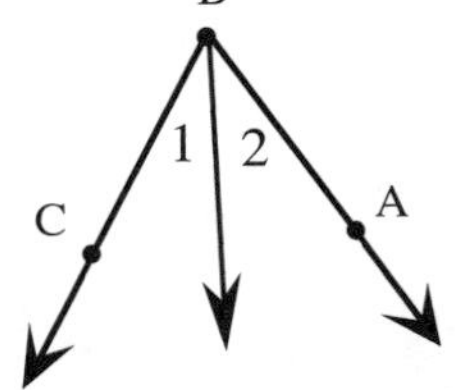

a) If $m\angle 1 = 30°$ and $m\angle 2 = 50°$, find $m\angle ABC$.

b) If $m\angle ABC = 67°$ and $m\angle 2 = 28°$, find $m\angle 1$.

PS-38. Calculate the measures requested in each separate problem below. Recall that any straight line, when considered as a straight angle, measures 180°.

a) Find $m\angle a$.

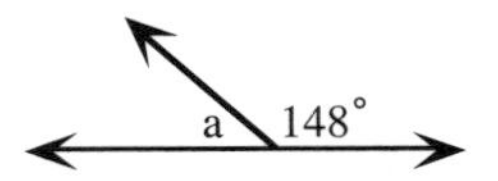


b) Find x, and the measure of the acute angle.

c) Find the measure of the obtuse angle.

PS-39. Find the measures of the following angles by calculation. In this problem, $\angle EBA$ is a straight angle. Note that parts (a), (b), and (c) are each separate problems.

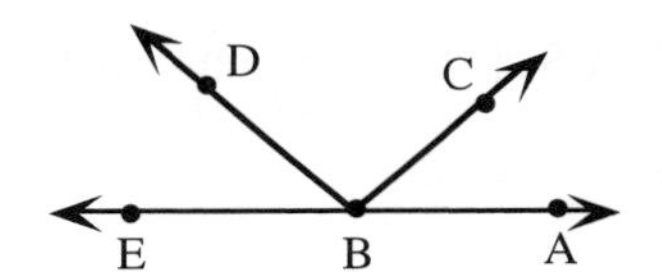


a) If $m\angle ABD = 133°$ and $m\angle CBD = 98°$, find $m\angle ABC$.

b) If $m\angle ABD = 119°$, find $m\angle DBE$.

c) If $m\angle CBD = 84°$, find $m\angle ABC + m\angle DBE$.

PS-40. **DIRECTIONS**: Sketch the next three figures in the pattern.

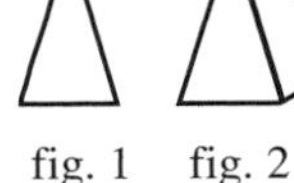

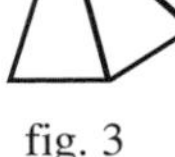

? ? ? ?

fig. 1 fig. 2 fig. 3 fig. 4 fig. 5 fig. 6 • • • fig. n

a) Make a table that includes the figure number and the number of segments in each figure.

b) Use the table to predict the number of segments in the 10th figure. How many segments will be in the 50th figure? Describe the rule for this pattern in a sentence or algebraically using n as the figure number.

c) How many segments are in figure n?

PS-41. What are the two possible slopes for the hypotenuse of each of these grid triangles? You may only rotate them, not flip them.

a)

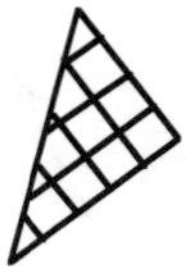

b)

c)

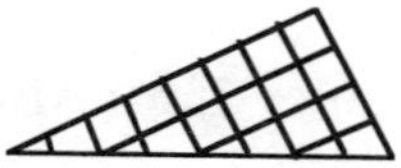

PS-42. Write two possible equations for the line containing the hypotenuse of the grid triangle in part (a) of the previous problem if the vertex of the right angle is at (0,0).

PS-43. Start with the number 6. Double it. Then double this result. Continue to double each new number until you find a pattern in the unit (one's) digits. Describe the pattern. You may want to make an organized table (list).

PS-44. Be sure that all of the new terms (vocabulary) introduced today have been added to your tool kit vocabulary grid.

## PARALLEL LINES AND ANGLE RELATIONSHIPS

PS-45. The box below has some reminders about notation from Unit 1. Read the information, add it to your tool kit if necessary, then go on with the next problem.

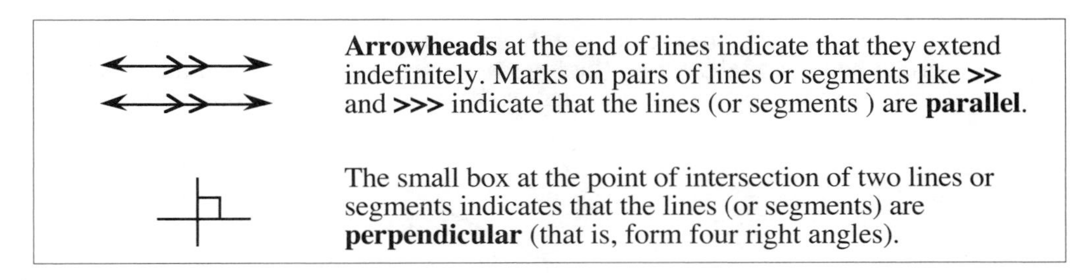

**Arrowheads** at the end of lines indicate that they extend indefinitely. Marks on pairs of lines or segments like **>>** and **>>>** indicate that the lines (or segments ) are **parallel**.

The small box at the point of intersection of two lines or segments indicates that the lines (or segments) are **perpendicular** (that is, form four right angles).

PS-46. In figure 1 below, s || t and r intersects (cuts) s and t. In figure 2, x and y are NOT parallel and w intersects x and y.

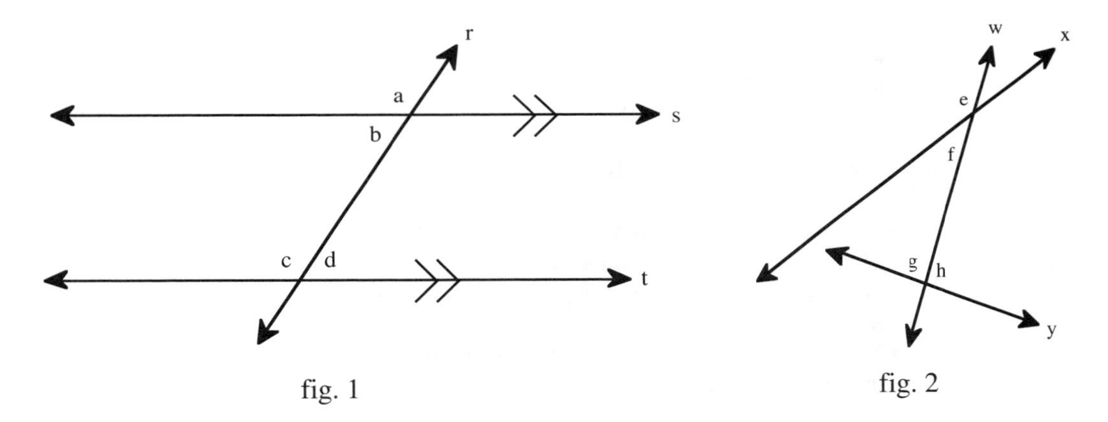


fig. 1

fig. 2

a) Take a sheet of plain paper (tracing paper is better, binder paper will work), place it over ∠a and, using a ruler as a guide, make a precise copy of the angle. Slide the copy of ∠a to ∠c and compare the sizes of the two angles. What do you observe?

b) Trace a copy of ∠b and try to fit it over ∠d. What do you observe?

c) Trace a copy of ∠e and try to fit it over ∠g. What do you observe?

d) Trace a copy of ∠f and try to fit it over ∠h. What do you observe?

e) Both figures show two lines that are cut (intersected) by a third line. Under what condition will certain pairs of angles be equal?

**PS-47.** **DIRECTIONS:** Your teacher will provide you with a resource page with the figures below. Keep it handy; you will use it in several problems over the next few days. Each figure presents a pair of parallel lines, *p* and *q*, which are intersected (cut) by a third line, *m*. Line *m*, often called a **transversal**, forms several angles at each point of intersection with *p* and *q*.

The experiments in the previous problem demonstrated the special relationship between certain pairs of angles **when the lines are parallel**. When given the measures of some of the angles, we can use what we know to confirm our observations.

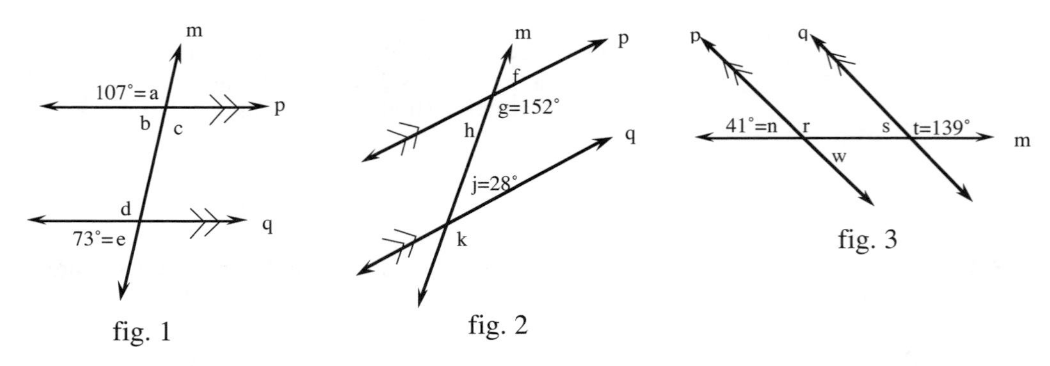


a) Use what you know about straight angles to calculate the measures of these angles: b, d, f, k, r, and s. When everyone in your team has completed the calculations and agrees with the results, check with your teacher to be sure your results are correct.

b) Keeping in mind that the condition we required of lines *p* and *q* is that they be parallel:

- In figure 1, compare the measures of angles a and d, then compare the measures of angles b and e.
- In figure 2, compare the measures of angles f and j, then compare the measures of angles g and k.
- In figure 3, make similar comparisons for $m\angle n$ and $m\angle s$, then $m\angle r$ and $m\angle t$.

c) Read the following definition and add it to your tool kit vocabulary grid.

Angles on the same side of two lines and on the same side of a third line (called a **transversal**) that intersects the two lines are called **CORRESPONDING ANGLES**. In the figure at right, angles 1 and 2 are corresponding angles, as are angles 3 and 4. Other examples of corresponding angles are on your resource page: angles a and d in figure 1, angles g and k in figure 2, and angles r and t in figure 3.

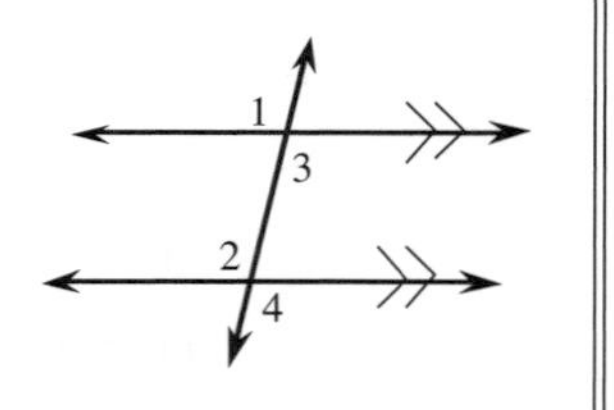


d) Use the definition above and your observations in part (b) to complete this conjecture:

**Conjecture:** If two parallel lines are cut by another line, then pairs of corresponding angles are ____.

PS-48. Use what you know about straight angles and your results from the previous problem to find the measures of angles c (fig. 1), h (fig. 2), and w (fig. 3).

a) Compare the measures of the following three pairs of angles:

• fig. 1: m∠c and m∠d • fig. 2: m∠h and m∠j • fig. 3: m∠w and m∠s.

b) How is each pair of angles related?

c) Read the following definition, then use it and your observation in part (a) to complete the conjecture about alternate interior angles.

Angles between a pair of lines that switch sides of a third line are called **ALTERNATE INTERIOR ANGLES**. In the figure at right, angles 1 and 2 and angles x and y are examples of pairs of alternate interior angles. Other examples of alternate interior angles are on your resource page: angles c and d in figure 1, angles h and j in figure 2, and angles w and s in figure 3.

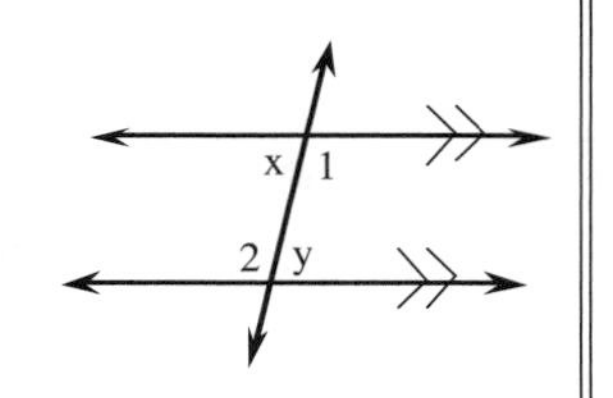


**Conjecture:** If parallel lines are cut by another line, then alternate interior angles are ___.

PS-49. Classify each of the following pairs of angles as corresponding, alternate interior, supplementary, or "none of these."

a)

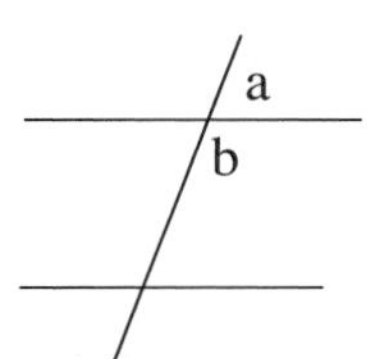


b)

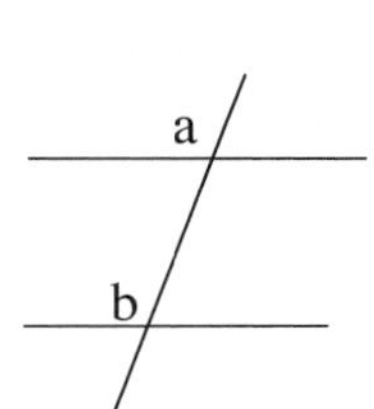


c)

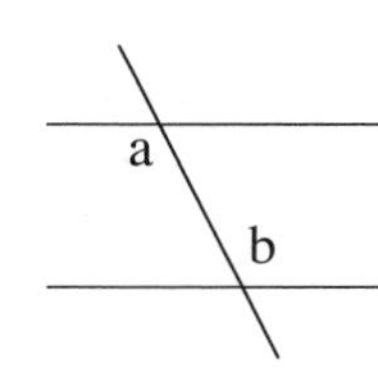


d)

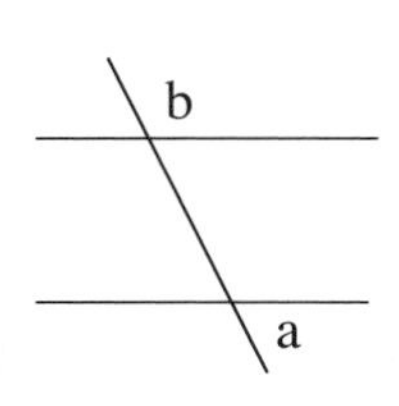


e)

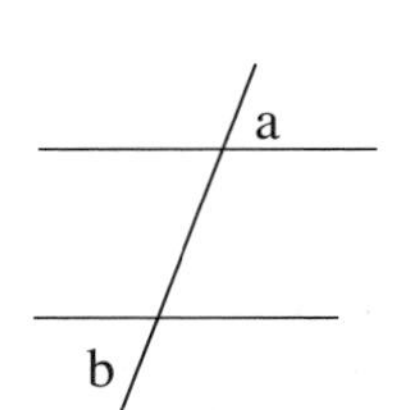


f)

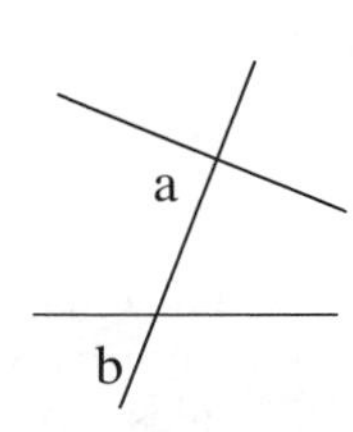


g)

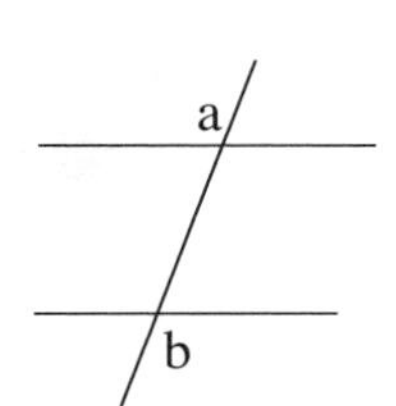


h)

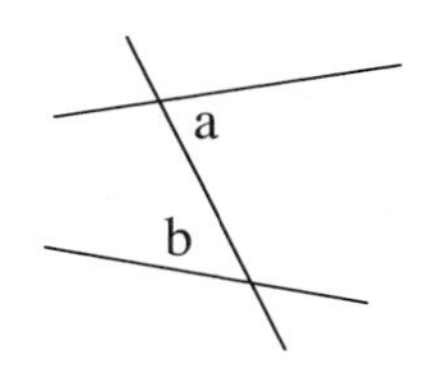


i)

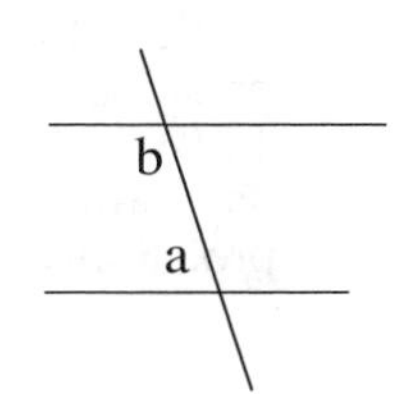


j) What condition is necessary to be able to say that pairs of corresponding angles or alternate interior angles are equal?

PS-50. Be sure that you have added all of today's definitions and conjectures to your tool kit vocabulary grid.

PS-51. Write equations for each problem based on the definition of supplementary angles. Note that each part is a separate problem.

a) If $m\angle 1 = 3x$ and $m\angle 2 = 2x$, find x and $m\angle 1$.

b) If $m\angle 1 = 5x$ and $m\angle 2 = 2x + 5°$, find x and the measure of each angle.

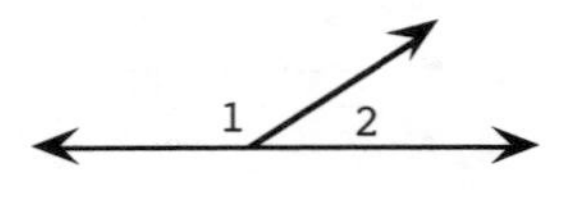


PS-52. Use the conjectures and definitions in your tool kit to solve parts (a) and (b). Each part is a separate problem.

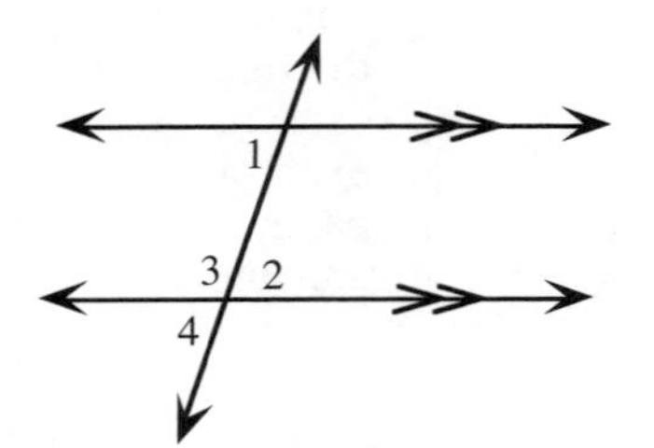


a) If $m\angle 1 = 63°$, find $m\angle 2$ and $m\angle 3$ by calculation.

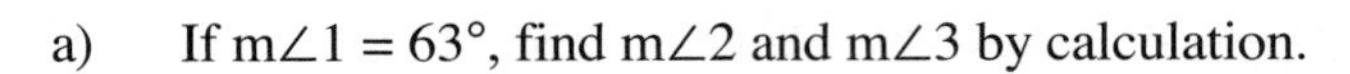

b) If $m\angle 1 = 74°$ and $m\angle 4 = 3x - 18°$, write an equation and find x.

c) If $m\angle 2 = 3x - 9°$ and $m\angle 1 = x + 25°$, write an equation to find x, then find $m\angle 2$.

PS-53. The ratio of the sides of a rectangle is 3 : 5 and its area is 60 square cm. Draw a diagram, write an equation, and find the dimensions of the rectangle.

PS-54. Suppose the area of the rectangle in the previous problem is 187 square cm. What are its dimensions?

PS-55. The Quality Cola Company is starting a new promotional gimmick which the company's president hopes will increase sales. Some cans and bottles containing Quality Cola will be painted blue on the inside, and if someone finds a can or bottle with a blue interior, that person will win $100. The president would like the probability of winning $100 to be 1.5%. If the company produces 10,000 cans of cola each day, how many should have blue interiors to achieve this probability?

PS-56. The graph below shows the average amount of rainfall (in inches) for an 11 year period. Based on this graph and any trends you may notice, predict the average rainfall for the next 5 years.

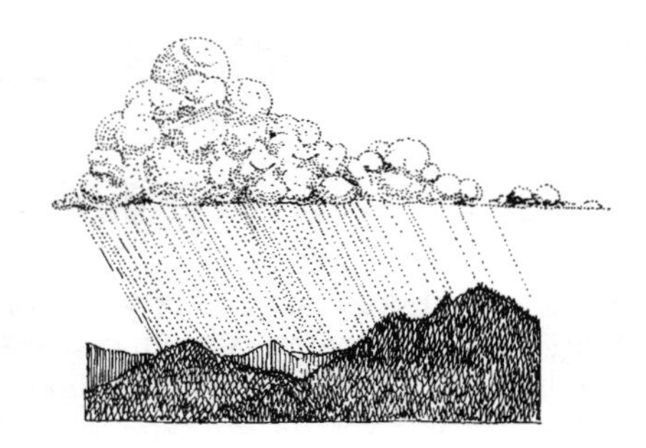

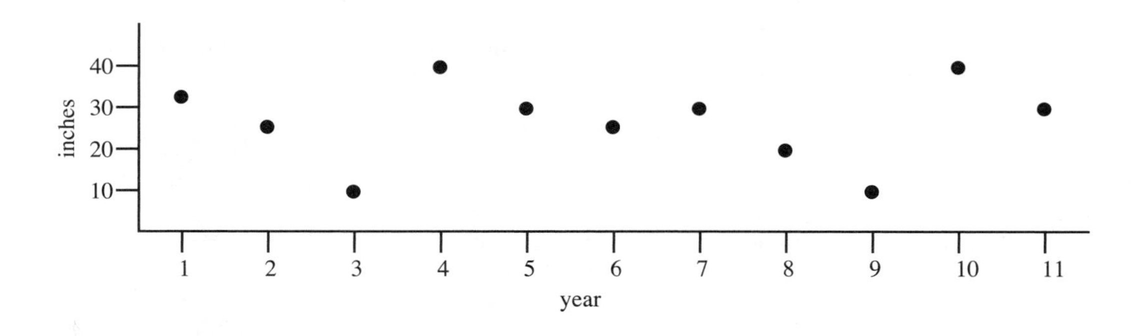


PS-57. A car rally started in Cowpoke Gulch and directed the participants to drive 4 miles north, 6 miles east, 3 miles north, and then 1 mile east to Big Horn Flat. Draw a diagram and calculate the direct distance (straight) from Cowpoke Gulch to Big Horn Flat.

# ADJACENT AND VERTICAL ANGLES

## UNDERSTANDING THE SPIRALING OF SKILLS AND CONCEPTS

The ideas and skills of geometry are spiraled throughout your text and over your additional years of any college preparatory math courses you take. What is spiraling? Put a dot on a piece of paper, then draw a curve that is somewhat like a circle but that keeps getting a little wider with each revolution. Suppose the dot represents the idea of "graphing." As you learn more and more about graphing, your understanding of it expands. Each new aspect of graphing could be placed along the expanding bands of the spiral. Each time you add a new band it is close to the previous band (what you already know) but takes the idea a little further. By the end of four years of college preparatory math you would have a spiral perhaps the size of a full sheet of paper with everything you learned about graphing. You would have several more pages for the other big ideas in the courses. You could even do a spiral that ties all the big ideas together.

The authors wrote this course so that you could build your understanding of the big ideas throughout the four years of your college preparatory math studies. **Clear understanding and mastery take time.** The authors wrote this text knowing you might not grasp all the information the first time you see it. So they give you an opportunity to use the ideas again and again, gradually adding more depth or something new to your understanding. If you do not fully understand a concept the first time, don't give up. **You will have the opportunity to see it and practice it again.** However, it is important to learn **as much as you can the first time** so you have a place to start the next time you see it.

**Some key pieces of advice:**

- Do not wait until the last minute. A full explanation of ideas does not occur later in the course.
- Do the spiraled practice problems consistently. The authors' goal is **mastery over time**, but you must practice regularly.
- Add information in boxes to your tool kit. Each new piece adds to your understanding.
- Keep a good notebook; many teachers structure their exams so that 60% of the material comes from prior units.

**MISSION POSSIBLE** (something to do)

Briefly discuss with your team how the following ideas have been spiraled through the course so far: area, perimeter, linear equations, and the Pythagorean Theorem. Then go on with the next problem.

PS-58. Use the resource page provided by your teacher or copy each of the nine figures below onto your paper. Determine which pairs of angles are adjacent and which are not by checking each figure to see if it has all three conditions--A, B, and C--described in the box below. Write "yes" or "no" next to each figure. Be sure to add this definition to your tool kit vocabulary grid.

We refer to angles that share the same vertex and one side as **ADJACENT ANGLES.** "Adjacent to" in the English language means "next to." For two angles to be next to each other we require that they satisfy these three conditions:

A. The two angles must have a common (shared or same) side.

B. They must have a common vertex (i.e., a common starting point for the sides).

C. They can have no interior points in common. The common side must be between the two angles (that is, no overlap is permitted).

Another way to express condition C is that one angle may not lie "on top of" the other angle.

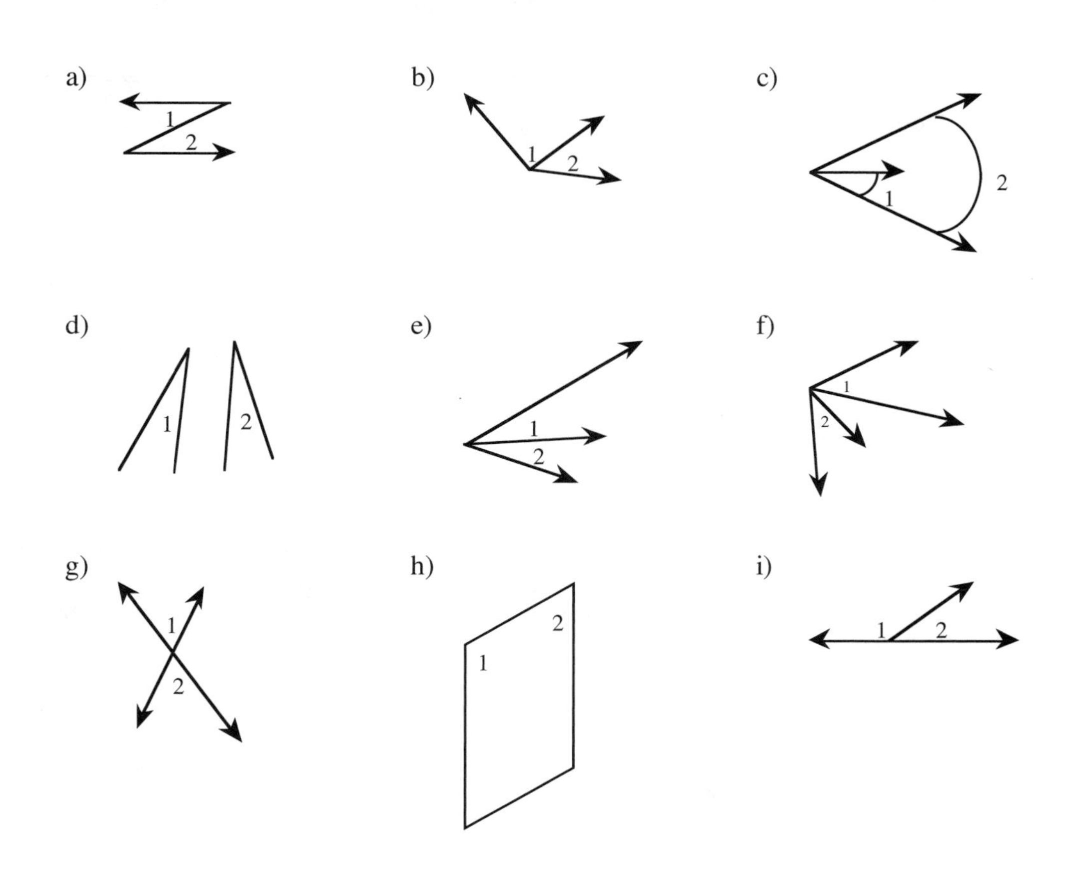

PS-59. Suppose two lines intersect as shown in the figure below, right. Complete parts (a) through (d), then add the definition in the box to your tool kit vocabulary grid.

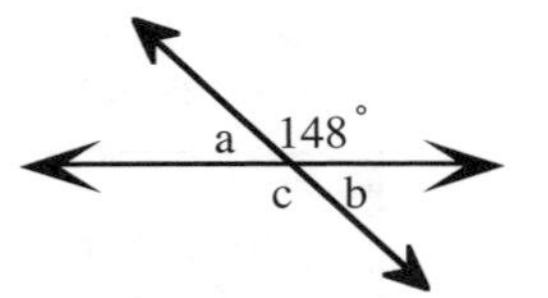


a) Calculate m∠b.

b) Compare your method for finding m∠b and m∠a. How do their measures compare?

c) Calculate m∠c and compare it to the angle to the right of ∠a and above ∠b.

d) Read the definition in the box below, then complete the conjecture below the box.

**VERTICAL ANGLES** are the two opposite (that is, non-adjacent) angles formed by two intersecting lines, such as angles 1 and 2 in the top figure at right. In the other figure, ∠3 by itself is not a vertical angle, nor is ∠4, although ∠3 and ∠4 together are vertical angles. Vertical is a relationship between PAIRS of angles, so you cannot call one angle a vertical angle.

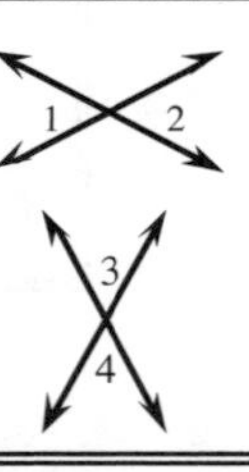


**Conjecture:** When two lines intersect, the resulting pairs of vertical angles are ___.

PS-60. Use the resource page from problems PS-47 and PS-48.

a) Compare the measures of the three pairs of angles below. How is each pair of angles related?

• m∠c to m∠a • m∠h to m∠f • m∠w to m∠n

b) Compare these results to your work in the previous problem. Your results here should confirm your conjecture about vertical angles.

PS-61. Use your conjectures about parallel lines and the angles formed with a third line to find the measures of the labeled angles. Show the step by step procedure you use AND name each angle conjecture (e.g., corresponding, alternate interior, vertical, or straight) you use that justifies your calculation.

a)

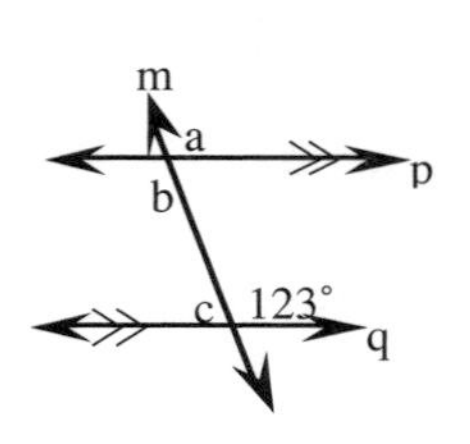


b)

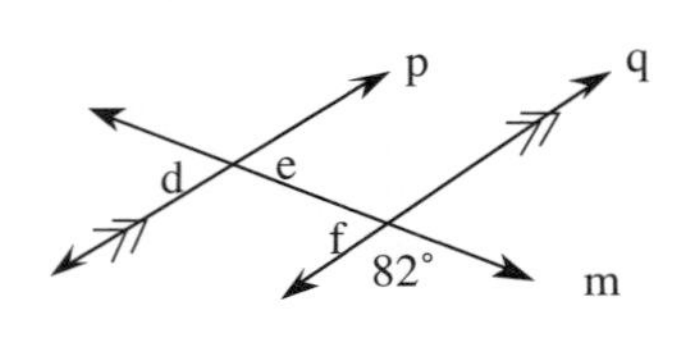


c)

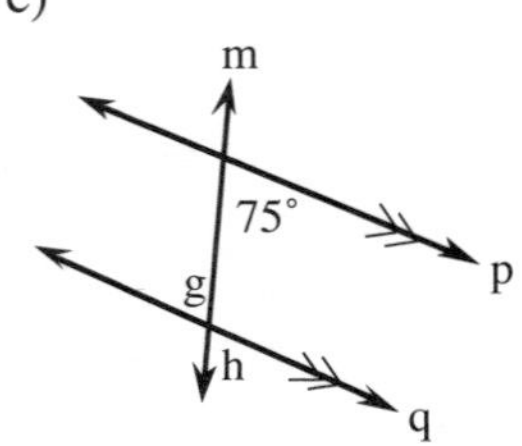

PS-62. If $m\angle 5 = 53^\circ$ and $m\angle 7 = 125^\circ$, find the measures of each numbered angle. Explain how you find each angle, citing definitions and conjectures that support your steps.

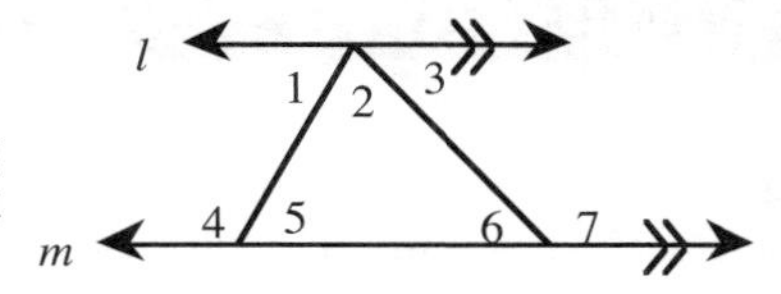

PS-63. Complete each part below based on the figure at right.

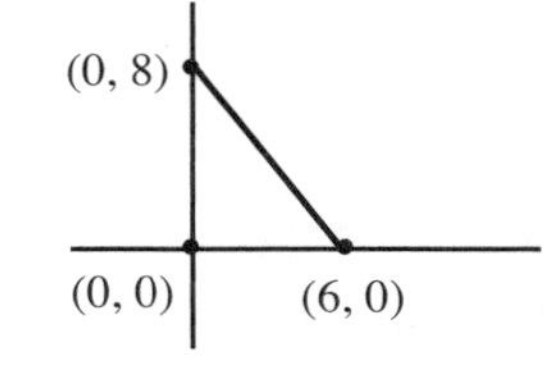


a) In the triangle at right, use the Pythagorean Theorem to find the length of the hypotenuse. Then find the midpoint of all the sides of the triangle.

b) Connect the midpoints of the segments in part (a) to form a new triangle. Is your new triangle also a right triangle? Discuss this and come to a conclusion that you can explain to the teacher or the class.

c) Calculate the area of both triangles.

d) What is the ratio of the length of a leg of the new triangle to the length of the corresponding leg of the big triangle?

e) What is the ratio of the area of the new triangle to the area of the original triangle?

PS-64. Find the measures of the angles requested and explain how you found them. Each letter is a separate problem.

a) If $m\angle 4 = 61^\circ$, find $m\angle 6$.

b) If $m\angle 1 = 48^\circ$, find $m\angle 8$.

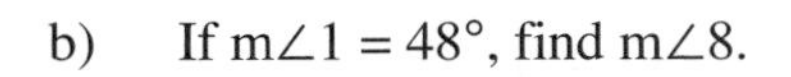

c) If $m\angle 2 = 137^\circ$, find $m\angle 8$.

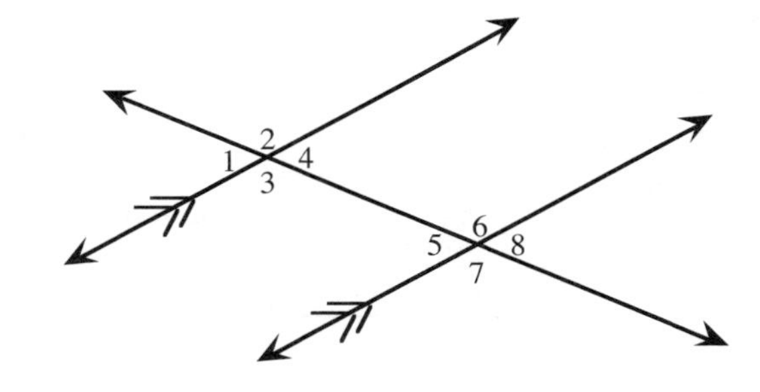


PS-65. If $x = \frac{1}{2}$ which will be larger, $(2x)^2$ or $2(x^2)$? Explain your choice.

PS-66. The length of the legs of a right triangle are consecutive integers (i.e., n and n + 1). The hypotenuse is 29 cm long. Draw a diagram, label the three sides, and find the lengths of the legs. Show the strategy or equation you use.

## PRACTICE WITH ANGLES

PS-67. Calculate the measure of each numbered angle and explain your reasoning in each case.

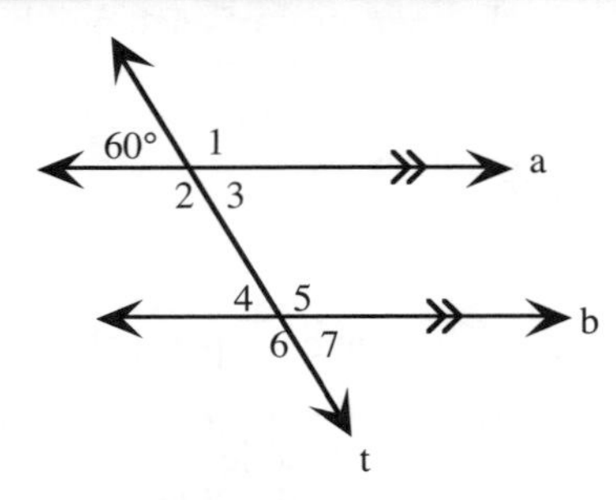


PS-68. Calculate the value of x for each part, (a) through (e), below. Write equations as necessary. Cite definitions and conjectures that justify your steps. Remember that each part is a separate problem!

> **EXAMPLE:** Suppose the measures of two angles in the figure below are stated algebraically:
>
> $$m\angle 1 = 8x + 14^\circ \text{ and } m\angle 6 = 2x - 3^\circ$$
>
> Since lines r and s are parallel, $m\angle 6 = m\angle 2$ because alternate interior angles are equal. Next, you can substitute $2x - 3^\circ$ for $m\angle 2$, and since $\angle 1$ and $\angle 2$ are supplementary, this means you can write the following equation:
>
> $$(8x + 14^\circ) + (2x - 3^\circ) = 180^\circ \quad (\text{from } m\angle 1 + m\angle 2 = 180^\circ)$$
>
> Another way to arrive at the same equation is to note that $m\angle 1 = m\angle 5$ (parallel lines, corresponding angles equal) and that $\angle 5$ and $\angle 6$ are supplementary. Solving either way, $x = 16.9^\circ$. You could now substitute for x and calculate the measures of $\angle 1$ and $\angle 6$.

a) $m\angle 2 = 5x - 30^\circ$, $m\angle 3 = 2x + 24^\circ$.

b) $m\angle 4 = 6x - 13^\circ$, $m\angle 5 = 4x + 33^\circ$.

c) $m\angle 3 = 5x + 10^\circ$, $m\angle 7 = 7x - 4^\circ$.

d) $m\angle 6 = 10x - 15^\circ$, $m\angle 8 = 15x$.

e) $m\angle 4 = x^2$, $m\angle 8 = 144^\circ$.

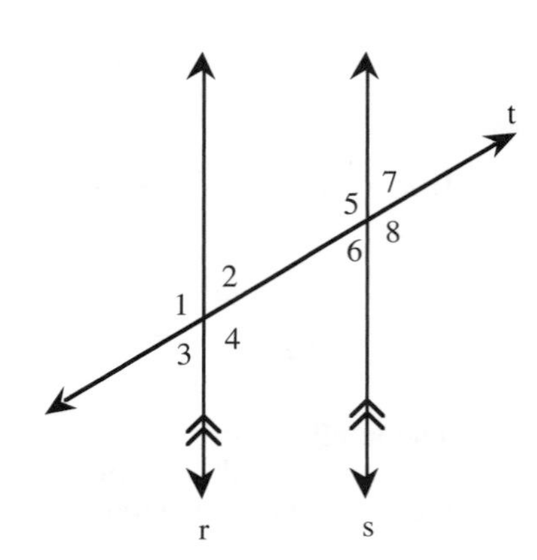

PS-69. Use your resource page from problems PS-47 and PS-48. Examine the pairs of angles b and d, g and j, and r and s on the resource page. If you add the measures of each <u>pair</u>, what do you observe?

a) Write a conjecture about two angles on the <u>same</u> <u>side</u> of a transversal which are <u>between</u> two parallel lines. Note: these are called interior angles with respect to the two parallel lines.

**Conjecture:** The sum of the measures of two interior angles on the <u>same</u> side of a third line (transversal) is ____.

Use your conjecture and the figure in the previous problem to answer parts (b) and (c).

b) If $m\angle 2 = 67^\circ$, what is $m\angle 5$?

c) If $m\angle 4 = 4x + 23^\circ$ and $m\angle 6 = 3x + 17^\circ$, find $m\angle 4$.

PS-70. Notice that both <u>pairs</u> of angles are called vertical, even though one pair (a and b) looks horizontal. Review the definition of vertical angles (PS-59) and decide with your team if this issue is a problem.

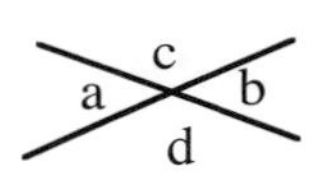


PS-71. Use the figure below to answer these questions.

a) If $m\angle 1$ is $37^\circ$, find the measures of the other three angles.

b) If $m\angle 1$ is $12^\circ$, find the measures of the other three angles.

c) If $m\angle 1$ is $y^\circ$, express the measures of the other three angles in terms of y.

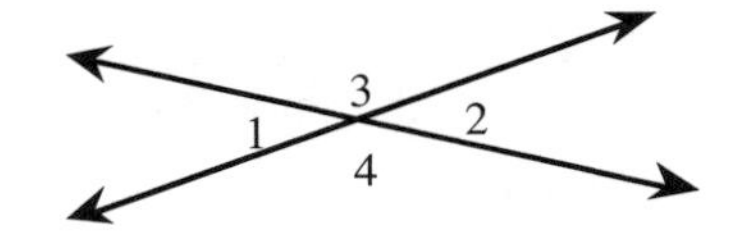


PS-72. Obtain a piece of graph paper and graph <u>both</u> equations on the <u>same</u> set of axes.

a) $y = \frac{3}{4}x + 4$

b) $y = -\frac{3}{4}x - 2$

c) Where do the lines cross?

d) Shade in the region bounded by (inside) these two lines and the y-axis. Find the area of this region. Did you have to approximate or is your answer exact?

PS-73. Calculate the measures of the indicated angles. Write equations as necessary. Each part is a separate problem.

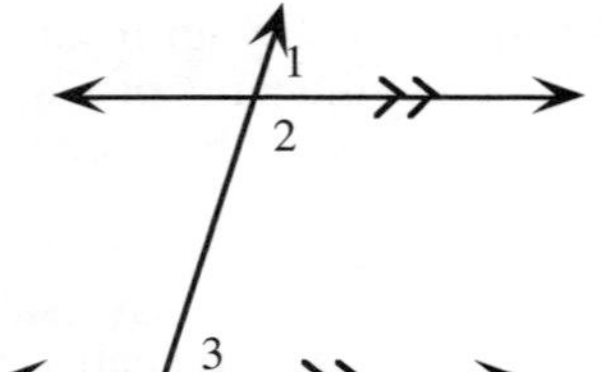
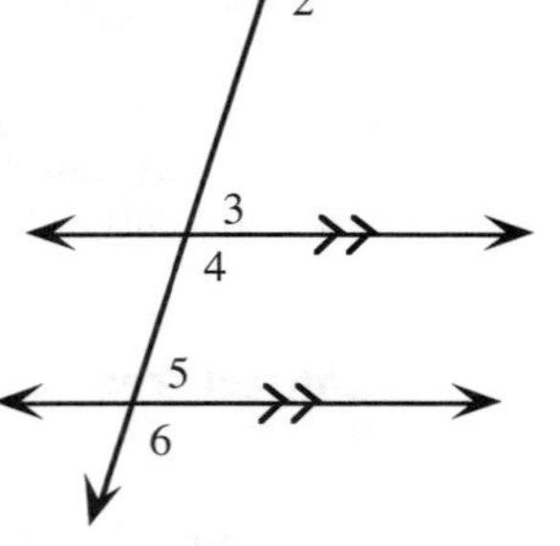


a) If $m\angle 1 = 73°$, find $m\angle 5$ and $m\angle 4$.

b) If $m\angle 4 = 2x - 35°$ and $m\angle 6 = x + 46°$, find x and $m\angle 6$.

c) If $m\angle 2 = 4x - 19°$ and $m\angle 3 = x + 44°$, find x and $m\angle 2$.

PS-74. Solve for y in terms of x, that is, each equation will become y = ... .

a) $6x + 5y = 20$

b) $4x - 8y = 16$

PS-75. If $4x - 3y < 9$, complete the following:

a) Replace x and y with 0 to show that the inequality will be true.

b) Show that if $x = 0$ and $y = 1$ the inequality will also be true.

c) Show that if $x = 4$ and $y = 2$ the inequality will be false.

PS-76. Demonstrate (prove) that the points A(1, 9), B(4, 5), and C(13, -7) are on the same line. A graph alone is insufficient evidence.

PS-77. Calculate the area of the shaded region. Use the appropriate units.

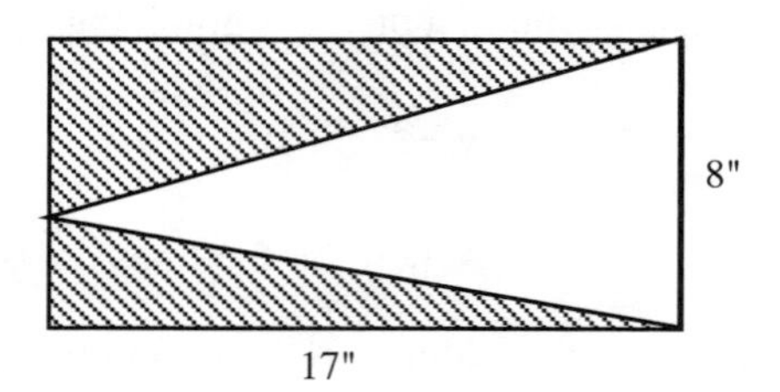

## GRAPHING LINEAR INEQUALITIES

PS-78.

Recall that the solution to a single variable equation (such as $2x + 3 = 9$), when graphed on the number line, can be thought of as a dividing point for the line. This point is useful for finding the solution to inequalities. Do parts (a) and (b) to verify that the solution graph at right matches your solution.

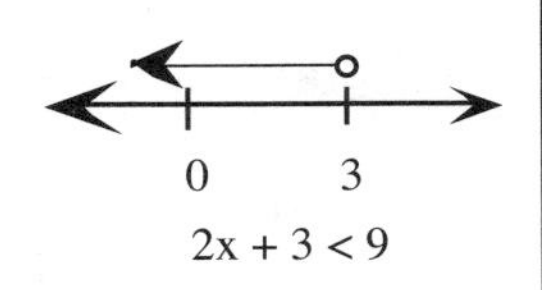


a) Solve $2x + 3 < 9$ by solving the inequality as if it were an equation.

b) Test the points $x = -2, 0, 1.5, 4$, and $6$ by substituting them, one at a time, into $2x + 3 < 9$ to see which ones result in a true statement. Draw a ray left or right from $x = 3$ to show the full solution.

PS-79.

a) Sketch the graph of $y = 2x + 3$ on a sheet of graph paper.

b) Suppose the linear equation $y = 2x + 3$ that you just graphed becomes the inequality $y < 2x + 3$. Then the line $y = 2x + 3$ can be thought of as the **DIVIDING LINE** for the two-dimensional graph since it divides the graph into two regions. **All of the points that make the inequality true lie on one side of the line.** We can test ANY point not on the line to see if it makes the inequality true.

c) Substitute the points (0, 0), (2, 2), and (3, -1), one at a time, into the original form of the inequality ($y < 2x + 3$). Is each result true or false? Where are these points located in relation to the dividing line?

d) Substitute the points (0, 5), (-2, 3), and (-3, 0), one at a time, into the original form of the inequality ($y < 2x + 3$). Is each result true or false? Where are these points located in relation to the dividing line?

e) Which side of the dividing line appears to contain values that make the inequality true? The **region** of the graph on this side of the line should be shaded to indicate that all the points in the region are true for the inequality. Shade it, and describe the region in relation to the line.

PS-80.

How many points do you need to test to find out which side of the dividing line is the solution? What point will be easiest to test for most linear inequalities?

PS-81.

Note that for $<$ and $>$, the line is dashed to show that it is not part of the solution; for $\leq$ and $\geq$ the line is solid. Should your graph from problem PS-79 look like the one at right? Why or why not?

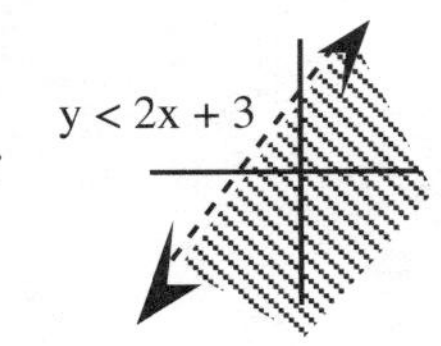


CAUTION: If the inequality is not in slope-intercept form and you have to solve it for y, always use the original inequality to test a point, NOT the solved form.

PS-82. Graph the following linear inequalities.

a) $y < \frac{2}{3}x - 1$

b) $4x + 3y \geq 9$

PS-83. Graph each equation on one set of axes on the upper half of a sheet of graph paper.

a) $y = \frac{1}{2}x - 1$

b) $y = \frac{1}{2}x + 1$

c) $y = \frac{1}{2}x + 3$

d) Now graph a fourth line on the same set of axes that fits the pattern of these three lines. Write the equation for your line.

PS-84. Graph both inequalities on the lower half of the graph paper you used in the previous problem.

a) $y \geq \frac{1}{2}x - 3$

b) $y \leq \frac{1}{2}x + 2$

PS-85. Do you think $y = \frac{3}{2}x + 5$ and $y = \frac{3}{2}x - 2$ would be parallel when graphed? Explain.

PS-86. Examine the equations and inequalities from the preceding three problems. The lines in these three problems are all parallel. Write a conjecture that explains how to use linear equations in slope-intercept form to determine whether or not lines are parallel.

PS-87. Examine the region common to both inequalities in problem PS-84. If possible, find the area of this region. If not, explain why not.

PS-88. Read the definition at the top of the box below, add it to your tool kit vocabulary grid, then make two lists, one headed complementary angles and the other supplementary angles. Name all the pairs of angles in the figures below that are complementary and those pairs that are supplementary.

**COMPLEMENTARY ANGLES** are two angles whose measures have the sum of 90°.

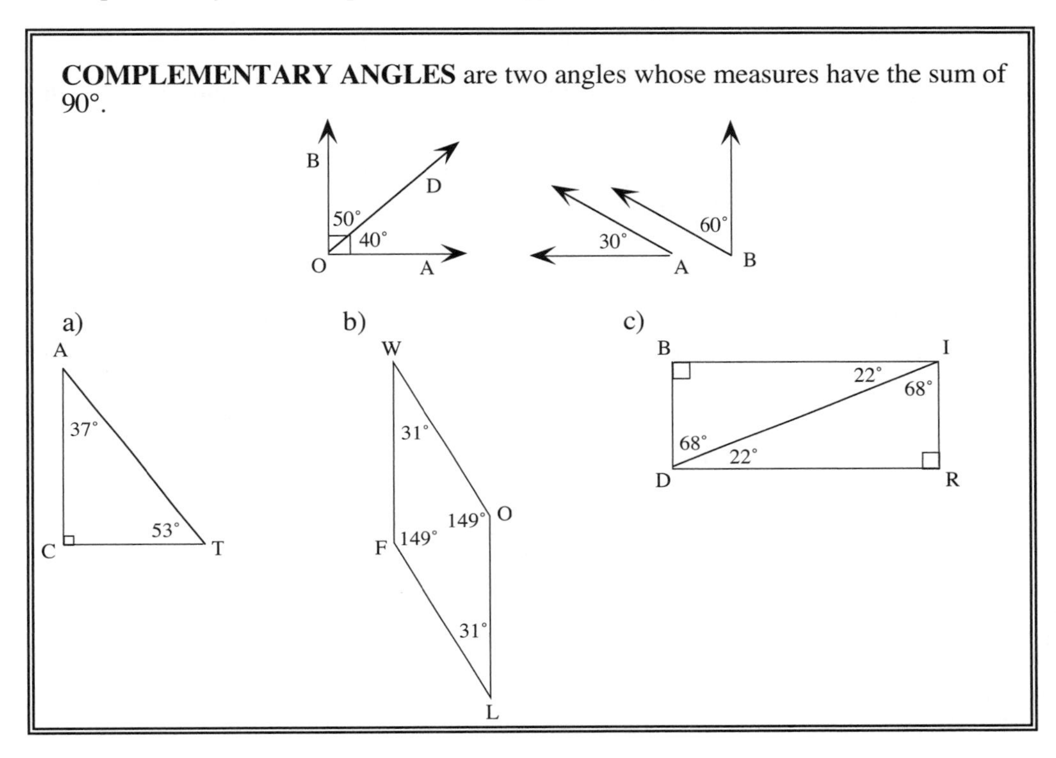


PS-89. Use your conjectures and definitions from recent problems to solve parts (a) through (d). Explain how you find each answer and name the definitions and/or conjectures you use. Write equations as necessary. Note that each part is a separate problem.

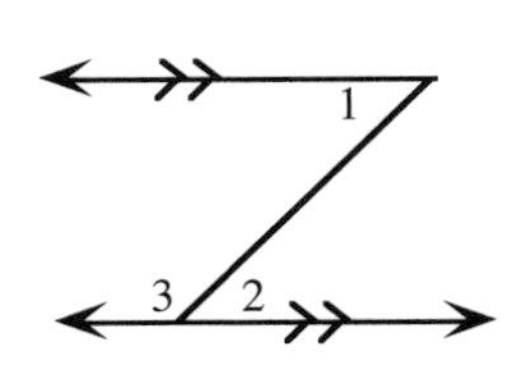


a) If $m\angle 1 = 59°$, find $m\angle 2$ and $m\angle 3$.

b) If $m\angle 1 = x + 27°$ and $m\angle 2 = 4x - 39°$, find x and $m\angle 1$.

c) If $m\angle 3 = 3x + 36°$ and $m\angle 2 = 4x - 52°$, find x and $m\angle 2$. Write an equation.

d) If $m\angle 1 = x + 16°$ and $m\angle 3 = 3x + 8°$, find x and $m\angle 1$.

PS-90. Vivian has $1.00 and she is thinking about buying a lottery ticket. As she is walking to the local Circle G store to buy a ticket, a girl scout approaches her and asks Vivian if she would like to buy a raffle ticket. "It only costs $1.00 per ticket and you can win $1000!"

"How many tickets have been sold?" Vivian asks.

"Only 500," replies the girl scout.

a) If Vivian's is the last ticket to be sold, what is the probability that she will win the $1000?

b) Suppose her state lottery payoff is $30,000,000. Which should Vivian buy: a lottery ticket or a raffle ticket?

PS-91. Of all the graphs below, which one best demonstrates the relationship between a person's I.Q. (that is, their basic intelligence) and their age? Justify your answer.

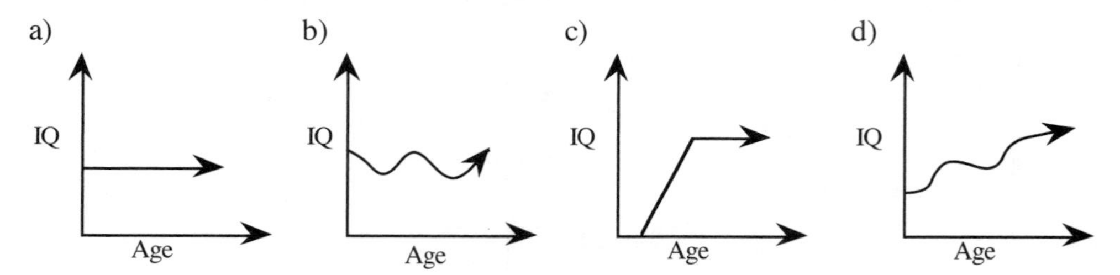

## ANGLES AND TRIANGLES

PS-92. Each person in your team should carefully draw a triangle that fills about a quarter of a sheet of paper. Each triangle should have a <u>different</u> shape.

a) Carefully cut out your triangle.

b) Draw a horizontal line across the lower portion of the remaining paper and mark point A somewhere near the middle.

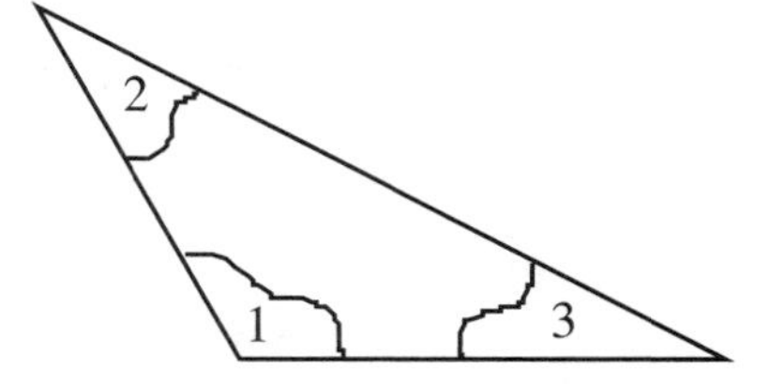


c) Mark fairly large pieces in each corner of your triangle as shown in the figure at right. Then number the corners 1, 2, and 3. Which corner gets which number is not important.

d) Separate each corner of the triangle by tearing it along the jagged line.

e) Place the three resulting angles on the upper side of the line with all three vertices at the labeled point A. Be sure that the angles do not overlap.

f) Compare the results of each person's work and write a conjecture about the sum of the three angles of any triangle.

PS-93. Answer the questions and be ready to explain the problem to your teacher or the class.

a) $m\angle 1 + m\angle 2 + m\angle 3 = 180^\circ$. Why?

b) $m\angle 4 = m\angle 1$. Why?

c) $m\angle 5 = m\angle 3$. Why?

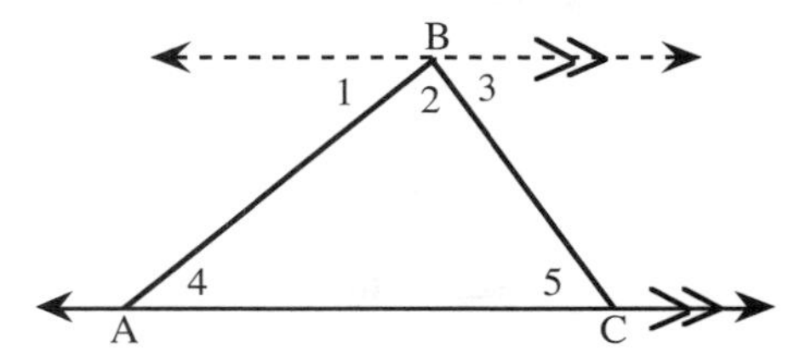


d) Explain how you can put the equations from parts (a), (b), and (c) together to form an equation that represents the statement in the box below.

> **TRIANGLE INTERIOR ANGLE SUM THEOREM**
>
> **The sum of the measures of the interior angles of a triangle is 180°.**
>
> This statement is no longer a conjecture. It is now a **theorem** because doing parts (a) - (d) produced a sequence of true, connected facts that verify the final statement. This process <u>proves</u> the conjecture, and a proven conjecture is known as a theorem.

PS-94. You already have a conjecture about alternate interior angles. Compare that conjecture to the statement below and write a short statement about how they differ. Then read the statement in the box below and use the new conjectures to do parts (a) through (f).

If a pair of alternate interior angles are equal, then the two lines are parallel.

> Conditional statements can be written in reverse. The reverse of the alternate interior angle conjecture is shown in the statement above. Next consider the corresponding angle conjecture:
>
> **Conjecture:** If two parallel lines are cut by another line, then pairs of corresponding angles are equal.
>
> **Reversal:** If a pair of corresponding angles are equal, then the two lines are parallel.
>
> The second statement (reversal) is called the **converse** of the first. We will study this in more detail in Unit 6. The two converses of our earlier conjectures tell us what condition (equal pairs of alternate interior or corresponding angles) is necessary to assure us that the pair of lines are parallel.

**DIRECTIONS:** On the resource page provided by your teacher, use the two converses in the box (above and on the resource page) and the information given in each part (a through g) below to determine **whether line *a* is parallel to line *b*.** Write yes or no and explain your reasons. You may want to use pencils, straws, or something like them to represent the lines to help you visualize the effects of having certain angles equal.

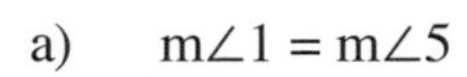

a) $m\angle 1 = m\angle 5$

b) $m\angle 3 = m\angle 6$

c) $m\angle 5 = m\angle 8$

d) $\angle 4$ and $\angle 6$ are right angles.

e) $m\angle 1 = m\angle 7$

f) $m\angle 3 = m\angle 5$

g) $m\angle 2 = m\angle 7$

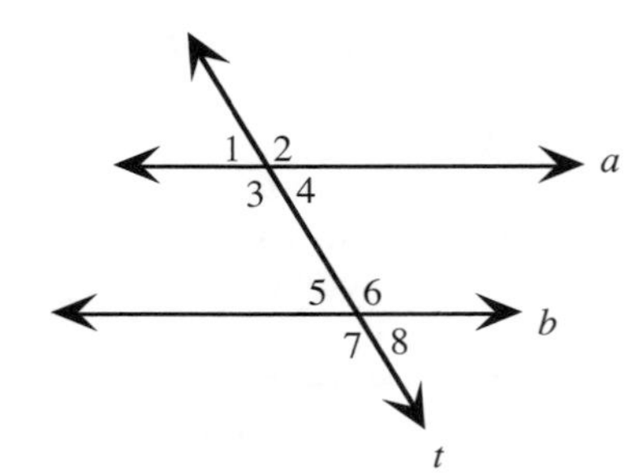

PS-95. Davis knows the following three facts are true:

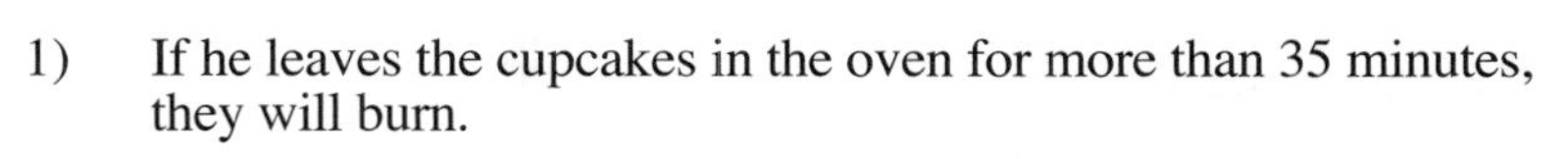

1) If he leaves the cupcakes in the oven for more than 35 minutes, they will burn.

2) His friends in Eagle Scout Troop 715 do not like burned cupcakes. In fact, burned cupcakes put them in a rather foul mood.

3) Being selected to represent the troop at the national jamboree is a popularity contest: if the troop members like you, they will choose you to attend.

The troop will meet today at 5:00 PM. At 3:00 PM, Davis puts some cupcakes into the oven and starts working on his homework. He falls asleep and does not wake up until 3:55 PM. What do you think will happen? Discuss this situation with your team. Write a clear description and support your statements with facts. If you express an opinion or make a guess, be sure to note that you are doing so (as opposed to stating a fact).

PS-96. Find the measures of the missing angles by using the theorem about the sum of the measures of the angles of a triangle.

a)

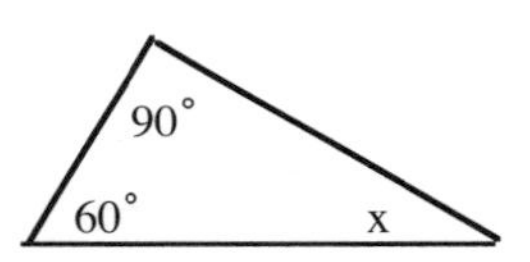


b)

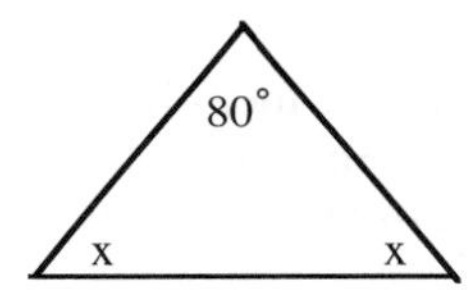


c)

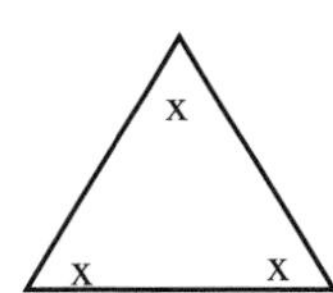


d)

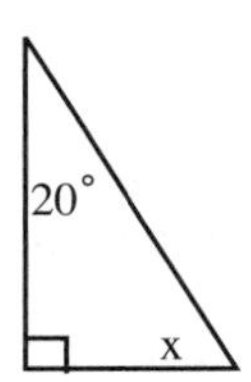


e)

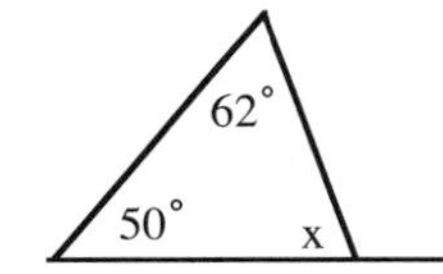


f)

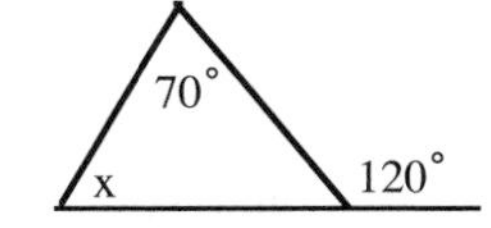

PS-97. Examine the figures at right.

a) What is happening to $m\angle CBD$ as the figure number increases?

b) What is happening to $m\angle ABC$ as the figure number increases?

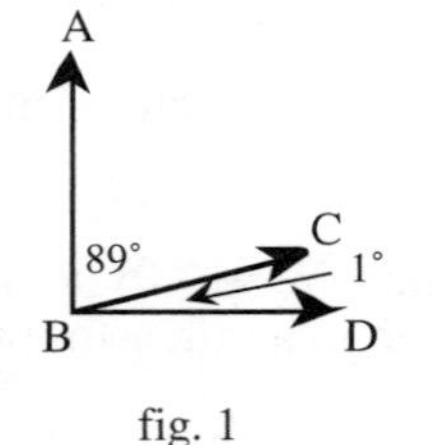

fig. 1

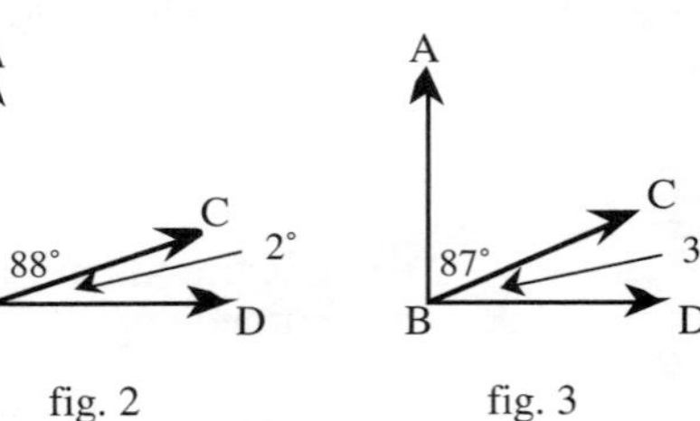

fig. 2 fig. 3

c) What will be $m\angle CBD$ in figure 47?

d) What will be $m\angle ABC$ in figure 47?

e) For <u>any</u> figure n, what will be $m\angle CBD$ and $m\angle ABC$?

f) Based on this pattern, what would be $m\angle CBD$ for figure 100? Do you think this pattern will go on forever? Explain.

PS-98. Use the conjectures and definitions about parallel lines to calculate the requested angle measures. Parts (a) and (b) are <u>separate</u> problems.

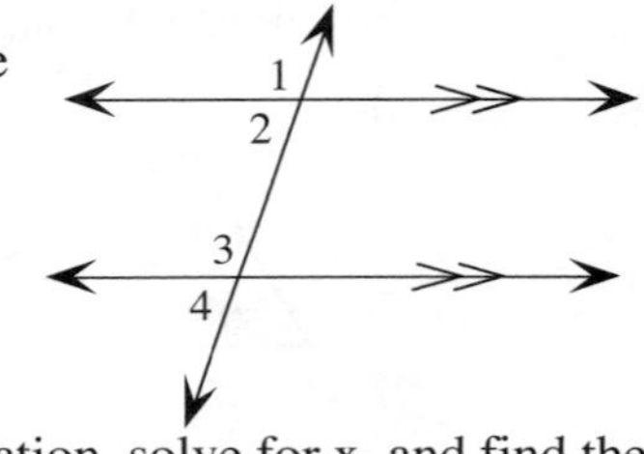


a) If $m\angle 3 = 102°$, find the measures of the other three angles. Explain your reasons using the definitions and conjectures from your vocabulary grid.

b) If $m\angle 2 = 2x + 66°$ and $m\angle 3 = 5x - 27°$, write an equation, solve for x, and find the measure of $\angle 3$. Explain your reasons.

PS-99. Draw two angles which fit each of the following descriptions. If no such pair of angles can be drawn, explain why not. Parts (a), (b), and (c) are <u>separate</u> problems.

a) Complementary and equal   b) Supplementary and vertical   c) Vertical and obtuse

PS-100. Solve $3(x + 8) = 2(x - 1)$. Check your solution.

PS-101. Show that $x = 12$ is <u>not</u> the solution to $6x + 3x = x + 100$.

PS-102. If $x = 6$, find each value.

a) $x^2$   b) $3(x^2)$   c) $(3x)^2$

## EXTERIOR ANGLES

PS-103. **DIRECTIONS**: Study the boxes below and add the definitions of exterior and remote interior angles to your tool kit vocabulary grid. Then, on the resource page provided by your teacher, calculate the missing angle measures in each figure and record them in the table. When you have completed the resource page, look for a pattern in the relationship between the measure of ∠BCD and the sum of the measures of ∠A and ∠B.

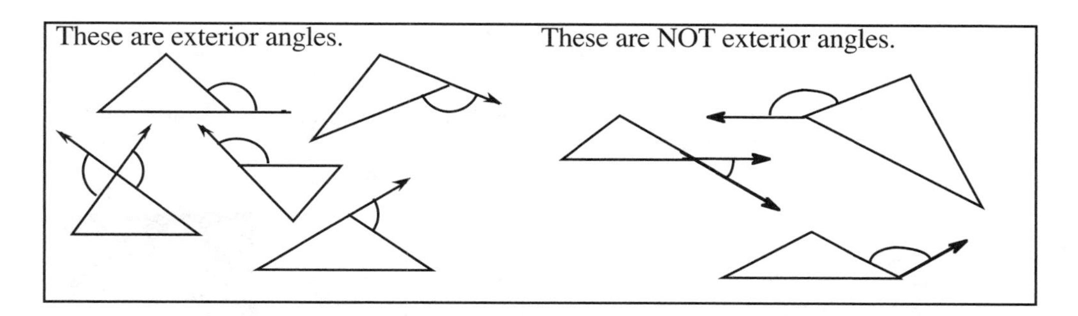


**EXTERIOR ANGLES** are formed by extending a side of the triangle. The two angles across the triangle from the exterior angle are called **REMOTE INTERIOR ANGLES.** In each figure below, ∠A and ∠B are remote interior angles with respect to exterior angle BCD.

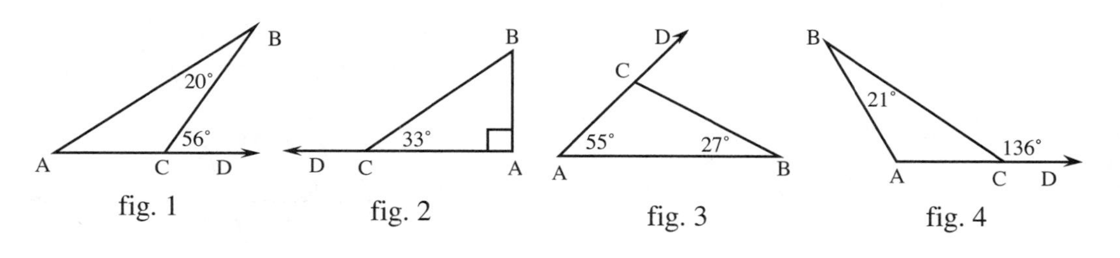


| Figure Number | m∠A | m∠B | m∠ACB | m∠BCD | m∠A + m∠B |
|---|---|---|---|---|---|
| 1 | | | | | |
| 2 | | | | | |
| 3 | | | | | |
| 4 | | | | | |

a) Compare your results for m∠BCD and the sum of m∠A and m∠B (the remote interior angles) for each figure. Discuss your observations with your team.

b) Write a conjecture about the relationship of an exterior angle to the two remote interior angles.

PS-104. Calculate the measures of the angles requested. Each part is a separate problem.

a) If $m\angle 1 = 53°$ and $m\angle 2 = 71°$, find $m\angle 4$.

b) If $m\angle 2 = 78°$ and $m\angle 4 = 127°$, find $m\angle 1$.

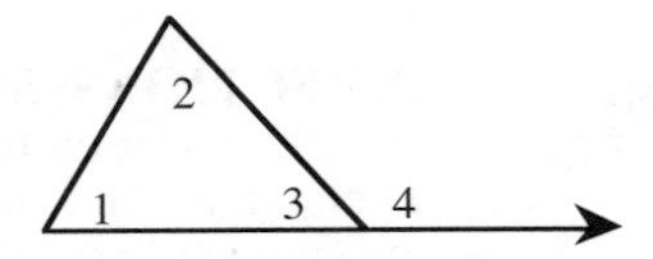


PS-105. Graph each inequality on the same set of axes by graphing the dividing line, testing a point and shading the solution.

a) $y \le -\frac{1}{3}x + 2$  b) $y > \frac{3}{4}x - 3$  c) $2x + y > 1$

d) Darken in the region where all the shadings overlap. What shape is this? Approximate its area.

PS-106. Based on the given information, determine which pairs of lines, if any, are parallel. If none are necessarily parallel, write "none." Straws, coffee stirrers, or dry pasta will help in making decisions by forming angles and holding them rigid.

a) $m\angle 2 = m\angle 7$

b) $m\angle 3 = m\angle 11$

c) $m\angle 1 = m\angle 12$

d) $m\angle 13 = m\angle 12$

e) $\angle 6$ and $\angle 7$ are supplementary

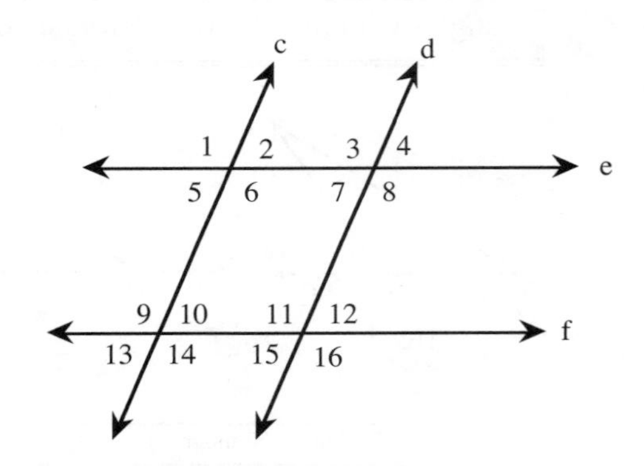


PS-107. In each figure below, the area of the triangle is 20 square units. Find the height of each triangle.

a)

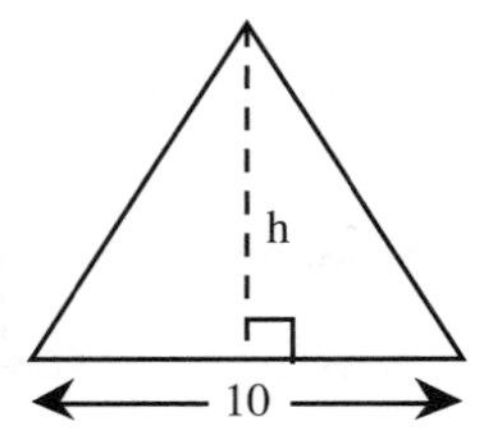


b)

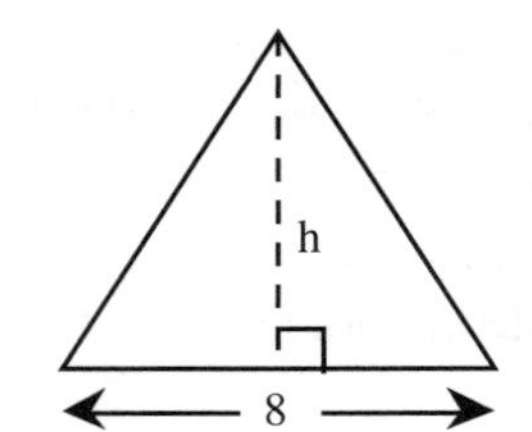


c)

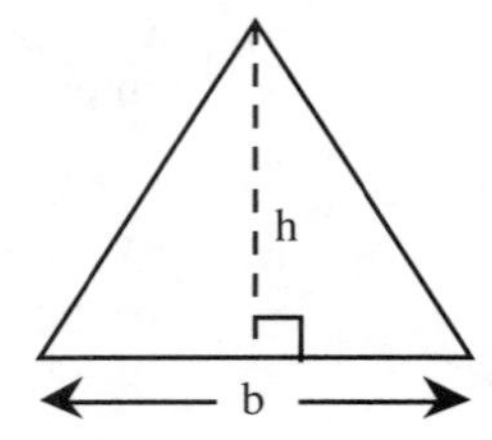

PS-108. Two angles are complementary and one of them is 32°. What must be the measure of the other? How do you know?

PS-109. Kim wants to build a 3-foot wide brick sidewalk around her rectangular pool which is 20 feet by 60 feet. If it takes 4.5 bricks every square foot, how many bricks will Kim need? Draw a diagram, write an equation, and make clear any assumptions you make.

PS-110. Given $y = \frac{2}{3}x + 3$. Use your conjecture about parallel lines and slope to write the equation of the line parallel to the given line with a y-intercept of $-1$.

PS-111. On a sheet of graph paper, scale axes for $-3 \le x \le 10$ and $-7 \le y \le 7$.

a) Graph the line $y = -\frac{4}{7}x + 4$.

b) Draw another line that will form an isosceles triangle with the line you drew in part (a) and the y-axis. Write the equation of that line.

c) Outline the triangle formed by the intersection of the two lines and the y-axis. Demonstrate (prove) whether this triangle is an equilateral triangle or not.

PS-112. Show that the origin (0, 0) is a solution to the inequality $y < -6x + 8$ by testing it.

## UNIT AND TOOL KIT REVIEW

PS-113. Calculate the difference of the squares of several different pairs of consecutive integers by making and completing a table like the one below.

| Integers | Squares | Difference of the Squares |
|---|---|---|
| 1, 2 | 1, 4 | 4 - 1 = 3 |
| 2, 3 | 4, 9 | |
| 3, 4 | | |
| 4, 5 | | |
| 5, 6 | | |
| ... | | |
| n, n + 1 | | |

a) Make a conjecture about the difference of consecutive integer squares by examining the pattern in the table that you have created above.

b) Use the last row of your table to show that your conjecture will work for <u>any</u> two consecutive integers.

PS-114. Make a table with two entries, the first titled odd counting numbers (1, 3, 5, ...) and the second the square of each number, minus 1.

a) Complete the table for the first seven odd numbers.

b) Write a conjecture for the results in your second column (or row and discuss it with your team.

c) What would be the result for the nth odd number, that is, 2n - 1?

d) Add a third row or column titled "Column (or Row) 2 divided by 8." Take each number in the second column (or row) and divide it by 8. (They <u>are</u> divisible by 8, right?) Describe the resulting sequence (familiar, isn't it!).

e) Write the result you got for the nth odd number in simpler form and use this form to help construct an argument that eight will always divide the result.

PS-115. Use the picture at right to answer each separate problem. State reasons that justify your answers.

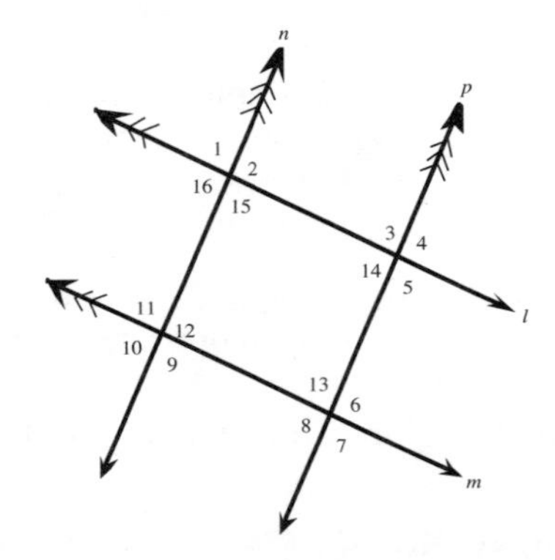

a) If $m\angle 1 = 50^\circ$, what is $m\angle 11$?

b) If $m\angle 14 = 100^\circ$, what is $m\angle 6$?

c) If $m\angle 2 = 105^\circ$, what is $m\angle 3$?

d) If $m\angle 9 = 60^\circ$, what is $m\angle 14$?

e) If $m\angle 16 = 95^\circ$, what is $m\angle 5$?

PS-116. Solve for x in each of the following figures.

a)

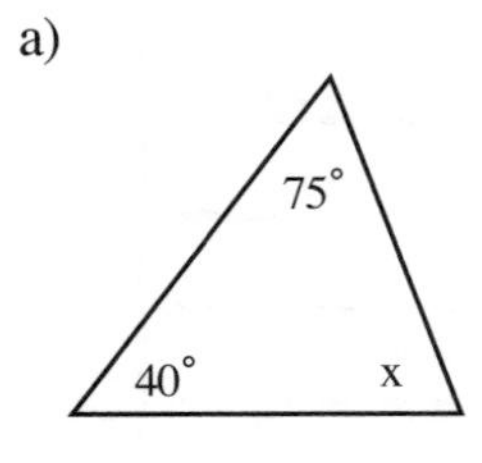


b)

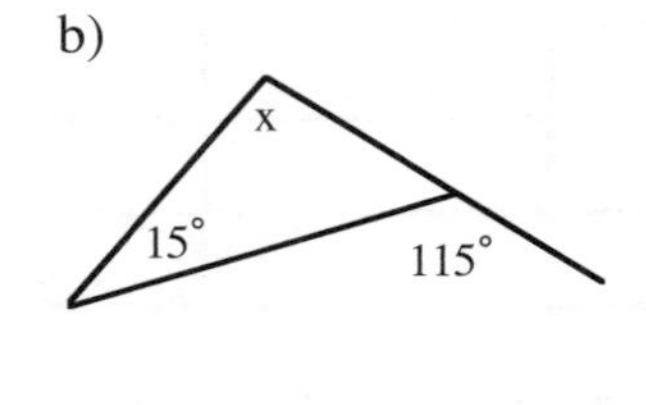


c)

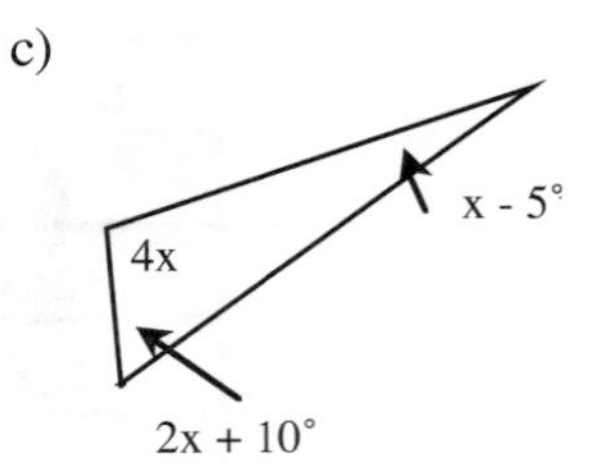


PS-117. Graph each of the following inequalities.

a) $y \geq -\frac{2}{3}x - 6$

b) $y < 2x + 5$

PS-118. How would the sum of the angles in a triangle be different if we agreed that a straight angle only had 100°?

PS-119. Start with the line $y = -\frac{5}{4}x - 2$.

a) Write the equation of the line parallel to it with a y-intercept of 3.

b) Write the equation of the line parallel to it with a y-intercept greater than 5.

PS-120. Is $y = -2x - \frac{5}{4}$ parallel to the line in the previous problem? Explain.

PS-121. VOCABULARY QUICK QUIZ

Here is a check to see how well you have learned the vocabulary of the unit. Copy each term below onto your homework paper. Next to each term draw a picture representing that term. Try to do as many as you can **without** looking back through the unit or your tool kit.

- bisect (segments, angles)
- vertex
- ray
- angle
- perpendicular lines
- parallel lines
- acute angle
- obtuse angle
- straight angle
- exterior angle
- vertical angles
- supplementary angles
- complementary angles
- alternate interior angles
- corresponding angles
- adjacent angles
- isosceles triangle
- equilateral triangle

PS-122. TOOL KIT CHECK-UP

Your took kit is your quick reference for the vocabulary and ideas of geometry, as well as any reminders you need about the algebra in the course.

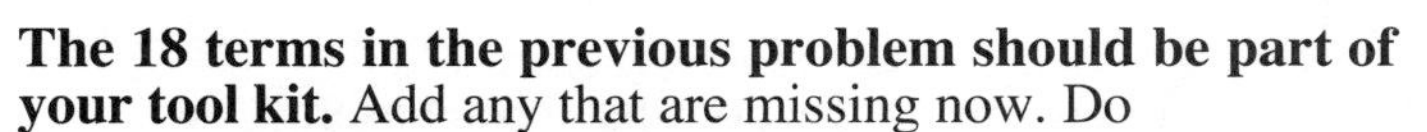

**The 18 terms in the previous problem should be part of your tool kit.** Add any that are missing now. Do the same for the terms and conjectures listed below. If you need help, skim through the unit looking for bold type and double line boxes. Also use the index in the back of the book.

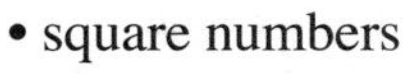

- square numbers
- triangular numbers
- alternate interior angle conjecture
- corresponding angle conjecture
- vertical angle conjecture
- how to name an angle
- linear inequalities
- exterior angle conjecture
- slopes of parallel lines conjecture
- sum of the measures of the angles in a triangle theorem

## PZL-15 HOW TO STUDY FOR MATH TESTS

Well, there isn't much more for me to say to help you succeed in this course, so this will be my last visit with you. Before I go, I thought we should talk about how to study for a test. Actually, studying for a test isn't much different than what you do each day IF you are following a consistent plan to learn geometry.

If I was sitting with your team and asked each of you how you study for a math test, I bet at least one person in the team would say, "I study the night before the test and memorize everything I can." This method does not work very well if you want to learn geometry. **The best way to study for an exam in this course is to participate in every class session**. In other words, you have to **learn the material as you go along**. Once you learn the material, then you go back and **review several days ahead of the exam**. You can practice the night before to smooth any rough edges. There *is no way you will be able to learn the material for the first time the night before an exam.*

I do understand that it is difficult to review several days ahead, but the authors have built in some ways for you to do it. That is what the tool kit assignments are for. Some teachers even give team quizzes and tests. That's another way to review. The real key is to **participate in every class session and ask good questions**. Then the night before a test is about polishing what you have learned, not a panicked cram session.

Here's a summary of the points I have made so far about studying for success in this geometry course. Call it the "Master Study Plan."

1. Do not fall behind.
2. Communicate with your partner, team, and teacher.
3. Build a complete tool kit.
4. Keep a good record of correct homework problems (notebook).
5. Know where to look for information (see PZL-6 in Unit 1).
6. Make the connections.

Good luck with the rest of the course. I hope my suggestions have been useful.

## GEOMETRY mini - PROJECT

PS-123. Discover a pattern that has both a geometrical representation <u>and</u> a numerical representation. Problem PS-32 is an example, but you will have to find another pattern.

a) Name your pattern and display the name and the pattern in a <u>pleasing</u> and <u>creative</u> way. This is to be done on a 12" x 9" sheet of paper, preferably card stock or construction paper.

b) Put your name and class period on the back. You will be graded on correct geometric pattern, correct numerical pattern, creativity, and clarity of presentation.

UNIT 4

# The TransAmerica Pyramid

SPATIAL VISUALIZATION

# *Unit 4 Objectives*

***The TransAmerica Pyramid:*** **SPATIAL VISUALIZATION**

This unit will help you develop your visualization skills. In particular, you will build some 3-dimensional solids and examine their characteristics, as well as practice drawing them on 2-dimensional surfaces.
You will extend your work with areas of polygons to areas of 3-dimensional surfaces and begin calculating volumes of solids.

The algebra focus in this unit parallels visualization by examining non-linear graphs, particularly parabolas and non-linear inequalities.

In this unit you will have the opportunity to:

- explore and discuss visualization activities.
- learn to draw isometric representations of cube stacks.
- build selected polyhedra, especially prisms and pyramids, to understand their properties and assist you with drawing them on 2-dimensional surfaces.
- calculate surface areas and volumes of polyhedra.
- extend graphing to non-linear relationships, especially parabolas.
- solve factored quadratic equations using the Zero Product Property.

*Read the problem below. Over the next several days you will learn what you need to solve it. Do not attempt to solve it now.*

SV-0. The TransAmerica building in San Francisco is built of concrete and is shaped like a square based pyramid. The building is periodically power-washed using one gallon of cleaning solution for each 250 square meters of surface. How much cleaning solution is needed to wash the TransAmerica building if an edge of the square base is 96 m and the height of the building (not the height of a sloping face) is 220 m? Drawing a sketch of the building will be helpful.

*UNIT* 4

ALGEBRA
GRAPHING
RATIOS
GEOMETRIC PROPERTIES
PROBLEM SOLVING
SPATIAL VISUALIZATION
CONJECTURE, EXPLANATION (PROOF)

# *Unit 4*

## *The TransAmerica Pyramid:* SPATIAL VISUALIZATION

## INTRODUCTION TO QUADRATIC FUNCTIONS

SV-1. **DIRECTIONS**: Your pet Pit Bull Babs has been digging up your neighbor's flower beds for some time, and now you must keep her restrained. You have 36 meters of fencing to enclose her in a rectangular pen and you would like her to have as much room as possible. How should you arrange the fencing?

To answer this question, draw rectangles on graph paper to represent the possible fencing arrangements. Share this work among team members. Remember that the perimeter must always be 36 units. Complete a table like the one below on the resource page provided by your teacher.

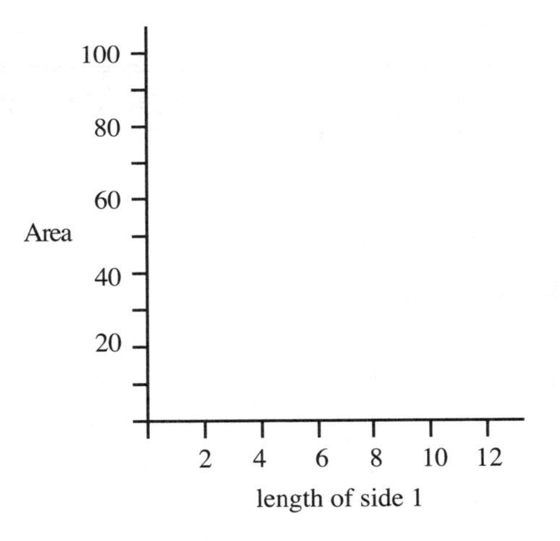


| Measure of 1st side | Measure of 2nd side | Area | Perimeter |
|---|---|---|---|
| 1 | 17 | 17 | 36 |
| 2 | 16 | 32 | 36 |
| 3 | | | |
| 4 | | | |
| 5 | | | |
| 6 | | | |
| 7 | | | |
| 8 | | | |
| 9 | | | |
| 10 | | | |
| 11 | | | |
| 12 | | | |

a) Graph your results. Does it make sense to connect the points you plotted?

b) Describe the graph.

c) What dimensions result in the smallest area? Is your answer the same if you use values other than integers?

d) What dimensions create the largest area?

e) Explain how you will set up your fencing in 2 - 3 complete sentences.

SV-2 Use your work from the previous problem to answer the questions below.

a) Write 2 - 3 sentences about the relationship between a rectangle's perimeter and its maximum area. Do you think this will always be the case? Explain.

b) What additions should be made to your table to complete the symmetric pattern (that is, same on both sides) of this graph? Make your graph symmetric.

SV-3. Here is one more line from the table of the previous problem. Fill in the two missing columns in terms of x. Begin by finding the measure of the other side in terms of x.

| Measure of 1st side | Measure of 2nd side | Area | Perimeter |
|---|---|---|---|
| x | | | 36 |

SV-4. The expression you wrote under AREA in the preceding problem can be written as $x(18 - x)$ or $18x - x^2$. How is the second expression different from other equations (functions) you graphed in the first three units?

Read the definitions and complete parts (a) and (b) in the box below.

A polynomial in which the largest exponent is two is known as a **QUADRATIC**. The graphs of **quadratic functions** are **PARABOLAS**.

These are quadratic functions:

$y = 3x^2 + 5x - 1$ $\quad y = x^2$ $\quad y = (x + 3)(x + 5)$ $\quad y = -5x^2 + 7$

a) Explain why the third example above is a quadratic.

These are **not** quadratic functions:

$y = 2x - 7$ $\quad y = x^3 + 8$ $\quad y = \frac{3}{x}$ $\quad x^2 + y^2 = 8$

b) Explain why these functions are not quadratics.

SV-5. To graph a quadratic function, we cannot just use the slope and the y-intercept as we did with linear functions. Discuss with your team why finding a number to represent the slope might not make sense for a quadratic function and record your reasons.

SV-6.

### How to Graph a Quadratic Function

There are many possible relationships between x and y. During the year we will explore several types of equations and their graphs, in addition to linear and quadratic functions. For now, we will write all of the equations in y-form so that it is easy to substitute values for x (input) and find each paired y-value (output).

Graph $y = x^2$ by completing the following steps:

a) On your paper, copy and complete the table for values of x and y.

| x | -3 | -2 | -1.3 | -1 | 0 | 0.8 | 1 | 2 | 3 |
|---|---|---|---|---|---|---|---|---|---|
| y | | | | | | | | | |

b) On graph paper, carefully plot the points and draw a smooth curve through them.

c) On the same set of axes, graph the following two quadratic functions. Use a table like the one above. Check with members of your team to be sure that everyone gets the same values. In the first equation, remember to square first, then change the sign of the result.

$y = -x^2$ $\qquad$ $y = (-x)^2$

SV-7. **DIRECTIONS:** Use the resource page provided by your teacher to put all three graphs on a single set of axes. When necessary, use your calculator to complete the tables for the functions using the given x-values. Graph the equation by plotting the resulting points and connecting them in a smooth curve. Work in pairs to reduce the workload as you make each table.

a) $y = \frac{2}{3}x + 2$

b) $y = x^2 + 7x + 10$

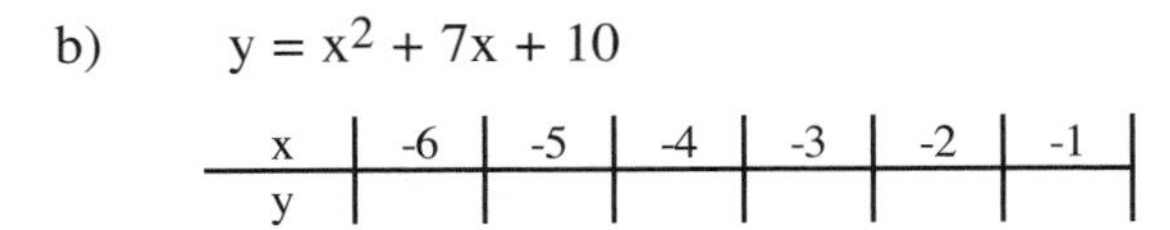

| x | -6 | -5 | -4 | -3 | -2 | -1 |
|---|---|---|---|---|---|---|
| y | | | | | | |

c) $y = \sqrt{x + 1}$

| x | -1 | -0.5 | 0 | 0.5 | 1 | 2 | 3 | 4 | 5 | 6 | 7 | 8 |
|---|---|---|---|---|---|---|---|---|---|---|---|---|
| y | | | | | | 1.7 | | | | | | |

d) Are your graphs in parts (a), (b), and (c) linear or curved? If a graph is linear, give its slope. If it is curved, is it a parabola?

SV-8. Write two or three sentences explaining how you can tell whether a graph will be curved or linear just by looking at the equation.

SV-9.

## THE ZERO PRODUCT PROPERTY

If $a \bullet b = 0$, then either $a = 0$ or $b = 0$.

Note that this property states that at least one of the factors MUST be zero. It is also possible that all of the factors are zero. This simple statement gives us a powerful result which is most often used with equations involving the products of binomials.

For example: Solve $(x + 5)(x - 2) = 0$.

By the Zero Product Property, since $(x + 5)(x - 2) = 0$, either $x + 5 = 0$ or $x - 2 = 0$.

a) Solve each equation for x.

b) Replace x in the original example above with each of your results, one at a time. Does each of these substitutions result in a true statement?

c) Add the Zero Product Property to your tool kit.

SV-10. Use the Zero Product Property to solve each of the following.

a) $(x - 2)(x + 1) = 0$

b) $(x + 7)(x + 2) = 0$

c) $x(x - 11.3) = 0$

d) $3x(x + 7.1)(x - 8) = 0$

e) $8(x - 3) = 0$

SV-11. Solve for x. Each part is a separate problem.

a) If $m\angle 1 = 3x - 18^\circ$ and $m\angle 5 = 2x + 12^\circ$, find x.

b) If $m\angle 3 = 4x - 27^\circ$ and $m\angle 6 = x + 39^\circ$, find x.

c) If $m\angle 4 = 49^\circ$ and $m\angle 6 = 5x + 41^\circ$, find x.

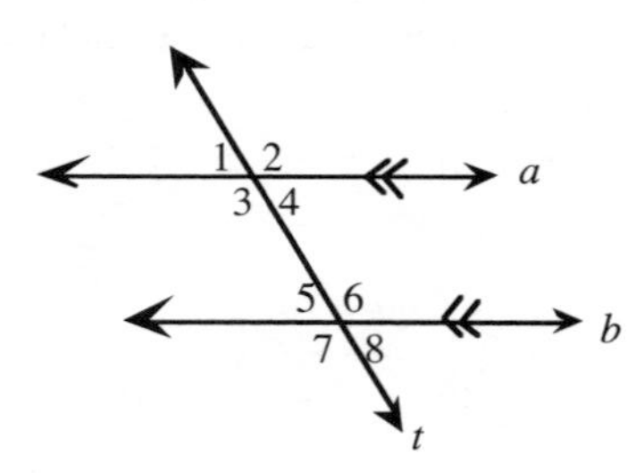


SV-12. Arrange the following statements in a logical order to make a convincing argument:
CONCLUSION: If this is my last class of the day, it will not be easy.

a) The lesson is confusing if you don't pay attention.

b) If this is my last class of the day then I'll be sleepy.

c) When I am sleepy it is difficult to pay attention.

d) If the lesson is confusing then the class will not be easy.

SV-13. Study the five figures and draw the next five figures in the pattern.

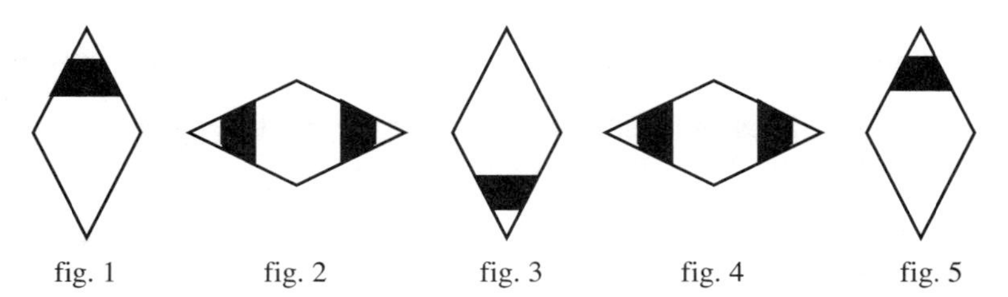


a) Predict what the 78th figure in the pattern will look like. Draw it and explain how you determined your answer.

b) Find the 79th figure in the pattern. Draw it and explain how you get your answer.

SV-14. Draw a pair of axes for $-3 \le x \le 5$ and $-5 \le y \le 10$.

a) Graph $y = (x - 2)(x + 1)$. Use $x = -2, -1, 0, 1, 2, 3$.

b) Where does the graph cross the x-axis?

c) Substitute each of the x-values that you found in part (b) into the equation in part (a), one at a time. Describe your results in a sentence or two.

d) Why does this make sense? Explain completely.

SV-15. Examine the graphs and their equations below.

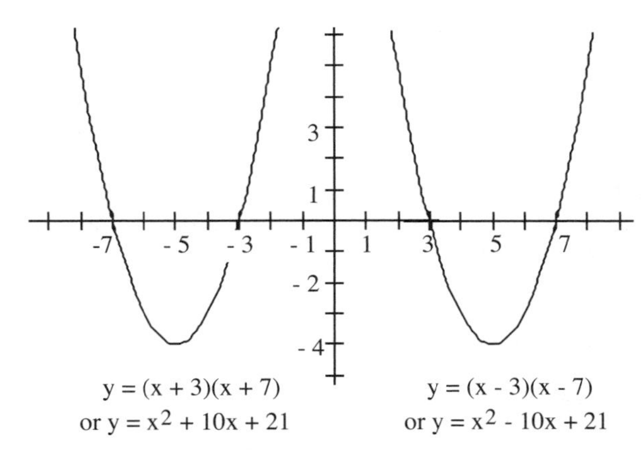


a) Discuss with your team how you know where the parabola crosses the x-axis by examining the first form of each equation. Reviewing the previous problem may help. Write your conclusion on your paper.

b) Briefly explain the relationship between the sign of each number in the binomials (first form of each equation) and the sign of the x-coordinates where the parabola crosses the x-axis.

c) Suppose you only had the second form of each equation, how could you transform it into the more useful product form?

SV-16. Consider the function $y = x^2 + 5x + 6$.

a) Will the graph of this function be straight or curved? Explain how you know.

b) How many times, at most, can it cross the x-axis? Use the graph in the previous problem and mentally move the parabolas around.

c) When it crosses the x-axis, what is the value of the y-coordinate at the point(s) of intersection?

d) If we know that $y = x^2 + 5x + 6$ can also be written as $y = (x + 3)(x + 2)$, write an explanation as to how we can figure out where the parabola crosses the x-axis without graphing it.

SV-17. Where would each of the following quadratic functions cross the x-axis? Look back over the two previous problems if you need help. You do not need to graph these equations to answer the questions.

a) $y = (x - 3)(x - 4)$ b) $y = x(x - 7)$ c) $y = (x - 4)(2x + 6)$

## GRAPHING NON-LINEAR INEQUALITIES

SV-18. Just as we used linear equations to understand linear inequalities, we can use functions whose graphs are curved to outline inequalities with curved boundaries.

EXAMPLE: Graph $y > x^2$

a) First graph $y = x^2$ by using the values in the table from SV-6. Connect the points with a smooth, **dashed** curve. Why is the curve dashed?

b) Pick a point not on the parabola and substitute it into the inequality to test if it satisfies the given inequality. Is one point enough? Why or why not?

For example, test point (1, 0) in $y \overset{?}{>} x^2$

c) Explain why your result in part (b) tells you to shade the interior region of the parabola.

d) Check! Does your answer look like the figure at right?

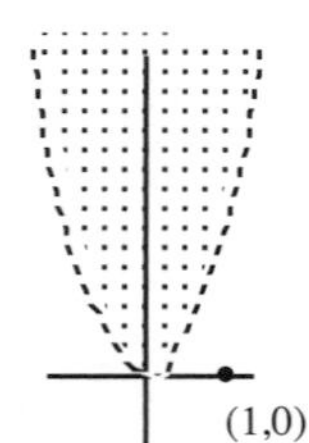


SV-19. Sketch (do not make a table and plot points) the graphs of the inequalities below on graph paper. You can do this quickly if you refer to your graphs in problem SV-6, part (c).

a) $y > -x^2$ b) $y \le (-x)^2$

SV-20. Do the following problem on the lower half of the resource page you used for SV-7, **OR** draw a pair of axes on a sheet of graph paper and scale them for $-9 \le x \le 9$ and $-9 \le y \le 9$.

a) For each inequality, graph the dividing line (solid or dashed?), using the slope-intercept method. Do the shading lightly. Using a different color for each inequality is helpful. Write the appropriate inequality along each line after you graph it.

1) $y \le x + 2$ 2) $y \ge \frac{1}{2}x - 2$ 3) $y \le -2x + 6$

b) Darken the region where the three shadings overlap (intersect). What is the name of the shape of the shaded region?

c) Explain why graphing three linear inequalities will often give you this kind of shape.

d) Describe or sketch a situation in which graphing three inequalities will not give you such a shape.

e) Approximate the area of the enclosed figure. Use a graph paper ruler as needed.

SV-21. Use the graphs you did for problem SV-7, parts (b) and (c). Test a point, then shade the appropriate regions so that they now represent graphs of the inequalities below. Just draw a quick sketch of what they become as inequalities; you do not have to carefully graph them again.

a) $y > x^2 + 7x + 10$ b) $y \le \sqrt{x + 1}$

SV-22. Describe what you see in each case. Try to see something in a three dimensional sense.

Case 1

Case 2

Case 3

SV-23. Armalene needs to make a die that will produce certain probabilities. She will use a normal cube, but rather than putting numbers on the faces, she will paint the faces different colors. She needs the probability of rolling the color blue to be $\frac{1}{2}$, the probability of rolling the color yellow to be $\frac{1}{3}$, and the probability of rolling the color red to be $\frac{1}{6}$. Help her out by describing how she should paint all six faces of the cube to produce these probabilities.

SV-24. Multiply the following.

a) $(x - 3)(x - 4)$ b) $x(x - 7)$ c) $3x(2x + 1)$

## VISUALIZING, PART 1

SV-25. **DIRECTIONS:** Today you will be working on an important skill in mathematics: visualization. Many of the following problems do not ask you to solve anything, but rather to describe a situation as you imagine it or see it in your mind's eye. For each of the problems, include a sketch AND write descriptions in complete sentences. Examine each figure below and explain how you know that what you decide is really true.

a) Is the diagonal line straight?

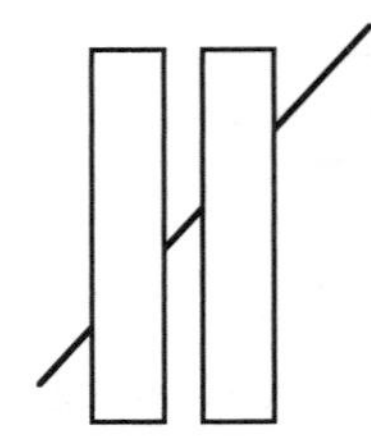

b) Are the horizontal slashes exactly in the middle of the center line of these triangles?

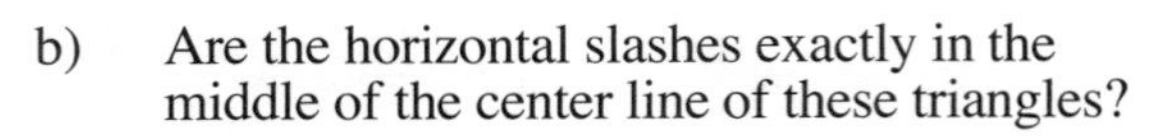

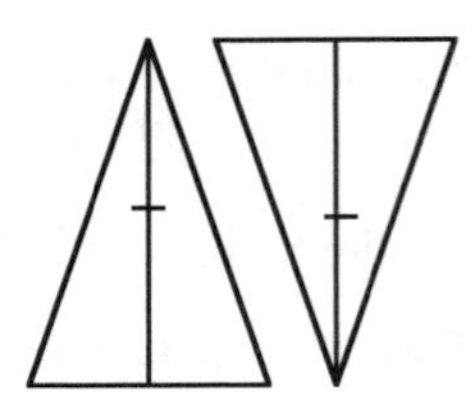

c) Which angle is larger, a or b?

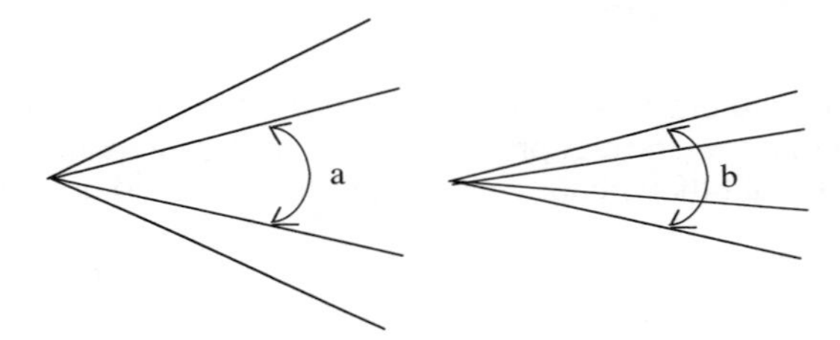


SV-26. The picture at right shows a line and a circle with no points in common on a flat surface. Can the line and the circle intersect in exactly one point? two points? three points? Explain each answer and illustrate it with an example where possible. Remember: the circle is the curve itself, not the region inside the curve.

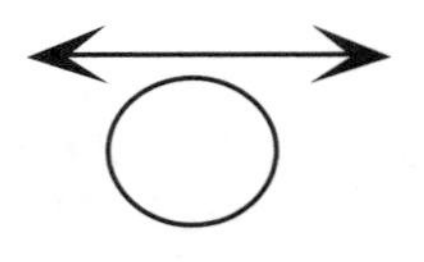

SV-27. One end of a ruler is nailed to the ground. What shape will the other end make as it moves freely?

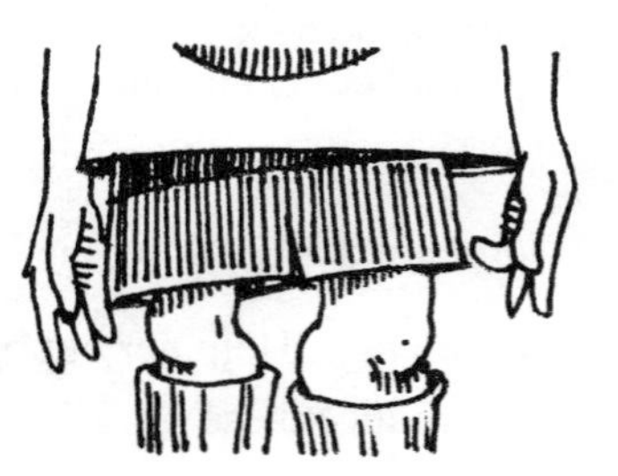

SV-28. A young child throws a ball onto a concrete playground and no one is there to catch or stop it as it bounces across the playground. Sketch and describe the path of the ball from the time it leaves the child's hand until it is no longer moving. A side view may help you in your description.

SV-29. Your best friend's birthday present is in this box.

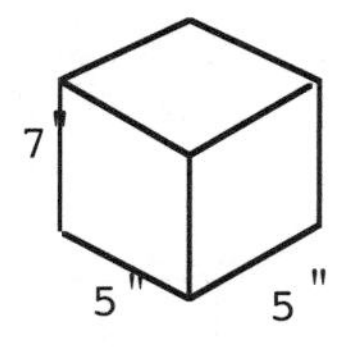


a) What are suitable dimensions for a sheet of wrapping paper to wrap the box completely (allowing for the necessary overlap)?

b) What is a suitable length of ribbon so that it wraps around all sides once and ties in a small bow?

SV-30. What shape is formed if the consecutive midpoints of the sides of a triangle are connected? What about the midpoints of a rectangle? A square? A hexagon? Draw a sketch of each figure.

SV-31. ABC Sports will be televising a big football game between the Philadelphia Eagles and the Washington Redskins. Based on the picture each camera broadcasts shown below, describe where in the stadium the camera is located.

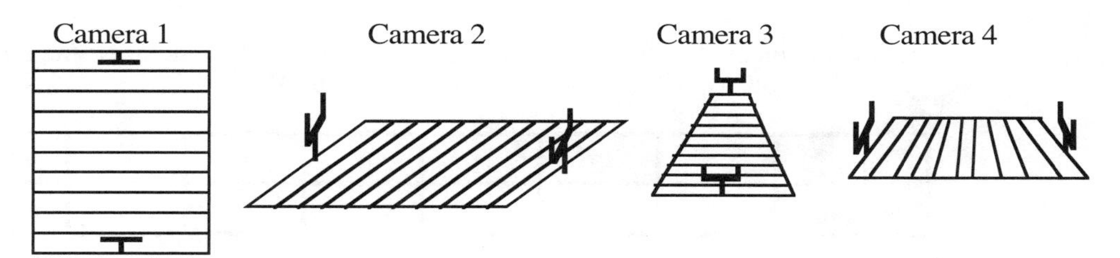


SV-32. Imagine that you have cut the shape at right out of paper and folded it along the dotted lines in the same direction until the sides come together. Describe what you get.

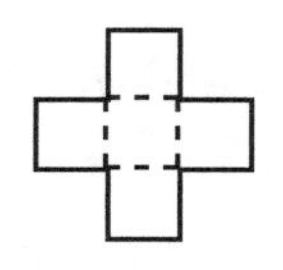

SV-33. Which of these figures (commonly called nets) will give you the same result as the preceding problem?

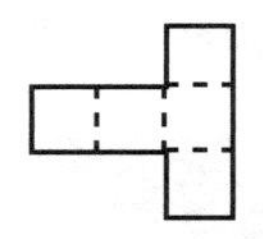

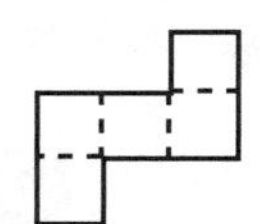

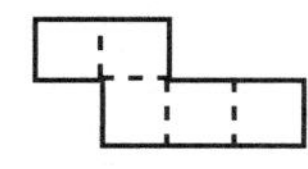

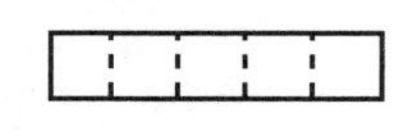

SV-34. If we add one more square to the net, we should be able to close the box. For instance, this net will do it:

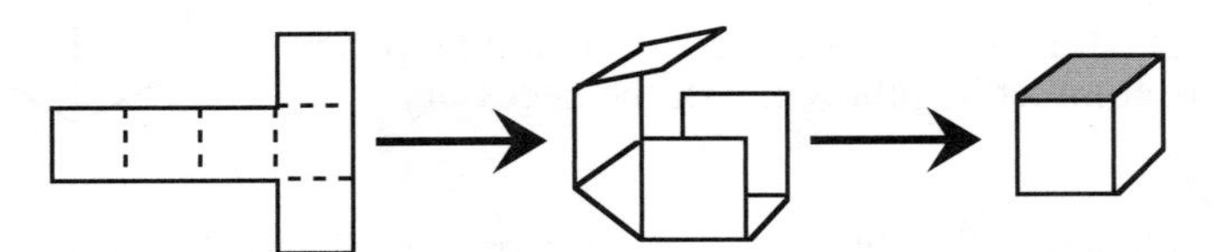

a) Draw a different net that will fold into a closed box.

b) The cube in the example above now has a hole drilled completely through it. Draw a picture of what the net looks like after it is unfolded (including the holes).

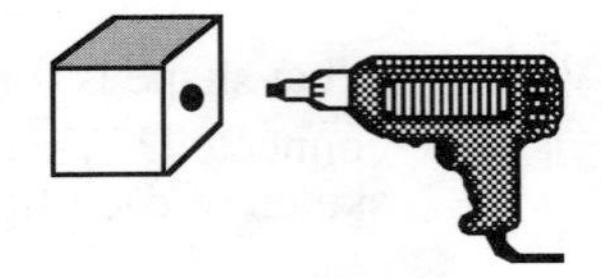

SV-35. On a sheet of graph paper, draw a pair of xy-coordinate axes for $-3 \le x \le 3$ and $-8 \le y \le 8$.

a) Graph the inequality $y \ge x^2 - 2$. Do the shading lightly! Remember: the result is a curved region. Why? Is the curve solid or dashed? Why?

| x(input) | -3 | -2 | -1 | 0 | 1 | 2 | 3 |
|---|---|---|---|---|---|---|---|
| y(output) | | | | | | | |

b) On the same set of axes, graph $y \le x$. Shade this region lightly also.

c) Darken the overlapping region. Estimate the coordinates of the points where the lines intersect (meet).

d) Estimate the area of the shaded region and explain your method.

e) Suppose we made an adjustment in one of the inequalities, changing $y \le x$ to $y \ge x$. Make a sketch of the new situation next to the first graph. Include the parabola.

f) Could you estimate the area of the new shaded region with this adjustment? Explain your answer completely.

SV-36. The graph below shows the amount of money in Hillary's savings account. Describe what is happening to Hillary's money, giving plausible reasons why the graph rises and falls at particular points.

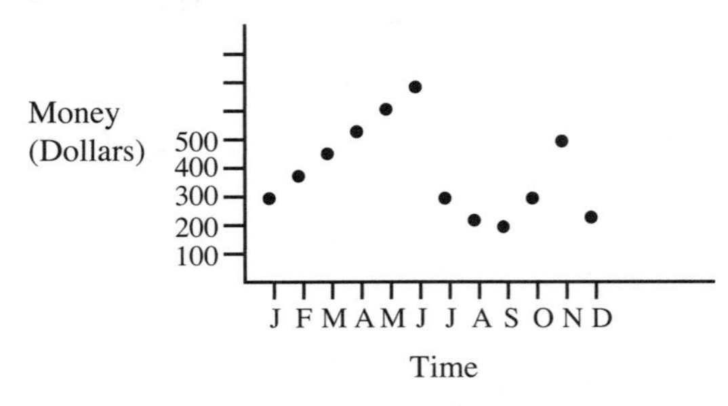

SV-37. Solve for x.

a) $(2x - 6)(x + 3) = 0$

b) $6x(5 - x)(x + 1) = 0$

SV-38. Where do each of the graphs of the following functions cross the x-axis? Solve without graphing.

a) $y = (x - 1)(x + 2)$

b) $y = 6x(5 - x)(x + 1)$

c) Compare your results for part (b) of the previous problem to part (b) of this problem. Explain what you see and enter the result in your tool kit.

## VISUALIZING, PART 2

SV-39. Examine each figure below and explain how you know that what you decide is really true.

a) Which line is the continuation of A?

b) Is the point midway between the top vertex and the base?

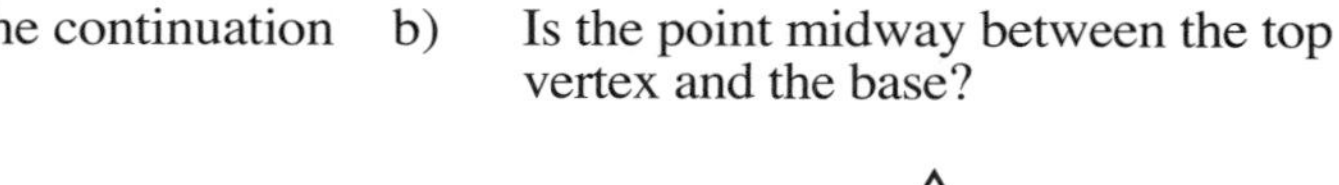

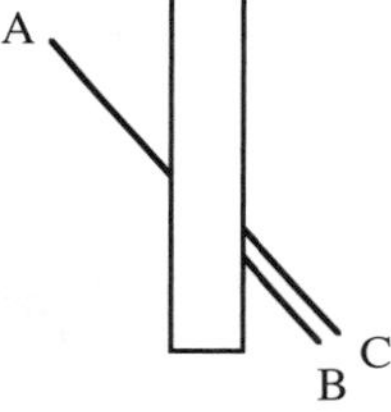


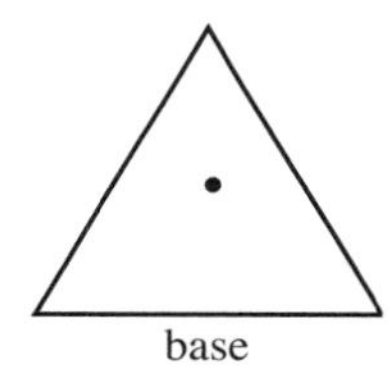


c) Which line segment is longer, $\overline{AB}$ or $\overline{BC}$?

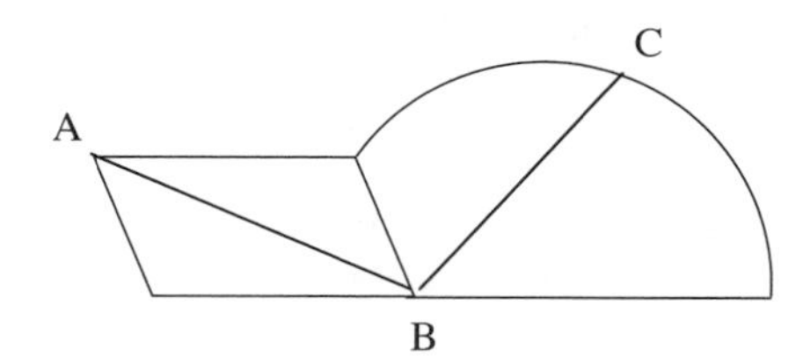


SV-40. Consider the figure at right.

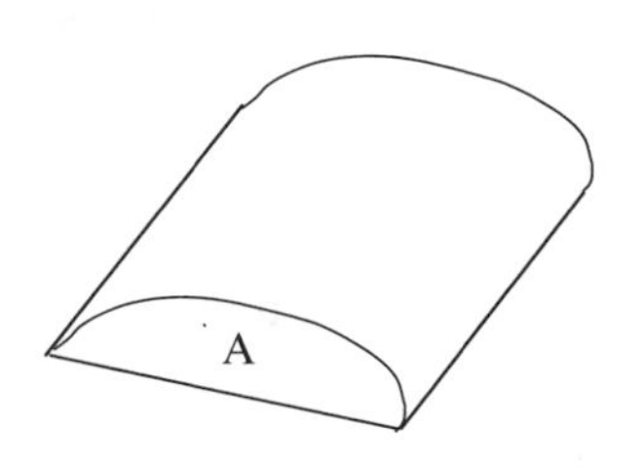


B

a) If you stood above this object and looked straight down from the top, what would you see?

b) If you looked straight toward the side labeled A, what would you see?

c) If you looked from the position marked B, that is, with the figure at your eye level, what would you see?

SV-41. Carol keeps her pet puppy on a rope that is tied to a stake. Describe the region in which her puppy may wander.

SV-42. Describe the region in which Carol's puppy in the previous problem may exercise if it is tied to a stake with a 20' rope at the corner of the building in the figure at right.

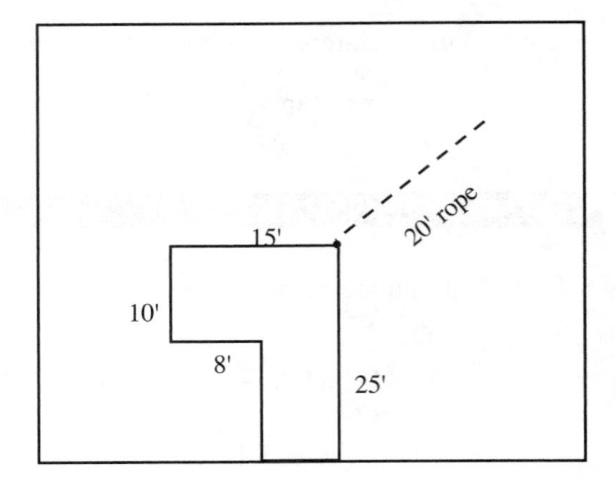


SV-43. George keeps his pet bee tied to a (much smaller) stake - a thumbtack as a matter of fact.

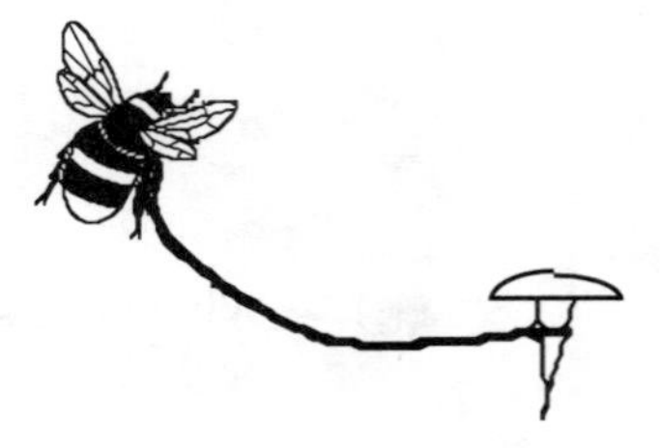

a) Describe the region in which the bee may move.

b) Suppose the tack is pushed into the corner between two walls halfway between the floor and ceiling. Now what is the bee's region of mobility?

c) Suppose the tack is pushed into the corner of a room, between two walls and the floor. Describe the region in which the bee may move.

SV-44. Imagine your classroom as a large box. How many corners does it have? How many edges?

SV-45. Graph $x^2 < y < x^2 + 1$. Be sure to shade the solution region. Note that this statement means the same thing as graphing the system of inequalities $y > x^2$ <u>and</u> $y < x^2 + 1$.

SV-46. Solve for x.

a) $(x + 6)(x + 1) = 0$

b) $x(x + 2)(x - 2) = 0$

c) Replace 0 with y in part (a) and make a quick <u>sketch</u> of the graph. You only need to show the basic <u>positioning</u> and <u>shape</u> of the curve based on the x-intercepts from your solution. Do <u>not</u> draw a point-plotted graph.

SV-47. A sheet of paper is folded in half, then folded in half again. The resulting rectangle is 6 inches by 5 inches. What is the area of the full sheet of paper?

SV-48. Calculate the value of x.

a)

b)

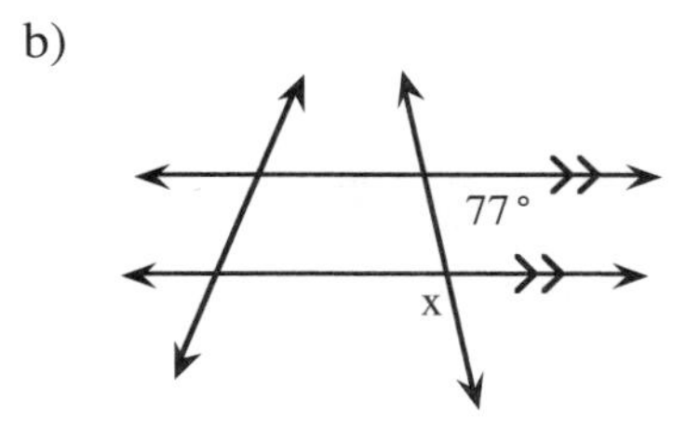


SV-49. Study the two graphs below, then complete parts (a) through (d).

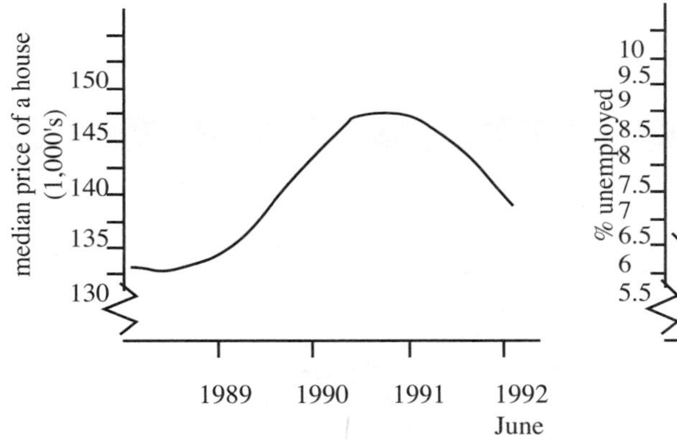


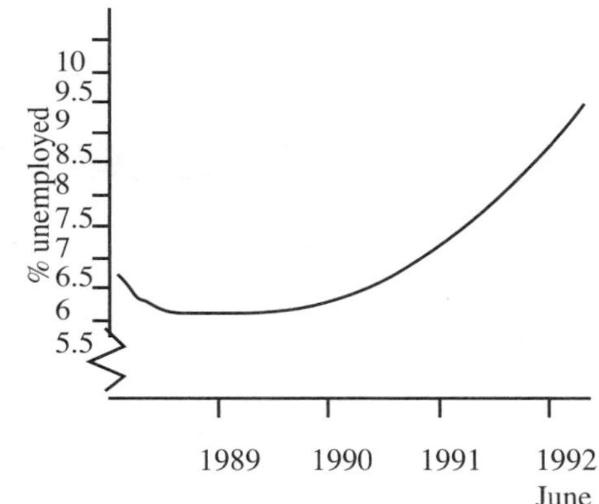


a) Write a short description of the information that each graph provides.

b) Write a short paragraph explaining what you conclude from the information in the two graphs.

c) Do the two graphs clearly and accurately report the data?

d) How might they be misinterpreted?

SV-50. Graph the points P(-3, -2), Q(7, -2), and R(2, 5) and connect them to form a triangle.

a) Find the perimeter of the triangle.

b) Find the area of the triangle.

c) What kind of triangle, scalene, isosceles, or equilateral, is $\Delta PQR$? What convinces you that your answer is correct?

SV-51. Write the equation of the line that:

a) is parallel to $y = \frac{2}{5}x + 3$ with y-intercept = -2.

b) passes through the points (0, 2) and (3, -5).

SV-52. Use the trapezoid shown at right.

a) Find the area.

b) Find the perimeter.

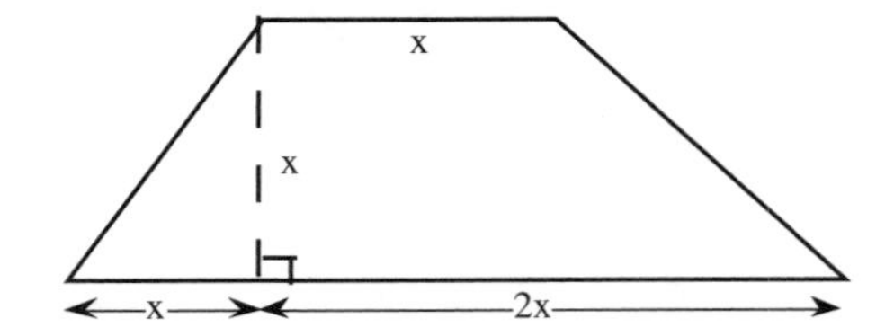


## ISOMETRIC DRAWINGS

SV-53. **DIRECTIONS**: Much of the focus of this course is two-dimensional geometry (figures on flat surfaces), but reality comes in at least three dimensions. An important and useful skill is to be able to draw 2-dimensional representations of 3-dimensional objects, and to be able to interpret 2-dimensional drawings as representing real-world objects. Research indicates that men tend to be better at doing this type of visual translation. These studies point out that in general girls and boys have different experiences as children and that boys have many more opportunities to do the type of visualization that later helps them in mathematics. Is this generalization true for the people in your team? Remember findings that are true for the general population do not necessarily apply to individuals.

a) Make a list of experiences you have had that might have helped you develop visualization skills. Perhaps they include the games, toys, and/or sports you played as a child.

b) Research also shows that people who have had relatively little experience with this kind of visualization catch up very quickly when they have the opportunities to practice. The following problems provide many such opportunities. Go on to the next problem now.

SV-54. **DIRECTIONS**: During the next several days, we will be discussing volume and surface area of solid 3-dimensional objects that are built from cubes. Read the information in the box below so that you become familiar with the basics of drawing three dimensional figures. When you have added the information to your tool kit, make four copies of a cube on isometric dot paper.

This is an **ISOMETRIC DOT PAPER GRID.**

IMPORTANT DETAIL: Isometric dot paper is not the same if it is turned sideways. To be sure that your paper is correctly lined up, draw a hexagon like the one shown below. If the <u>sides are vertical</u>, then the paper is turned correctly; if the sides have points, then turn the paper 90°. Another way to check that your paper is turned correctly is to look at the left margin of the paper. The dots should make a straight, vertical column. If the dots zigzag in and out between rows, the paper is turned the wrong way.

This is turned correctly:

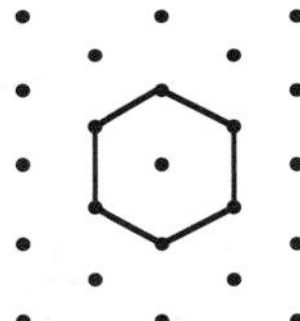

this is not:

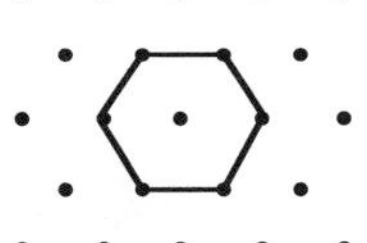

This is a **3-D ISOMETRIC** drawing of a cube.

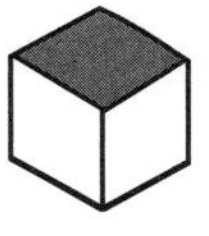

By stacking and pushing cubes together, we can create other solids. Two cubes can be joined to form two isometric solids as shown at right:

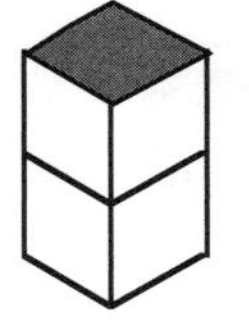

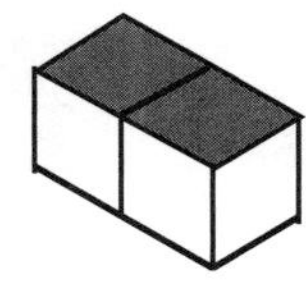

<u>Keep real cubes handy as you are working on these problems.</u> Setting up what is pictured using real cubes can be very helpful when interpreting the drawings.

When we look at an isometric drawing we will ALWAYS view it from the <u>front</u>, <u>right</u>, and <u>top</u> parts of the solid as shown at right. Building models on a 3" x 5" index card marked "front," "back," "left," and "right" is helpful, because you can turn the model and view it from several directions without it falling apart or having to rebuild it.

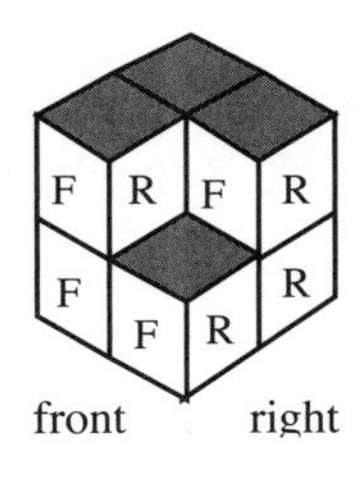


Do not forget to make four copies of a cube.

SV-55. One example of a 3-cube solid is shown at right. Copy it onto isometric paper and draw two different 3-cube solids that give different isometric drawings.

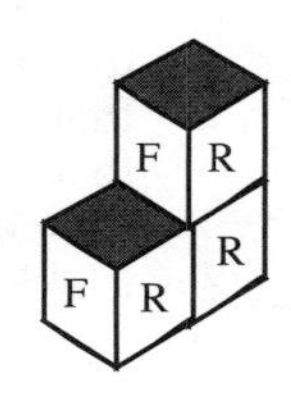


SV-56. The figures at right are examples of a 4-cube solid and a 5-cube solid. Copy each onto isometric dot paper.

SV-57. Copy each of the following shapes onto isometric dot paper.

a)

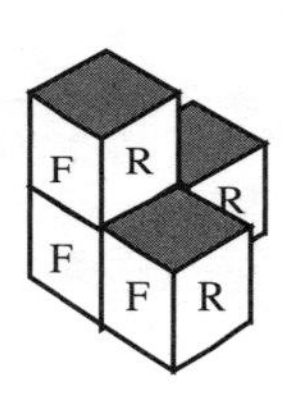


b)

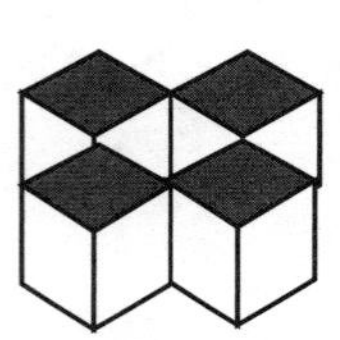

c)

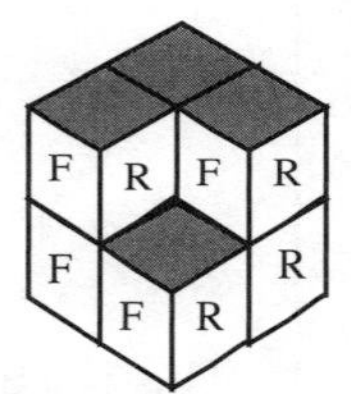


SV-58. Create an isometric drawing for a solid that is made up of 6 cubes.

SV-59. Create an isometric drawing for a solid that is made up of 8 cubes.

SV-60. State whether the graphs of each of the following functions will be straight or curved. If you know the name of the curve (parabola, for example), give it. You do not need to graph them. If you are unsure, make a quick sketch by plotting a few points.

a) $y = 3x^2 - 5$

b) $y = 7x + 4$

c) $3x - 2y = 8$

d) $y = 7x^4 - 3x^3 + 1$

e) $y = \sqrt{16 - x^2}$

f) $y = \frac{12}{x}$

SV-61. Draw a pair of xy-coordinate axes.

a) Will the boundary of the graph of $y \geq x^3$ be curved or straight? Explain how you know.

b) Graph $y \geq x^3$ for $-4 \leq x \leq 4$. Also include $\pm 0.5$ and $\pm 1.5$ in your table of values. Scale your axes appropriately. ($\pm$ means use both $+0.5$ and $-0.5$, etc.)

SV-62. Solve for the variable(s) in each part of the problem. Write the names of the definition(s) and conjecture(s) that justify the steps in your solution.

a)

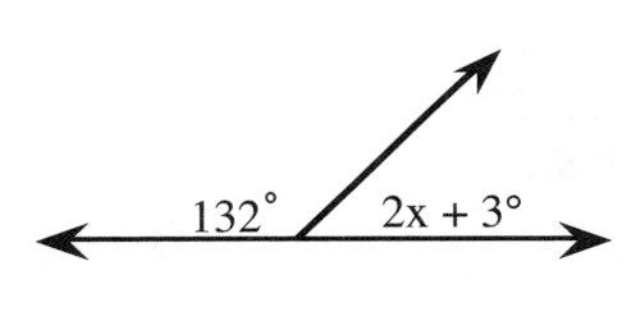


b)

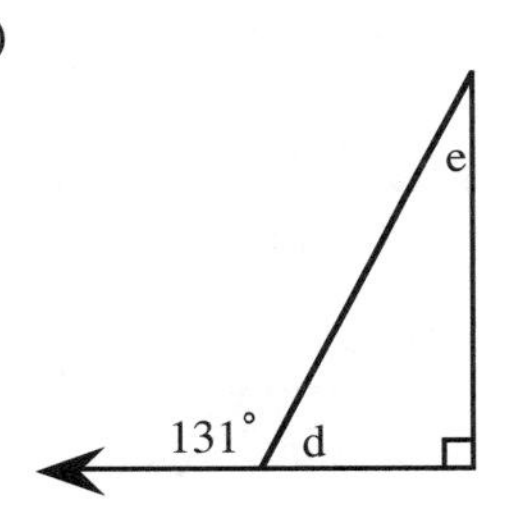


c)

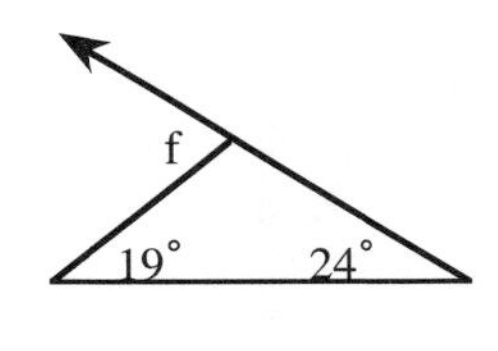


SV-63. A friend of yours offers this proposition to you: "Either flip a quarter or roll a die. If you flip three tails in a row, I'll give you the quarter. If you choose to roll the die, I'll give you the quarter if you get three 5's in a row." If you have nothing to lose, which will you do? Justify your answer.

SV-64. Solve for y.

a) $3x + 2y = 6$

b) $x - 5y = 12$

## MAT PLANS FOR ISOMETRIC DRAWINGS

SV-65. Read the box below and add the definitions, with illustrations, to your tool kit.

**DIRECTIONS**: The easiest way to represent multiple cube shapes on square grid paper is with a Mat Plan. A **MAT PLAN** is a top (or bottom) view of a multiple cube solid. The following is an example of a Mat Plan:

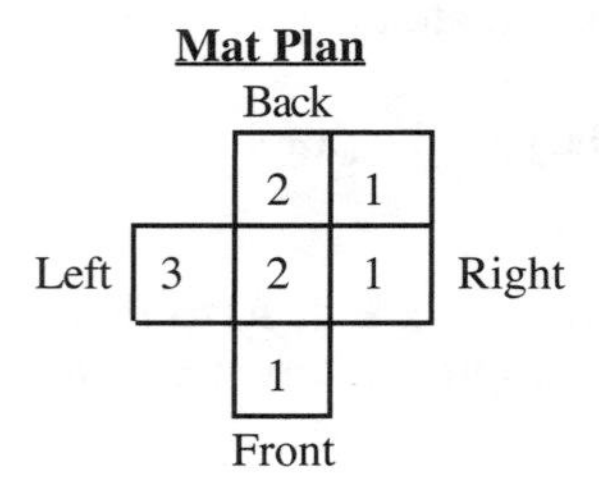


The number in each square represents the number of cubes in that stack. Enough information is given for you to be able to reconstruct the multiple cube shape.

a) How many cubes will it take to build this solid?

This is an **ISOMETRIC VIEW** of the solid represented by this Mat Plan. The shaded areas are the tops of each stack.

b) If you did not already know the Mat Plan for this isometric view, could this picture be hiding a cube?

c) What would be the Mat Plan if there were a hidden cube?

REMEMBER: So that we can always recognize what each of us draws, that is, so that we draw figures in a consistent orientation (position), we will always draw isometric views so that the <u>right</u> face, the <u>front</u>, and the <u>top</u> are the three represented views. Another way to remember this is to think of the base of the isometric view as the Mat Plan rotated 45° clockwise (see the figures in the next problem).

SV-66. **DIRECTIONS:** Read the note about isometric drawing in the box below. Then use isometric paper to draw an isometric view for each of the Mat Plans in parts (a) and (b):

A helpful technique for isometric drawing is to first transform the outline of the Mat Plan to isometric paper. The middle figure shows the Mat Plan rotated 45° clockwise. This view can now act as a floor plan and you can build the picture next to it as shown.

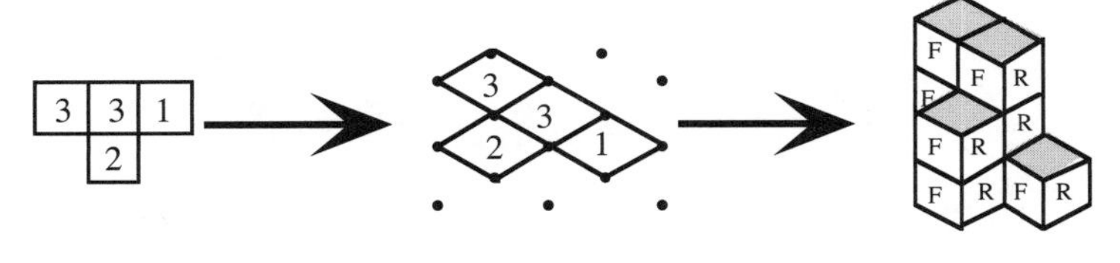


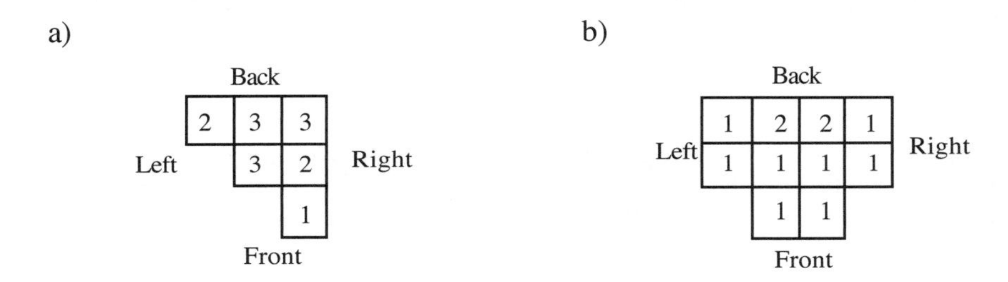


SV-67. Read the box below and add the definition to your tool kit. Then count the number of cubes you used to build each solid in the previous problem. What is the volume of each?

**VOLUME** is the number of 1 x 1 x 1 cubes, or parts of cubes, that fit inside a three dimensional figure or object. For example, the solid in the box above has a volume of nine cubic units.

SV-68. Create Mat Plans from the following isometric views and find the volume of each.

a)

b)

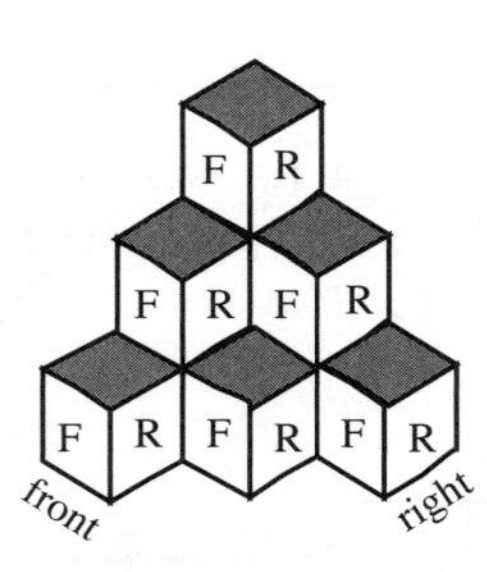


c) Did you have to make any assumptions about hidden cubes when you drew the Mat Plans? If so, what assumptions did you make in each case and why?

d) What other view would you need to see to be sure how many cubes there are? Explain.

SV-69. Eleanor has the Mat Plan at right and she must find the volume of the solid it represents. She first decides to build the solid. How many cubes will she need to build it? What is its volume?

SV-70. Explain to Eleanor in one or two complete sentences how she could get the volume for a solid directly from its Mat Plan, that is, without building it.

SV-71. Read the box below and add the definitions, with illustrations, to your tool kit.

A Mat Plan shows the top (or bottom) view of a solid and identifies the height of each cube stack in the solid.

Any cube stack that has the same shape for its top and bottom is a **PRISM.**

The following cube stacks are all examples of prisms made with cubes.

   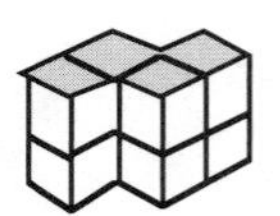

What is the volume of each prism?

SV-72. Draw a Mat Plan for each of the following cube stacks and use the Mat Plan to find the volume of each stack. First find the area of the top of each stack.

a)

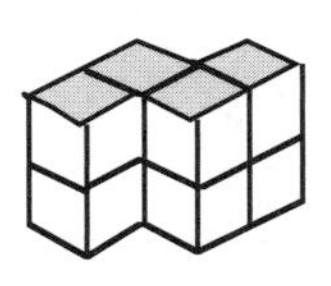

b)

c)

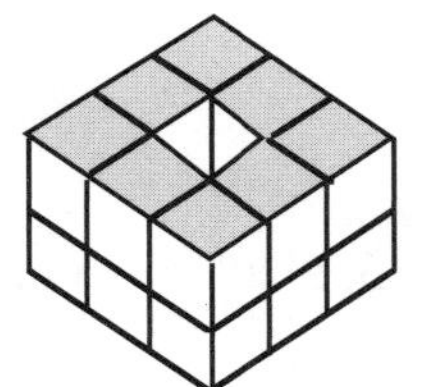

d)

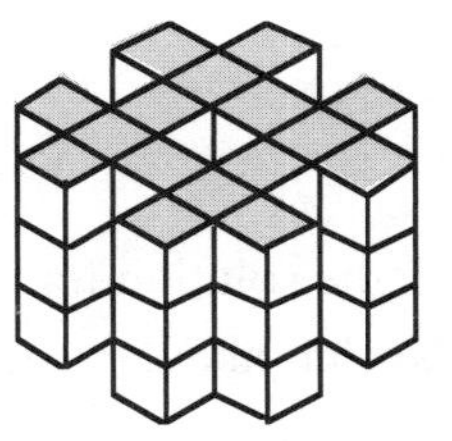

SV-73. Maria claims that there is a relationship between the area of the bottom of a prism, the height of that prism, and the volume of the prism.

a) Look over your answers from the previous problem and discuss with your team what Maria might be thinking.

b) Write a one or two sentence conjecture for how to calculate the volume of a prism based on your discussion.

c) Explain to a fifth grader an easy way to calculate the volume of a prism when you know the area of the base and the height. Be sure to tell him or her why this works.

SV-74. Using the figure at right, decide which of the individual statements below (if any) have <u>enough information to conclude</u> that $\overline{AT}$ is perpendicular to $\overline{CT}$. <u>Explain</u> your reason for each separate problem.

a) $m\angle 1 = 40^\circ$ and $m\angle 4 = 50^\circ$. Is $\overline{AT} \perp \overline{CT}$?

b) $m\angle 2 = m\angle 3$. Is $\overline{AT} \perp \overline{CT}$?

c) $\angle 2$ and $\angle 3$ are complementary. Is $\overline{AT} \perp \overline{CT}$?

d) $m\angle 1 = m\angle 2$ and $m\angle 3 = m\angle 4$. Is $\overline{AT} \perp \overline{CT}$?

e) $m\angle 1 = m\angle 4$ and $m\angle 2 = m\angle 3$. Is $\overline{AT} \perp \overline{CT}$?

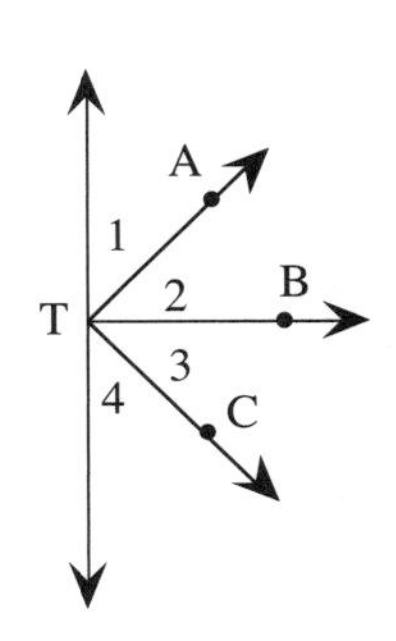

SV-75. Find x.

a)

85°
l
x
21°
m

b)

$x^2$
$x^2$
98°

SV-76. Graph the solution for $y \le \frac{3}{5}x + 2$ and $y > -2x - 3$. Shade each inequality differently and be alert about dashed and solid lines.

SV-77. For each of the following polygons, write an algebraic expression to represent its area.

a) Square

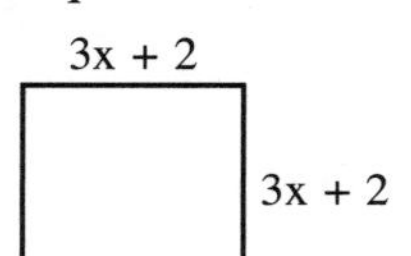


b) Rectangle

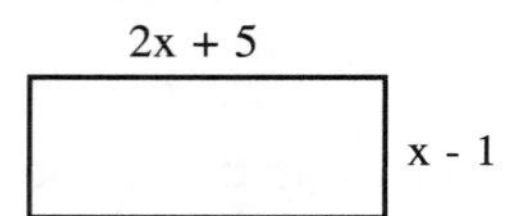


c) Triangle

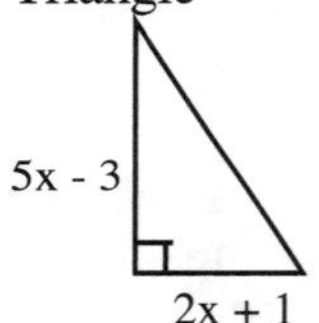


SV-78. Suppose you know that two adjacent, supplementary angles are equal as shown in the figure. Write a conjecture about the relationship between the two lines. Hint: what is the measure of each marked angle? How do you know?

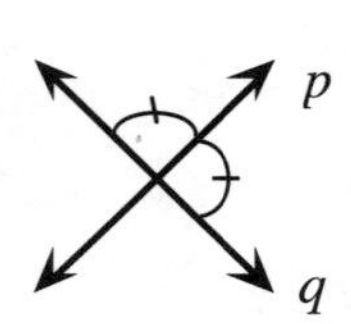

## POLYHEDRA AND SURFACE AREA

SV-79. Read the box below and add the definitions, with illustrations, to your tool kit. Then obtain 18 gum drops and 30 toothpicks from your teacher and work with a partner to complete the problem below.

> In the figures below, each polygon formed by toothpicks on the outer surface of a solid represents a **FACE**. Each toothpick represents an **EDGE** (a common side of two polygonal faces), and each gum drop represents a **VERTEX** (plural: vertices; a point where three or more sides of faces meet).
>
> Note: these figures, called **polyhedra**, by definition may not have holes in them. We use the "open" toothpick figures below to illustrate their parts.

a) Use gum drops and toothpicks to build the solid shown at right, which is called a **regular tetrahedron**. Then count the number of vertices and edges in the solid and record your results in the appropriate sections of the table on the resource page.

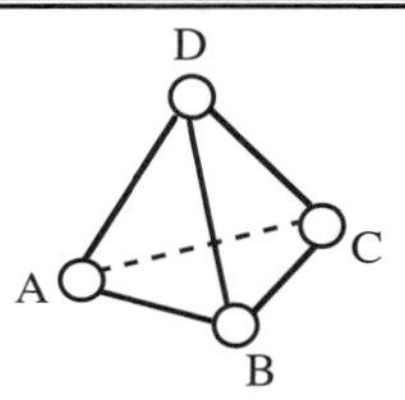


b) Use some more gum drops and toothpicks to build a **cube**. Draw a sketch of your cube and label its vertices. Record the number of vertices and edges in the same table.

c) Use the remaining gum drops and toothpicks to build the solid at right, which is called a **regular octahedron**. Record the number of vertices and edges in the table.

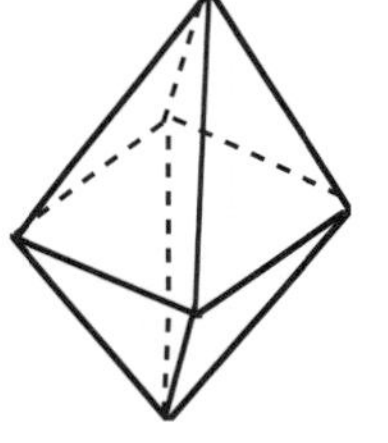

d) Without using gum drops and toothpicks, complete the rest of the table for the number of vertices and edges of the remaining solids.

e) Cut out the paper tetrahedron template (provided by your teacher) along the solid lines. Fold along the dotted lines, creasing well. Glue or tape together the resulting solid with the shaded flaps tucked inside. Name the kind of polygon that is each face (on each side) of the solid.

f) Count the number of faces on the tetrahedron and record your results in the appropriate section of the table. How many faces does a cube have? Record your answer in the table on the resource page, then read problem SV-80 and complete the rest of the table for the number of faces for each of the remaining solids.

SV-80. Let VR = number of vertices, E = number of edges, and F = number of faces. In 1736, the great Swiss mathematician Euler (pronounced Oiler) found a relationship among VR, E, and F.

a) For each row, calculate VR + F.

b) Write an equation relating VR, F, and E.

SV-81. Is it possible to make a tetrahedron with non-equilateral faces? If not, explain why not. If so, draw a sketch.

SV-82. How many edges does the solid at right have? (Don't forget the ones you can't see.) How many vertices does it have?

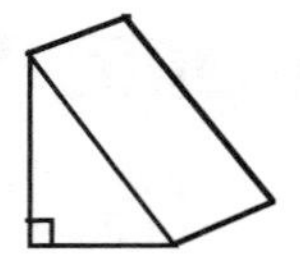

SV-83. Read the following definition of a **polyhedron**, study the figures below, and discuss their characteristics. Then add the definition, with illustrations, to your tool kit. Note that the cylinder (can) and sphere (ball) are not polyhedra. Why not?

The solids you have been building and analyzing are called **polyhedra** (singular: polyhedron). A **POLYHEDRON** is a 3-dimensional surface formed by polygonal regions that has no holes in it. The polygonal regions are joined at their sides, forming **edges** of the polyhedron. Each polygon is a **face** of the polyhedron. For example, a shoe box with its lid closed (ignore the overlapping lip of the top) is a rectangular prism. It is not considered a polyhedron when its lid is off (it has a hole in it).

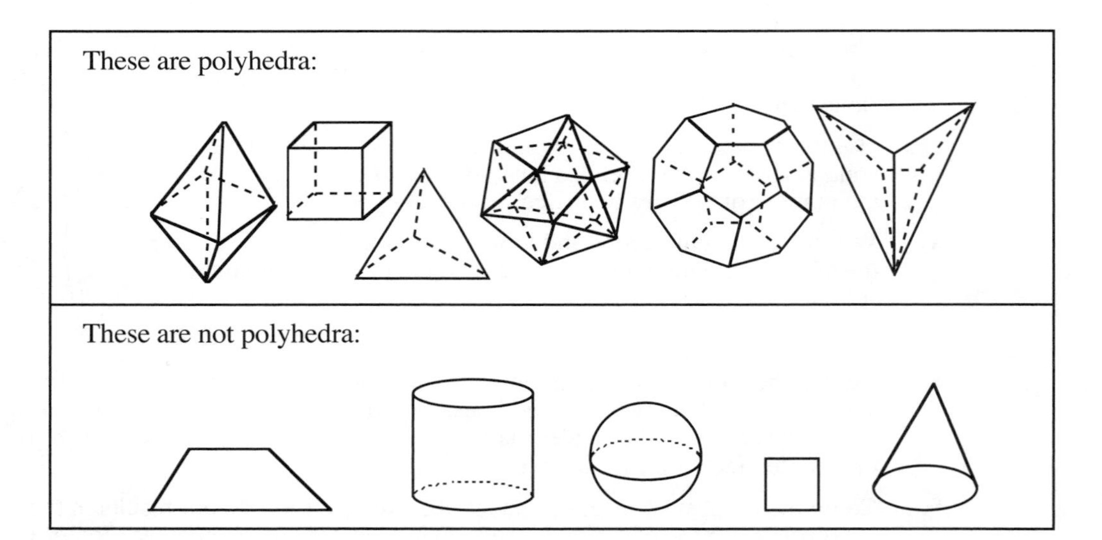

SV-84. Classify the following as a polyhedron or not a polyhedron, then read the box below them.

a) 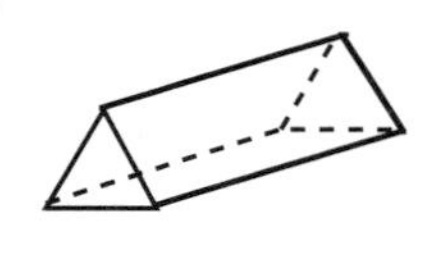

b) 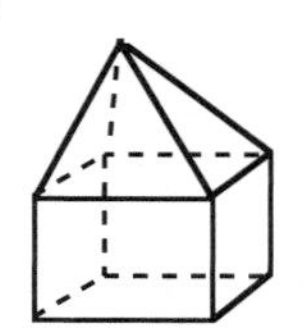

c) 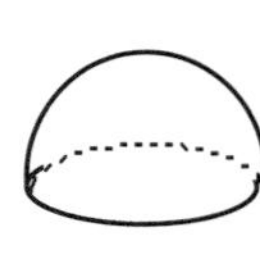

d) 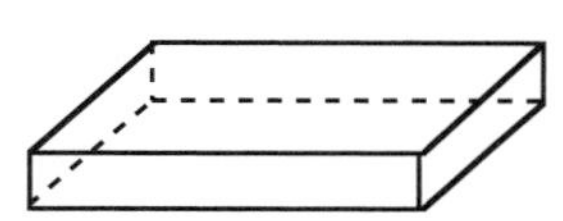

e) 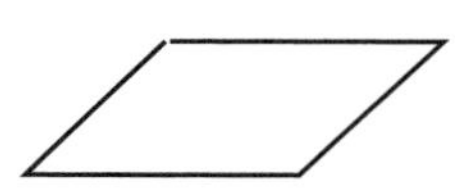

f) 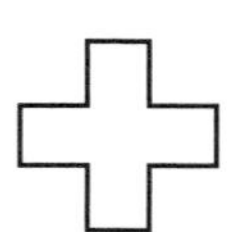

g) 

h) 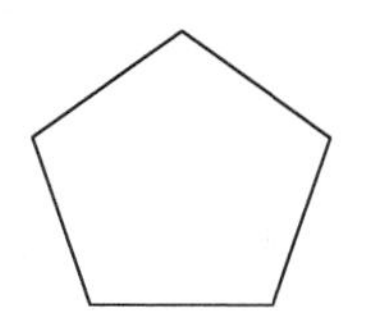

i) 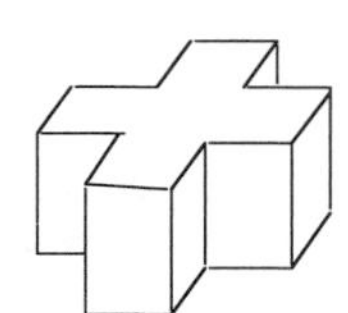

Polyhedra are classified by the number of faces they have. Here are some of their names:

| | | | |
|---|---|---|---|
| **4 faces** | Tetrahedron | **9 faces** | Nonahedron |
| **5 faces** | Pentahedron | **10 faces** | Decahedron |
| **6 faces** | Hexahedron | **11 faces** | Undecahedron |
| **7 faces** | Heptahedron | **12 faces** | Dodecahedron |
| **8 faces** | Octahedron | **20 faces** | Icosahedron |

SV-85. Name the idea(s) you use to solve for x and y in each problem below.

a) 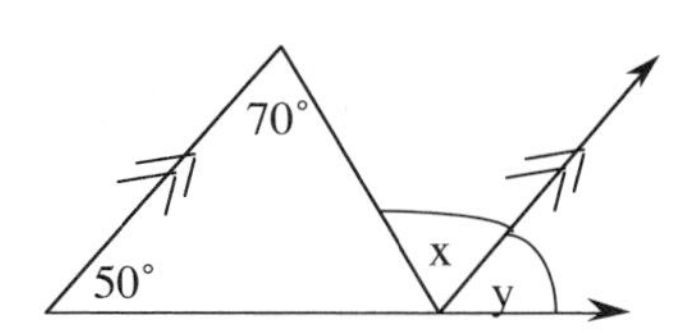


b) 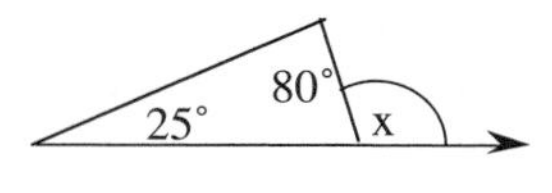


SV-86. Write the equation of the line with the given slope and y-intercept.

a) $m = \frac{2}{3}$; (0, 6)

b) $m = 0$; (0, - 2)

c) $m = 1.1$; $(0, \frac{2}{3})$

d) $m = 6$; (0, 0)

SV-87. Svend drew the prism at right but forgot to mark the unit cube lines on its surfaces. Glenda took a cube and determined that the area of the top is 18 square inches and that the prism is 4 cubes high. What is the volume of the prism?

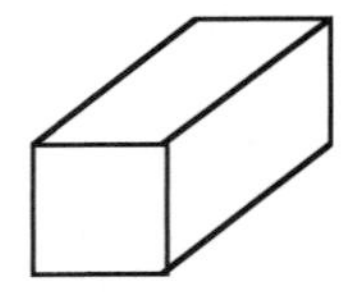

SV-88. The length of a rectangle is three meters longer than its width. Its area is 40 square meters. What are the rectangle's dimensions?

SV-89. Solve for x.

a) $(x + 7)(x - 3) = 0$

b) $6x(3 - x)(x + 2) = 0$

SV-90. Complete the color square, if possible. Explain your steps in a logical, step by step paragraph.

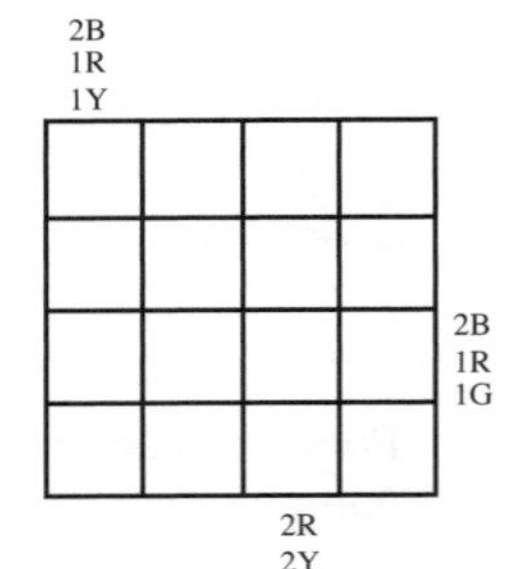


## DRAWING POLYHEDRA

SV-91. In the **pyramid** example at right, just the vertices of the solid are shown on the left. The open dot represents a hidden vertex. The picture on the right shows the solid with the edges drawn in. The dashed segments represent edges that cannot be seen. Notice that any segment (edge) originating from the open dot is a dashed segment.

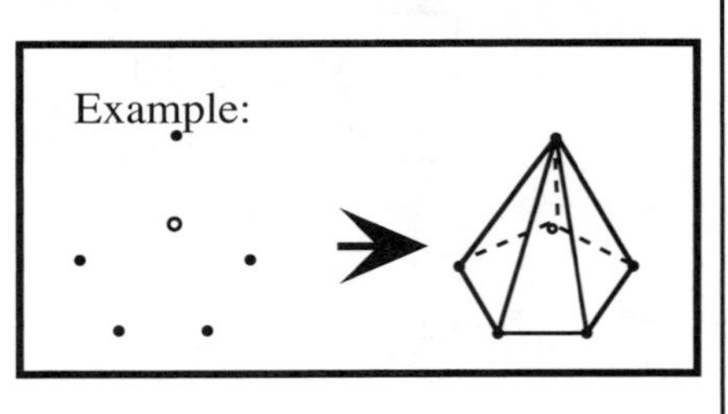


On the resource page provided by your teacher, draw in the edges to make pyramids from the three sets of points below. Show hidden edges as dashed segments.

a) b) c)

SV-92.

The example at right shows how to make a **prism** with hidden vertices and edges. Each face in this example and most faces in the problems that follow are quadrilaterals. Look for patterns that form trapezoids, parallelograms, and rectangles before you start connecting dots.

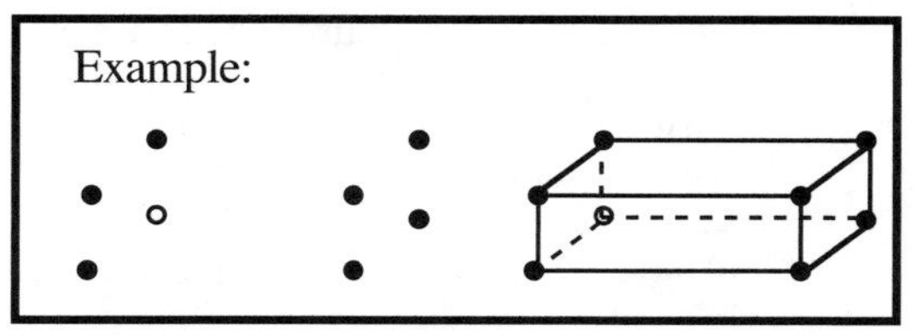


Using the dots on the resource page, draw in the edges to make prisms from the sets of points. Show hidden edges as dotted lines. Remember: look for a repeated arrangement before you begin!

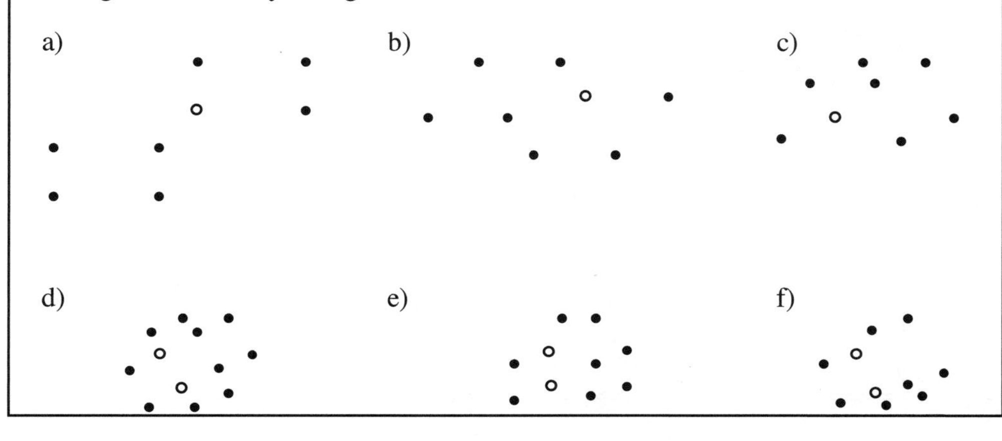


SV-93. Here are the names of the six prisms you drew in the previous problem. Choose a correct name for each picture on the resource page:

parallelogram-based prism    hexagonal-based prism    trapezoidal-based prism

rectangular prism    pentagonal-based prism

**SV-94.** **DIRECTIONS:** Working with a partner, obtain the resource templates from your teacher. Each of you build one of the solids. After you have built them, do parts (a), (b), and (c), read the following two boxes and add the definition, with illustrations, to your tool kit.

a) List at least two or three similarities between a prism and a pyramid.

b) List at least two or three differences between these two solids.

c) Why do you suppose the pyramid you built (and the one below) is called a hexagonal pyramid? What name would you give to the prism?

A **PYRAMID** is

1) a polyhedron
2) with a base that is a polygon and
3) lateral faces that are formed by connecting each vertex of the base to a single given point (the vertex of the pyramid) that is above or below the surface that contains the base. The lateral faces do NOT need to be congruent.

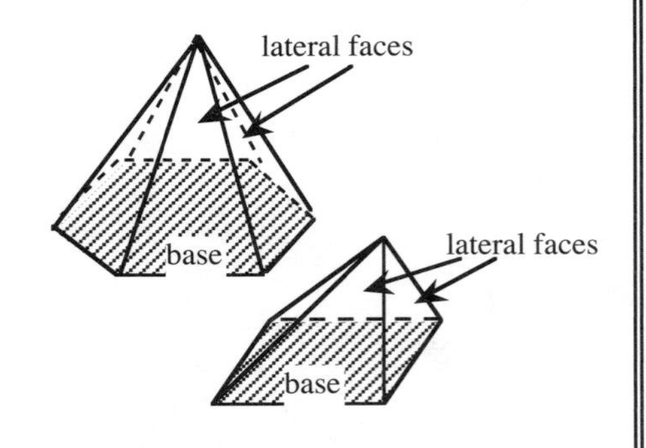


A **PRISM** is:

1) a polyhedron
2) with two **congruent** (same size and shape) parallel bases that are polygons, and
3) **lateral faces** (faces on the sides) that are parallelograms formed by connecting the corresponding vertices of the two bases. Lateral faces may also be rectangles, rhombi, or squares.

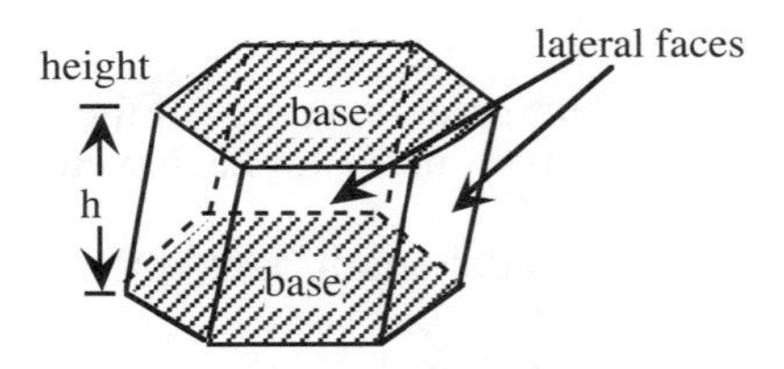


The **VOLUME** of a prism is the product of the **base area** (B) and **height** (h) (perpendicular distance between the bases)**:** **V = Bh**. Volume is expressed in **cubic** units.

SV-95. Prisms and pyramids do not always sit on their bases. Sometimes they rest on one of the other faces.

a) Draw pictures of a prism sitting on its base and the same prism on one of its faces.

b) Draw pictures of a pyramid sitting on its base and the same pyramid on one of its faces.

SV-96. On your paper, sketch the figure that is the base for each of the following solids (do not sketch the entire figure), then name the solid using the name of its polygonal base and either "prism" or "pyramid."

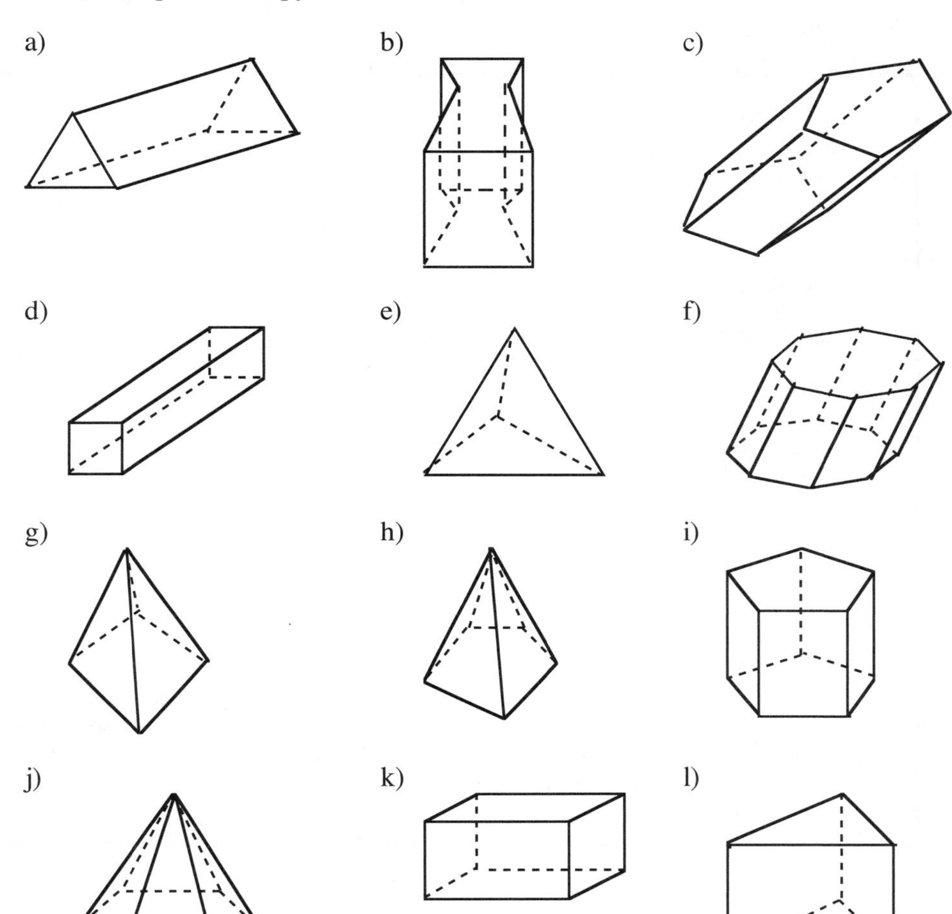

SV-97. Graph the following points using $0 \le x \le 23$ and $-19 \le y \le 11$ for axes, and connect them as you would in a dot to dot puzzle (that is, connect points 1 and 2, 2 and 3, 3 and 4, etc.). Where you see ---- , connect the points with dotted or broken lines. (For example in (a), 3 and 4 would be connected with a solid line, but 4 and 5 would be connected with a dotted line.) Name each solid you draw.

| a) | b) |
|---|---|
| 1. (0, 4.5) | 1. (12, 0) |
| 2. (3.5, 11) | 2. (22, -11) |
| 3. (5, 1) | 3. (16, -19) |
| 4. (0, 4.5) | 4. (12 , 0) |
| --------------- | 5. (2, -19) |
| 5. (18, 4.5) | -------------- |
| --------------- | 6. (8, -11) |
| 6. (23, 1) | -------------- |
| 7. (21.5, 11) | 7. (12, 0) |
| --------------- | |
| 8. (18, 4.5) | Next connect (2, -19) to (16, -19) with a solid segment and (8, -11) to (22, -11) with a dotted segment. Your second picture is complete. |
| Next connect (5, 1) to (23, 1) and connect (3.5, 11) to (21.5, 11) with solid segments. Your first picture is complete. | |

SV-98. Multiply and express as a sum:

a) $(x + 2)(x - 4)$  b) $(x^2 + 1)(x + 2)$

c) $(x^2 + 2)(x - 4)$  d) $(2x^2 - 1)(3x + x)$

SV-99. Graph $y = \frac{4}{3}x + 2$ on a sheet of graph paper.

a) Find the slope of the line.

b) Neatly and accurately draw a line perpendicular to this line through its y-intercept and find its slope. The perpendicular line should contain the point (4, -1).

c) How are the slopes of the two lines similar? different?

SV-100. Find the volume of each prism below.

a)

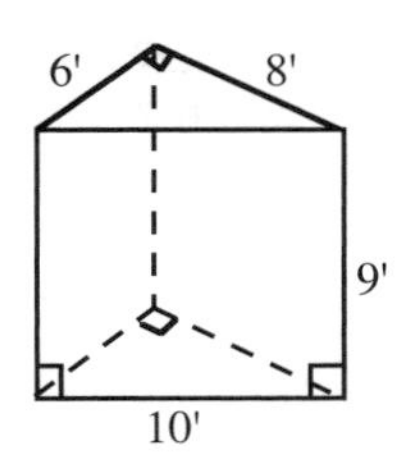


b) All surfaces are parallelograms.

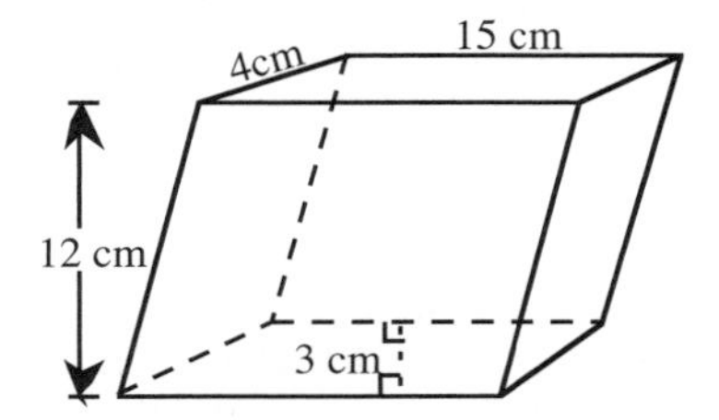


## SURFACE AREA

SV-101. Obtain a resource template from your teacher. Each member of the team should build one of the four solids. Before you start cutting, find the area of each polygon on your template (steps (a) and (b)). Then complete the remaining steps:

a) MEASURE the length of the segment(s) labeled x and the sides of the polygon(s) to the nearest mm. Write each of the lengths along the sides and heights. Note that no segments in the rectangular prism are labeled x. Why not?

b) Find the area of the faces of the solid (each polygon) and write each area on its corresponding face. Be certain that you DO NOT include the shaded glue flaps in your area calculations!

c) Cut out and build the solid by gluing or taping the flaps and edges together.

d) For each solid, find the sum of the areas of the faces (including the base). This sum represents the **total surface area** of the polyhedron.

e) Complete the table from your data, then share your results with your team. Explain to the others in your team how you did the problem.

| Name | Picture of the parts | Number and names of polygons added to get surface area | Total surface area (TSA) |
|---|---|---|---|
| Square Pyramid | □ + △ | | |
| Triangular Pyramid | | | |
| Rectangular Prism | | | |
| Octahedron | | | |

SV-102. The TransAmerica building in San Francisco is built of concrete and is shaped like a square based pyramid. The building is periodically power-washed using one gallon of cleaning solution for each 250 square meters of surface. How much cleaning solution is needed to wash the TransAmerica building if an edge of the square base is 96 m and the height of the building (<u>not</u> the height of a sloping face) is 220 m? Drawing a sketch of the building will be helpful.

SV-103. Calculate the <u>total</u> <u>surface</u> area of the following figures by using subproblems to first draw a picture of what each face looks like as a single polygon. Part (a) has been done as an example for you. Don't forget the bases!

a) The base of the pyramid is a square.

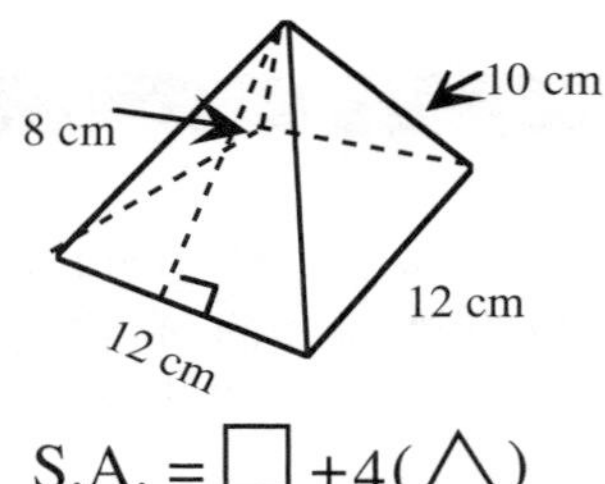


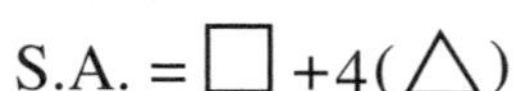


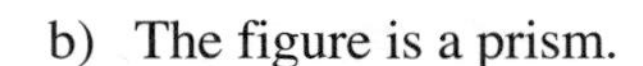


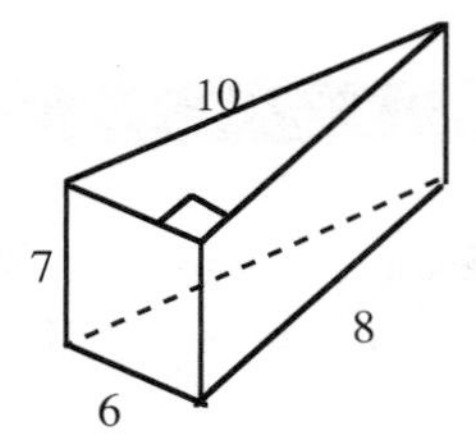


SV-104. A box is in the form of a cube 3 feet on a side. Will a rigid five foot stick fit inside the box? Draw a diagram and solve the subproblems.

SV-105. A box has a rectangular base which is 3 feet by 4 feet. The height of the box is 5 feet. Will an 8 foot stick fit into it? Stop! Can you make this problem easier by generalizing from the previous one?

SV-106. The figure at right is the front face of a rectangular prism. Use what you have learned about drawing solids to draw the rest of the prism so that its depth is greater than either of the dimensions of the face. Be sure to use dashed lines where appropriate.

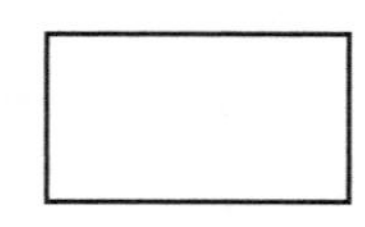

SV-107. Examine the figure at right and determine everything you know about the lines and angles.

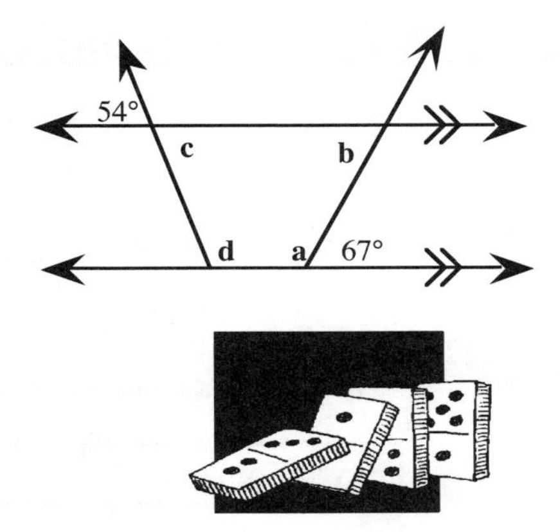


a) Find the measures of the four lettered angles. Explain how you found each angle and name the idea(s) you used to do so.

b) Calculate the sum of the four lettered angles. The result <u>should</u> suggest a conjecture for the sum of the four angles of a trapezoid. Use what you know about parallel lines and their related angles to write a convincing argument (a proof) that this result must ALWAYS be true for a trapezoid.

SV-108. Solve each system of equations.

a) $x + y = 10$
$x - y = 2$

b) $y = 3x - 5$
$2x + y = 10$

c) Demonstrate (prove) that your answers are correct.

## UNIT AND TOOL KIT REVIEW

SV-109. Consider each case at right.

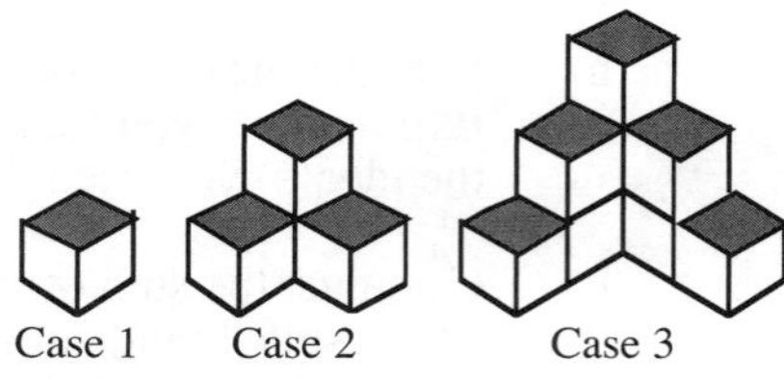


a) Draw case 4.

b) Make a table on your paper and complete it for each case.

| Case Number | Number of Cubes Needed to Build It |
|---|---|
| 1 | |
| 2 | |
| 3 | |
| 4 | |
| 5 | |
| ... | |
| 100 | |

c) How many cubes are needed to build case 5?

d) How many are needed to build case 100?

e) What is the volume of the solid in case 100?

f) What is the volume for case n?

SV-110. Suppose the cube stack in case 5 above is assembled securely and then dipped into a bucket of blue paint so that it is completely submerged. How many cubes have exactly (that is, only):

a) zero faces painted?

b) 1 face painted?

c) 2 faces painted?

d) 3 faces painted?

e) 4 faces painted?

f) 5 faces painted?

g) 6 faces painted?

h) Compare the sum of your answers to parts (a) through (g) to the result for case 5 in the previous problem. Do they agree? Should they agree? Why or why not?

SV-111. Find the surface area of the figures below.

a)

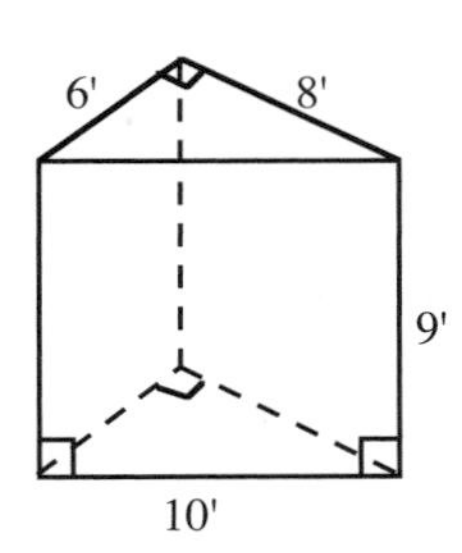


b) The front and back faces are parallelograms; the other four faces are rectangles.

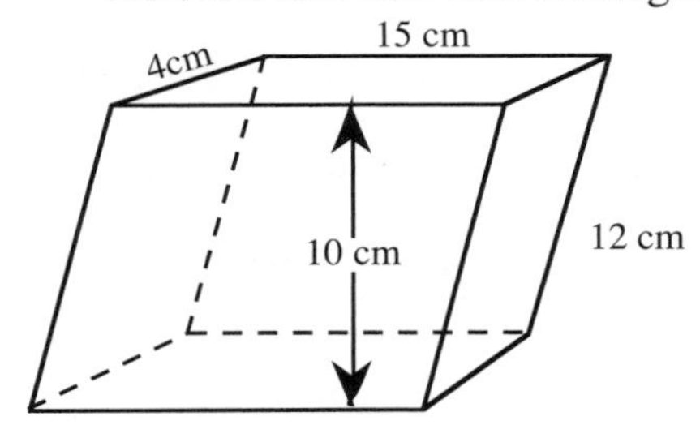


SV-112. You are given two squares, one of side length 3 and one of side length 2.5. The smaller square lies completely inside the larger and is tilted so that its vertices lie on the edges of the larger square.

a) Sketch the situation described.

b) Find the area between the two squares.

c) Find the length of a side of the smallest possible square that would fit so each vertex was on a side of the square of side length 3. What is the area between these two squares?

SV-113. Solve for x.

a) $(x + 1)(x - 2) = 0$

b) $x(x + 1)(x - 1) = 0$

c) $3(x - 5)(x + 2) = 0$

SV-114. Draw three pairs of xy-coordinate axes on graph paper. Scale the axes for $-8 \le x \le 8$ and $-10 \le y \le 10$. Graph the following inequalities by first graphing the curve then using a test point to see where to shade. Share the work on each graph by having each team member calculate results for two or three x-values.

a) $y \le (x + 2)^2$

| x | -5 | -4 | -3.5 | -3 | -2.5 | -2 | -1.5 | -1 | -0.5 | 0 | 1 |
|---|---|---|---|---|---|---|---|---|---|---|---|
| y | | 4 | | | | | 0.25 | | | | |

In part (b), subtract first, then square, then change signs.

b) $y < -(x - 2)^2$

| x | -1 | 0 | 1 | 1.5 | 2 | 2.5 | 3 | 4 | 5 |
|---|---|---|---|---|---|---|---|---|---|
| y | -9 | | | | | -0.25 | | | |

c) $y \ge x^3$

| x | -2 | -1.5 | -1 | -0.75 | -0.5 | 0 | 0.5 | 0.75 | 1 | 1.5 | 2 |
|---|---|---|---|---|---|---|---|---|---|---|---|
| y | | -3.38 | | | | | 0.13 | | | | |

SV-115. Identical cubes are stacked in the corner of a room as shown at right. Note that you may want to redraw the figure to help you visualize the changes described in the following questions.

a) How many cubes are not visible?

b) Remove the visible cubes. How many will be <u>visible</u> now?

c) If you continue to remove each set of visible cubes, howmany more steps will it take to remove all of the cubes?

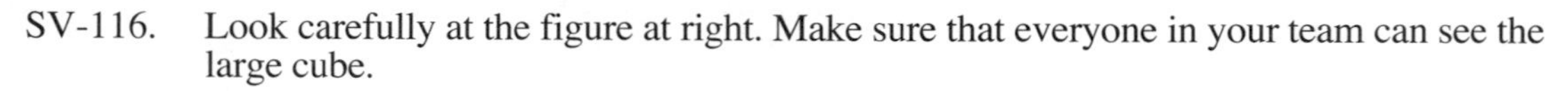

SV-116. Look carefully at the figure at right. Make sure that everyone in your team can see the large cube.

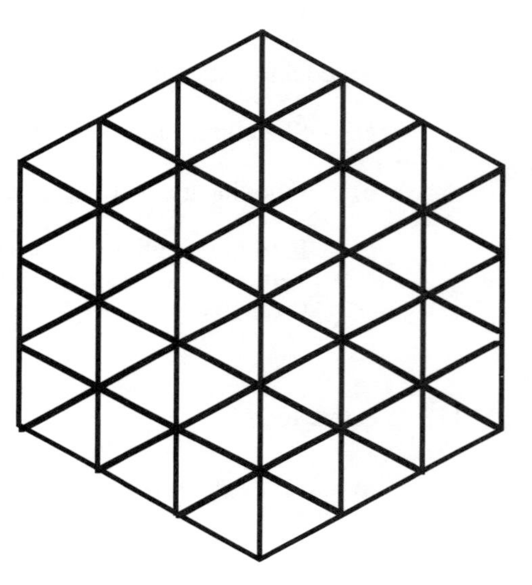

a) Trace one of the small cubes onto your paper. How many small cubes make up the large one?

b) Shift your focus to a 1 x 1 x 1 cube. What kind of two-dimensional geometric shape is formed by the border of the single cube? How many triangles do you see in this cube?

c) How many triangles do you see in a 2 x 2 x 2 cube?

d) How many small triangles are in the figure?

e) How many triangles are in an n x n x n cube, that is, a cube with each side length n?

SV-117. TOOL KIT CHECK-UP

Your tool kit contains reference tools for geometry. Return to your tool kit entries. You may need to revise or add entries.

Examine the list below. Each item should be included in your tool kit; add any that are missing now. You will find them in your text by skimming through the unit or by using the index in the back of the book.

- quadratic
- parabola
- isometric view
- mat plan
- polyhedron
- prism, pyramid
- face, edge, vertex
- base area
- Zero Product Property

# UNIT 5 Congruence and Triangles

# *Unit 5 Objectives*
## CONGRUENCE AND TRIANGLES

In this unit you will explore the idea of congruence, namely, the conditions under which figures have the same size and shape. First you will examine polygons on flat surfaces and xy-coordinate graphs and consider turning, sliding, and flipping them. These processes are known as transformations. Next you will study congruence between parts of triangles to determine when you can be certain they are the same size and shape. The algebra focus in this unit completes the work with solving quadratic equations that you started in Unit 4. You will factor and solve quadratic equations, then learn how to use the Quadratic Formula to solve any quadratic equation, factorable or not.

In this unit you will have the opportunity to:

- learn to turn, slide, and flip polygons on a flat surface while preserving their size and shape.
- discover and use the conditions under which pairs of triangles must be congruent.
- prove that parts of triangles and quadrilaterals are congruent to corresponding parts of other triangles and quadrilaterals.
- determine the limits to the length of the third side of a triangle.
- extend your knowledge of linear equations and slope to include perpendicular, horizontal, and vertical lines.
- review factoring and use it to solve quadratic equations.
- use the Quadratic Formula to solve quadratic equations.

*Read the problem below, then go on to the first problem. We will come back and solve it later.*

CG-0. Maria wondered if $\overline{LR} \parallel \overline{FE}$ in the figure at right. She knew that CR = AE, CL = AF, and $\overline{CL} \parallel \overline{AF}$. Her friend Paula said she thought they were parallel, so Maria challenged her to prove it. Decide whether or not Paula will be successful and explain why or why not.

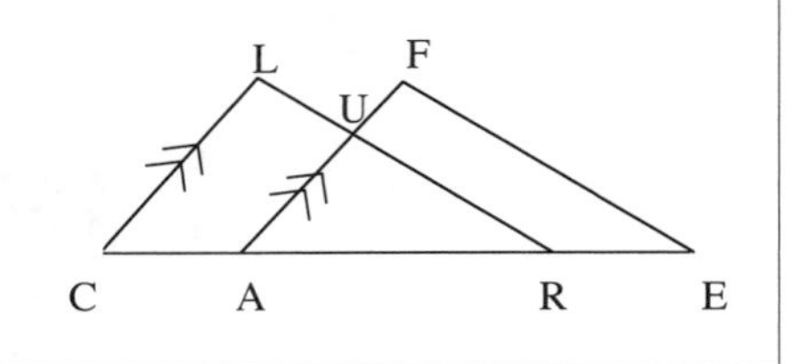


*UNIT* 5

ALGEBRA
GRAPHING
RATIOS
GEOMETRIC PROPERTIES
PROBLEM SOLVING
SPATIAL VISUALIZATION
CONJECTURE, EXPLANATION (PROOF)

# Unit 5

## CONGRUENCE AND TRIANGLES

### SYMMETRY AND REFLECTIONS

CG-1. On the resource page provided by your teacher, find as many lines as you can for each figure that, when folded along the line you draw, the two parts of the figure (that is, their segments and angles) fit together exactly. Some figures have more than one such line and some may have no lines. Draw the lines lightly on the figure.

Finally, read the definition and add it to your tool kit.

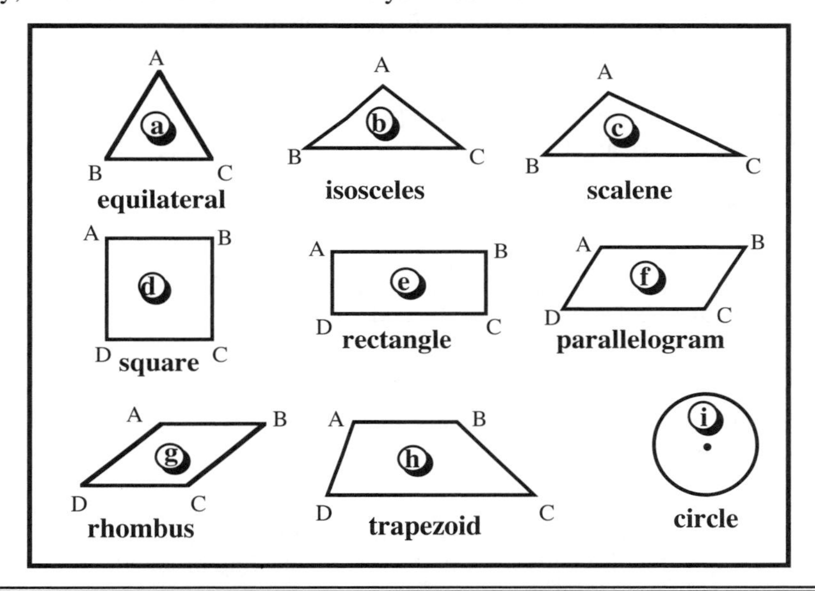


> Any line which divides a figure so that each side folds over the line to fit the other side exactly is called a **LINE OF SYMMETRY.** The figure is said to have **line symmetry.**

CG-2. There are several ways to move figures around on a flat surface. You could take a triangle, for example, and slide it across a table, or turn it 50°, or flip it over. Such movements are known as **transformations**, or **rigid motions**. Imagine that line $l$ on your resource page is a mirror.

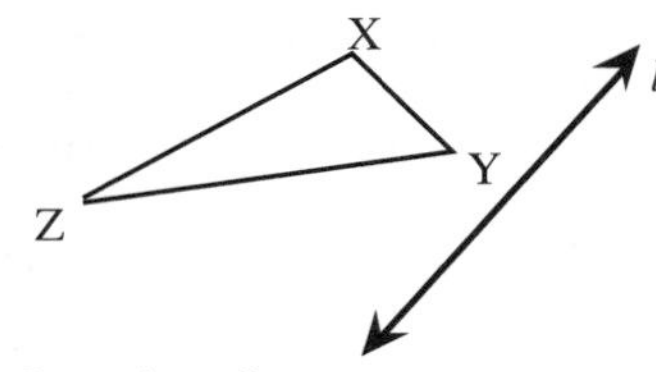


a) What would the reflection of ΔXYZ look like in the mirror?

b) Use the resource page to draw a reflection of ΔXYZ across line $l$. Label it ΔX'Y'Z' to show that point X corresponds to point X' (read X' as "x prime," etc.), Y to Y', and Z to Z'. If you need help getting started, notice that, in the mirror, point Z' will be straight across line $l$ and the same distance from $l$ as the original point Z.

c) Describe the change in position of ΔX'Y'Z' compared to that of ΔXYZ.

CG-3. On the resource page, reflect Quadrilateral ABCD across line $m$. Label the image A'B'C'D'. Use a mirror to help you get started.

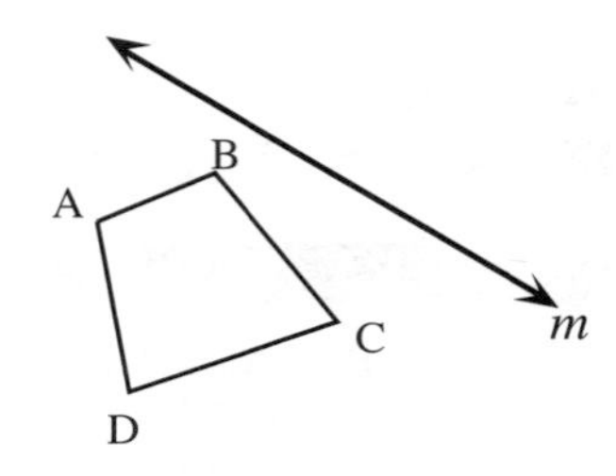


a) Use your ruler to draw line segments connecting corresponding points (for example, A should be connected to A').

b) What do you notice about the segments $\overline{AA'}$, $\overline{BB'}$, $\overline{CC'}$, and $\overline{DD'}$? Do they have anything in common?

c) How do these same four segments relate to line $m$?

CG-4. Draw a pair of xy-axes on graph paper, each scaled from -8 to 8. Plot the points A(5, 2), B(2, 4), and C(4, 6) and connect them to form ΔABC.

a) Reflect ΔABC across the x-axis. Label it ΔPQR and give the coordinates of each point.

b) Will you get the same result if you reflect ΔABC across the y-axis? Draw it and label its vertices A', B', and C'.

c) What do you notice about the vertices?

d) How does the size and shape of each reflected triangle compare to the original ΔABC?

CG-5. Read the information in the box below, add the definitions to your tool kit, and revise your work in the three preceding problems if it does not fit the definitions.

Transformations like those in the three preceding problems that preserve the size and shape of a figure across a line in a mirror image are known as **REFLECTIONS** or **flips**.

In problem CG-2, X'Y'Z' is read "X prime, Y prime, Z prime." This notation is often used to indicate that X' is the **image** of X, etc., thus making the correspondence between points clear.

All **image** points are connected to their original points (called **preimage** points) by segments that are perpendicular to and bisected by the line of symmetry. Thus, in problem CG-2 segments $\overline{XX'}$, $\overline{YY'}$, and $\overline{ZZ'}$ are each perpendicular to line $l$. Line $l$ in turn bisects each of these segments. Any point A reflected across a line $n$ satisfies both of these conditions as shown in the figure at right.

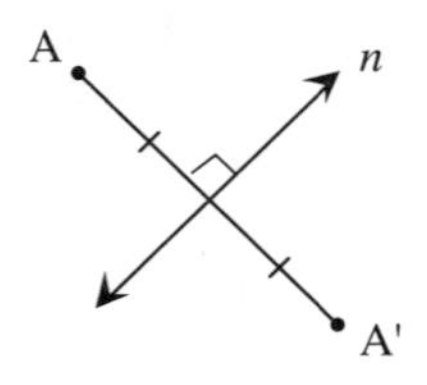


Sometimes people say "ΔXYZ is **mapped** to ΔX'Y'Z'."

**CG-6.** Read the review of factoring below, add anything you need from it to your tool kit, then practice factoring in the next problem.

## A QUICK REVIEW OF FACTORING

You know how to multiply two polynomials, such as
$(x + 2)(2x + 3) = 2x^2 + 3x + 4x + 6 = 2x^2 + 7x + 6$. The geometric interpretation of the polynomial, called **factoring with rectangles,** is

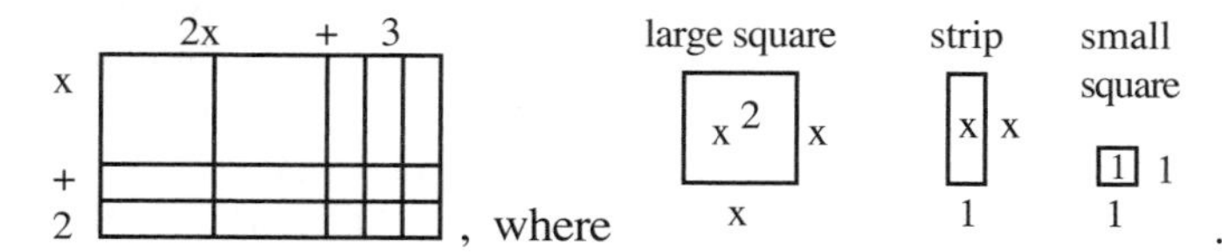


In the figure above left, there are two regions of area $x^2$, seven regions of area $x$, and six regions of area 1. In the figure above right, the linear dimensions are the outside of the basic figures, the area of each figure is shown inside its perimeter.

Often we want to un-multiply or **FACTOR** a polynomial. For example, to factor $x^2 + 5x + 4$ we can use one large square (for $x^2$), five strips (for 5x) and four small squares (for 4) and see if we can arrange them to form a rectangle. In this case, it is possible and we see that

$$x^2 + 5x + 4 = (x + 4)(x + 1).$$

Another technique to factor trinomials is called the **sum and product pattern**. If you took CPM Algebra 1, you practiced this method in diamond problems.

Multiplying $(x + 4)(x + 1)$ gives $x^2 + 5x + 4$. The 5x is the result of adding 4x and 1x; the 4 is the product of 4 and 1, so to go backwards--that is, to factor--we simply need to find two integers whose sum is 5 and whose product is 4.

Some important types of quadratics to know how to factor are (for example):

| | |
|---|---|
| $x^2 + 7x = x(x + 7)$ | common term |
| $x^2 - 25 = (x + 5)(x - 5)$ | difference of two squares |
| $x^2 - 6x + 9 = (x - 3)^2$ | binomial squares |

Factoring is often useful for solving quadratic equations and writing complicated expressions in simpler ways.

CG-7. In parts (a) and (b), find the dimensions of the rectangles in terms of x. Next, complete the guided sum and product problem in part (c). Finally, factor the trinomials in parts (d), (e), and (f) using the method of your choice.

a)

| $x^2$ | $5x$ |
|---|---|
| $2x$ | $10$ |

b)

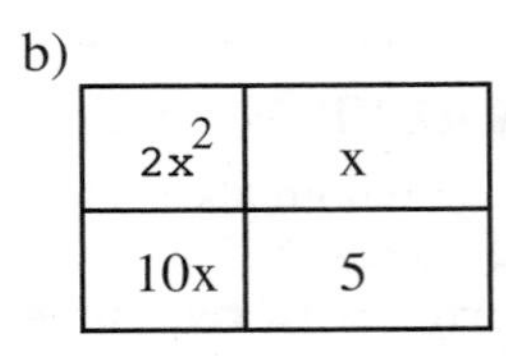


c) Factor $x^2 - 2x - 8$:

1) What are the possible integer factors of 8?

2) Test these pairs, with various sign combinations, to find a sum of -2. Then write the factored form of the trinomial.

d) $x^2 + 10x + 16$ e) $x^2 + 15x + 54$ f) $x^2 + 11x + 30$

CG-8. Solve for all variables and state the reason(s) from your tool kit that justify your work.

a) b) c)

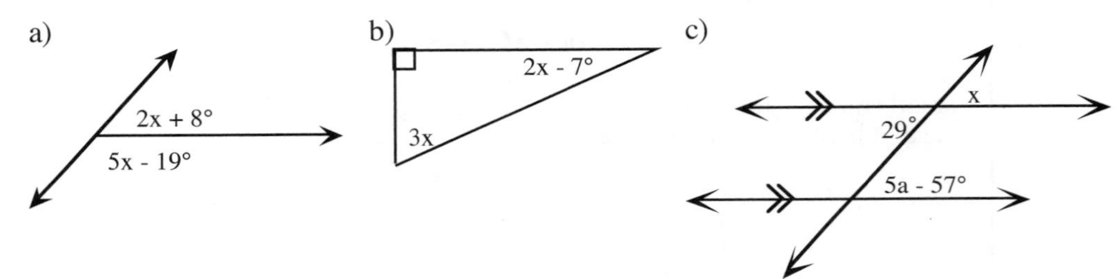


CG-9. Draw two segments that satisfy all of the following conditions simultaneously, that is, one drawing with all three conditions:

1) The segments are equal in length.
2) Only one of them is bisected.
3) They are perpendicular.

CG-10. Use the figure at right.

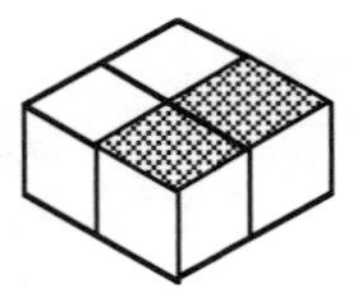

a) Add a cube to each shaded face and then draw the new solid on isometric dot paper.

b) Calculate the volume and total surface area of the solid you drew in part (a).

## TRANSLATIONS AND ROTATIONS

CG-11. **DIRECTIONS:** Take a sheet of graph paper and draw a set of xy-axes with $-6 < x < 10$ and $-7 < y < 7$. Plot the points A(- 5, 2), B(2, - 4), and C(1, 1) on the graph and connect them to form ΔABC.

a) Draw ΔA'B'C' by adding 4 to each x-coordinate and - 2 to each y-coordinate to get three new points, A', B', and C'. For example, the coordinates of A' are (- 5 + 4, 2 - 2) = (- 1, 0).

b) Describe what you notice about the two triangles in a sentence or two.

c) Using the same graph, add 7 to each original x-coordinate and 3 to each original y-coordinate, and connect the three new ordered pairs to form a third triangle. Label it A", B", and C". What do you notice?

d) Read the information in the box below and add the definition to your tool kit.

> **Transformations** like the ones above that preserve the size, shape, and orientation of a figure while sliding it to a new location are known as **TRANSLATIONS** or **slides.**

**CG-12.** **DIRECTIONS:** Suppose the points D(2, 0), E(6, 0), and F(3, 4) are plotted and connected to form a triangle as shown on the grid below right. Copy them onto a sheet of graph paper (include the axes). ΔD'E'F' is formed by **rotating** (turning) ΔDEF 90° counter-clockwise using the origin as the center of rotation. The vertices of the new triangle are D'(0, 2), E'(0, 6) and F'(-4, 3). Use a piece of tracing paper to outline the x- and y-axes, then trace a triangle that is the same size and shape as ΔDEF and is in the same position along the x-axis. Use it to help visualize the rotations. Complete parts (a) and (b) on your graph paper.

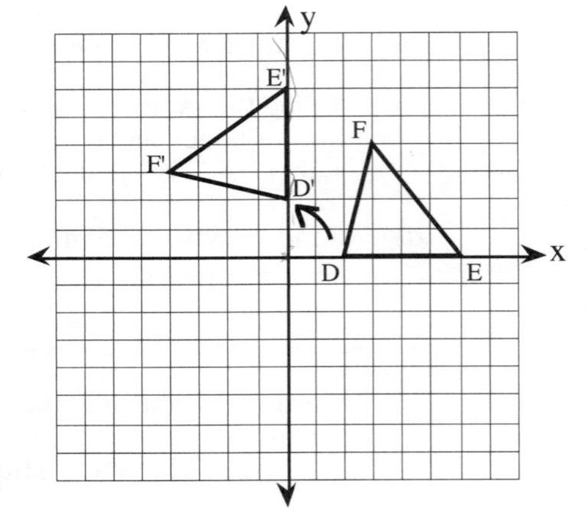


a) Place the point of a pen or pencil at the origin (0, 0) and rotate the traced copy of ΔDEF 180° counter-clockwise about the origin to get ΔD"E"F" and draw the new triangle on your graph beneath the tracing paper.

b) Put the traced copy of ΔDEF back in its original position on the x-axis, then rotate it 270° counter-clockwise about the origin to get ΔGHI and draw the new triangle on your graph beneath the tracing paper.

c) In what ways are the triangles in parts (a) and (b) the same? In what ways are they different? Write your observations in a few sentences.

d) Read the information in the box below and add the definition to your tool kit.

**Transformations** like the ones above that preserve size and shape while turning all the points in the original figure the same number of degrees in relation to a fixed center point (like the origin) are known as **ROTATIONS** or **turns.**

Recall that we labeled the image of ΔDEF after a transformation as ΔD'E'F' to show exactly which points corresponded to which other points. Here, D → D', E → E', and F → F'. When primes are not used, the order of the vertices determines the correspondences (matches). For example, if we know that ΔPQR is the image of ΔXYZ, we know that P corresponds to X, Q corresponds to Y, and R corresponds to Z.

CG-13. Examine the grid triangles at right.

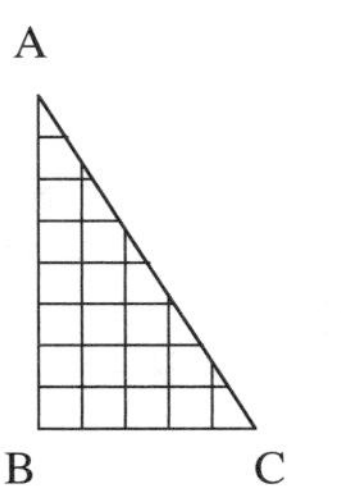

a) What transformation(s) were used to move $\Delta ABC$ to show the positive slope of the hypotenuse (CC')?

b) Find both possible slopes for the hypotenuse of the grid triangle.

c) Compare the slopes: what is the same? what is different?

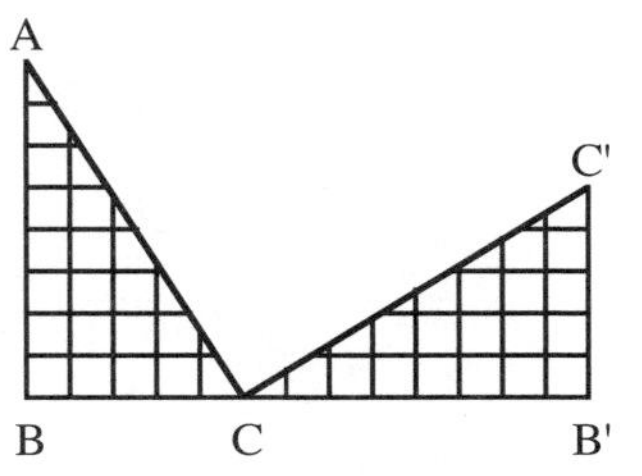


d) Recall that the acute angles of a right triangle are complementary. Explain why $\angle ACC'$ must be 90°.

e) Based on your result in part (d), what is the relationship between segments $\overline{AC}$ and $\overline{CC'}$?

f) Write a conjecture about the relationship between the slopes of two perpendicular lines.

CG-14. In each part, name the transformation or set of transformations you would use to move the figure on the left to coincide with the one on the right.

a)

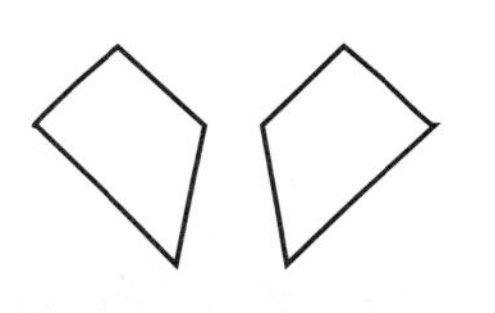

b)

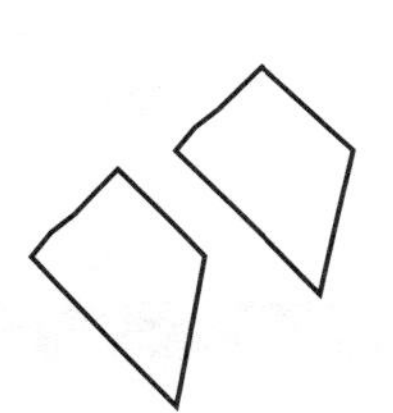

c)

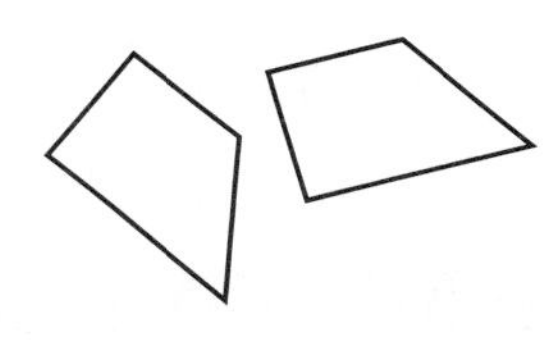

d)

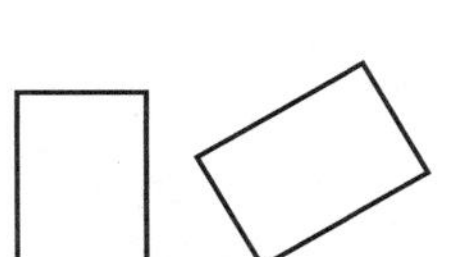

e)

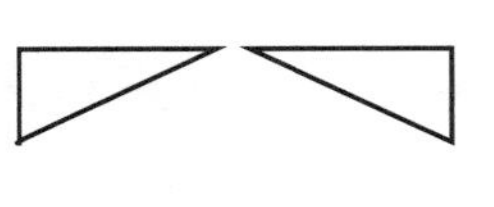

f)

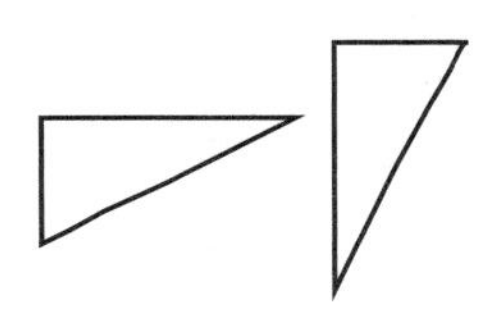

CG-15. Draw a pair of xy-axes on graph paper, each scaled from -8 to 8. Plot the points R(-5, 2), C(-4, 3), H(-1, 4) and S(-3, 1) and connect them to form a quadrilateral.

a) First, reflect Quad RCHS across the y-axis to create R'C'H'S'. Can you predict the coordinates before you draw the reflection?

b) Reflect the image (R'C'H'S') across the x-axis. Label this image R"C"H"S" (read "R double prime, C double prime," etc.).

c) Describe what you notice in a sentence or two.

CG-16. Find the measures of $\angle B$ and $\angle BCA$ in each triangle below. Show your subproblems and state the reason(s) (from your tool kit) that justify your work.

a)

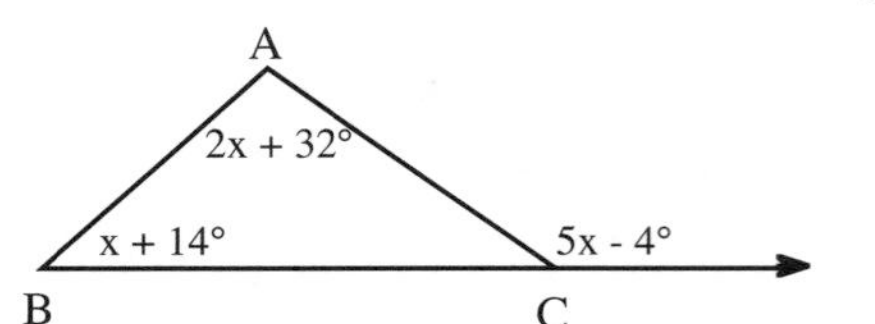


b)

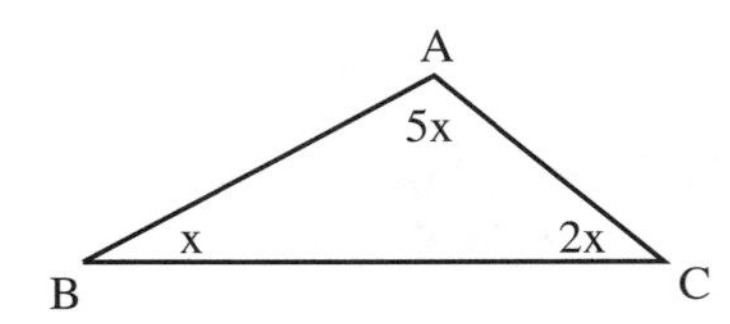


CG-17. Draw two segments that satisfy all of the following conditions simultaneously, that is, one drawing with all three conditions:

1) The segments are equal in length.
2) They bisect each other.
3) They are perpendicular.

CG-18. Find the area of a rectangle which has one side 8 cm and whose <u>diagonal</u> is 10 cm. Show all subproblems.

CG-19. Use your experience with rectangles or sums and products of integers to help you factor these quadratics:

a) $x^2 - 6x + 8$    b) $x^2 - 11x + 18$    c) $x^2 - x - 20$

## CONGRUENCE

CG-20. Draw pictures to help answer these questions. Tracing paper may be helpful.

a) Suppose you are given two lines. Would it always be possible to pick up one line and move it to coincide, that is, fit exactly, with the other line? Explain completely.

b) Suppose you are given any two rays. Would it always be possible to pick up one ray and move it to coincide (fit exactly) with the other ray? Explain completely.

c) Suppose you are given any two line segments. Would it always be possible to pick up one line segment and move it to coincide (fit exactly) with the other line segment? Explain completely.

d) Write a conjecture stating the conditions under which two segments could coincide exactly.

CG-21. Compare the three angles below.

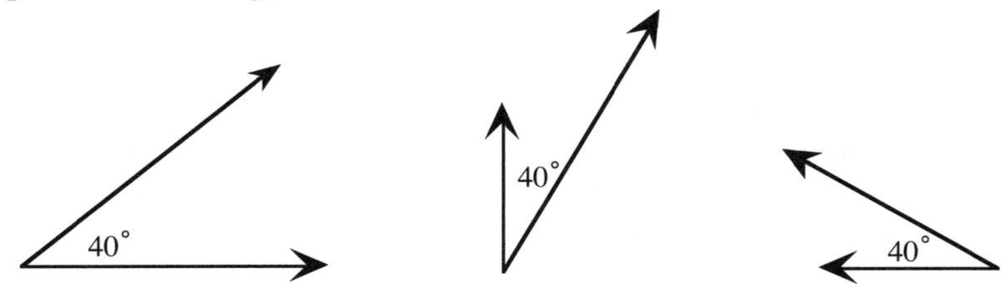


a) Discuss with your team whether the angles are the same size and shape. Write the conclusions of the team in a short paragraph.

b) Write a statement describing a way to tell whether angles are the same size.

CG-22. Earlier in the year we used the term **congruent** when we talked about figures that had exactly the same size and same shape. To get a better feeling for what this idea means, your teacher will give you a sheet of waxed paper, tracing paper, or plain paper and a resource page with shapes on it.

a) On the paper, trace figures (a), (b), (c), and (d). USE YOUR RULER for shapes (a), (b), and (d). Make your drawings neat and accurate!

b) Slide, turn, and/or flip the sheet of paper with the traced figures on it to see if you can get your copy of figure (a) to precisely match any of the other figures in the box on the resource page. List the figures which match figure (a).

c) Repeat part (b) for figures (b), (c), and (d); list the figures which can be matched to each of them.

CG-23. Read the information in the box below and add the definitions and notation to your tool kit. Then answer the questions below the box.

Two figures are **CONGRUENT** if they have exactly the same size and shape. This means that one figure can be slid, turned, and/or flipped so that it fits exactly on the other figure. The symbol for congruent is ≅.

When two figures are congruent, we always list **CORRESPONDING LABELS** in the same order. For example, in the figures below, it is proper to say that $\Delta ABC \cong \Delta DEF$ because A **corresponds** to D, B **corresponds** to E, and C **corresponds** to F. However, it would be <u>incorrect</u> to say $\Delta ABC \cong \Delta DFE$. We can show the corresponding parts easily with arrows as follows:

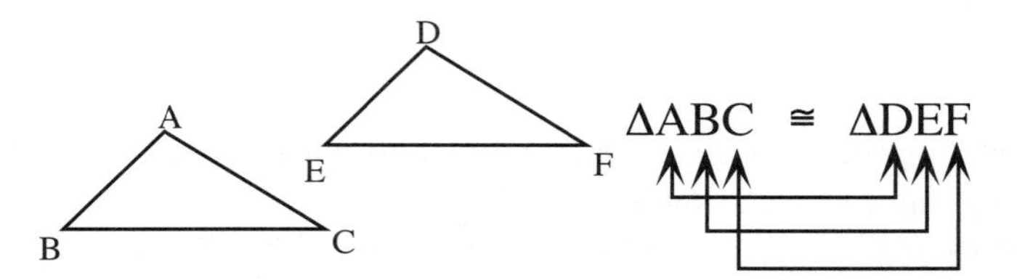


Reminder: when we want to indicate that two or more segments or angles are equal in a figure, we use an equal number of slash marks as illustrated in the three figures below.

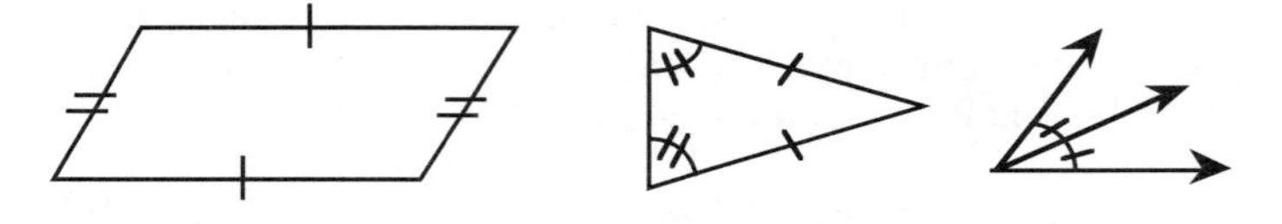

The two triangles below are congruent. Which of the following statements are correctly written and which are not? Explain why and why not. Discuss your answers with your team.

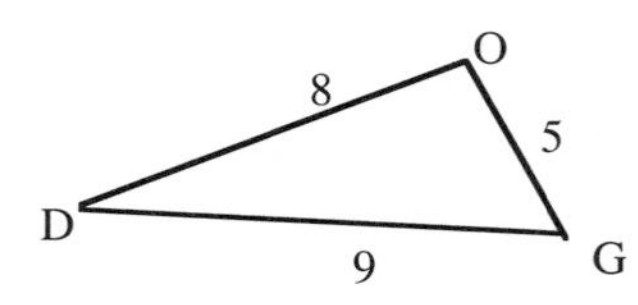


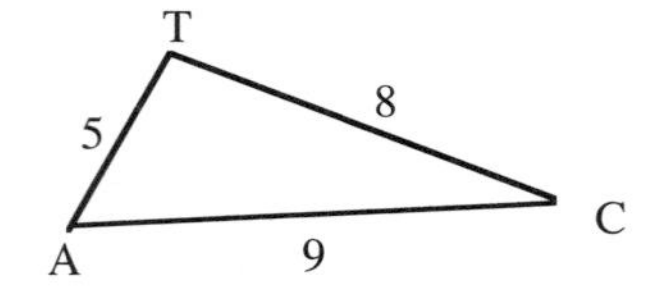


a) $\Delta DOG \cong \Delta CAT$

b) $\Delta DOG \cong \Delta CTA$

c) $\Delta OGD \cong \Delta CTA$

d) $\Delta CAT \cong \Delta DGO$

CG-24. Draw a horizontal line on your paper and draw $\Delta A'B'C'$ so that it is congruent to $\Delta ABC$ below. Explain (prove) exactly how you did it so that a student who is absent today would be convinced you are correct.

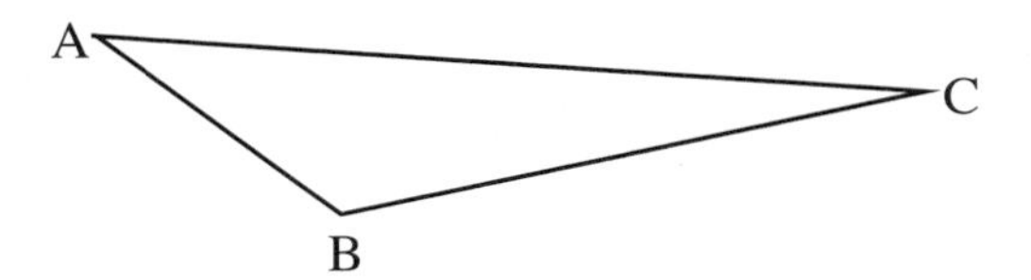


CG-25. Draw $\Delta PQR$ using $\angle P$, $\angle Q$, and $\overline{PQ}$ below. Place your paper over each angle and trace it to get an accurate copy. Start by drawing a horizontal ray with its endpoint labeled P. Notice that you now have no choice where to place $\overline{PQ}$. Next place $\angle P$ at endpoint P with one of its sides coinciding with (on top of) $\overline{PQ}$. Explain why, when you next place $\angle Q$ in your figure, the shape of $\Delta PQR$ is determined (set).

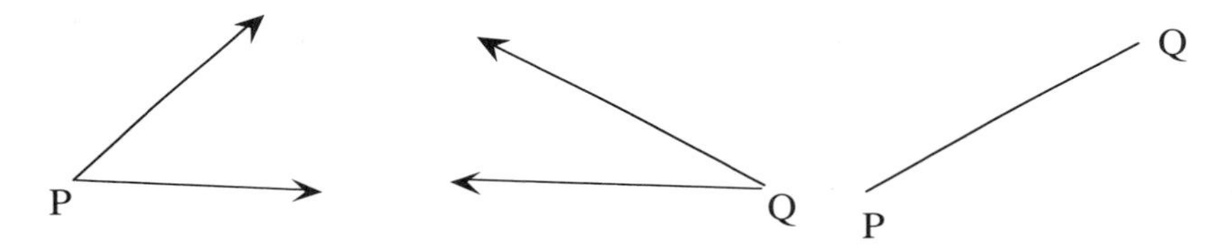


CG-26. Sammy drew a careless picture of two triangles even though his textbook said that $\Delta ABC \cong \Delta QPR$. Fortunately you know how to use the congruence statement to figure out which pairs of sides and angles are equal. Name all six pairs of corresponding, congruent parts.

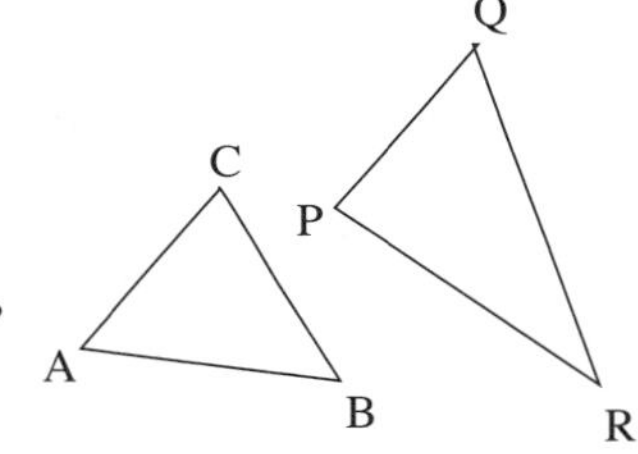


CG-27. Draw $\overrightarrow{AB}$ on your paper. Find two rays, $\overrightarrow{AC}$ and $\overrightarrow{AD}$, such that $\angle BAC \cong \angle PQR$ and $\angle BAD \cong \angle PQR$.

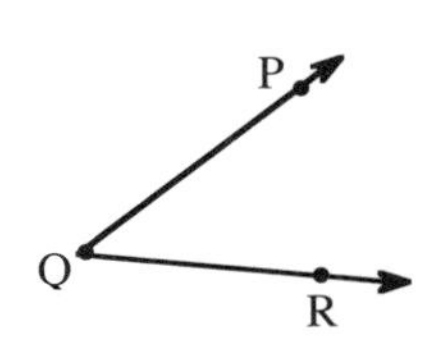

CG-28. Assume that each picture shows a pair of congruent figures. Complete a correct congruence statement to illustrate which parts correspond. Remember: letter order is important!

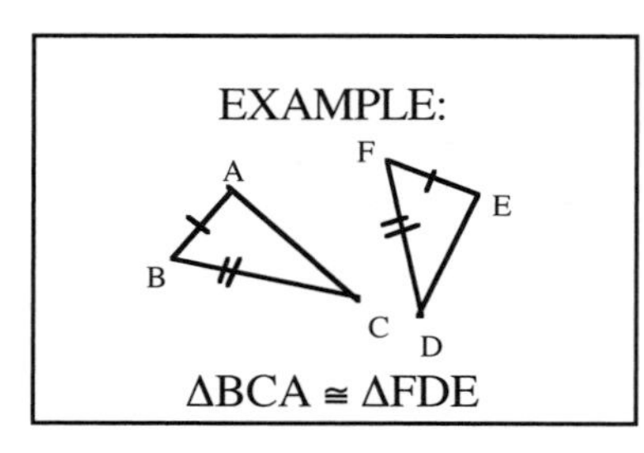


a) Trapezoid ABCD ≅ ?

b) Pentagon RIGHT ≅ ?

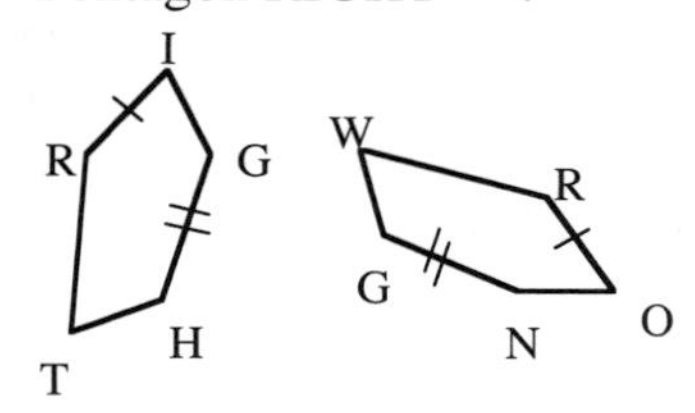


c) Δ ? ≅ Δ ?

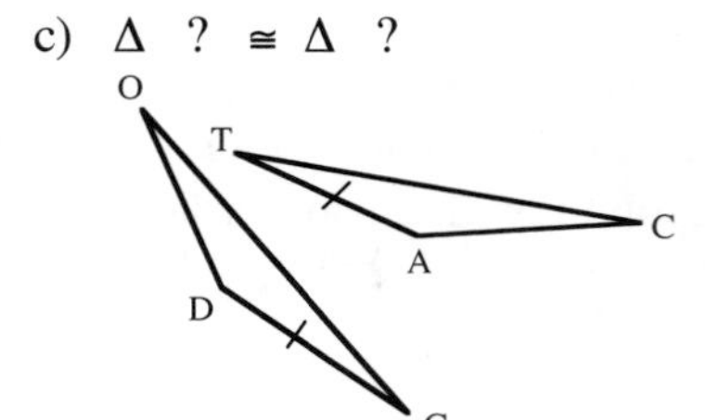


CG-29. If $y = \frac{5}{3}x - 3$, what is the slope of any line perpendicular to it? parallel to it?

CG-30. Factor to solve these equations.

a) $x^2 - 9x + 18 = 0$ b) $x^2 - 3x - 10 = 0$ c) $x^2 - 49 = 0$

# CONGRUENT TRIANGLES

CG-31. Read the following information and record it in your tool kit. Then go on to the next problem.

We know that a triangle has six parts: three sides and three angles.

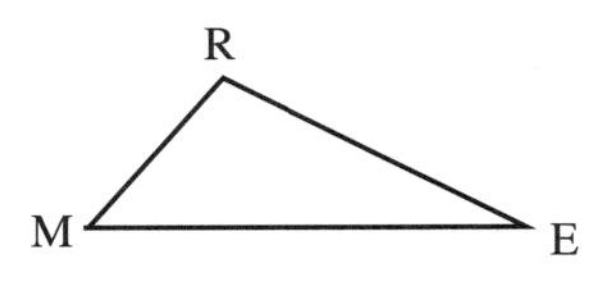


$\overline{MR}$ $\angle M$

$\overline{RE}$ $\angle R$

$\overline{EM}$ $\angle E$

If two triangles are congruent, then the **corresponding six parts are congruent,** and vice versa.

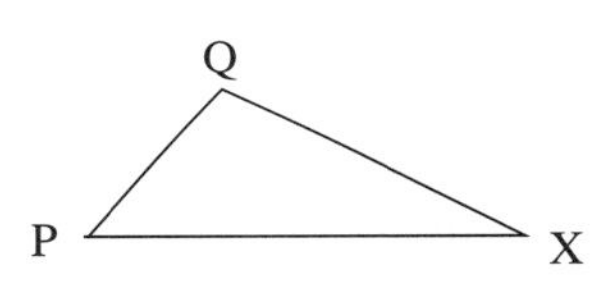


$\Delta MRE \cong \Delta PQX$ means that:

$\overline{MR} \cong \overline{PQ}$ $\angle M \cong \angle P$

$\overline{RE} \cong \overline{QX}$ $\angle R \cong \angle Q$

$\overline{ME} \cong \overline{PX}$ $\angle E \cong \angle X$

So whenever you know that two triangles have **all six** corresponding parts congruent, you know the triangles are congruent. But do we need to know about <u>all six</u> parts to make that conclusion? Would five, four, three, or two be enough?

The goal of the next several problems--especially the next two problems--is to determine the <u>minimum</u> number of congruent correspondences between parts of the two triangles that are needed to ASSURE that the two triangles are congruent.

**CG-32.** **DIRECTIONS:** Your teacher will provide you with a packet of plastic angles and line segments. For each part below each person in your team OR each team starts with the same two pieces specified and uses them to make a triangle. Can you make different triangles using only two of a triangle's six parts--or does everyone get the same size and shape? Sketch and label each triangle your team makes.

a) First try one side, LS3, and one angle, ∠6. Are the triangles different or congruent?

b) Next try two angles, ∠7 and ∠8. Are the triangles different or congruent?

c) Now use two sides, LS3 and LS4. Are the triangles different or congruent?

d) Discuss with your team whether or not knowing two of the six parts of any triangle is sufficient information to draw exactly the same triangle without looking at each other's paper. Write a summary of your conclusions and your reasons for them.

**CG-33.** Follow the same general instructions for the parts below as you did in the preceding problem including sketching and labeling the figures you make. This time, however, the question is whether you can make different triangles using only three of a triangle's six parts OR whether everyone gets the same triangle. Be sure to write your decision in each part as to whether the triangles are all congruent or not.

a) Use three sides, LS1, LS2, and LS3. Is everyone's triangle congruent?

b) Use three angles, ∠7, ∠8, and ∠9. Is everyone's triangle congruent?

c) Use two sides and the angle BETWEEN them. Place LS2 and LS4 along the sides of ∠9. Is everyone's triangle congruent? Note that ∠9 is called the **included angle** since it is between two specified sides.

d) Use two sides and one angle but the angle is not between the two sides. Use LS1, LS3, and ∠5. Is everyone's triangle congruent?

e) Use two angles and the side that connects their vertices. Slide one side of each angle--∠7 and ∠8--along side LS3 until an endpoint of LS3 is at the vertex of each angle. Is everyone's triangle congruent? Note that LS3 is called the **included side** since it is between two specified angles.

f) Discuss with your team whether or not knowing three of the six parts of any triangle is sufficient information to draw exactly that triangle. Write a summary of your conclusions and your reasons for them.

**CG-34.** **DIRECTIONS:** In part (a) of the previous problem you found that if you are given the lengths of the three sides of a triangle, all the triangles you could make were congruent to each other. Suppose you start with TWO triangles knowing that their three pairs of corresponding sides are congruent:

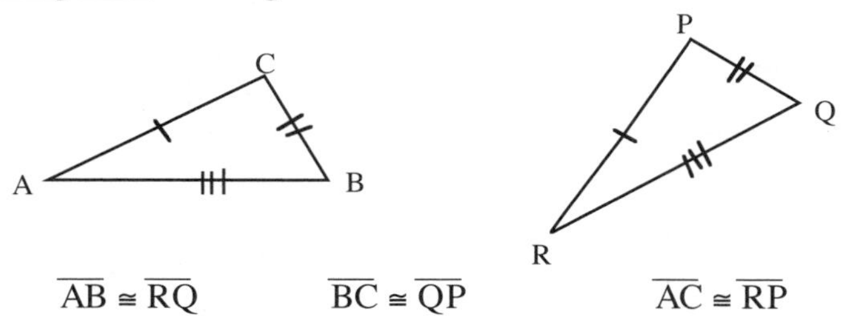


$\overline{AB} \cong \overline{RQ}$ $\overline{BC} \cong \overline{QP}$ $\overline{AC} \cong \overline{RP}$

a) What can you say about the two triangles and why? What can you say about their angles? Why?

We can abbreviate your conclusion as **SSS** (side--side--side).

b) The other parts that you experimented with in the preceding problem can be abbreviated similarly, such as SAS (side--angle--side, part (c)) and ASA (angle--side--angle, part (e)). Which of the remaining four arrangements in that problem will allow you to conclude that a pair of triangles are congruent? For example, if two triangles have all three pair of corresponding angles equal (and that is all you know for sure), can you conclude that the triangles are congruent? For each of the four other parts (b) through (e), if your answer is yes, draw a diagram like the one above and specify which other parts of the triangle are congruent. For the arrangements that do not guarantee congruence, explain why not.

c) Once your team decides which arrangements work and which do not, compare your results to those of one or two other teams in the class. Be prepared to give explanations of your results to the class. It is important that the whole class reach agreement on which properties work and which do not before you continue with the unit.

**CG-35.** Refer to the triangles below and the statements at right and determine which of the statements are correct and which are not.

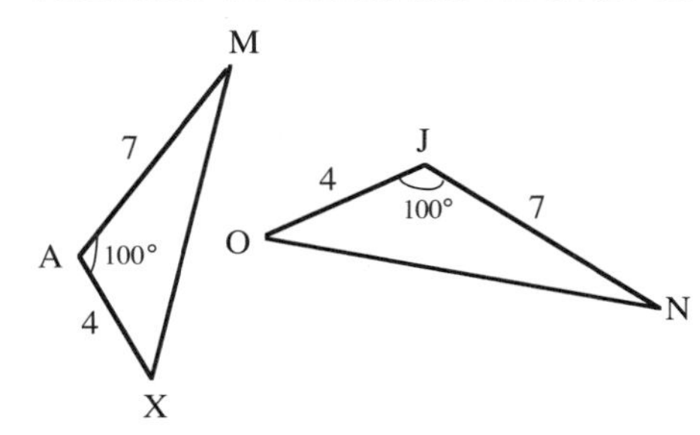


a) $\Delta MAX \cong \Delta JON$

b) $\Delta MAX \cong \Delta NJO$

c) $\Delta JON \cong \Delta AMX$

d) $\overline{MX} \cong \overline{NO}$

e) The triangles are not congruent.

CG-36. Find the total surface area and volume of each figure. Note the information about each base.

a) Pentagonal prism with $B = 49$ sq. cm.

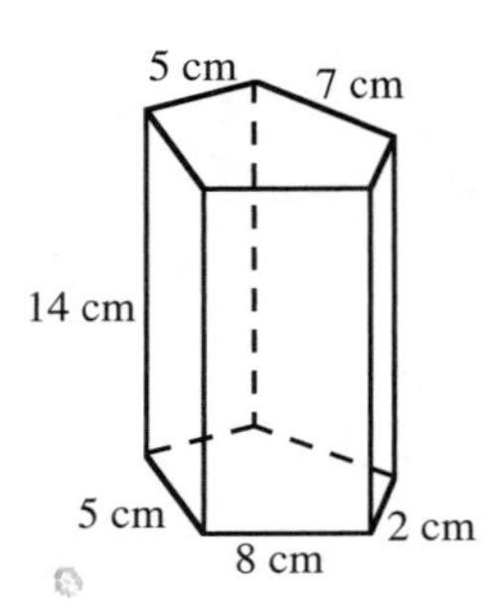


b) Rectangular prism with a 2' by 3' cut out.

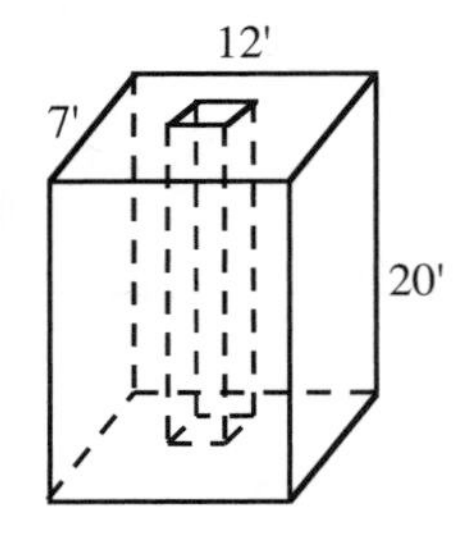


CG-37. Find the equation of the line with a y-intercept of (0, 7) that is:

a) perpendicular to $y = \frac{1}{2}x - 4$.

b) parallel to $y = \frac{1}{2}x - 4$.

CG-38. Given a right triangle with hypotenuse 16 cm and one leg of length 9 cm.

a) Find the length of the other leg.

b) Suppose you know that one of the angles of the triangle in part (a) is 34.23°. What are the sizes of the other angles?

CG-39. Factor to solve these equations. Remember the Zero Product Property from the last unit.

a) $x^2 + 2x - 24 = 0$

b) $x^2 + 11x + 24 = 0$

c) $x^2 - 10x + 24 = 0$

d) $x^2 + 5x + 4 = 28$

# USING CONGRUENT TRIANGLES

CG-40. **DIRECTIONS:** You have had the opportunity to see for yourself that when certain combinations of <u>three parts</u> of triangles are congruent, everyone must build the same triangle. This observation--and the conditions under which congruence is assured--will be useful when you compare the <u>corresponding parts</u> of <u>two</u> triangles to see if the two triangles are congruent. Read the conclusions about triangle congruence in the box below and add the properties to your tool kit.

In order to conclude that two triangles are congruent (without showing that all six corresponding parts of the triangles are congruent), you need only show that certain combinations of three pairs of corresponding parts are congruent. These combinations, called **TRIANGLE CONGRUENCE PROPERTIES**, are:

- **SSS** This represents side--side--side, which means all three pairs of corresponding sides are congruent.
- **SAS** This represents side--angle--side, which means two pairs of corresponding sides AND the angle they form (the included angle) are congruent.
- **ASA** This represents angle--side--angle, which means two pairs of corresponding angles AND the side they share (the included side) are congruent.

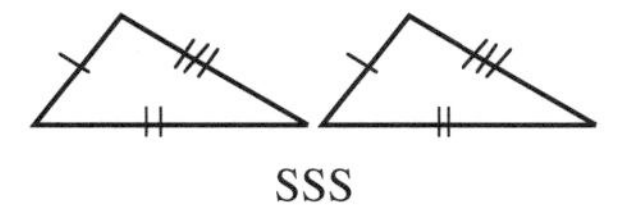

SSS

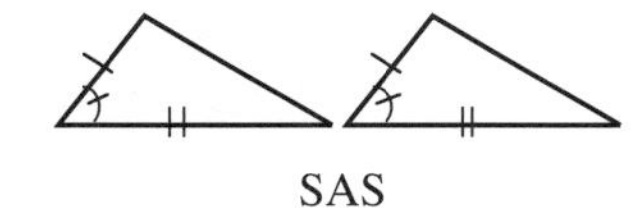

SAS

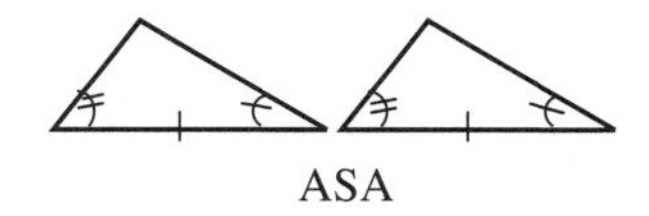

ASA

Once you have demonstrated that two triangles are congruent, you may state that any of the other pairs of corresponding parts are congruent.

CG-41. **DIRECTIONS:** Use the three triangle congruence properties and the information about each pair of triangles below to decide whether the two triangles are congruent. If they are, write a clear explanation justifying your response similar to the example below. Then write the correct sequence of vertices for the second triangle. If the triangles are not congruent, briefly explain why not. Discuss each problem with your teammates.

**Example:** $\Delta ABC \cong \Delta$ __?__.

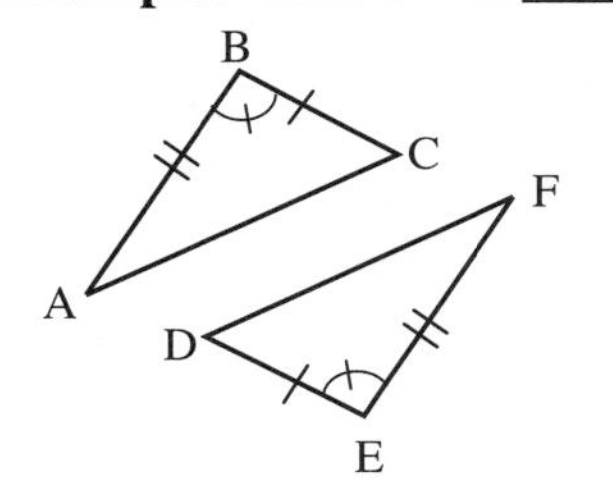


**Solution:** Yes, $\Delta ABC \cong \Delta FED$, because

- congruent segment $\overline{AB}$ corresponds to $\overline{FE}$;
- congruent segment $\overline{BC}$ corresponds to $\overline{ED}$; and,
- congruent angles $\angle B$ and $\angle E$ are included angles.

Thus, the SAS congruence property applies.

Note: <u>only</u> the order FED is acceptable--why?

**>>>This problem continues on the next page.>>>**

Note: in some of the figures below you will be able to use tool kit definitions and conjectures to find additional information about parts of the figures. Also state these properties in your answer to justify the additional information or relationships you use. For example, in part (b), you also know that ∠BEA ≅ ∠DEC because they are vertical angles

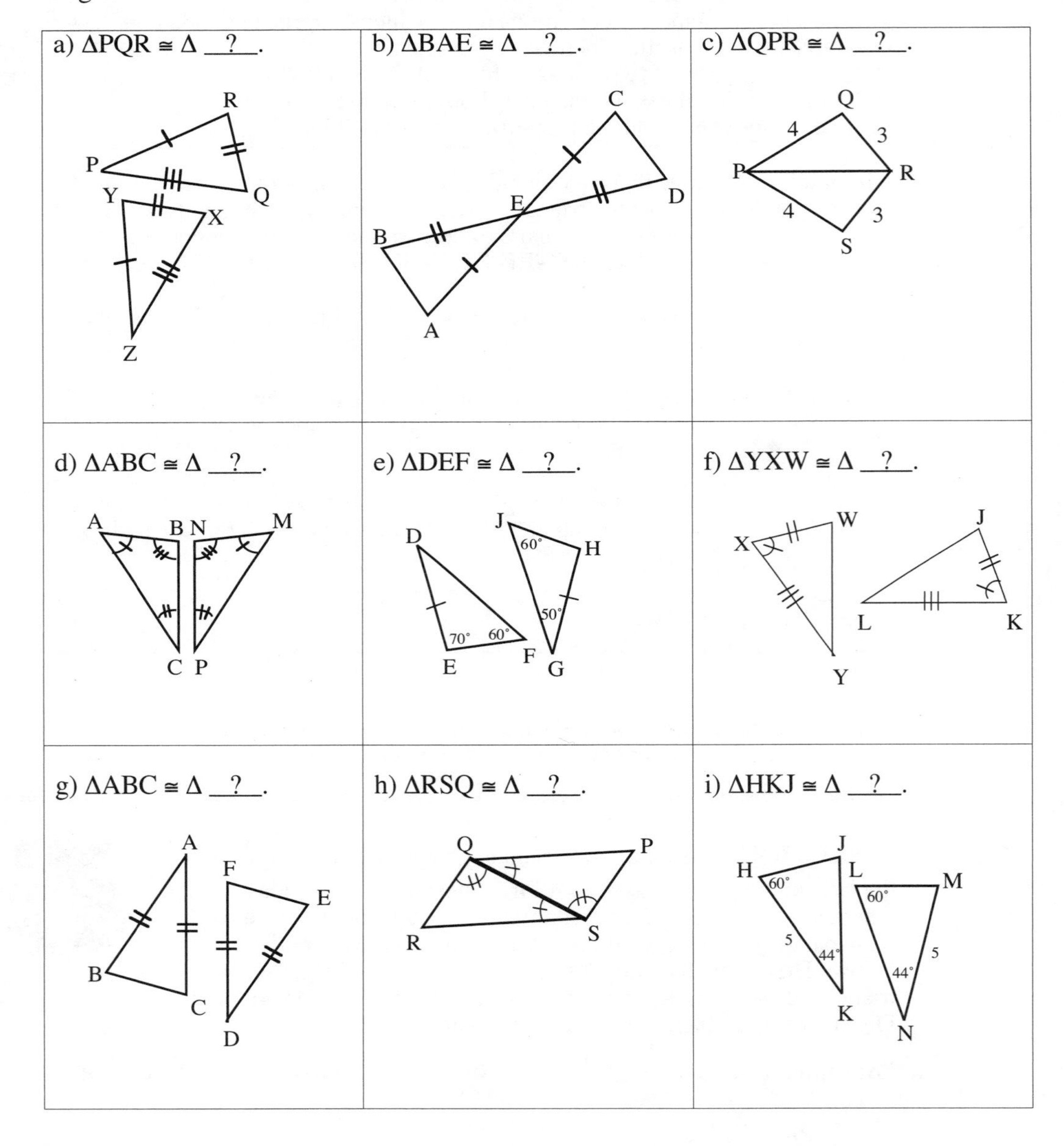

CG-42. Use the resource page provided by your teacher or draw this figure on graph paper and be sure to label it as shown.

a) Translate (slide) $\Delta ABC$ so that the point B is at the origin. Label the new triangle $\Delta A'B'C'$ and give the coordinates of each NEW point.

b) Is $\Delta ABC \cong \Delta A'B'C'$? Justify your answer.

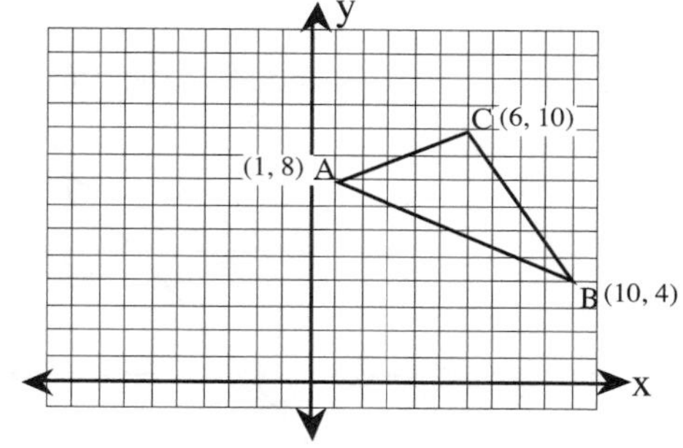


c) Calculate the slopes of each of the following segments.

slope of $\overline{AB}$ = slope of $\overline{A'B'}$ =

slope of $\overline{AC}$ = slope of $\overline{A'C'}$ =

slope of $\overline{BC}$ = slope of $\overline{B'C'}$ =

d) Write a conjecture about the slopes of the sides of translated triangles in conditional (if-then) form. Explain why it will always be true.

CG-43. Use the upper portion of the resource page (or your graph) from the previous problem.

a) Graph $y = \frac{1}{8}x + 12$ and describe the steepness of the line.

b) Write the equation of the line that has a y-intercept of 12 and is absolutely flat (horizontal). Hint: look at the slope of the line in part (a).

c) Describe the equation of any horizontal line.

d) Graph $y = 3$ on the same axes as part (a).

e) Extend this idea to vertical lines. Write the equation of the vertical line that passes through (5, 0), then graph it.

CG-44. Multiply.

a) $(x + 1)(3x - 2)$ b) $(3x + 1)(x + 2)$ c) $(2x - 5)(x + 1)$

CG-45. Where would each graph cross the x-axis?

a) $y = x(x - 4)$ b) $y = (x + 3)(x - 5)$ c) $y = (x - 9)(x + 9)$

CG-46. Solve each equation by factoring. Remember the zero product rule from the last unit.

a) $x^2 + 6x + 8 = 0$ b) $x^2 + 8x + 15 = 0$ c) $x^2 - 6x + 8 = 0$

CG-47. Solve for x and state the reason(s) from your tool kit that justify your work.

a)

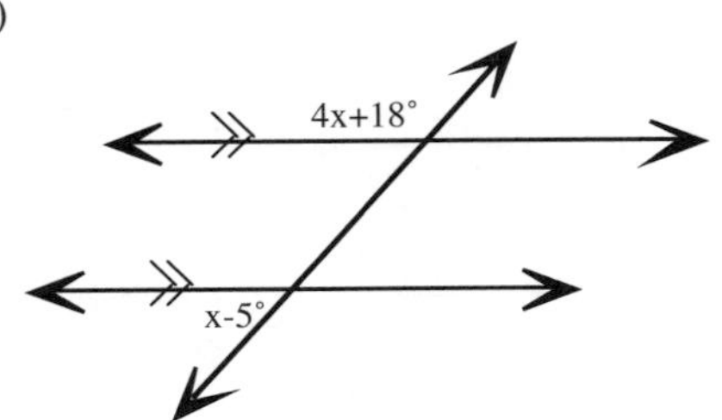


b)

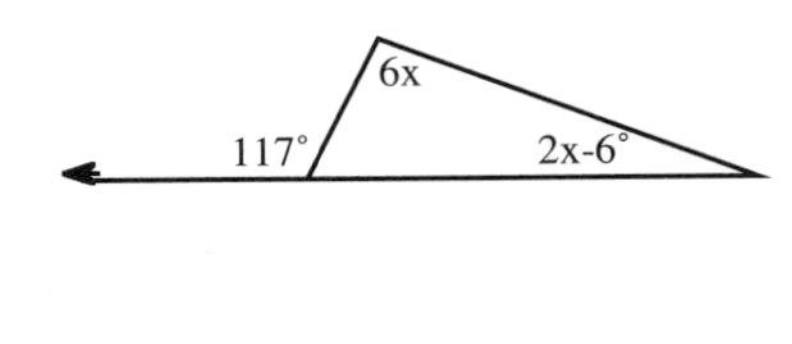


CG-48. Each face of this pyramid is an equilateral triangle. If AD = 5, what is BC? Explain completely.

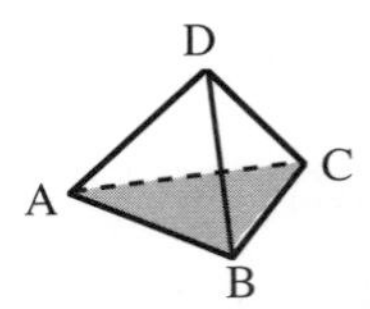


CG-49. What would the solid at right look like if the figure were rotated so that the left side became the front? Draw the resulting solid on isometric dot paper.

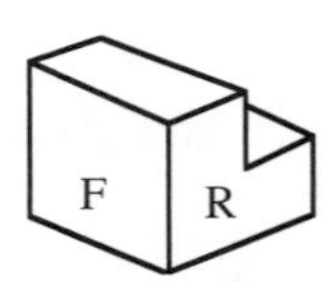


## TRIANGLE CONGRUENCE PROOFS

CG-50. Given the parallelogram ABCD, identify two congruent triangles and explain why they are congruent using one of the triangle congruence properties. Remember that arrow marks and slash marks do not mean the same thing!

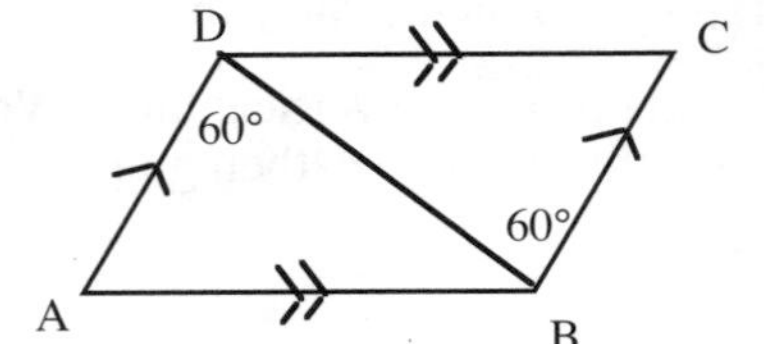

**CG-51.** In the previous problem you probably found that $\Delta ADB \cong \Delta CBD$ because you knew $\angle ADB \cong \angle CBD$, $\overline{DB} \cong \overline{DB}$ and $\angle DBA \cong \angle BDC$. Now you know three more facts because you have demonstrated that the triangles are congruent.

a) Name the other three pairs of parts that are congruent.

b) Read the information in the box below and add the abbreviation to your tool kit.

> Once you have demonstrated that two triangles are congruent, you may state that <u>any</u> of the other pairs of corresponding parts are congruent. When you need a reason to justify such a statement, use the following phrase:
>
> **≅ Δs give us ≅ parts**
>
> Remember: you will only use this statement AFTER you have shown (or been told) that two triangles are congruent.
>
> One further point: in the preceding problem, we could say $\overline{AD} \cong \overline{BC}$ OR $AD = BC$ and $\angle A \cong \angle C$ OR $m\angle A = m\angle C$. What is the difference? The first comparison using congruence (≅) means the <u>physical</u> segments or angles are the same size and shape. The second comparison using equal (=) means the <u>length</u> of the segments or the <u>measures</u> of the angles are the same.

**CG-52.** In the figure at right, $\overline{AB} \parallel \overline{DE}$ and C is the midpoint of $\overline{AE}$. Copy the figure on your paper.

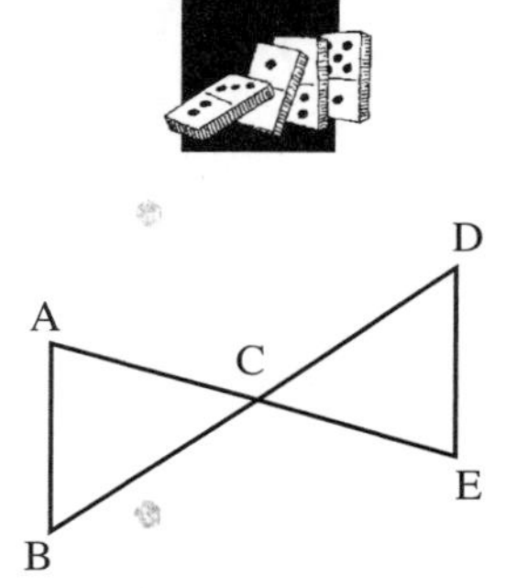


a) Identify <u>pairs</u> of angles and sides that are congruent and use slash marks to show which parts are congruent. Then state the tool kit item that supports your markings.

b) Explain clearly and completely why the two triangles are congruent and write a correct triangle congruence statement using the letters that name the vertices.

c) Why does the information from part (b) assure you that $\overline{AB} \cong \overline{DE}$?

d) What other parts are congruent? Why?

CG-53. For each figure on the resource page, locate and label two additional points so that QRST is the same size and shape as MNOP. Be sure that you map M to Q, N to R, etc.

CG-54. In each case, use the information given to find the value of x. You must justify why you can say that certain parts of the figure are congruent or equal. In other words, you will need to prove that the triangles in each problem are congruent.

a)

A, D, 6, B, x, E, C

b)

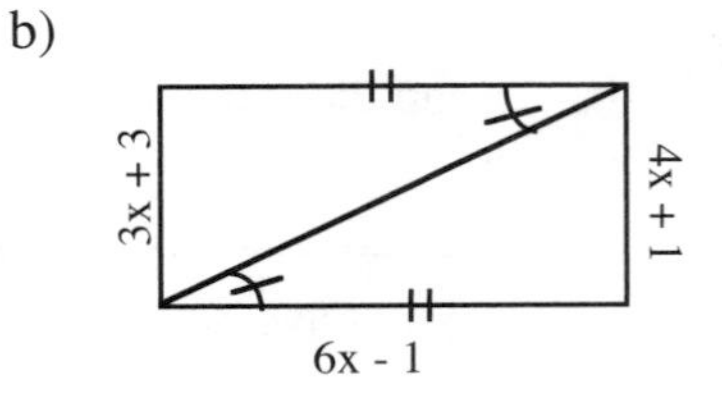


c)

3x-1, 2x, 6x + 2, 4x + 8

d) The height of an equilateral triangle bisects the opposite side.

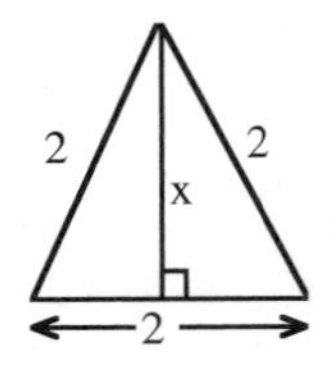


CG-55. ABCD is a square lying on a flat surface, E is directly above the point D. Does AE = EC? Explain clearly why or why not, that is, prove your answer.

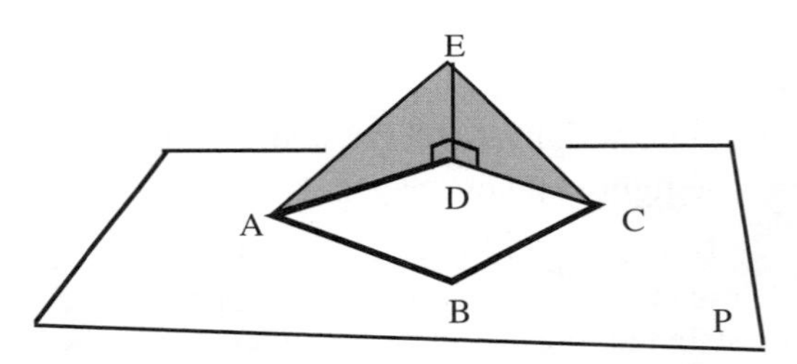

CG-56. Find the area of each polygon.

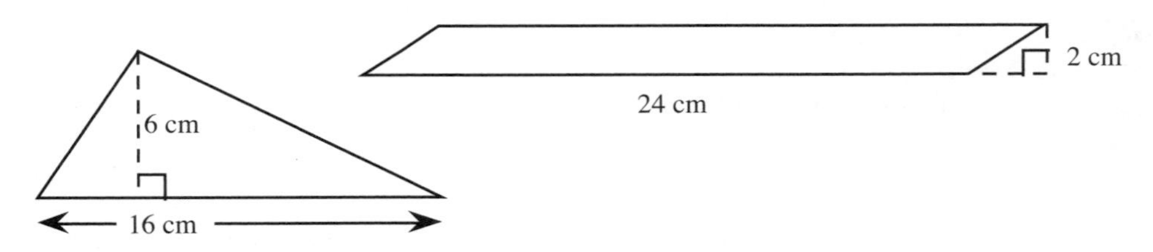


a) Compare the areas of the two figures above.

b) Do your results mean that the two figures are congruent? Explain in detail why or why not.

CG-57. Suppose two triangles are congruent. Do you know that they have the same area? Explain why or why not.

CG-58. Multiply:

a) $(x - 5)(x + 5)$  b) $(x + 8)(x - 8)$

c) $(x + 6)(x + 6)$  d) $(x - 3)(x - 3)$

CG-59. Factor these quadratics. Parts (a) through (c) are called perfect square trinomials. After you complete the problem, explain why you think they are called that.

a) $x^2 + 14x + 49$  b) $x^2 - 18x + 81$  c) $x^2 - 22x + 121$

d) $x^2 - 49$  e) $y^2 - 81$

## MORE CONGRUENCE PROPERTIES

CG-60. $\Delta ABC$ is a right triangle. If you know the lengths of any two sides, do you also know (or can you determine) the length of the third side? Explain clearly why or why not.

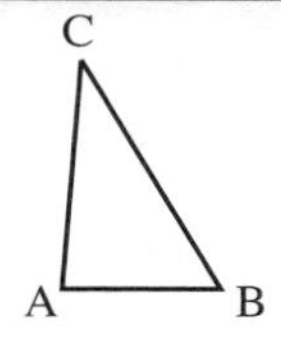


CG-61. Draw a right triangle with legs of length 3 cm and 5 cm (use a ruler). Did everyone in your team get a triangle congruent to yours? Explain why or why not.

CG-62. **DIRECTIONS**: Figures 1 and 2 below are an example of why SSA (two sides and a non-included angle) is NOT a pattern that ALWAYS results in congruent triangles. Figure 3 also illustrates why SSA is unreliable. In figure 3, $\Delta A'CB$ and $\Delta ACB$ both have two pairs of corresponding sides and one pair of corresponding angles congruent. HOWEVER, their pair of corresponding angles at vertex C, namely, $\angle BCA'$ and $\angle BCA$ are clearly **not** congruent.

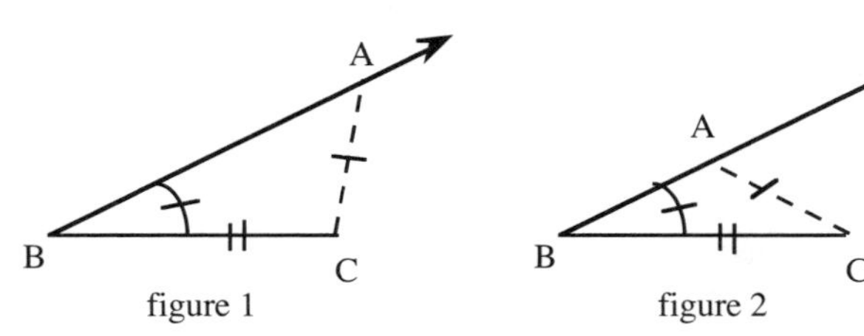

figure 1 figure 2

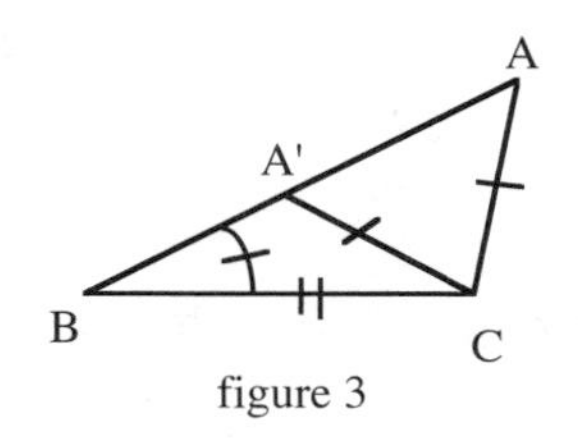

figure 3

However, we want to see if SSA might be true in certain cases.

a) Have each person in your team build a right triangle with everyone using the same length for the hypotenuse and the same length for one leg. Use an index card or the corner of your paper to be sure you have a right angle.

b) Compare your triangles. What do you notice?

c) State a conjecture about pairs of right triangles that have one pair of corresponding legs congruent and each hypotenuse congruent. Give a convincing argument that your conjecture is true. Hint: see if you can use your work in CG-60 and CG-61 to help show that this special case in a right triangle can be explained in terms of one of the three basic triangle congruence properties.

CG-63. SSA does not usually guarantee congruent triangles, EXCEPT when working with a right triangle. Read the box below and add the triangle congruence property to your tool kit. Then answer the question below the box.

> For pairs of right triangles, when one pair of corresponding legs are congruent and each hypotenuse is the same length, the two triangles are congruent by the **HYPOTENUSE-LEG triangle congruence property**, abbreviated **HL.**

Explain why H (hypotenuse) and L (leg), rather than SSA, are the correct letters to use to abbreviate your conjecture in the previous problem.

CG-64. Consider the two triangles at right. You have seen that if two angles and the included side of one triangle are equal to the corresponding side and angles of another, then the two triangles are congruent. We have labeled this relationship ASA. Notice that in ΔDUB, we have information about two angles and the included side (namely ∠D, ∠U, and $\overline{DU}$), but in ΔWHS we have two angles and a side that is NOT included between the two known angles.

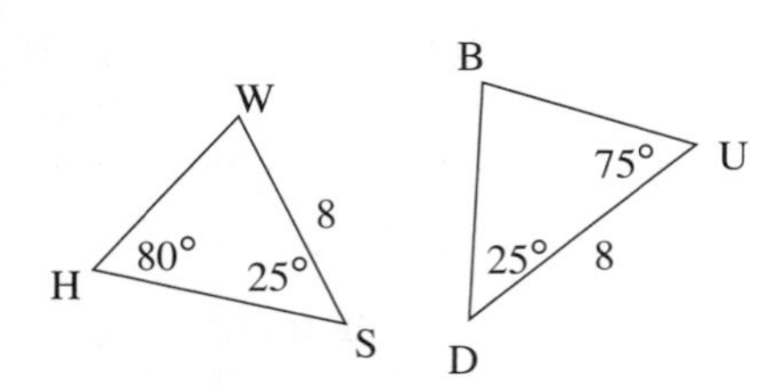


a) If you knew that m∠W = 75°, would the triangles be congruent? Why or why not?

b) Calculate the measure of ∠W.

c) Do you think the order of corresponding congruent parts matters for two angles and a side? Explain.

CG-65. Be sure that all of the congruence properties for triangles are listed in your tool kit with a marked diagram of a pair of triangles that illustrates each property. You should have:

**SSS, SAS, ASA (AAS), and HL**

a) In which one(s) is the order important?

b) What other patterns have you seen that DO NOT work in all cases? Explain why they do not assure congruence between two triangles.

CG-66. **DIRECTIONS:** Use the triangle congruence properties and the information about each pair of triangles below to decide whether the two triangles are congruent. If they are, explain how you know you can use one of the triangle congruence properties, then write its abbreviation and the correct sequence of vertices for the second triangle. If the triangles are not congruent, briefly explain why not.

a) ΔABC ≅ Δ __?__.

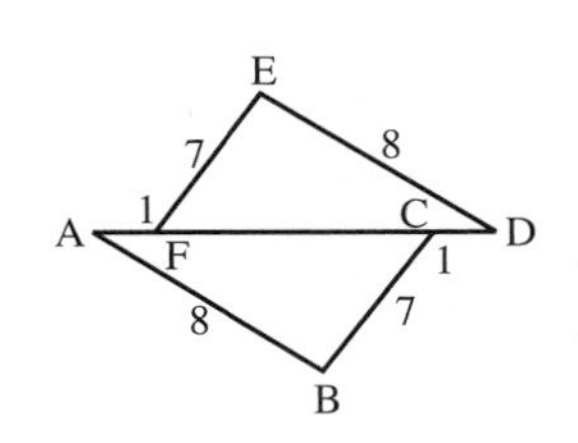

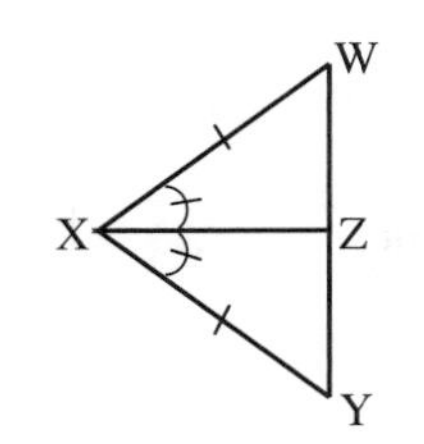

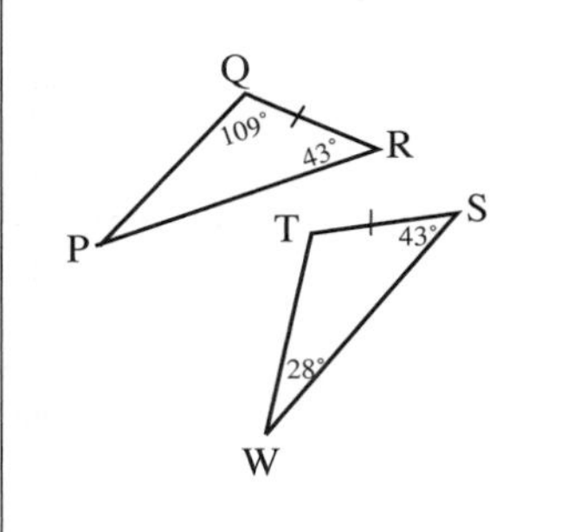


d) ΔGHJ ≅ Δ __?__.

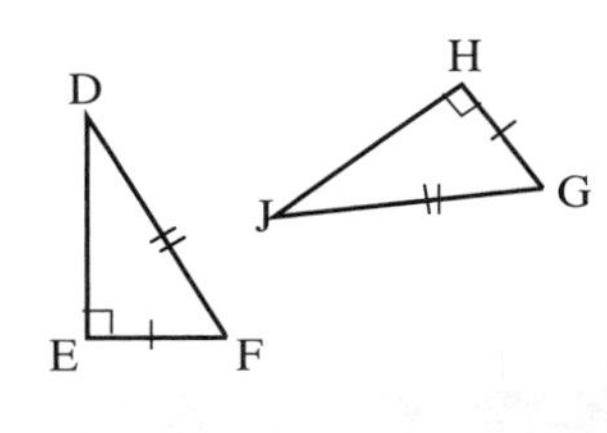

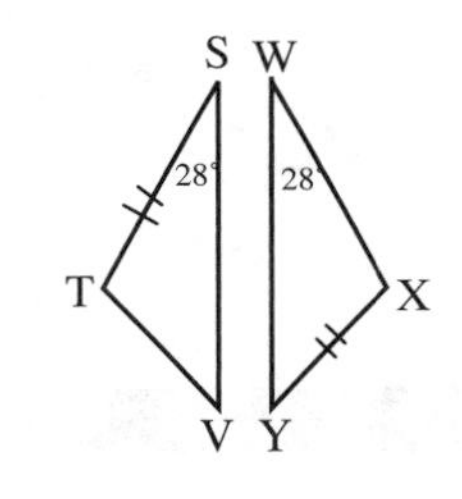

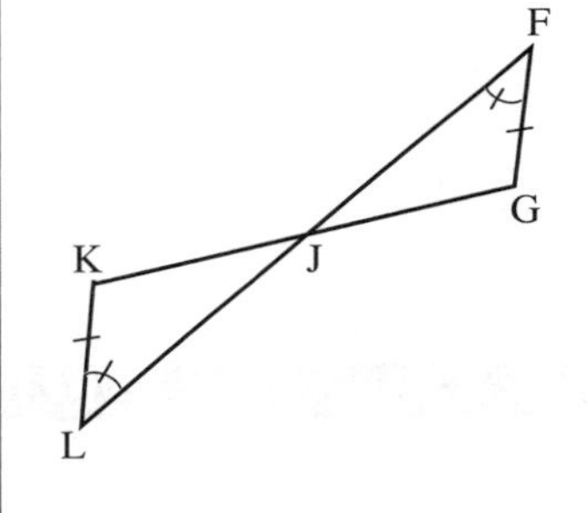


CG-67. When the line $y = -\frac{3}{5}x + 3$ crosses the x- and y-axes, it forms a triangle in the first quadrant.

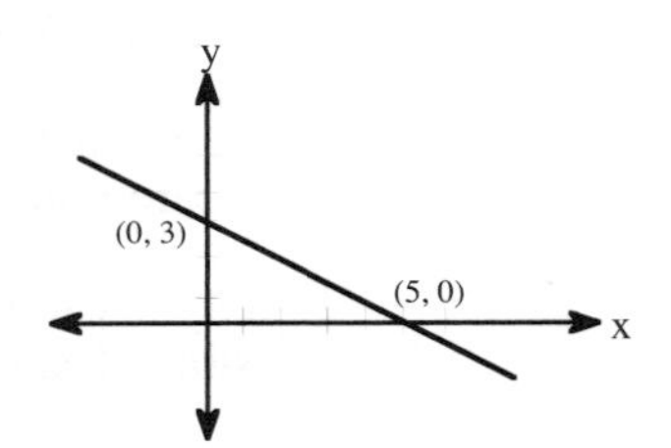


Give the equation of a line which together with the x- and y-axes would create a congruent triangle in the:

a) second quadrant.

b) third quadrant.

CG-68. Solve each of the following equations for x.

a) $x^2 - 140 = 4$

b) $x(x + 1)(x + 2) = 0$

c) $x^2 - 6x + 5 = 0$

d) $2x + 3y = 5$

CG-69. From the given Mat Plans, create an isometric drawing of ONE of the solids below.

a)

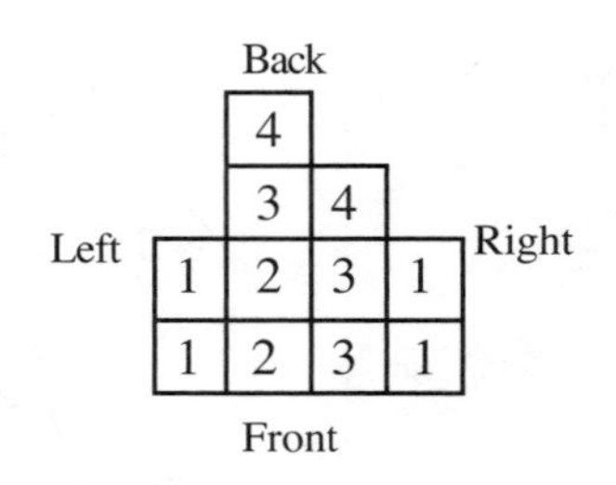


b)

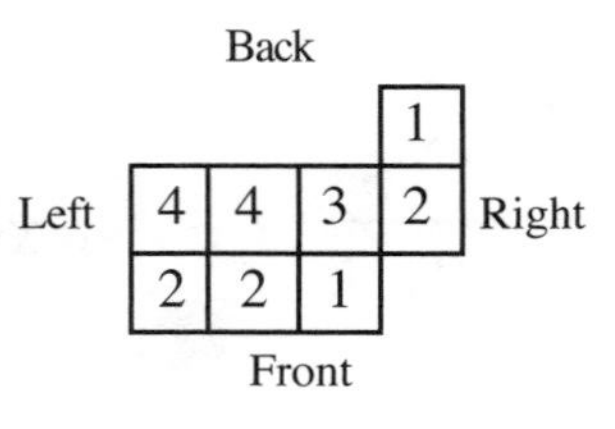


## THE TRIANGLE INEQUALITY

CG-70. Carefully cut the manipulative provided by your teacher into three lengths: 4 cm, 5 cm, and 10 cm. Now build a triangle using the three pieces for its sides.

a) Compare your triangle with other triangles made in your team. Are they congruent? What happened?

b) Next, cut manipulatives to build a triangle with sides of 5 in., 2 in., and 9 in. Compare your triangles. Are they congruent? What do you notice?

c) Finally, try to build a triangle with sides of length 5 cm, 6 cm, and 9 cm. Is it possible? Explain.

d) Based on parts (a), (b), and (c), can <u>any</u> three lengths be used for the sides of a triangle? If not, which combinations will work? Write your conclusion as clearly as you can.

CG-71. Suppose you had two segments of lengths 7' and 12' and you wanted to draw a third segment to form a triangle. The previous problem suggests that there are some restrictions on the lengths you may choose.

a) In figure 1 below, if you push the 7' segment over to the 12' segment (imagine that vertex B is hinged), how much of the 12' segment sticks out when C touches $\overline{AB}$?

b) If you push ∠B down as shown in figure 2, how long will the segment $\overline{AC}$ be when it becomes straight (that is, when ∠ABC becomes 180°)?

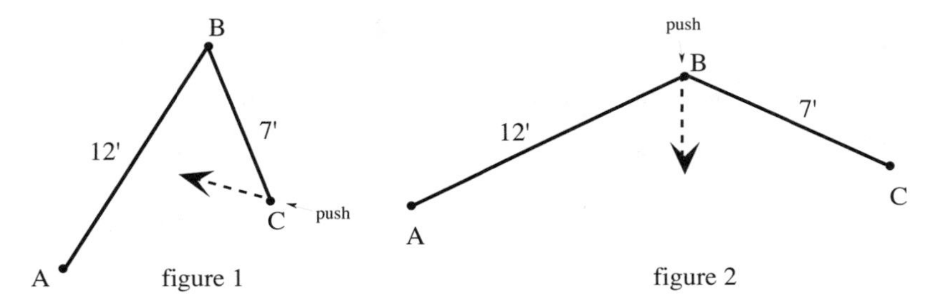


figure 1 figure 2

c) The values you found in parts (a) and (b) give you the **minimum** and **maximum limits** for the length of the third segment in any triangle with two sides of lengths 7' and 12'.

1) What are the limits on the length of the third side, $\overline{AC}$?

2) Write the limits for the length of $\overline{WY}$ in the triangle at right.

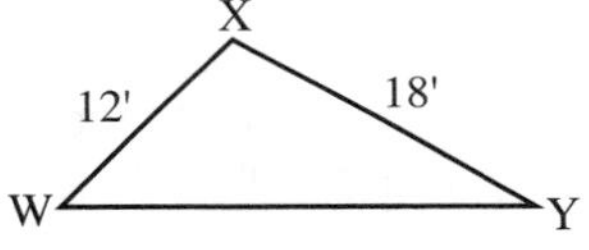


3) In the triangle at right, with m > n, what are the limits for the length of $\overline{DF}$ using m and n?

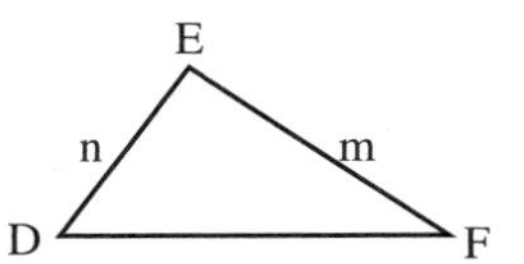


Read the summary of the Triangle Inequality in the box below and add it to your tool kit.

The property that states **maximum and minimum limits** for the length of the third side of any triangle is known as the **TRIANGLE INEQUALITY**. For any triangle, the length of each side must be less than the sum of the lengths of the other two sides. Stated algebraically based on the figure below:

$PQ < PR + QR$
$PR < PQ + QR$
$QR < PQ + PR$

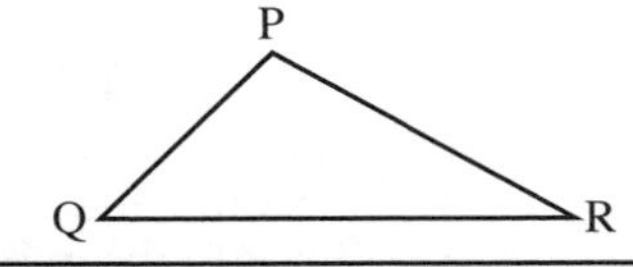

CG-72. Find the quadratic(s) that cannot be factored using integers and show why it (they) are not factorable. Factor the other(s).

a) $x^2 + 9x + 7$  b) $x^2 - x - 42$  c) $x^2 - 5x + 24$

CG-73. Read the information about the Quadratic Formula in the box below, add it to your tool kit, then use it to solve the next set of problems.

## THE QUADRATIC FORMULA

You have used factoring to solve quadratic equations. You might have wondered what you would do if the trinomial was not factorable! You can solve any quadratic equation by using the **QUADRATIC FORMULA.**

**If $ax^2 + bx + c = 0$, then $x = \dfrac{-b \pm \sqrt{4^2 - 4ac}}{2a}$.**

For example, suppose $3x^2 + 7x - 6 = 0$. Here $a = 3$, $b = 7$, and $c = -6$.

Substituting these values into the formula results in:

$$x = \frac{-(7) \pm \sqrt{7^2 - 4(3)(-6)}}{2(3)} \Rightarrow x = \frac{-7 \pm \sqrt{121}}{6} \Rightarrow x = \frac{-7 \pm 11}{6}$$

Remember that square numbers have both a positive and negative square root. The sign $\pm$ represents this fact for the square root in the formula and allows us to write the equation once (representing two possible solutions) until later in the solution process.

Split the numerator into the two values: $x = \dfrac{-7 + 11}{6}$ or $x = \dfrac{-7 - 11}{6}$

Thus the solution for the quadratic equation is: $x = \frac{2}{3}$ or $-3$.

Generally you will not have a result that has integer or fraction factors. Your answer will contain a square root.

The power of the quadratic formula is that it always works to find the solution to a quadratic equation, whether the trinomial is factorable (using integers) or not.

CG-74. Solve the following problems with the Quadratic Formula. Part (a) summarizes the main steps.

a) Solve $2x^2 - 5x + 2 = 0$ by using the quadratic formula.
  1) What are a, b, and c in this equation?
  2) Substitute the values of a, b, and c into the quadratic formula.
  3) Simplify the expression to find the solution.

b) $x^2 + 3x - 10 = 0$  c) $x^2 + 9x + 7 = 0$

CG-75. Make a table and graph $y = x^2 + 3x - 10$ for $-6 \le x \le 3$. Explain how the graph and the solution to $x^2 + 3x - 10 = 0$ in part (b) of the previous problem are related.

CG-76. Without graphing, explain the relationship between the solution to problem CG-74 (c) and its equation $y = x^2 + 9x + 7$.

CG-77. Explain what is wrong in each picture. There are several different ways to explain these problems. Most important is to be sure that your explanation is clear and well-supported. If a line looks straight, you may assume that it is.

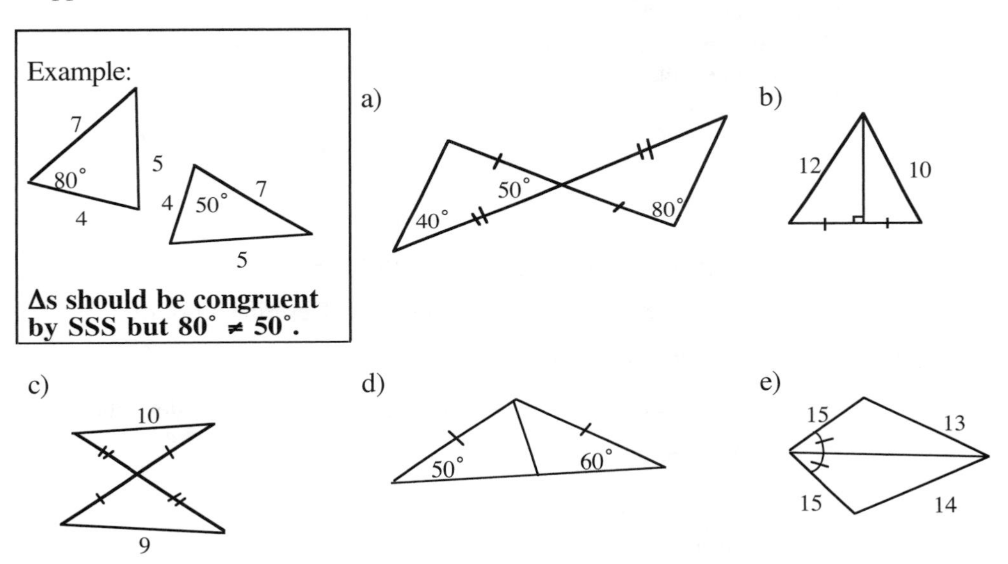


CG-78. Find the minimum and maximum limits for the third side of a triangle if the other two sides are 8" and 13".

CG-79. Graph $y < \frac{1}{3}x + 4$ AND $y > (x - 1)^2$ on the same set of axes. Remember to shade your solution regions. Then approximate the area of their common region.

CG-80. Find the area of the shaded portion of the square. Note that the white region is also a square and its vertices are midpoints of the sides of the large square.

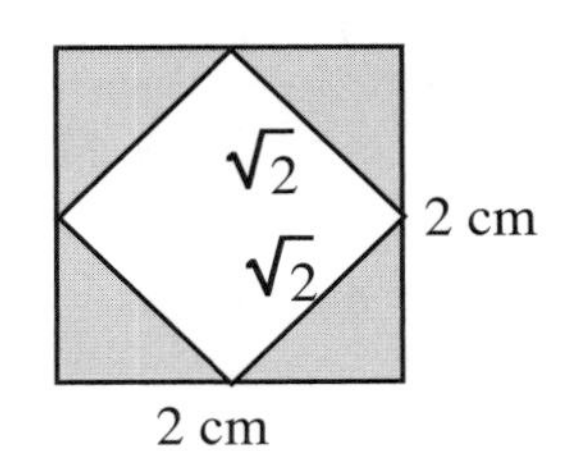

## UNIT AND TOOL KIT REVIEW

CG-81. Use the resource page or draw $\Delta ABC$ on your graph paper, then reflect, translate, or rotate as directed. Give the coordinates for the vertices of each new triangle. For <u>each</u> part, start with the <u>**original**</u> triangle.

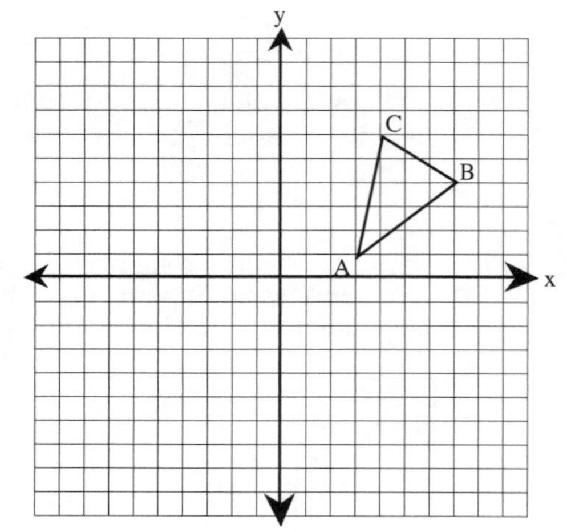

a) Reflect (flip) $\Delta ABC$ across the x-axis.

b) Translate (slide) $\Delta ABC$ to the third quadrant so that vertex C has coordinates (-5, -1). c) Rotate (turn) $\Delta ABC$ 90° counter-clockwise about the origin, that is, into the second quadrant. The coordinates of B' will be (-4, 7).

CG-82. On a sheet of graph paper, graph $y = x^2 - 4$ for x = -4, -3, ..., 4.

a) Highlight the line of symmetry for the curve. Describe it briefly.

b) Where does the curve cross the x-axis?

c) Where does $y = (x - 3)(x - 5)$ cross the x-axis? State and describe its line of symmetry.

CG-83. Find the total surface area of a square-based pyramid with isosceles triangular faces. Start by finding the height of one of the faces.

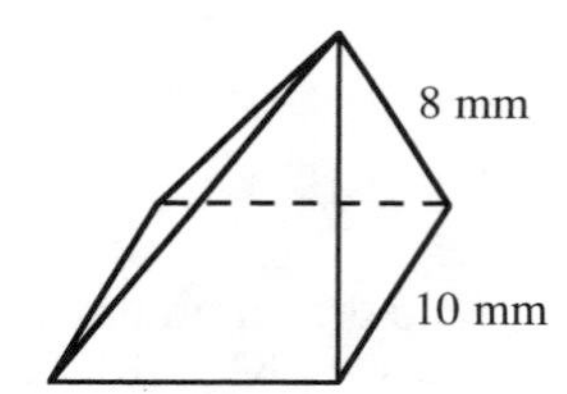


CG-84. Solve the following equations using the Quadratic Formula. Be sure to reduce all fractions completely.

a) $24x^2 - 2x - 15 = 0$

b) $-2x^2 - 3x + 8 = 0$

c) $3x^2 + 11x - 20 = 0$

d) $x^2 + 13x - 2 = 0$

CG-85. Solve part (a) numerically, then write the general dimensions for <u>any</u> length diagonal as directed in part (b).

a) A rectangle is twice as long as it is wide and has a diagonal of length 25 cm. Draw a diagram and find its dimensions.

b) A rectangle is twice as long as it is wide and has a diagonal of length D cm. Find its dimensions.

CG-86. Find the minimum and maximum limits for the third boundary (side) of a triangular region if the lengths of two sides are 800 yards and 364 yards respectively.

CG-87. If the pair of triangles in each problem below are congruent, write the correct correspondence, state the congruence property and any other ideas you use that make your conclusion true. Otherwise, explain why you cannot conclude that the triangles are congruent. Note that the figures are not necessarily drawn to scale.

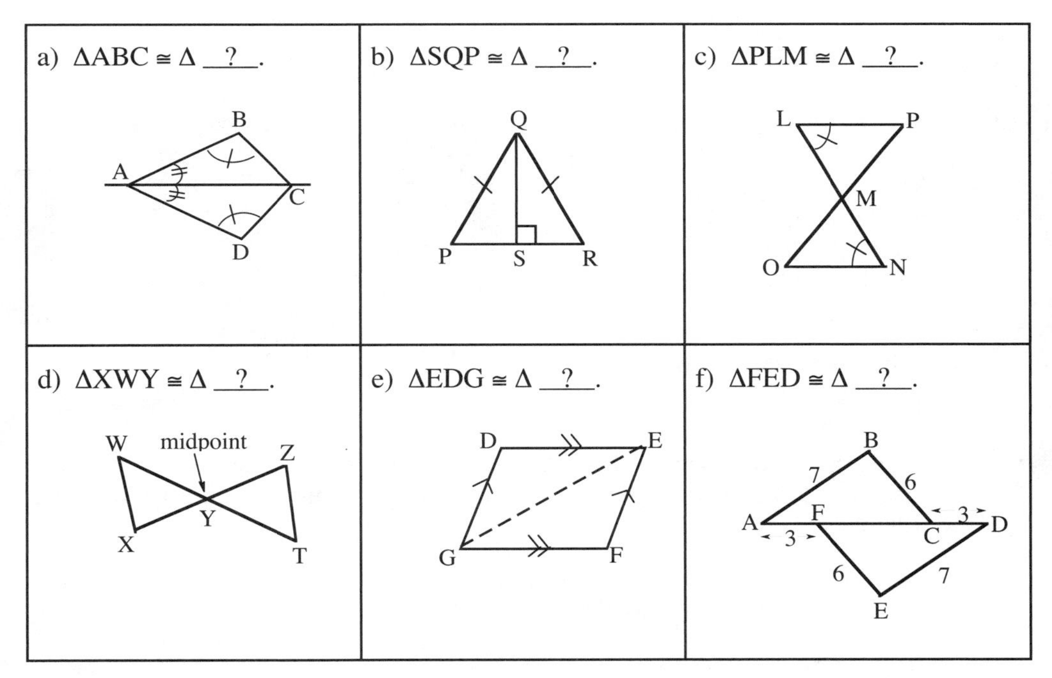


CG-88. Factor.

a) $x^2 - 64$

b) $4x^2 + 28x$

c) $x^2 - 12x + 36$

CG-89. Use the Quadratic Formula to determine approximately where the parabolas cross the x-axis.

a) $y = -3x^2 + 2x + 2$

b) $y = x^2 + 2x - 1$

CG-90. On a sheet of graph paper, draw a pair of xy-axes, plot the points A(-2, 1), B(-5, 2), and C(-3, 4), and connect them to form $\Delta ABC$.

a) Reflect (flip) $\Delta ABC$ across the y-axis and label its vertices X, Y, and Z.

b) Translate (slide) $\Delta ABC$ so that A' has coordinates (-4, -3) and label it $\Delta A'B'C'$.

c) Rotate (turn) $\Delta ABC$ 180° counter-clockwise about the origin (0, 0) and label it $\Delta A''B''C''$.

d) How are all four triangles related? Why?

CG-91. Paula knows that in the diagram below, $\overline{CR} \cong \overline{AE}$, $\overline{CL} \cong \overline{AF}$, and $\overline{CL} \parallel \overline{AF}$. Her friend Maria asks her whether or not $\overline{EF} \parallel \overline{RL}$. Paula responds, "Well, I think so--they sure look like it!"

"But the figure may be distorted!" Maria says. "Convince me."

Paula thinks to herself, "If I get $\angle REF \cong \angle ARL$, then the lines will be parallel. But how do I do that? ... Aha! The angles are corresponding parts of two congruent triangles!"

a) Which triangles does Paula know are congruent? Why are they congruent?

b) Why do equal angles make the lines parallel?

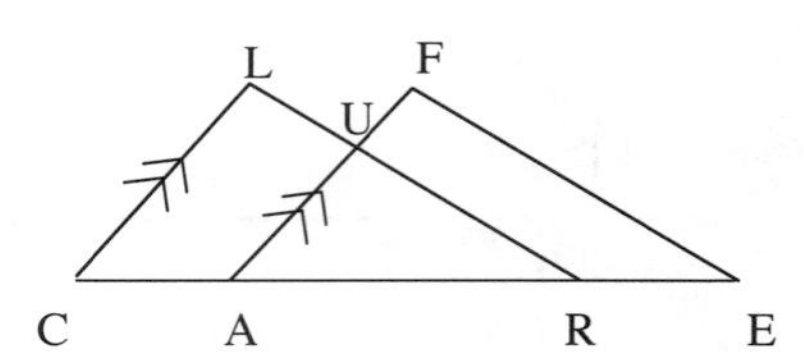

CG-92. Write the correct correspondences, state the congruence property that justifies why the triangles are congruent, and include the tool kit definitions and conjectures you use to get the information you need. If you cannot conclude that the triangles are congruent, explain why not.

a) $\Delta CAB \cong \Delta$ __?__.

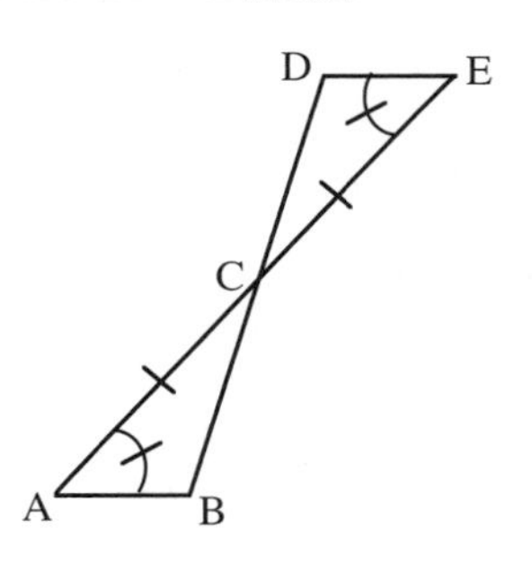


b) $\Delta CBD \cong \Delta$ __?__.

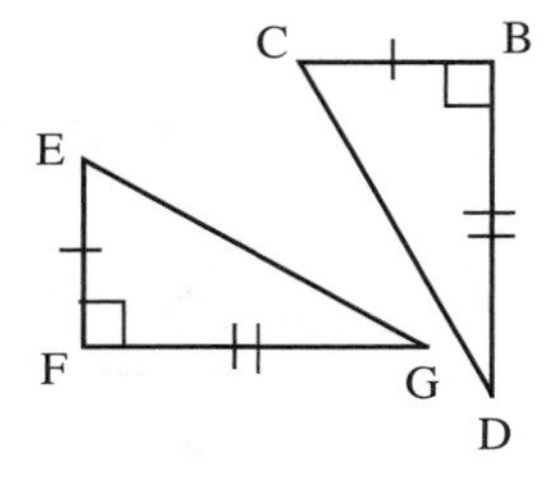


c) Consider $\Delta LJI$ and $\Delta HJK$. Are they congruent?

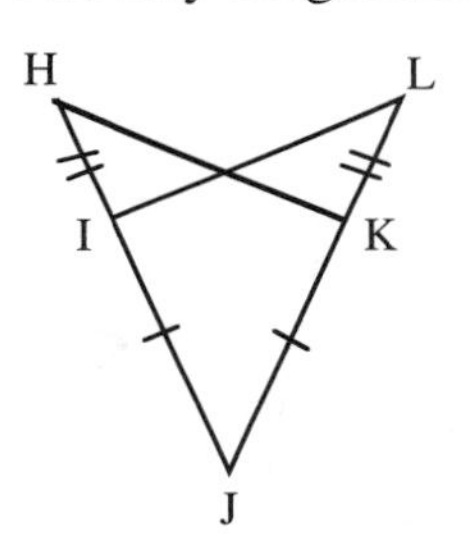


d) $\Delta PRQ \cong \Delta$ __?__.

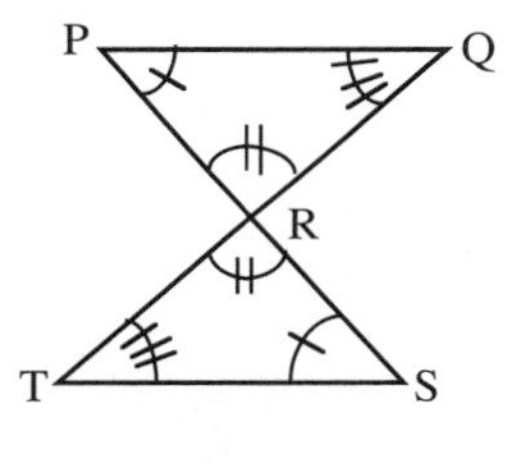


CG-93. Factor these quadratics. From your results, state where the curve crosses the x-axis, state the line of symmetry, and draw a rough sketch of the graph without making a table of values.

a) $y = x^2 + 4x - 21$

b) $y = x^2 - 2x - 24$

CG-94. Use the Quadratic Formula to solve each quadratic equation.

a) $10x^2 - x - 2 = 0$

b) $x^2 + 2x - 5 = 0$.

CG-95. Suppose you know that $\Delta TAP$ is congruent to $\Delta DOG$ and that $TA = 14$, $AP = 18$, $TP = 21$, and $DG = 2y + 7$.

a) Draw an approximate sketch of $\Delta TAP$ and $\Delta DOG$.

b) Find y.

CG-96. A surveyor knows that when she sights two sides of a triangular-shaped parcel of land, the lengths are 957 meters and 638 meters. Without surveying the third side, what are the maximum and minimum limits on its lengths?

CG-97. For each figure below, determine whether or not the two smaller triangles in each figure are congruent. If so, explain why and solve for x. If not, explain why not.

a) b)

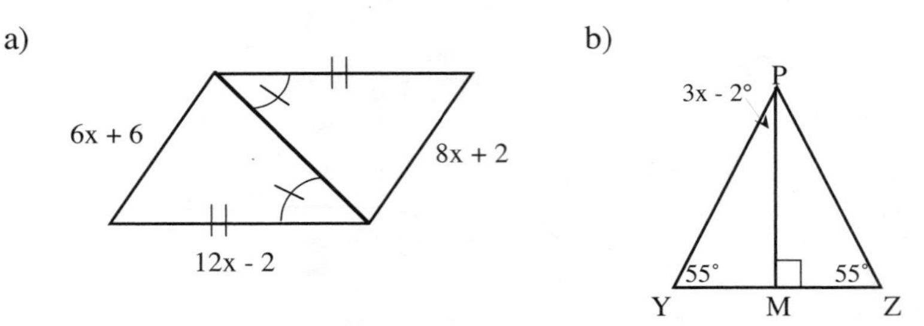


CG-98. State which triangles are congruent and justify your statement. Some figures are not drawn accurately. Be sure to base your replies on what you <u>know</u> from looking at the markings.

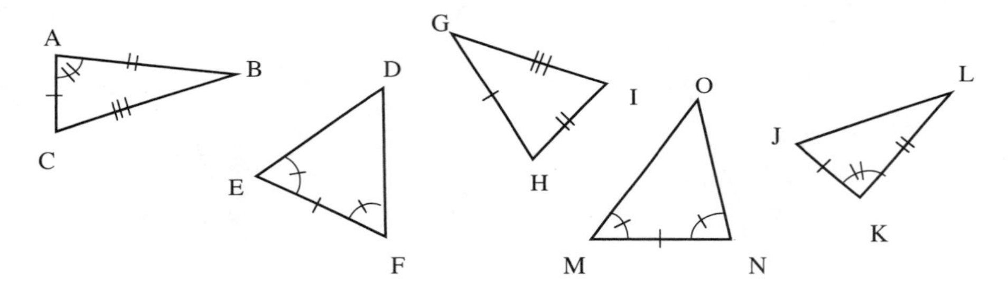


CG-99. TOOL KIT CHECK-UP

Your tool kit contains reference tools for geometry. Return to your tool kit entries. You may need to revise or add entries.

Examine the list below. Each item should be included in your tool kit; add any that are missing now. You will find them in your text by skimming through the unit or by using the index in the back of the book.

- line of symmetry
- transformation
- reflection (flip)
- translation (slide)
- rotation (turn)
- congruent
- corresponding (parts)
- Triangle congruence properties: SSS, SAS, ASA (AAS), HL
- Triangle Inequality
- factoring
- slopes of perpendicular, vertical, and horizontal lines
- Quadratic Formula.

## GEOMETRY mini-PROJECT

CG-100. The diagram below shows n x n x n cubes divided up into 1 x 1 x 1 cubes. The little cubes have either zero, one, two, or three outside faces.

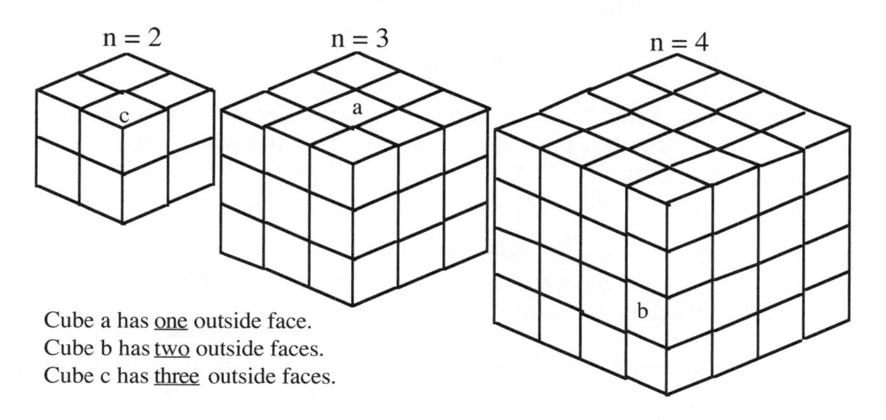


| n | # of small cubes with 3 outside faces | # of small cubes with 2 outside faces | # of small cubes with 1 outside face | # of small cubes with 0 outside faces |
|---|---|---|---|---|
| 2 | 8 | 0 | 0 | 0 |
| 3 | 8 | 12 | 6 | 1 |
| 4 | | | | |
| 5 | | | | |
| ... | | | | |
| 10 | | | | |
| n | | | | |

**>>>This problem continues on the next page.>>>**

Use real cubes or use the pictures that precede the table to help with the first three rows.

a) Complete the table. As you fill in each section, check with another student or your team to make sure that there is agreement on the data. An important part of this activity is to analyze how you can get each entry in the table, so discuss your methods with other students.

b) Choose either the 10 row or the n row and select one entry you or your team found. Discuss the approach you used to find the entry. Was the approach you used one of these: reasoning from a pattern, reasoning logically, or using a combination of the two?

d) After you have filled in each row, check the numbers horizontally. This means add up the numbers across the row. For example, in the row n = 4, there are 64 small cubes to be accounted for.

e) The last row, which is all algebraic, can also be checked horizontally. You will have to use some algebra skills to simplify the expression to make sure it checks. Show your check algebraically.

CG-101. A rectangular prism 3 cm by 4 cm by 3 cm, such as the one pictured below, is painted red on all six faces. It is then cut into 36 cubes that are 1 cm on a side and all 36 of these cubes are placed in a bag. What is the probability of randomly selecting a cube from the bag that is painted red on exactly:

a) 3 faces?

b) 1 face?

c) 4 faces?

d) 2 faces?

e) no faces?

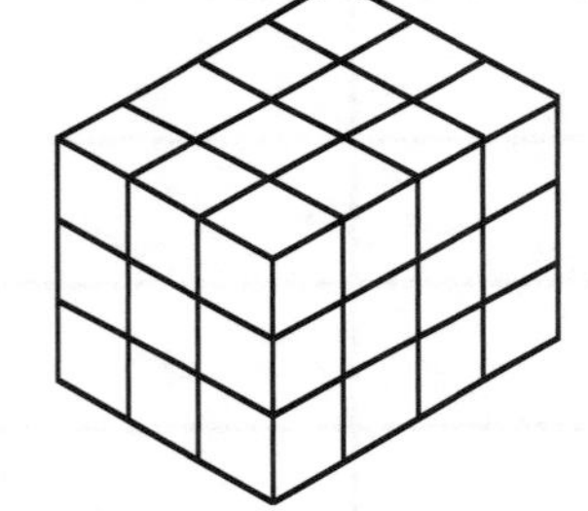

f) If you take the total surface area of all cubes, what percentage would be painted red? Careful here!

UNIT 6

# Tool Kit

## WRITING PROOFS

# Unit 6 Objectives

## *Tool Kit:* WRITING PROOFS

So far in this course we have concentrated on learning the basic vocabulary, properties, and relationships of geometry. In this unit, as we complete the first half of the course, we will introduce several styles of presenting logical arguments--known as proofs--and practice these styles by proving several of your conjectures from the previous units. While this unit's main focus is proof, the mathematical content will be familiar. This will give you a good chance to review and consolidate what you have studied so far.

The algebra focus will be laws of exponents, although linear and quadratic equations and inequalities, including their graphs, are included. You will also continue to solve various types of equations.

In this unit you will have the opportunity to:

- prove several of the conjectures you made in previous units, including the Pythagorean Theorem, parallel relationships, and triangle properties.
- practice various proof formats--paragraph, flowchart, and two column--and decide which style is best for you.
- learn how to use indirect proof (proof by contradiction) to justify conclusions.
- review most of the ideas from the first six units while practicing logical argumentation.
- review and practice the laws of exponents.

*Read the problem below, then go on to the first problem on the next page. We will come back and solve TK-0 later.*

TK-0. Given the square at right, prove that the shaded figure must be a square, then use the figure as a basis to prove the Pythagorean Theorem:

$$a^2 + b^2 = c^2$$

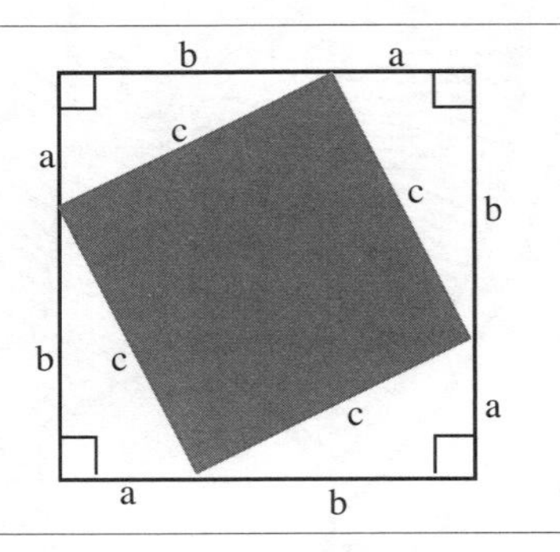


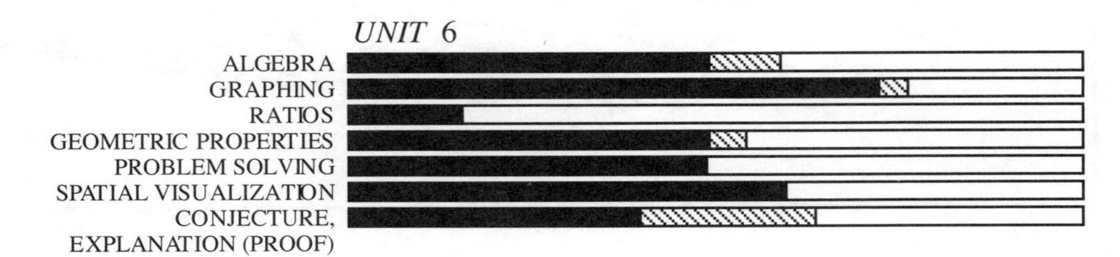

# *Unit 6*

## *Tool Kit:* WRITING PROOFS

### TOOL KIT PRACTICE AND ARROW DIAGRAMS

TK-1. Over the first five units you have studied many of the fundamental notions, terms, and processes that are necessary to do more interesting and challenging problems. Hopefully you have been building a personalized tool kit to use as you work on geometry. However, your tool kit is incomplete. You will be adding to it as you continue through the course. This unit will give you an opportunity to learn more about how, when and where to use your tool kit. In particular, you will continue to use the information in your tool kit to help you write good, clear explanations of your work. You will continue to develop your skill of writing justifications for mathematical relationships so that you can prove to your reader or audience that what you (or the problem) claims to be true must be true. Before you begin this unit, see how much you know by completing the matching problems provided by your teacher. This is just a self-test, so relax and do your best.

TK-2. Take a moment to review the substitution property below. You have used this property previously for equations, expressions, and linear systems. Add the property to your tool kit, then go on with the next problem.

**SUBSTITUTION PROPERTY**

1) Expressions: If $x = 4$, evaluate $2x^2 - 3x$. $2(4)^2 - 3(4) = 2(16) - 12 = 20$.

2) Linear systems: If $y = 2x - 4$ and $y = 3x + 6$, then $2x - 4 = 3x + 6$.

3) Angles: If $m\angle b = m\angle c$ in the figures at right,
$m\angle a = 180° - m\angle b$ (straight angle)
$m\angle d = 180° - m\angle c$ (straight angle)
Then $m\angle a = m\angle d$.

a b

c d

In the first figure, solve for y. In the second figure, solve for x. What can you conclude about x and y and what property did you use?

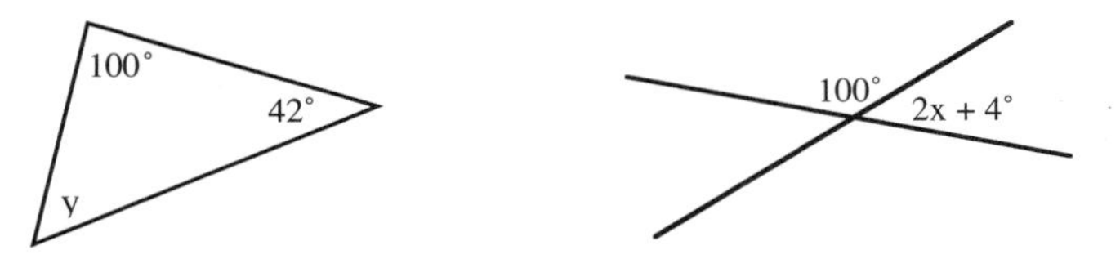


In the following problems, solve them as directed, but use your tool kit to explain and/or justify the steps in your solution process. Some problems are simple applications of a single idea, so a sentence or phrase using the appropriate term will be sufficient. However, when problems involve several ideas and multiple steps in their solutions, order your explanation so that it has a beginning, middle, and end just as you would in an essay. Often there is more than one way or one order in which to do a problem, so doing a problem differently does not necessarily make you wrong!

TK-3. Find x and the measure of both angles. Give a reason (from your tool kit) for what you do to solve this problem.

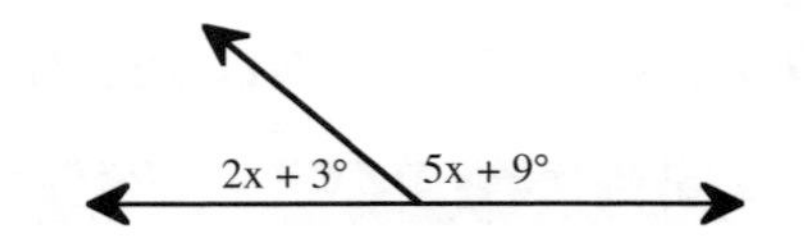


TK-4. What statements from your tool kit could you use to persuade someone in your team that $\overline{QP} \cong \overline{SP}$ ?

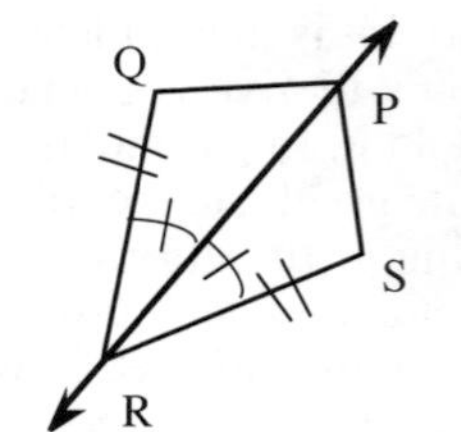

TK-5. Find the measures of the numbered angles. Use your tool kit to name the ideas you use.

a)

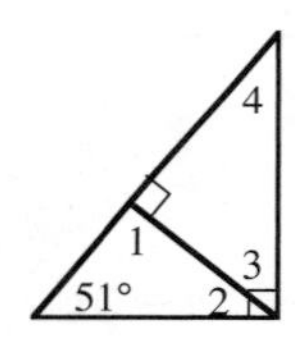


b)

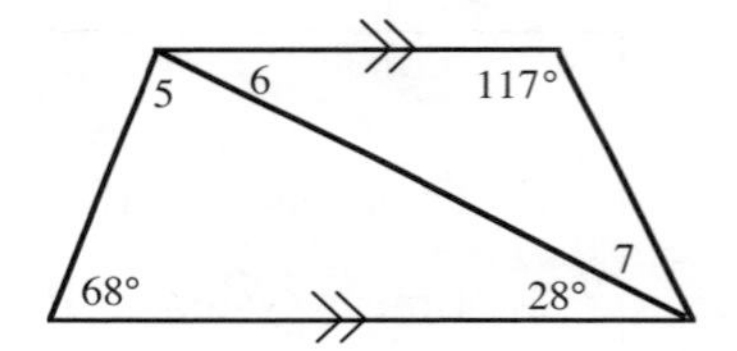


TK-6. Find the m∠x. Use your tool kit to write the names of the ideas you use in your solution.

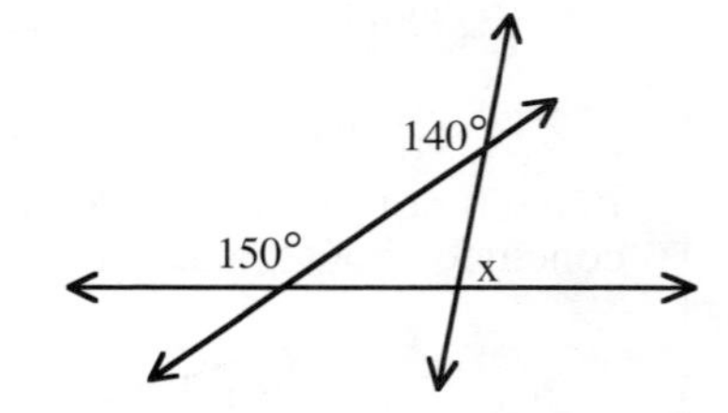

TK-7. Read the following definition, add it to your tool kit, then complete parts (a), (b), and (c) following the directions located below the examples.

Justifying your steps, as you did in the previous problems, takes time. To shorten your reasons that support your explanations, often all you need to do is to draw an **ARROW DIAGRAM**. Arrow diagrams can represent definitions and conditional statements.

**EXAMPLES**

| **Arrow Diagrams** | **Definitions/Conditional Statements** |
|---|---|
| 1) | Straight angles equal 180°. |
| 2) | An acute angle measures less than 90°. |
| 3) | If a triangle is a right triangle, then the sum of the squares of the lengths of the legs is equal to the square of the length of the hypotenuse (The Pythagorean Theorem). |

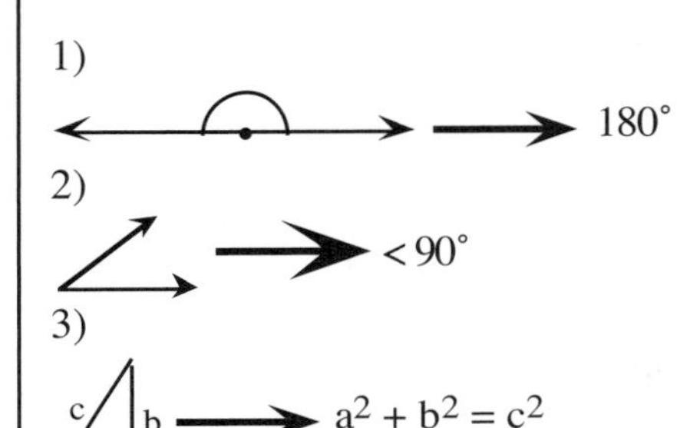


**DIRECTIONS**: Write the statements in parts (a) and (c) as conditional statements (if-then) using the arrow diagram style. In part (b), translate the arrow diagram into a conditional statement.

a) If both pairs of sides in a quadrilateral are parallel, then the quadrilateral is a parallelogram.

b)

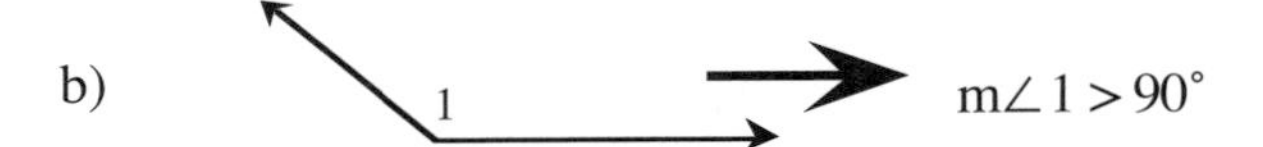


c) If two triangles are congruent, then the corresponding parts of the two triangles are congruent.

TK-8. In Unit 3 you wrote two useful conjectures:

1) If two parallel lines are cut by a third line, then the measures of pairs of corresponding angles are equal.

2) If two parallel lines are cut by a third line, then the measures of pairs of alternate interior angles are equal.

Transform each of these conjectures into an arrow diagram.

TK-9. For the two acute angles shown, $m\angle 1 = 6x - 3°$ and $m\angle 2 = x + 2°$. Solve for x <u>and</u> the measure of each angle. State the ideas from the tool kit that justify the procedure you use.

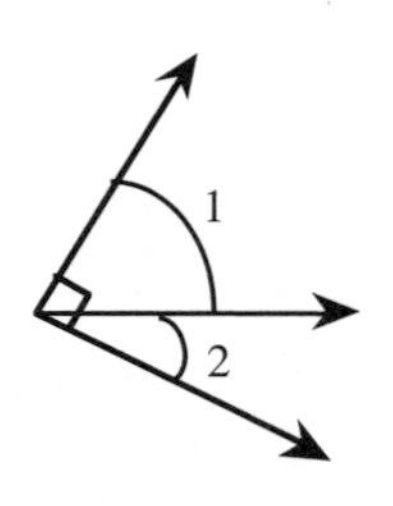


TK-10. Find the minimum and maximum limits for the length of the third side of a triangle if the other two sides and 83' and 117'.

TK-11. Use the figure at right to answer the questions below.

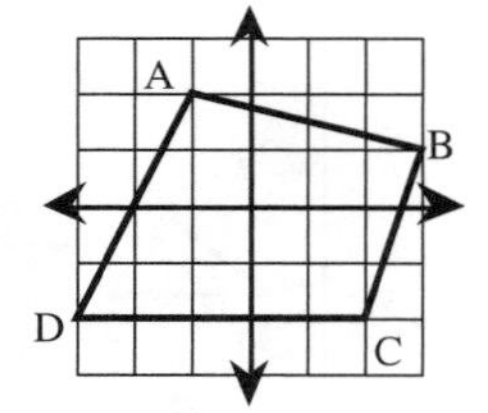


a) Calculate the perimeter and area of quadrilateral ABCD.

b) What is the slope of $\overleftrightarrow{AD}$?

c) What are the coordinates of the midpoint of $\overline{BC}$?

TK-12. Factor:

a) $x^2 + 2x - 35$

b) $4x^2 + 24$

TK-13. Solve for x. Remember the Quadratic Formula means that for any $ax^2 + bx + c = 0$,

$$x = \frac{-b \pm \sqrt{b^2 - 4ac}}{2a}.$$

a) $5x^2 - 3x - 10 = 0$

b) $4x^2 - 24 = 0$

## PROOFS OF FAMILIAR IDEAS

TK-14. In July of 1969, the United States sent Apollo XI to the moon. The lunar lander, named Eagle, settled gently on the moon's surface and Neil Armstrong became the first person to step onto the moon. A few skeptics have claimed that the entire lunar program was faked by the government. They claim that all of the TV and radio transmissions were done in large airplane hangers with stages and special effects. Take a few minutes to convince your partner or team that humans really have walked on the moon. Write down your reasons so that you can present your proof logically to the class.

TK-15. Suppose Harry is doing his homework and must solve the equation 3(9 - 2x) = 4x - 7. He checks the back of the book and finds x = 3.4, and writes it on his paper. The next day Harry's teacher gives a quiz and says, "Prove that the answer to 3(9 - 2x) = 4x - 7 is 3.4." Use your algebra skills to solve the problem, and write a brief justification for each step.

TK-16. Earlier in the year you made the conjecture that "Vertical angles are equal." Are you convinced that this is always true? Maybe there is a way to position two lines so that the measures of a pair of vertical angles are just a little different. Would you bet $20 that no one has ever found a pair of vertical angles that were not equal?

a) Discuss this possibility with your partner or team. Try to find an example of a pair of vertical angles that are not equal.

b) If you think you have found such a situation, write an explanation of why your discovery is correct. If you still believe that vertical angles are equal, write a convincing argument that this is so. Refer to the figure at right. If you cannot get started, part (c) has some hints.

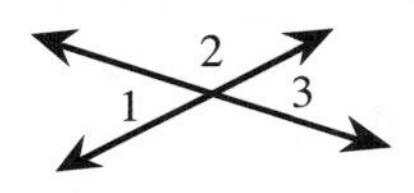


c) Write two equations using each pair of supplementary angles, one pair at a time. Use substitution to combine the two equations into one equation.

d) Finally, carefully read the box on the next page and add the information to your tool kit.

**>>>This problem continues on the next page.>>>**

Hereafter we will refer to your conjecture above as the **VERTICAL ANGLE THEOREM**: If two lines (or segments) intersect, then the measures of each pair of vertical angles are equal.

The previous three problems raise the issue of what it means to <u>prove</u> something. Simply stated, a **PROOF** convinces an audience that a **conjecture** is true for ALL cases (situations) that fit the conditions of the conjecture. For example, "If a figure is a triangle, then the sum of the measures of the angles is 180°." Because you proved this conjecture near the end of Unit 3, you know that it is ALWAYS true, at least as long as the triangle is flat!

In one sense, an algebra course proves that the rules of arithmetic you believed and used in elementary school are always true for all numbers. In many problems such as Digit Place and Color Square games, you actually wrote proofs using logical arguments based on the rules of each game to convince yourself and your teammates that the solution to the game WAS correct. Likewise, you have proved a few characteristics of odd and even integers. However, you also made several mathematical conjectures based on patterns that have only been verified experimentally. In this unit you will use your tool kit to prove some of your previous conjectures, as well as a few new ones.

TK-17. You have explored the idea that the measure of an exterior angle of a triangle ($\angle 4$) is equal to the sum of the measures of the two remote (farthest away from $\angle 4$) interior angles ($\angle 1$ and $\angle 2$). While you may believe that this conjecture is true, we have not <u>proved</u> it! An argument to prove the conjecture is outlined below. Use the outline to complete a <u>paragraph</u> which will be your proof.

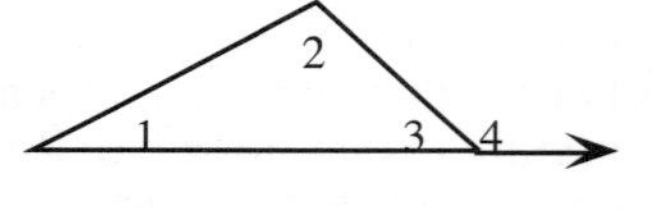

Theorem: An exterior angle of a triangle is equal to the sum of the two remote interior angles. (Hereafter known as the **EXTERIOR ANGLE THEOREM.**)

$m\angle 1 + m\angle 2 + m\angle 3$ = _______ because _________________________.

$m\angle 3 + m\angle 4$ = _______ because ____________________________.

$m\angle 1 + m\angle 2 + m\angle 3 = m\angle 3 + m\angle 4$ because ___________________________________.

Now subtract ___________ to get $m\angle 1 + m\angle 2$ = ________________ .

TK-18. Use your tool kit to name the idea(s) you use to solve the problem below.

a) Find the measures of the numbered angles.

b) Add the results for the three numbered angles and write a conjecture about the sum of the exterior angles of a triangle.

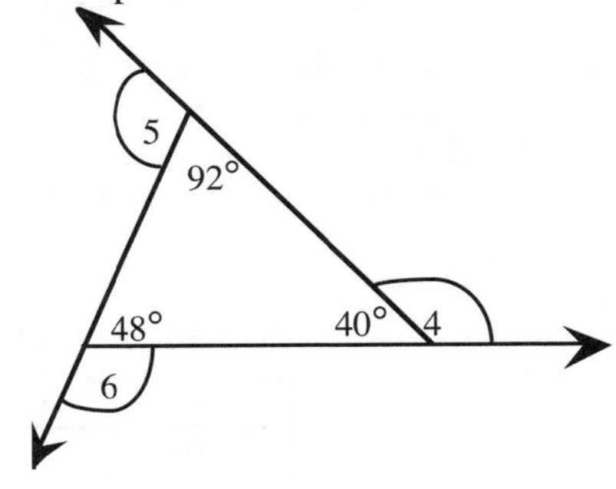


TK-19. In each problem, solve for x and then find m∠B. Do not forget the tool kit part!

a)

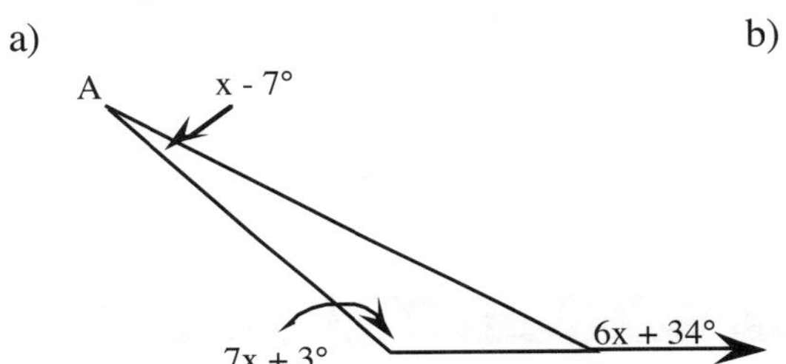


b)

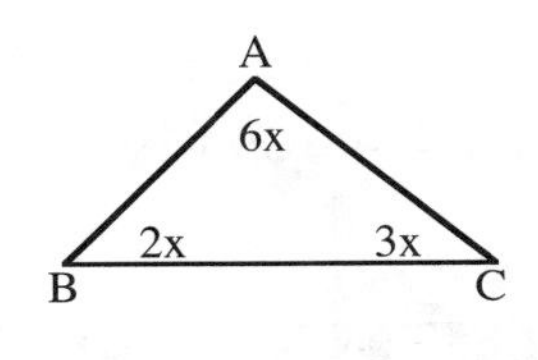


TK-20. Below is an argument to show that if $\overline{AD} \parallel \overline{EH}$ and $\overline{BF} \parallel \overline{CG}$, then m∠1 = m∠4. It is scrambled. Reorder the argument to make it logical and use your tool kit to write a reason that justifies each step.

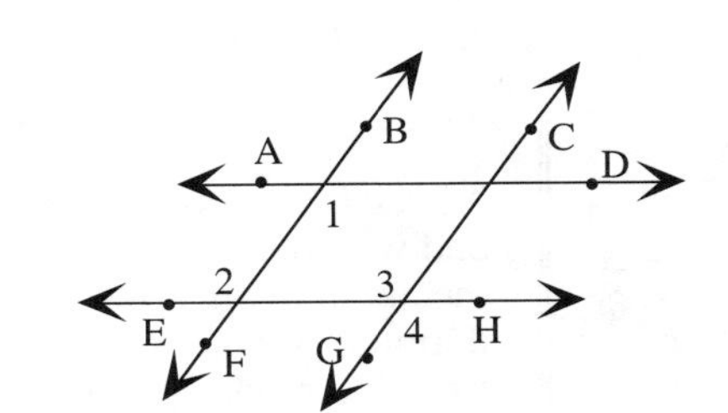


1) m∠2 = m∠3
2) $\overline{AD} \parallel \overline{EH}$ and $\overline{BF} \parallel \overline{CG}$
3) m∠1 = m∠4
4) m∠1 = m∠2
5) m∠3 = m∠4
6) m∠1 = m∠3

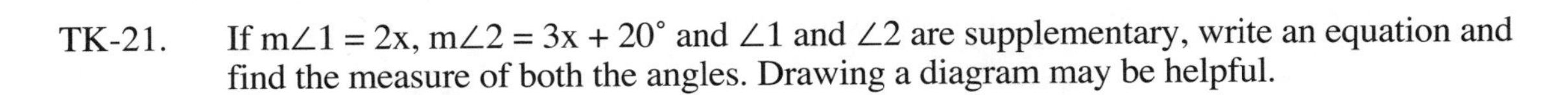

TK-21. If m∠1 = 2x, m∠2 = 3x + 20° and ∠1 and ∠2 are supplementary, write an equation and find the measure of both the angles. Drawing a diagram may be helpful.

TK-22. Write a conditional (if-then) statement that represents the arrow diagram at right.

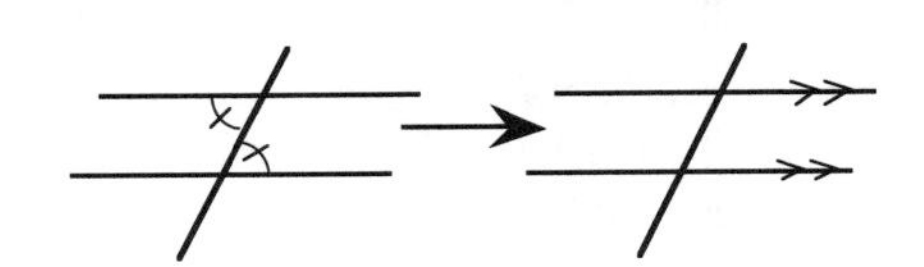

TK-23. As Kim begins to get in shape, she records in a notebook the date and distance she cycled, and the time it took her. Below are some selected dates from her records. Based on the information in Kim's log, approximately how far would she need to ride in order for her ride on July 4th to last about two hours?

| Date | Distance | Time |
|---|---|---|
| 6/14 | 12 miles | 1 hour |
| 6/19 | 15 miles | 55 minutes |
| 6/25 | 18 miles | 1 hour, 10 minutes |
| 7/2 | 22 miles | 1 hour, 25 minutes |

TK-24. Solve for x.

a) $3(2x - 5) = 38 - 2x$ b) $x^2 - 4x - 12 = 0$ c) $\frac{4x + 3}{2x} = \frac{2}{5}$

## THEOREMS AND MORE PROOFS OF FAMILIAR IDEAS

TK-25. Read the box below, add the information to your tool kit, then write an explanation of the difference between a conjecture and a theorem.

Notice that proofs usually start with the first part of the conditional statement and end with the conclusion (second part) of the conditional statement. Doing a proof means showing how the condition and related, previously accepted properties, definitions, and proved information lead logically to the conclusion. It is often useful to represent conditional statements two ways.

Example of two ways to represent a conditional statement:

1) In words:

"If two sides of a triangle are congruent, then the two angles of the triangle opposite those sides are congruent."

**OR**

2) As an arrow-diagram:

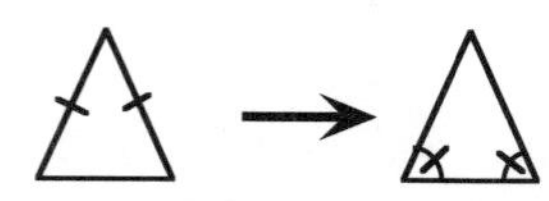

The arrow diagram makes the words of the conditional statement visual. Note that the triangle at left above represents ANY isosceles triangle, so once you prove this conjecture, you may use it anytime you have an isosceles triangle. This is the power of proof. Once a conditional statement is proven, its conclusion may be applied in all circumstances that meet the condition(s). Any conjecture that we prove is referred to as a **THEOREM**. Thus, the example above will be known as the **Isosceles Triangle Theorem** once you finish the next problem.

TK-26. By definition, an isosceles triangle has two sides of equal length. An interesting and extremely useful property of isosceles triangles is that the angles opposite the equal sides are also equal. This property is written as a conditional statement in the box below. Read the information, add it to your tool kit, then prove the theorem.

> If $\Delta ABC$ is isosceles with $\overline{BA} \cong \overline{BC}$, then the angles opposite these sides are congruent, that is, $\angle A \cong \angle C$. This statement is known as the **ISOSCELES TRIANGLE THEOREM.** The information in the conditional ("if") part of the statement is sometimes written as, "**Given** $\Delta ABC$ is isosceles with $\overline{BA} \cong \overline{BC}$, **prove**..." When you need a reason to justify a statement in a proof from the conditional statement, writing "given" or "stated as true" is sufficient.

Suppose you start with the triangle shown with $\overline{BA} \cong \overline{BC}$ and D as the midpoint of $\overline{AC}$. By drawing $\overline{BD}$, you form two more triangles. This suggests that if you could prove that $\Delta ABD \cong \Delta CBD$, you would know that $\angle A \cong \angle C$. (Why?) Use all of this information and your tool kit to write a short, logically sequenced paragraph that proves the conjecture. If you would prefer to set up your explanation as an ordered list, that is okay, too.

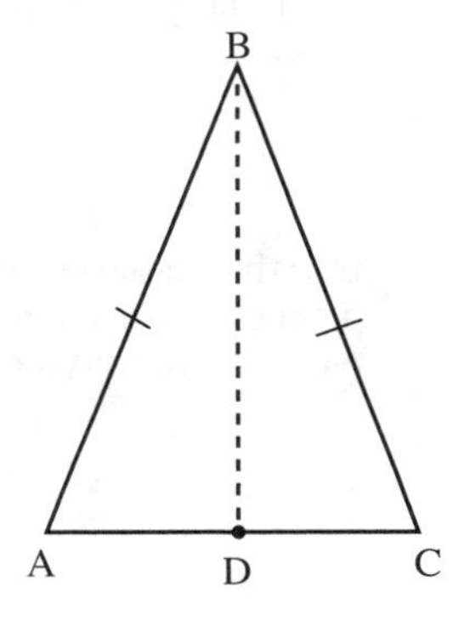


TK-27. If two angles of one triangle are congruent to two angles in another triangle, are the third pair of angles congruent? Write a paragraph or a line by line presentation that completely supports your conclusion.

If $\angle A \cong \angle D$ and $\angle B \cong \angle E$, is it true that $\angle C \cong \angle F$?

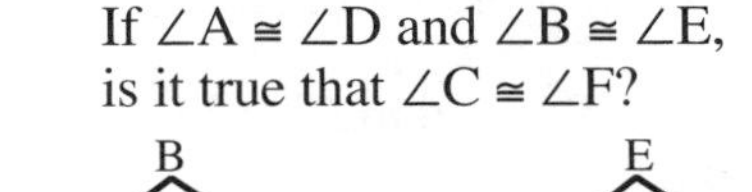


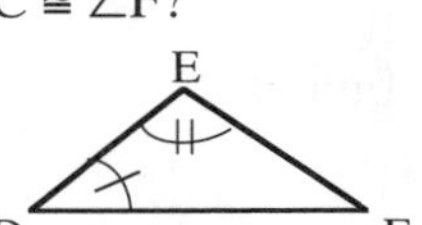

TK-28. When Alfred looked at the previous problem he said, "This is easy! Anyone can see that $\angle C$ is the same as $\angle F$ because they look equal in the figures." Discuss this issue with your team, then write a short paragraph that explains to Alfred why this does not prove anything!

## A Note about Proof Format in This Course

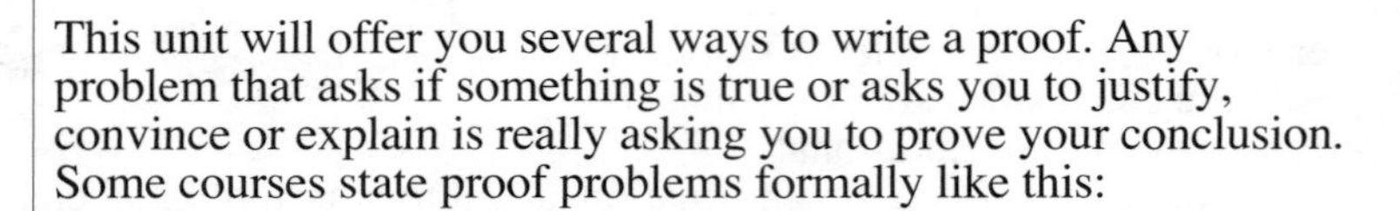

This unit will offer you several ways to write a proof. Any problem that asks if something is true or asks you to justify, convince or explain is really asking you to prove your conclusion. Some courses state proof problems formally like this:

**GIVEN:** any triangle **PROVE:** the sum of the three interior angles is 180°.

In this course we will not usually do this for two reasons. First, we want you to concentrate on writing logical arguments that show how connected pieces of information lead to a valid (true) conclusion. In most cases, you will be free to present your argument in the form that is easiest and clearest for you. Second, the "Given...Prove..." format already tells you that the conclusion is true. In some problems it is appropriate to state this, then have you confirm the statement with a proof. However, real life--whether academics or daily living--is not so certain. Consequently, we will often ask, "If you have a triangle, is it true that..." This approach asks you to <u>think</u> <u>through</u> a situation and convince your audience (teacher, teammates, ...) that your conclusion is valid.

TK-29. Use the resource page provided by your teacher. As you complete each part of this problem, use your tool kit to explain how you solved it. Each part is a separate problem. Note: there may be more than one way to do each problem.

a) If $m\angle 3 = 78°$, find $m\angle 18$.

b) If $m\angle 13 = 113°$, find $m\angle 18$.

c) If $m\angle 12 = 74°$, find $m\angle 21$.

d) If $m\angle 6 = 132°$, find $m\angle 15$.

e) If $m\angle 11 = 98°$ and $m\angle 7 = 53°$, find $m\angle 4$.

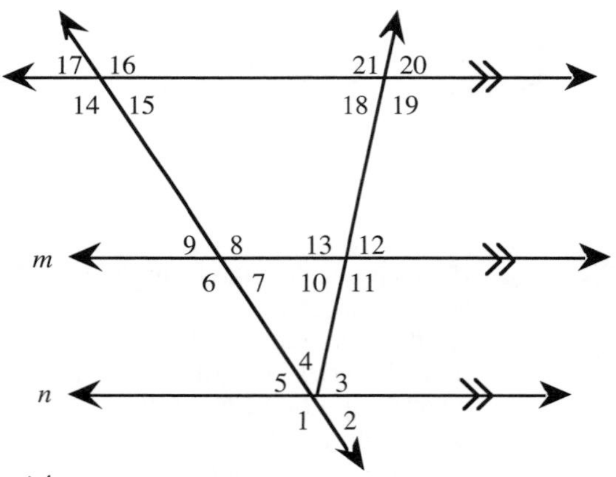


f) If $m\angle 20 = 81°$ and $m\angle 9 = 57°$, find $m\angle 4$.

g) Challenge: find a second way to do part (f) that is significantly different from the way you just did it.

TK-30. Here is a proof with the last conditional statement missing. Fill it in. Be sure that you respond with a conditional statement.

PROVE: If the electricity goes off after midnight, then I will be late for school.

Proof: If the electricity goes off after midnight, then the timer on the coffee maker will shut down. If the timer shuts down and restarts, it automatically resets for midnight. If the timer resets to midnight, then the coffee will not be ready when my mother is supposed to get up in the morning. If the coffee is not ready, my mother will not get up. If my mother does not get up then she will not wake me up.

If ...______________________________________________________________.

TK-31. If $m\angle 1 = m\angle 3$, does $m\angle NRP = m\angle ORQ$? Show all your work leading to your conclusion, giving a reason (from your tool kit) for each statement you make.

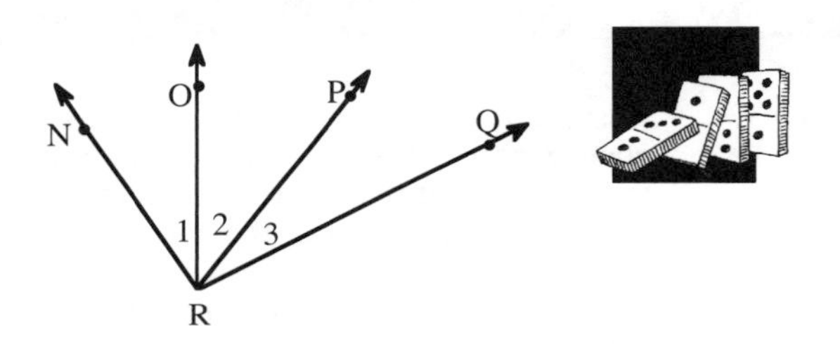


TK-32. If $m\angle 1 = m\angle 4$, does $m\angle ABE = m\angle CBE$? Give an argument justifying your answer. Support your argument by giving reasons from your tool kit.

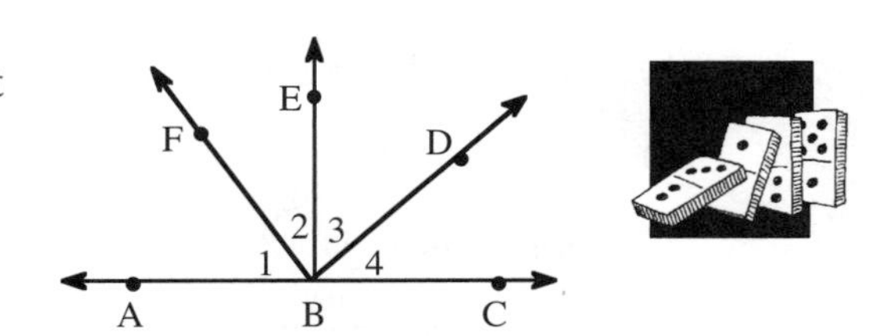


TK-33. Solve for x. Remember the Quadratic Formula.

a) $2x^2 + 5x - 3 = 0$  b) $3x^2 - 6x + 2 = 0$

TK-34. The sum of an integer and three times the square of the same integer is 114. Find the integer.

TK-35. Draw a sketch of the graph representing the relationship between the age of a person and the average speed at which that person drives.

TK-36. Is the graph you drew in the previous problem true for all drivers at each age? To what extent does your graph reflect stereotyping?

## FLOWCHART PROOFS

TK-37. Assume these rules: The box contains cheese, and the box of cheese is kept in the freezer, but the crust is not in the freezer. Put the letter of the appropriate statement into a flowchart oval to create a logical sequence that tells how to make a cheese pizza.

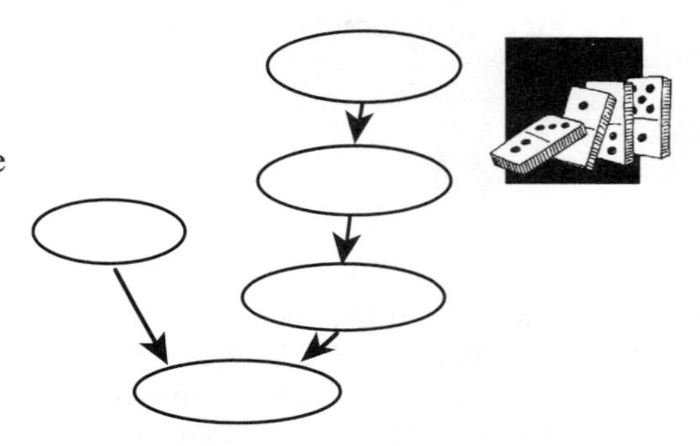

a) Get the box.

b) Put the cheese on the crust.

c) Get the crust.

d) Open the box.

e) Go to the freezer.

TK-38. Read the information and example in the box below and add it to your tool kit, then solve the Digit-Place problem below the box using a flow-chart.

In the first five units, you have been writing explanations in sentence, list, and paragraph forms. As problems become more complicated, you will need ways to see and organize information. This process is much like how you approach writing an essay or a research paper: when writing a paper you list ideas and information, cluster ideas around some central topics, then organize raw material in a manner that will accomplish your purpose. Some students like to write ideas randomly on paper as they pop into their heads, others prefer to list them, some simply write unorganized paragraphs. In this course we expect you to use the method most natural for you unless you are practicing a particular style of proof. The next style of proof is **FLOWCHART PROOFS** where statements are written in ovals. We return to the Digit Place Game to introduce them.

**Example:** first, write each clue in an oval, then see how they can be used together to reach a conclusion. Notice how the connecting lines show the logic of the proof.

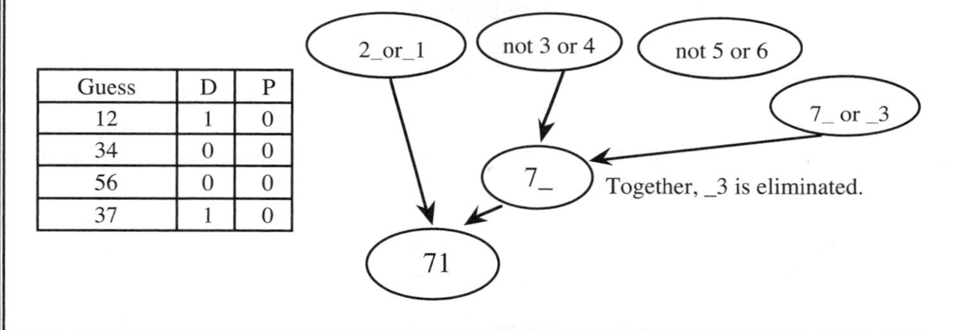

| Guess | D | P |
|---|---|---|
| 12 | 1 | 0 |
| 34 | 0 | 0 |
| 56 | 0 | 0 |
| 37 | 1 | 0 |



Flowchart proof practice:

Solve the Digit Place problem by rewriting each clue in an oval, then organizing a flowchart proof of your conclusion. Note that this is a different game than above.

| Guess | D | P |
|---|---|---|
| 12 | 0 | 0 |
| 34 | 1 | 1 |
| 56 | 1 | 0 |
| 37 | 0 | 0 |

TK-39. Solve the Digit Place problem by rewriting each clue in an oval, then organizing a flowchart proof of your conclusion.

| Guess | D | P |
|---|---|---|
| 34 | 1 | 0 |
| 56 | 0 | 0 |
| 78 | 1 | 1 |
| 18 | 0 | 0 |

TK-40. Use a flowchart proof to support your answer to the question: If $m\angle 2 = m\angle 3$, does $m\angle 1 = m\angle 3$?

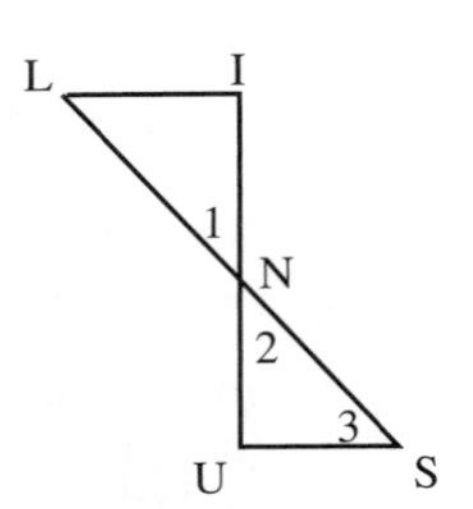


TK-41. Here is Susan's proof of the Exterior Angle Theorem:

"The sum of the measures of angles 1, 2, and 3 is 180° and so is the sum of the measures of angles 3 and 4. So angles 1 and 2 must equal 4."

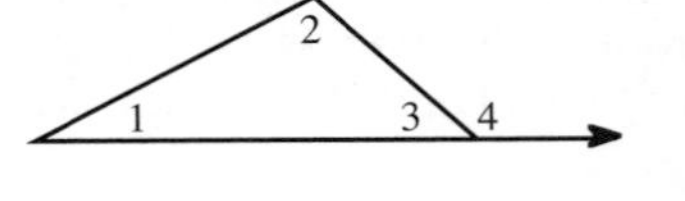

Susan's argument is not wrong, but it is incomplete. Why? Write the necessary improvements to make her proof complete and convincing. If you need help getting started, compare Susan's argument to problem TK-17.

TK-42. Here is an example of a flowchart proof. Read it, then answer the question at the end. You did this proof in paragraph form in problem TK-27.

*If two angles of one triangle are equal to two angles of a second triangle, then their third angles are equal.*

You know that $m\angle u = m\angle x$ and $m\angle v = m\angle y$. You want to show how these two statements lead to the conclusion that $m\angle w = m\angle z$.

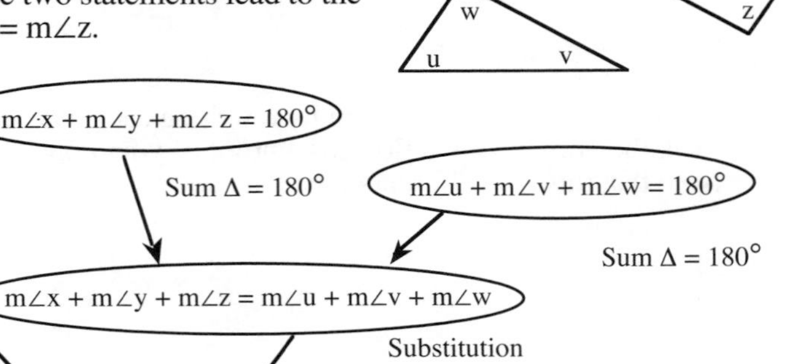


Explain completely what happened to create the last oval.

TK-43. Solve the Digit Place problem by rewriting each clue in an oval, then organizing a flowchart proof of your conclusion.

| Guess | D | P |
|---|---|---|
| 37 | 1 | 1 |
| 35 | 0 | 0 |
| 42 | 0 | 0 |
| 86 | 1 | 0 |

TK-44. Solve each linear system.

a) $x = 2y + 4$
$x = 10 - 4y$

b) $6x - 2y = 1$
$3x + 10y = 6$

TK-45. In the figures below, are the two triangles congruent? Why or why not? Support your conclusion by using the tool kit. Remember: the figures are not drawn to scale.

a)

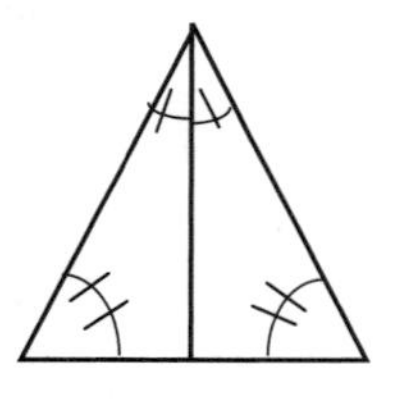

b)

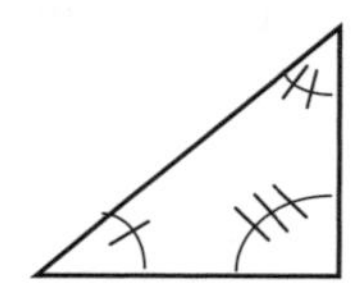

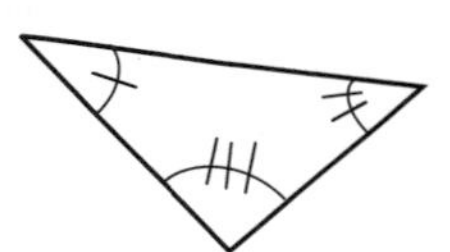

c)

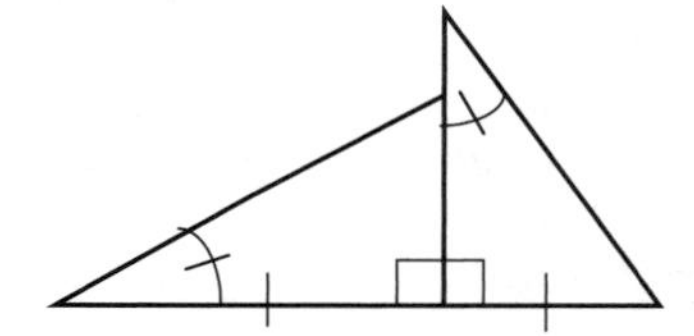

d)

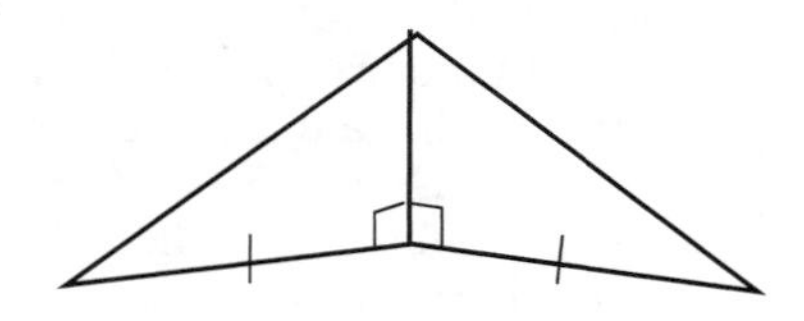

TK-46. In this diagram, find the measure of each labeled angle. Explain your process by using your tool kit.

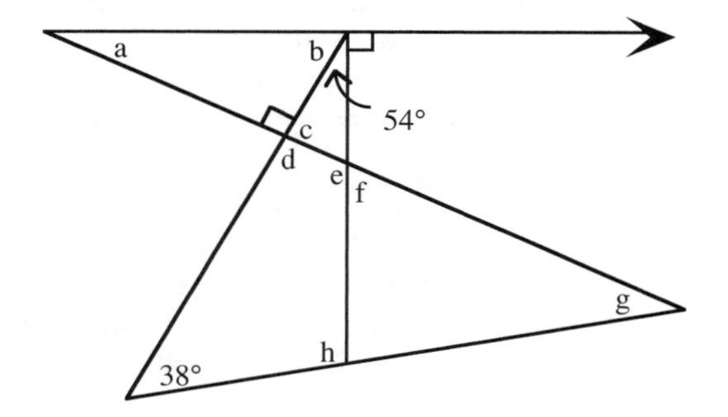

TK-47. Find the total surface area and volume of the prism at right. Show all subproblems.

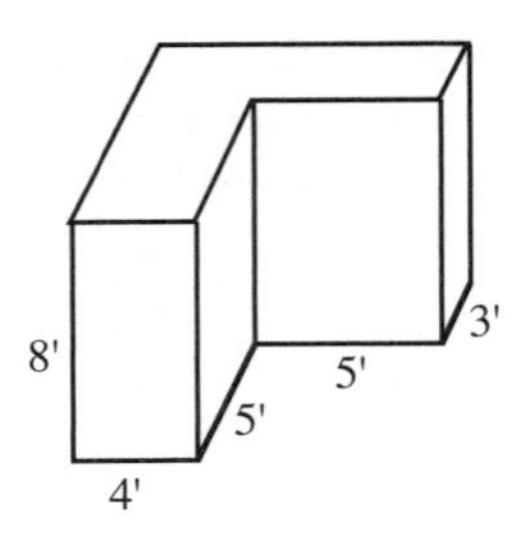


## FLOWCHART PROOFS OF PARALLELISM CONJECTURES

TK-48. Create a flowchart to indicate the steps needed to find the area of the shaded region. Note that you are <u>not</u> going to find the actual area! Begin by writing down all the things you need to do without worrying about order. Some students find this easier to do by drawing pictures of the subproblems they would use. After you think you have them all, put them into order in a flowchart.

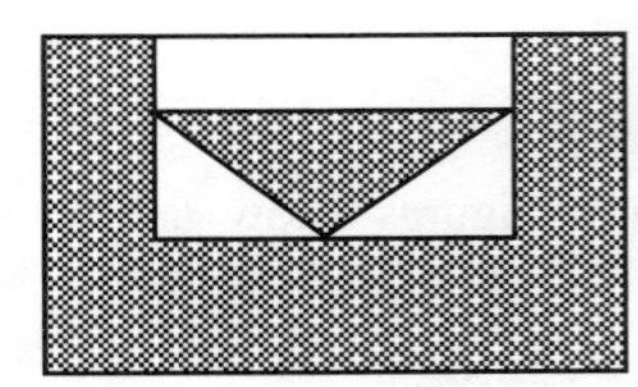

TK-49. Read the information in the box below, add it to your tool kit, then prove the theorem below the box.

> Earlier in this unit you considered the differences between a **conjecture** and a **proof**. In order to prove observations, there needs to be a starting point or basic properties that are accepted as true. In mathematical systems, these basic truths are known as **AXIOMS** or **POSTULATES**. In Algebra 1, you accepted properties such as "adding the same value to both sides of an equation preserves equality" and the Distributive Property as true, then learned how to apply them in various situations. In Geometry, we accept facts like straight angles measure 180° and the distance between two points is unique, that is, there is only one distance and one line between two points on a flat surface. One important postulate we assumed in Unit 3--although we discovered it by exploration and called it a conjecture--was that when two parallel lines on a flat surface are cut by a third line, pairs of **corresponding angles** are equal. We will accept this statement as true in this course and use it to prove several of our conjectures so that they become theorems. Use it now to prove the theorem below.

In the figure at right, if $l \parallel m$ and cut by line $n$, then $m\angle 1 = m\angle 2$. This statement is known as the **ALTERNATE INTERIOR ANGLE THEOREM**. Write a flowchart proof to confirm that this previously developed conjecture is valid.

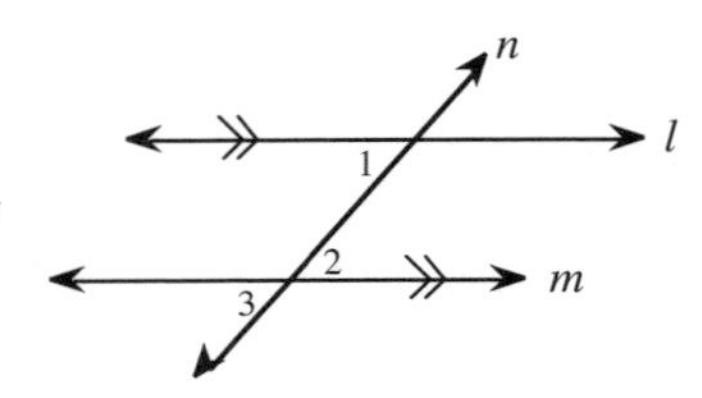

TK-50. Another conjecture from Unit 3 said, "If $l \parallel m$ and cut by line $n$, then $m\angle 2 + m\angle 3 = 180°$." Write a flowchart proof to confirm that this previously developed conjecture is valid.

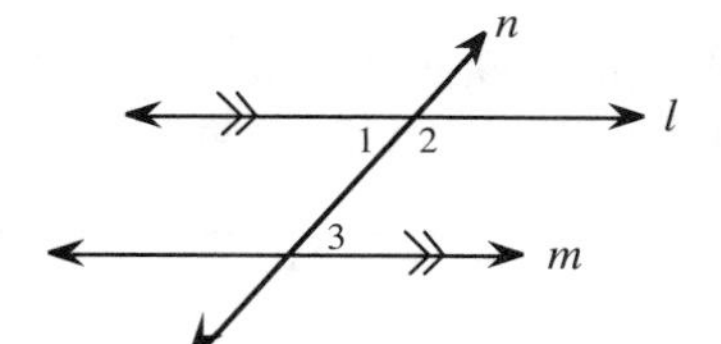

TK-51. In the figure for the preceding problem, label the angle that is the supplement of $\angle 2$ above line $l$ "angle 4." Briefly explain how you could prove the previous problem using angles 2, 3, and 4.

TK-52. If BD = CD and $m\angle ADB = m\angle ADC$, is $m\angle ACD = m\angle DAB$? Justify your conclusion with any style of proof. Justify any statements you make with the tool kit. A good start is to mark all of the parts that you know are or can justify to be congruent.

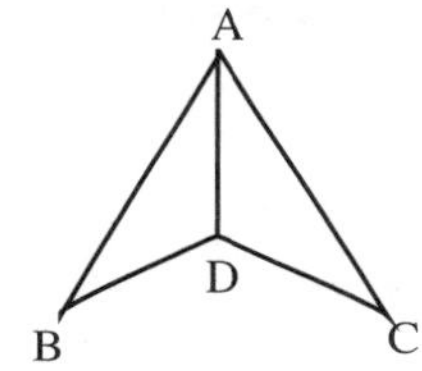

TK-53. A few days ago Tuan read the instructions that asked him to prove the Isosceles Triangle Theorem. Then he said that he did not have to prove it because, "If it is in the book it must be true." If you were in Tuan's team, how would you respond? Discuss your views with your team then write your response in a short paragraph.

TK-54. Read the information that follows, add it to your tool kit, then do the problems in the last box below.

## LAWS OF exponents

Exponents are used to represent repeated multiplication.

$3^4$ says to multiply 3 by itself four times, that is, $3^4 = 3 \cdot 3 \cdot 3 \cdot 3$.

As long as an exponent is a reasonably small integer, you can write the base repeatedly to see what it represents. However, by using some general laws you can work much more efficiently with exponents. Think of the laws, and all of the previous algebra review, as parts of your algebra tool kit.

For $x^a$, x is the **BASE** and a is the **EXPONENT**.

| Examples: | Exponent Laws |
|---|---|
| $x^3x^4 = x^{3+4} = x^7$ <br> $x^3x^4 = (x \cdot x \cdot x) \cdot (x \cdot x \cdot x \cdot x)$ <br> $= x \cdot x \cdot x \cdot x \cdot x \cdot x \cdot x$ <br> $= x^7$ | Law 1: For any numbers a and b, $x^a \cdot x^b = x^{a+b}$ <br> Proof: $x^a \cdot x^b = \underbrace{(x \cdot x \cdot \ldots \cdot x)}_{\text{a of these}} \cdot \underbrace{(x \cdot x \cdot \ldots \cdot x)}_{\text{b of these}}$ <br> $= \underbrace{x \cdot x \cdot \ldots \cdot x \cdot x \cdot x \cdot \ldots \cdot x}_{\text{a + b of these}}$ <br> $= x^{a+b}$ |
| $\frac{x^8}{x^3} = x^{8-3} = x^5$ | Law 2: $\frac{x^a}{x^b} = x^{a-b}$ |
| $(x^3)^4 = x^{3 \cdot 4} = x^{12}$ | Law 3: $(x^a)^b = x^{ab}$ |
| $(x^2y^3)^4 = x^{2 \cdot 4}y^{3 \cdot 4} = x^8y^{12}$ | Law 4: $(xy)^a = x^ay^a$ and $(x^ay^b)^c = x^{ac}y^{bc}$ |
| $\left(\frac{x^2}{y^4}\right)^3 = \frac{x^{2 \cdot 3}}{y^{4 \cdot 3}} = \frac{x^6}{y^{12}}$ | Law 5: $\left(\frac{x}{y}\right)^a = \frac{x^a}{y^a}$ |

Use the first two laws of exponents to simplify the following:

a) $x^3x^5$

b) $x^7x^3x^2$

c) $x^6x^{2.5}$

d) $\frac{x^{12}}{x^7}$

e) $\frac{x^{21}}{x^8}$

f) $\frac{x^{10}}{x^9}$

TK-55. Write a proof for Law 2 for $a > b$ similar to the one given in the example for Law 1.

TK-56. Solve the Digit Place problem and write a proof of your conclusion.

| Guess | D | P |
|---|---|---|
| 95 | 1 | 0 |
| 63 | 0 | 0 |
| 78 | 0 | 0 |
| 10 | 1 | 0 |

TK-57. Prove your conclusion to the following question by using a flowchart, list or paragraph. Use your tool kit to justify your statements. Start by drawing the figure on your paper and marking the equal parts.

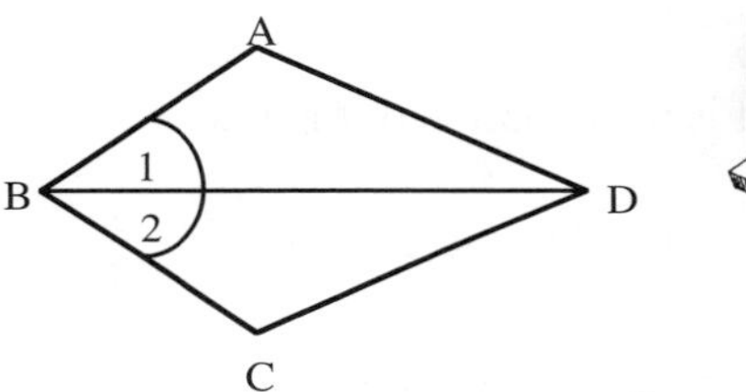

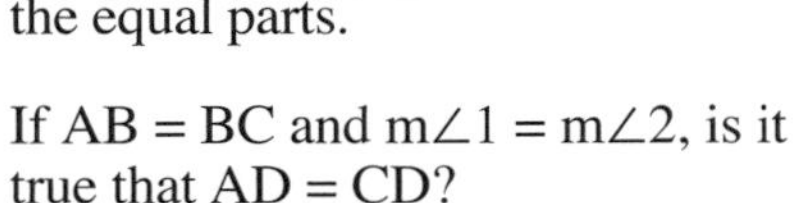

If $AB = BC$ and $m\angle 1 = m\angle 2$, is it true that $AD = CD$?

TK-58. If $m\angle A = 5x + 5^\circ$ and $m\angle B = x + 13^\circ$, find $x$ and $m\angle A$.

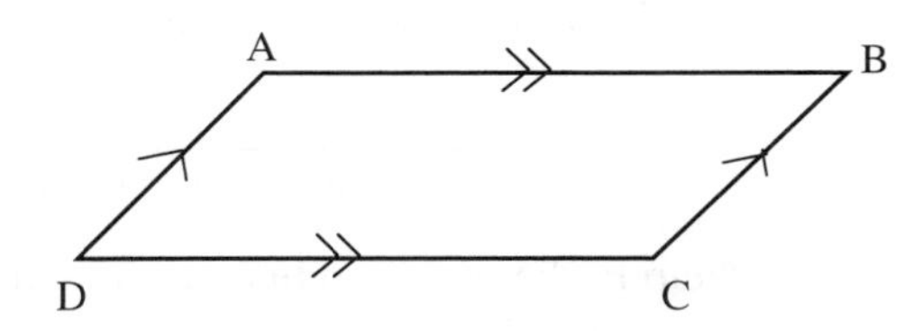

TK-59. A ball 12" in diameter rolls around inside the bottom of a rectangular box that is 14" wide and 30" long. The ball always touches a side of the box. To help you answer the questions below, get a ball and a box to help you visualize the problem. They do not have to be the right dimensions to give you a visual image of the diagram.

a) Imagine that the ball is covered in wet paint, so that as it rolls, it leaves a trail. Draw a diagram of the bottom of the box and the path the ball leaves on it.

b) What is the name of the shape of the ball's path? What are its dimensions?

c) Find the length of the ball's path.

TK-60. What is the equation of the line passing through (0, 15) and:

a) Parallel to the graph of $y = -\frac{3}{4}x + \frac{2}{3}$?

b) Perpendicular to the graph of $y = -\frac{3}{4}x + \frac{2}{3}$?

TK-61. Eddie needs to graph a line on his homework assignment. This always gives him trouble so he decides he better make an input-output table, find the values, and plot all the points. He shows you the table below, and asks for your opinion.

| x (input) | -3 | -2 | -1 | 0 | 1 | 2 | 3 |
|---|---|---|---|---|---|---|---|
| y (output) | -17 | -12 | -7 | 17 | 3 | 8 | 31 |

Help Eddie out. Find his mistakes, and replace them with the correct y-values. Explain how you know which ones to correct.

TK-62.

## FLOWCHART EXAMPLE

Below is an example of how to write a flowchart proof to justify a conclusion using the idea of congruent triangles. **Notice that the tool kit reasons appear below each oval to justify the statement inside.** The ovals are connected in a logical sequence to reach the desired conclusion. You can use this example as a model when you use a flowchart for a proof.

If E is the midpoint of $\overline{AD}$ and $\angle A$ and $\angle D$ are both right angles, prove that $\overline{AB} \cong \overline{DC}$.

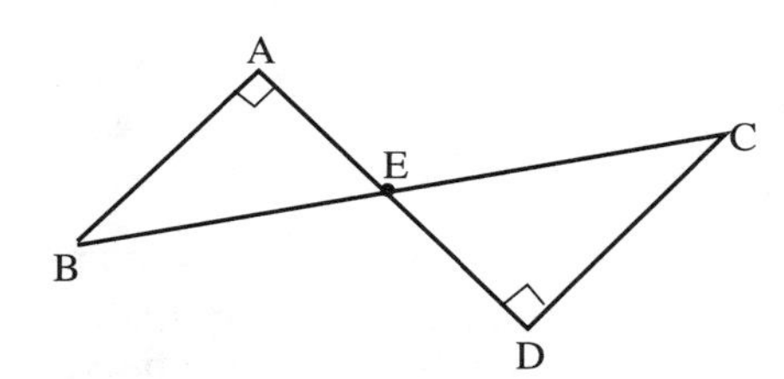


Mark the parts of the two triangles that you know are congruent.

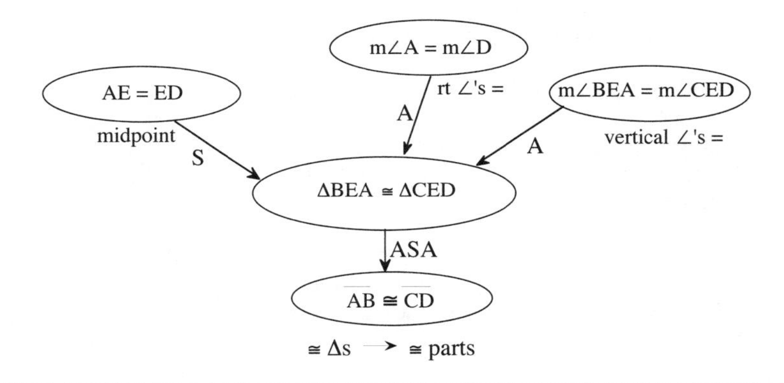

**CAUTION!** In problem TK-28, Alfred wanted to conclude that a pair of corresponding angles between two triangles were equal because they looked equal. Remember that figures are not presumed to be drawn to scale unless you are told that they are. The only assumptions that you may make about figures are:

- lines that appear to be straight are straight.
- pairs of vertical angles are always congruent.
- common sides of two triangles (or polygons) are congruent.

TK-63. If DC = AC and $m\angle 1 = m\angle 2$, is it true that $m\angle 3 = m\angle 5$? Prove your answer using a flowchart and support your reasons with the tool kit. First draw the figure on your paper, and mark what you know. Review problem TK-57 if you need help getting started.

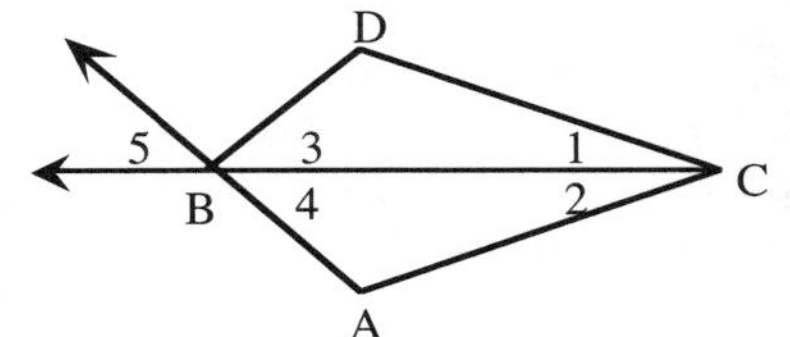

## PROVING THE PYTHAGOREAN THEOREM

TK-64. If the figure at right is drawn as marked, prove that the shaded region is a square, that is, prove that:

1) all four sides are congruent; and,

2) all four angles marked "z" measure 90°.

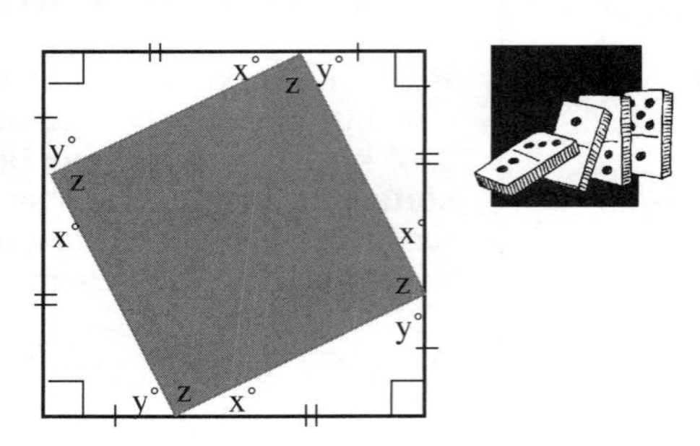


TK-65. You can use the preceding problem as the basis to prove the **Pythagorean Theorem.** Since you <u>know</u> that the shaded figure is a square (you proved it above):

a) Find the area of the large square two ways:

1) Using the dimensions of the large square.

2) Using subproblems, that is, finding the sum of the areas of the five figures inside the large square.

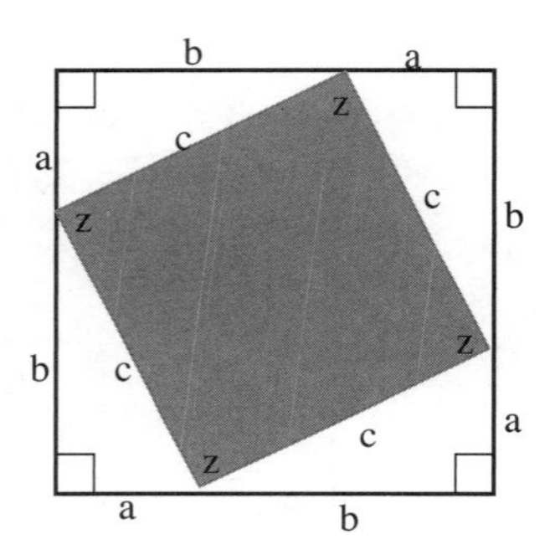


b) Since both areas (1) and (2) represent the area of the same figure, the algebraic expressions of the area are equal. Write an equation and subtract common terms from both sides.

c) Your result should confirm the Pythagorean Theorem for each of the four triangles in the figure. Why does the work in this problem <u>prove</u> the Pythagorean Theorem?

TK-66. In the following diagram, $m\angle 4 = 114°$, $m\angle 1 = m\angle 3$. Draw the diagram on your paper so that you can mark it and find $m\angle 6$.

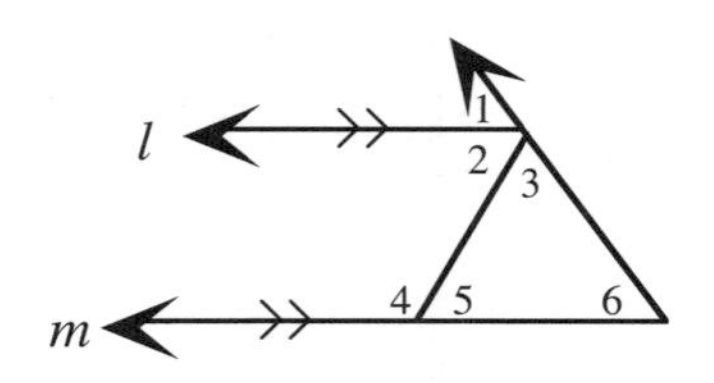


TK-67. If $\overline{AB} \parallel \overrightarrow{CE}$, is it true that $m\angle 2 = m\angle BCD$ (the sum of angles 3 and 4)? Use the style of your choice and your tool kit to write the proof.

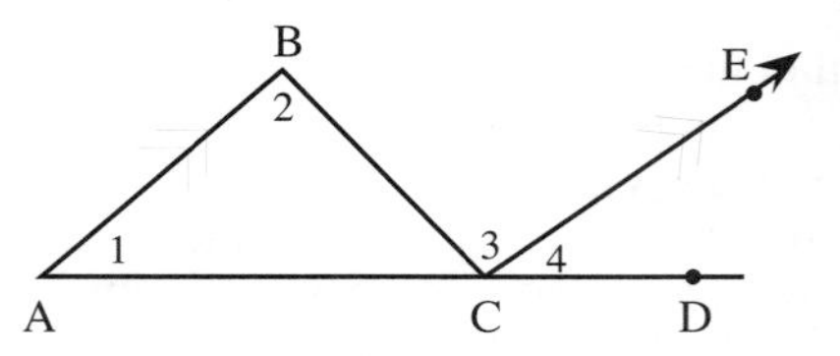

TK-68. Use the third and fourth laws of exponents to simplify the following:

a) $(x^3)^4$   b) $(x^2)^6$   c) $(x^3y^4)^3$   d) $(x^5y^3)^2$

TK-69. Use your tool kit to name the ideas you use when finding the value (in degrees) of each variable.

a)

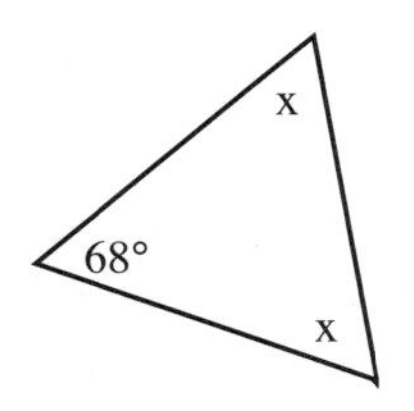


b)

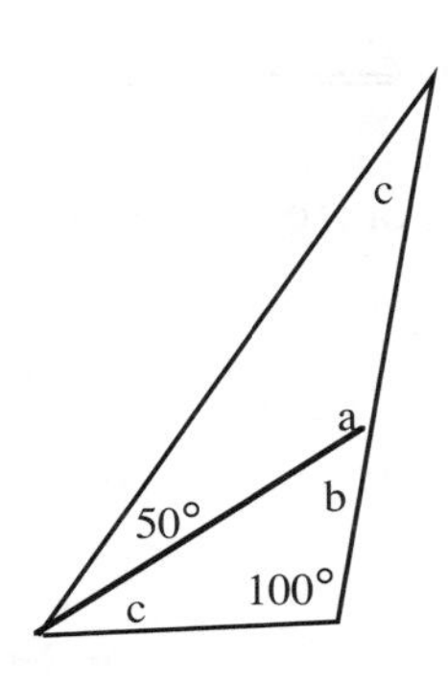


TK-70. Write an equation to solve the following problem. Start by drawing and labeling a diagram for the rectangle and the square.

One side of a rectangle is nine inches long and the other side is two inches shorter than the side of a square. The perimeter of the rectangle is two inches longer than the perimeter of the square. Find the dimensions of the rectangle.

TK-71. Calculate the area of this region. All angles which appear to be right angles are right angles.

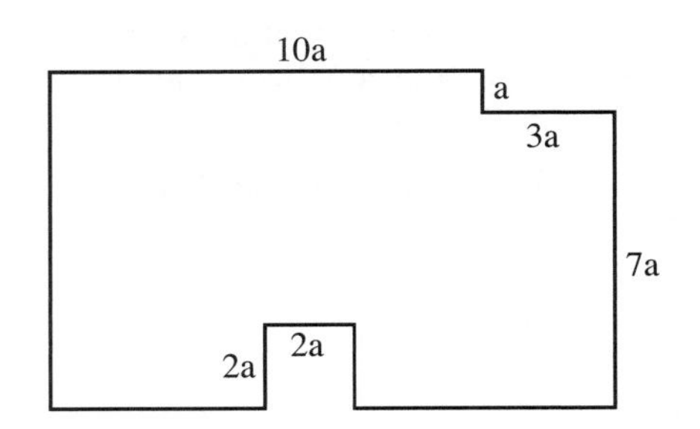


TK-72. What is the probability of rolling an ordinary die and seeing:

a) a one?

b) an even number?

c) a multiple of three?

d) a number greater than or equal to 2?

e) a single-digit number?

f) a two-digit number?

## CONVERSES AND COUNTEREXAMPLES

TK-73. Look closely at the **converses** in the examples below and decide whether or not each one is always true. If it is not always true give an example which shows that it is not. Then add this definition to your tool kit.

The **CONVERSE** of a conditional statement is formed by switching the if and then parts of the statement.

Statement: A: If a triangle is equilateral, then it is equiangular.
Converse A: If a triangle is equiangular, then it is equilateral.

Statement: B: If $x = 4$, then $x^2 = 16$.
Converse B: If $x^2 = 16$, then $x = 4$.

Statement C:

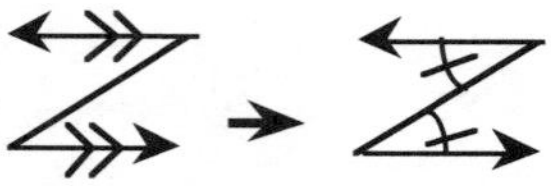

Converse C:

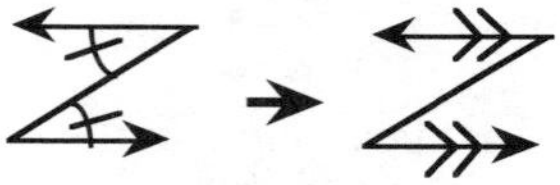

Statement: D: ABCD is a square → ABCD is a parallelogram.
Converse D: ABCD is a parallelogram → ABCD is a square.

Notice that the words if and then stay in the same place. The condition becomes the conclusion and the conclusion becomes the condition.

TK-74. Read the box below and add the definition to your tool kit. Then answer this question: "Does LaShaun need to find <u>more</u> dark-haired people who have blue eyes to prove that the generalization is incorrect? Explain."

> An example which shows that a generalization has at least one exception, that is, a situation in which the statement is false, is called a **COUNTEREXAMPLE**. For example: Oscar says, "All dark haired people have brown eyes." LaShaun does not believe him and she finds Aaron who has dark hair and <u>blue</u> eyes. Aaron is a <u>counterexample</u> to Oscar's generalization. Therefore, Oscar's statement is false.

TK-75. Write the converse of each statement below. Explain whether or not the converse is <u>always</u> true. If the converse is not always true, support your claim by drawing or writing a counterexample.

a) If $x > 7$, then $x > 5$.

b) If two distinct lines are not parallel, then they must intersect.

c) If m and n are the slopes of any two distinct lines and $m = n$, then the lines are parallel.

TK-76. Write the <u>converse</u> of the Isosceles Triangle Theorem. Decide whether or not you believe the converse to be true. If you believe it is true, prove it in the style of your choice. If you believe it is false, produce a counterexample. It may be helpful to review problem TK-26 (The Isosceles Triangle Theorem). If you use this problem as a model for your proof, start with $\overline{BD}$ bisecting $\angle ABC$ <u>instead of</u> $\overline{BD}$ bisecting $\overline{AC}$. However, you may not use both conditions in your proof.

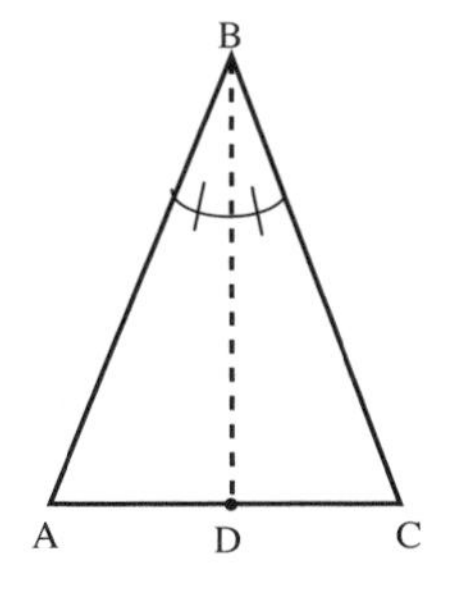


TK-77. If $m\angle A = m\angle D$ and $AB = CD$,

a) is $AE = DE$? Prove your answer.

b) is $m\angle D = 90°$? Prove your answer.

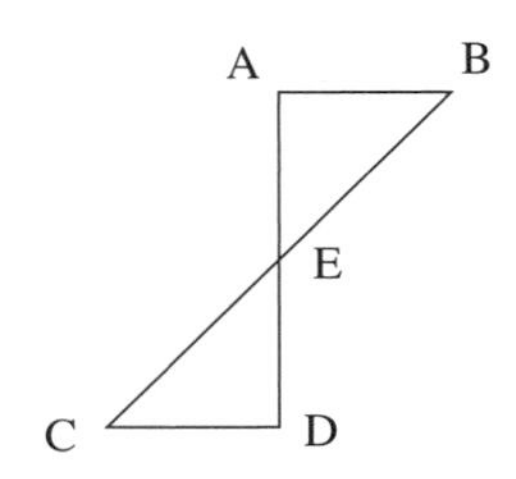

TK-78. If $\angle PSQ \cong \angle RSQ$, is it true that $\overline{SP} \cong \overline{SR}$? Mark a copy of each figure with what you <u>know</u> (not what you might assume!). Then prove your answer in the style of your choice and support your reasons with the tool kit.

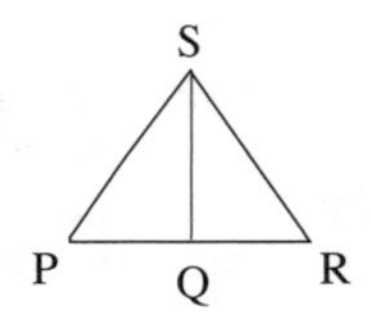

TK-79. Here is a Digit Place game with the first few guesses:

| Guess | D | P |
|---|---|---|
| 123 | 0 | 0 |
| 451 | 1 | 1 |

The teacher asks: "Explain what you know so far." Some sample student responses are shown. Your job is to evaluate the quality of these responses. Discuss each person's answer with your team, then write your evaluation on your paper.

Fred: There must either be a 4 or a 5 in the first or second place.

Sam: From the first guess we know that 1, 2, and 3 are not in the number. The second guess tells us that either the 4 or 5 is a correct digit, but not both, and that the number is either _5_ or 4_ _ .

Judy: The first guess tells us that 1, 2, and 3 are not in the number. My third guess would be 678.

TK-80. If $\overrightarrow{RO}$ bisects $\angle PWE$ and $\angle P \cong \angle E$, is $\angle 1 \cong \angle 3$? Prove your answer.

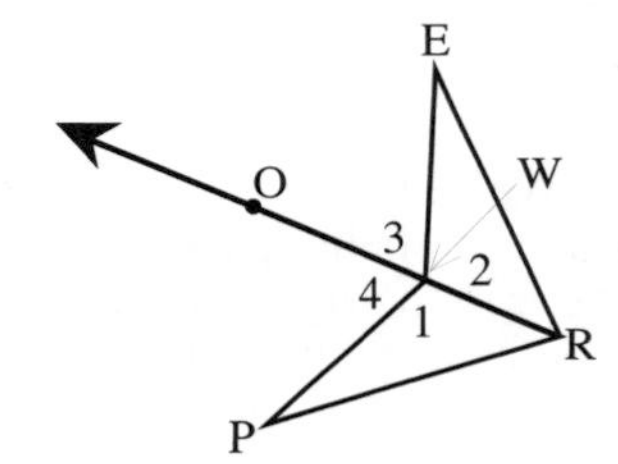

TK-81. Using the same given information and diagram as above, is $\overline{RE} \cong \overline{RP}$? Prove your answer.

TK-82. Simplify the following:

a) $x^7x^8x^2$ b) $(x^{12})^4$ c) $\left(\frac{x^2}{y^3}\right)^4$ d) $(3x^4y^5)^2$

TK-83. Assume each initial statement is true. Write the converse of each statement. Decide if the converse is always true or not. You may have to draw a picture. If it is not always true, give a counterexample to demonstrate that the converse is a false statement.

a) If $\angle A$ and $\angle B$ are vertical angles, then $\angle A \cong \angle B$.

b) If $\Delta ABC$ is an obtuse triangle, then $m\angle A + m\angle B + m\angle C = 180°$.

c) If n is an odd integer and m is an even integer then mn is even.

TK-84. Draw the graphs of the inequalities below, estimate the coordinates of their point(s) of intersection from the graph, then estimate the area of the overlapping shaded regions.

$$y \geq x^2 - 2 \qquad\qquad y - x < 4$$

TK-85. Find the volume of the trapezoidal prism at right. Show all subproblems.

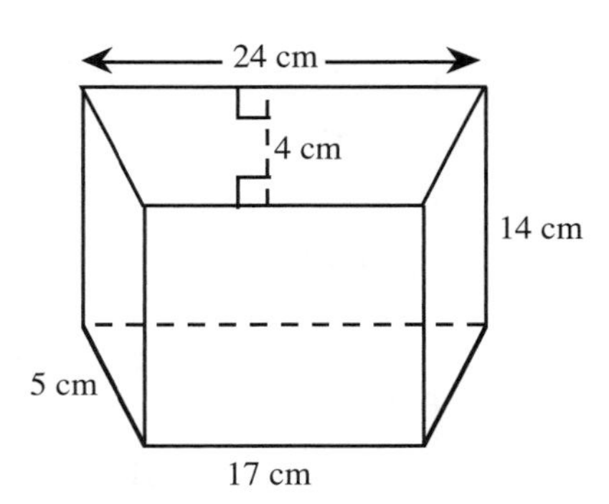


TK-86. Solve for x.

a) $3x^2 + 7x + 1 = 0$

b) $-x^2 - 5x + 2 = 0$

## TWO COLUMN PROOF

TK-87.

**INTRODUCTION:** Here is a model of how some people and many geometry books do proofs in the list style. They are often called **T-proofs** or **double (two) column proofs** because of their layout.

**DIRECTIONS:** Study the model, then do parts (a) and (b) below the box. Note that all you know is that WASH is a parallelogram. This problem proves facts about parallelograms that we have assumed and used in previous problems.

If WASH is a parallelogram, then $\overline{WA} \cong \overline{SH}$.

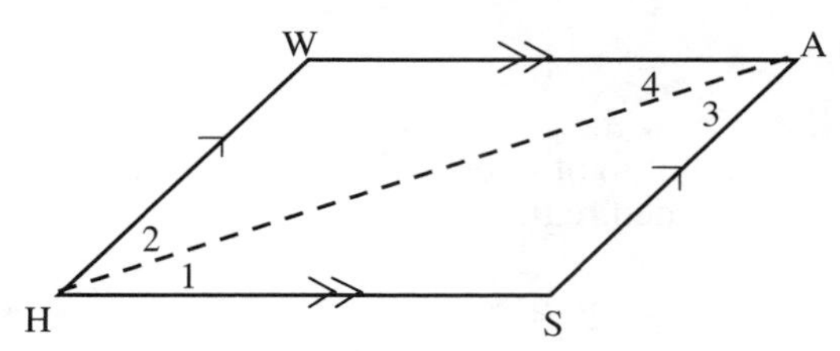


| STATEMENT | REASON (This statement is true because…) |
|---|---|
| $\overline{HS} \parallel \overline{WA} \longrightarrow m\angle 1 = m\angle 4$ | |
| $\overline{HW} \parallel \overline{SA} \longrightarrow m\angle 2 = m\angle 3$ | |
| $\overline{HA} \cong \overline{HA}$ | same segment |
| $\Delta WHA \cong \Delta SAH$ | ASA |
| $\therefore \overline{WA} \cong \overline{SH}$ | **$\cong \Delta$s $\rightarrow \cong$ parts** |

Note: $\therefore$ is the symbol for the word "**therefore**."

a) Is $\overline{WH} \cong \overline{SA}$? Why or why not?

b) Is $\angle W \cong \angle S$? Why or why not?

TK-88. The following problem is not the same problem as the example above. It is similar in content to the model EXCEPT that you are now trying to prove that the figure is a parallelogram rather than starting with this fact as your condition (given). Do this proof in the two column style modeled in the previous problem. Recall that to have a parallelogram, you need to know that BOTH pairs of opposite sides are parallel

If $\overline{AB} \parallel \overline{CD}$ and $\overline{AB} \cong \overline{CD}$, prove that ABCD is a parallelogram. Stated another way, under the conditions given, prove that $\overline{BC} \parallel \overline{AD}$.

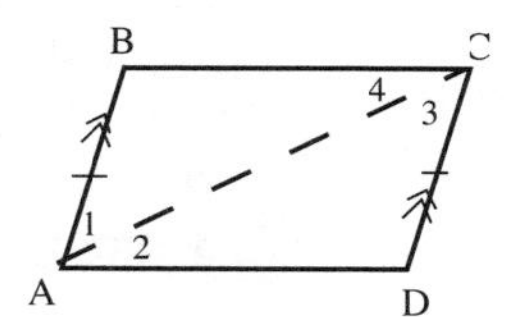


TK-89. Examine the diagram at right. E is the midpoint of $\overline{AB}$; F is the midpoint of $\overline{AC}$. $\overline{EF}$ is known as a **mid-segment.** Use a ruler and tracing paper to investigate the figure as directed below.

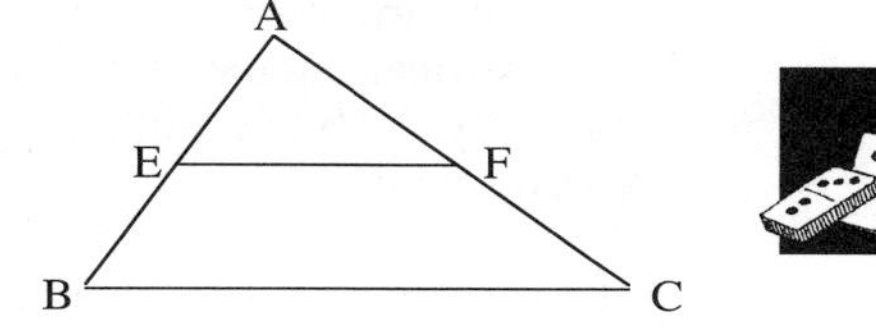


a) Which angles appear to be equal? Test your conjecture with tracing paper or plain paper.

b) Which segments appear to be parallel? Use your work in part (a) to justify your observation.

c) Measure EF and BC. What appears to be true? Do you think connecting the midpoints will always produce this result? Draw some other types and shapes of triangles, add a mid-segment, and measure the two lengths.

d) Write a conjecture stating two different ways that mid-segment $\overline{EF}$ is related to $\overline{BC}$.

TK-90. We hope your conjecture in the previous problem included statements that mid-segment $\overline{EF}$ is parallel to $\overline{BC}$ and is one-half the length of $\overline{BC}$. What we want to do now is prove that this is in fact always true for every triangle. We can use subproblems to prove what seems to be a fairly complicated theorem. Before we start, we will add two segments to the figure as shown above. $\overline{GE}$ is an extension of $\overline{EF}$ and congruent to $\overline{EF}$. $\overline{GB}$ simply connects points G and B.

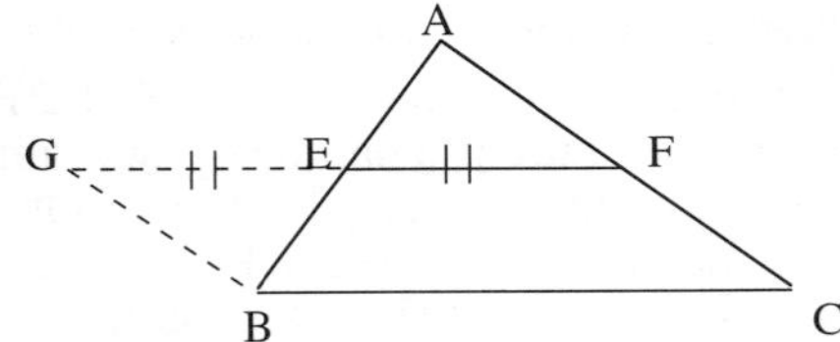

REMEMBER: E and F are midpoints.

a) If we copy only a part of the figure as shown at right, it looks more familiar. Known information is marked. Write a brief justification, using tool kit vocabulary, about why the two triangles are congruent.

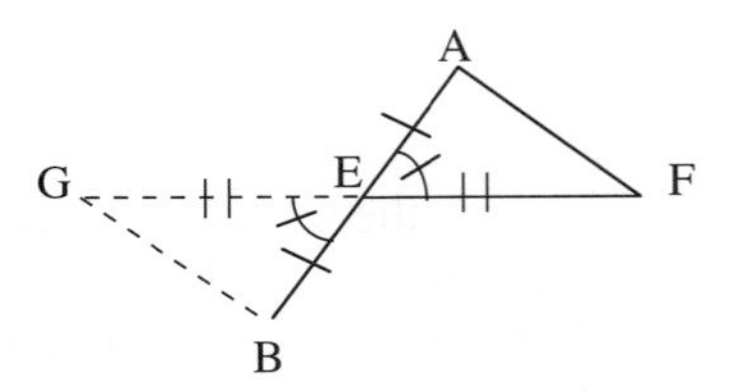


b) If the triangles are congruent, is $\overline{GB} \cong \overline{FA}$? Explain. Then explain why $\overline{GB} \parallel \overline{FA}$.

c) Use the information we know so far to explain why we can now conclude that $\overline{GB} \cong \overline{FC}$ in the original figure.

d) Use parts (b) and (c) and problem TK-88 to explain why quadrilateral GFCB must be a parallelogram.

TK-91. Copy the theorem and definition in the box below into your tool kit. Be sure to include a diagram. Write a brief summary of why you know the theorem is true using your work in the previous two problems.

**THE MID-SEGMENT THEOREM**: The segment that connects the midpoints of any two sides of a triangle is half the length of and parallel to the third side of the triangle. In the figure at right, $\overline{EF} \parallel \overline{BC}$ and $EF = \frac{1}{2}BC$. $\overline{EF}$ is known as a **mid-segment.**

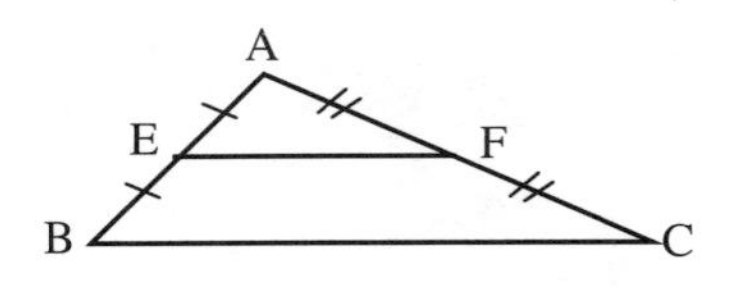


Segments and lines that are added to existing figures are called **AUXILIARY LINES**. $\overline{GE}$ and $\overline{GB}$ above are examples of auxiliary lines, as is the dashed line in the figures for the Isosceles Triangle Theorem (TK-26). Auxiliary lines are often used to create more triangles. You can draw several auxiliary lines in a figure. The key is to recognize the one(s) that will help you solve the problem or write your proof.

TK-92. In the three figures, D and E are the midpoints of the sides of the triangles. Find $x$ in inches and $y$ in degrees for problems (a), (b), and (c). For each problem, write down the tool kit ideas you used. Your newest addition should be very useful.

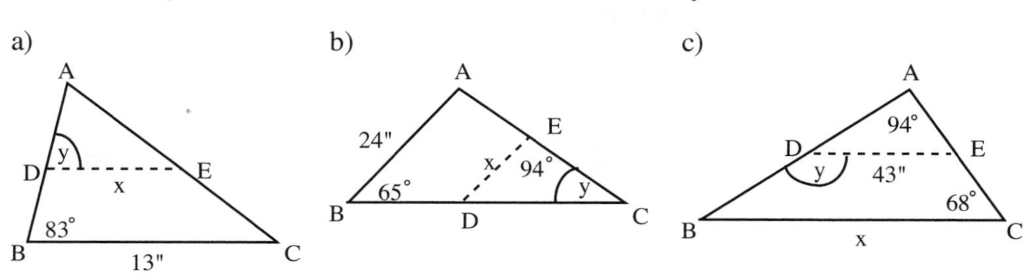


TK-93. Use the laws of exponents to simplify the following:

a) $(2x^3)^2$

b) $3^2 3^x$

c) $(3x^2)^3(2x^3y^4)^2$

d) $\frac{x^3}{x^{11}}$

e) $\left(\frac{x^2}{y^3z^5}\right)^3$

TK-94. Find the measure of each angle identified with a variable.

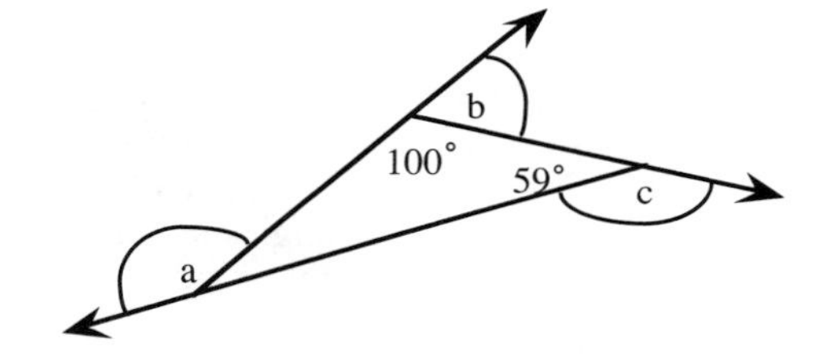


TK-95. Find the value of a and m∠x. Be sure to justify your solution with ideas from the tool kit.

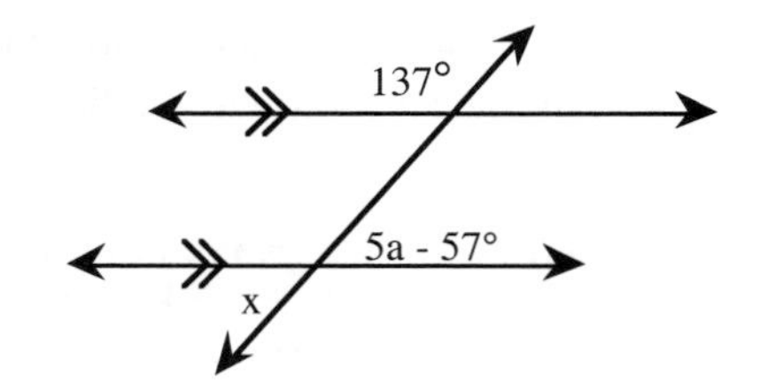


TK-96. Use the figure at right to complete parts (a) and (b) below.

a) If m∠1 is 37°, find the measures of the other three angles.

b) If m∠1 is y°, express the measures of the other three angles in terms of y.

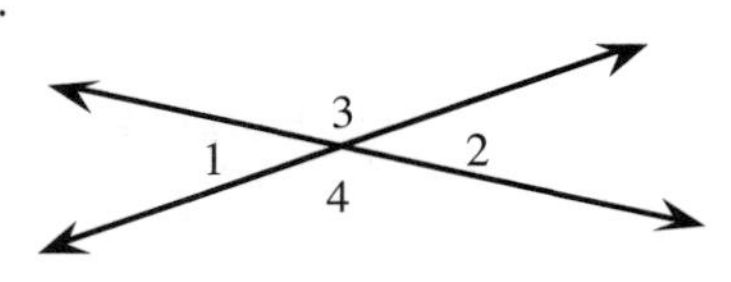

TK-97. Assume the figure at right has no hidden cubes.

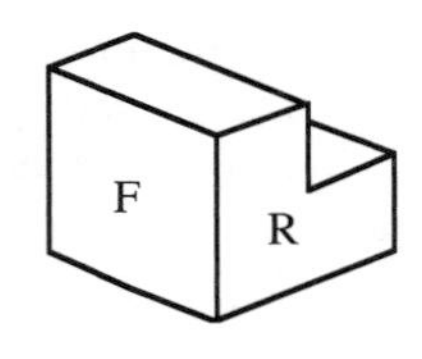


a) What will the solid look like if the figure at right were rotated so that the right side becomes the front? Draw the rotated solid on isometric grid paper.

b) What will the solid look like if the figure at right were rotated so that the back becomes the front? Draw the rotated solid on isometric grid paper.

TK-98. Find the total surface area and volume of the rectangular solid at right.

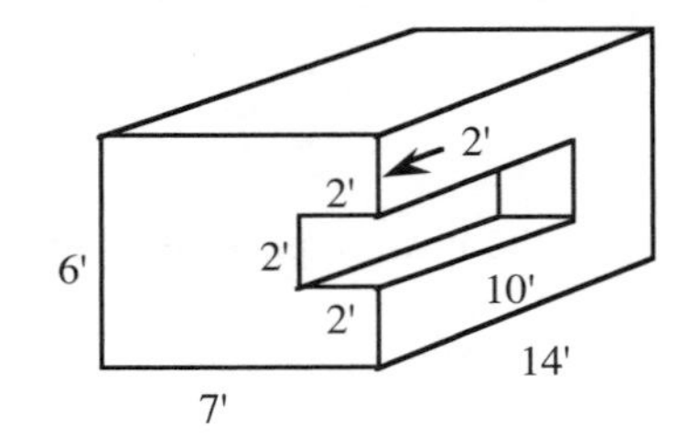


## PROOF BY CONTRADICTION

TK-99. Something is wrong with this diagram. Explain what fact(s) from your tool kit is (are) violated and how you know this to be true. You could start with 88° and find the measures of angles a, b, and c, or start with 67° and find the same angle measures.

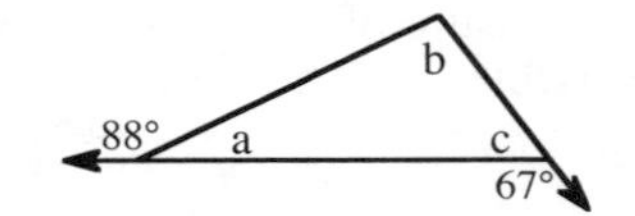

TK-100. Melvin looked at the figure at right and claimed that the center square in the first row is blue.

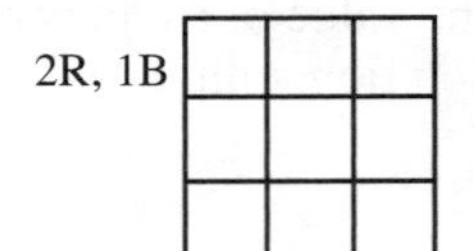

Help him out by convincing him that this cannot be true. However, your argument must begin as follows: "SUPPOSE the middle square in the first row **is** blue. Then...."

The two problems you have just done are examples of **INDIRECT PROOF** or **PROOF BY CONTRADICTION**. You used this kind of argument frequently in Unit 2 whenever you did Digit Place or Color Square problems. In general, this type of proof begins by assuming that something is true and then showing that such an assumption eventually leads to a contradiction of a known fact (usually something in your algebra or geometry tool kit).

TK-101. Given the equation $y = \frac{1}{2}x - 3$, Bert claims that the point (4, 1) lies on the line. Use an indirect argument to show him that it does not. Start with, "Assume that the point (4, 1) lies on the graph of $y = \frac{1}{2}x - 3$. Then... (Do not graph it!).

TK-102. Why is this figure impossible on a flat surface? List as many reasons (2 minimum) as possible. If you could bend the surface so that it was no longer flat, might it be possible to draw a triangle with two 90° angles? How about three 90° angles? Explain. Include a sketch of any figures you describe.

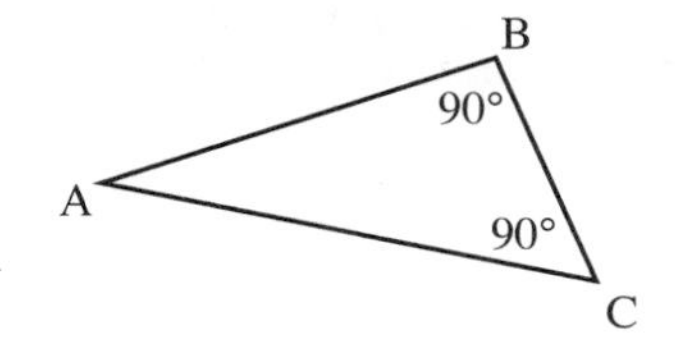

TK-103. A crime suspect claimed to be deaf. Detective Trick Tracy stood behind the suspect and shouted, "Hands up!" The suspect was startled and jumped. Use an indirect proof (proof by contradiction) to prove that the suspect is <u>not</u> deaf. Start by supposing that the suspect <u>is</u> deaf.

TK-104. Use indirect reasoning (indirect proof) to prove that a triangle has <u>at most</u> one obtuse angle. Start by supposing that it has two.

TK-105. Maryanne claims that A(-3, 2), B(2, -1), and C(5, - 4) lie on the same line. Assume that they do and show that this assumption contradicts a well-known fact about straight lines.

TK-106. Find each product or quotient.

a) $x^2(3x^4 - 5x^2 - 8)$

b) $(3x^4 - 8y)(x^5 + 2y^5)$

c) $\frac{54x^8y^3}{12x^4y^{14}}$

d) $(3x^2y^5)^2(2x^5y)^3$

TK-107. Examine the graph at right.

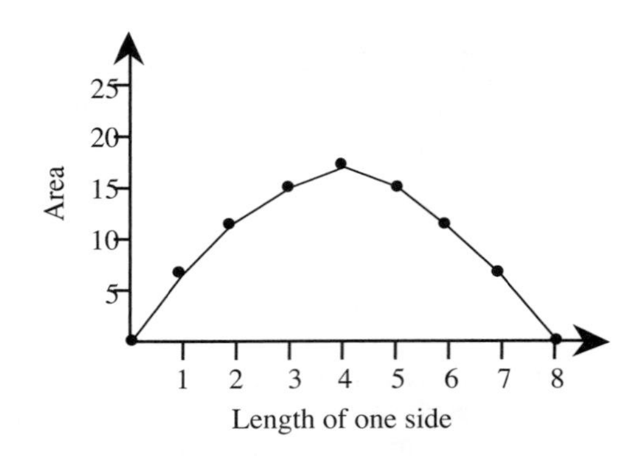


a) What information can you obtain from the graph?

b) When does this graph represent the shape with the greatest area?

c) At the point of greatest area, tell everything you can about the shape.

TK-108. Reuben is planning to do 50 graphing problems tonight for homework. His teacher has told him that the probability of him getting any one graphing problem correct is $\frac{3}{4}$. Based on his teacher's prediction, how many of the 50 graphing problems do you think Reuben will get correct? Justify your answer.

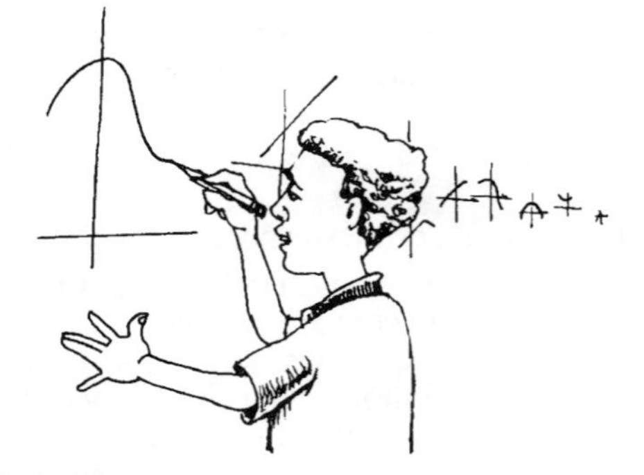

TK-109. On a sheet of graph paper, set up a pair of coordinate axes and plot A(2, 3), B(3, 7), C(8, 4), and D(4, 5).

a) Reflect quadrilateral ABCD across the y-axis. Label the new quadrilateral A'B'C'D'. List the coordinates of the vertices.

b) Rotate quadrilateral ABCD 270° counterclockwise about the origin. Tracing paper is helpful. Label the new quadrilateral A"B"C"D". List the coordinates of the vertices.

c) Translate ABCD so the point A maps to (-4, -6). Label the new quadrilateral A'''B'''C'''D'''. List the coordinates of the vertices.

d) Are all three quadrilaterals congruent? Justify your answer.

TK-110. Solve for x.

a) $-2x^2 + 6x + 7 = 0$

b) $x^2 - 10x + 25 = 0$

# UNIT AND TOOL KIT REVIEW

TK-111. Write the converse of each statement below. Decide whether the converse is true or false and justify your answer. REMEMBER, one counterexample proves that a statement is false.

a) If $x \leq 3$ then $x < 4$.

b) If two triangles are congruent, then they have the same area.

TK-112. Gabrielle has three segments of lengths 5, 12, and 13. She wonders if these three segments will form a right triangle. She decides that they do not. Start with this assumption and write a proof by contradiction to show that these three lengths <u>do</u> form a right triangle.

TK-113. Write the conditional form for the arrow diagram at right. Prove the conditional statement in the style of your choice. If you need help getting started, consider how an isosceles triangle compares to an equilateral triangle.

TK-114. In the figure at right, if $m\angle X = m\angle A$ and $\overline{XY} \cong \overline{AB}$,

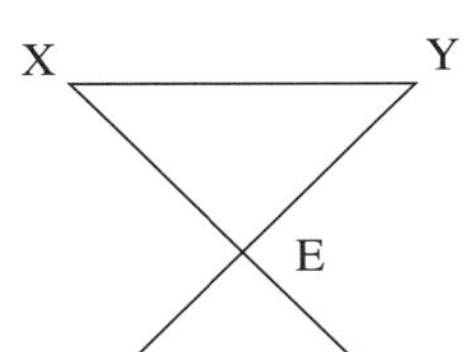

a) Is $m\angle Y = m\angle B$? Justify your answer.

b) Is $\Delta XEY$ isosceles? Justify your answer.

c) Is $\overline{YE} \cong \overline{BE}$ ? Justify your answer.

TK-115. On a sheet of graph paper, set up a pair of coordinate axes with $-5 \leq x \leq 20$ and $-5 \leq y \leq 15$. <u>Carefully</u> graph each of the following inequalities on this set of axes.

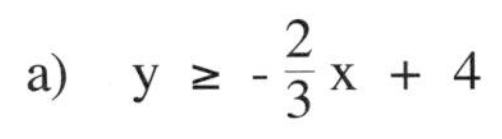

a) $y \geq -\frac{2}{3}x + 4$

b) $y \leq 2x + 12$

c) $y \geq \frac{1}{2}x - 3$

d) $y \leq -\frac{7}{16}x + 12$

e) Darken the common region and find its area by calculation. Show your subproblems.

TK-116. Find the measure of each numbered angle. Justify your answers.

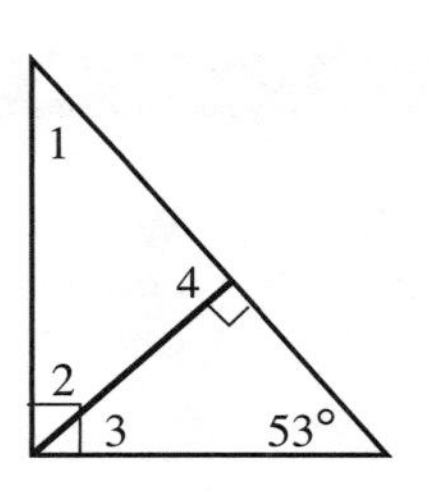


TK-117. Are the points (1, 2), (2, 4), and (4, 8) all on a straight line? Justify your answer completely.

TK-118. If Q is the midpoint of $\overline{PR}$, is Q the midpoint of $\overline{XY}$? Justify your answer completely.

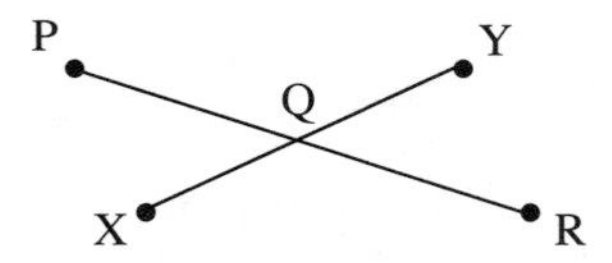

TK-119. Points D and E are midpoints of sides $\overline{AB}$ and $\overline{BC}$. $m\angle B = 38°$, $m\angle BDE = 115°$, and AC = 14 cm.

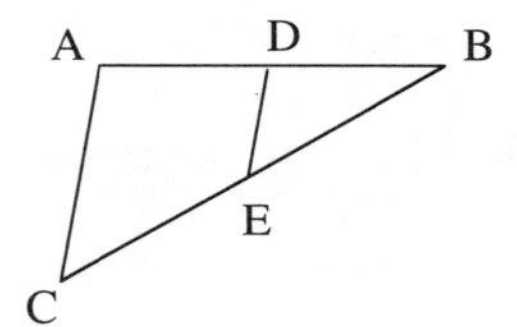


a) Find $m\angle C$.

b) Find DE.

TK-120. Simplify.

a) $x^7x^3x^2$

b) $\frac{x^{24}}{x^6}$

c) $(p^3)^2$

d) $\frac{y^4}{y^4}$

TK-121. Start with the equation $y = -2x + 5$. Find the equation of the line passing through (0, -2) and:

a) perpendicular to $y = -2x + 5$.

b) parallel to $y = -2x + 5$.

TK-122. TOOL KIT CLEAN-UP

Examine the elements in your tool kit from Units 0 - 6. You may need to create a new, fresh, consolidated tool kit representing the first semester of geometry and algebra review.

Include items you value the most in your tool kit. Spend time on this, as this is your tool kit for next semester!

a) Examine the list of tool kit entries from this unit. Check to be sure you have all of these entries (see below). Add any you are missing.

b) Identify which concepts in your complete tool kit (Units 0 - 6) you understand well.

c) Identify the concepts you still need to work on to master.

d) Choose entries to create a Unit 0 - 3 tool kit that is shorter, clear, and useful. You may want to consolidate or shorten some entries. Use this shorter version with your tool kit entries for Units 4 - 6.

e) Have you used your tool kit regularly? Why or why not?

- arrow diagrams
- proof
- Isosceles Triangle Theorem
- flowchart proofs
- two column proofs
- indirect proof
- converse
- counterexample
- mid-segment
- auxiliary lines
- substitution
- laws of exponents

TK-123. Your friend, Eniko, assures you that since SSS made pairs of triangles congruent, SSSS must make pairs of quadrilaterals congruent.

a) Suppose that you know that all four sides of a quadrilateral have the same length. Does this mean that it automatically must be a square? If you say yes, PROVE IT! If you say no, give a counterexample.

b) Do you think that Emiko's suggestion that SSSS is a congruence theorem for quadrilaterals is correct? Justify your belief in one or two sentences.

TK-124. When you tell Eniko she is wrong about SSSS, she replies that you misunderstood. Maybe those particular four measurements do not work, but some combination of <u>four consecutive</u> <u>parts</u>, sides and angles, will. She just cannot remember which four you need. Help her out by listing all the possible combinations of four sides and angles which could occur. Compare answers with the other people in your team to be sure you have all of the possibilities. You now have a list of conjectures which may or may not be true congruence theorems for quadrilaterals.

TK-125. Each team should select one member to post the team's list of conjectures for the class.

a) Are all of the lists the same?

b) Which of the combinations which look different are really the same? For example, SASS is really the same as SSAS going around the quadrilateral the other way.

TK-126. Each team should pick (or your teacher will assign) a different one of the conjectures from the teams and try to decide whether it is true. If your team believes that the conjecture is true, give a convincing argument. If not, give an example of two quadrilaterals which satisfy the properties, but are <u>no</u>t congruent. This will be your counterexample.

TK-127. The figure at right is a trapezoid based on two congruent right triangles. Find the area of the trapezoid two ways: first using the bases and height of the trapezoid with the trapezoid area formula, then by dissection using the three triangles. Use the two results to demonstrate that $a^2 + b^2 = c^2$. Explain why this result proves the Pythagorean Theorem.

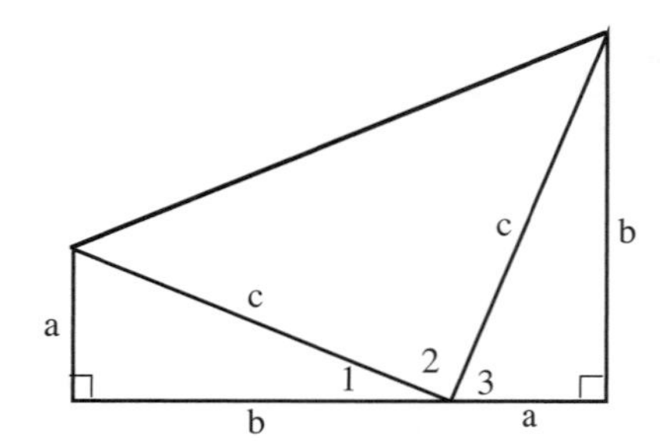

TK-128. The two squares at right have the lengths of a and b respectively.

The third figure is a polygon formed by placing the two squares side by side so that the bottom edge has length b + a.

The fourth figure shows how to cut the third figure into three pieces, two of which are congruent right triangles.

Draw larger versions of these figures on your paper or use the resource page provided by your teacher. Cut figure 4 along the dashed line and reassemble the three pieces to form a square. Then explain how your result proves the Pythagorean Theorem.

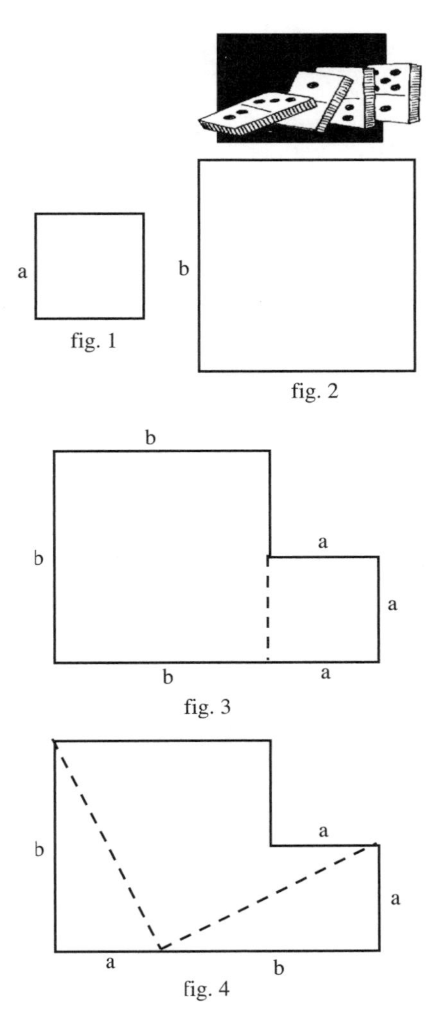

UNIT

# The Height of Red Hill

TRIGONOMETRY

# Unit 7 Objectives
## *The Height of Red Hill:* TRIGONOMETRY

This unit introduces the basic right triangle trigonometry ratios for use in area problems and applications. You will also build an angle measuring device to use with the trigonometry to measure physical items around your campus. Near the end of the unit you will learn how to use trigonometry in any kind of triangle. You will also discover the relationship between slope and one of the trigonometric ratios.

The algebra focus in this unit teaches you how to eliminate fractions from equations. You will also review how to simplify radicals.

In this unit you will have the opportunity to:

- learn how to find the lengths of sides and the measures of angles in right triangles when you only know two of the six measurements.
- solve interesting application problems.
- use radicals and the Pythagorean Theorem to write constant ratios between the sides of special right triangles.
- eliminate fractions in equations prior to solving them.
- review the simplification process for radicals.

*Read the following problem. Over the next several days you will learn what is needed to solve it. Do not attempt to solve it now.*

T-0. From the parking lot at the Red Hill Shopping Center, the angle to the top of the hill is about 25°. From the base of the hill you can also sight the top but at an angle of 55°. The horizontal distance between sightings is 740 feet. How high is Red Hill?

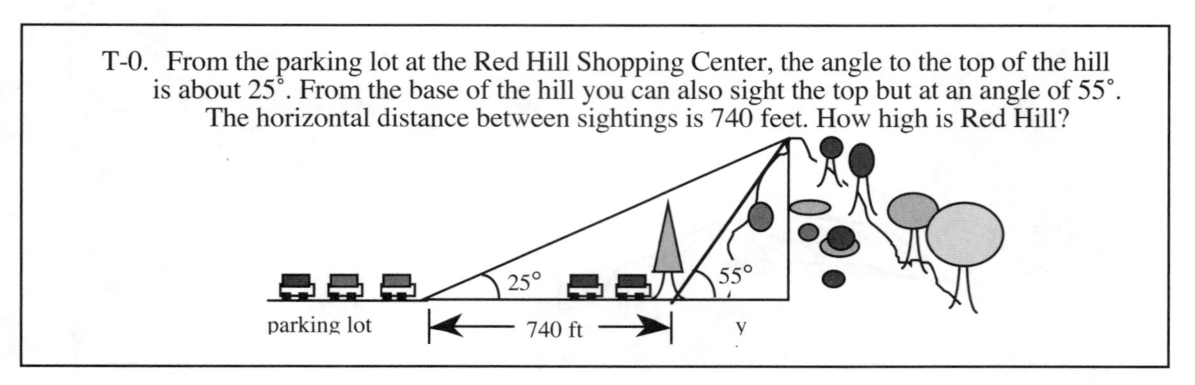


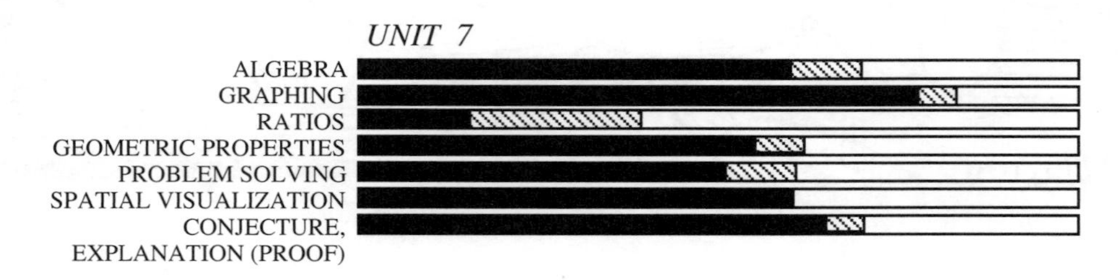

# *Unit 7*

## *The Height of Red Hill:* TRIGONOMETRY

## RIGHT TRIANGLE TRIGONOMETRY

T-1. **DIRECTIONS:** So far this year we have looked at several types of functions, among them linear ($y = mx + b$), quadratic ($y = ax^2 + bx + c$), and cubic functions ($y = x^3$). With these functions, inputs and corresponding outputs have both been numbers.

Today we will look at three more functions: the tangent, sine (pronounced "sign"), and cosine (pronounced "co-sign) functions. These functions take angle measures as inputs and have decimal answers as outputs. We will be investigating these functions and their properties in this unit. They are very useful for solving a wide variety of problems.

Complete the table on the resource page that your teacher gives you. Use your calculator to fill in the missing values in the table. Each value should be to four decimal places. Your purpose in completing the table is to learn how to use your calculator to input an angle measure and find its decimal value. You will learn what these three functions mean over the next few days. As you work, record your results and be looking for patterns and relationships, both down columns and across rows.

| Angle | Sine | Cosine | Tangent |
|---|---|---|---|
| 0° | | | |
| 1° | | | |
| 5° | | | |
| 10° | | | |
| 15° | | | |
| 20° | | | |
| . | | | |
| . | | | |
| . | | | |
| 85° | | | |
| 89° | | | |

a) When you have completed the table, make sure that each person has a good copy to keep.

b) What relationships do you see between the three functions? Describe at least two.

c) In the sine column, what happens to the numbers as the angle measure increases?

d) Similarly, in the cosine column, what happens to the numbers?

e) What do you notice about the range of values for each function: sine, cosine, and tangent?

T-2. You stopped your table at 89°. Use a calculator to find the value of each function at 89.5°. What do you think will be the value for each at 90°? Check it. Explain what your answer for the tangent function means. Add these values to your table.

T-3. Use your calculator to find sin 32°. Which of the following values do you believe is the same as sin 32°? Use your table and its patterns to make an educated guess. Check your answer. How are these two angles related?

a) cos 32°   b) tan 58°   c) cos 58°   d) tan 32°

T-4. From your table, estimate each of the following values, then use your calculator to get more accurate results. Write down both your estimate and your calculator result.

a) tan 4°   sin 4°   cos 4°

b) tan 17°   sin 17°   cos 17°

c) tan 49°   sin 49°   cos 49°

T-5. Read the following information about eliminating fractions in equations, add it to your tool kit, then do the practice problems at the bottom of the box.

**DIRECTIONS:** Take a survey of the people in your team: how many of them enjoy doing problems involving fractions? Often people do not like working with fractions either because they are not used to them or because they are afraid of making mistakes when using them. Here is the perfect solution, called **FRACTION BUSTERS**. Suppose you are given an equation that has a fraction in it like $\frac{2}{3}x + 6 = 10$. To rewrite this equation without any fractions, we ask ourselves, "How can we eliminate the three?" (Although the fraction is $\frac{2}{3}$, the part we would like to eliminate is the denominator.) Fraction Busting uses the Multiplication Property of Equality (that is, the fact that multiplying both sides of an equation by the same number changes its <u>form</u> but not its balance (equality)) to rearrange the equation. Start by multiplying <u>everything</u> on <u>both</u> sides of the equation by three.

$\mathbf{3}(\frac{2}{3}x) + \mathbf{3}(6) = \mathbf{3}(10)$   multiply all parts by 3, then divide common factors to reduce

$2x + 18 = 30$   solve as usual

Notice that we now have an equation with no fractions.

Solve the following equations using the Fraction Busters method.

a) $\frac{3}{4}x + 2 = \frac{29}{4}$   b) $\frac{4}{5}x - 4 = 6$   c) $3x + \frac{2}{7} = 5$

T-6. On a sheet of graph paper, scale your axes five squares per unit for $-4 \le x \le 4$ and $-4 \le y \le 4$. Copy and complete the table below for $y = \frac{1}{x}$. Note that this function is known as a "hyperbolic function." You will study this family of curves in Algebra 2.

| x | -4 | -3 | -2 | -1 | -0.5 | -0.3 | 0 | 0.3 | 0.5 | 1 | 3 | 4 |
|---|---|---|---|---|---|---|---|---|---|---|---|---|
| y | | -0.33 | | | | | | 3.33 | | | | |

a) What does your result for $x = 0$ tell you about this value?

b) Describe the curve you drew.

c) Will this graph ever have any points in the second or fourth quadrants? Explain.

T-7. Draw an arrow diagram for the statement below and write a proof in the style of your choice.

If $\Delta ABC$ is isosceles with $AB = BC$, then the height from the vertex angle ($\angle B$) bisects the side opposite the vertex angle ($\overline{AC}$).

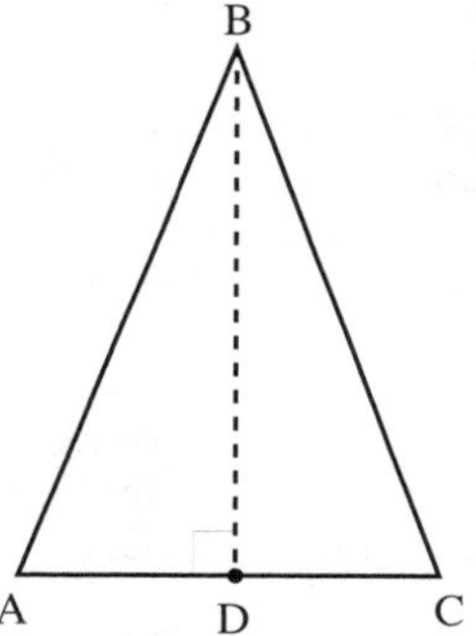

T-8. Solve for x, show all subproblems, and check your solutions.

a) $3(x - 5) > 16 - 4x$

b) $x(2x - 5)(5x + 2) = 0$

c) $-3x^2 - 4x + 5 = 0$

d) Find x and y if: $2x - y = 7$
$3x + y = 13$

T-9. For each of the following pairs of triangles, decide if the two triangles are congruent. Give a reason justifying your answer. Remember drawings may not be accurate. Use only the measurements given.

a)

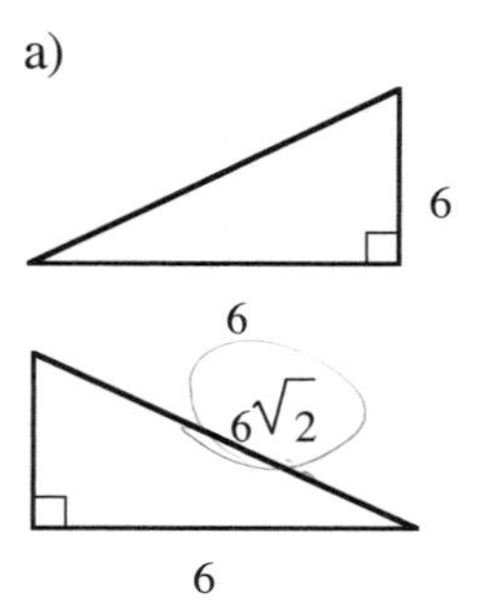


b)

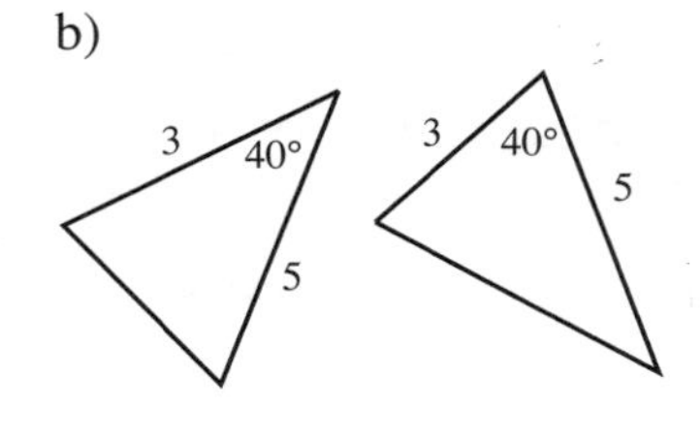


c)

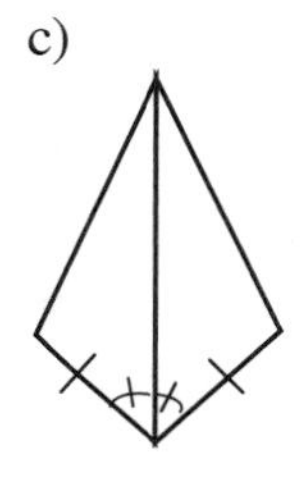

T-10. The figure at right is called a **truncated pyramid**. Both bases are parallel and the three lateral faces are trapezoids. The height of each face is 9 cm. Find the total surface area of the figure. Show and number all subproblems.

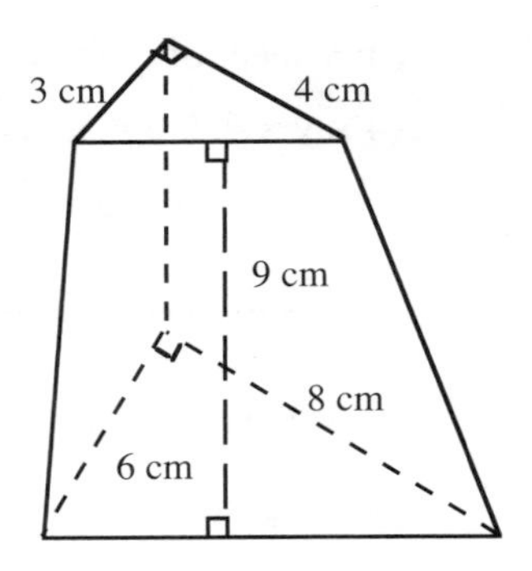


## THE TANGENT RATIO

T-11. A support cable is attached to the top of a 120 foot pole. The cable will be pulled tight and fastened to the ground forming an angle of 60° with the cable and the flat ground. How far from the pole should the cable be attached?

a) Now that you have read the problem, draw a picture of the situation labeling what you KNOW and what you want to FIND.

b) To solve this problem, we need more information. Explain in a sentence why the problem above <u>cannot</u> be solved using just the Pythagorean Theorem.

T-12. In the figure in the box below, we know that the hypotenuse is the longest side. If you were standing at ∠A, the leg <u>closest</u> to you would be $\overline{AC}$ and the leg <u>farthest</u> away would be $\overline{BC}$. Read the definition, add it to your tool kit, then go on to problem T-13.

The leg <u>nearest</u> ∠A is called the **ADJACENT LEG**. The leg <u>farthest</u> from ∠A is called the **OPPOSITE LEG**. Remember that the **hypotenuse** is always opposite the right angle and is always the longest side.

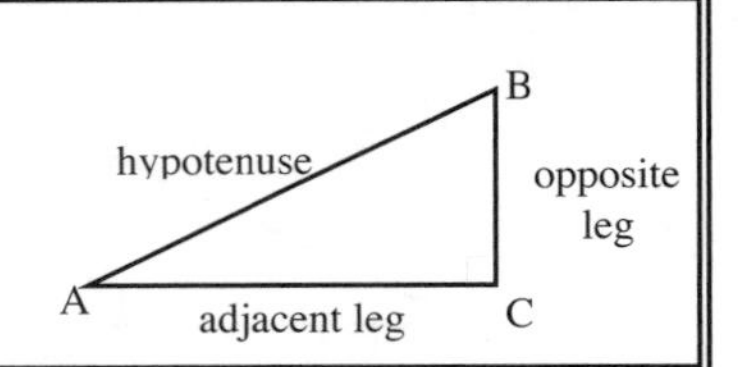

T-13. In the triangle in problem T-12 we used $\angle A$ as our reference point to label the opposite and adjacent legs. Suppose $\angle B$ is our reference point. Copy the triangle on your paper and label the hypotenuse, adjacent leg, and opposite leg with respect to $\angle B$, that is, as if you were standing at $\angle B$.

T-14. Recall that a **ratio** is a comparison of two numbers written as a fraction. There are six different ratios of pairs of sides in relation to $\angle A$ in the triangle in problem T-12. List them all, then read the box below and add the definition to your tool kit.

EXAMPLES: one of them is $\frac{\text{opposite leg}}{\text{adjacent leg}}$, another is $\frac{\text{opposite leg}}{\text{hypotenuse}}$.

Certain **ratios** of sides of right triangles are given specific names. The first ratio of the last example, $\frac{\text{opposite leg}}{\text{adjacent leg}}$, is known as the **TANGENT RATIO** of an angle (abbreviated **tan**). The relationship written as an equation is:

$$\tan A = \frac{\text{opposite leg}}{\text{adjacent leg}}$$

T-15. In the picture below, the second triangle is just an enlargement of the first. (The sides were multiplied by two.)

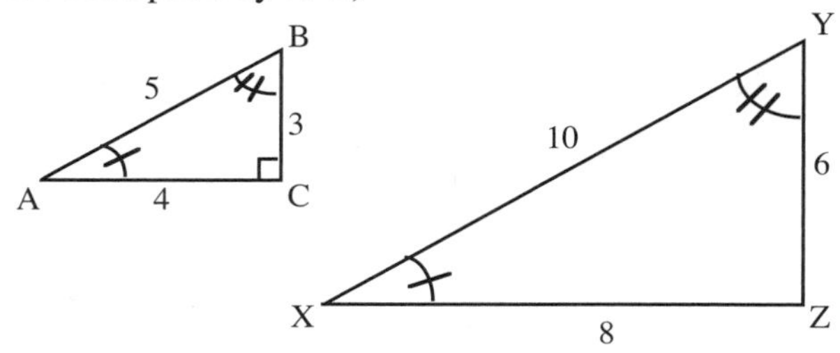


From the first triangle, $\tan A = \frac{\text{opposite leg}}{\text{adjacent leg}} = \frac{3}{4}$

a) Using the second triangle, write the ratio for tan X. Reduce it. Compare the tangent values for angles A and X. What do you notice?

b) What is tan B? Be careful! You are using a different reference angle. Rotate the page if you need to see the ratio clearly. Compare tan B to tan Y and write your observation in a sentence.

c) What would you expect the tangent of $\angle X$ to be if the length of the sides of $\Delta XYZ$ were triple that of $\Delta ABC$? Why?

T-16. Use your calculator to find tan 59°. Use the resource page provided by your teacher and follow the instructions below to help you find out why that value appeared.

a) Each of you will work with a different size right triangle with a 59° angle at A. What are the measures of the other two angles?

b) Carefully measure the lengths of the two legs in your triangle in millimeters and write each length along the side. Measure accurately! What are the lengths?

c) Using the lengths of the two legs you found in part (b), write the ratio of the side length opposite ∠A to the side length adjacent to ∠A. Use your calculator to find the value of the ratio in decimal form. What did you get?

d) Explain your results in light of the fact that all of the triangles were a different size.

T-17. Write a conjecture about tangents of angles that are equal based on your results in the previous two problems.

T-18.

**EXAMPLE**

Consider the triangle at right. Suppose we want to find the length of the opposite leg when we know the length of the adjacent leg is 20. The tangent value of a specific angle (in this case, 30°) is always constant (that is, it does not change if the side lengths increase or decrease). Thus, regardless of the lengths of the sides of the triangle, the value "tan 30°" is the same and **simply represents a number** in the equation below.

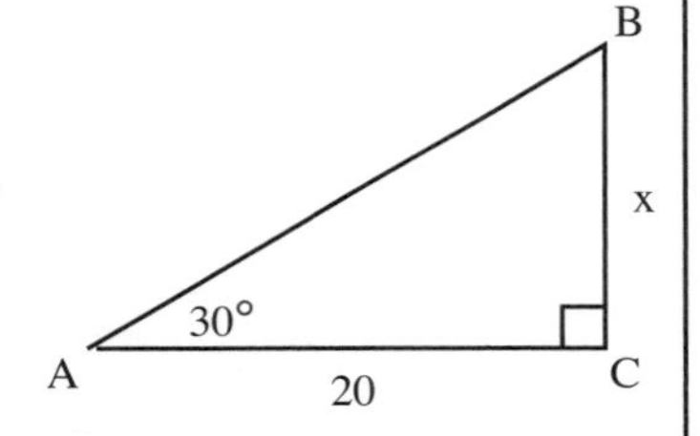


a) $\text{Tan A} = \frac{\text{opposite leg}}{\text{adjacent leg}}$. Complete the equation using this definition with the x and 20 from the figure above: tan 30° = ?

b) Solve the equation for x. Do not find the value of tan 30° until the end of the problem. Use Fraction Busters as necessary.

c) Suppose you use ∠B to solve for x. Complete the following equation and solve for x. Your result should be the same as in part (b). tan 60° = ?

T-19. Find the length of the unknown side in each of the following right triangles. Be sure to write down the equation you will solve in each case.

a) 44°, 28, x

b) y, 32°, 150

c) 47, 32°, x

T-20. Use the Fraction Busters method to solve for x.

a) $\frac{3}{4}x - 2 = 8$

b) $\frac{12}{x} + 2 = 7$

c) $\frac{18}{x} = 3$

d) $\frac{9}{2x} + 2 = \frac{8}{x}$

T-21. Draw an arrow diagram for the statement below, then write a proof in the style of your choice.

If $\Delta ABC$ is isosceles with $AB = AC$, and $\overline{AD}$ bisects $\angle BAC$, then $\overline{AD}$ also bisects $\overline{BC}$.

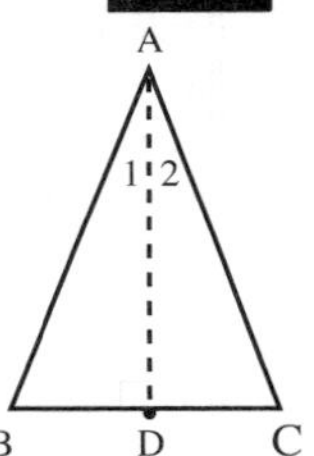


T-22. Find the tangent ratio for $\angle A$ in each triangle shown below. Convert the ratio to a decimal and estimate the measure of the angle by using your table of tangent values from problem T-1.

a)

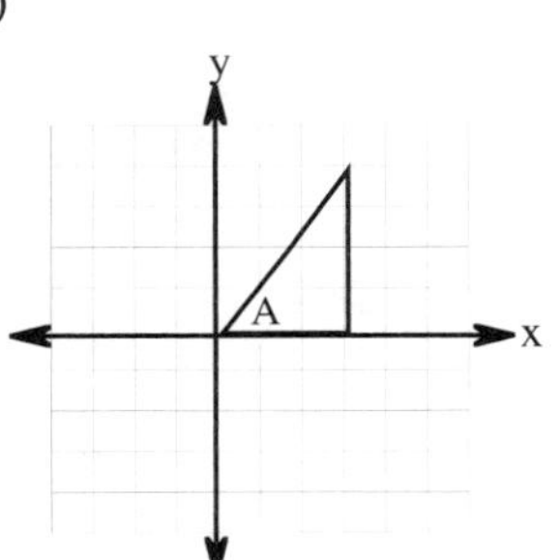


b)

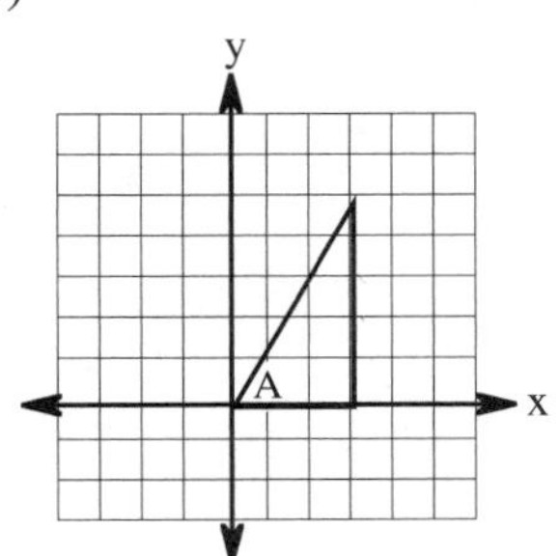


c)

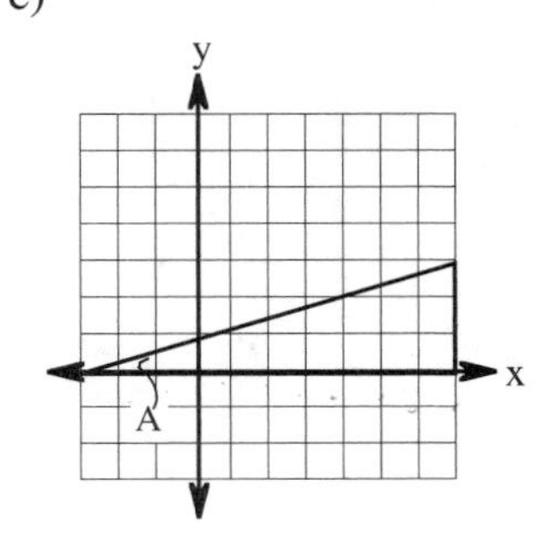


T-23. Use graph paper to <u>carefully</u> graph the two parabolic inequalities below, then estimate the area of their intersection.

$y \geq x^2 - 3$ $\qquad$ $y \leq 9 + 2x - x^2$

T-24. Use substitution for systems of equations to algebraically find the exact points of intersection of the two curves in the preceding problem.

T-25. Just a reminder to be sure that you are keeping a tool kit AND that you are keeping it up to date. You should also have your tool kit from the first six units handy, especially for the homework problems.

## THE SINE AND COSINE RATIOS

T-26. Read the next definition in the box below and add it to your tool kit. Then complete parts (a), (b), and (c) that follow the example.

The next special ratio of sides is called the **SINE RATIO** (abbreviated **sin** and pronounced "sign"). In this right triangle, the sine of ∠A is:

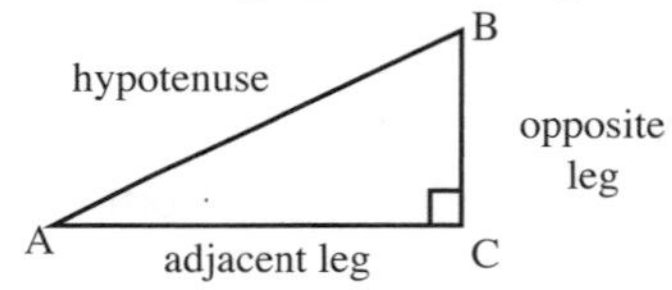


$$\sin A = \frac{\text{opposite leg}}{\text{hypotenuse}}$$

We will use the sine ratio in the same way we used the tangent ratio to find lengths of sides of right triangles. In the first example below we know the angle measure is 33° and the length of the hypotenuse is 20 cm. We want to find the length of the opposite leg.

$$\sin 33^\circ = \frac{x}{20}$$

$$20(\sin 33^\circ) = x$$

$$20(0.5446) = x$$

$$10.89 \text{ cm} \approx x$$

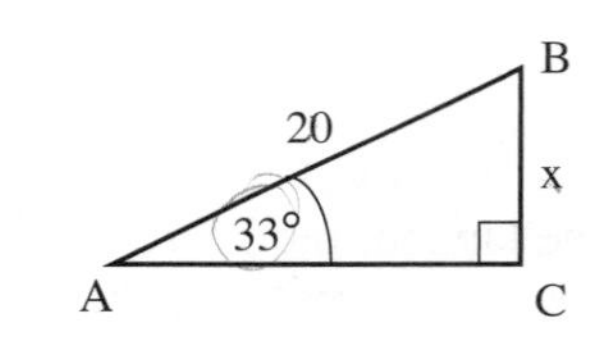


a) Explain to your team what happened in each step in the example above.

b) Suppose the hypotenuse was unknown:

$$\sin 33^\circ = \frac{15}{x}$$

$$x(\sin 33^\circ) = 15$$

$$x = \frac{15}{\sin 33^\circ}$$

$$x \approx 27.54$$

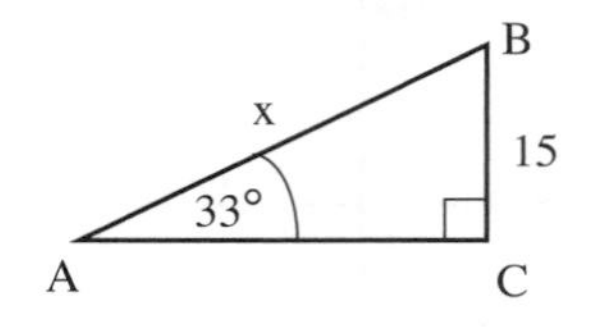


c) Explain to the members of your team what happened in each step in the example above.

T-27. Use the sine ratio to find the length of the side XY.

T-28. Read the next definition in the box below and add it to your tool kit. Then complete parts (a) and (b) that follow the example.

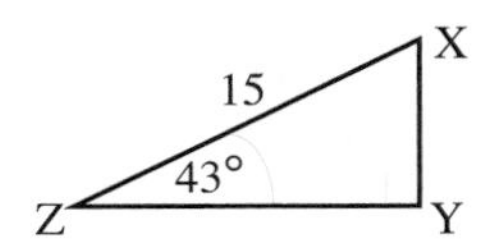

The third trigonometric ratio we will use is called the **COSINE RATIO** (abbreviated **cos** and pronounced "co-sign").

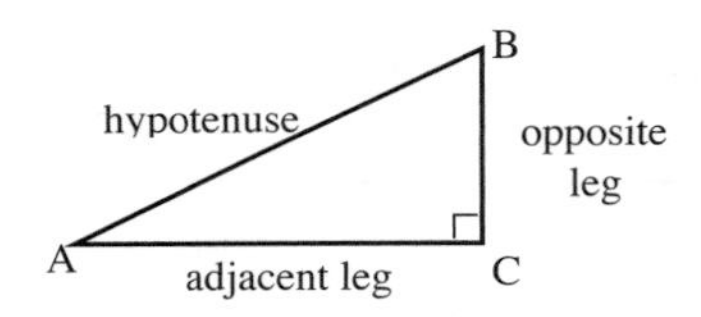


$$\cos A = \frac{\text{adjacent leg}}{\text{hypotenuse}}$$

In the example below we know the angle and the adjacent leg. We want to find the length of the hypotenuse.

$$\cos 30^\circ = \frac{20}{x} \quad \Rightarrow \quad (x)(\cos 30^\circ) \approx \frac{20}{x}(x)$$

$$\Rightarrow \; x(\cos 30^\circ) = 20 \quad \Rightarrow x = \frac{20}{\cos 30^\circ}$$

$$\Rightarrow x \approx \frac{20}{0.8660} \quad \Rightarrow \quad x \approx 23.09 \text{ cm}$$

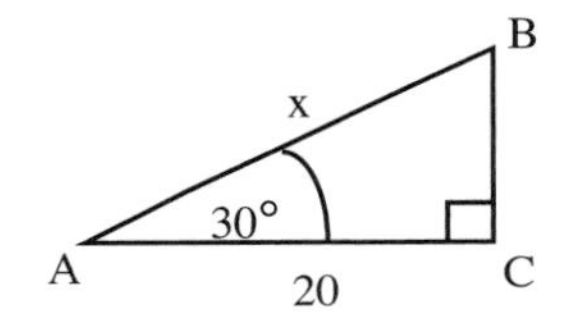


a) Explain what happened in each step in the example above.

b) Draw a copy of the figure on your paper and label side $\overline{AC}$ "x" and side $\overline{AB}$ "42." Show a step by step example that solves for x.

T-29. Use the cosine ratio to find the length of the adjacent leg. Show the steps you used.

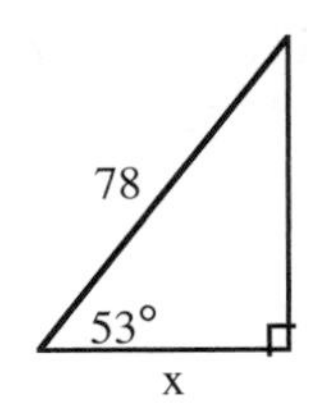

T-30. You now have three **TRIGONOMETRIC RATIOS** you can use in relation to $\angle A$ in any right triangle ABC:

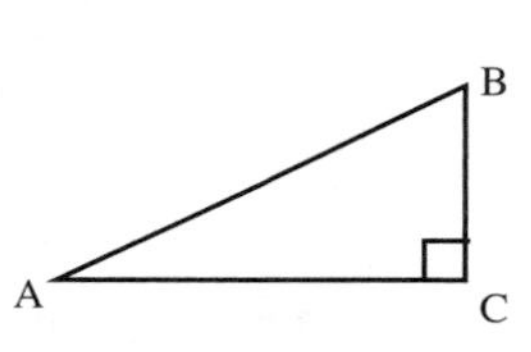


$$\tan A = \frac{\text{opposite leg}}{\text{adjacent leg}} = \frac{BC}{AC}$$

$$\sin A = \frac{\text{opposite leg}}{\text{hypotenuse}} = \frac{BC}{AB}$$

$$\cos A = \frac{\text{adjacent leg}}{\text{hypotenuse}} = \frac{AC}{AB}$$

The other acute angle in the right triangle above ($\angle B$) can also be used with the three trig ratios.

a) Write the segment ratios for tan B, sin B and cos B.

b) Compare the ratios for sin A and cos B as well as those for sin B and cos A. State your observations in a sentence or two.

T-31. For each triangle below, state which trigonometric ratio you would use to solve the problem. Then write the equation **BUT DO NOT SOLVE IT!**

a)

b)

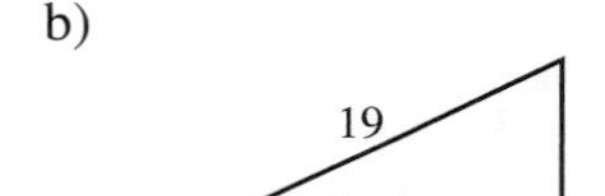


c)

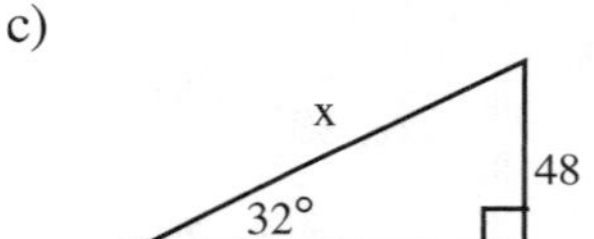


d)

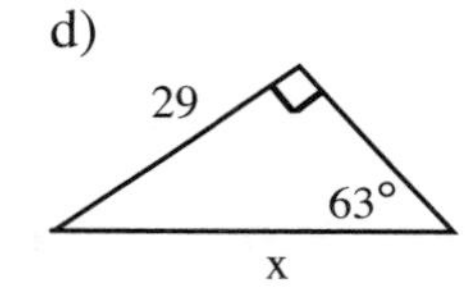


e)

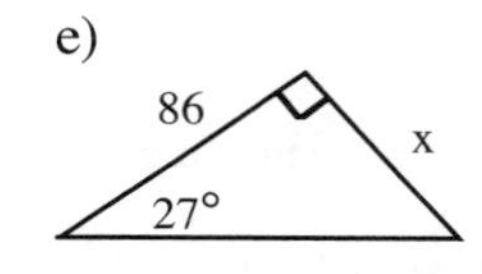


f)

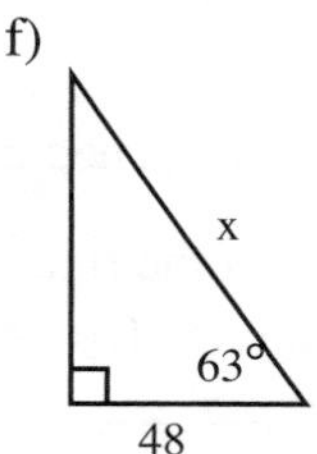


g)

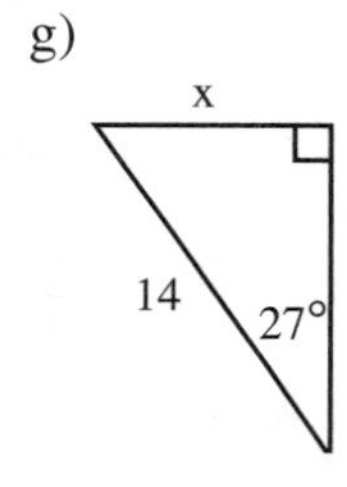


T-32. Use the appropriate trigonometric ratio to solve for the indicated lengths. State the ratio you use.

a)

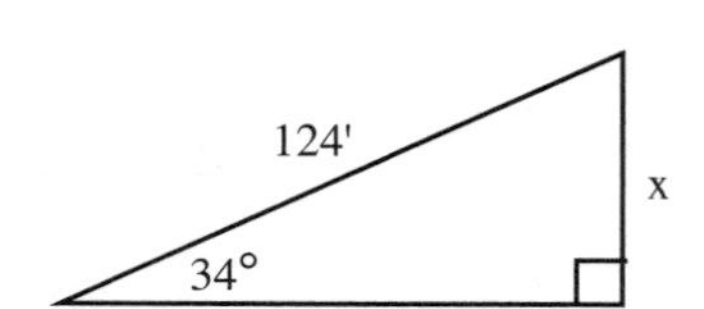


b)

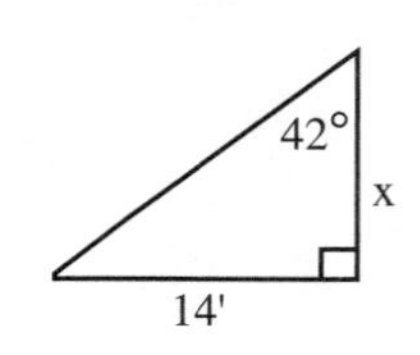


T-33. If the hypotenuse of any right triangle ABC has a length of 1 cm, can the length of the leg opposite either acute angle ever be more than 1 cm? Explain and draw a picture to support your answer.

T-34.

## MORE FRACTION BUSTERS

**DIRECTIONS:** Suppose you wanted to solve this equation which has two different denominators:

$$\frac{3}{4}x + \frac{1}{6} = 3.$$

We can still use the idea of **fraction busters** to solve this equation. We want to multiply everything in the equation by some number that will cancel <u>all</u> the denominators. (Note: "cancel" means that the number you choose can be divided by each denominator so that the result for each denominator's value is 1.) In this particular equation, we need to cancel both a 4 and a 6. What number will cancel both of these numbers? Certainly their product, 24, will cancel both of them, but can you think of another <u>smaller</u> number that will work? This number is called the **least common multiple**. For 4 and 6, the least common multiple is 12.

| | |
|---|---|
| $12\left(\frac{3}{4}x + \frac{1}{6}\right) = 12(3)$ | multiply by 12 |
| $12\left(\frac{3}{4}x\right) + 12\left(\frac{1}{6}\right) = 36$ | distribute the 12 |
| ${}^{3}\cancel{12}\left(\frac{3}{\cancel{4}}x\right) + {}^{2}\cancel{12}\left(\frac{1}{\cancel{6}}\right) = 36$ | simplify (cancel/divide) |
| $9x + 2 = 36$ | multiply what is left and solve |

If you had multiplied by 24 you would have the equation $18x + 4 = 72$, which is also easy to solve. The important thing is to find <u>some</u> number that will eliminate the fractions. If it is the smallest number, you will save some time at the end, but do not worry too much about it.

In either case, the process results in NO MORE FRACTIONS!

Solve the following equations using the Fraction Busters method.

a) $\frac{2}{3}x = \frac{1}{2} + 4$

b) $\frac{5}{6} - 2x = \frac{x}{3}$

**T-35.** Use the appropriate trig ratio to write an equation and solve for the indicated lengths (one at a time, of course!). You can use Fraction Busters to help solve the equations. Remember that "sin 25°" and "tan 25°" represent constant ratios. As such, they are simply another way to write a number and can be manipulated in equations just like any number. Use your calculator when you need to know the decimal approximation for trig values, usually at the END of the problem.

a)

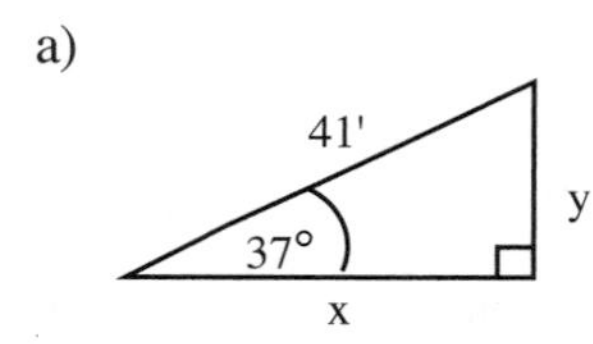


b)

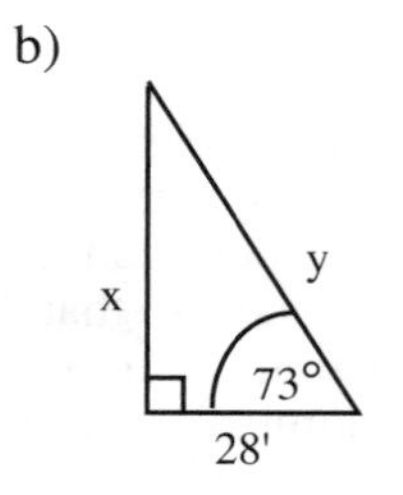


c)

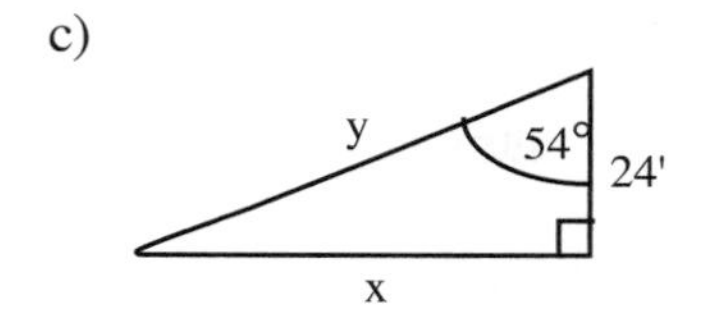


d)

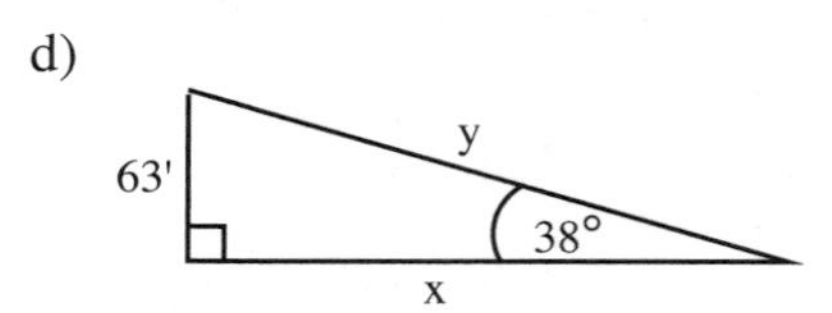


**T-36.** If ΔABC is equilateral, prove that the height bisects the opposite side (the triangle's base).

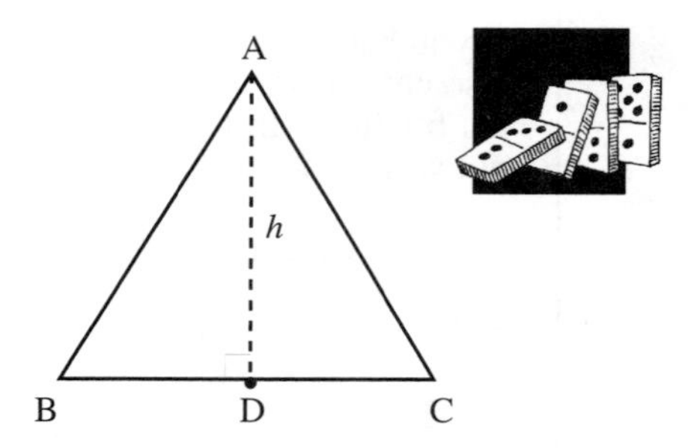


**T-37.** Joe wonders if ΔABC at right is isosceles. Suppose he presented this proof to your class:

Draw auxiliary line segment $\overline{AD}$ so that it bisects $\overline{BC}$ and is perpendicular to $\overline{BC}$. Then ΔADB ≅ ΔADC by SAS and AB = AC by ≅ Δs → ≅ parts.

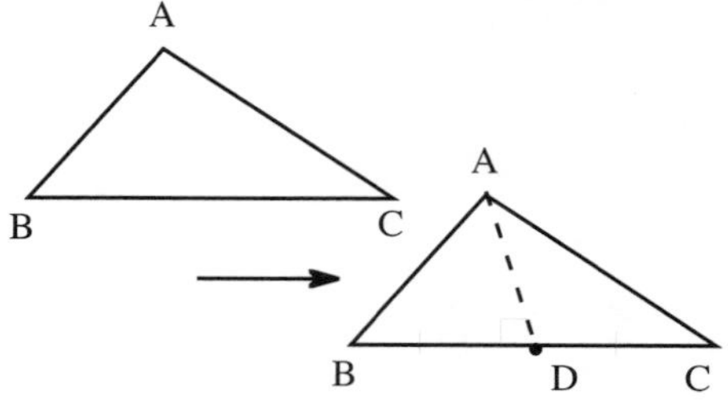


Write a brief critique of his effort. Explain why he is correct or point out the flaw(s) in his logic and /or process. If you think he is wrong, a counterexample will support your argument.

## USING TRIGONOMETRY TO SOLVE PROBLEMS

T-38. DIRECTIONS: We will be looking at some real world situations in which the trigonometric ratios allow us to get more information, or to solve the problem completely. In each case, the actual computation is not very difficult with a calculator, so put your time into making sure you are drawing an accurate diagram, using the correct trig ratio, and writing an accurate equation. Read the example below, then go on to the next problem.

**EXAMPLE**

One criteria used in judging kite-flying competitions is the size of the angle formed by the kite string and the ground. This angle can be used to find the height of the kite. Suppose the length of the string is 600 feet and the angle at which the kite is flying measures 40°. Calculate the height, h, of the kite.

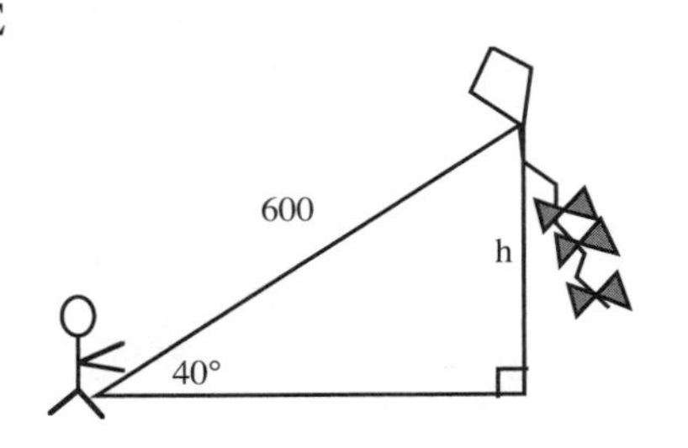


If the judges wish to figure out the height or altitude of the kite, they can do so using the sine ratio by writing:

$$\sin 40^\circ = \frac{h}{600} \Rightarrow 600(\sin 40^\circ) = h \Rightarrow 600(0.6428) = h \Rightarrow 385.7' \approx h.$$

***REMEMBER:*** ***THE FIRST STEP IN EACH PROBLEM IS TO DRAW AN ACCURATE DIAGRAM.***

T-39. You are flying a kite in a competition and the length of the string is 725 feet and the angle at which the kite is flying measures 35° with the ground.

a) How high is your kite flying?

b) Suppose your friend is standing directly under the kite. How far is your friend from you?

T-40. Now you have the tools to solve the problem you read earlier in the unit.

A support cable is attached to a 120 foot pole. The cable will be fastened to the ground so that an angle of 60° is formed with the cable and the ground. Draw a diagram and solve the following:

a) How far from the pole should the cable be attached? Show your work.

b) How long must the cable be?

**T-41.** The trig table resource page you used at the beginning of the unit told you the sine, cosine, and tangent values of various angles between 0° and 89°. You can also use the table to answer this question:

What is the angle whose cosine value is 0.5736?

A quick check of your table would reveal that the angle is 55°.

Most of the time you will find the decimal trig values by writing the ratio of two sides of a right triangle, then converting the fraction to a decimal. Suppose you were doing a problem and were asked, "Find the angle a handicapped access ramp to Mary's front porch makes with the flat ground under it if her porch is four feet high and the ramp is 18' long." You could write the following equation:

$$\sin A = \frac{\text{porch height}}{\text{ramp length}} = \frac{4}{18} \approx 0.2222$$

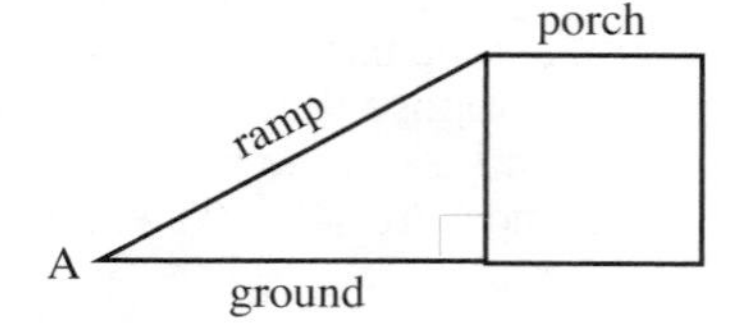


Unfortunately, this value is not listed in your table. You could estimate m∠A from the table--perhaps 14°.

There is, of course, an efficient way to find angle measures using your calculator. In problems T-16 and T-17 you found the size of a triangle with a 59° angle did not affect the value of the tangent ratio. Consequently, ANY time you need the value of "tan 59°" you always get 1.664279482. Thus, if you find that, in a right triangle, the ratio of the $\frac{\text{opposite}}{\text{adjacent}}$ sides is 1.664279482, you would expect to find that the angle in the triangle is 59°.

The preceding paragraph illustrates that you can "undo" the trig ratios just like you can undo addition by subtraction, multiplication by division, and squaring by square root. Finding the measure of an angle when you know the trig ratio is the **inverse** of finding the trig ratio from a known angle.

When you use a calculator to do this, you must press the "inv" or "2nd" key first, then the sin, cos, or tan key. Exactly how you do this on your calculator depends on the brand and model you have. Practice finding angle measures from known trig ratios below. Remember, first change the ratio (fraction) into a decimal.

a) $\sin A = \frac{13}{21}$

b) $\cos A = \frac{59}{71}$

c) $\tan A = \frac{51}{37}$

d) $\tan A = \frac{8}{13}$

e) What angle does the ramp make with the porch in the example above?

T-42. Read the information in the box below about simplifying radicals and add it to your tool kit, then practice the procedure with the problems at the bottom of the box.

Sometimes it is more convenient to leave square roots in radical form than to find their calculator approximations (decimal values). The idea is to look for perfect squares (i.e., 4, 9, 16, 25, 36, 49, ...) within the number (that is, as **factors** of the number) that are inside the radical sign (**radicand**) and take out the square roots of any perfect squares. For example:

$$\sqrt{9} = 3$$

$$\sqrt{18} = \sqrt{9 \bullet 2} = \sqrt{9} \bullet \sqrt{2} = 3\sqrt{2}$$

$$\sqrt{12} = \sqrt{4 \bullet 3} = \sqrt{4} \bullet \sqrt{3} = 2\sqrt{3}$$

$$\sqrt{80} = \sqrt{4 \bullet 20} = \sqrt{4 \bullet 4 \bullet 5} = \sqrt{4} \bullet \sqrt{4} \bullet \sqrt{5} = 2 \bullet 2 \bullet \sqrt{5} = 4\sqrt{5}$$

or $\sqrt{16} \bullet \sqrt{5} = 4\sqrt{5}$

When we cannot find any more perfect square factors inside the radical sign, The product of the integer (or fraction) and remaining radical is said to be in **SIMPLE RADICAL FORM**.

Simplify the following radicals WITHOUT using your calculator or decimal approximations. Express your answers in simple radical form.

a) $\sqrt{25}$

b) $\sqrt{49}$

c) $\sqrt{32}$

d) $\sqrt{75}$

T-43. Match each of the radicals on the left with its simplified radical form on the right. Use the technique of finding perfect square factors in the choices in the left column. DO NOT use your calculator.

| | |
|---|---|
| a) $\sqrt{12}$ | 1) $2\sqrt{5}$ |
| b) $\sqrt{20}$ | 2) $3\sqrt{2}$ |
| c) $\sqrt{24}$ | 3) $3\sqrt{3}$ |
| d) $\sqrt{18}$ | 4) $2\sqrt{3}$ |
| e) $\sqrt{27}$ | 5) $2\sqrt{6}$ |

T-44. A 45' ramp leading to a building forms an angle with the ground of 9°.

a) What is the height that the ramp will reach?

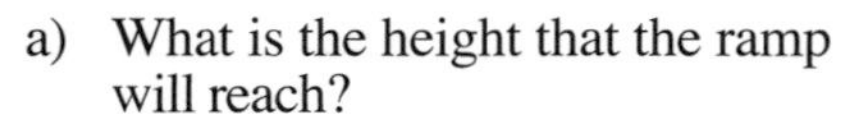

b) How far away from the base of the building does the ramp extend?

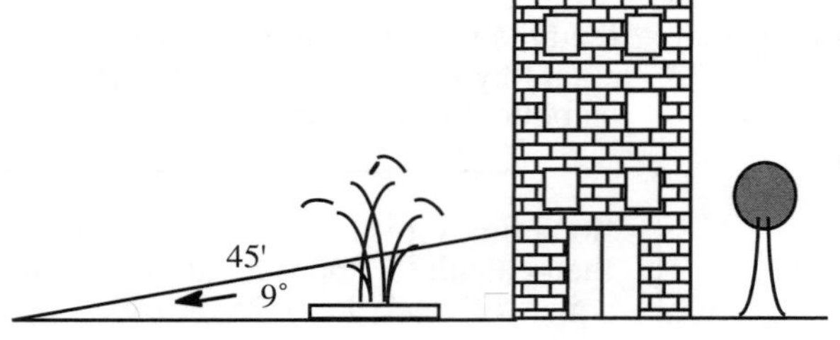


T-45. Find the area of the triangle at right. Show the subproblems you use.

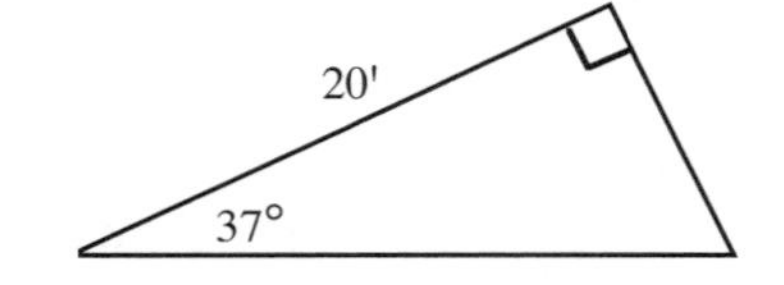


T-46. A ladder leaning against a vertical wall makes an angle of 75° with the flat ground. If the foot (bottom) of the ladder is 4 feet from the base of the wall, would this ladder be long enough for you to retrieve the suction dart that your friend stuck 17 feet up the wall?

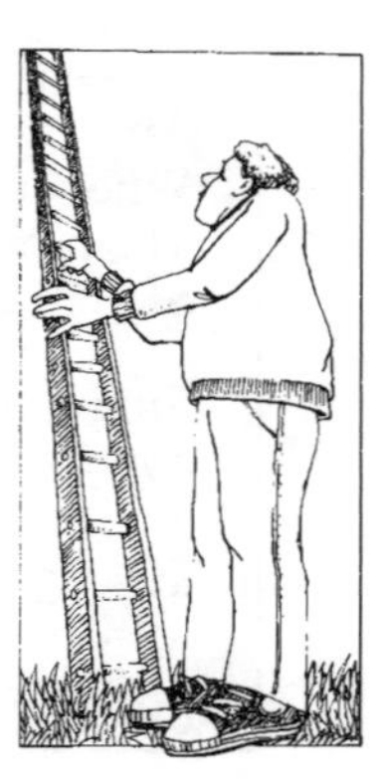

T-47. Express the following in simplest radical form, that is, simplify the following radicals WITHOUT using your calculator or any decimal approximation. Just find the square(s), take out square root(s), and write them as a product.

a) $\sqrt{96}$

b) $\sqrt{72}$

c) $\sqrt{40}$

d) $\sqrt{32}$

T-48. Find the area of the shaded region in the figure at right. Show all subproblems.

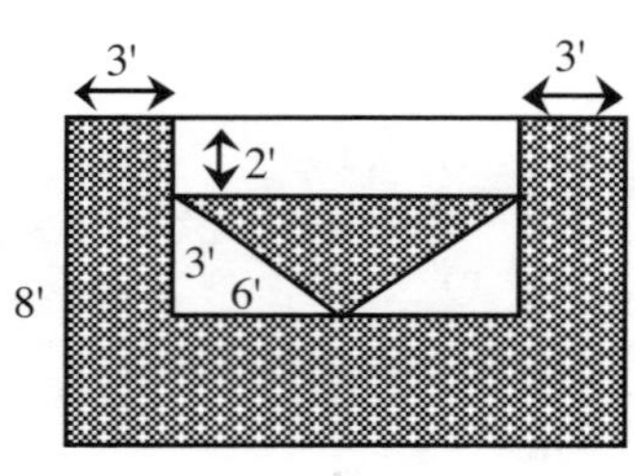

## PROOFS AND MORE APPLICATIONS

T-49. You know that the sine value for angles between 0° and 90° ranges between 0 and 1. Explain why 1 is the <u>maximum</u> value possible. Use the definition of sine and the two figures below to prove your response.

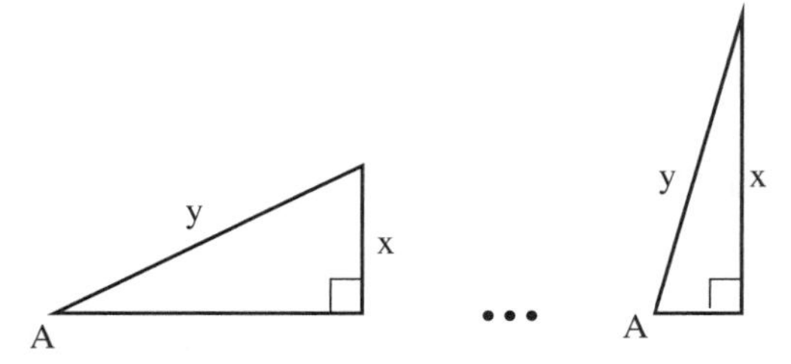


T-50. In any right $\Delta ABC$, we know that $\angle A$ and $\angle B$ are complementary. Use the trig definitions to prove that $\sin A = \cos B$?

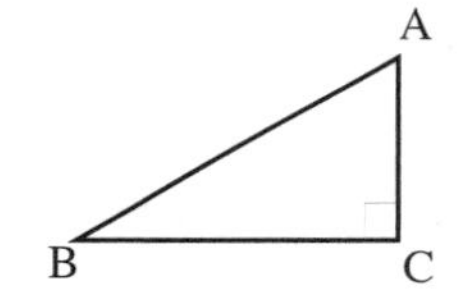

T-51. At 2:00 PM, Deion observed a caterpillar crawling up a barn wall. He estimated his distance to the wall as 8 feet and his angle of sight up to the caterpillar as 10°.

a) If Deion's eyes are about 6 feet above the floor, about how high is the caterpillar on the wall at 2:00?

b) At 2:05 the caterpillar has moved up the wall and Deion's angle of sight is now 25°. <u>How far</u> has the caterpillar crawled in the 5 minutes?

c) If the caterpillar is 15 feet up the wall at 2:17, at what angle is Deion observing the caterpillar?

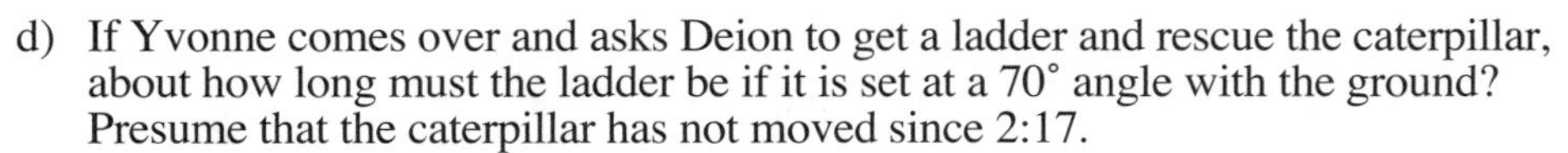

d) If Yvonne comes over and asks Deion to get a ladder and rescue the caterpillar, about how long must the ladder be if it is set at a 70° angle with the ground? Presume that the caterpillar has not moved since 2:17.

T-52. $\Delta ABC$ is equilateral, and $\overline{AD}$ is one of its heights.

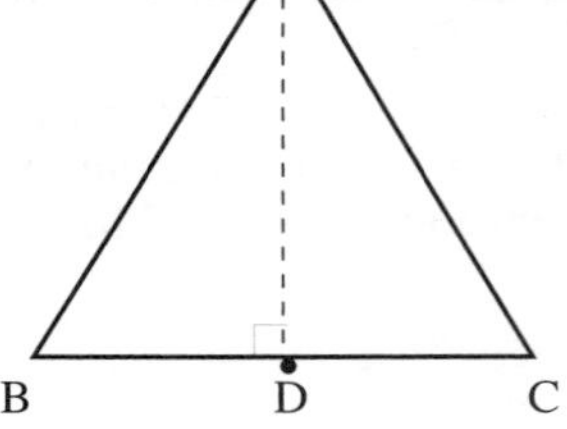


a) Copy the figure on your paper and write the measures of all the acute angles on the figure.

b) Is $\Delta ADB \cong \Delta ADC$? Prove it.

c) If AB = 2, find BD and AD. Leave your results in simple radical form.

d) If AB = 10, find BD and AD. Leave your results in simple radical form.

e) If AB = 7, find BD and AD. Leave your results in simple radical form.

f) $\Delta ABD$ and $\Delta ACD$ are called "30°- 60°- 90° triangles." Why do you think they are called that? What properties do such triangles have?

T-53. A twenty foot ladder leans against the top edge of a wall and makes an angle of 65° with the ground.

a) Find the distance between the foot of the ladder and the building. (Remember to make a drawing.)

b) Show three different ways you could use to calculate the height of the wall.

c) Which method is most accurate? Why?

T-54. On a sheet of graph paper, graph $y = -x + 6$. Then find:

a) The area of the triangle bounded by the line and both axes.

b) The length of the hypotenuse.

c) The height of the triangle from the origin (0, 0) to the hypotenuse.

d) The slope of the height in part (c).

T-55. In the figure at right, C is the midpoint of $\overline{AD}$, $\angle A \cong \angle D$ and $\angle BCA \cong \angle ECD$. Prove that BC = EC.

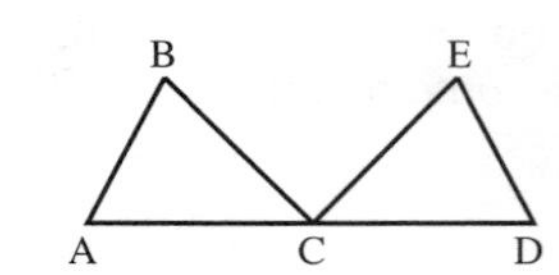

T-56. A train reaches the foothills of the Sage Brush Mountains and begins its climb at an angle of $5^\circ$ with the horizontal. After it travels 1000 yards along the track, how much elevation has it gained? (Remember to draw a picture.)

T-57. Solve for x if:

a) $\frac{x}{6} + x = 5$

b) $\frac{3}{4}x + \frac{2}{3} = x - \frac{5}{2}$

c) $\frac{12}{x} - \frac{9}{2x} = x - \frac{1}{2}$

T-58. Write in simple radical form.

a) $\sqrt{98}$

b) $\sqrt{180}$

c) $\sqrt{54}$

T-59. In any $\Delta ABC$ with AB = 23 cm and AC = 17 cm., what are the minimum and maximum limits of the length of side $\overline{BC}$?

T-60. Jake is playing in a high stakes poker game. Four cards in the suit of hearts are showing (face up) on the table. Twenty cards have been dealt. Jake has three hearts in his hand (that is, not showing). Jake is considering how much to wager before each person is dealt two more cards. Jake could try for a flush (five cards in the same suit). If no other hearts beyond those previously mentioned were dealt in the initial deal and Jake will receive the next card dealt, what is the probability that his next card will be a heart? Remember: 52 cards in the deck, 4 suits of 13 cards.

T-61. **INTRODUCTION:** Most of the problems in this unit have taken you a lot less time and effort than it would have a few years ago. Before scientific calculators were available at a reasonable price, people had to search through books of tables to find the values of the trig ratios. Often people relied on knowing the relationships for a few special angles in certain triangles. They used a table in the back of a textbook or a book of tables--like your resource page for problem T-1 but much more detailed--for other angles.

Sometimes knowing about the special triangles is still useful because they can help to solve some problems quickly, sometimes even in your head. Some teachers, tests, and textbooks you may encounter will expect you to know about them.

For the first three isosceles right triangles (45°- 45°- 90°) below, we used the Pythagorean Theorem to find the length of the hypotenuse and expressed it in simple radical form, NOT as a decimal approximation.

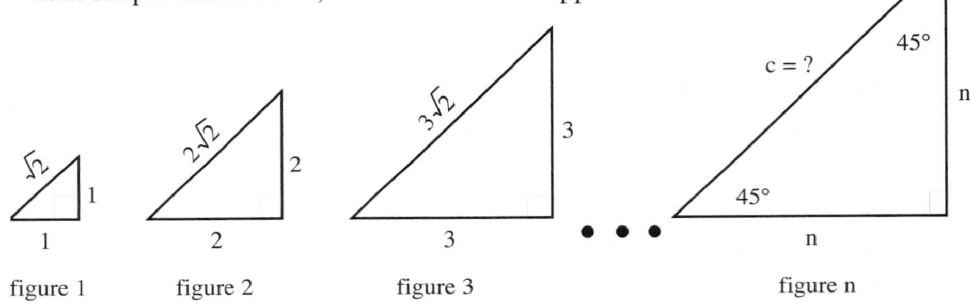


a) Look for a pattern in the first three figures. You may need to make a table labeled legs and hypotenuse. Draw the next two diagrams, figures 4 and 5, and solve for the length of the hypotenuse. Test your pattern or draw more figures to get more data to find one. State your observations.

b) Use your results for the first several triangles to express the length of c in terms of n (figure n). Write a conjecture about how to find the length of the hypotenuse of any isosceles right triangle without using the Pythagorean Theorem.

**T-62.** Another way to understand the isosceles right triangle relationship is to see the triangle as half of a square. On a piece of paper carefully draw a square.

a) Draw in one diagonal and shade in one of the right triangles formed.

b) Label the angle measures in the other right triangle.

c) If one leg has a length of x, what is the length of the other leg? Why?

d) What is the length of the hypotenuse in terms of x? Use the Pythagorean Theorem and write it in simple radical form.

**T-63.** For the first three $30° - 60° - 90°$ triangles below, we used the Pythagorean Theorem to find the length of the longer leg and expressed it in simple radical form, not as a decimal approximation.

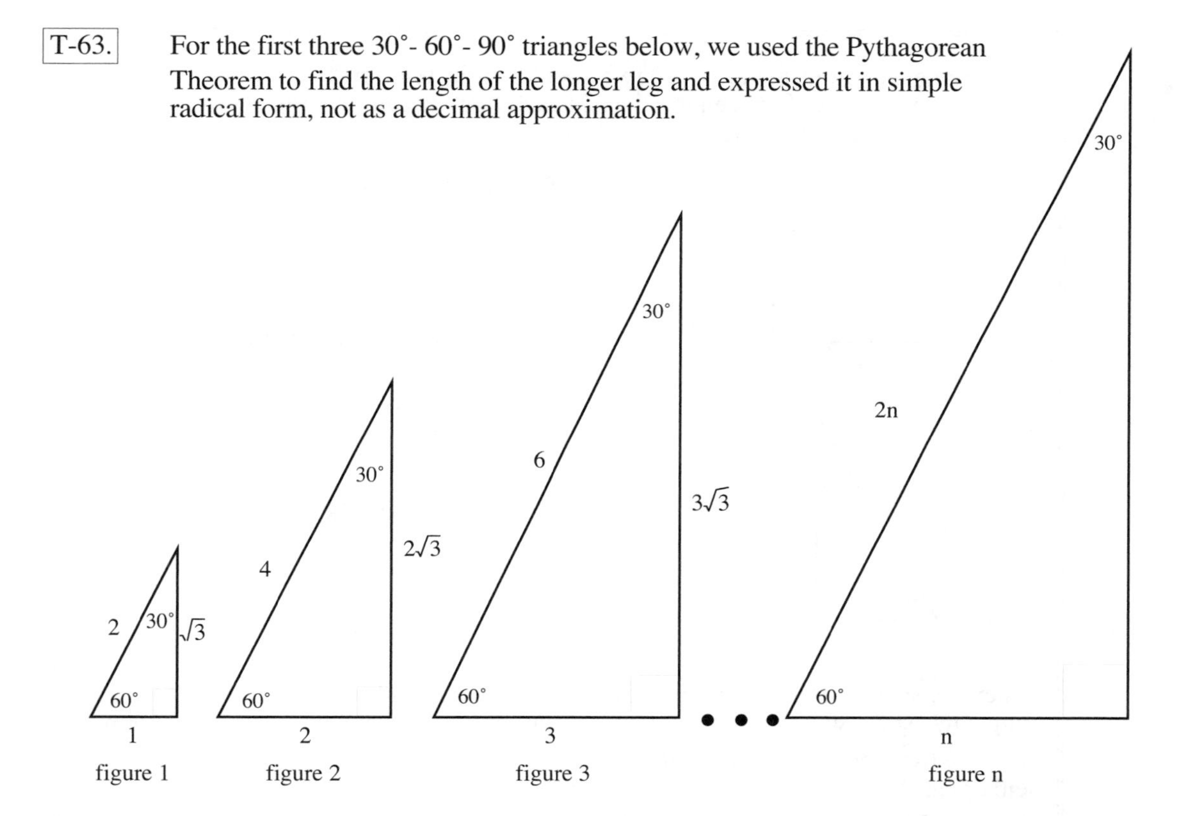


a) Look for a pattern in the first three figures. You may need to make a table labeled short leg, long leg, and hypotenuse. If necessary, draw and solve the next two triangles, figures 4 and 5, to get more data. State your observations.

b) Use your results for the first several triangles to express the length of the long leg in terms of n (figure n). Write a conjecture about how to find the length of the long leg for <u>any</u> $30° - 60° - 90°$ triangle <u>without</u> using the Pythagorean Theorem.

T-64. Another way to understand the 30°- 60°- 90° triangle relationship is to see the triangle as half of an equilateral triangle like you did in problem T-52. On a piece of paper carefully draw an equilateral triangle.

a) Draw in a height, dividing the equilateral triangle into two right triangles. Shade in one of the right triangles.

b) Label the angle measures in the other right triangle.

c) If the hypotenuse has a length of 2x, what is the length of the short leg? Why?

d) What is the length of the long leg in terms of x? Use the Pythagorean Theorem and write it in simple radical form.

The triangles you just studied are two **special right triangles**. Knowing these relationships (ratios) between the sides in each triangle can save time doing problems in this course and on tests like the SAT.

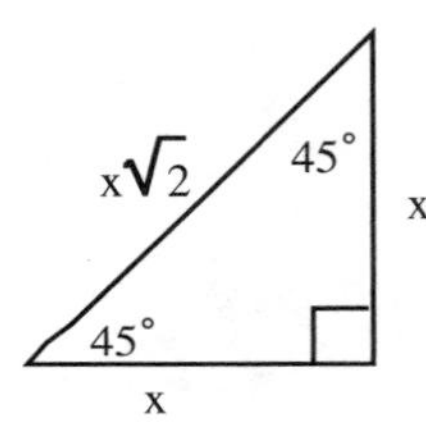


**THE ISOSCELES RIGHT**

or

**45°- 45°- 90° TRIANGLE:**

"half of a square" with side ratios

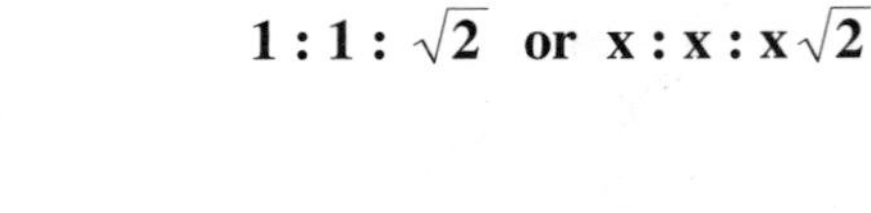

$\mathbf{1 : 1 : \sqrt{2}}$ **or** $\mathbf{x : x : x\sqrt{2}}$

2x
30°
x√3
60°
x

**THE 30°- 60°- 90° TRIANGLE:**

"half of an equilateral triangle" with side ratios

$\mathbf{1 : \sqrt{3} : 2}$ **or** $\mathbf{x : x\sqrt{3} : 2x}$

They are special triangles because the **ratios** of the sides for triangles with these angle measures stay the same no matter what the actual side lengths are. You know that the hypotenuse of an isosceles right triangle is always $\sqrt{2}$ times the length of either leg. You know that the hypotenuse of a 30°- 60°- 90° triangle is always twice the length of the short leg and that the long leg is $\sqrt{3}$ times the length of the short leg.

T-65. Find values for x and y in each triangle without using your calculator.

a)

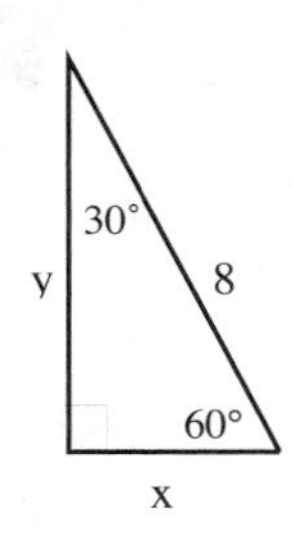


b)

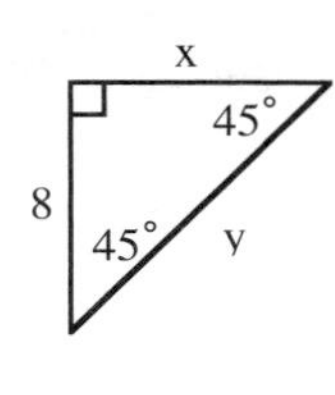


T-66. Find the perimeter of a square which has a diagonal of length $5\sqrt{2}$ and show how you obtained your answer.

T-67. Find values for x and y in the triangle at right. Show your subproblems.

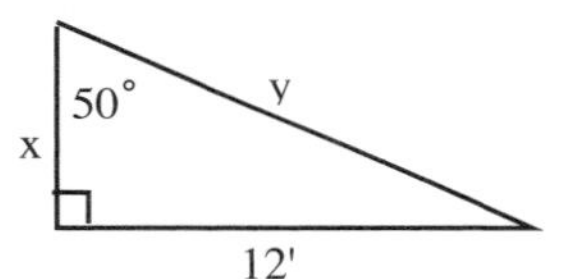


T-68. Rapunzel is looking down from the top of the Leaning Tower of Pisa. The distance from the top of the tower wall to the bottom is 184 feet, and the tower leans to make an angle of 83° with the ground. How long is Rapunzel's hair if the ends just touch the ground?

T-69. On your paper draw a square ABCD and the diagonal $\overline{AC}$. Without using your calculator,

a) if the side of the square is 3 cm, what is the length of diagonal AC?

b) if the side of the square is 5 cm, what is the length of the diagonal $\overline{AC}$?

T-70. Prove the statement below.

If the height from one angle of a triangle bisects the opposite side, then the triangle is isosceles.

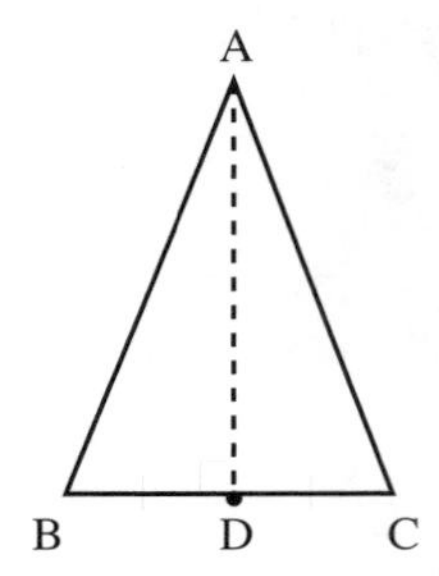

T-71. Simplify the following exponential expressions.

a) $(x^3y^5)^3$

b) $(3x^2y^2)(5x^3y^6)$

c) $\left(\frac{x^3}{y^2}\right)^4$

d) $\frac{(4x^2y^4)^3}{8x^4y^5}$

T-72. Write the equation of the line that passes through (0, 4) and is:

a) perpendicular to the graph of $y = \frac{2}{3}x - 1$.

b) parallel to the graph of the same equation.

T-73. Solve for all variables.

a) $3x + 2y = 8$
$x - 4y = -2$

b) $x^2 - 11x + 8 = 0$

T-74. Write each of the following in simple radical form.

a) $\sqrt{50}$

b) $\sqrt{200}$

c) $\sqrt{289}$

d) $\sqrt{344}$

## CLINOMETER CONSTRUCTION

T-75. **HOW TO MAKE A CLINOMETER:** Working in pairs, cut out <u>and</u> tape or glue a protractor scale to an index card as shown. **Attach the top edge $\overline{XY}$ along the top edge of the card.** Tape on a soda straw as shown. Put the string into the straw at X, pull it through the straw, and staple it to the card near Y. Tie a weight to the other end of the string so that it will hang vertically a little below the bottom of the card.

When using the clinometer, you should stand at a convenient distance from an object whose height you wish to measure. Sight the top of the object through the straw. Mark the point E where the string intersects the protractor scale. Your partner can help you with this. Carefully study the diagrams which follow so that you will be ready to use your clinometer tomorrow.

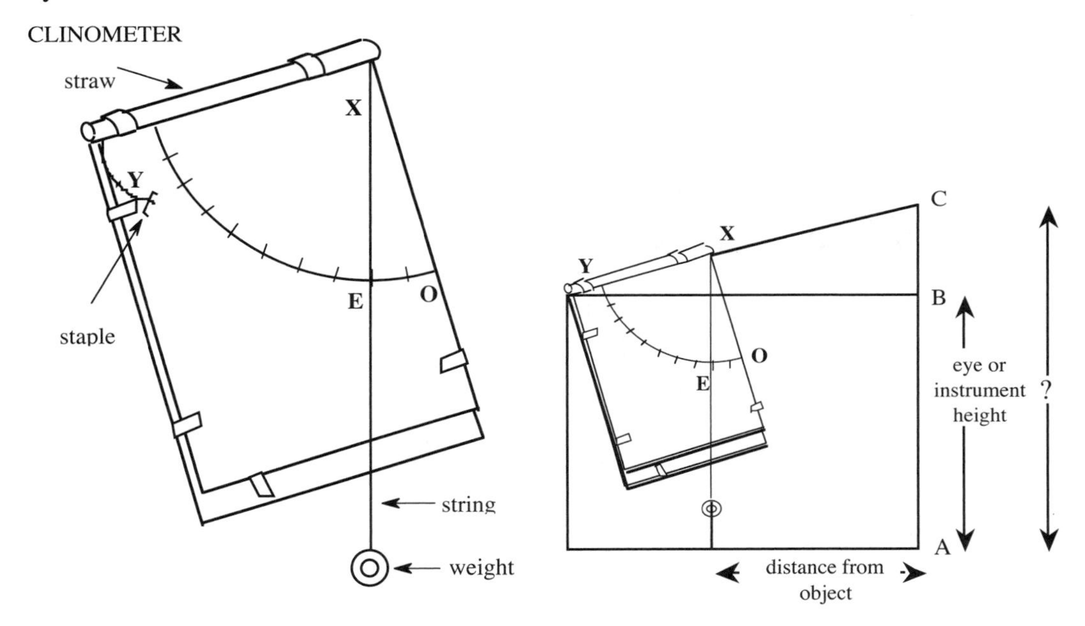


* The Clinometer activity is based on a presentation given by Micheal Palmer at the 1987 Invitational Summer Institute of the Northern California Mathematics Project.

T-76.

**DIRECTIONS:** Take another survey of your team. Which of the next two problems would you rather do without a calculator?

$1.234 \overline{)5.00000}$ or $5 \overline{)1.234}$

Without a calculator, the second division problem is much easier than the first. Before scientific calculators were readily available, people had to devise ways to simplify division problems such as the one above at left. This type of division problem occurs frequently when working with radicals. The radical expression $\frac{1}{\sqrt{2}}$, for instance, is very difficult to write as a decimal if you cannot use a calculator. First, you would need to find the decimal approximation of $\sqrt{2}$ (which, by the way, never ends or repeats) and then divide. However, the problem $\frac{\sqrt{2}}{2}$, which we will soon see is equivalent to $\frac{1}{\sqrt{2}}$, is much easier, as shown in the comparison below.

$1.414 \ldots \overline{)1.000}$ $2 \overline{)1.414 \ldots}$

Rewriting expressions which have radicals in the denominator so that they no longer have those radicals in the denominator is called **RATIONALIZING THE DENOMINATOR**. Here is an example of how it is done.

In the number, $\frac{2}{\sqrt{3}}$, we find the $\sqrt{3}$ in the denominator inconvenient. We multiply both the numerator and the denominator by the radical in the denominator.

$$\frac{2}{\sqrt{3}} = \frac{2}{\sqrt{3}} \bullet \frac{\sqrt{3}}{\sqrt{3}} = \frac{2 \bullet \sqrt{3}}{\sqrt{3} \bullet \sqrt{3}} = \frac{2\sqrt{3}}{\sqrt{9}} = \frac{2\sqrt{3}}{3}$$

And look! No more radicals in the denominator! It is somewhat similar to Fraction Busters in that we multiply by the radical we want to remove written in the convenient form of $\frac{\sqrt{3}}{\sqrt{3}} = 1$. Multiplying by 1 does not change the value of the original number. Using $\frac{\sqrt{3}}{\sqrt{3}}$ as a form of 1 does change the form of the number, which is our purpose. Radicals in the denominator are hard to grasp in a way that makes sense like halves and thirds. For $\frac{1}{2}$, the 2 in the denominator means to break a whole into two equal parts. But what does it mean to break a whole into $\sqrt{2}$ parts? Rationalizing the denominator returns an expression with an irrational denominator to more familiar terms.

Some other rules for radicals: $\sqrt{a} \cdot \sqrt{b} = \sqrt{a \cdot b}$ and $\sqrt{\frac{a}{b}} = \frac{\sqrt{a}}{\sqrt{b}}$

Rationalize the denominator in each of the following expressions.

a) $\frac{1}{\sqrt{5}}$

b) $\frac{3}{\sqrt{3}}$

c) $\frac{\sqrt{2}}{3\sqrt{5}}$

T-77. ABCD is a trapezoid in which AB = 16', BC = 10', and m∠BCD = 40°. Redraw the diagram on your paper and put in a useful auxiliary line.

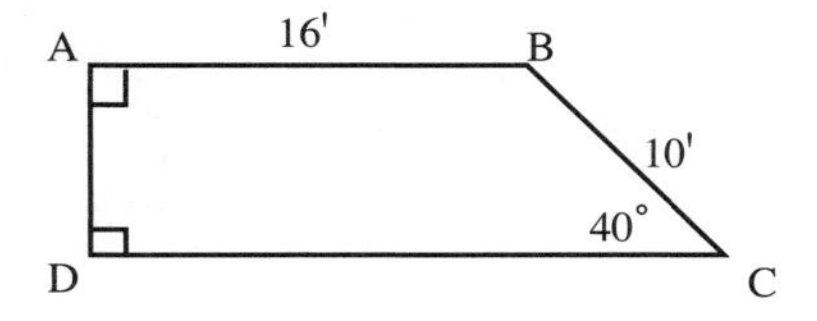


a) Find AD and CD.

b) Find the area of ABCD.

T-78. Multiply as indicated and then write in simple radical form.

a) $\sqrt{2} \bullet \sqrt{3}$

b) $\sqrt{6} \bullet \sqrt{8}$

c) $\sqrt{15} \bullet \sqrt{45}$

d) $\sqrt{10} \bullet \sqrt{10}$

T-79. ΔABC is a right triangle with ∠A and ∠B as the acute angles. Is it true that (tan A)(tan B) = 1? Prove your answer.

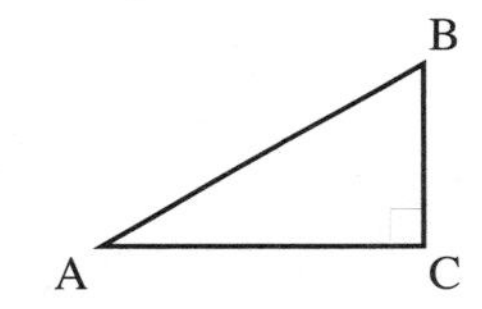

T-80. If a jet climbs at an angle of 8°, what is the minimum distance between its take-off point and a 120 foot tower, so the jet will clear the tower by at least 50 feet?

T-81. Sketch a triangle with a 90° and two 45° angles. Let one leg be 8 units long.

a) What is the length of the other leg?

b) Find tan 45° by using the appropriate sides. Does this value agree with your calculator?

T-82. Solve for x and y without using a calculator.

a)

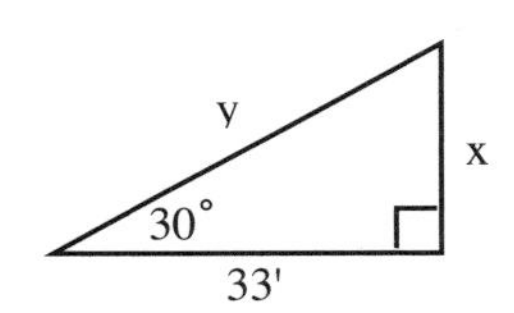


b)

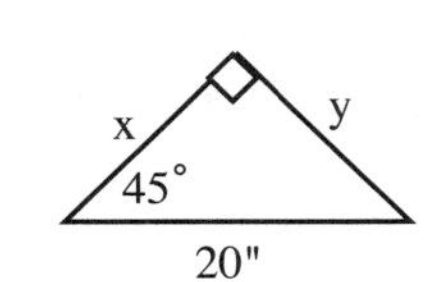

T-83. Solve for x.

a) $\frac{1}{5} + \frac{1}{4}x = \frac{7}{5} - \frac{1}{20}x$

b) $x^2 + 3x - 40 = 0$

T-84. To complete this problem you will need your clinometer that you made yesterday, the resource page, Measuring With a Clinometer, and possibly a tape measure.

T-85. If you knew the height of an object, explain how you could use a clinometer to determine your distance from the object. Use diagrams.

T-86. In $\Delta RST$ find ST and RS if $m\angle R = 55°$, $m\angle S = 75°$, and RT = 30 cm. Show all subproblems.

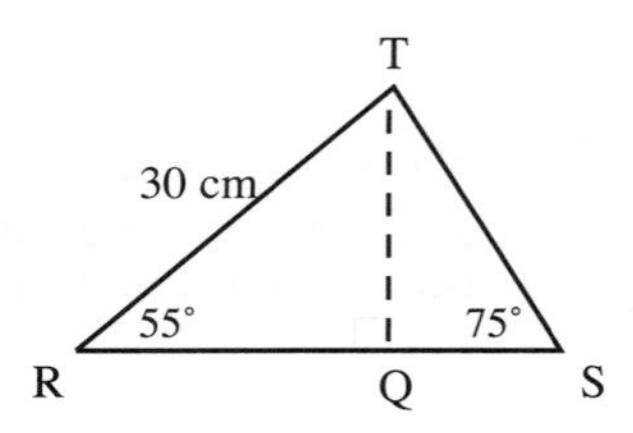


T-87. Find the area of each figure. Show all subproblems.

a)

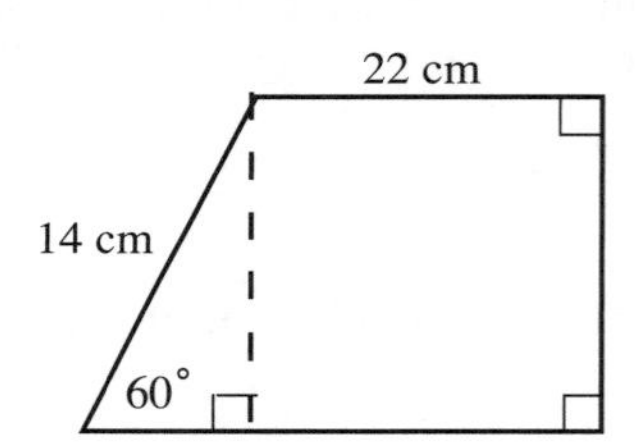


b)

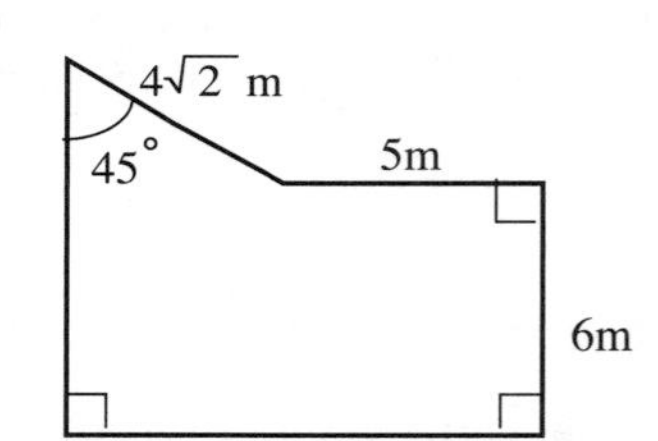


T-88. Find the area of the rectangle.

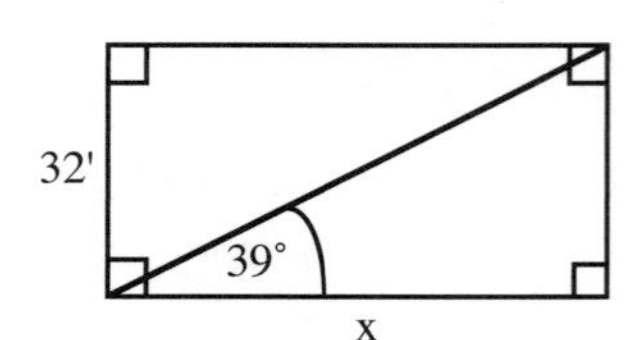

T-89. The maximum grade for an interstate highway is 7%, that is, 7 feet of increased height for each 100 feet of horizontal distance covered.

a) What is the measure of the angle of inclination for a 7% grade?

b) In Kensington County the maximum grade allowed for driveways is 25%. How large an angle is that?

T-90. Find the lengths of both the hypotenuse and the long leg <u>without</u> using a calculator.

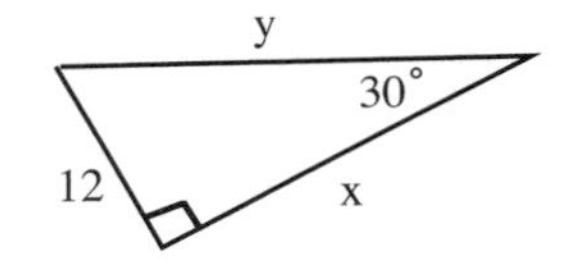


T-91. In each part below, show how you arrive at your answer.

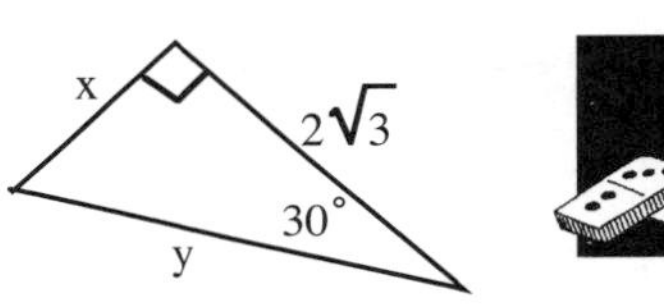


a) Find the lengths of both the short leg and the hypotenuse.

b) Use the Pythagorean Theorem to prove that your answers are correct.

T-92. Solve for x. Remember Fraction Busters.

a) $\frac{3}{\sqrt{2}}x + \frac{1}{\sqrt{2}} = \sqrt{2}$

b) $\frac{4}{9} - \frac{6}{13}x = 9x + \frac{1}{9}$

T-93. Rationalize the denominator and write the expression in simple radical form. If you need help, see problem T-76 for an example.

a) $\frac{5}{\sqrt{3}}$

b) $\frac{12}{\sqrt{6}}$

c) $\frac{6}{\sqrt{8}}$

T-94. Read the following information about a special kind of fraction busting, add it to your tool kit, then do the practice problems at the bottom of the box.

**DIRECTIONS:** Some equations that involve fractions can be simplified in one step. When two fractions (or ratios) are written as an equation, it is called a **proportion**. For example, $\frac{x}{3} = \frac{7}{8}$ is a proportion.

Solving algebraically, the steps are:

$$24\left(\frac{x}{3}\right) = \left(\frac{7}{8}\right)24 \quad \text{Fraction Busting}$$

$$\overset{8}{\cancel{24}}\left(\frac{x}{\underset{1}{\cancel{3}}}\right) = \left(\frac{7}{\underset{1}{\cancel{8}}}\right)\overset{3}{\cancel{24}} \quad \Rightarrow \quad 8x = 21 \quad \Rightarrow \quad x = \frac{21}{8} = 2.625$$

Some people use a short cut to go directly to the third equation above. They call it **cross-multiplying**. Why does this work?

$$\frac{x}{3} = \frac{7}{8} \Rightarrow (x)(8) = (3)(7) \Rightarrow 8x = 21$$

CAUTION! If you choose to save some work by simply **cross-multiplying** the ratios and solving the resulting one-step equation, BE SURE you have a proportion to start with.

**NOT ALLOWED:** $\frac{2x}{3} - 5 = \frac{4}{7}$ **PERMITTED:** $\frac{2x}{3} = \frac{4}{7}$

Use cross-multiplication to solve the following proportions:

a) $\frac{3x}{5} = \frac{6}{11}$

b) $\frac{12}{x} = \frac{37}{18}$

c) $\frac{2x + 1}{4} = \frac{3}{8}$

d) $\frac{\sin 51^{\circ}}{14} = \frac{\sin 32^{\circ}}{x}$

## LAW OF SINES

T-95. Be sure that everyone in your team understands how to solve problem T-86. In particular, everyone should have the subproblems clearly sequenced.

T-96. This problem is essentially the same as T-86. Here we seek to generalize the process. Use the triangle at right with side lengths a, b and c to:

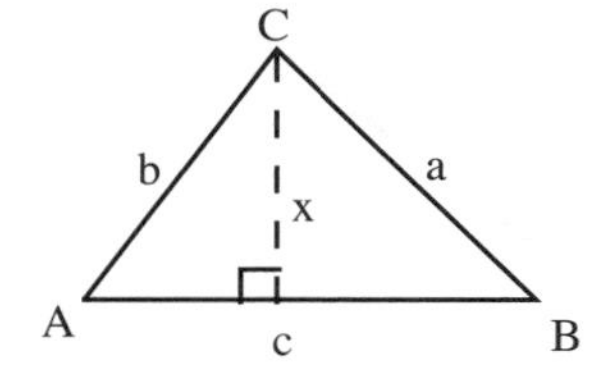

a) Write a trigonometric equation using x, a, and $\angle B$ that you could solve for x if you knew the value of a and $m\angle B$.

b) Write a trigonometric equation using x, b, and $\angle A$ that you could solve for x if you knew the value of b and $m\angle A$.

c) Solve both equations for x.

d) Use part (c) to show that $\frac{\sin A}{a} = \frac{\sin B}{b}$.

e) What do you think is probably true of $\frac{\sin C}{c}$?

f) The relationship for sine ratios in $\Delta ABC$ is summarized in the box below. Read the information and add it to your tool kit.

In the previous problem you demonstrated the **LAW OF SINES**. For any $\Delta ABC$, it is always true that:

$$\frac{\sin A}{a} = \frac{\sin B}{b} \qquad \frac{\sin A}{a} = \frac{\sin C}{c} \qquad \frac{\sin B}{b} = \frac{\sin C}{c}$$

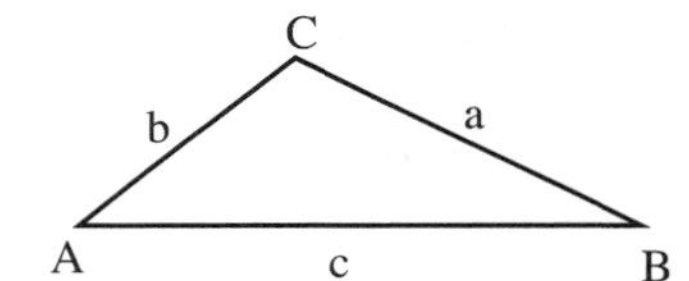


This law is a powerful tool because it allows you to use the sine ratio to solve for measures of angles and lengths of sides of any triangle, not just right triangles. The law also works for angle measures between 0° and 180°. For example, if $m\angle A = 53°$, a = 18', and $m\angle B = 39°$, you can solve for b by substituting the values into the law's ratios and solving the proportion. Remember that the sine values represent numbers in this example.

| | |
|---|---|
| $\frac{\sin 53°}{18} = \frac{\sin 39°}{b}$ | substitute using the Law of Sines |
| $b(\sin 53°) = 18(\sin 39°)$ | cross-multiplying |
| $b = \frac{18(\sin 39°)}{\sin 53°}$ | solving for b |
| $b \approx 14.18'$ | calculations |

Note that you can use your calculator to convert the sine values into decimal approximations at any time during the solution. However, it is usually most convenient to do so at the end of the problem.

T-97. Use the Law of Sines to find the labeled length in each triangle.

a)

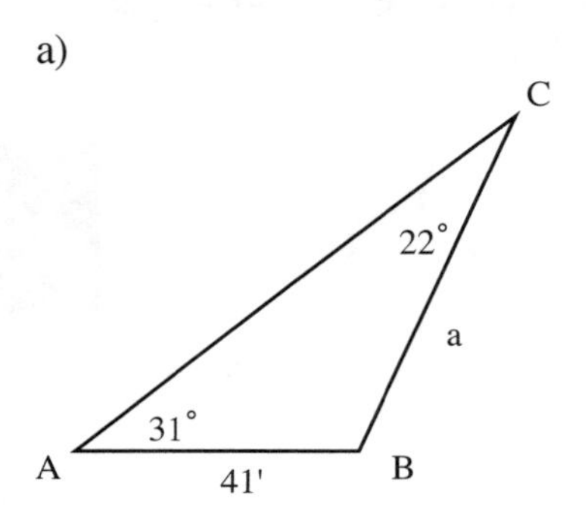


b)

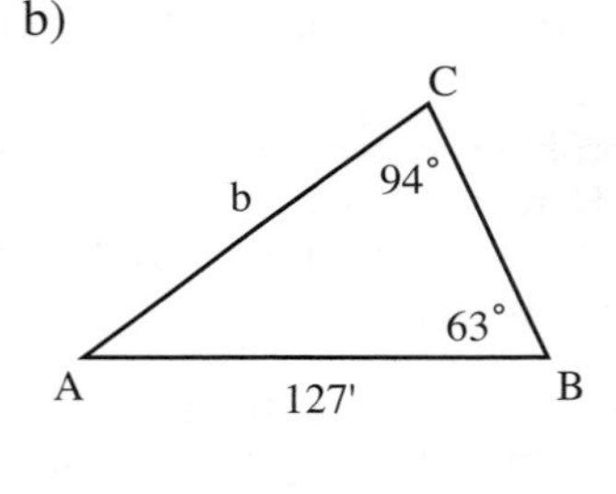


T-98. Sammy Scout is assigned to find the widest part of Polecat Pond. He drives stakes at A, B, and C since he believes that $\overline{AC}$ is the widest part of the pond. He measures $\overline{AB}$ and finds that its length is 684'. ∠B measures 79° and ∠C measures 53°. How wide is the pond between points A and C?

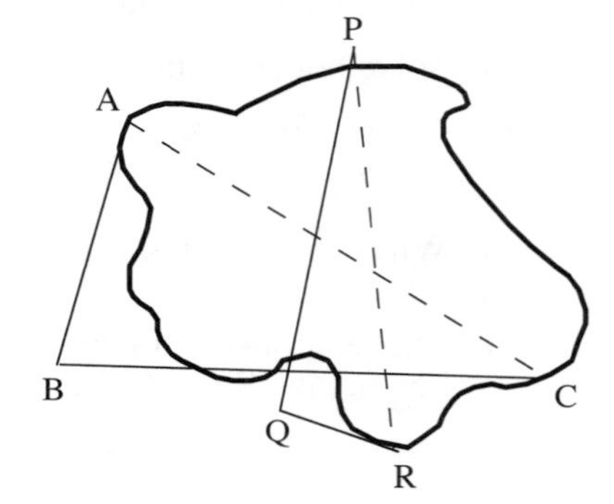


T-99. Barbara Bluebird thought that the widest part of the pond was from P to R. She measured $\overline{QR}$ to be 321', ∠Q as 108°, and ∠P as 18°. How wide is the pond from P to R? Who is correct?

T-100. The square pyramid with base ABCD shown at right has sides 16 cm long and its triangular faces each have two equal edges 24 cm long. Point E is the top vertex and point P is the center of the base.

a) Find the distance from A to P.

b) Explain why ∠APE is a right angle.

c) Find the distance from P to E.

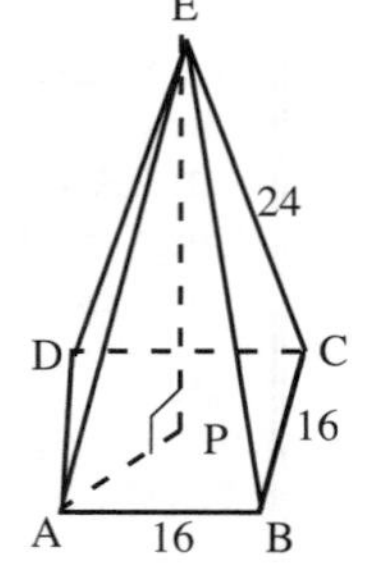


T-101. Find the area of the triangle at right. Show all subproblems.

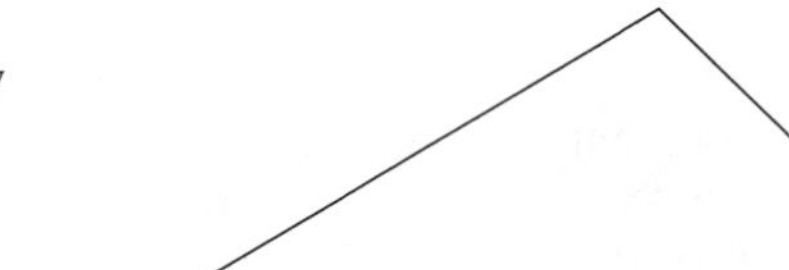

T-102. Use the Law of Sines to solve the following problems.

a)

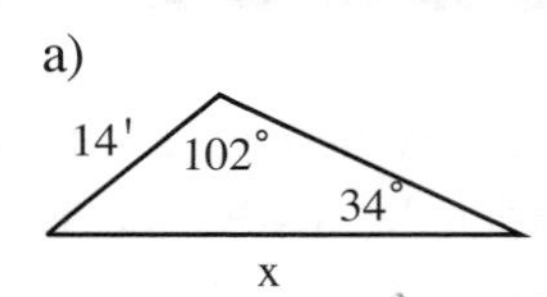


b)

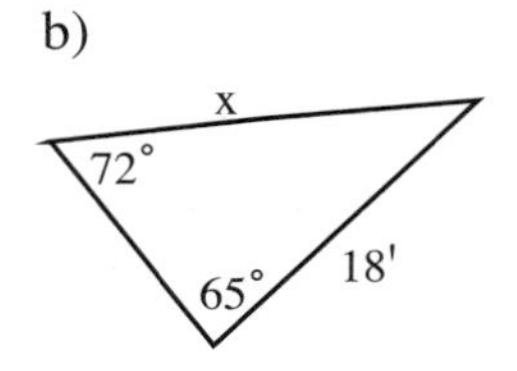


c)

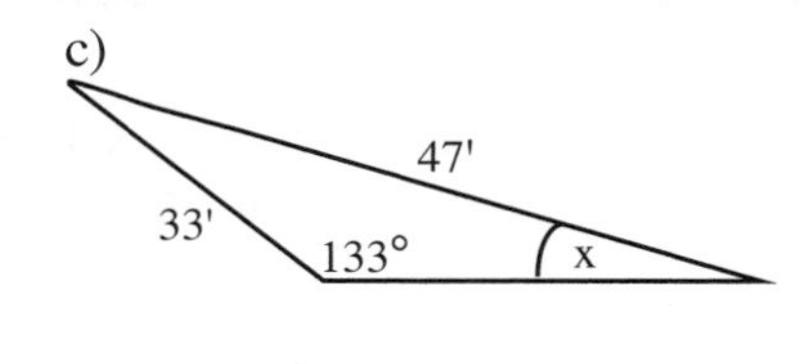


T-103. Write each expression in simple radical form.

a) $\sqrt{54}$ b) $\sqrt{6} \bullet \sqrt{12}$ c) $\frac{3}{\sqrt{12}}$

T-104. Each red stick is 3 cm longer than two blue sticks placed end to end. Four red sticks and three blue sticks together extend to a distance of 111 cm. How long is a blue stick?

T-105. Find the area of the parallelogram. Show all subproblems.

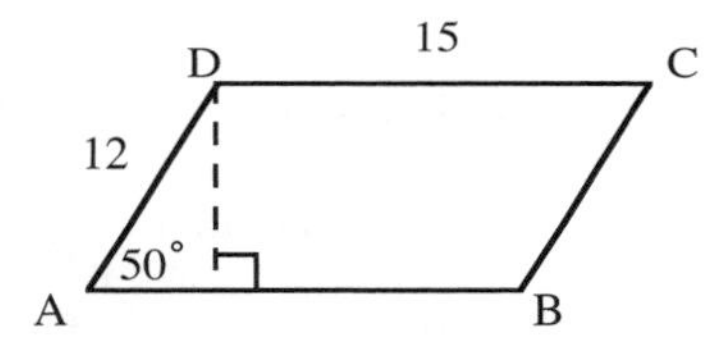


T-106. In problem T-102 you should have found the following values for x: part (a) ≈ 24.49', part (b) ≈ 17.15', and part (c) ≈ 30.9°.

a) Label the figures below with these values, then find the length of the third side and the measure of the third angle for each triangle. Write these values on the figures.

b) Make a two-column table for each figure. Put the measure of each angle (smallest to largest) in one column and the length of the angle's opposite side in the second column.

c) Write a conjecture about the size of an angle of any triangle (i.e., smallest angle, largest angle) and its relationship to the length of the side opposite the angle.

1)

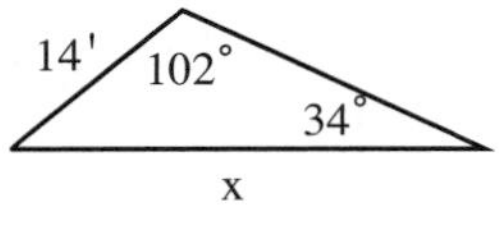

2)

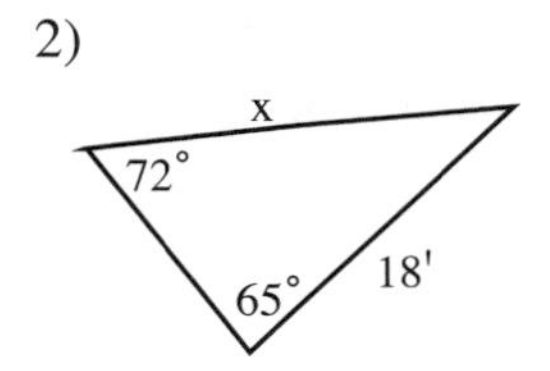


3)

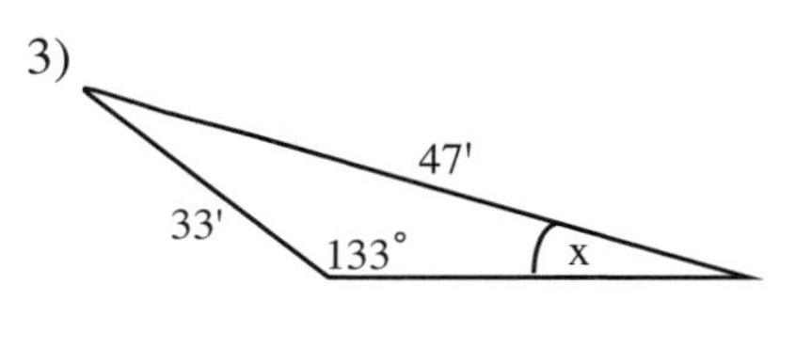

## TANGENT AND SLOPE

T-107. Graph the line $y = \frac{3}{4}x$. Label the point where the line intersects the y-axis A. Label the points B(4, 3) and C(4, 0). You should now have a right triangle, ΔABC, that is attached to the line. As a matter of fact, it is a slope triangle.

a) What is the slope of the line?

b) In ΔABC, what is tan A?

c) Compare your answers for parts (a) and (b) and describe what you observe.

d) Find m∠BAC to the nearest degree.

T-108. Graph the line $y = \frac{2}{3}x - 2$ **on the <u>same</u> set of xy-coordinate axes you used for problem T-107**. The following parts are one way to help you find the angle that this line forms with the x-axis.

a) Draw right ΔTEL such that T(3, 0), E(6, 0), and L(6, 2).

b) What is tan T?

c) Is tan T the same as the slope of the line?

d) Find m∠T to the nearest degree.

e) Use the same process to explore ΔMOT for M(0, -2), O(3, -2), and T(3, 0). What is tan M?

T-109. Write a conjecture stating the relationship between the slope of a line and the tangent of the angle that the line makes with the x-axis. Write another sentence stating how you can find the measure of that angle.

T-110. A friend you are working with has graphed a line and tells you that the line has a slope of $\frac{1}{2}$. She asks you to graph the line yourself.

a) Can you be sure that the line you graph is the <u>same</u> line? Explain.

b) She tells you that she has labeled the angle the line makes with the x-axis ∠A. Can you tell her the measure of ∠A? If so, what is it? If not, why not?

T-111. Find the measure of the angle formed between each line and the x-axis <u>without graphing them</u>.

a) $y = 3x$ b) $y = x + 1$ c) $y = \frac{1}{3}x - 1$

T-112. The angle measures listed below correspond to the angle that is formed when each line crosses the x-axis. Assuming that each line passes through the origin (0, 0), give the equation of each line.

a) 64° b) 14.04°

c) What would each equation be if the y-intercept were -3 in each case?

T-113. Two 30°-60°-90° triangles each have a <u>hypotenuse</u> measuring 8 cm.

a) Are the two triangles congruent? Prove your answer.

b) A <u>different</u> pair of 30°-60°-90° triangles each have a <u>leg</u> measuring 8 cm. Are the two triangles congruent? Prove your answer.

T-114. $\Delta JAN$ and $\Delta FEB$ are congruent right triangles with $m\angle J = 90°$.

a) Draw a diagram of the two triangles.

b) Explain why $\angle N \cong \angle B$.

c) Show why $\frac{JN}{AN} = \frac{FB}{EB}$ by using the definition of the cosine ratio.

T-115. Find x and y in each figure. Using the special triangle ratios is fastest, but you may also use trig ratios.

a)

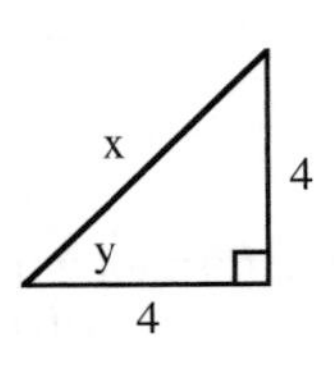


b)

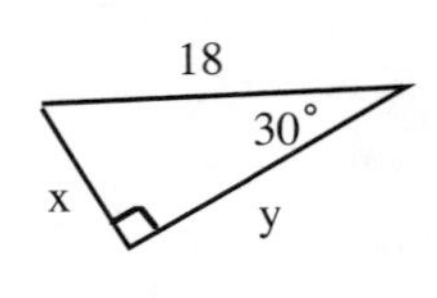


c)

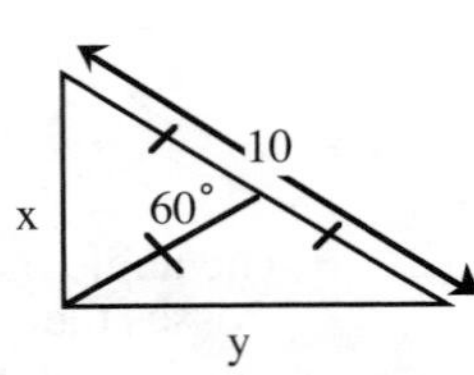


d)

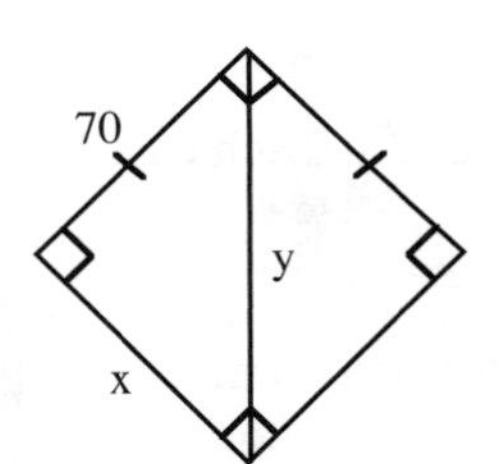


e)

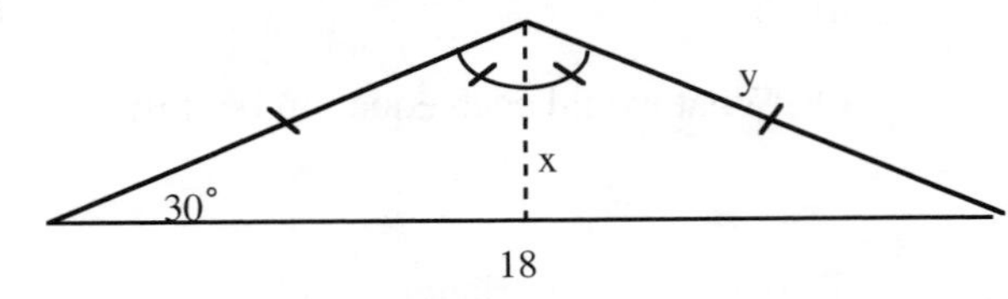


T-116. Find m∠C, then use the Law of Sines to find a and b.

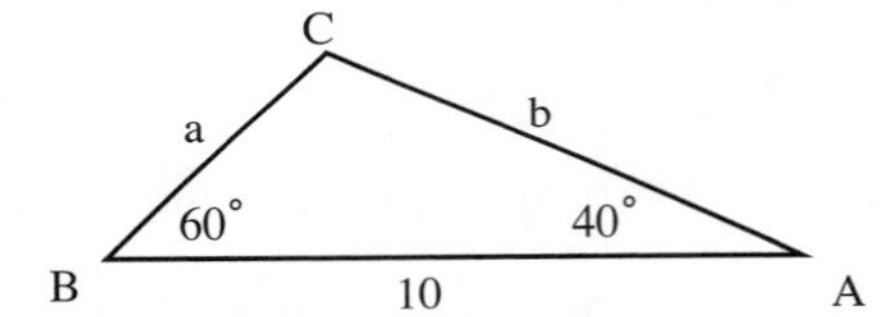


T-117. The sum of two numbers is 40 and the larger is 9 more than the smaller. Find the numbers.

T-118. Rationalize the denominator and write the expression in simple radical form.

a) $\frac{\sqrt{5}}{\sqrt{8}}$ b) $\frac{3\sqrt{7}}{2\sqrt{12}}$ c) $\frac{\sqrt{4}}{\sqrt{32}}$

T-119. Solve for x.

a) $4x + \frac{2}{3}(x + 1) = \frac{1}{9}(2x - 3)$ b) $3(x + \frac{2}{5}) = \frac{4}{5} - 3x$

## UNIT AND TOOL KIT REVIEW

T-120. **The Height of Red Hill:** From the parking lot at the Red Hill Shopping Center, the angle of sight (elevation) to the top of the hill is about 25°. From the base of the hill you can also sight the top but at an angle of 55°. The horizontal distance between sightings is 740 feet. How high is Red Hill? Show your subproblems.

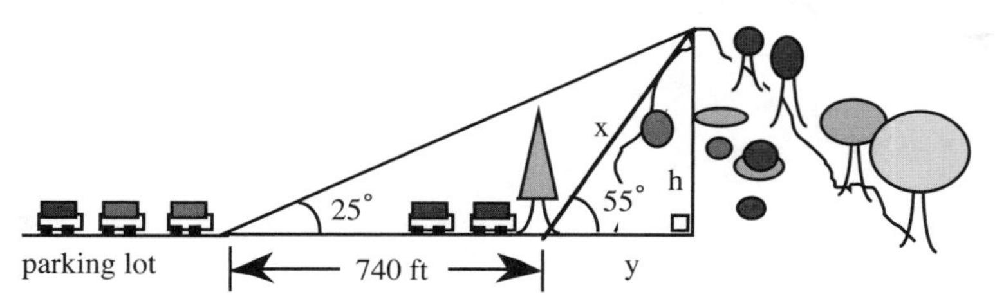


T-121. Use the grid triangle at right to find:

a) the length of $\overline{AC}$.

b) the slope of $\overleftrightarrow{AC}$.

c) the measure of ∠C.

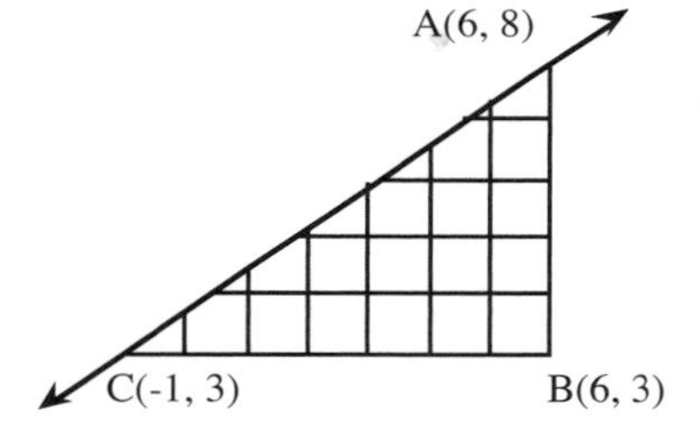


T-122. ∠A has a tangent of 0.5095 and ∠B has a tangent of 0.6009. Which is the larger angle? Explain how you arrived at your conclusion. If you used your calculator, explain how you could reach the same conclusion without a calculator or a table.

T-123. An angle of a right triangle has a cosine of 0.9375. Which is longer--the leg adjacent to the angle or the leg opposite the angle? What is the measure of the angle? Justify your conclusion.

T-124. An angle of a right triangle has a tangent of 1.0000. Describe the triangle. Include as much as you know from the information given. Does the length of the sides of the triangle affect how you answer the question? Explain.

T-125. If you graph the line $y = \frac{2}{5}x + 1$, what is the measure of the angle it makes with the x-axis? Show and explain your process.

T-126. Solve each of the following for x. If you need help getting started, rotate the figure until it looks more like those you worked with earlier in the Unit.

a)

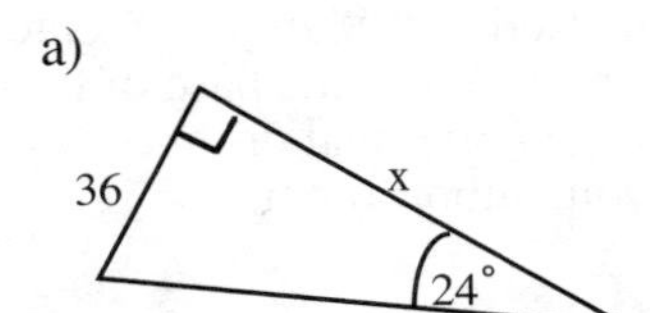


b)

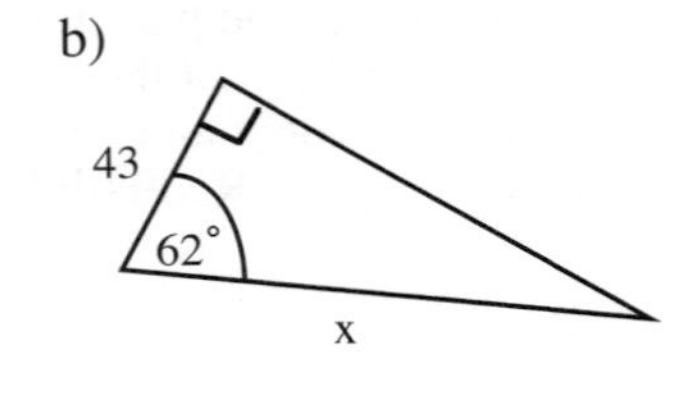


c)

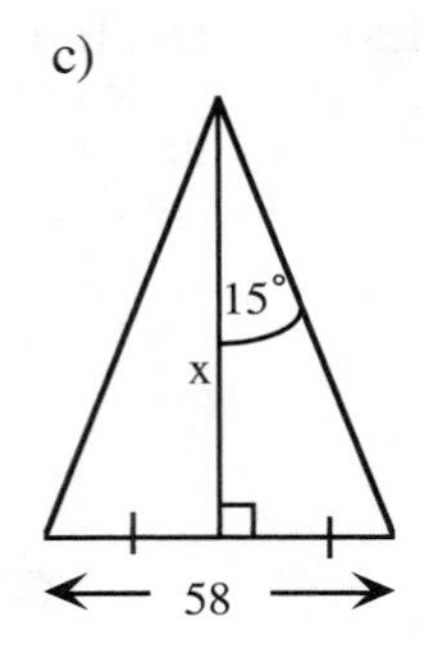


T-127. A famous mathematical spider, Dr. Web, has attached his web to the mushroom. He figures the distance from the top of the mushroom to the ground along his web is 9 cm and the angle it makes with the ground is 49°.

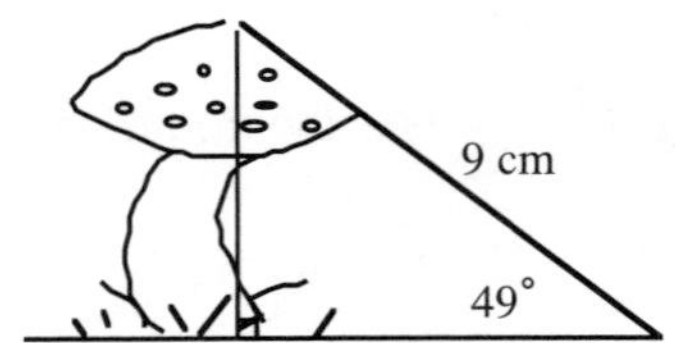


a) How tall is the mushroom?

b) How far is the bottom of the web from the base of the mushroom stem?

T-128. Δ PQR is an isosceles triangle with base angles of 46° and equal sides of length 26 cm. Draw a diagram and use the Law of Sines to solve the following.

a) Find the length of its base.

b) Find its perimeter.

c) Find its area.

T-129. Solve for x:

a) $\frac{x}{6} - \frac{1}{3} = \frac{x}{8}$

b) $x^2 - 5x + 2 = 0$

T-130. Write in simple radical form.

a) $\sqrt{108}$

b) $2\sqrt{3} \bullet 3\sqrt{6}$

c) $\frac{6}{\sqrt{10}}$

d) $\frac{2\sqrt{3}}{\sqrt{8}}$

T-131. TOOL KIT CHECK-UP

Your tool kit contains reference tools for geometry and algebra. Return to your tool kit entries. You may need to revise or add entries.

Be sure that your tool kit contains entries for all of the items listed below. Add any topics that are missing to your tool kit NOW, as well as any other items that will help you in your study of this course.

- adjacent, opposite legs (in right triangle)
- sine, cosine, and tangent ratios
- Law of Sines
- special right triangles:
  - 45°- 45°- 90°
  - 30°- 60°- 90°
- tangent and slope ratios
- fraction busters
- simple radical form
- rationalizing the denominator
- triangle side and angle size (T-106(c))

# The Trekee Clubhouse Logo

**SIMILARITY**

# *Unit 8 Objectives*

## *The Trekee's Clubhouse Logo:* SIMILARITY

This unit focuses on figures that have the same shape but not necessarily the same size. You will explore this concept, known as similarity, in polygons as well as scale drawings. The idea will be extended to two and three dimensions, specifically, studying the relationship between linear measurements, area, and volume of prisms and pyramids. You will have the opportunity to prove some similarity properties as well as to continue your work with trigonometry and practicing your algebra skills.

In this unit you will have the opportunity to:

- explore properties of figures with the same shape.
- discover the basic theorems for similarity and apply them to triangles.
- learn efficient ways to calculate lengths, areas, and volumes between similar two and three dimensional figures, especially when all of the dimensions of one or both of the figures or solids are not known.
- continue to practice algebra topics.

*Read the problem below. Over the next few days you will learn what is needed to solve it. Do not try to solve it now.*

S-0. **The Trekee's Clubhouse Logo Problem.** A group of Star Trek fans plan to build a new meeting area with an enlarged sign. The T on the old sign was outlined with a curved piece of neon tubing 13.4 feet long and the T itself was covered with gold leaf. On their current building, the circular border of the sign has a radius of 2 feet, while the new sign will have a circular border with a 5 foot radius. How long a piece of neon tubing will they need? How much gold leaf will they need to cover the new T if the old one needed 3.6 ounces? If the current model of the Enterprise weights 58 pounds, how much will the new one weigh?

*UNIT* 8

ALGEBRA
GRAPHING
RATIOS
GEOMETRIC PROPERTIES
PROBLEM SOLVING
SPATIAL VISUALIZATION
CONJECTURE AND EXPLANATION (PROOF)

# *Unit 8*

## *The Trekee Clubhouse Logo:* SIMILARITY

S-1. **DIRECTIONS:** Draw a pair of xy-axes centered on a full sheet of graph paper. Plot, label, and connect the following points to form ΔABC. (Note: it might be wise to do each of the following problems in a different color.)

A(1, 6), B(3, 1), and C(4, 3)

a) Add 3 to each x- and y-coordinate of the points A, B, and C and plot the new triangle: ΔDEF.

b) Add -5 to each x-coordinate and 2 to each y-coordinate of the points A, B, and C and plot the new triangle: ΔGHI.

c) Multiply each x- and y-coordinate of the points A, B, and C by 2 and plot the new triangle: ΔJKL.

d) Multiply each x- and y-coordinate of the points A, B, and C by -3 and plot the new triangle: ΔMNO.

e) Divide each x- and y-coordinate of the points A, B, and C by 2 and plot the new triangle: ΔPQR.

f) What do you notice about the collection of triangles?

g) Separate the triangles into two groups such that all the triangles in a given group are alike in some way. How did you separate them and why did you separate them that way?

S-2. Suppose Eulalia uses a map of Pennsylvania to determine that Valley Forge is 14 miles from downtown Philadelphia. Did she really measure 14 miles? Explain how she probably determined the distance.

Maps are examples of **SCALE DRAWINGS**. They represent the land accurately and are a manageable size. They are, of course, a reduced version of the original, which is **similar** to the original in that it has the same shape, and the distance between points on the map is proportional to the same distance on land. Thus, maps conveniently allow users to determine distances between two points. An example is shown at right.

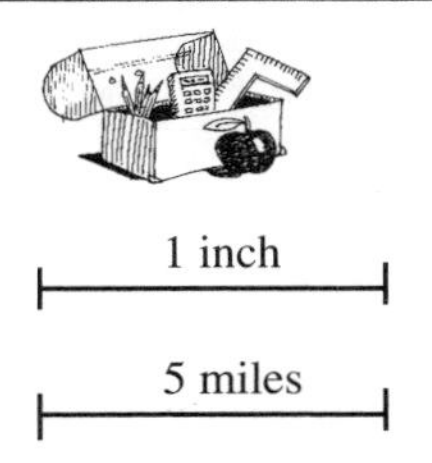


In a scale drawing, it is important to decide on the unit of measure. American maps usually represent distances in miles, but they certainly cannot use actual miles as the unit of measure. Otherwise, a map of Pennsylvania would be over 250 miles long and 450 miles wide! So, a map will include a **scale** which shows the units in which the map is drawn and how that converts to miles.

If a scale drawing is on graph paper, then each side of a square will represent a unit: a foot, a mile, a kilometer, etc.

S-3. Guillermo needs a scale drawing of his house on its suburban lot. The lot is 56' x 110'. The garage is 20 feet back from the street. He has the sketch of his house--not drawn to scale--with the measurements shown at right.

Draw a scale drawing of Guillermo's house on its lot. Use graph paper and be sure to state your scale.

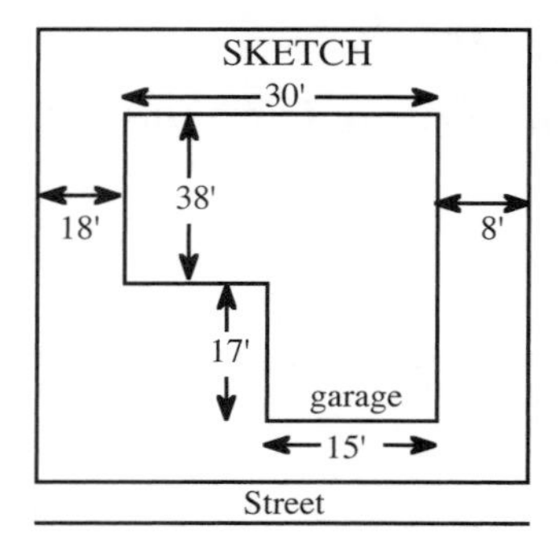


S-4. **DIRECTIONS:** Read the box below, then answer the questions that follow. Use your figures from problem S-1 as models to help visualize each situation.

> If two figures, like ΔDEF and ΔJKL in problem S-1, have the same shape, but are not necessarily the same size, we say that they are **SIMILAR**. We write **ΔDEF ~ ΔJKL**. You can create a figure like ΔDEF similar to ΔJKL by enlarging (or reducing) ΔJKL and then rotating, translating, or reflecting the enlargement (reduction). This definition applies to all figures, not just triangles.

a) If two figures, like ΔABC and ΔDEF in problem S-1, are congruent, are they always, sometimes, or never similar?

b) If two figures are similar, are they always, sometimes, or never congruent?

c) If ΔABC ≅ ΔDEF and ΔDEF ~ ΔJKL, are ΔABC and ΔJKL always, sometimes, or never similar?

d) If ΔPQR ~ ΔGHI and ΔGHI ~ ΔMNO, are ΔPQR and ΔMNO always, sometimes, or never similar?

e) If ΔABC ≅ ΔDEF and ΔDEF ~ ΔJKL, are ΔABC and ΔJKL always, sometimes, or never congruent?

f) Are all of the triangles in problem S-1 similar? Justify your response.

g) Photographic enlargements made from the same negative are similar. Why?

S-5. Figure ABCD at right is a parallelogram. In Unit 6 you proved that each diagonal cuts the parallelogram into two congruent triangles and that opposite sides and opposite angles in the parallelogram are congruent. Use these facts to help prove that AE = CE.

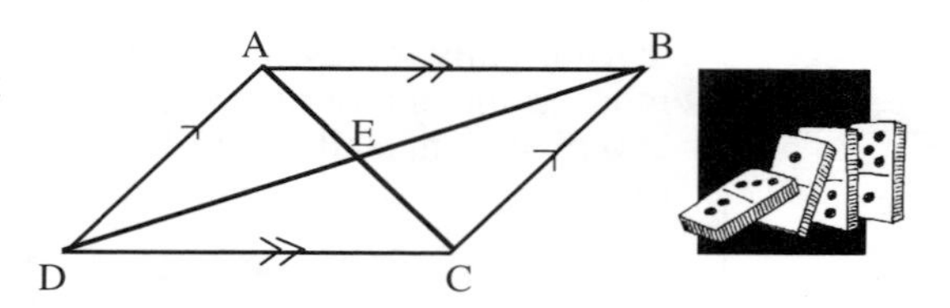

S-6. An isosceles right triangle has an area of 72 square centimeters.

a) Show how to put two of these triangles together to form a square.

b) Find the lengths of each side of the original triangle.

S-7. Solve the following problems for all unknowns.

a)

b) $\frac{2}{3}x - 18 = \frac{7}{4}(2x + 5)$

c)

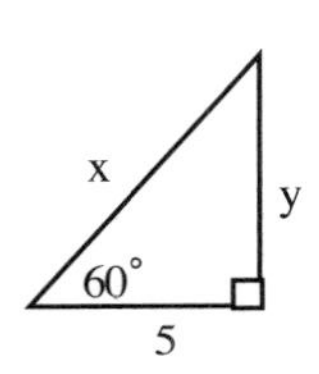


d)

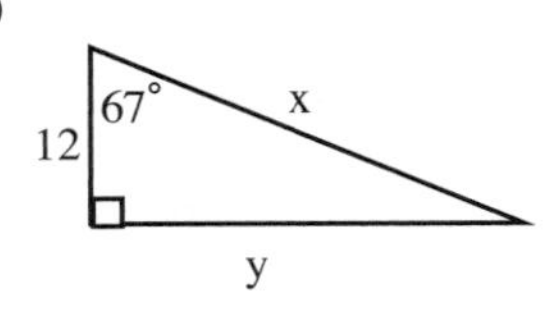


S-8. Solve for x:

a) $\frac{13}{38} = \frac{x}{110}$

b) $\frac{\sqrt{46}}{82} = \frac{x}{9}$

c) $(\sqrt{2})x + \sqrt{8} = \sqrt{18}$

d) $\frac{x}{(\sqrt{2})} + \sqrt{8} = \sqrt{72}$

S-9. Determine whether the two triangles at right are similar by completing the following:

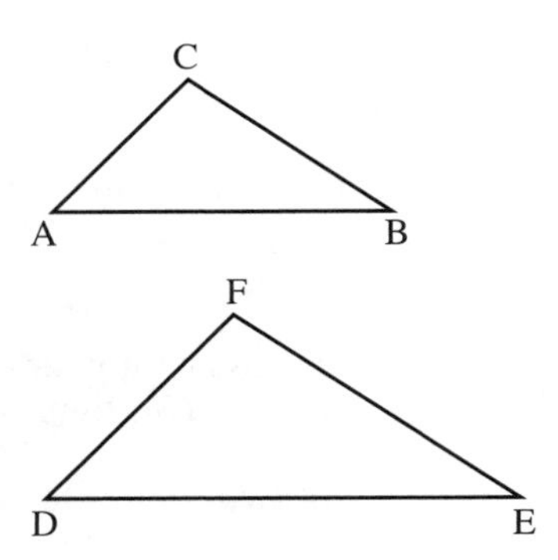


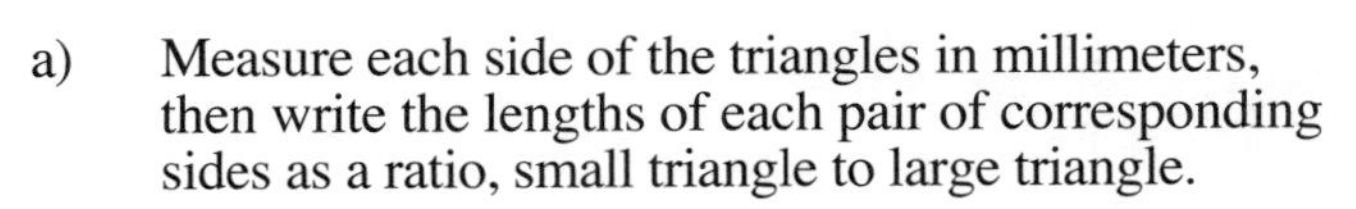

a) Measure each side of the triangles in millimeters, then write the lengths of each pair of corresponding sides as a ratio, small triangle to large triangle.

b) Convert the ratios into decimals, compare the results, then record your observations and decide whether the triangles are similar.

S-10. Sketch a figure similar but NOT congruent to each of the following:

a) a circle of radius 4 cm

b) a square of side 6 cm

c) a rectangle with dimensions 3 cm by 6 cm

d) a right angle

S-11. Discuss with your teammates if the figures described in each part below are similar or not. Be sure to write a reason for and/or draw a diagram that illustrates each answer.

a) all equilateral triangles

b) all right triangles

c) all cubes

d) all spheres

e) all tetrahedra (this is the plural of "tetrahedron")

f) all right isosceles triangles

S-12. **DIRECTIONS:** Set up a pair of xy-axes and plot the following points: A(5, 2), B(1, 2), C(5, 6), D(3, -2), E(6, -5), and F(2, -6). Do not connect any of them yet.

a) Connect points A, B, and C to form ΔABC. Calculate the length of each side of the triangle.

b) What kind of triangle is ΔABC? Justify your answer.

c) Connect points D, E, and F. Calculate the lengths of the sides of the triangle.

d) What kind of triangle is ΔDEF? Justify your answer.

e) Are the triangles similar? Explain.

S-13. Midori took some measurements of a wall with two windows which she plans to wallpaper. She made a quick sketch and wrote in some dimensions.

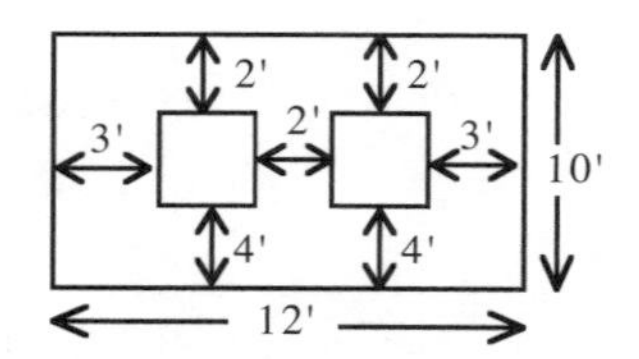


a) Do you need to make a scale drawing before you can find the surface area of the wall? Explain completely.

b) How much wallpaper will she need?

S-14. Discuss with your partner or team how to determine whether two figures are similar. Then write a sentence or two that summarizes what is necessary for two figures to be similar so that a friend who is not taking geometry would understand similarity.

S-15. For each picture below, draw a figure on your graph paper that is similar but not congruent to the one given.

a)

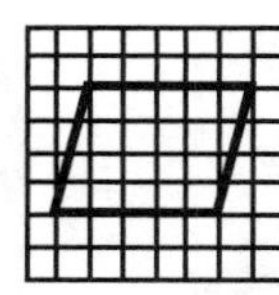

b)

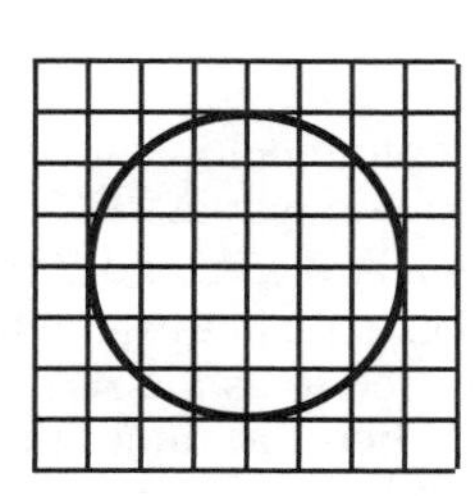

S-16. A triangle has sides of lengths 10 cm, 20 cm, and $10\sqrt{3}$ cm. Sketch the triangle, then find the measures of its angles. Justify your answer.

S-17. A 20' rope is attached to the top of a 16' flagpole.

a) If the rope could be attached <u>anywhere</u> on the ground (limited, of course, by the length of the rope), describe the region containing all the points on the ground where the rope could be attached.

b) What angle does the full length of the rope make with the ground when it is pulled tight?

S-18. Solve the following for x:

a) $x^2 + 5x - 6 = 0$

b) $x^2 - 14x = 2$

S-19. Write in simple radical form.

a) $\sqrt{32}$

b) $\dfrac{8}{\sqrt{28}}$

c) $\sqrt{54} + \sqrt{384}$

## RATIOS OF CORRESPONDING SIDES OF SIMILAR TRIANGLES

S-20. Read the theorem below and add it to your tool kit. Then explain why $\Delta ABC \sim \Delta A'B'C'$ if all we know is $\angle A \cong \angle A'$ and $\angle B \cong \angle B'$. Note that once you explain this, the theorem below may be referred to as the **AA Triangle Similarity Theorem** rather than the AAA Triangle Similarity Theorem.

**AAA Triangle Similarity Theorem**

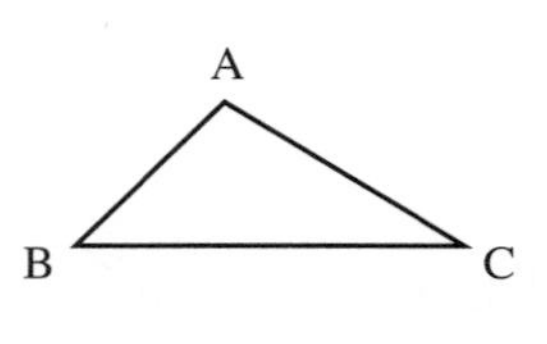


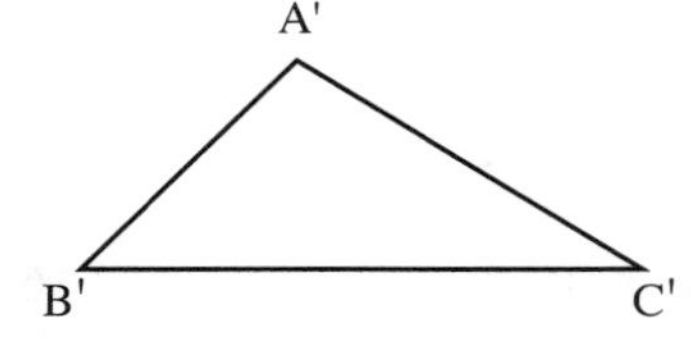


If $\Delta ABC$ and $\Delta A'B'C'$, with $\angle A \cong \angle A'$, $\angle B \cong \angle B'$, and $\angle C \cong \angle C'$, then $\Delta ABC \sim \Delta A'B'C'$.

S-21. The triangles at right are drawn to scale.

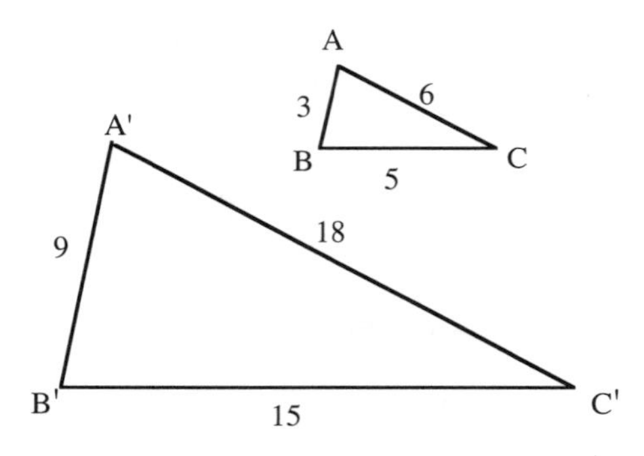


a) Are the triangles similar? Explain.

b) Calculate $\frac{AB}{A'B'}$ and reduce.

c) Calculate $\frac{BC}{B'C'}$ and $\frac{AC}{A'C'}$ and reduce.

d) What do you notice about these ratios?

S-22. Calculate and reduce these ratios for the sides of the triangles in the previous problem.

a) $\frac{AB}{BC} =$ $\frac{A'B'}{B'C'} =$

b) $\frac{BC}{AC} =$ $\frac{B'C'}{A'C'} =$

c) $\frac{AB}{AC} =$ $\frac{A'B'}{A'C'} =$

d) What do you notice?

S-23. **DIRECTIONS:** Discuss part (a) with your team members and agree on a conjecture, then complete the other parts of the problem.

a) Write a conjecture relating similar triangles and side lengths.

b) Next calculate these ratios for the same triangles (problem S-21), reduced of course.

$\frac{AB}{A'B'} =$ $\frac{AB}{B'C'} =$

$\frac{AC}{B'C'} =$ $\frac{BC}{A'B'} =$

c) Should any of these ratios be the same? Explain.

d) Refine your conjecture in part (a) if necessary. What must be stressed about the sides you use?

S-24. On graph paper, outline and shade the interior of a rectangle with a length of 1 and a width of 3.

a) Outline and shade the interior of a new rectangle whose length and width are 8 units longer than the first. What are the dimensions of this new rectangle?

b) What is the ratio of the lengths of the two rectangles?

c) What is the ratio of their widths?

d) Does adding the same amount to each side result in two similar figures, that is, is the new rectangle a uniform enlargement of the original? Explain why or why not.

S-25. **Carefully** read the box below, add the information to your tool kit, and then use it in the following problem.

The **RATIO OF SIMILARITY** between any two similar figures is the ratio of any pair of corresponding sides. Simply stated, once it is determined that two figures are similar, all of their pairs of corresponding sides have the same ratio.

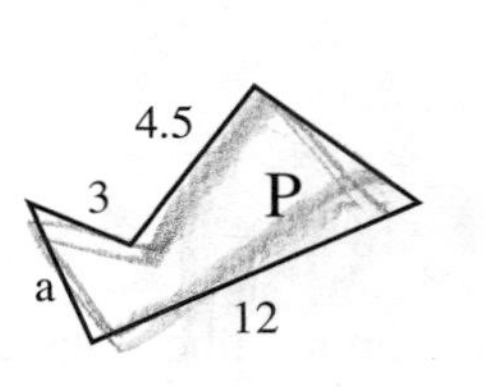


~

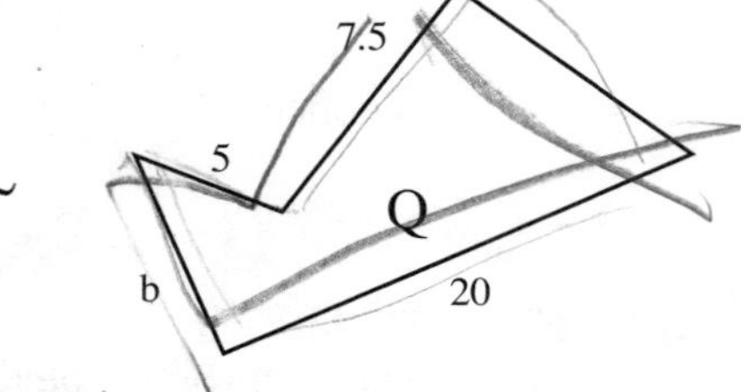


The ratio of similarity of figure P to figure Q, written P : Q, is $\frac{3}{5}$.

a) Test $\frac{4.5}{7.5}$ as well as $\frac{12}{20}$. Do they have the same ratio?

b) What does $\frac{a}{b}$ equal?

Note that the ratio of similarity is always expressed in lowest possible terms. Also, the order of the statement, P : Q or Q : P, determines which order to state and use the ratios between pairs of corresponding sides.

S-26. Find the ratio of similarity, P to Q, for the following similar figures:

a)

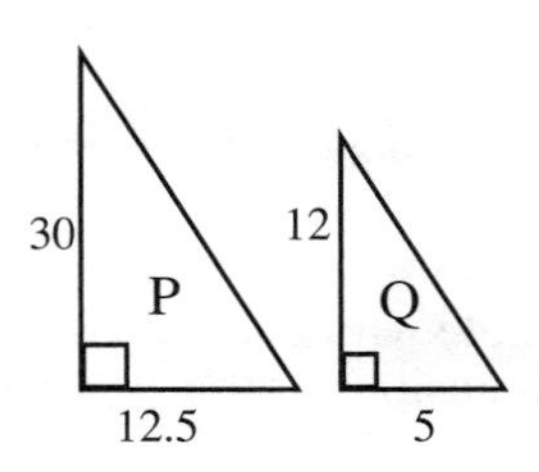


b)

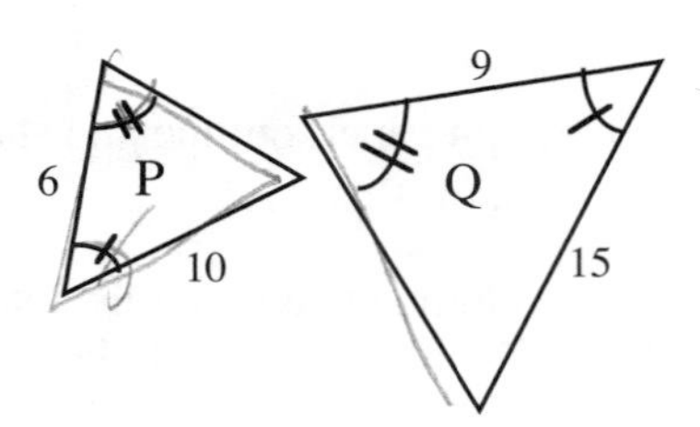


c)

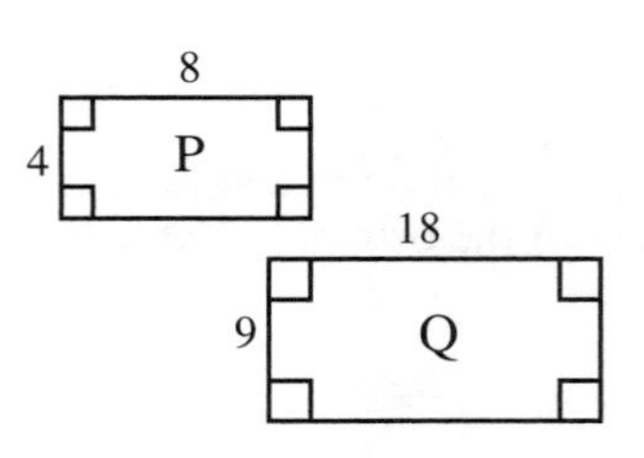


d)

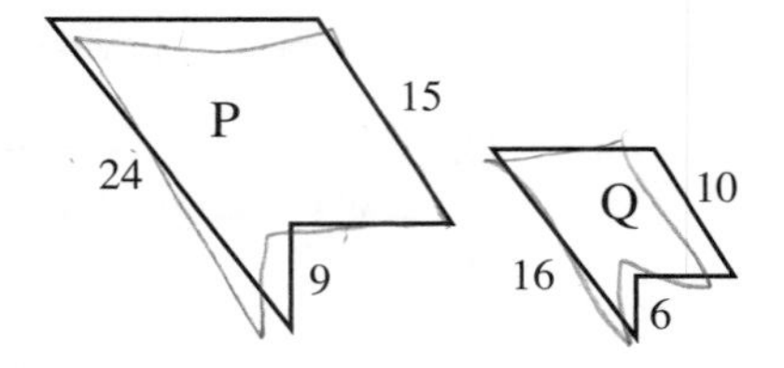

S-27. $\Delta$DEF and $\Delta$D'E'F' are similar.

a) Find E'F' by solving the equation (proportion) below.

$$\frac{6}{15} = \frac{8}{E'F'}$$

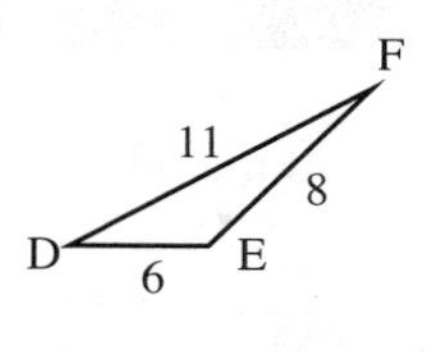


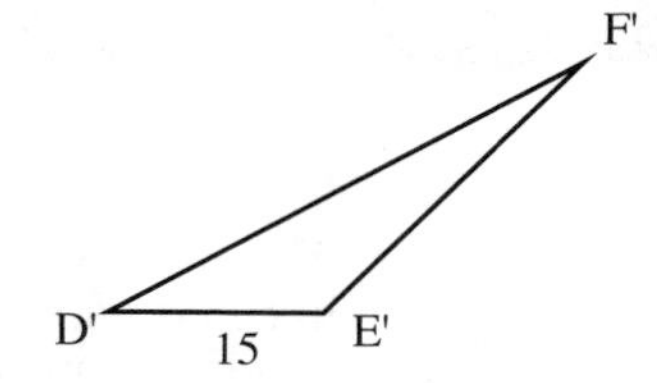


b) Write an equation like the one in part (a) to solve for D'F'.

S-28. Use the figure at right to answer the questions below.

a) Is $\Delta ABC \sim \Delta EDC$? Prove your answer.

b) If AB = 10 cm, BC = 8 cm, and CD = 6 cm, find DE. Be sure that you match corresponding sides.

c) Under the same conditions as part (b), find CE.

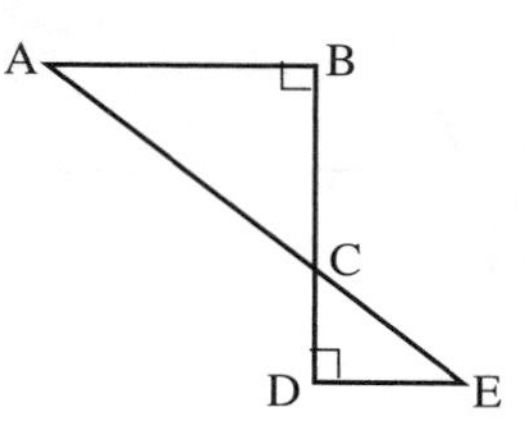

S-29. In trapezoid ABCD at right:

a) find the length of $\overline{AD}$.

b) find the area of ABCD. .

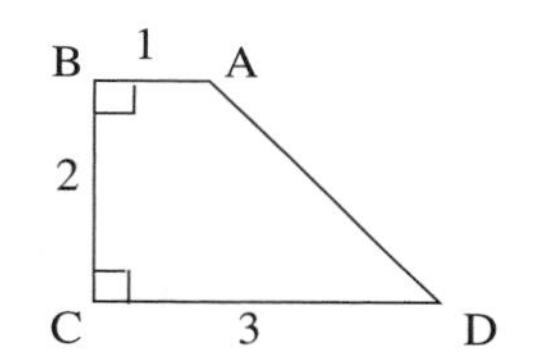


S-30. Use ratios of similar triangles to solve for x and y.

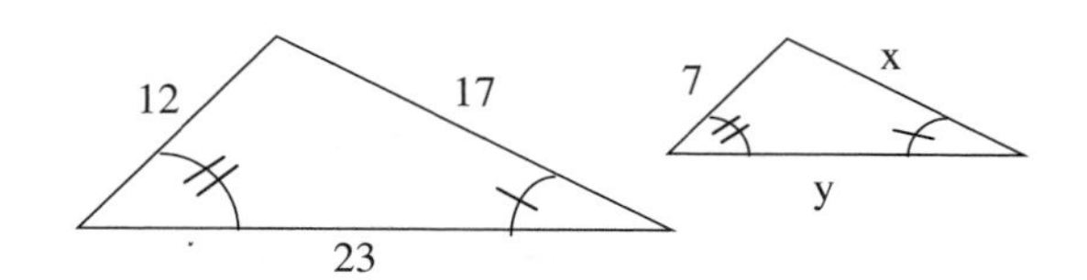


S-31. The diagonal of a rectangle is three times the length of its shortest side. Find the area of the rectangle if the length of its longest side is 20 cm.

S-32. Sarah and Armida drove 380 miles to Los Angeles together. Armida drove 50 miles more than Sarah. How far did each drive?

S-33. Simplify the following:

a) $(4x^3y^4)^2(3x^2y^2)$

b) $3x^2(4x^4 - x^3 + 13x - 8)$

c) $(2x - 5)(0.85x + 9)$

d) $(3x - 2)^2$

## SIMILAR TRIANGLES

S-34. Eleanor and John are working on a geometry problem together. They know that line $m$ is parallel to $\overline{BC}$ and they first want to show that $\Delta AED \sim \Delta ABC$.

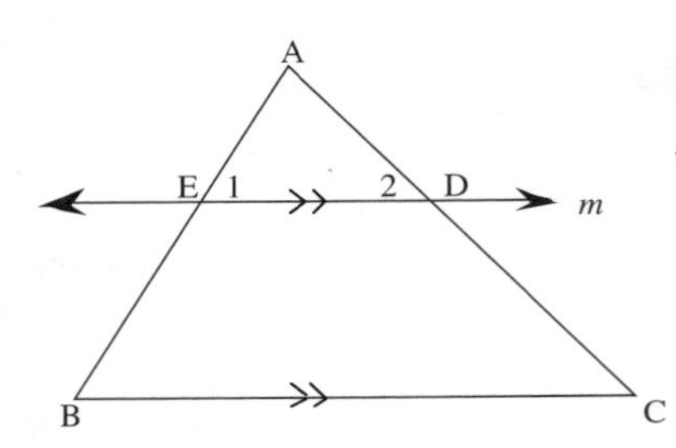


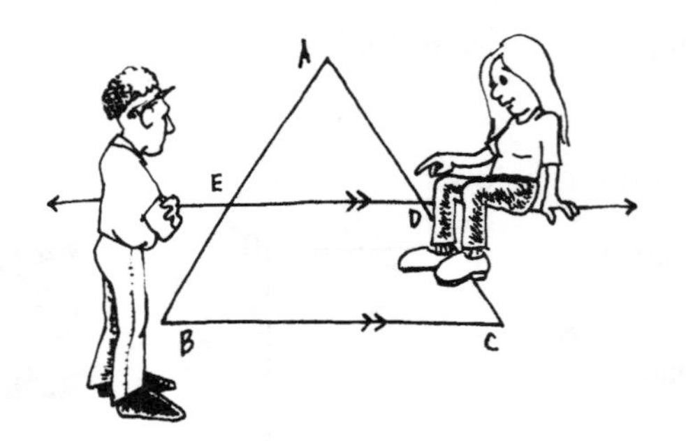


Eleanor says, "This is easy. We have parallel lines so the triangles are similar by AA." "Hold on a minute!" John replies. "What angles are equal?" "Corresponding angles!" she says.

a) Name the pairs of corresponding angles Eleanor sees. Are they equal? Why or why not?

b) Are the triangles ($\Delta AED$ and $\Delta ABC$) similar? Explain.

When John sees that the triangles are similar, he suggests redrawing them as shown at right. "Look," he says, "Now you just write an equation like we did in some of the earlier problems." He suggests the following equation:

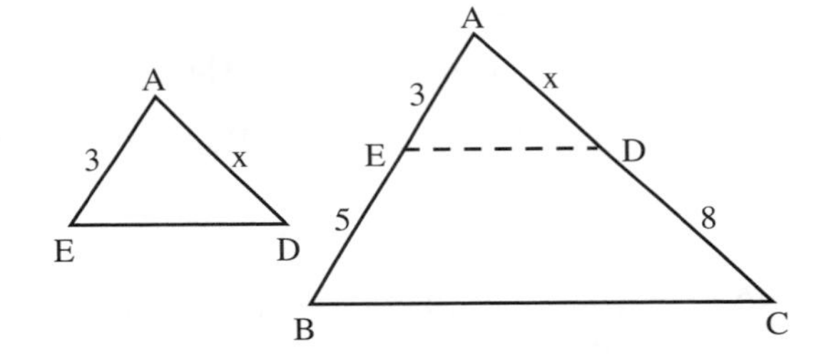


$$\frac{3}{3+5} = \frac{x}{x+8}$$

c) Solve the equation for x.

Eleanor asks what would happen if x was the length of a lower piece of the large triangle? So they change the values of the segments as shown at right, then write the equation below:

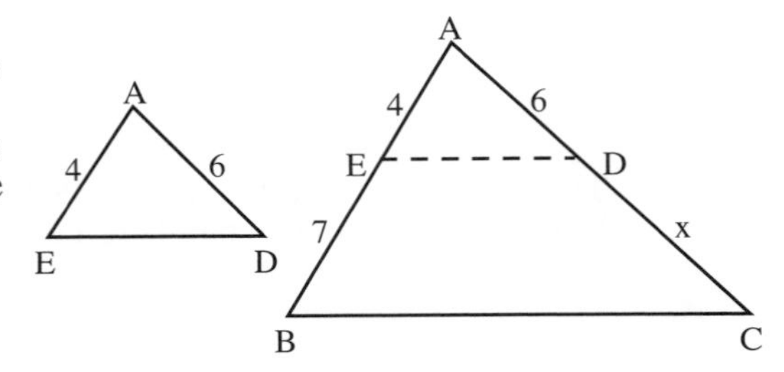


$$\frac{4}{4+7} = \frac{6}{6+x}$$

d) Solve the equation for x.

e) In the original figure above, suppose AE = 6, AD = 8, and DC = 16. Draw figures, write an equation, and solve for EB. Then add the following definition to your tool kit.

An equation stating that two ratios are equal is called a **PROPORTION**. The equations in this problem are examples of proportions.

S-35. Find x in each figure. Be consistent in matching corresponding parts of similar figures.

a)

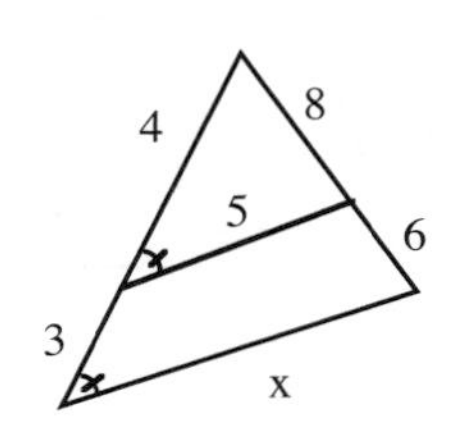


b)

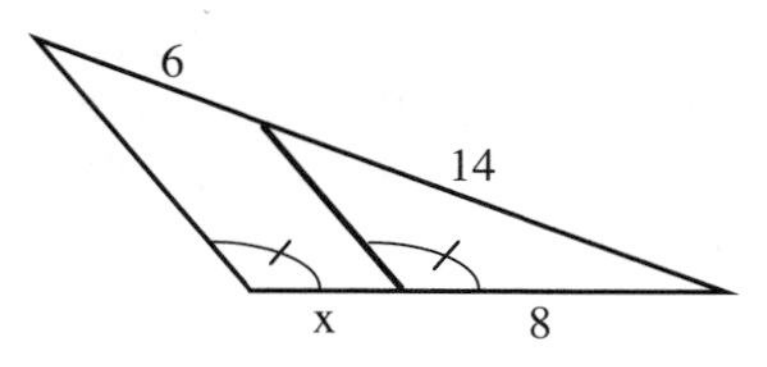


c) Careful!

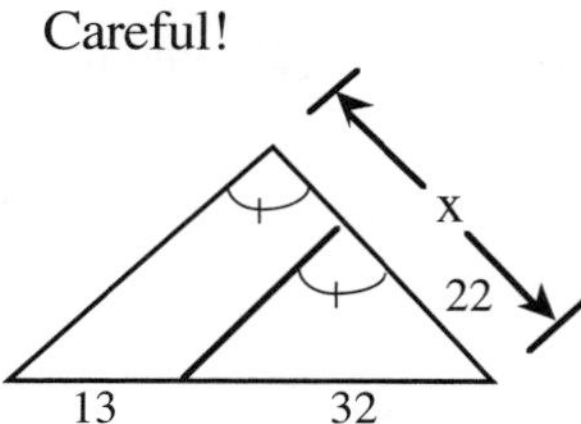


d)

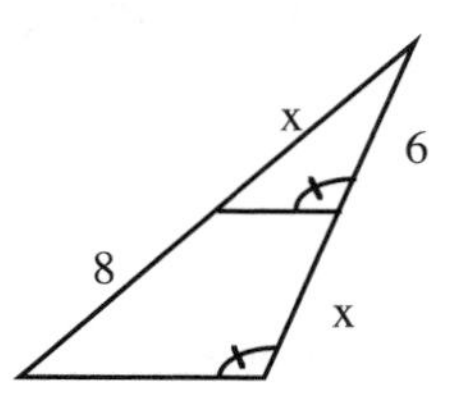


S-36. To find the distance AB across the swamp, sights were taken as shown with $\overline{CD}$ parallel to $\overline{AB}$. Find the length of AB. Show your work. Which segment lengths did you NOT need?

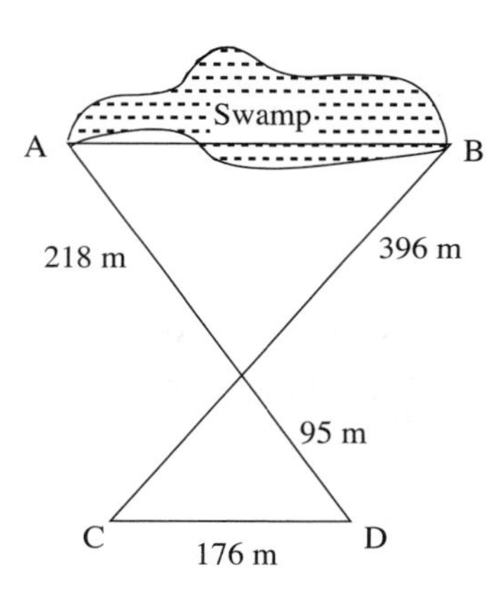

S-37. Shalley starts with a 1 x 2 rectangle. She is going to draw a sequence of 100 different rectangles by taking her original 1 x 2 rectangle and adding one to both the length and width. Her second rectangle is 2 x 3, her third is 3 x 4, and so on.

a) What are the dimensions of the 15th rectangle?

b) Are all the rectangles similar? If not, explain why not. If so, what is the ratio of similarity?

S-38. The figure at right has a smaller isosceles triangle ($\Delta PRS$) drawn inside a larger isosceles triangle ($\Delta PQT$). Both triangles have a common vertex ($\angle P$). Is it true that $\Delta PQS \cong \Delta PTR$? Write a proof or explain why they are not necessarily congruent.

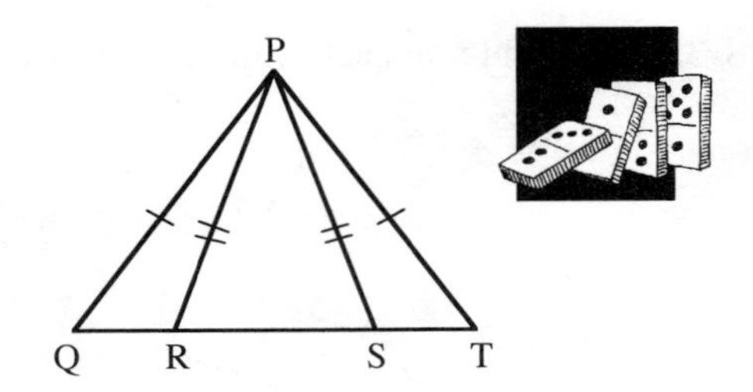


S-39. At the same time the shadow of a six foot fence post is five feet long, the shadow of a flag pole is 20 feet long. How high is the flag pole?

S-40. Find p and q from the given information.

a)

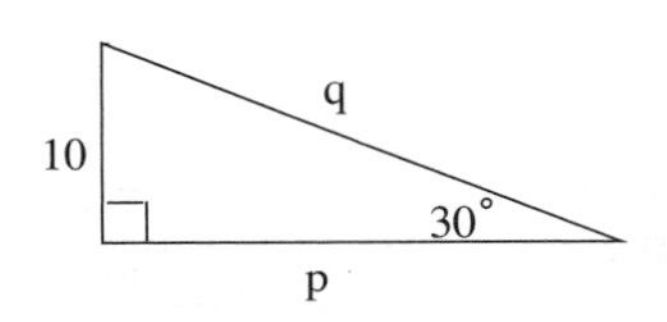


b)

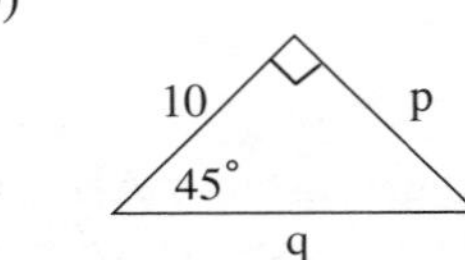


S-41. Graph $y > \frac{1}{3}x - 3$ and $y \geq -x^2 - 2x$ (for $-4 \leq x \leq 2$) on the same set of coordinate axes. Describe the region of the graph common to <u>both</u> inequalities. Reminder: do not ignore the negative sign that precedes the $x^2$ term--square first, <u>then</u> change the sign.

a) How would the graph change if the **first** inequality were changed from > to < ? Make a sketch of what it would look like.

b) Change the inequality signs above so that the graph becomes an enclosed region. Estimate the area of the region.

S-42. Solve for x and y:

a) $2x - 3y = 17$
$3x + y = 31$

b) $y = 2x - 1$
$4x - 5y = 8$

S-43. Simplify the following.

a) $(a^4b^6)^3$

b) $\dfrac{x^5y^7}{x^2y^6}$

c) $(a^8b^2)^5(a^2b^4)^3$

## APPLICATIONS OF SIMILARITY IN TRIANGLES

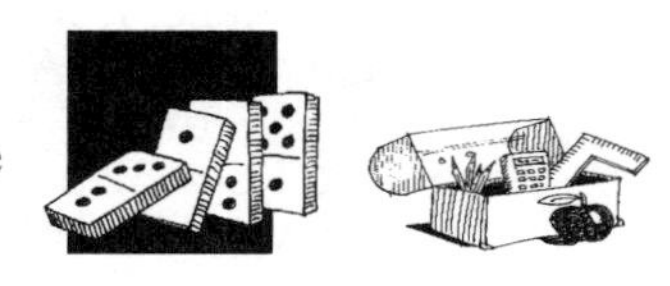

S-44. In Unit 5 you studied transformations and congruence. The figure at right shows a pool table and seven balls. Suppose you wanted to shoot the cue ball (C) and hit the ball labeled A. Since you cannot hit it directly (it is blocked by several other balls), you decide you will hit the cue ball into the right side of the table so that it rebounds and continues on to hit A. How can you determine where to hit the side of the table so that ball C will hit ball A?

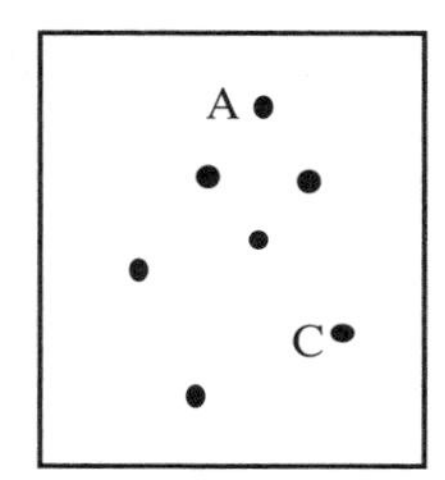


a) Trace a copy of the table on your paper and reflect (flip) ball A across the right side of the table and label the point A'. An easy way to reflect A is to fold the paper along the right side of the pool table. Label the point where $\overline{AA'}$ crosses the side of the pool table D.

b) You have a clear line of sight from C to A'. Draw the segment $\overline{CA'}$, label the point where it crossed the side of the pool table B, then draw $\overline{BA}$.

c) Use the properties of a reflection to prove that $\Delta ADB \cong \Delta A'DB$.

d) Label the lower right corner of the table E. Prove that $m\angle EBC = m\angle DBA$.

e) Read the following definition and add it to your tool kit.

In the figure at right, $\angle a$ is the **ANGLE OF INCIDENCE** and $\angle b$ is the **ANGLE OF REFLECTION**. In the pool table example, $\angle a$ is the angle of approach and $\angle b$ is the angle of rebound. $\mathbf{m\angle a = m\angle b}$. In this figure the arrow marks indicate direction, not parallelism.

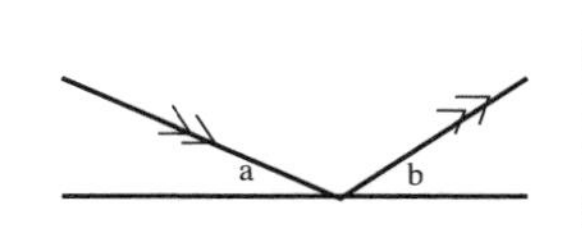


S-45. Latoya looks down into a mirror that is lying on the ground between her and a flagpole. As she looks into the mirror, she can see the top of the pole. Her eye level is five feet from the ground and she is standing 4 feet from the mirror, which is 30 feet away from the base of the pole.

a) Draw a picture of the situation, filling in any measurements you know, and prove that the triangles are similar.

b) How tall is the pole?

S-46. Latoya forgot her clinometer, so she devised another method to find the height of the flagpole. Her eye is at A, 5 feet above the ground. A friend holds an 8 foot stick vertically 32.5 feet from the pole (so FD = 32.5). If D'A = 3.31 feet, how tall would Latoya measure the flagpole to be if she can just see the top of the flagpole over the stick? Show how she would do this based on the given data.

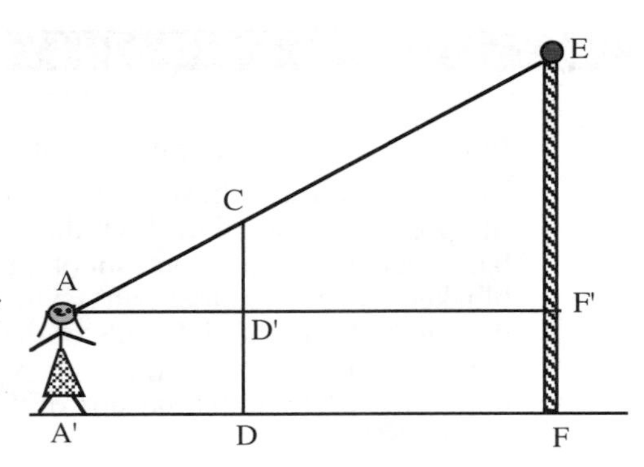


S-47. In the previous problem we assumed that the ground is level. What if the pole is at the top of a hill which makes a 20° angle with the horizontal? If Latoya stands up straight, the stick, $\overline{CD}$, is vertical, and line AF' is parallel to the hillside, will this method still work? Explain why or why not.

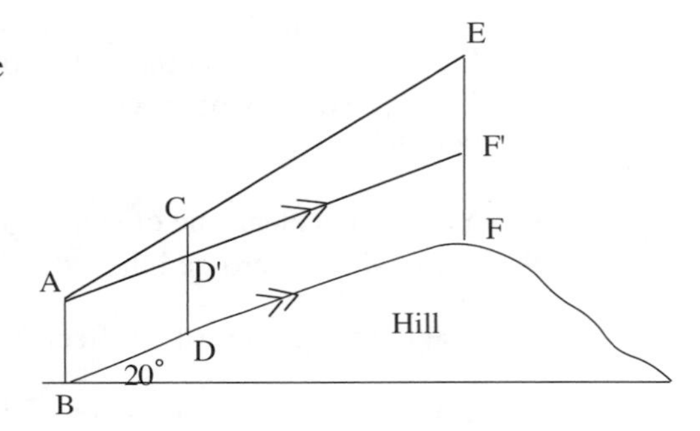


S-48. Use the information in each diagram to solve for x.

a) $\Delta ABC \sim \Delta DEF$

b) Careful!

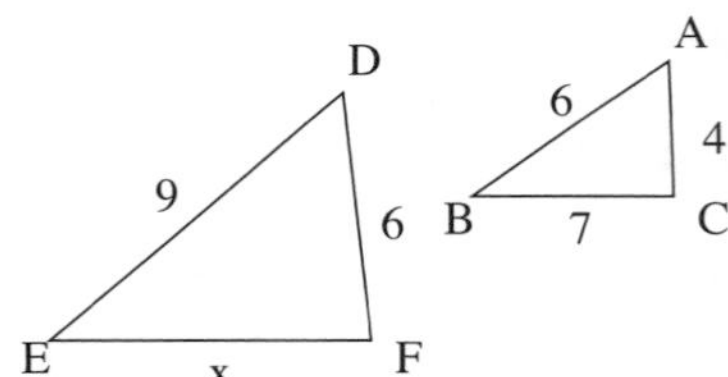


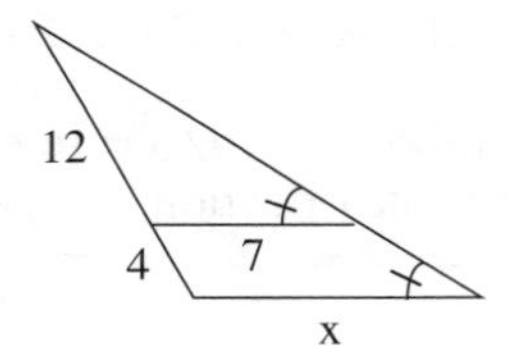


S-49. On dot paper or graph paper, draw $\Delta DON$ so that the ratio of sides of $\Delta DON$ to $\Delta TIM$ is 1:1.

a) Are the triangles similar?

b) Are the triangles congruent?

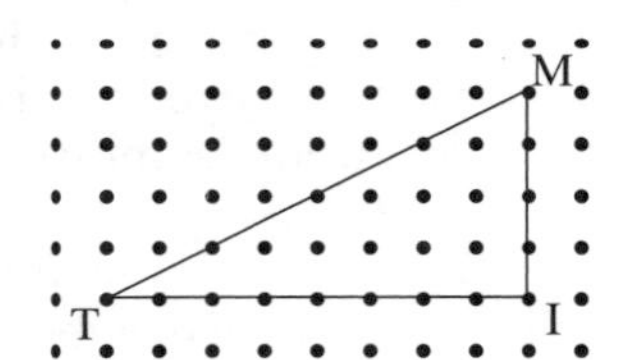

S-50. To measure the distance MT across a river, points R and P were lined up with T, a tree. Then distances MP, PS, and SR were staked out as shown.

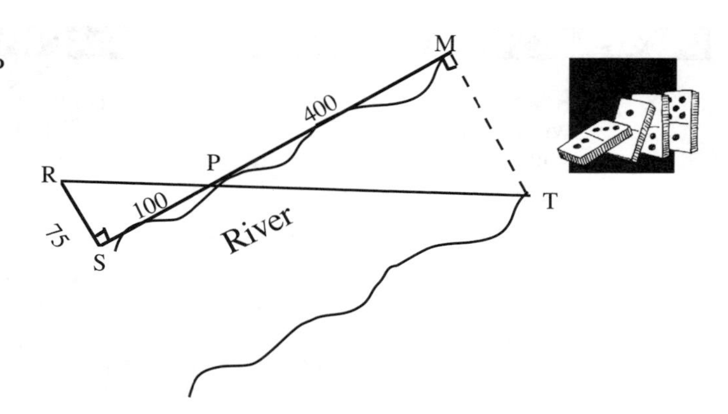


a) Why are the triangles similar?

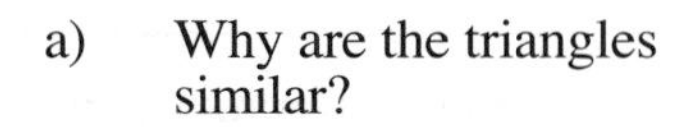

b) How wide is the river?

S-51. The sum of a certain integer and the square of the same integer is 992. Find the integer. Guess and check with a calculator might be helpful.

S-52. Nebio is 6'2" tall. When his shadow is 7'6" long, what is the angle made by the ground and his line of sight to the sun?

S-53. Consider the graphs of these four equations.

$y = x$ $\qquad$ $y = x^2$ $\qquad$ $y = x^3$ $\qquad$ $y = x^4$

a) If we started substituting for values of x, with $x \geq 1$, which of these equations will give larger results sooner?

b) If we substitute for values of x within the interval $0 \leq x \leq 1$, which of these equations will give the largest values for y? Test 0.5 for each equation to verify your conjecture.

S-54. Solve for x.

a) $\frac{x}{2} - 3 = \frac{1}{7}$

b) $0.03724x + 0.2214 = 3.9715$

c) $x^2 - 7x - 19 = 0$

d) $(x - 1)(x + 1) + (x - 1)(x + 2) = 0$

## COMPARING THE PERIMETER AND AREA OF SIMILAR SHAPES

S-55. Read problem S-0 at the beginning of this unit. Over the next few days you will learn what is needed to solve it. DO NOT ATTEMPT TO SOLVE IT RIGHT NOW. After you have a sense of what the problem is asking, reread the definition in problem S-25, then go on to problem S-56.

S-56. **DIRECTIONS:** Figure a is a 1 x 1 square. Each subsequent figure is formed by multiplying the side length of the first figure, which is 1, by a **magnifying factor** (constant multiplier). For example, the sides of figure c are 3 times the length of figure a (and thus a 3 x 3 square). Make two tables on your paper like the ones below and fill in the rest of the perimeters and areas in the first table, then the ratios in part (a).

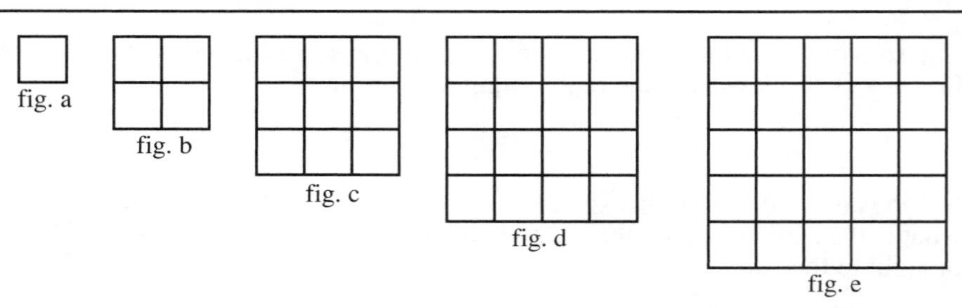


| | side length | perimeter | area |
|---|---|---|---|
| figure a | 1 | 4 | 1 |
| figure b | 2(1) = 2 | 8 | 4 |
| figure c | 3(1) = 3 | | |
| figure d | 4(1) = 4 | | |
| figure e | 5(1) = 5 | | |

a) Write the following ratio names, calculate them, and REDUCE WHEN POSSIBLE:

| Comparing figure c to figure a | Comparing figure d to figure a |
|---|---|
| $\frac{\text{length of side of fig. c}}{\text{length of side of fig. a}}$ | $\frac{\text{length of side of fig. d}}{\text{length of side of fig. a}}$ |
| $\frac{\text{perimeter of fig. c}}{\text{perimeter of fig. a}}$ | $\frac{\text{perimeter of fig. d}}{\text{perimeter of fig. a}}$ |
| $\frac{\text{area of fig. c}}{\text{area of fig. a}}$ | $\frac{\text{area of fig. d}}{\text{area of fig. a}}$ |

b) Use the results from part (a) to write a summary of your observations about the relationships between side ratios, perimeter ratios, and area ratios in relation to the magnification factors. Look for patterns. You can check your ideas by writing the same three ratios comparing figure b to a and e to a. Formulate your ideas into a conjecture.

c) The magnification factor between figure b and figure d is two. Write the three ratios (side, perimeter, and area) comparing figure d to b as in part (a) and test your conjecture from part (b). Is it true? If not, revise your conjecture.

S-57. **DIRECTIONS:** For EACH pair of right triangles below, do ALL four tasks.

1) Calculate the area of each figure.
2) Write the ratio of the areas, small to large, and reduce if possible. Note: this is easier if you leave answers as fractions, even if the denominator is 1. Do not write decimals.
3) Write the ratio of the shortest pair of corresponding sides, small to large, reduce if possible, and then square it.
4) Compare your results in parts (2) and (3). What do you notice?

a)

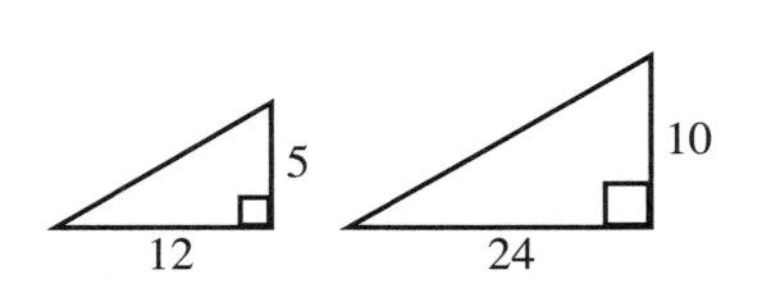


b)

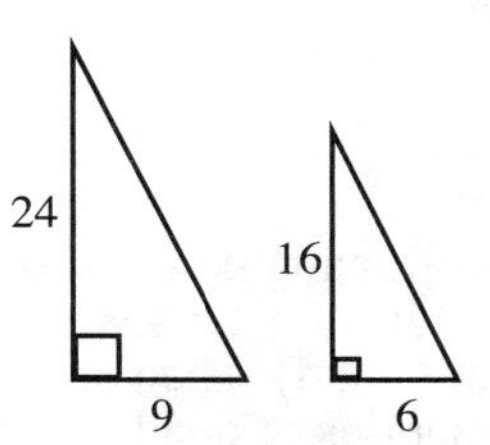


S-58. **DIRECTIONS:** For EACH problem below, do ALL four tasks.

1) Calculate the area of each figure.
2) Write the ratio of the areas, small to large, and reduce if possible.
3) Write the ratio of the shortest pair of corresponding lengths, small to large, reduce if possible, and square it.
4) Compare your results in parts (2) and (3). What do you notice?

a)

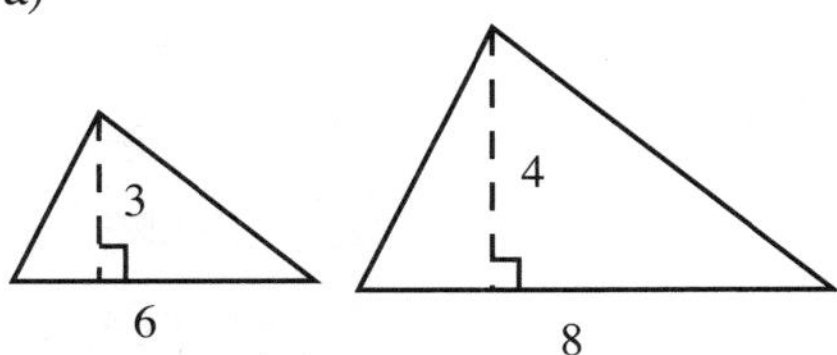


b)

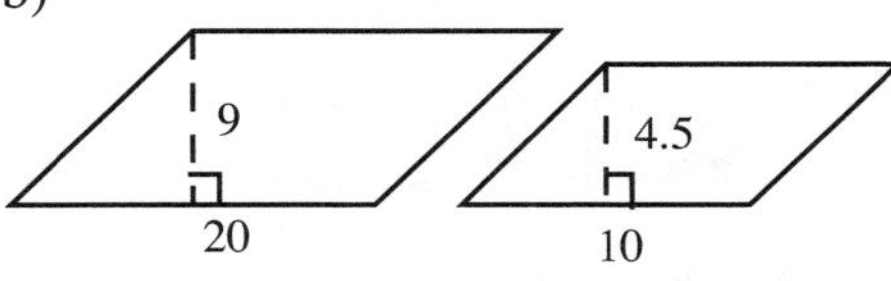


c)

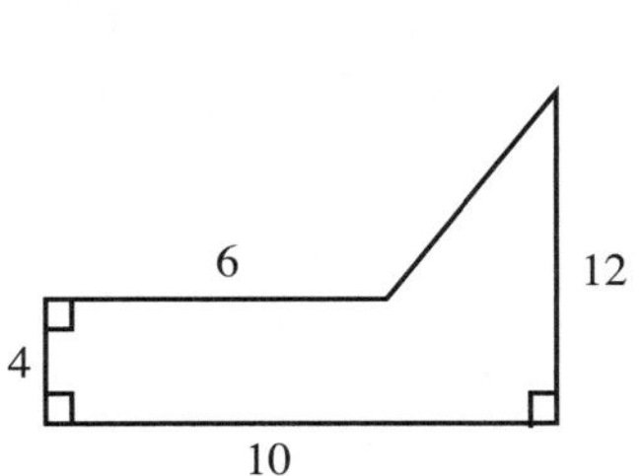


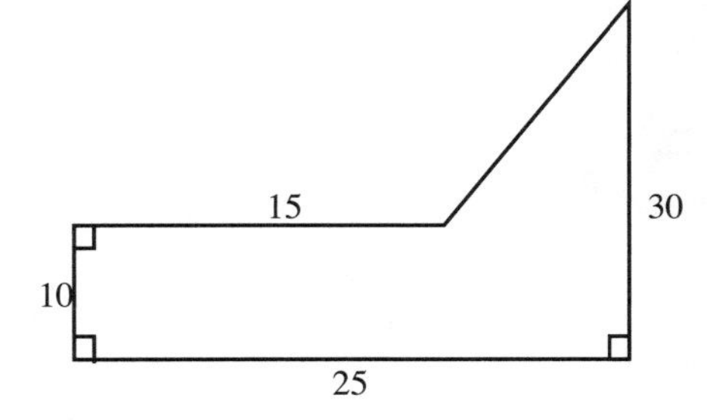

S-59. Given $\Delta ABC$ is similar to $\Delta A'B'C'$ as shown below.

a) Find the ratio of similarity for the triangles.

b) Find the unknown lengths A'B' and B'C'.

c) Find the ratio of their areas.

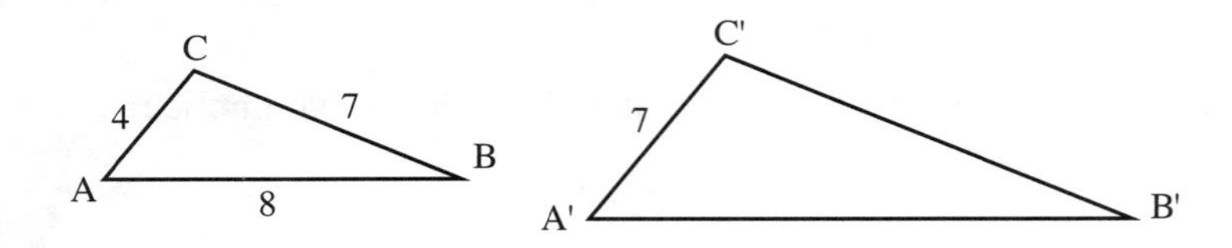


S-60. Plot the following points and connect them to form quadrilateral TEAM: T(-6, 8), E(-2, 10), A(-1, 2), M(-4, 4). Then complete the following transformations of quadrilateral TEAM.

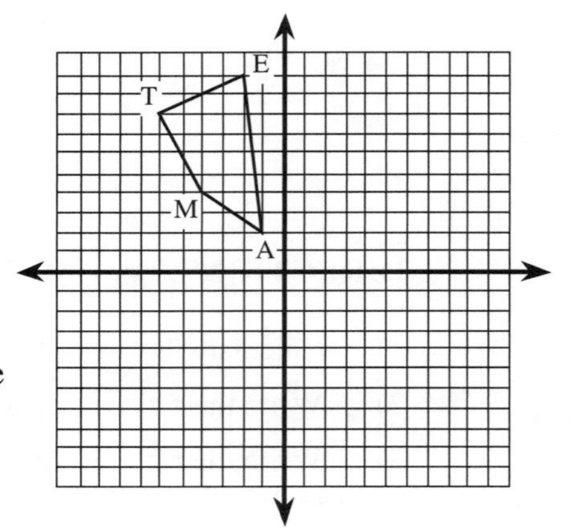


a) Reflect (flip) TEAM across the y-axis.

b) Translate (slide) TEAM into the third quadrant so that T' has coordinates (-9, -4).

c) Rotate (turn) TEAM 180° counter-clockwise about the origin. Here's a check: Point M' will have coordinates (4, -4).

S-61. A cable 100 feet long is attached 70 feet up the side of a building. If it is pulled taut,

a) what angle does the cable make with the ground?

b) what angle does it make with the building?

S-62. A rectangle's length is 60% greater that its width. Find its dimensions if its perimeter is 400 cm.

S-63. To determine whether or not his homework will be graded, Marshall is given this choice: either he can roll a die or flip a coin. If he rolls a die and an even number comes up, the teacher will grade his homework. If an odd number comes up, he does not have to turn in his homework but he will get full credit. If he flips the coin, his homework is graded if heads comes up. If tails comes up, he gets full credit. What do you think he should do? Justify your answer.

S-64. Sketch a graph showing the relationship of the height of a football after it has been punted and the time, in seconds, since it has been punted.

## LENGTH, AREA, AND VOLUME RATIOS FOR SIMILAR FIGURES

S-65. Complete the number sequence patterns for the first ten terms. Look for a pattern that is NOT additive.

a) 1, 4, 9, 16,…

b) 1, 8, 27, 64, 125, ...

c) For any number, n, write a general term that describes each sequence. Also write the name of each set of numbers.

**S-66.** **DIRECTIONS:** So far we have examined similarity ratios for flat (two-dimensional) figures. If we extend the table developed in problem S-56 with squares by adding a column for volume, we can examine cubes and the relationship between volumes of similar solids (three-dimensional figures). Use the resource page provided by your teacher.

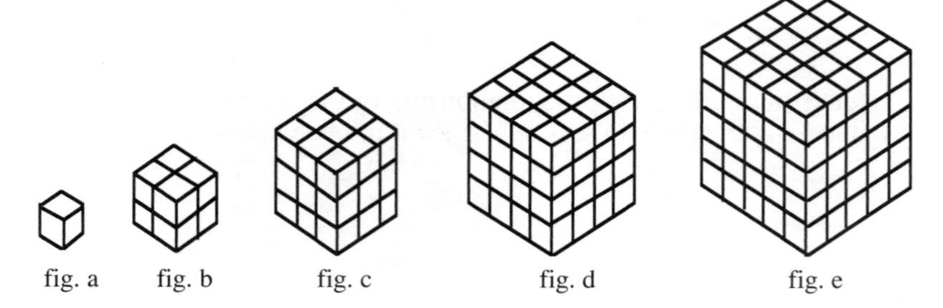


a) In our new table, notice that the first three columns now refer to edges and faces. Complete the fourth column for the volumes of the cubes on the resource page.

| | edge (side) length | face perimeter | face area | volume |
|---|---|---|---|---|
| figure a | 1 | 4 | 1 | 1 |
| figure b | 2(1) = 2 | 8 | 4 | 8 |
| figure c | 3(1) = 3 | 12 | 9 | |
| figure d | 4(1) = 4 | 16 | 16 | |
| figure e | 5(1) = 5 | 20 | 25 | |

b) Use the table in part (a) to complete the following table of ratios on your resource page. Reduce perimeter ratios to lowest terms. Avoid decimals here; the fractions make the relationships easier to see. Note that the ratios compare the LARGER FIGURE to the SMALLER FIGURE. Review **magnification factor** in the directions for problem S-56 before you begin.

| ratio | magnification factor | edge length (side) | perimeter of a face | area of a face | volume |
|---|---|---|---|---|---|
| $\frac{\text{fig. b}}{\text{fig. a}}$ | | | | | |
| $\frac{\text{fig. c}}{\text{fig. a}}$ | $\frac{3}{1}$ | $\frac{3}{1}$ | $\frac{12}{4}=\frac{3}{1}$ | $\frac{9}{1}$ | $\frac{27}{1}$ |
| $\frac{\text{fig. d}}{\text{fig. a}}$ | | | | | |
| $\frac{\text{fig. e}}{\text{fig. a}}$ | | | | | |
| $\frac{\text{fig. c}}{\text{fig. b}}$ | | | | | |
| $\frac{\text{fig. d}}{\text{fig. c}}$ | $\frac{4}{3}$ | $\frac{4}{3}$ | $\frac{16}{12}=\frac{4}{3}$ | $\frac{16}{9}$ | $\frac{64}{27}$ |
| $\frac{\text{fig. e}}{\text{fig. b}}$ | | | | | |

**S-67.** **INTRODUCTION:** Each figure in the previous problem is similar to the others. If you choose any figure as your original figure, the others are either magnifications or reductions of the original (the new sides or edges are formed by multiplying the original length by a constant value). For example, if figure b is chosen as the original figure, figure e is a 2.5 magnification and figure a is a $\frac{1}{2}$ reduction.

a) Compare the edge and perimeter ratios in each row of the table. Length and perimeter are linear (one-dimensional). Briefly state the relationship between the edge (side) ratios and the perimeter ratios.

b) Compare the area ratios in each row to the edge and perimeter ratios. Notice that the values in area ratios are all square numbers. Area is measured in square units: it is two-dimensional. Briefly state the relationship between the edge (side) ratios and the area ratios.

c) Compare the volume ratios in each row to the edge and perimeter ratios. Notice that the values in area ratios are all cubic numbers. Volume is measured in cubic units: it is three-dimensional. Briefly state the relationship between the edge (side) ratios and the volume ratios.

S-68.

## THE $r:r^2:r^3$ THEOREM

Once you know two figures are similar with a ratio of similarity $\frac{a}{b}$, the following proportions for the SMALL (sm) and LARGE (lg) figures (which are enlargements or reductions of each other) are true:

$$\frac{side_{sm}}{side_{lg}} = \frac{a}{b} \qquad \frac{P_{sm}}{P_{lg}} = \frac{a}{b} \qquad \frac{A_{sm}}{A_{lg}} = \frac{a^2}{b^2} \qquad \frac{V_{sm}}{V_{lg}} = \frac{a^3}{b^3}$$

a small ~ b large

a small ~ b large

a small ~ b large

a small ~ b large

For two-dimensional figures, the theorem refers to the ratios of sides, perimeters, and areas. For three-dimensional figures, the theorem refers to the ratios of edges, areas of faces or total surface area of the solids, and volume.

### EXAMPLES

a) Suppose the ratio of the sides of the tetrahedra (plural of tetrahedron) above is 3:5. Then the ratios of their surface areas is $\left(\frac{3}{5}\right)^2 = \frac{9}{25}$

b) If the total surface area of the large tetrahedron is 240 square units, you can use ratios to solve for the total surface area of the small tetrahedron.

$$\frac{A_{sm}}{A_{lg}} = \frac{a^2}{b^2} \Rightarrow \frac{A_{sm}}{240} = \frac{3^2}{5^2} \Rightarrow \frac{A_{sm}}{240} = \frac{9}{25} \Rightarrow 25(A_{sm}) = 240(9) \Rightarrow A_{sm} = 86.4$$

c) If the volume of the small tetrahedron is 65 cubic units, then the volume of the large tetrahedron can be found by:

$$\frac{V_{sm}}{V_{lg}} = \frac{a^3}{b^3} \Rightarrow \frac{65}{V_{lg}} = \frac{3^3}{5^3} \Rightarrow \frac{65}{V_{lg}} = \frac{27}{125} \Rightarrow 65(125) = V_{lg}(27) \Rightarrow V_{lg} = 300.93$$

**NOTES:** First, always square or cube the ratio of similarity, NOT the areas or volumes themselves. Second, the theorem is also true if you use the ratio of large to small; just be consistent in writing both large values in the numerators and both small values in the denominators. Third, **weight** is sometimes used interchangeably with **volume** in application problems in this course.

S-69. The two rectangular prisms at right are similar. The ratios of their vertical edges is 4:7. Use the $r:r^2:r^3$ Theorem to find the following <u>without</u> knowing the dimensions of the prisms.

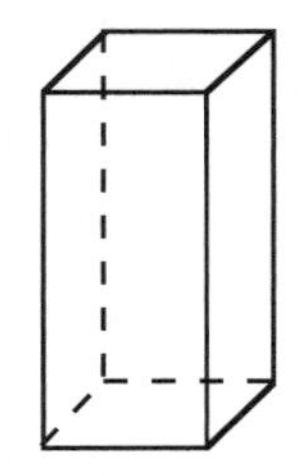

a) Find the ratio of their surface areas.

b) Find the ratio of their volumes.

c) Suppose the perimeter of the front face of the large prism is 18 units. Find the perimeter of the front face of the small prism.

d) Suppose the area of the front face of the large prism is 15 square units. Find the area of the front face of the small prism.

e) Suppose the volume of the small prism is 21 cubic units. Find the volume of the large prism.

S-70. A cube has edge length 5. Another cube has edge length triple the original. Compute $\frac{\text{volume of new cube}}{\text{volume of original cube}}$ and reduce this ratio. Briefly describe the numerical relationship between the resulting ratio and the magnification factor of 3.

S-71. Each red stick is 4 cm longer than each blue stick. Five red sticks and two blue sticks together measure exactly 100 cm. How long is a blue stick?

S-72. A construction worker needs to know the angle to cut a board that will stretch diagonally across a rectangular wall from floor to ceiling. He measures the wall and finds it is 7.67' high and the distance along the wall's base is 10.5'. At what angle should he cut the board so that it goes from the lower left corner to the upper right corner of the wall?

S-73. A triangle has a side of length 6 cm included between angles of 43° and 67°. Draw a diagram and find the lengths of the other two sides.

S-74. Quadrilateral DEFG is similar to quadrilateral D'E'F'G'. The side lengths of the first quadrilateral are DE = 12 cm, EF = 15 cm, FG = 20 cm, and DG = 18 cm. Draw and label the figures, then find the side lengths of D'E'F'G' if F'G' = 24 cm.

S-75. Express in simplest radical form (i.e. NO decimals).

a) $\sqrt{320}$

b) $\frac{6\sqrt{6}}{\sqrt{10}}$

c) $6\sqrt{3} + 2\sqrt{18} - 4\sqrt{48}$

## APPLICATIONS OF $r:r^2:r^3$, PART 1

S-76. **The Trekee Clubhouse Logo Problem.** The local group of Star Trek fans has just inherited $500,000 from their late founder and president. The will specifies that they are to spend the money to build a new headquarters for their club meetings. They want to put up a sign outside the new headquarters with their organization's logo that looks exactly like the old sign, only larger. The T on the old sign was outlined with a curved piece of neon tubing 13.4 feet long and the T itself was covered with gold leaf. On the top of the sign was a model of the Starship Enterprise which was three feet long. On their current building, the circular portion of the sign has a radius of 2 feet, while on the new sign the circular portion will have a 5 foot radius.

a) How long a piece of neon tubing will they need?

b) If they had to buy 3.6 (Troy) ounces of gold to cover the T in their old logo, how much gold will they need for their new logo?

c) If the current model of the Enterprise weighs 58 pounds, how much will the new enlarged model weigh if it is made from the same material? Note that weight is often used interchangeably with volume.

S-77. Suppose that the Trekees in the previous problem also wanted to make a small indoor sign similar to the old outside sign but with a radius of 1 foot.

a) Find the length of neon tubing needed.

b) Find the amount of gold needed.

c) Find the weight of this Enterprise model.

S-78. The result which you used to solve the Trekee's Sign Problem we called the $r:r^2:r^3$ Theorem for similar figures.

a) One of the local eighth grade classes will be studying this theorem in a few weeks. Explain the theorem (<u>why</u> it is named this and <u>how</u> it works) on a poster so that when it is posted in the eighth grade classroom, those students can use it and understand it.

b) The magnification factor is r. Explain what r is and how it is related to the ratio of similarity.

S-79. Rectangle KLMN has a diagonal 32% longer than its longest sides. Rectangle K'L'M'N' is similar to KLMN. Is it true that the diagonal of K'L'M'N' is 32% longer than its longest sides? Justify your answer.

S-80. For each picture below, make a copy of the figure on your graph paper, then draw a second figure that is similar, but not congruent, to the first one.

a)

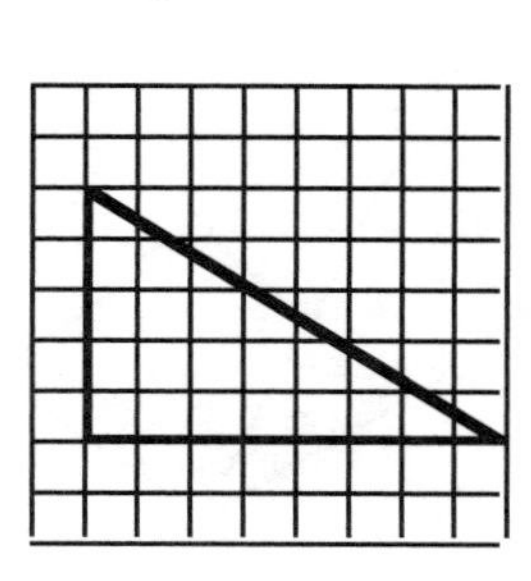

b)

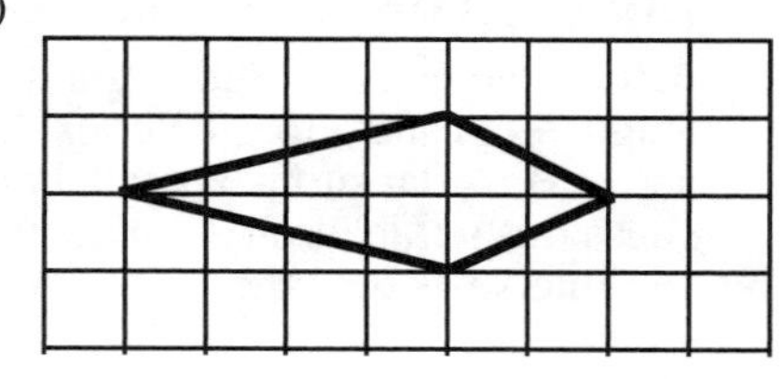

S-81. Label the original figures you copied in the previous problem F and G. Label the two new figures F' and G'.

a) Compute the following ratios.

$$\frac{\text{area (F')}}{\text{area (F)}}, \quad \frac{\text{perimeter (F')}}{\text{perimeter (F)}}, \quad \frac{\text{area (G')}}{\text{area (G)}}, \text{ and } \frac{\text{perimeter (G')}}{\text{perimeter (G)}}.$$

b) Compare the ratio of the perimeters to the square root of the corresponding ratio of the areas for each figure. State what you notice.

S-82. The ratio of similarity for two similar polygons is 2:9.

a) If the perimeter of the larger polygon is 86, what is the perimeter of the smaller polygon?

b) If the area of the smaller polygon is 13, what is the area of the larger polygon?

S-83. An equilateral triangle has side length of 84 cm. Draw an equilateral triangle and:

a) find the height of the triangle.

b) find the area of the triangle.

S-84. The two triangles are similar. Solve for x and y.

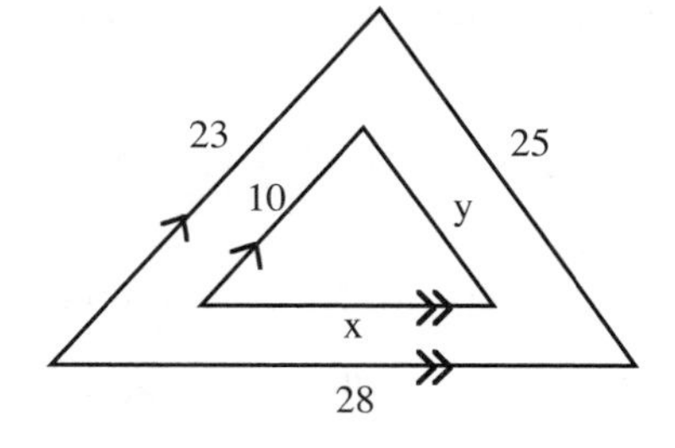


S-85. Solve for x. Remember to treat the trig statements algebraically just like the numbers they represent.

a) $x + \cos 60° = 3$

b) $(\sin 45°)x + \cos 60° = 3$

c) $(\sin 45°)x = 3$

d) $\frac{\sin 45°}{\sqrt{2}} x + \cos 60° = 3$

## APPLICATIONS OF $r:r^2:r^3$, PART 2

S-86. **DIRECTIONS:** Tetrahedron ABCD has a base, ΔBCD, which is an equilateral triangle of side length 9 cm and a fourth vertex A, 4 cm directly above B. A flat surface parallel to the base triangle slices this tetrahedron 2 cm above the base (at E) and intersects the other edges at F and G.

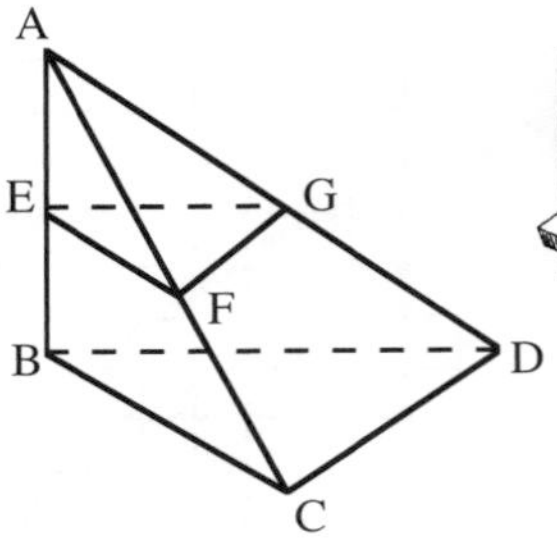

A
E G
F
B D
C

a) Prove that the faces of the new (small) tetrahedron above the slicing plane are similar to the faces of the original (large) tetrahedron.

b) What is the ratio of similarity for the edges of the smaller tetrahedron to the edges of the larger tetrahedron? Check the data in the directions.

c) What are the side lengths of the base of the smaller tetrahedron? Explain.

d) What will the ratio, small to large, of the surface areas be for the tetrahedrons?

e) What will the ratio, small to large, of the volumes be for the tetrahedrons?

f) If the volume of the large tetrahedron is $27\sqrt{3}$ cubic cm, what is the volume of the smaller tetrahedron?

S-87. Find the volume of the truncated tetrahedron of the previous problem. The truncated tetrahedron is what is left of the larger solid when the top, smaller, pyramid is removed.

S-88. Suppose the height of tetrahedron ABCD in problem S-86 is 6 cm. Find the ratios of the small tetrahedron to the large tetrahedron for their edges, surface areas, and volumes if the flat parallel slicing surface is:

a) 4 cm above the base b) 2 cm above the base c) 1 cm above the base

Note that each of these problems will have three parts to it.

S-89. Eight congruent square-based pyramids made of lead are melted down and poured into a mold to make a single pyramid similar to each of the first eight.

a) How many times larger is the new volume compared to the volume of one of the original eight?

b) What is the ratio of the volumes, small to large, of these similar solids?

c) What is the ratio of similarity (that is, lengths), small to large, for these solids?

d) If the base edges of the original pyramids were 12 cm long and the slant edges (those that go from the base vertices to the top of the pyramid) are 10 cm long, how long are the corresponding edges of the new one? Explain.

e) Answer the same problem as in (c) if the base edges are b cm and the slant edges are s cm long.

S-90. Imagine drawing a diagonal on two faces of a cube so that the two diagonals have a common vertex.

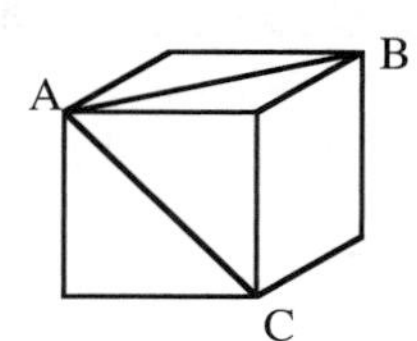


a) If you connect B and C:

1) What two-dimensional figure do the three diagonals form? Hint: visualize the figure by slicing the cube through all three diagonals.

2) What three-dimensional figure do the three diagonals (with the edges of the cube) form?

b) What is the angle between each pair of diagonals?

c) If the cube has an edge length of 10 cm, what is the area of the two-dimensional figure formed by the diagonals? Hint: think about special triangles.

S-91. For the triangle at right,

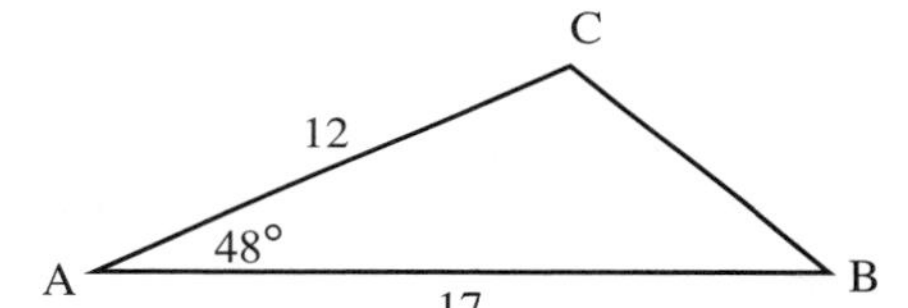


a) draw in the height from the vertex and find its length.

b) find the area of the triangle.

S-92. Graph $y \leq x^2 - 2$.

a) Estimate where the graph crosses the x-axis.

b) Calculate where the graph crosses the x-axis exactly.

S-93. A vertical pole rises above the south end of a football field.

a) In early September, when the sun is directly west of the pole, the sun is 26° above the horizon and the pole casts a shadow 82 feet long. About how tall is the pole?

b) Later in September, the sun is only 14° above the horizon when it is directly west of the pole. About how long a shadow does the pole cast then?

S-94. Solve for x and y:

$$y = x^2 - 7$$
$$3x + y = 11.$$

## TRIANGLE SIMILARITY THEOREMS

**S-95.** In the triangles at right, both have one 80° angle. We will explore whether they are similar or not.

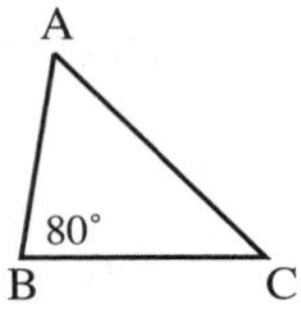


a) Since $\angle B \cong \angle B'$, does this alone guarantee that $\Delta ABC \sim \Delta A'B'C'$? Explain.

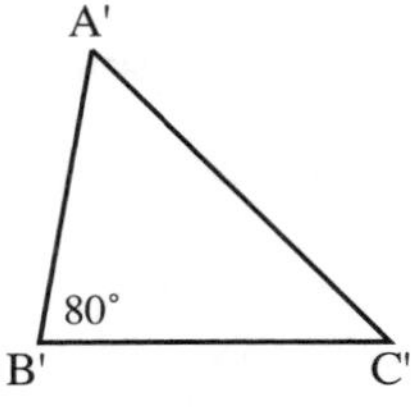


b) Measure the side lengths in millimeters and compute the following ratios. Write them as decimals.

$\frac{AB}{A'B'} =$ $\qquad$ $\frac{BC}{B'C'} =$

c) What appears to be true about the ratios?

d) If the triangles are similar, $\frac{AC}{A'C'}$ should also have the same ratio. Measure these lengths and calculate the ratio.

e) Do you think there is a SAS theorem for similarity? If so, write it as a conjecture (this means using complete sentences). If not, explain completely why not.

**S-96.** Suppose $\Delta WHS$ has side lengths WH = 3 cm, HS = 4 cm, and WS = 6 cm. What would be the side lengths of $\Delta W'H'S'$ with sides twice as long in $\Delta WHS$?

a) Are $\Delta WHS$ and $\Delta W'H'S'$ similar? Explain.

b) Do you think there is an SSS theorem for similarity? If so, write it as a conjecture (this means using complete sentences). If not, explain completely why not.

S-97. Read the box below, add the theorems to your tool kit, then complete parts (a), (b), and (c) below the box.

> In problem S-95 , the triangles have two pairs of corresponding sides proportional, and the included angles congruent. We call these conditions the **SAS~ Theorem**. In problem S-96, one triangle is a magnification of the other, so all pairs of corresponding sides are proportional. We call this condition the **SSS~ Theorem**.

Recall the congruence properties from Unit 5, SSS, SAS, ASA, and HL.

a) Do we need an ASA Theorem for similarity? Explain.b) How is the SAS~ theorem different from SAS in Unit 5?

c) How is the SSS~ theorem different from SSS in Unit 5?

S-98. Are the two triangles at right similar? Explain why or why not.

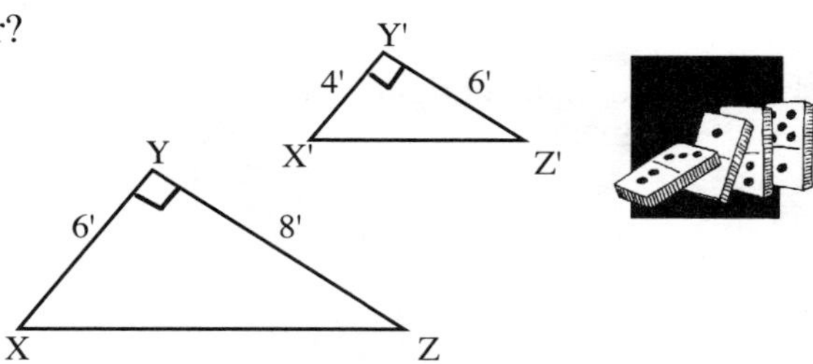


S-99. The ratio of the areas of two similar polygons is 16:81.

a) What is the ratio of similarity (the length ratio) for the two polygons?

b) If the perimeter of the smaller polygon is 112, what is the perimeter of the larger polygon?

S-100. Using a ruler, draw a large $\Delta ABC$ on your paper like the one at right. Use D and E to label the midpoints of sides $\overline{AC}$ and $\overline{BC}$ respectively. Draw $\overleftrightarrow{DE}$.

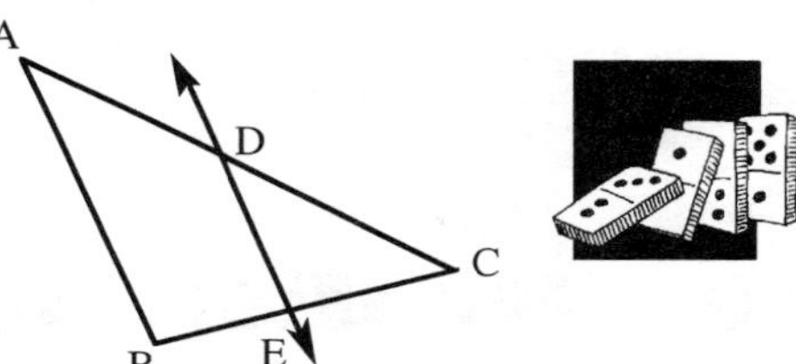


a) Find all segments that are in the ratio 1:2. You might need to use your ruler!

b) Explain why the two triangles are similar.

c) What do you notice about the lengths of $\overline{AB}$ and $\overline{DE}$? Justify that it is true using the ideas from this unit.

S-101. For the figure at right:

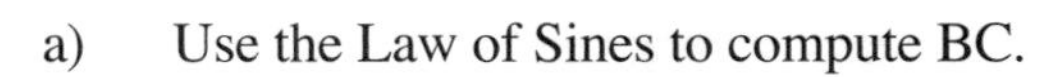

a) Use the Law of Sines to compute BC.

b) Calculate the length of CD.

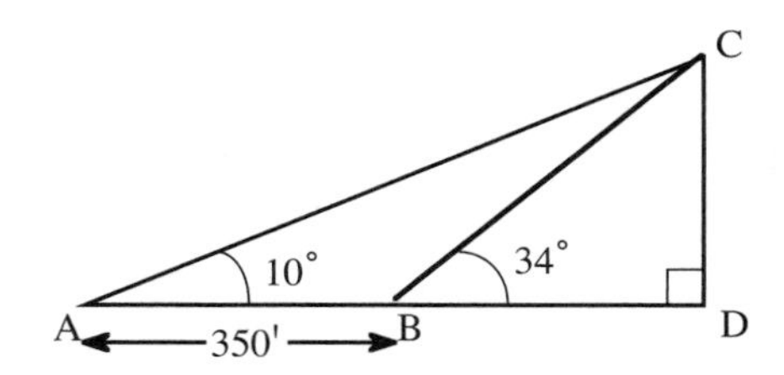


S-102. The ratio of the areas of two similar polygons is 3:16.

a) Find the ratio of similarity for the polygons.

b) If one side of the smaller polygon is 6, find the length of the corresponding side of the larger polygon.

S-103. A flagpole 50 feet high casts a shadow 70 feet long. At the same moment, Tabatha casts a 6' shadow. How tall, to the nearest inch, is Tabatha?

S-104. In the figure at right, find all lengths not given.

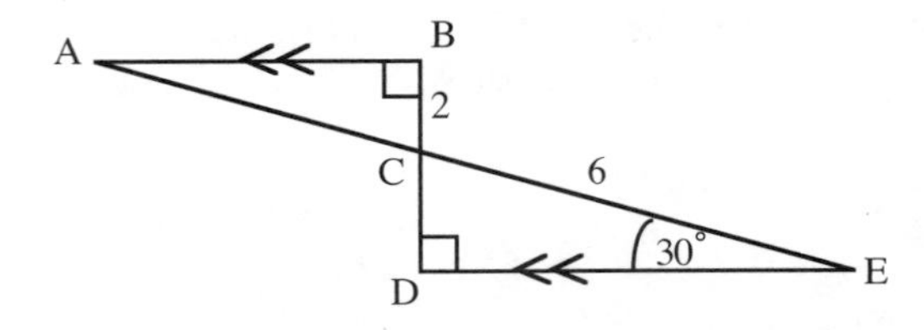


S-105. Sketch the line $y = \frac{2}{3}x + 4$.

a) What are the x- and y-values where the line crosses the y-axis?

b) What are the x- and y-values where the line crosses the x-axis?

S-106. The actual roof of the Pyramid Power of the Ancients shop is a square based pyramid with base edges of length 15 meters and slanted edges of length 9.6 meters. The model of this roof has base edges of length 25 cm and slanted edge of length 16 cm. Keri thinks that the model and the roof are similar to each other, while Hitomi thinks they are not. With whom do you agree? Why? Write your conclusion in one or two complete sentences.

## UNIT AND TOOL KIT REVIEW

S-107. Figure 1 is a regular hexagon. This means that all sides and interior angles are equal. Each interior angle measures 120°. Figure 2 shows all of its diagonals that go through its center. Prove that the six triangles are congruent and equilateral.

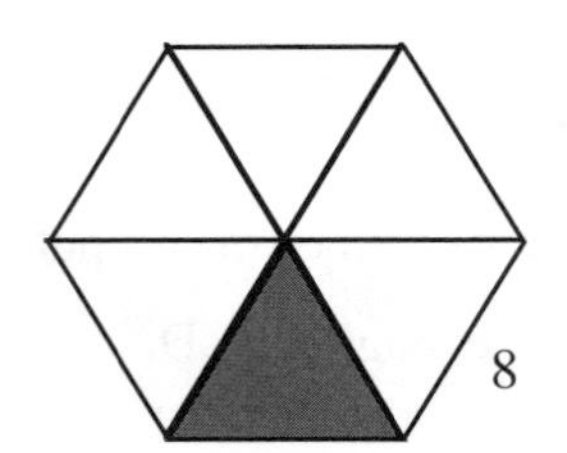


figure 1 figure 2

S-108. Use the hexagons in the preceding problem.

a) Find the area of the shaded triangle in figure 2.

b) Use part (a) to help find the area of the hexagon.

c) Find an expression that can be used to find the area of an equilateral triangle with a side length of s. Hint: start by finding the height in terms of s.

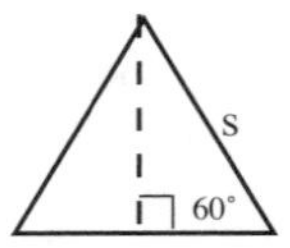

S-109. A, B, and C are midpoints of the sides of scalene $\Delta DEF$. Mark the figure to show what you know, then prove that all four small triangles are congruent.

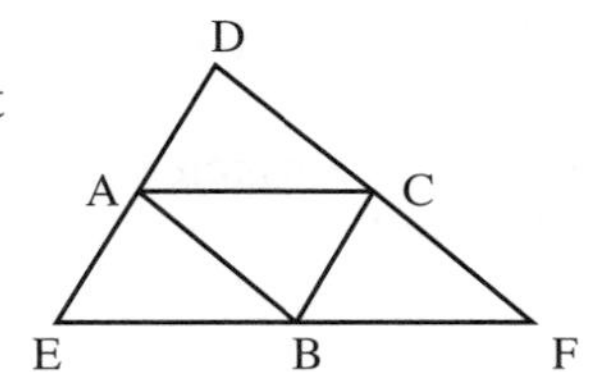

S-110. Solve for the indicated variables in parts (a) and (b) below.

a)

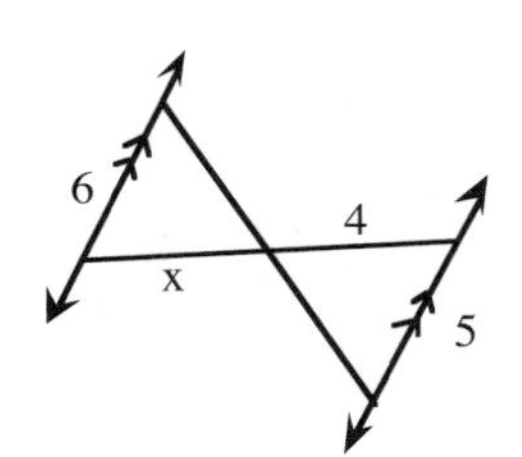


b)

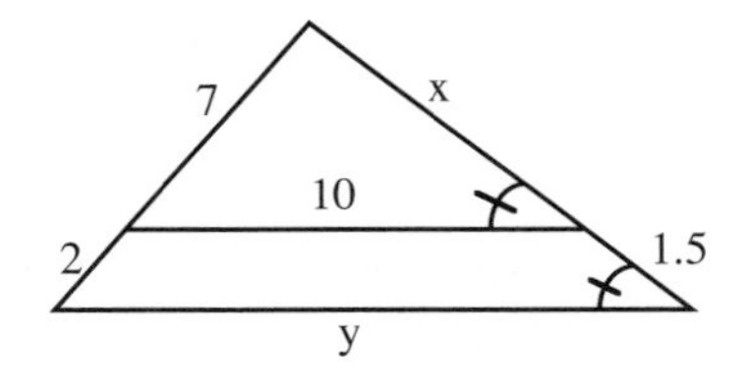


S-111. Two pyramids are similar. Their ratio of similarity is 5:8.

a) If an edge of the small pyramid is 14 units, find the length of the corresponding edge of the large pyramid.

b) If the perimeter of the base of the large pyramid is 48 units, find the perimeter of the base of the small pyramid.

c) If the area of a face of the small pyramid is 70 square units, find the area of the corresponding face of the large pyramid.

d) If the volume of the small pyramid is 117 cubic units, find the volume of the large pyramid.

S-112. You are assigned to measure the height of the dome section of the Capitol Building in Washington, D.C. Describe two ways you can do this.

S-113. The figures and table in problem S-56 can be used to help solve the following problem:

"How many squares are on a checker/chess board?"

The answer is not 64! Use your problem solving skills of making an organized table and looking for patterns to solve this problem. Be sure to discuss your ideas with a partner or your team.

S-114. TOOL KIT CHECK-UP

Your tool kit contains reference tools for geometry. Return to your tool kit entries. You may need to revise or add entries.

Be sure that your tool kit contains entries for all of the items listed below. Add any topics that are missing to your tool kit NOW. You may want to include a sample problem for each of the main ideas in this unit

- similar figures (similarity)
- ratio of similarity
- proportions
- angle of incidence and angle of reflection
- magnification factor
- $r : r^2 : r^3$ Theorem
- triangle similarity theorems
  AA~, SAS~, and SSS~

S-115. When John and Eleanor were completing problem S-34, Harry asked if there was a way to write simpler proportions.

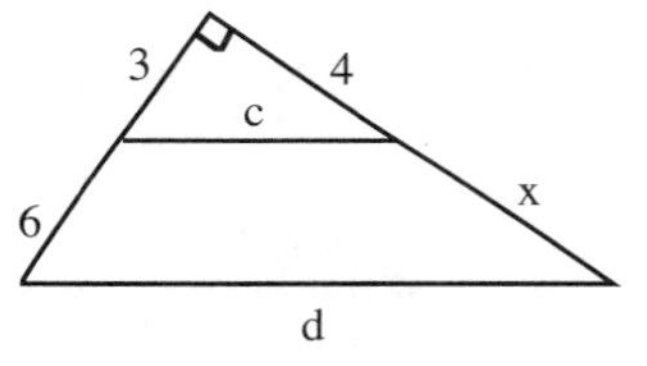


"Hey! Instead of writing $\frac{3}{3+6} = \frac{4}{4+x}$, could I just write $\frac{3}{6} = \frac{4}{x}$ and solve that?" John and Eleanor thought this was a great idea, but wanted to be sure that it would always work.

a) Solve both of John's equations. Does his idea seem to work?

b) Solve parts (c) and (d) from problem S-34 using the shortcut. Does it work?

c) Next redraw the figure above as two separate triangles like Eleanor and John did in the previous problem. Solve for c and d.

d) Now solve for d using Harry's method for $\frac{3}{6} = \frac{5}{d}$. Compare your result to your answer in part (c) and explain.

**SUMMARY:** Success with writing and solving proportions depends in part on writing ratios between corresponding parts of similar figures. The parallel segments in figures like the ones in this and problem S-34 are whole sides of each triangle. When you use the shortcut, you are using "broken" sides for the nonparallel segments, which are not the same as whole segments. Thus, Harry's idea works fine for ratios between the "broken sides," but cannot be used when the parallel sides are involved. In this case, whole sides must be matched with whole sides.

# UNIT 9 Urban Sprawl

## POLYGONS, AREA, AND PROOF

# *Unit 9 Objectives*

## *Urban Sprawl:* POLYGONS, AREA, AND PROOF

This unit explores the properties of polygons with particular attention given to quadrilaterals. The first part of the unit focuses on the interior and exterior angles of polygons. This study is followed by work with figure dissections. In Unit 1 you worked mostly with simple triangles and special quadrilaterals like rectangles and trapezoids. In this unit you will combine your experience with figure dissection and trigonometric ratios to calculate the areas of complex two-dimensional figures, prisms, and pyramids. The remainder of the unit looks at families of quadrilaterals--trapezoids, kites, parallelograms, rectangles, rhombi, and squares--and proves properties of the figures and their diagonals.

In this unit you will have the opportunity to:

- discover the basic relationships and formulas for interior and exterior angles of polygons (n-gons).
- use figure dissections and trigonometry to find the area of complex two- and three-dimensional figures.
- practice and extend your ability to present a mathematical proof with special emphasis on proving properties of quadrilaterals.

*Read the problem below. Over the course of the unit you will learn what is needed to solve it. Do not try to solve it now.*

US-0. As American cities expanded during the Twentieth Century, there were often few controls on how the land was divided. In the town of Dry Creek, one particular tract of land had the shape and dimensions shown in the figure at right. The developer planned to build five homes per acre. Determine the number of homes that can be built on this tract of land. Show all dissections and subproblems.

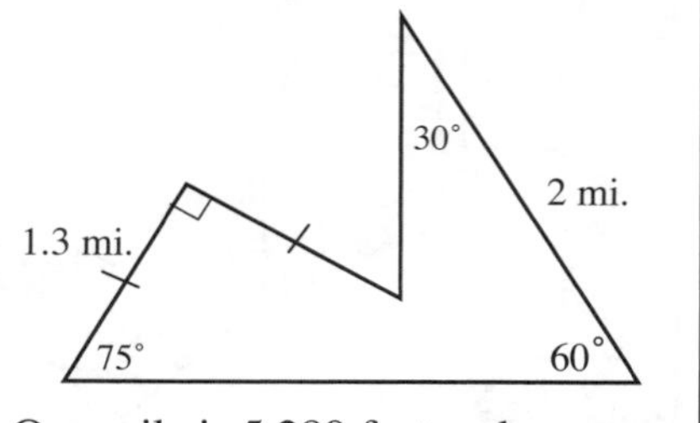


One mile is 5,280 feet and one acre contains 43,560 sq. ft.

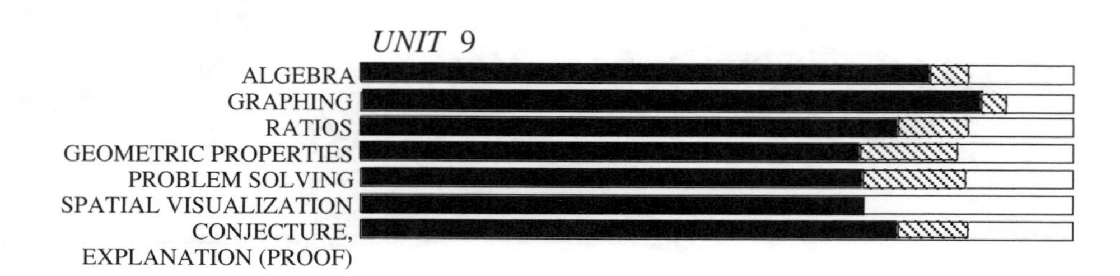

# Unit 9

## *Urban Sprawl:* POLYGONS, AREA, AND PROOF

US-1. Study the figures in the two boxes below. Then use your observations to determine which of the figures--A through H--are polygons. For each figure that you reject, explain why it is not a polygon. Then read the information in the double-line boxes and add it to your tool kit.

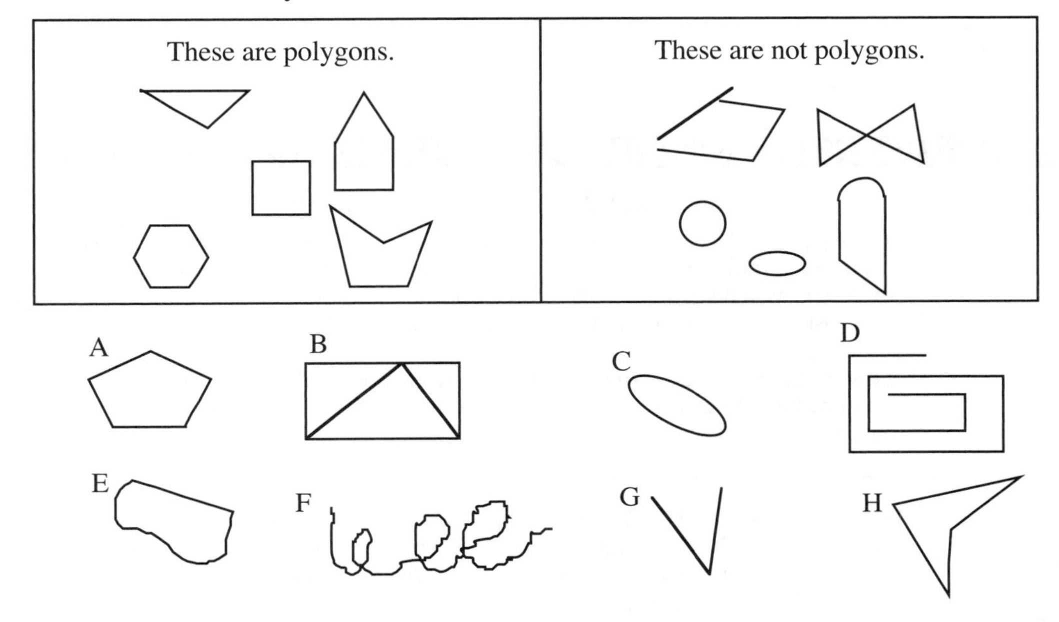


In Unit 1 a **POLYGON** was defined as a two-dimensional closed figure made up of straight line segments connected end to end. These segments may not cross (intersect) at any other points. The term polygon does not give you any specific characteristics about a figure. Knowing a shape is a triangle, however, tells you the shape is a polygon <u>and</u> it has three sides. Some terms for types of polygons are:

| Name of Polygon | Number of Sides |
|---|---|
| Triangle | 3 |
| Quadrilateral | 4 |
| Pentagon | 5 |
| Hexagon | 6 |
| Heptagon | 7 |

| Name of Polygon | Number of Sides |
|---|---|
| Octagon | 8 |
| Nonagon | 9 |
| Decagon | 10 |
| 11-gon | 11 |
| n-gon | n |

The names for polygons with more than eight sides are not commonly used. Usually people refer to a seven-sided polygon as a 7-gon, an eleven-sided polygon as an 11-gon, etc. The only names you need to remember are the first four and the octagon.

Any segment which connects two vertices (and is not a side) is called a **DIAGONAL** of a polygon.

$\overline{AD}$, $\overline{BE}$, and $\overline{AC}$ are three of the diagonals in this polygon.

$\overline{AB}$ is a **SIDE**, not a diagonal.

$\angle F$ is one of the six **INTERIOR ANGLES** of the polygon.

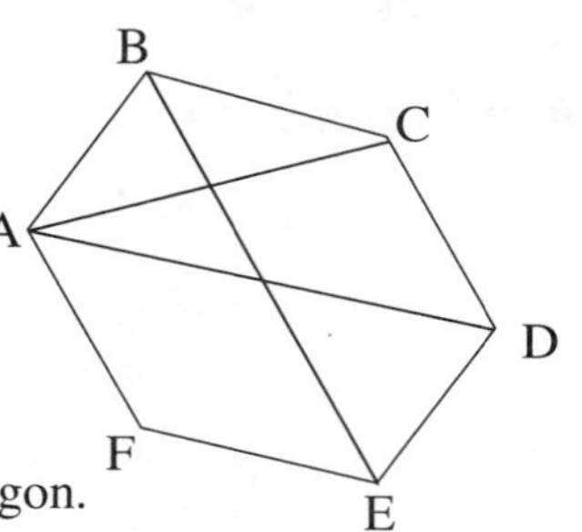


## INTERIOR ANGLES OF POLYGONS

US-2. On the resource page provided for this problem, draw all the diagonals possible in each polygon **from the highlighted vertex**. Complete the table on your resource page and answer the questions. Remember--you are not drawing every possible diagonal in each figure, just the ones from the vertex with the heavy dot.

Once you have answered the questions, discuss the results with your team and write a conjecture that states how to find the sum of the measures of the interior angles of any polygon.

**DISSECTION PRINCIPLE**: Every polygon can be dissected (or broken up) into triangles which have no interior points in common. This principle is an example of the problem solving strategy of **subproblems**. Finding simpler problems will help us solve the larger problem. We used this principle in the preceding problem and earlier in the year for area problems.

US-3. Rudolph dissected pentagon ABCDE at right into five triangles by using point Z in the interior.

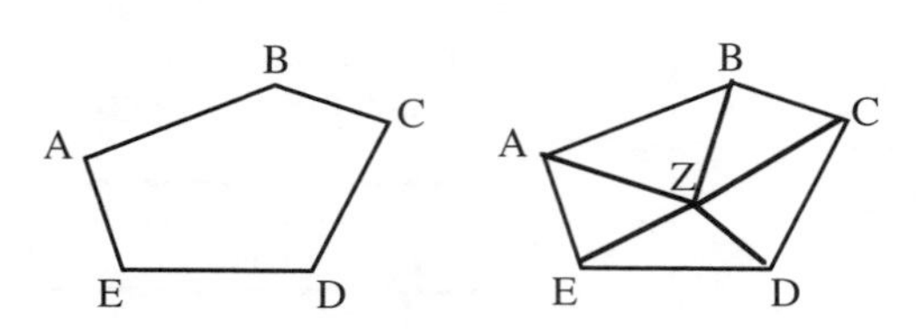


a) When he added the measures of all the angles in the five triangles, what sum did he get?

b) What is the sum of the measures of the five angles around point Z? c) Use your results from parts (a) and (b) to find the sum of the interior angles--that is, angles A, B, C, D, and E--of the pentagon.

US-4. Ngoc saw what Rudolph had done with his pentagon (above) and dissected the hexagon at right in the same way. Draw and dissect a hexagon on your own paper, then repeat parts (a), (b), and (c) from Rudolph's problem for Ngoc.

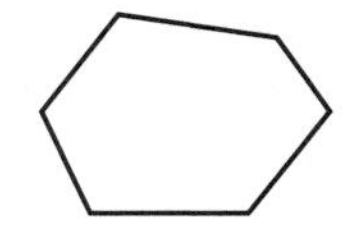

US-5. Jabari is trying to find the sum of the measures of the interior angles of a 100-gon. You notice that he is trying to draw a 100-gon! When you ask him why he is trying to draw it, he says, "I need to draw in all the triangles just like I did on the resource page."

a) If Jabari does draw in all the triangles, as you did on the resource page, how many triangles will he have?

b) Explain to Jabari how to do this problem **without** drawing a 100-gon and without drawing in all the triangles.

US-6. The figure at right has been dissected and every angle labeled. Use the three triangles to prove that the sum of the interior angles of ANY pentagon will ALWAYS be 540°.

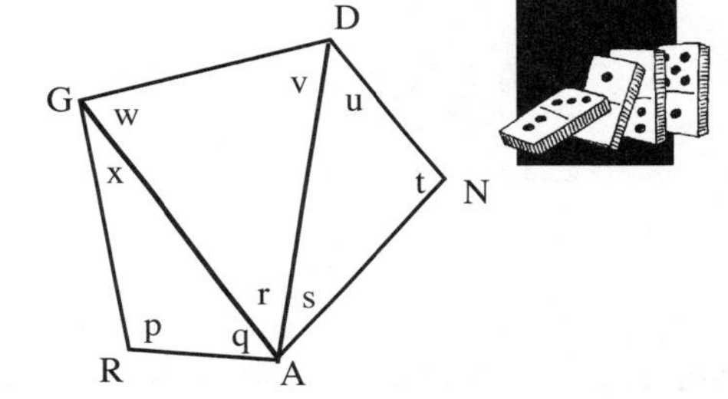


a) Write three equations, one each for the sums of the angles in each triangle.

b) Add the three equations to write one long equation for the sum of all nine angles.

c) Substitute the three letter name for each angle of the pentagon for the lower case letter(s) at each vertex of the pentagon. For example, $m\angle NDG = u + v$.

US-7. Find the value of x. Use the work you have already done today to reduce the number of calculations you have to do.

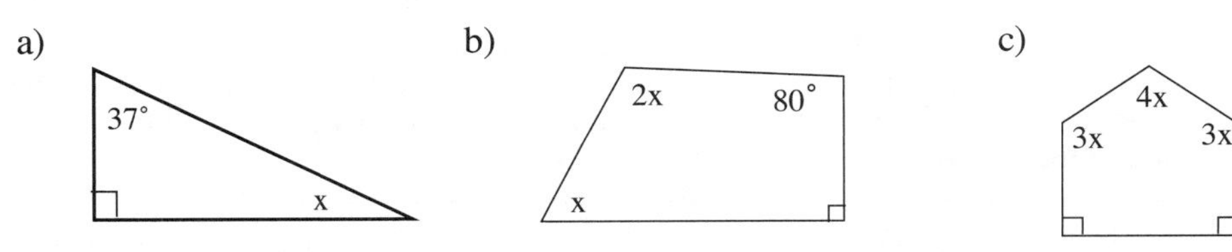


US-8. Solve for x and y.

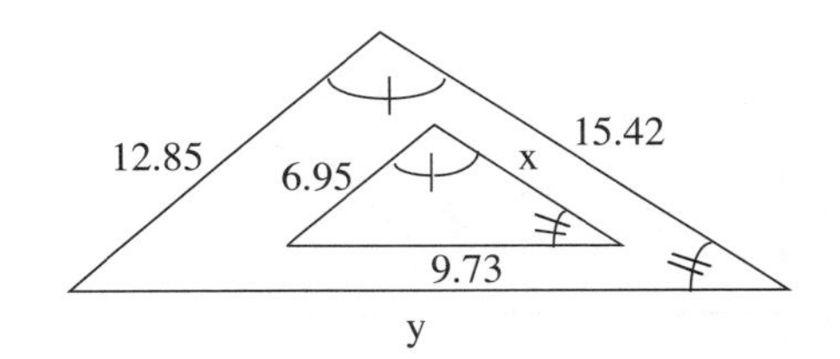

US-9. Find m∠z. Explain how the problem solving strategy of working backward could be useful here.

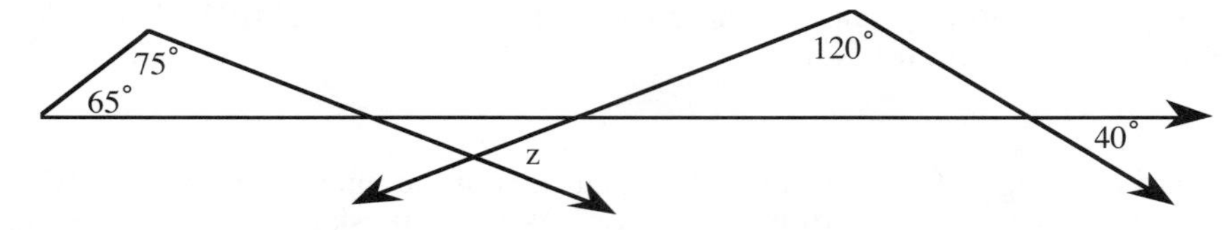


US-10. A rectangle on a sheet of paper is duplicated using an enlarging copy machine so that the sides of the enlargement are twice as long as those of the original. Write the ratio of the areas of the two rectangles. Explain how you arrived at your answer.

US-11. Simplify these expressions so that no parentheses or brackets remain.

a) 2x[3x + (x - 1)6]

b) 5[3x - 2(x - 5)] - 4(x - 2)

c) 4x + [x - 5(x + 3)]

US-12. Calculate the total surface area and volume of the figure at right.

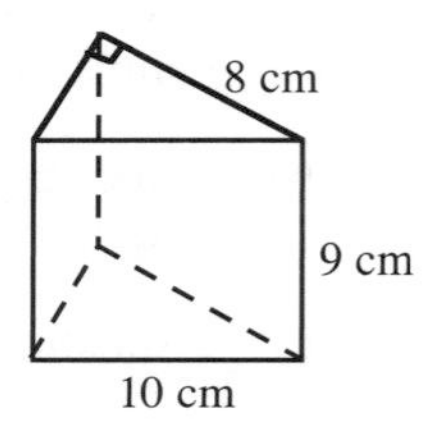


## REGULAR POLYGONS

US-13. Study the following figures. Compare and contrast convex and non-convex polygons. Discuss their similarities and differences with your team members. Then write a definition of a convex polygon and add it and the definition below to your tool kit.

| These are non-convex polygons. | These are convex polygons. |
|---|---|
|  | 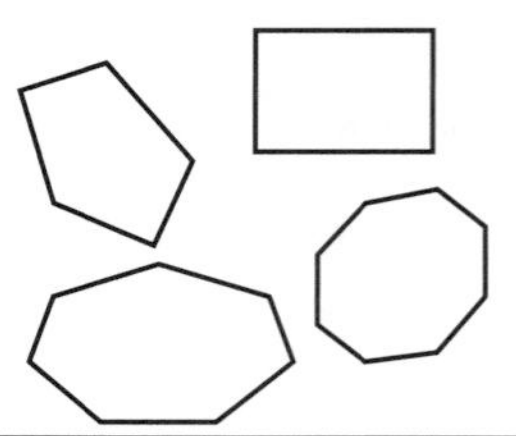 |

A **REGULAR POLYGON** is a convex polygon with all angles congruent and all sides congruent. By **convex** we mean each pair of interior points can be connected by a segment without leaving the interior of the polygon.

**US-14.** In problem US-2 you found that the sum of the interior angles of an n-gon is (n - 2)180°. Look over your resource page from that problem. How could you represent the number of degrees in <u>one</u> interior angle of a regular n-gon? How about a pentagon?

US-15. Remember Jabari? He finally understood what you were saying about the sum of the measures of the angles of a 100-gon, and he has calculated the sum to be 17,640°.

a) Is he correct? Justify your answer.

b) Now he would like to know the measure of <u>one</u> angle of the 100-gon if the 100-gon is regular. Explain to him how he can find it and why the method works.

US-16. Just to be sure Jabari **really** understands and can use your results, carefully explain to him how to find the sum of the measures of the angles of an n-gon <u>and</u> how to find the measure of a single angle in a regular n-gon. This is important! Make sure everyone in your team understands this and has it written down correctly on his or her paper. (Put this in your tool kit for this unit.)

US-17. The figures below are regular polygons. Solve for x and y. Show all steps leading to your solutions.

a)

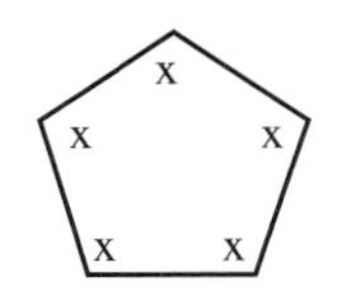


b)

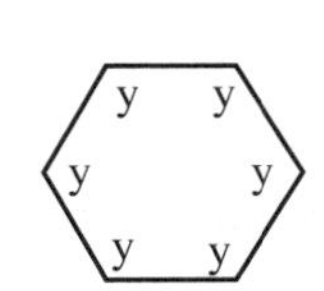


c)

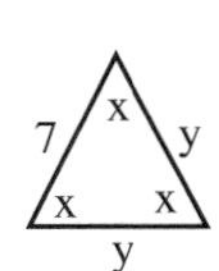


US-18. Just when Jabari understood everything you taught him, his teacher tells him the sum of all the interior angles of a polygon and expects him to find the number of sides in this polygon. The first problem his teacher gives him says the sum of the angles is 720°. Jabari thinks he should draw polygons until he finds one with 720° in it!

a) Help him out by calculating how many sides the polygon has. How did you do it?

b) Explain to Jabari how he can calculate the number of sides of a polygon himself just by knowing the sum of the measures of all the angles.

US-19. How many sides does a polygon have if the sum of the measures of the interior angles is:

a) 1440°?

b) 900°?

c) 1980°?

d) 1860°?

US-20. In the figure at right, if PQ = RS and PR = SQ, prove that $\angle P \cong \angle S$.

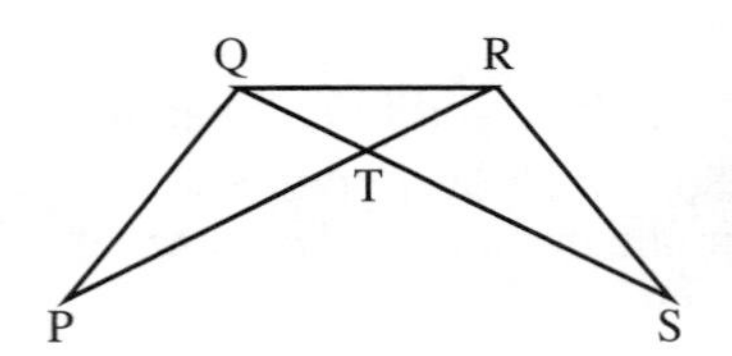

US-21. Solve for x and y:

a)

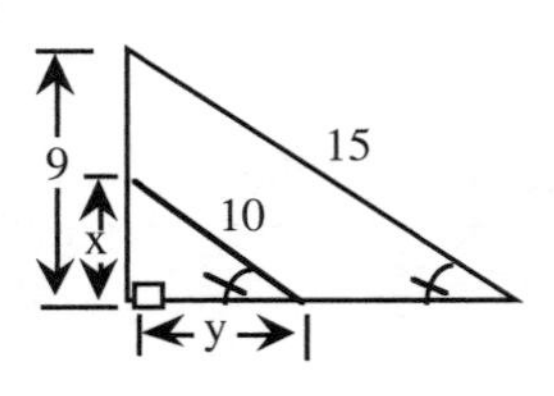


b)

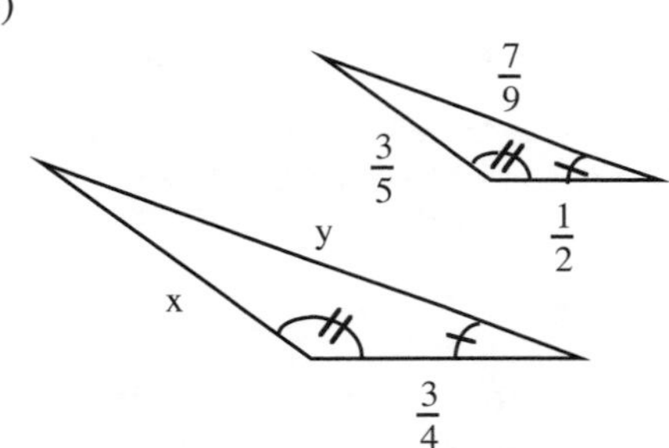


US-22. A square 3 feet on a side has a logo for a soft drink inside which has a total area of 4 square feet. A similar square 7 feet on a side has a similar logo inside. What area does the larger logo cover?

US-23. Solve for x and y. Show all steps leading to your solution.

a) $x + y = 1$
$3x - 2y = 13$

b) $3x - 2y = 8$
$2x - 3y = -13$

US-24. Find $x$ $(0^\circ \le x \le 90^\circ)$ if:

a) $\sin x = \frac{3}{7}$

b) $\sin x = 0.24$

c) $3\sin x = 0.24$

d) $3\sin x - 1 = 0.24$

US-25. Write each of the following as simply as possible.

a) $\frac{x^{20}y^{10}}{x^5y^7}$

b) $(x^3y^2z)^5$

c) $(x^5y^2)^2 - (x^4y)^3$

## EXTERIOR ANGLES

US-26. **DIRECTIONS:** In problem US-2, you found the sum of the <u>interior angles</u> for any polygon. Now we investigate a corresponding problem for <u>exterior angles</u>.Use the resource page provided by your teacher. Each member of your team should select <u>one</u> of the convex quadrilaterals and complete parts (a) through (c). If you are working with a partner, each of you should do two figures.

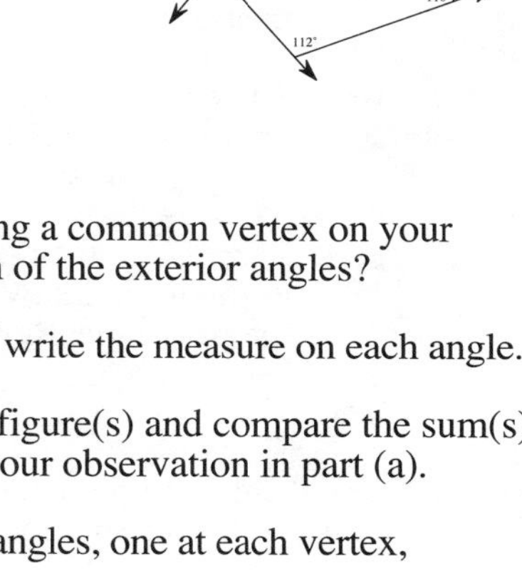

a) Start with any exterior angle in your figure and trace it on a sheet of paper. Next, slide the vertex of the copied angle to the next exterior angle's vertex in your figure and trace this angle so that it is <u>adjacent to</u> the first angle you traced. Continue this process until you have a copy of every exterior angle in your figure sharing a common vertex on your tracing paper. What do you notice about the sum of the exterior angles?

b) Calculate the measure of each exterior angle and write the measure on each angle.

c) Add the measures of the exterior angles in your figure(s) and compare the sum(s) to the results of your teammates or partner and your observation in part (a).

d) Write a conjecture about the sum of the exterior angles, one at each vertex, of n-gons.

US-27. Let us see if we can prove your conjecture for a pentagon.

a) What is the sum of the measures of one exterior angle and its adjacent interior angle? (e.g., what is $m\angle a + m\angle b$?)

Write five equations illustrating this, one for each angle of the pentagon.

b) What is the sum of the interior angles of a pentagon?

c) Add the five equations you wrote in part (a) and use your answer to part (b) to prove that the sum of the exterior angles of a pentagon is 360°. Refer to your solution to problem US-6 if you need hints for how to proceed.

US-28. The measures of four of the exterior angles of a pentagon are 57°, 74,° 56°, and 66°. What is the measure of the remaining exterior angle?

US-29. Jabari's teacher next asks him to find the number of sides of a regular n-gon given only the measure of one of its interior angles. Debbie says to use the equation and solve for n:

$$m\angle = \frac{(n - 2)180}{n}$$

Then Bertha says that the process is much simpler than doing all the algebra necessary to solve Debbie's equation. Follow her process:

a) Suppose one angle of a regular n-gon is 120°. Extend one side of the n-gon to create exterior angle a. What is its measure?

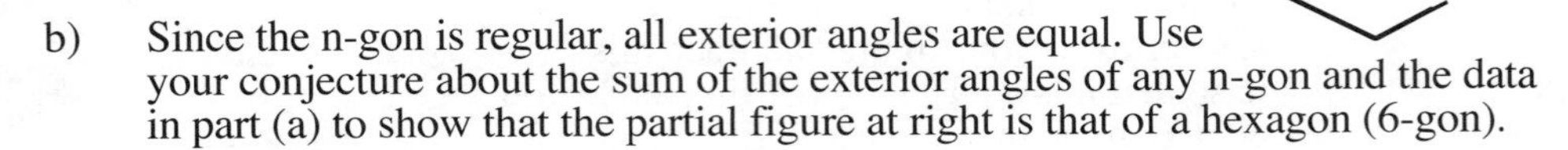


b) Since the n-gon is regular, all exterior angles are equal. Use your conjecture about the sum of the exterior angles of any n-gon and the data in part (a) to show that the partial figure at right is that of a hexagon (6-gon).

c) Summarize the method of using the measure of an exterior angle of a regular n-gon to find how many sides it has when all you know is the measure of an interior angle. Remember--the n-gon must be regular to use this method.

US-30. Use the procedure in the previous problem to find the number of sides in a **regular** polygon if each interior angle has a measure of:

a) 60°? b) 156°? c) 90°? d) 140°?

US-31. Be sure that the results of your polygon angle investigations in this unit match the information in the box below. Read it, make any adjustments to your tool kit that are necessary, then do the next problem.

The properties of interior and exterior angles in polygons, where n represents the number of sides in the polygon (n-gon), can be summarized as follows:

1) The sum of the measures of the **INTERIOR ANGLES** of an n-gon is sum = $(n - 2)180°$;

2) The measure of each angle in a regular n-gon is $m\angle = \frac{(n - 2)180°}{n}$;

3) The sum of the **EXTERIOR ANGLES** of any n-gon is always 360°.

**US-32.** Two surveyors are measuring distances from Emmet's Peak to some nearby points. Carissa is 5 miles from Emmet's Peak and Fernando is 3 miles away.

a) Based on the information given, how close together could Carissa and Fernando be?

b) How far apart could they be?

c) Fernando sights the top of Emmet's Peak, and then he sights Carissa. He determines that the measure of the angle between these sightings is 104°. Make an accurate sketch and label it with all the data, then use the Law of Sines to find the angle Carissa would measure between Emmet's Peak and Fernando.

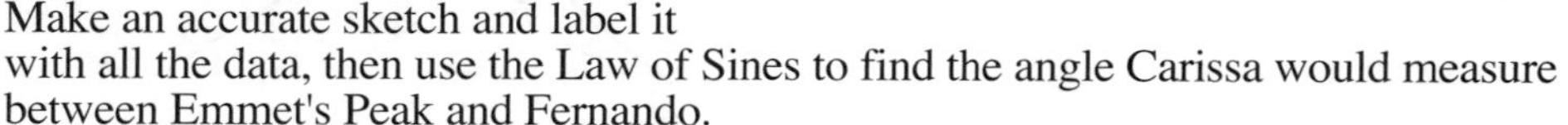

d) What is the measure of the angle formed by their respective lines of sight to the top of Emmet's Peak, that is, the angle whose vertex is at the peak?

e) Find the actual distance between Carissa and Fernando.

US-33. Both of these figures are regular polygons. Solve for x and y. Show the steps in your solution.

a)

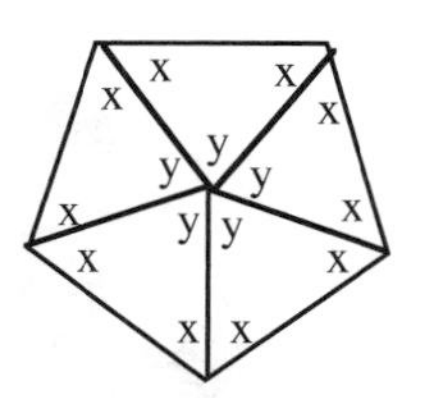

b)

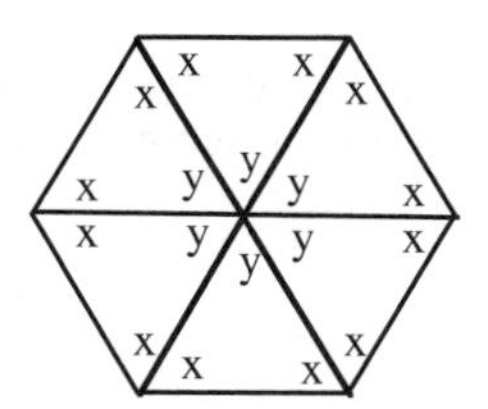

US-34. Show all subproblems leading to your solution.

a) Find the value of x.

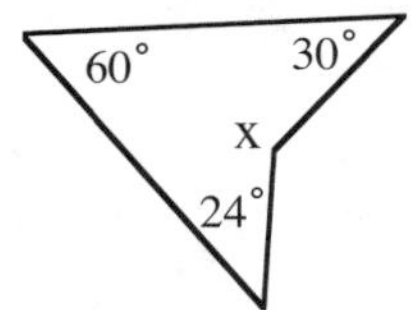


b) Find the measure of each angle.

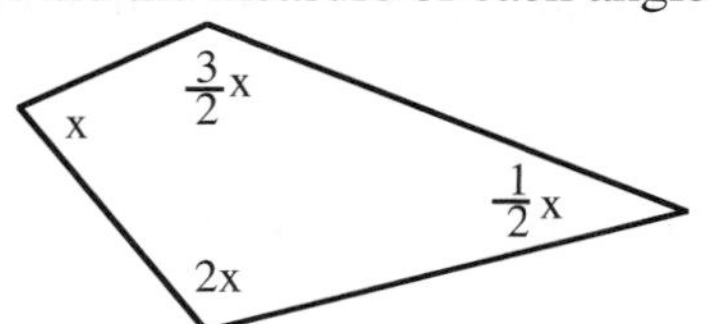

US-35. At right is a scale drawing of the floor plan for Nzinga's playhouse. The actual dimensions of the playhouse will be 15 times the size of the plan at right. How many square cm will her playhouse contain?

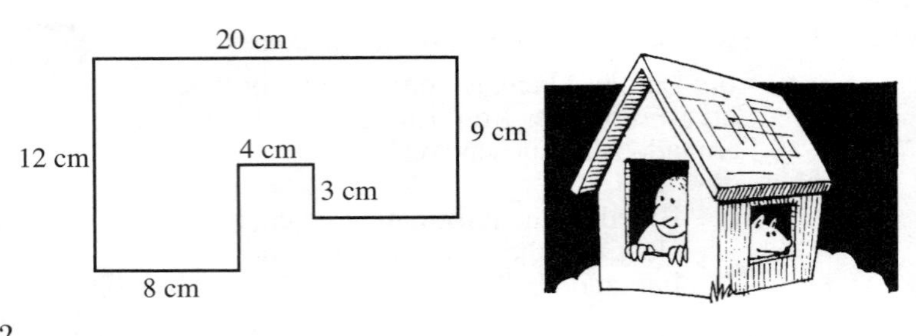


US-36. Solve for x and y.

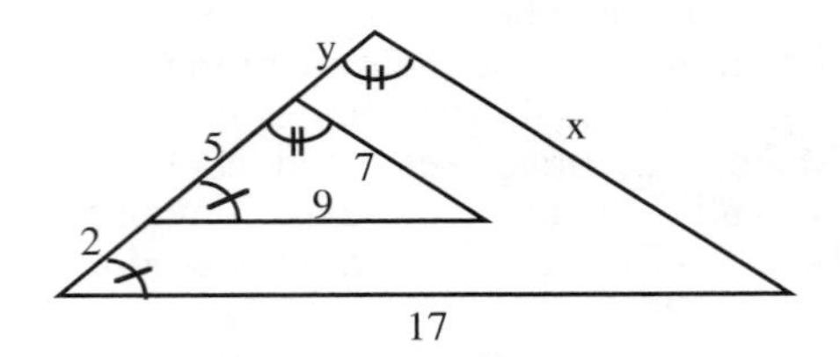


US-37. These hexagons are in a bag:

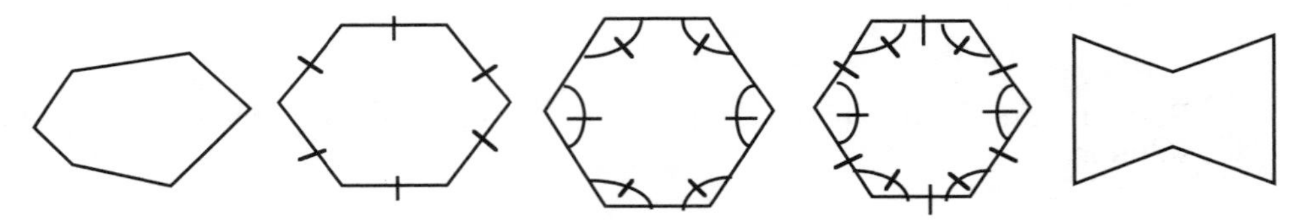

If Bret is going to reach into the bag and pull out a hexagon at random, what is the probability he pulls out a regular hexagon?

US-38. Consider a 30°- 60°- 90° triangle.

a) If the shortest side is 6 cm, find its area.

b) If the triangle has an area of 20 sq. cm., find the length of the shortest side.

## FINDING AREAS OF POLYGONS BY DISSECTION

US-39. **DIRECTIONS:** Sometimes it is easier to place a complicated polygon inside another simpler polygon for which you can find the area. Study the example below and then find the areas of the following polygons. Show your subproblems. All lengths are in centimeters.

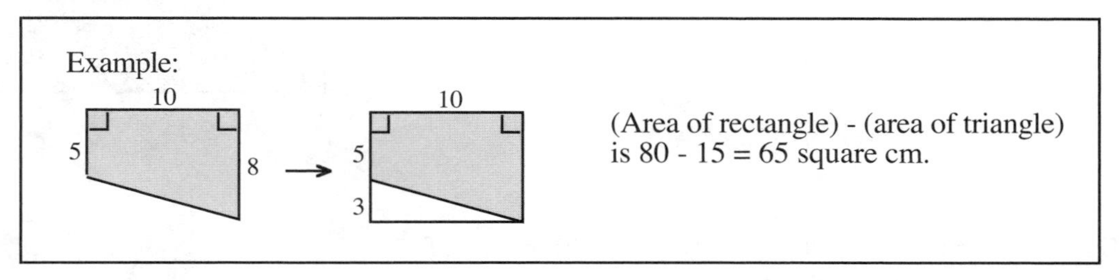


a) In the example above, how do you know the short leg of the right triangle has a length of 3? Discuss this with your team and write your consensus.

b)

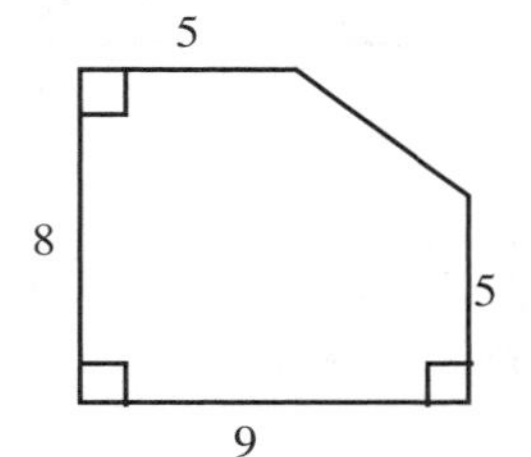


c)

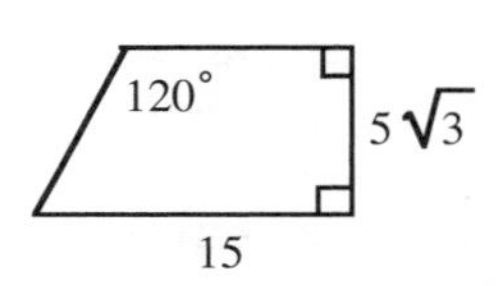


d)

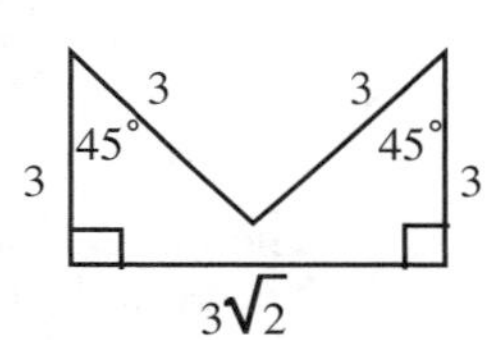


US-40. **DIRECTIONS:** Another way to compute the area of complicated polygons is by internally breaking them into simpler polygons whose area you can find. For example, in part (a) you could draw a diagonal to form two triangles. It is up to you, however, to decide where to add diagonals and/or auxiliary lines. Find the area of the following polygons. Show your subproblems.

a)

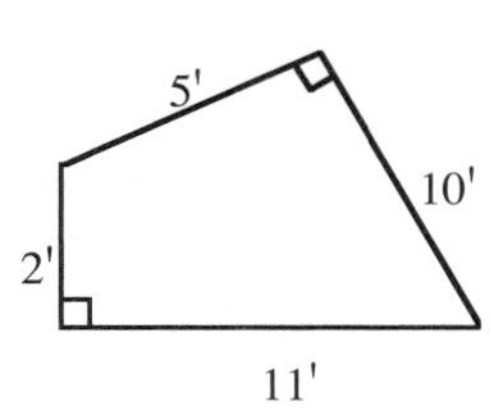


b)

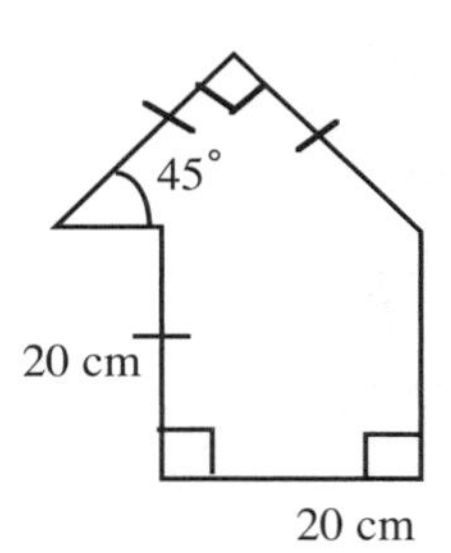


c)

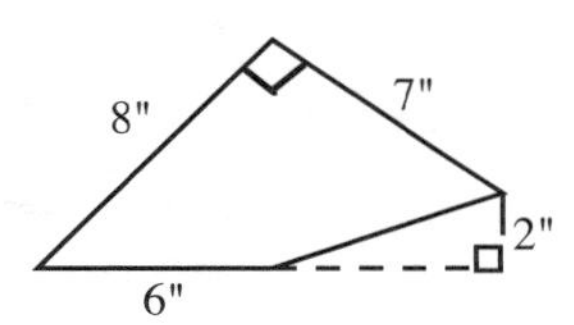


d) a regular hexagon

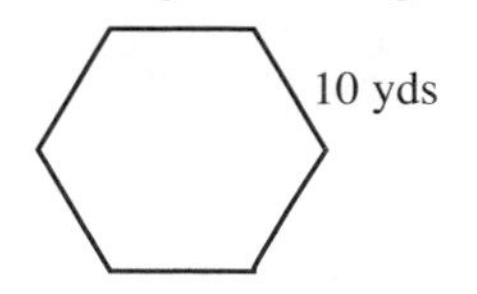

US-41. In the figure at right, a square with side length 1 and a square with side length 4 are placed as shown. What is the area of the shaded region? Caution: Redraw the figure on graph paper to see that the segment from A to C through B is not a straight line.

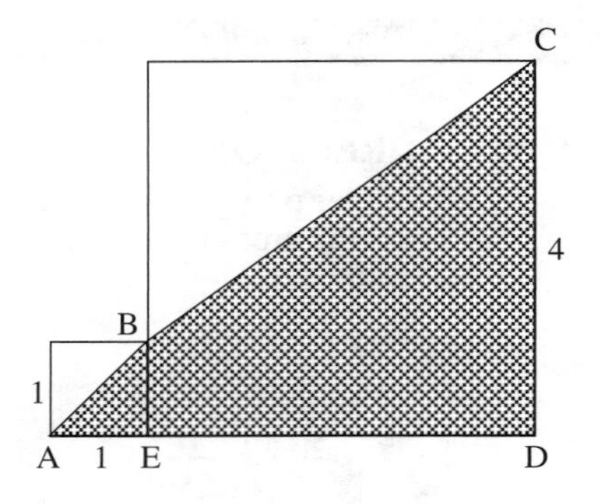


US-42. In the figure at right, CB = CD, $\overline{CB} \perp \overrightarrow{AB}$, $\overline{CD} \perp \overrightarrow{AD}$. Prove that $\overrightarrow{AC}$ bisects $\angle BAD$.

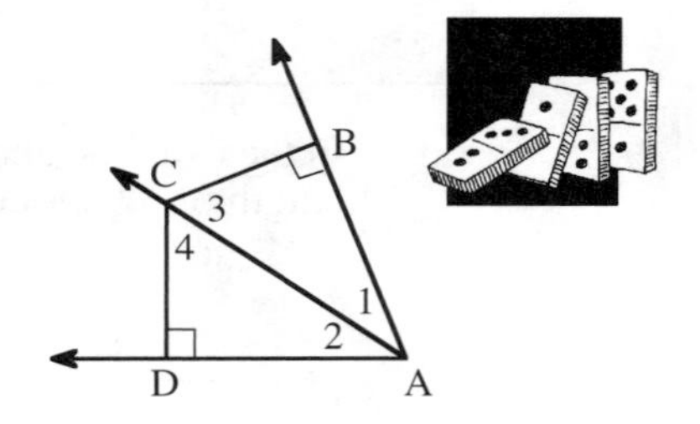


US-43. The dimensions of Hector's bedroom are 1.25 times larger than his sister Adele's bedroom. If it took 13.5 square yards of carpet to cover Adele's bedroom, how many square yards of carpet will be needed to cover Hector's bedroom floor?

US-44. Decide whether the following are always congruent, sometimes congruent, or never congruent. Justify each answer.

a) Equilateral triangles.

b) Regular hexagons with one side of length 5 cm.

c) Circles with an area of $21\pi$.

d) Squares.

e) Regular 17-gons.

f) Right triangles whose sides are in the ratio 3 : 4 : 5.

US-45. A jacket priced at \$69.75 is sold at 30% off. What is the sale price?

US-46. The following tables of values are points from separate graphs. Without plotting them, decide whether or not they would form straight lines or curves if they were plotted. Support your answer with one or two complete sentences explaining the basis for your decision.

a)

| x | 150 | 151 | 152 | 153 | 154 | 155 |
|---|---|---|---|---|---|---|
| y | 322 | 326 | 330 | 334 | 338 | 342 |

b)

| x | -35 | -34 | -33 | -32 | -31 | -30 |
|---|---|---|---|---|---|---|
| y | 52 | 54 | 58 | 64 | 72 | 82 |

US-47. Fraction bust, then solve for a.

a) $\frac{20}{a} = 5$  b) $\frac{20}{a} + \frac{8}{2a} = 6$  c) $\frac{9a}{4} - \frac{5}{a} = 2$

US-48. Calculate the area of each shape. Show the dissections and subproblems that you use.

a)

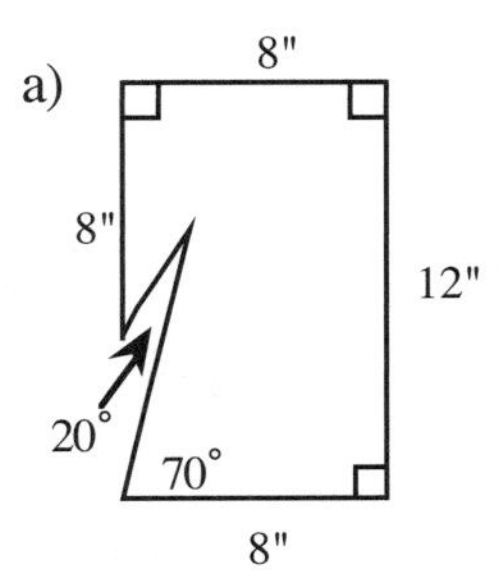


b)

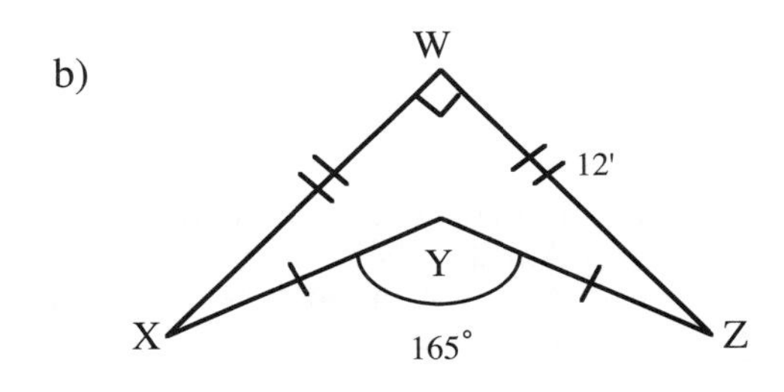


US-49. Calculate the total surface area of each solid. Show any subproblems you use. In particular, draw the two-dimensional figures you use for each part of the surface area of the three-dimensional figure.

a) The base is a square.
The triangular faces are all congruent.
The lateral edges are all 6 m.

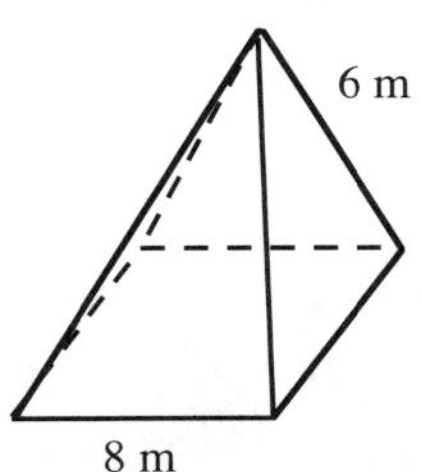


b)

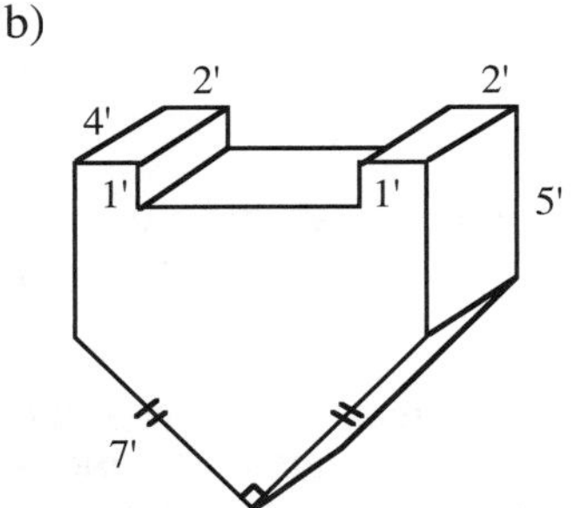


US-50. Calculate the volume of the prism in part (b) of the previous problem. First determine which face is the base.

US-51. A highway patrol car (H) is sitting 0.2 miles directly north of an intersection (I) when the officer observes a Jaguar (J) speeding through the intersection heading east. Fifteen seconds later, as the officer begins pursuit, the angle between their cars (∠IHJ) is 70°. The streets run perpendicular to each other, so the officer must first drive 0.2 mile south, then turn east. Draw a diagram and use it to help answer the following questions.

a) How far did the Jaguar travel in 15 seconds?

b) How fast is the Jaguar traveling?

US-52. In the figure at right, ∠A ≅ ∠ABD and ∠C ≅ ∠CBD. Is AD = CD? If yes, prove it; if not, show why not.

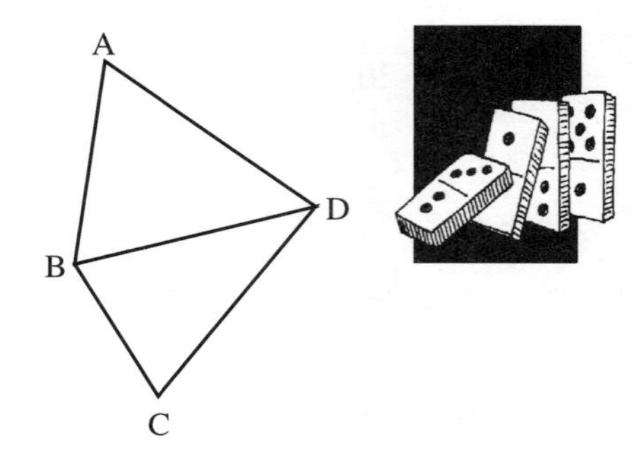


US-53. Benny says he has a regular polygon with one exterior angle measuring 22.5°. Benny tells Anisa, "There is no way you can figure out how many sides my polygon has with just this information!" "Au contraire!" Anisa replies. "I can <u>easily</u> figure out the number of sides it has!"

Decide who is correct. If it is Benny, what other information is needed to determine the number of sides? If it is Anisa, calculate the number of sides.

US-54. A right triangle has area 18 sq. cm. Find the area of a similar right triangle if it has dimensions which are:

a) twice as large.

b) $\frac{2}{3}$ as large.

c) 1.47 times as large.

US-55. Find all points (x, y) that are on the graph of $3x + 2y \geq 6$.

US-56. If ABCD is a rectangle, what is the relationship of the areas of regions 2 and 3 to the area of region 1? Justify your answer.

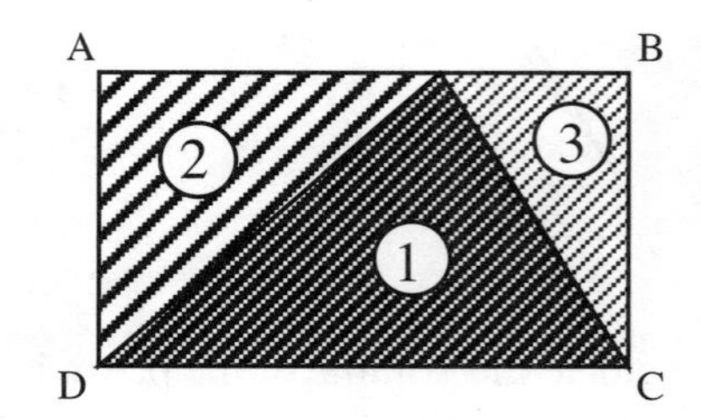

## AREA OF A REGULAR POLYGON

US-57. **DIRECTIONS:** You have learned how to find the area of triangles, various quadrilaterals, and regular hexagons in previous problems in Units 1-8. Today we would like to explore the area of regular n-gons to find a method for calculating their areas. Use the resource page provided by your teacher for the next several problems.

O is the center of the pentagon with $\overline{OB} \perp \overline{XY}$.

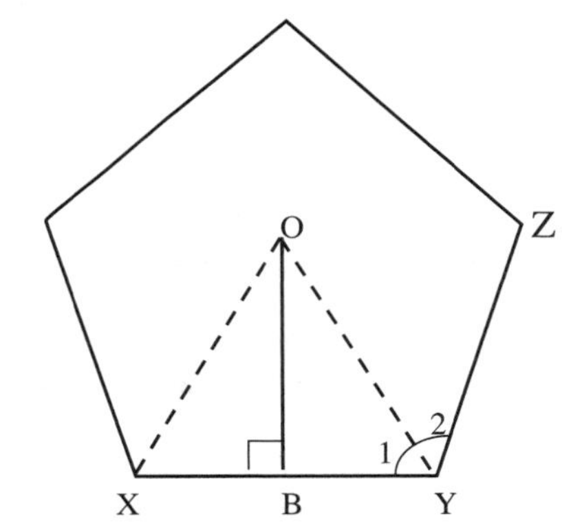


a) Explain why $\overline{OB}$ is the bisector of $\overline{XY}$.

b) Explain why $\overline{OY}$ bisects $\angle XYZ$.

c) What is $m\angle XYZ$? $m\angle 1$?

d) If XY = 12", find OB.

e) Find the area of $\Delta OXY$, and use your result to find the area of the regular pentagon. Explain your method.

US-58. The figure at right is a regular 7-gon with center at P and $\overline{PB} \perp \overline{QR}$. If QR = 12", find the area of the 7-gon. If you need help, some of the subproblems are listed below and details are modeled in the preceding problem.

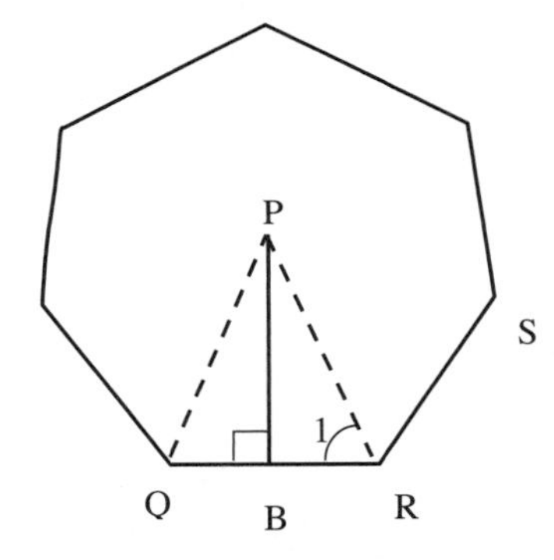


What is $m\angle QRS$? $m\angle 1$?

What is PB?
What is the area of $\Delta PQR$?

US-59. Use the process developed in the previous two problems to find the area of a regular 10-gon with a side length of 12" and center at A. Part of the figure is shown at right.

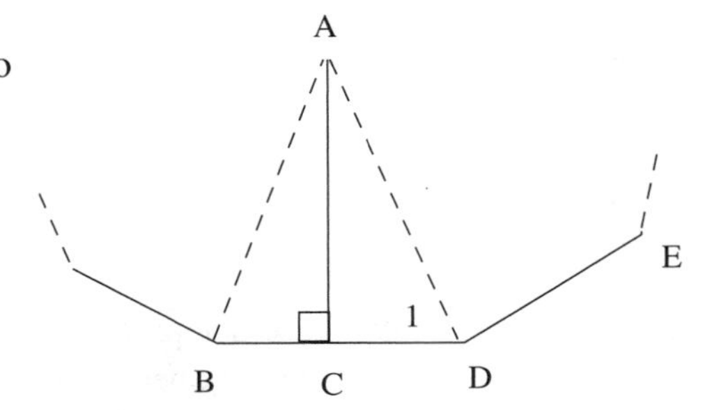


US-60. Use your work from the previous three problems to describe a general method (that is, a sequence of subproblems) for calculating the area of any regular n-gon.

US-61. **DIRECTIONS:** In this problem, each member of the team should use the process described in the preceding problem to do one of the n-gons below, then share the result with the team and go on to part (a). Find the areas of the following regular n-gons if the length of a side in all cases is 12":

3-gon 4-gon 6-gon 12-gon

a) Plot the above areas and those for regular 5-, 7-, and 10-gons from the previous problems on the resource page provided by your teacher.

b) Based on your graph, make predictions for the areas of a regular 8-gon, 9-gon, and 11-gon with side lengths of 12. Mark these points (preferably with a different color) on your graph.

c) Does it make sense to connect the points? Explain.

d) Describe the effect that increasing the number of sides of a regular n-gon has on The area of the n-gon when the length of the side remains the same.

US-62. Is it true that if an n-gon is non-convex (as in the figure at right), then the sum of the interior angles is still found by using (n - 2)180°? Either prove it for the given figure or else give a counterexample.

US-63. Calculate the total surface area and the volume of the prism at right. The bases are parallelograms. All vertical edges are perpendicular to the bases. Show all subproblems.

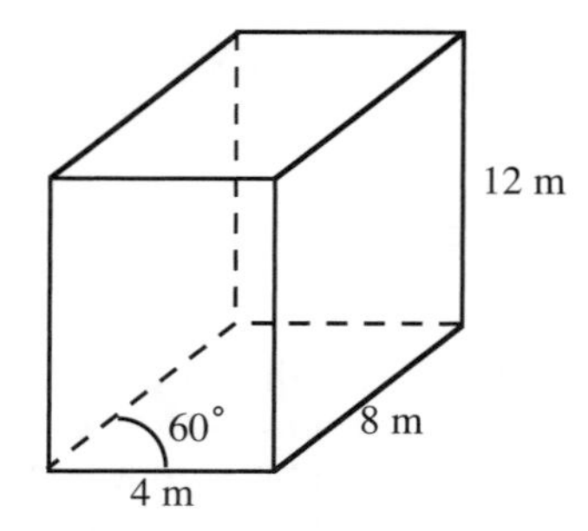


US-64. As Winter Recess approaches, you and some friends plan to construct a kid's toboggan slope on a patch of open land owned by your uncle. Kristin built a scale model. The model required 11 feet of rope to represent the fencing and 1.65 cubic feet of clay to represent the snow hill. If the scale of the model is 1:32, how much fencing and snow will be needed for the toboggan run?

US-65. Margaret has just received a letter from her pen pal Clarissa in Boston. Clarissa's letter describes their backyard by saying:

> "It is a triangle with one side 60 feet long, one side 44 feet long, and the third side 36 feet long."

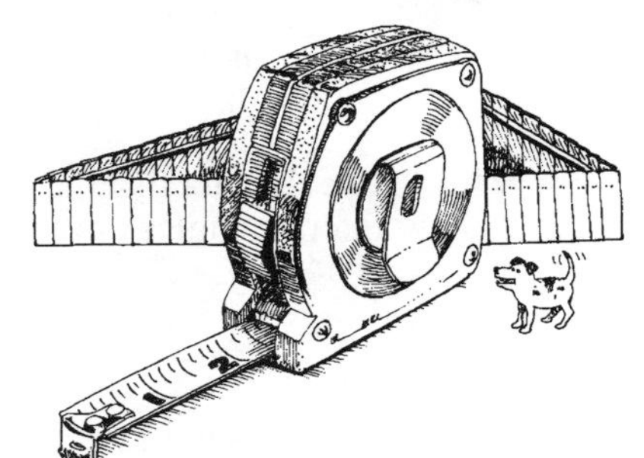

Margaret measures her own backyard which is also a triangle and finds its sides are also 60 feet, 44 feet, and 36 feet.

a) Are the two yards congruent in shape? Explain why or why not.

b) If Clarissa's 60 foot fence is to her left as she steps into the yard, must Margaret's left fence also be 60 feet?

US-66. The solid represented by the Mat Plan at right is built and glued together.

Back

| | | | |
|---|---|---|---|
| | 3 | 2 | |
| Left | 2 | 1 | Right |
| | 1 | 3 | |

Front

a) Draw the isometric view of this solid.

b) What is the surface area of the solid?

c) The solid is dipped in red paint and then left to dry. Once dried, the cubes are broken apart and each cube is dropped into a bag. If you are going to reach into the bag and pull out a cube at random, what is the probability that the cube has at least one red face?

## PROOF WITH POLYGONS

US-67. Given that ABCDE is a regular pentagon, discuss with your team as many different ways to find m∠F as possible. List them.

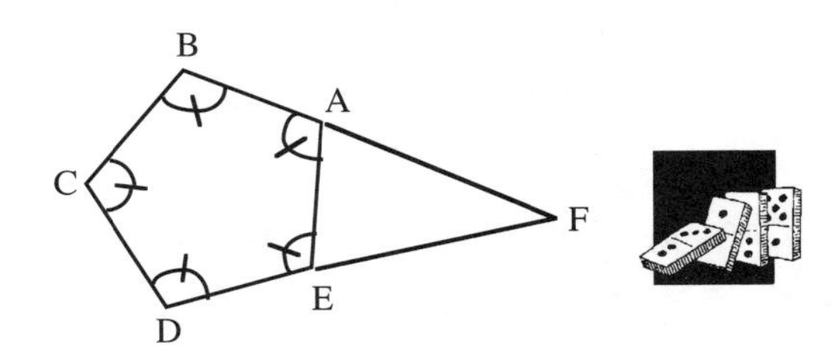

**US-68.** Two ways are suggested to find $m\angle F$ below. Solve for $m\angle F$ using <u>each</u> of these two ways and justify each step of your solution.

a) Use quadrilateral BCDF and the fact that the sum of the measures of the interior angles of a quadrilateral is 360°. Redraw the figure without $\overline{AE}$ to help visualize this approach.

b) Use the fact that the exterior angles of $\Delta AFE$ at A and E are supplements of interior angles of the triangle, and that the sum of the angles of a pentagon is 540°.

**US-69.** Look at the diagram for the preceding two problems. Prove that $\Delta AFE$ is isosceles or give a counterexample that shows why it is not.

**US-70.** Draw diagonal $\overline{BD}$ in the figure for problems US-67 and US-68. Note that we will use it as an auxiliary line for this problem.

a) Is $\Delta BDF$ isosceles? Prove it or explain why not.

b) Is $\overline{AE}$ parallel to $\overline{BD}$? Prove it or explain why not.

c) Is it true that $\angle C \cong \angle F$? Prove it or explain why not.

**US-71.** A regular pentagon with two diagonals is shown. Find the measures of $\angle 1$, $\angle 2$, and $\angle 3$. Justify your results.

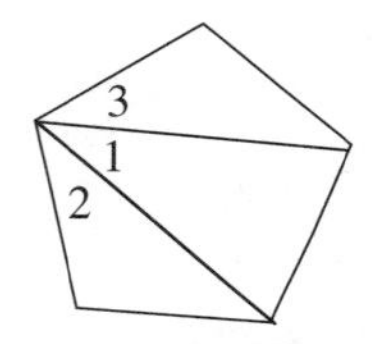


**US-72.** In the figure at right, $\overline{AB} \cong \overline{DC}$ and $\angle ABC \cong \angle DCB$.

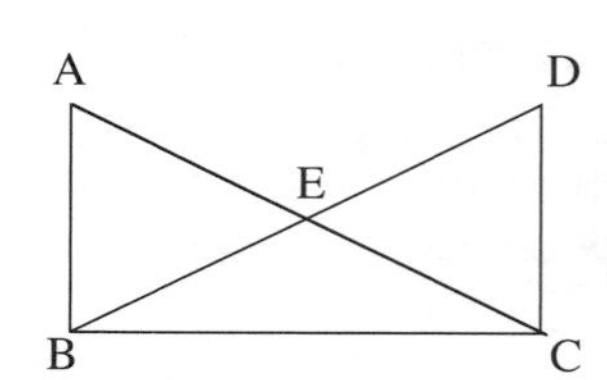

a) Is $\overline{AC} \cong \overline{DB}$? Prove your answer.

b) Does the size of $\angle ABC$ and $\angle DCB$ make any difference in your solution to part (a)? Explain why or why not.

US-73. A statue to honor Benjamin Franklin will be placed outside the entry to the Liberty Bell exhibit hall. The designers decide that a smaller version will be placed on a table inside the building. The dimensions of the life-size statue will be four times those of the smaller statue. Planners estimate they will need $1\frac{1}{2}$ pints of anti-tarnishing fluid to coat the small statue. They also know that the small statue will weigh 14 pounds.

a) How many pints of anti-tarnishing fluid will be needed to coat the life-size statue?

b) How much will the life-size statue weigh?

US-74. A quadrilateral has vertices at (0, 0), (100, 0), (25, 14), and (63, -26). In any way that you choose, find the sum of the interior angles of this quadrilateral.

US-75. Mario has $700 and is saving $40 per month. Camille has $450 and is saving $65 per month. Write an equation for each of them that states how much money they have (y) after x months, then answer the following questions.

a) When will Camille have as much money as Mario?

b) When will Camille and Mario together have $3,000?

c) When will Camille have twice as much money as Mario?

US-76. Find the values of x and y. Show all steps and state the reasons which justify your solutions.

a)

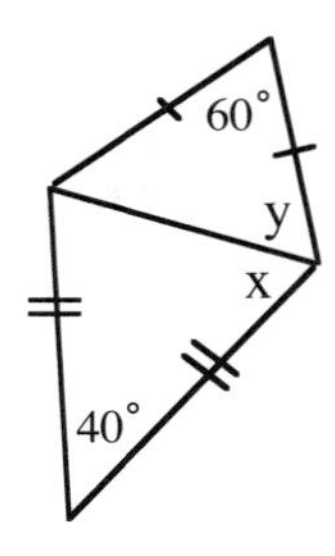


b)

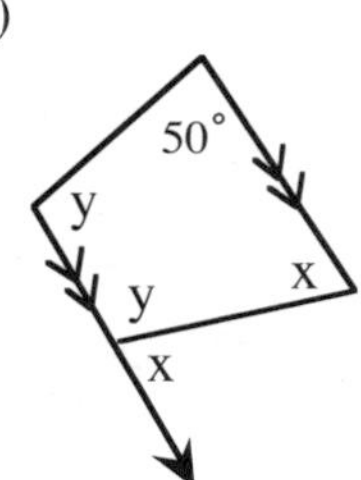


US-77. In the figure below, D and E are midpoints of their sides. Answer each part below and explain how you solve each part.

a) Find DE.

b) Find m∠ADE.

c) Find the ratio of AD to AB.

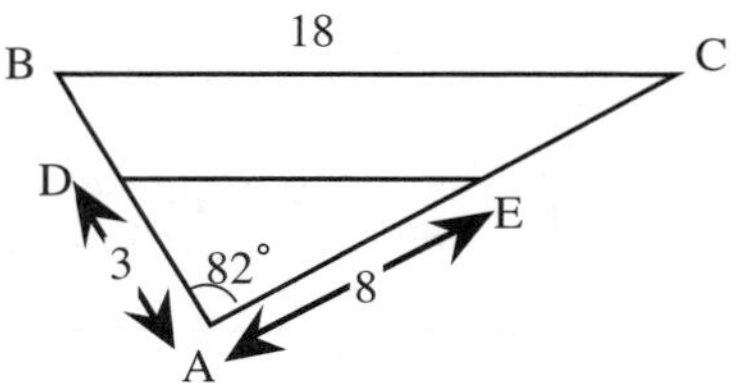

US-78. If AB = 3 cm and BC = 5 cm,

a) What is the smallest AC can be? Justify your answer.

b) What is the largest AC can be? Justify your answer.

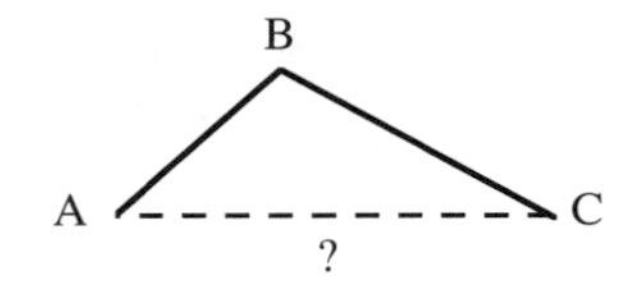


US-79. Solve for x. Write all answers in simplest form without decimals.

a) $\frac{x}{3} + \frac{5}{\sqrt{2}} = 10$

b) $\frac{x^2}{2} + \frac{7x}{4} + \frac{1}{2} = 0$

## PROPERTIES OF QUADRILATERALS

**US-80.** Your teacher will give you a resource page for quadrilaterals. Over the next few days you will be exploring various properties of specific quadrilaterals. As you discover more information about each kind of quadrilateral, you should add it to the vocabulary and properties resource page. This page will be part of your tool kit for this unit. **First you will need to add the definitions below next to the figures on the resource page.**

| | |
|---|---|
| **QUADRILATERAL** | polygon with exactly 4 sides. |
| **KITE** | quadrilateral with 2 pairs of consecutive, equal sides. |
| **TRAPEZOID** | quadrilateral with 1 pair of parallel sides. |
| **ISOSCELES TRAPEZOID** | trapezoid with a pair of equal base angles (from the same base). |
| **PARALLELOGRAM** | quadrilateral with 2 pairs of parallel sides. |
| **RECTANGLE** | quadrilateral with 4 right angles. |
| **RHOMBUS** | quadrilateral with 4 congruent sides. |
| **SQUARE** | quadrilateral with 4 right angles **and** 4 congruent sides. |

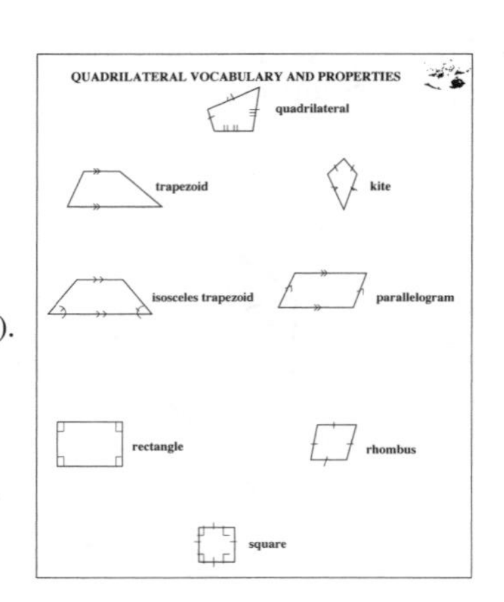


So far in this course you have discovered several properties of quadrilaterals. Illustrate (mark) any of the properties you might remember--like congruent parts--on the appropriate pictures and list them below the figure on your resource page. Once you have copied the definitions on your resource page, start the next problem.

**US-81.** In Unit 6 you proved three properties of parallelograms (refer to the figure below, right):

- diagonal $\overline{PR}$ creates two congruent triangles (as does $\overline{QS}$);
- pairs of opposite sides are equal (QR = PS and PQ = SR)
- pairs of opposite angles are equal ($m\angle Q = m\angle S$ and $m\angle QPS = m\angle SRQ$).

List these properties below the parallelogram on your vocabulary resource page and mark them on the figure there as well.

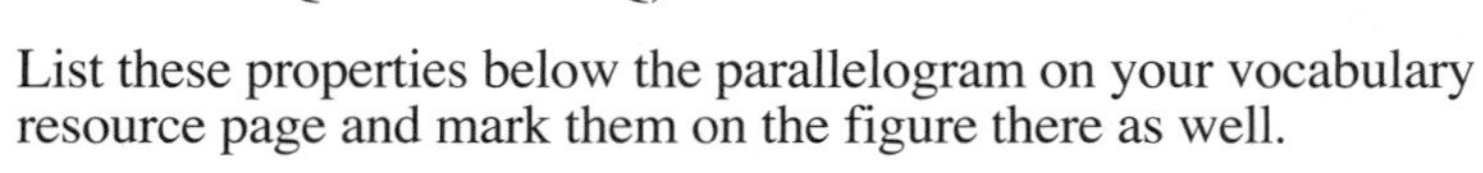

a) Prove that the diagonals of a parallelogram bisect each other. Start by copying the figure at right and drawing diagonal $\overline{QS}$. Label the point where the diagonals intersect T. Remember: refer to the properties of parallelograms and mark everything you know about the parts of the parallelogram and triangles from previous definitions and the theorems you proved. Then analyze the information and plan your proof.

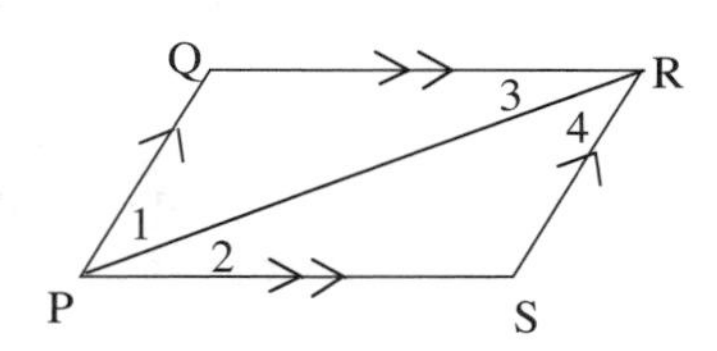


b) Draw a counterexample that <u>disproves</u> the generalization that PR = QS. <u>Could</u> PR = QS in some cases? Explain why or why not?

**US-82.** Trapezoid ABCD is isosceles with $\angle B \cong \angle C$. Prove that AB = DC.

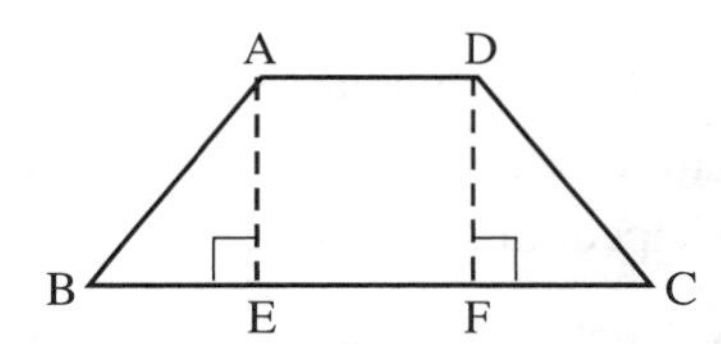

US-83. Name each figure below. Be as specific as you can, but do not be fooled by how the figure looks. Draw your conclusions based on what you <u>know</u> <u>for</u> <u>sure</u>.

a)

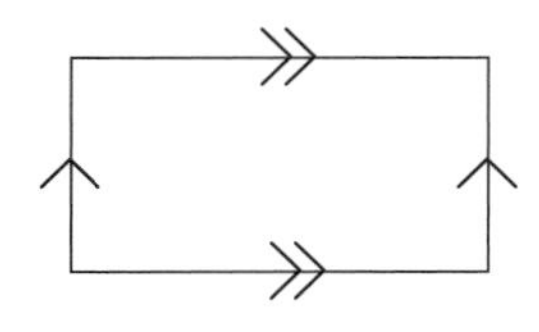

b)

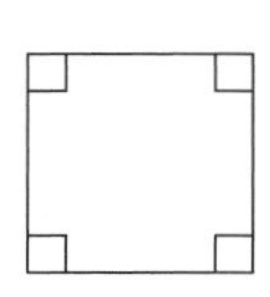

c)

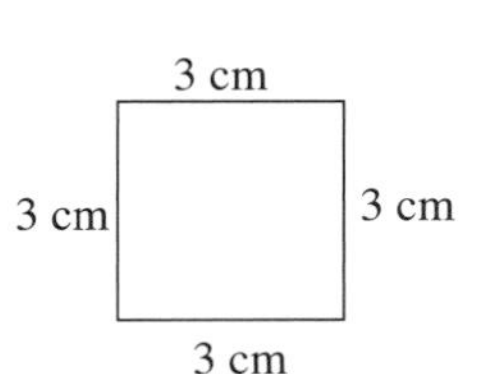


d)

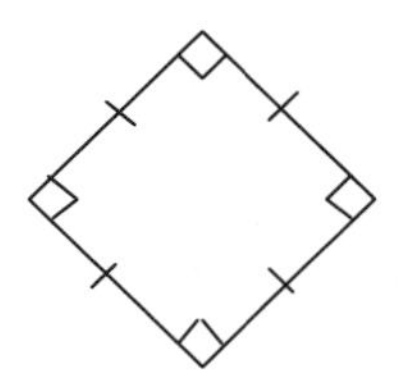

US-84. **DIRECTIONS:** For each part of the following problem, try to build a QUADRILATERAL (but not a trapezoid, rectangle, rhombus, or square) that satisfies the given characteristic(s) with the manipulatives provided by your teacher. If you are successful in building such a quadrilateral, make a sketch of it on your paper. If you believe the quadrilateral is impossible to build, explain why in one or two complete sentences. Be sure to discuss your conclusions with your team.

Draw a quadrilateral that has:

a) exactly one right angle.

b) exactly two right angles.

c) exactly three right angles.

d) one acute angle, one right angle, and two obtuse angles.

US-85. Draw and label the following 4 different PARALLELOGRAMS. Write a specific name next to each one. Refer to your vocabulary resource page for definitions.

a) ABCD is equilateral.

b) EFGH is equiangular.

c) IJKL is equilateral and equiangular.

d) MNOP is neither equilateral nor equiangular.

US-86. The Blackbird Oil Company is considering the purchase of 20 new oil storage tanks. The standard model holds 12,000 gallons. Its dimensions are $\frac{4}{5}$ the size of the similarly shaped jumbo model, that is, the ratio of the dimensions is 4:5.

a) How much more storage capacity would the purchase of the twenty jumbo models give Blackbird Oil?

b) If jumbo tanks cost 50% more than standard tanks, which tank is a better buy?

US-87. Find the value of z. Show all steps leading to each solution.

a)

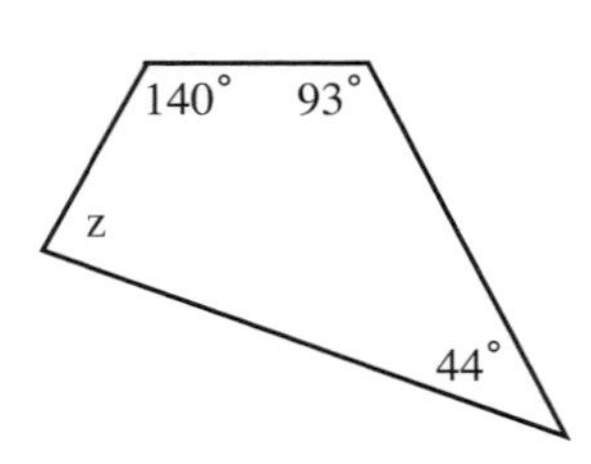


b)

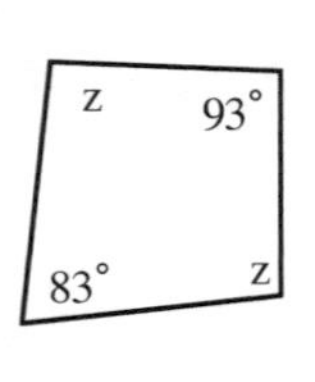


c)

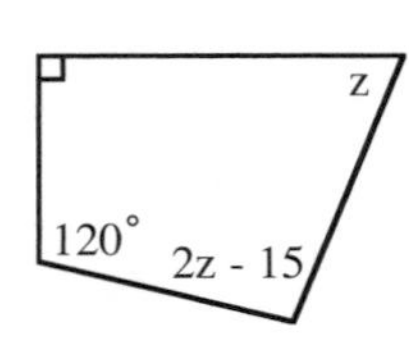


US-88. Find the value of x and redraw this diagram to fit what you found.

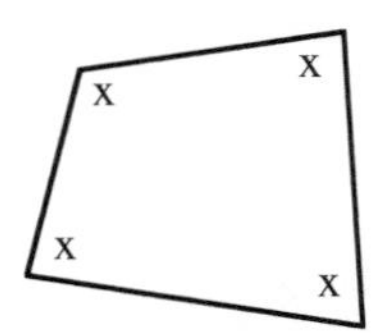

US-89. Patrick and his friend Patti are given the graph at right to analyze.

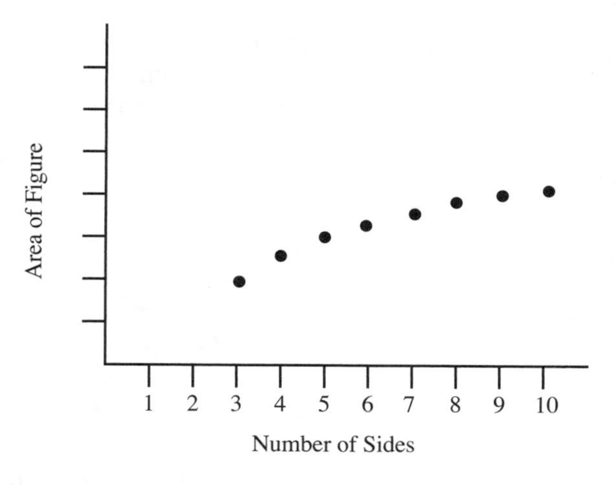


Patti exclaims "This is ridiculous! The graph is all messed up! It doesn't even have points plotted until x = 3 and none of the points are connected! We can't do ANYTHING with this problem!"

"Don't be silly!" Patrick comments. "I know why it starts at 3 and I know why the points are not connected."

Examine the graph closely. Why do the points start at three and why aren't they connected?

US-90. Solve for w if:

a) $5w^2 = 17$ b) $5w^2 - 3w - 17 = 0$ c) $2w^2 = -3$

## QUADRILATERAL PROOFS

US-91. Each member in the team should draw a rectangle on their paper. Be sure that each rectangle is a different size. Label the vertices as shown in the figure, then draw in the diagonals.

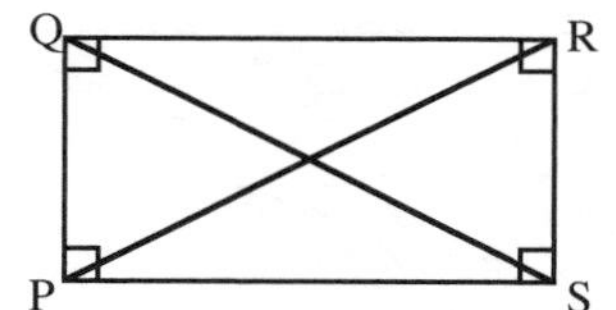

a) Measure the lengths of the diagonals in your rectangle and make a conjecture based on your result.

b) Prove your conjecture and add this information below the rectangle on your vocabulary resource page.

US-92. Figure DOVE is a rhombus. Use the fact that a rhombus is a special kind of parallelogram to prove the following:

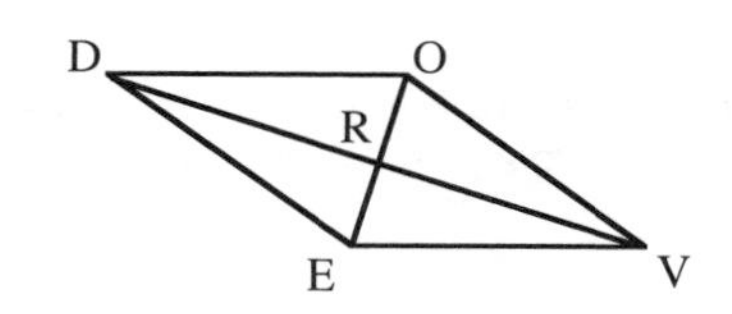

a) The diagonals of a rhombus are perpendicular.

b) The diagonals of a rhombus bisect the angles of the rhombus.

US-93. The length of each side of a rhombus is 10 cm and $m\angle A$ 60°. Find the length of the longer diagonal, $\overline{AC}$. You will need to use the properties of a rhombus to get all the data you need.

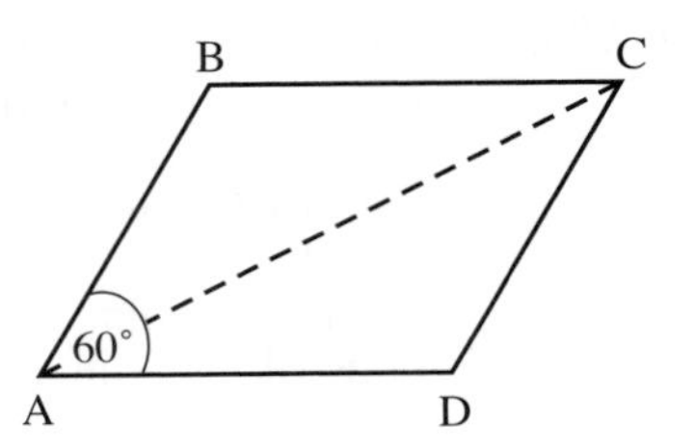


US-94. Graph each of the following equations on the same set of axes. Darken in the region bounded by all four lines.

$$y = \frac{4}{3}x \qquad y = \frac{4}{3}x + 5 \qquad x = 0 \qquad x = 3$$

a) Lightly shade the region bordered by these lines, then label the vertices of the resulting quadrilateral with the letters M, A, T, and H.

b) Find the length of each of the sides of MATH.

c) What type of quadrilateral is MATH? Why? Be as specific as you can.

d) Draw the diagonals of MATH. Write the equation for each diagonal.

US-95. Calculate the surface area of the following figures. First draw pictures of the area subproblems.

a) The base is a trapezoid.

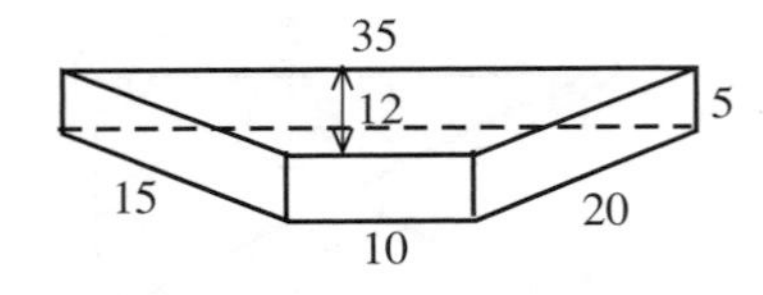


b)

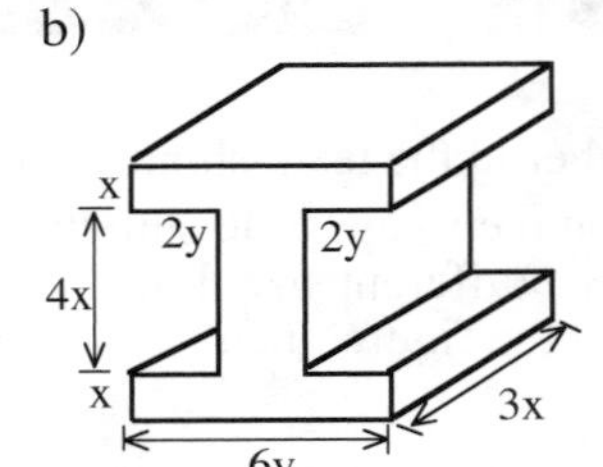


US-96. Show all steps leading to your solutions below.

a) Find x and the measure of each angle.

b) Must this pentagon be a regular pentagon? Give a convincing argument why or why not.

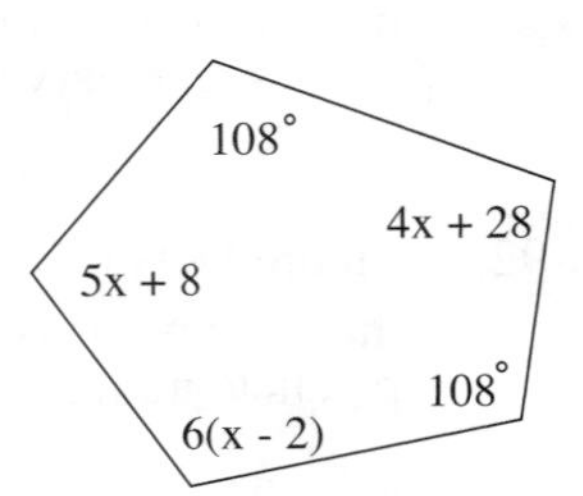

US-97. Recently, Randy took his SAT tests. One problem, which gave him a lot of trouble, asked him for the value of sin 60°. He did not have a calculator, so he guessed at the answer from these choices:

1) -3   2) $\frac{1}{2}$   3) $\frac{1}{3}$   4) $\frac{\sqrt{3}}{2}$   5) 6

a) What is the probability that by guessing he selected the correct answer?

b) Could Randy have increased his chances of getting the correct answer if he knew some facts about the sine ratio? Explain completely.

US-98. Solve these equations for x. Show all steps leading to your solution. Answers will contain a, b, and/or c.

a) $cx - a = b$   b) $\frac{x}{a} - b = c$   c) $(x - a)(x - b) = 0$

d) $ax^2 - acx = 0$   e) $\frac{x}{a + b} = \frac{1}{c}$

US-99. Consider the quadratic function $y = (x - \frac{2}{3})(x + \frac{3}{4})$.

a) Where does the graph of $y = (x - \frac{2}{3})(x + \frac{3}{4})$ cross the x-axis?

b) Multiply $(x - \frac{2}{3})(x + \frac{3}{4})$. Express your answer using fractions, not decimals.

c) Use Fraction Busters on your result in part (b) to write the equation $(x - \frac{2}{3})(x + \frac{3}{4}) = 0$ without fractions.

## UNIT AND TOOL KIT REVIEW

US-100. As American cities expanded during the Twentieth Century, there were often few controls on how the land was divided. In the town of Dry Creek, one particular tract of land had the shape and dimensions shown in the figure at right. The developer planned to build five homes per acre. Determine the number of homes that can be built on this tract of land. Show all dissections and subproblems.

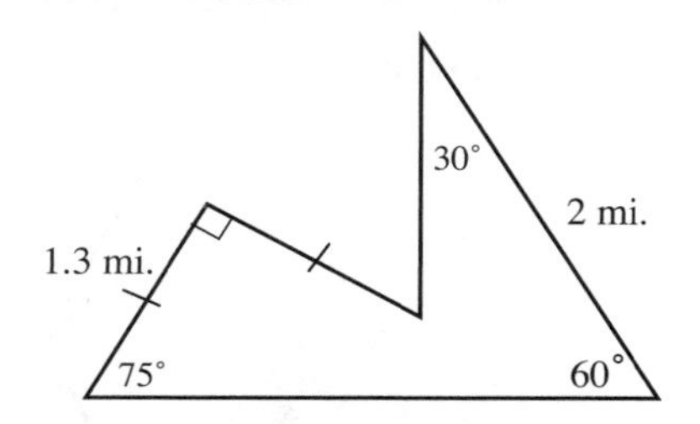


One mile is 5,280 feet and one acre contains 43,560 sq. ft.

US-101. In a certain parallelogram ABCD, the bisectors of two consecutive angles (A and D) meet at a point (P) on a non-adjacent side.

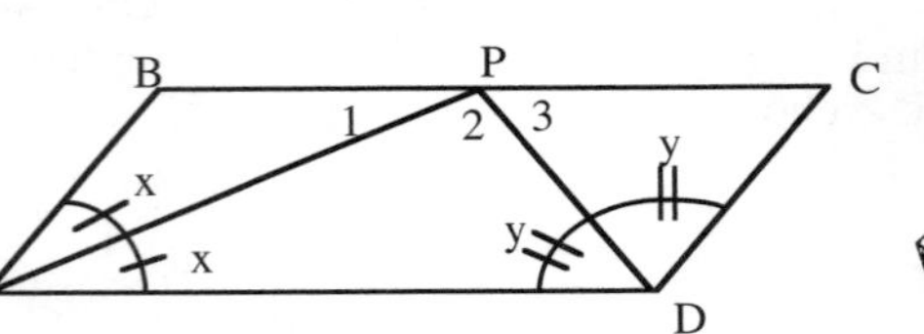


a) Keeping in mind that ABCD is a <u>parallelogram</u>, write down anything you and your team can determine about the figure. In particular, look for all of the segments and angles that a re equal to each other and mark them appropriately.

b) Prove that what you discovered in part (a) is true.

c) Are ΔABP and ΔPCD isosceles? Prove it or show why they are not.

d) Find the following area ratios AND explain how you got your answer:

1) $\frac{\text{area } \Delta APD}{\text{area } ABCD}$ 2) $\frac{\text{area } \Delta PBA}{\text{area } ABCD}$ 3) $\frac{\text{area } \Delta PBA}{\text{area } \Delta PCD}$

e) Does the ratio in part (d3) mean that ΔPBA ≅ ΔPCD? Explain.

US-102. Calculate the area in part (a) and the total surface area in part (b). Review problems US-57 through US-61 if you need help getting started.

a) A regular 8-gon.

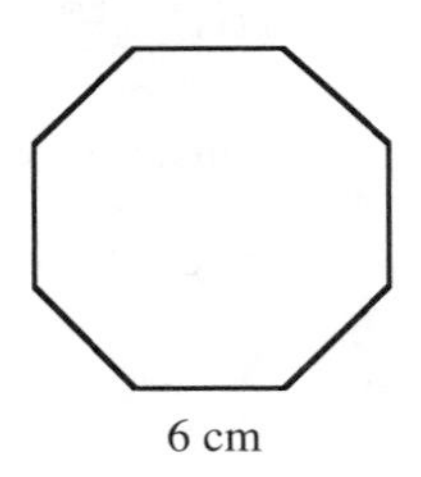


b) The base of the prism is a regular 8-gon.

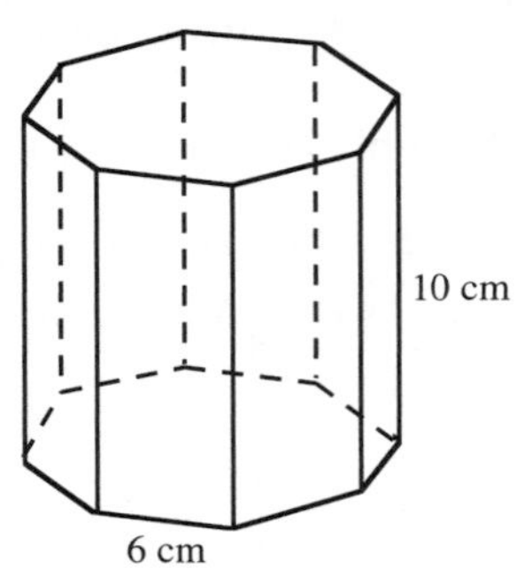


US-103. Find the value of x.

a)

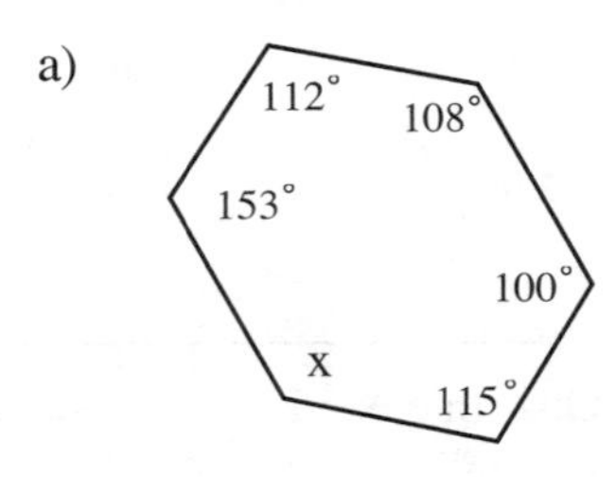


b)

156° x 140° 95° 110° 131° 115°

US-104. ABCD is a trapezoid. If AB = 16, BC = 10, and $m\angle C = 40°$, find AD and CD.

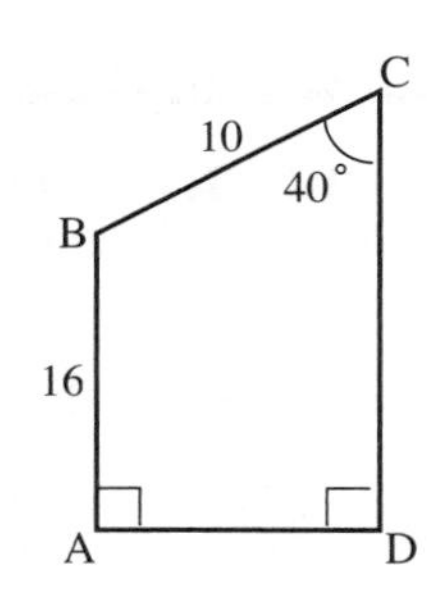


US-105. The figure at right is a square-based pyramid with congruent faces and h = 8. Calculate the total surface area of the figure. First draw pictures of the subproblems you will need.

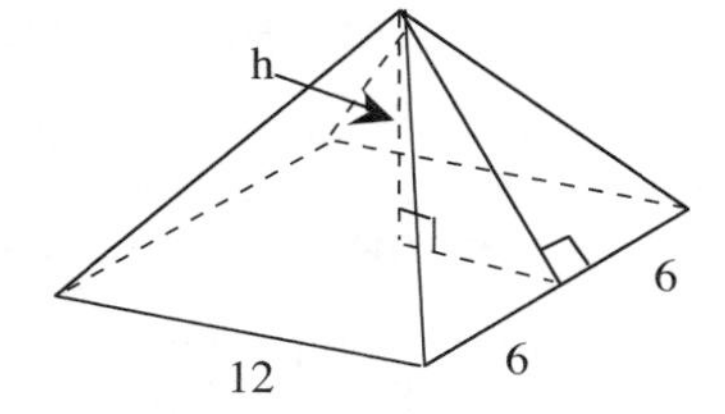


US-106. TOOL KIT CHECK-UP

Your tool kit contains reference tools for geometry and algebra. Return to your tool kit entries. You may need to revise or add entries.

Be sure that your tool kit contains entries for all of the items listed below. In particular, be sure that your quadrilateral grid is complete. Add any topics that are missing to your tool kit NOW, as well as any other items that will help you in your study of this course.

- polygon (n-gon)
  - diagonal
  - side
  - interior angle
  - exterior angle
  - regular
  - convex
- polygon angle formulas
  - $(n - 2)180°$ (sum of interior angles)
  - $m\angle = \dfrac{(n - 2)180°}{n}$ ($m\angle$ regular n-gon)
  - sum of exterior angles of an n-gon = 360°
- figure dissection (area)
- quadrilaterals and their properties
  - kite
  - trapezoid
  - isosceles trapezoid
  - parallelogram
  - rectangle
  - rhombus
  - square

## DESCRIBING QUADRILATERALS FROM THEIR DIAGONALS

US-107. **DIRECTIONS:** In this problem you will explore properties of the diagonals of various quadrilaterals. As you discover new properties, add them to your quadrilateral vocabulary grid. Read the instructions, 1 - 5, below, before you begin the problem.

**INSTRUCTIONS:** For each part of the problem, (a) through (g) below:

1) Build the two diagonals described (using strips of paper, uncooked linguini, drinking straws, coffee stirrers, etc.).

2) Hold your manipulatives in place and mark the endpoints of the diagonals on your paper. Draw segments connecting the endpoints to form a quadrilateral. Use a straightedge to draw your figures.

3) Identify what kind of quadrilateral you drew: a parallelogram? a rectangle? a rhombus? a square? a plain quadrilateral? Write a brief justification of why these characteristics of the diagonals MUST result in this type of figure.

4) Compare your quadrilateral to those made by the other people in your team. Did all of you draw the same kind of quadrilateral?

5) Organize your work by making a table with the headings shown below. Have one row for each part, (a) through (g).

**Be sure to add the properties you discover about diagonals by the appropriate figure on your quadrilateral vocabulary grid.**

| Properties of Diagonals | My Quadrilateral (sketch) | Specific Name of Quadrilateral |
|---|---|---|
| | | |

WHAT KIND OF QUADRILATERAL DO YOU GET FROM DIAGONALS WHICH ARE:

a) Not equal in length, bisect each other, and not perpendicular?

b) Not equal in length, bisect each other, and are perpendicular?

c) Equal in length, bisect each other, and perpendicular?

d) Equal in length, bisect each other, and are not perpendicular?

e) Equal in length, only one is bisected, and are not perpendicular?

f) Equal in length, neither is bisected, and are perpendicular?

g) Equal in length and intersect at a point one-third of the way along each one?

US-108. 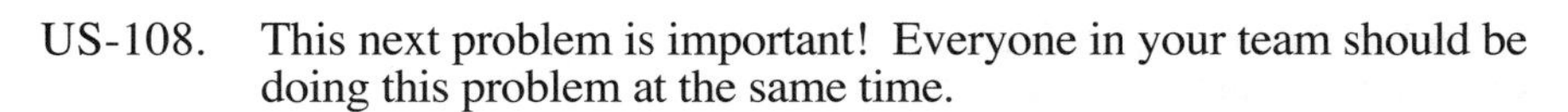
This next problem is important! Everyone in your team should be doing this problem at the same time.

a) Look back at parts (a), (b), (c), and (d) of the previous problem. Does everyone in your team agree with your answer? Come to a team consensus.

b) Each person should choose a different one of these four parts and write the conjecture as a theorem using careful "if-then" language. For example, you might say, "If ABCD is a quadrilateral and the diagonals are ..., then...". Share your conjecture with your team and make sure everyone agrees with the phrasing of your statement.

c) Prove your conjecture. Make sure you draw a diagram!

US-109. The following miscellaneous statements are all about a quadrilateral ABCD. Draw a diagram for each statement. If the statement is TRUE, give a brief explanation (proof) as to why it is true. If it is FALSE, use your diagram as a counterexample and explain as necessary.

a) If AB = CD, then ABCD is a rectangle.

b) If $\angle A \cong \angle B \cong \angle C$, then ABCD is a rectangle.

c) If ABCD is a rectangle, then $\angle A \cong \angle B \cong \angle C$.

d) If AC = BD, then ABCD is a rectangle.

e) If ABCD is a rectangle, then ABCD is a parallelogram.

f) If ABCD is a parallelogram, then $\angle A \cong \angle C$.

g) If $\angle A \cong \angle C$, then ABCD is a parallelogram.

h) If AB = CD and $\overline{AB} \parallel \overline{CD}$, then ABCD is a parallelogram. Drawing a diagonal will help.

i) If AB = CD and AD = BC, then ABCD is a parallelogram. Again, a diagonal will help.

US-110. For each of the following problems, try to build a quadrilateral that satisfies the given characteristics. If you are successful in building the quadrilateral, draw its sketch. If you believe the quadrilateral is impossible to build, explain why as clearly as you can.

a) Diagonals are perpendicular and the figure is not a square.

b) Diagonals are congruent and the figure is not a rectangle.

c) Diagonals bisect each other and the figure is not a parallelogram.

d) One diagonal is bisected, the other is not.

US-111. Palmer and Teller have the two quadrilaterals shown at right. They know that all pairs of corresponding angles are congruent and that $\overline{ME} \cong \overline{OJ}$ and $\overline{EK} \cong \overline{JN}$. They now need to know if the two quadrilaterals are congruent.

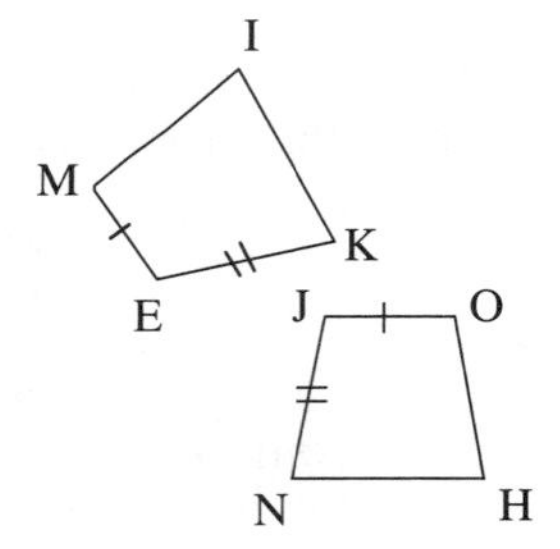


"They sure look congruent!" Teller exclaims.

"Haven't you learned anything?" Palmer responds. "You can't assume they are congruent just because they look congruent. The only hope we have is to make some congruent triangles."

"Well that's easy!" Teller yells. "I can see some already."

Palmer quickly reacts, "How can you say that? There are NO triangles in the picture!"

"Well they are not there yet, but I can easily make some."

a) Copy the quadrilaterals on your paper and use auxiliary lines to create appropriate triangles.

b) Which triangles are congruent and why?

c) What other triangles are congruent? Why?

d) Now that those triangles are congruent, what other parts of the figures are congruent?

e) What do you know about the quadrilaterals? Why?

US-112. An equilateral triangle has two vertices at (0, 0) and (20, 0) and its third vertex has y-coordinate greater than 0. Sketch all parts of the problem below.

a) Find the coordinates of the third vertex.

b) Find the area of the triangle.

c) One side of the triangle (not the base which is on the x-axis) is used as the side of a square. Draw the square and notice that the entire figure is now a pentagon. What is the area of the pentagon?

US-113. On a sheet of graph paper, set up a pair of xy-axes with $0 \le x \le 10$ and $-16 \le y \le 16$. Graph the three equations below on this one pair of axes.

$$y = \frac{2}{3}x + 5 \qquad y = -\frac{2}{3}x + 9 \qquad y = -\frac{5}{3}x + 15$$

a) Darken the interior region bordered by these three lines and the x- and y-axes (the region that lies within the first quadrant).

b) Reflect the darkened figure in (a) across the x-axis.

c) What kind of polygon is formed by the combination of both figures?

d) What is the area of this polygon?

US-114. Imagine the original shape in the preceding problem reflected across the y-axis rather than the x-axis.

a) Describe the shape that would be formed.

b) Without graphing it, what is the new shape's area?

US-115.

During an experiment with a weather balloon (A), some high school students took sightings and measurements resulting in the figure and data below. Find the height (h) of the balloon above the ground. Show all subproblems.

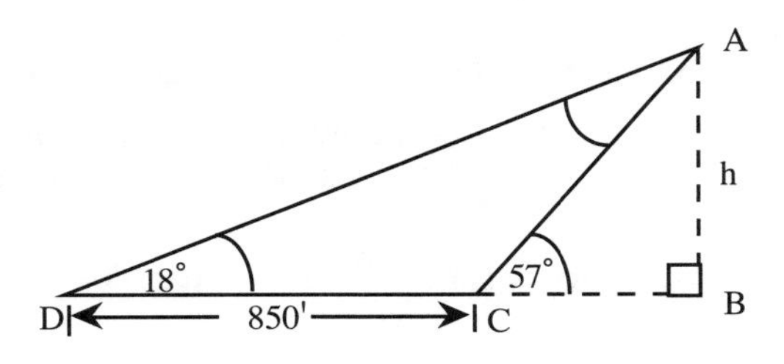


US-116. Given isosceles right ΔBFE overlaps square ABCD with AB = 1 and EB = 2, what is the area of the shaded region CDEF? Show all the subproblems leading to your solution.

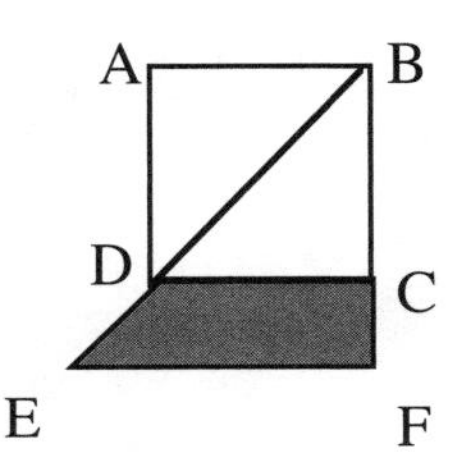

US-117. Graph and label the points A(0, 0) and C(6, 8). Locate two other points B and D such that:

a) ABCD is a square.

b) ABCD is a rhombus, but not a square.

c) Find all points (x, y) such that if B has coordinates (x, y) then AB and BC could be two edges of a rhombus. (You can do this either with algebra or with geometry.)

US-118. Use a whole sheet of large grid graph paper and place a pair of axes with the origin at the center of the paper. Carefully draw the graph of each of the following lines. Be neat. You will need this graph for the next problem as well. If you have been neat and careful, the four vertices of the resulting figure should have integer coordinates.

$$y = -\frac{3}{2}x - 8 \qquad y = -\frac{3}{4}x + 10$$

$$y = \frac{3}{4}x + 10 \qquad y = \frac{1}{2}x$$

a) Lightly shade the interior region bounded by the segments of the lines. What is the name of the region's boundary?

b) Find the length of each side of the figure formed in part (a), and find the perimeter.

c) Find its area.

US-119. Find and label the midpoint of each side of the figure formed above.

a) Form a new figure by connecting the consecutive midpoints you found. What does the new figure look like?

b) Prove that the figure is a parallelogram.

# The One-Eyed Jack Mine

CIRCLES AND SOLIDS

# Unit 10

## *The One-Eyed Jack Mine:* CIRCLES AND SOLIDS

This unit completes the introduction of topics that are usually covered in a geometry course. There are quite a few vocabulary terms for circles and three-dimensional figures in the unit. You will study angles, arcs, and line segments in relation to the circle. Then you will extend the area of polygons and circles to prisms, pyramids, cylinders, and cones. You will also consolidate previous work you did with volume and extend it to new solids.

Be sure to record each idea in your tool kit when it is introduced. In fact, we recommend that you make a Unit 10 tool kit vocabulary grid like the one you used in Unit 3 to help keep track of all the new terms.

In this unit you will have the opportunity to:

- investigate the number $\pi$ (pi).
- learn and apply the fundamental properties and relationships for circles:
  - --arcs and angles
  - --tangents, secants, chords, diameters, and radii.
  - --circumference and arc length.
  - --area of circles and sectors.
- review surface area and volume for prisms and extend these ideas to pyramids, cylinders, and cones.

*Read the following problem and make sure everyone in your team understands it. Do not attempt to solve it now. We will come back to it later.*

| | |
|---|---|
| CS-0 | The abandoned One-Eyed Jack Mine is about $3\frac{1}{2}$ miles off the main road adjacent to the Salmon River Wilderness area. Near the mine's entrance are the remains of a circular vat about 18 feet in diameter that was connected by a huge belt to a smaller circular drive wheel 10 feet in diameter. The distance between the wheel and the vat is 8 feet. Calculate the length of belt needed to go around the drive wheel and the vat. |

# *Unit 10*

## *The One-Eyed Jack Mine:* CIRCLES AND SOLIDS

### PI AND CIRCUMFERENCE

CS-1. Take a piece of dental floss or string and a circular disk. Mark off the length of the circle's diameter (distance across the disk at its widest point) on the dental floss. Wrap the floss around the circumference (perimeter) of the disk, marking on the disk each time you have gone the length of the diameter. Be accurate! About how many diameters are there in the circumference of the disk? Does the size of the circle seem to affect your answer?

CS-2. Use at least four different sized circular disks.

a) Copy a table like the one below on your paper and record the disk names or descriptions in the table. Measure the diameters and circumferences (use the dental floss) to the nearest millimeter and record this information in the table. Be as accurate as you can!

b) Calculate the ratio of the circumference to the diameter for each entry. Write your answer to 4 decimal places. Record this in the column headed $\frac{c}{d}$.

| Disk (description) | Diameter (d) | Circumference (c) | $\frac{c}{d}$ |
|---|---|---|---|
| | | | |
| | | | |
| | | | |
| | | | |

c) Calculate the average ratio, $\frac{c}{d}$, for your table. Now compare the average ratio for your team and report that value to the class.

d) Calculate the class's average.

e) Based on the results you saw in (b) and (c), write a conjecture about the ratio of the circumference of a circle to its diameter.

CS-3. **DIRECTIONS:** As in Unit 3, this unit will contain many vocabulary terms. Most of them you have heard and used before. To keep track of them all, start a vocabulary grid similar to the one you used in Unit 3. Each time you come to a new term, or a term in capital, bold letters, draw a picture illustrating the term in one of the boxes on your grid and write the term next to it. Be sure you include abbreviations and helpful markings on the figures you draw. Do this neatly. Read the box below and add the definitions to your tool kit.

**Caution! Read this information carefully!**

A **CIRCLE** is the set of all points that are the same distance from a fixed point, G. We use **⊙G** as the **symbol to represent a circle**, in this case circle G, where G is the center. The fixed point is called the **CENTER** of the circle and the distance from the center to the points on the circle is called the **RADIUS** (usually denoted r). A line segment drawn through the center of the circle with both endpoints on the circle is called a **DIAMETER** (denoted d). Note: d = 2r.

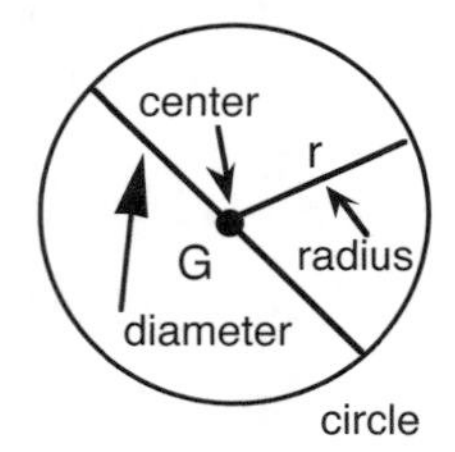


You can think of a circle as the rim of a bicycle wheel. The center of the circle is the hub where the wheel is bolted to the bicycle's frame. The radius is a spoke of the wheel.

The **CIRCUMFERENCE** of a circle (C) is its perimeter, or distance around the circle. The previous problem demonstrated that pi (pronounced "pie" and written as π) is a constant value that represents the ratio of the circumference to the diameter of any circle regardless of the size of the circle. **AREA** describes the region inside the circle itself.

$$\pi = \frac{\text{circumference}}{\text{diameter}} = \frac{C}{d} \qquad C = \pi d = 2\pi r \qquad A = \pi r^2$$

Note that by solving the equation $\pi = \frac{C}{d}$, above left, for C, we arrive at the familiar means of calculating the circumference of any circle, above middle.

As a decimal, π never ends and never repeats. The first twenty-eight digits of π are:

**3.141592653589793238462643338**.

It is impossible to use all the digits of pi in calculations (since the digits go on forever and π cannot be written as a fraction), so often 3.14 or $\frac{22}{7}$ are used as an approximation of π. Unless you are told otherwise in a problem or by your teacher, use the π key on your calculator to do circle problems in this course.

By now you should have six terms illustrated with a picture, or pictures, on your vocabulary grid.

CS-4. Write all answers to 5 decimal places.

a) Write $\frac{22}{7}$ as a decimal.

b) Which is a better approximation for $\pi$, 3.14 or $\frac{22}{7}$? Why?

c) The Egyptians used $(4)\left(\frac{8}{9}\right)\left(\frac{8}{9}\right)$ for $\pi$. Write this as a decimal.

d) Some other fractions used to approximate $\pi$ are $\frac{3925}{1250}$ and $\frac{355}{113}$. Write these fractions as decimals.

e) Of all the approximations for $\pi$ you have seen, which do you think is the best and why?

CS-5. In parts (a) and (b), use the circumference and area formulas in the box in problem CS-3 to find:

a) the circumference and area of a circle with a diameter of 2.5 inches.

b) the circumference and area of a circle with a radius of 6 inches.

c) the diameter of a circle with a circumference of $21\pi$ inches.

d) the radius of a circular pond with a circumference of 31.4 meters.

CS-6. Graph the following inequalities for $-3 \le x \le 3$ on the same set of axes.

$$y \le -x^2 + 3 \qquad\qquad y \ge x^2 - 3$$

Darken in the overlapping region. Is the region circular? Justify your answer.

CS-7. While the earth's orbit is actually slightly elliptical, it can be approximated by a circle with a radius of 93,000,000 miles.

a) What is the approximate circumference of the earth's orbit?

b) About how fast is the earth traveling in its orbit in miles per hour?

CS-8. If $\angle C$ and $\angle A$ are right angles and $AB = BC$, is ABCD a square? Prove your answer.

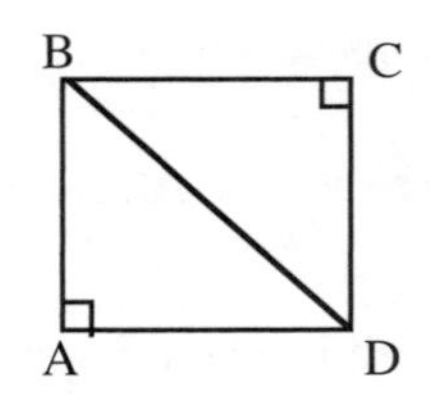

CS-9. Write the following in simplest radical form.

a) $\sqrt{128}$

b) $\frac{6}{\sqrt{18}}$

c) $\frac{3\sqrt{2}}{2\sqrt{3}}$

d) $\sqrt{75} + 2\sqrt{2} - 3\sqrt{3}$

CS-10. Use the figures at right.

a) Calculate the area of figure 1.

b) Suppose points A and B were stretched away from the figure until the acute angles were decreased to 30°, as shown in figure 2. Would the total area increase or decrease? Prove your answer.

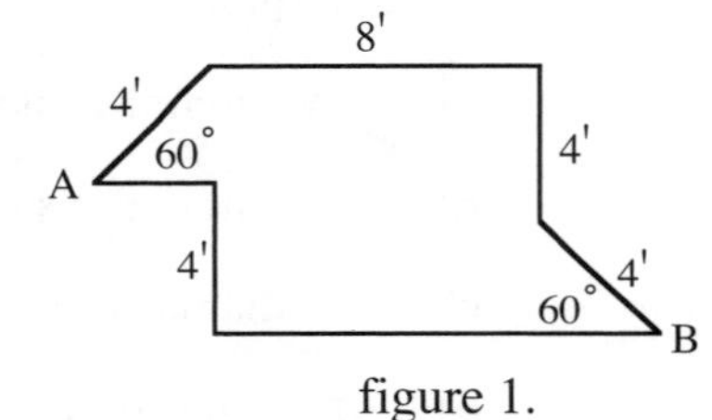

figure 1.

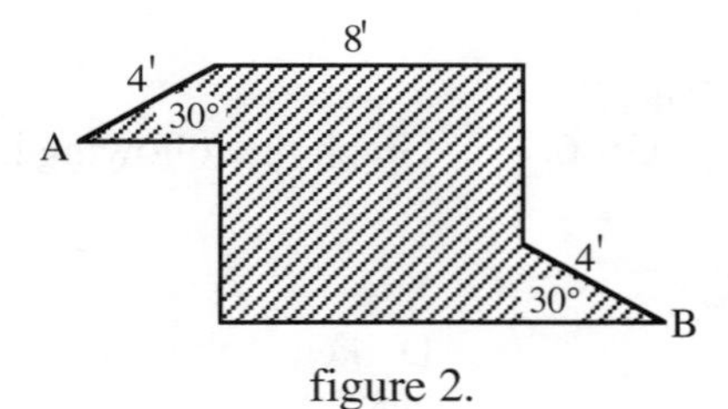

figure 2.

## ANGLES AND ARCS

CS-11. Add the definitions in the box below to your tool kit, then complete parts (a) through (e) below the box.

**Caution! Read this information carefully!**

A **CENTRAL ANGLE** has its vertex at the center of a circle.

An **ARC** is a part of a circle. Remember that a circle does not contain its interior. A bicycle tire is an example of a circle. The piece of the tire between any two spokes of the bicycle wheel is an example of an arc.

One way to discuss an arc is to consider it as a fraction of 360°, that is, as a part of a full circle. When speaking about an arc using degrees, we call this the **ARC MEASURE**. The arc between the endpoints of the sides of a central angle has the same measure (in degrees) as its corresponding central angle.

When we want to know how far it is from one point to another as we travel along the arc, we call this the **ARC LENGTH** and measure it in feet, inches, centimeters, etc.

Example: O is the **center** of the circle. $\angle AOB$ is a **central angle**. The sides of the angle intersect the circle and cut off arc AB. We use the symbol of an arc drawn over the points, $\overset{\frown}{AB}$, as an abbreviation. In this case the measure of the arc is 60°. However, there are really two arcs, one large one (a major arc) on the exterior of the angle and one small one (the minor arc). When you give the measure of $\overset{\frown}{AB}$, you should focus on the **MINOR ARC**. To refer to a **MAJOR ARC**, a third point on the major arc would be used to clearly name the arc, such as $\overset{\frown}{ACB}$. Its measure, abbreviated $m\overset{\frown}{ACB}$, is 300°.

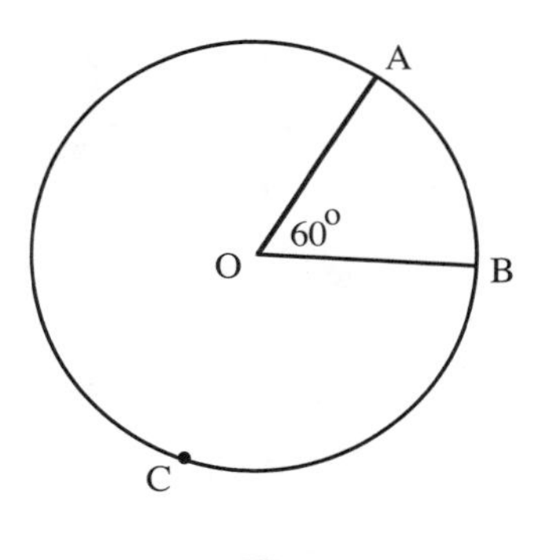


The three circles shown at right are called **CONCENTRIC** circles. This means they all have the same center.

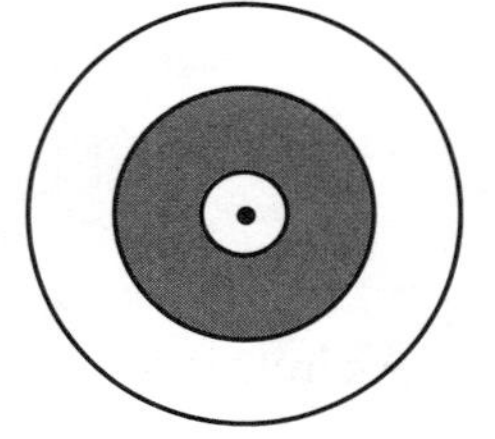

a) Find another real-life example of an arc (remember that it has to be part of a circle, not just any curve).

b) You can describe an arc as a fraction of a circle. Make a sketch of an arc that is $\frac{1}{4}$ of a circle.

**>>> Problem continues on the next page.>>>**

c) Make an accurate drawing of an arc that is $\frac{1}{6}$ of a circle whose radius is 10 cm.

d) Using the arc you drew in part (c), draw the radius from each endpoint of the arc to the center of the circle. What is the degree measure of the angle between the two radii?

e) Arcs are also measured in degrees. What do you think is the degree measure of this arc? Why?

CS-12. Use the figures at right to answer the following questions. O is the center of each circle.

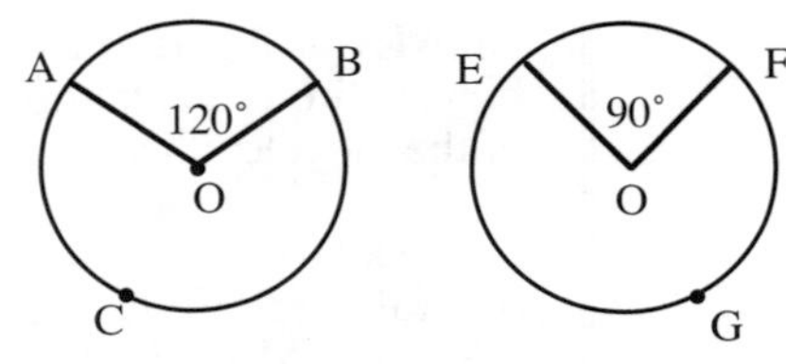


a) How many degrees are needed for an arc to go completely around a circle?

b) In the diagram shown, what is the measure of the minor arc $\overset{\frown}{AB}$? the <u>major</u> arc $\overset{\frown}{ACB}$?

c) What is the measure of central angle EOF?

d) What is the measure of $\overset{\frown}{EGF}$?

CS-13. The figure at right shows two **concentric** circles.

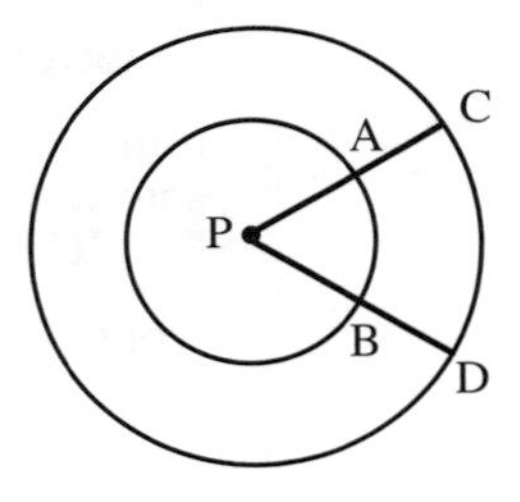


a) Which arc has the larger <u>measure</u>: $\overset{\frown}{AB}$ or $\overset{\frown}{CD}$? Explain.

b) Which arc has the greater <u>length</u>? Explain.

c) How does increasing the length of the radius of a circle affect the arc measure and the arc length? Explain.

CS-14. A circle is divided into n <u>congruent</u> arcs. What is the measure of each arc if :

a) n = 4?

b) n = 27?

c) n = 100?

d) n is any positive integer?

**CS-15.** In each circle, the vertex of the angle is at the center. Find the <u>measure</u> of x for each of the following:

a)

b)

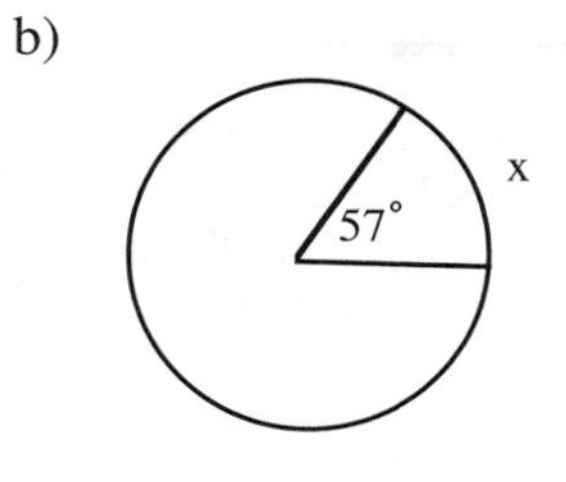


c)

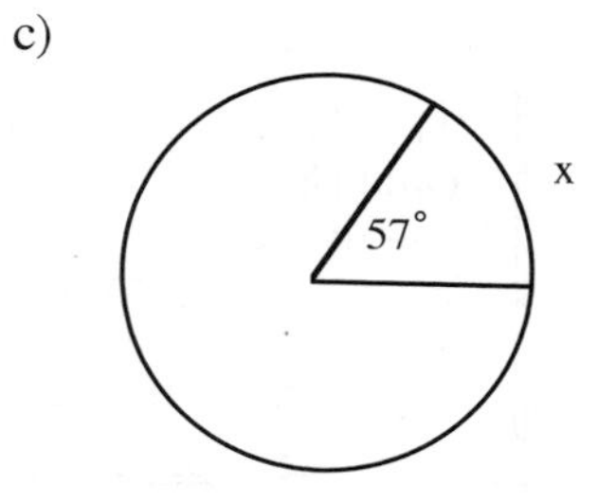


d)

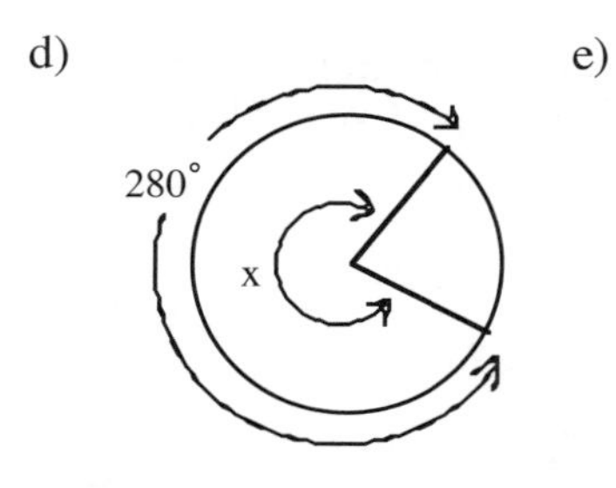


e)

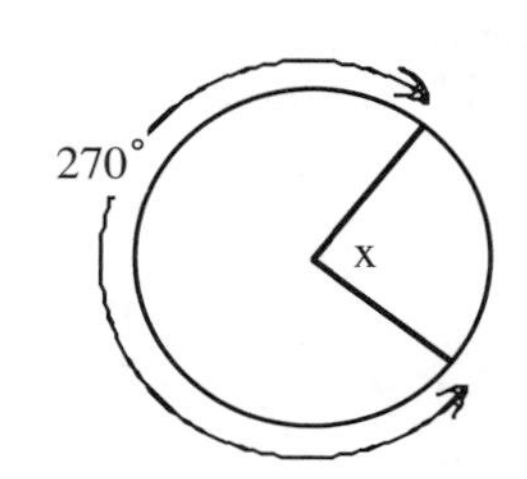


f) $m\angle 1 = m\angle 2$

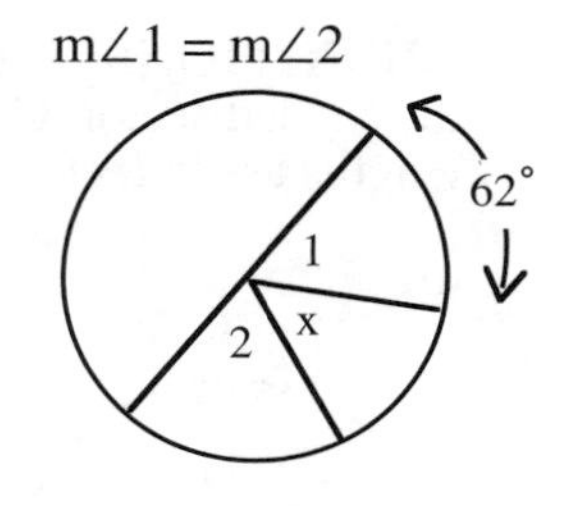

**CS-16.** Read the definition in the box below, add it to your tool kit, then do the problems below the box.

> The shaded regions in circles that resemble pieces of pie in the next problem are called **SECTORS** of a circle. **Sectors** are formed by the two radii of a central angle and the arc between their endpoints on the circle. It is often convenient to leave your answers **in terms of $\pi$** when calculating areas of circles or sectors. This means the symbol $\pi$ will often be in your answer. You will not substitute 3.14 or any other approximation of $\pi$. For example, the area of a circle with r = 6 is $36\pi$.

**DIRECTIONS:** Use your knowledge of arcs and areas of circles to find the area of the shaded sector. C, F, and U are the centers of their respective circles. All radii lengths are in feet.

a)

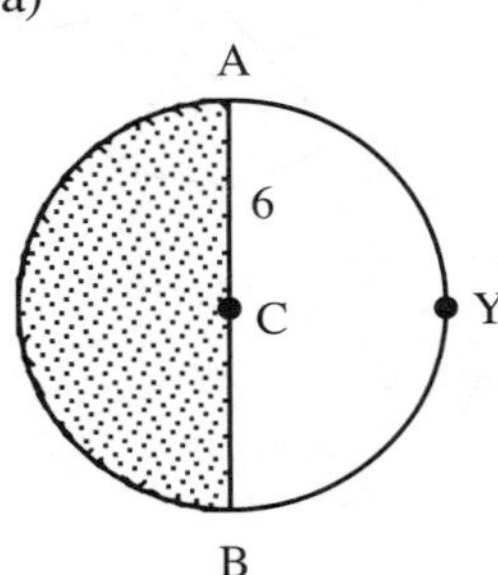


$\frac{m\overset{\frown}{AB}}{360^\circ} =$

area of ⊙C =

area of the semi-circle (shaded region) =

b)

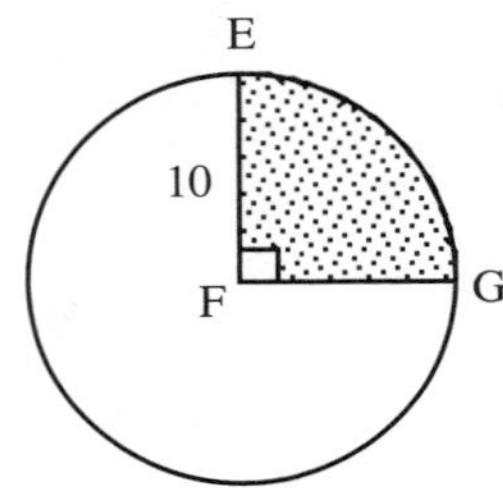


$\frac{m\overset{\frown}{EG}}{360^\circ} =$

area of ⊙F =

area of sector EFG (shaded region) =

c)

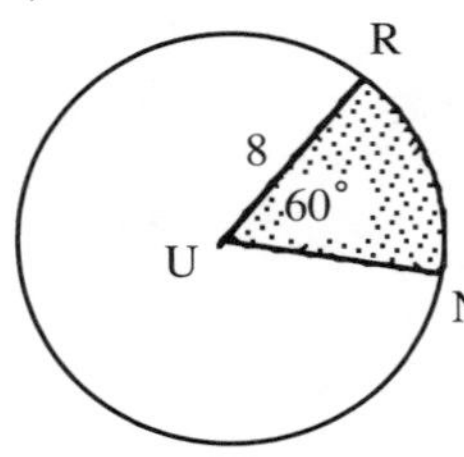


$\frac{m\overset{\frown}{RN}}{360^\circ} =$

area of ⊙U =

area of sector RUN (shaded region) =

**CS-17.** Flipper can swim around the edge of his circular pool in 20 seconds. How long would it take him to swim straight across the pool along a diameter?

**CS-18.** A regular polygon has an exterior angle measuring 20°.

a) If one side of the polygon is 6 cm, what is the perimeter of the polygon?

b) What is the area of the polygon?

**CS-19.** Solve for the variable(s).

a) $\frac{8}{5}x - \frac{13}{3} = \frac{1}{6}x$

b) $\frac{x}{3} - \frac{3}{2x} = 2$

c) $3x + 2y = 6$
$y = -x - 4$

d) $3\sin 30^\circ + 6z = -2$

CS-20. A scientist was working on a device to shrink objects. One day she tested it on her car and shrunk it to $\frac{1}{12}$ its original size. If her car was 16 feet long and weighed 3,200 pounds, how long is the shrunken car? How much does it weigh?

## INSCRIBED ANGLES AND INTERCEPTED ARCS

CS-21. Read the definition in the box below, add it to your tool kit, obtain a resource page from your teacher, then complete parts (a), (b), and (c) below the box. Colored pens or pencils are useful with the figures.

An **INSCRIBED ANGLE** is an angle with its vertex on the circle and whose sides intersect the circle. The arc formed by the intersection of the two sides of the angle and the circle is called an **INTERCEPTED ARC**. $\angle AOB$ is an inscribed angle, $\overset{\frown}{AB}$ is an intercepted arc.

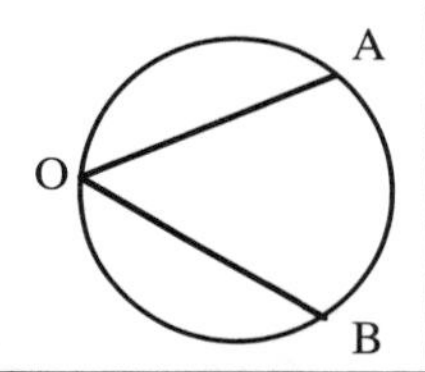


a) In Circle 1, outline each of the angles C, D, and E with a different color so that each angle intercepts an arc. Find the intercepted arc for each angle and outline it with the same color as the angle.

b) Use a ruler or other straightedge to carefully trace $\angle C$ on tracing paper, then place it over $\angle D$ and compare their sizes. Do the same for $\angle C$ and $\angle E$. What appears to be true about the measures of these three angles?

c) Make another copy of $\angle C$ on your tracing paper so that it is adjacent to the first copy, that is, the two copies have the same vertex and share a common side. What appears to be the sum of the two angles? What, then, is the measure of each copy of $\angle C$?

d) Name each angle's intercepted arc and compare the arc's measure to that of each angle you traced in part (b). How are they related?

e) Test your results in Circle 1 using Circle 2. First trace $\angle F$ and compare its size to $\angle J$. Next locate the arc these two angles intercept. What should the measures of the two angles be, based on the measure of $\overset{\frown}{GH}$ and your conjecture?

f) $\Delta XYZ$ is equilateral. What is the measure of each angle in the triangle? If your conjecture in part (d) is correct, two copies of $\angle F$ should fit exactly in $\angle Y$. Use the copy of $\angle F$ on your tracing paper to trace two copies of it in $\angle Y$. What do you observe?

g) Discuss your results with your team, refine your conjecture in part (d) if necessary, and add it to your tool kit.

CS-22. Read the theorem in the box below and modify your conjecture from the previous problem if necessary. Then find the measures of the indicated angles and arcs in parts (a) through (f) below the box. Where marked, C is the center of the circle.

**INSCRIBED ANGLE THEOREM**

The measure of any inscribed angle is half the measure of its intercepted arc. Likewise, any intercepted arc is twice the measure of any inscribed angle whose sides pass through the endpoints of the arc.

$$\mathbf{m\angle ADB = \frac{1}{2}\overset{\frown}{AB} \text{ and } \overset{\frown}{AB} = 2m\angle ADB}$$

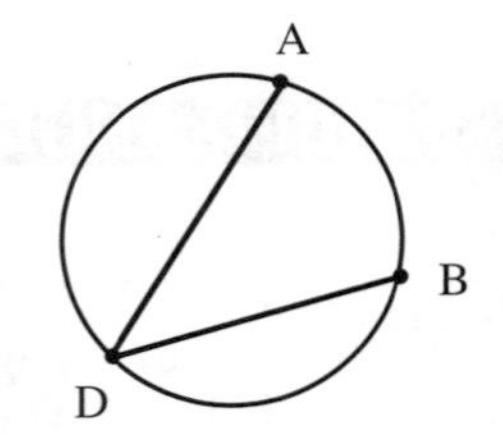


a)

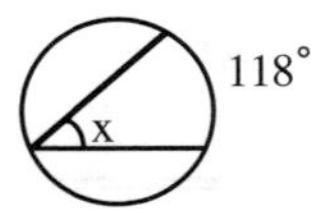


b)

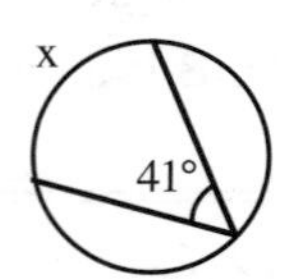


c)

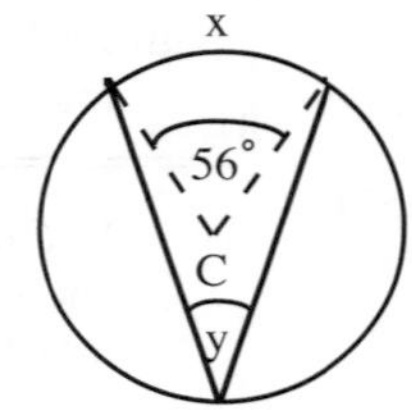


d)

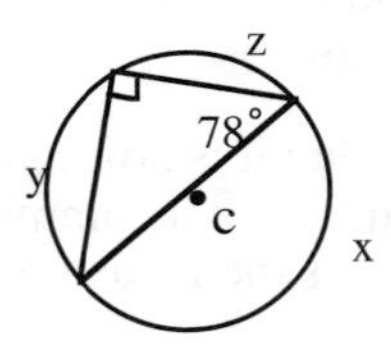


e)

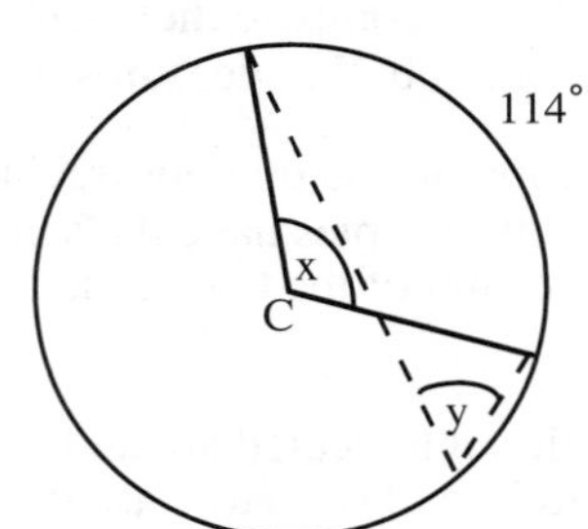


f)

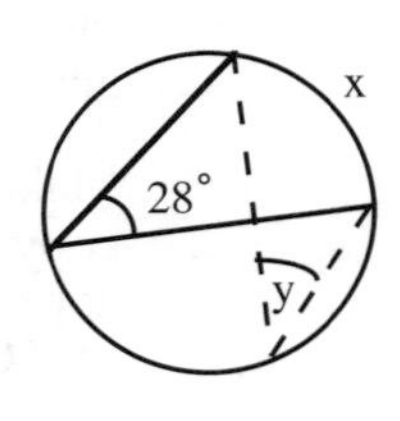

CS-23. Read the following definition, add it to your tool kit, then use it to do parts (a) and (b) below the box.

The endpoints of any diameter divide a circle into two congruent arcs. Each arc is called a **SEMICIRCLE.**

semicircle

a) Find m∠A.

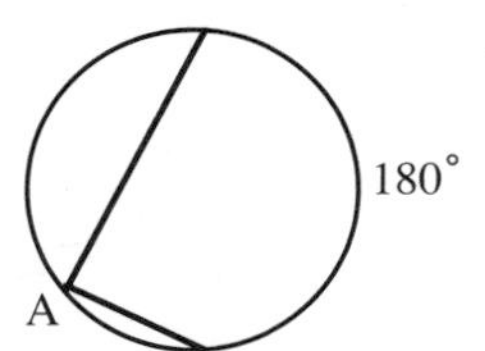


b) Find m$\overset{\frown}{ABC}$ .

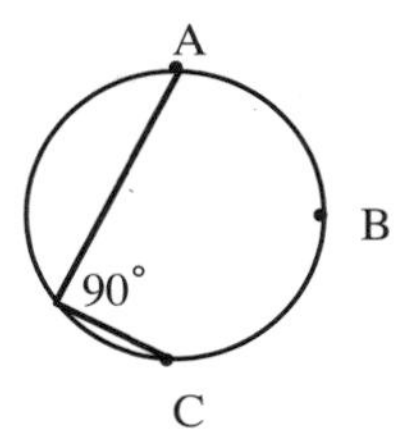


CS-24. In the figure at right, ⊙U has diameter $\overline{GR}$.

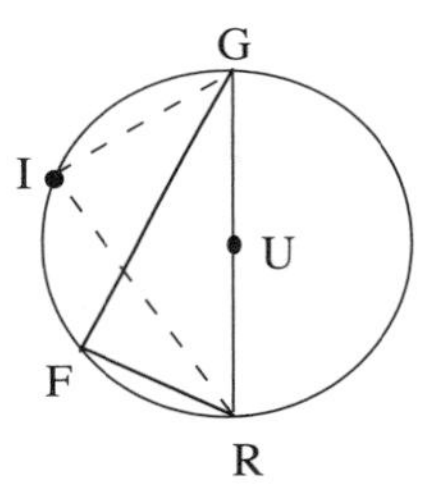


a) Find m∠F and m∠RIG.

b) What conclusion can you draw about an angle inscribed in a semicircle? Write your conjecture in one or two complete sentences.

CS-25. Find the following measures for circle H:

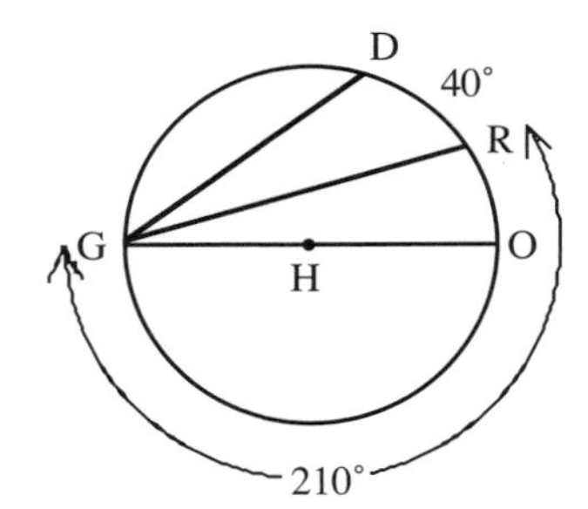


a) m$\overset{\frown}{GD}$.

b) m$\overset{\frown}{OR}$.

c) m∠RGO.

CS-26. A mountain bike tire has a diameter of 22 inches.

a) If a rider can get it to go 240 revolutions in one minute, how far will the bike have traveled in that minute?

b) Does this seem like a fast pace? Why or why not?

CS-27. Solve for a and b.

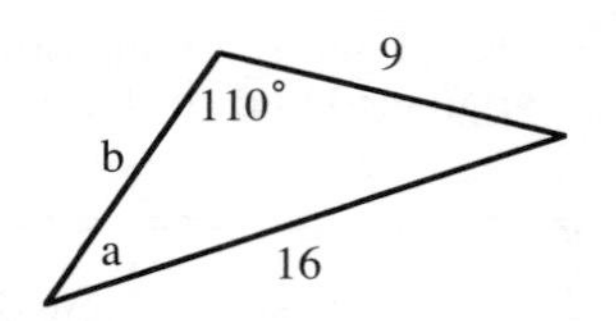


CS-28. In circle O, which triangle has the least area? Prove your answer.

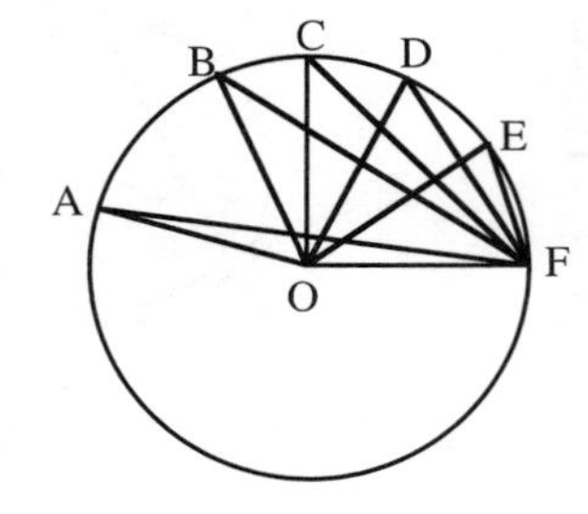

CS-29. Solve for x.

a) $3x^2 - 5x - 7 = 0$

b) $\frac{23 \sin 28^\circ}{3x} = \frac{x}{\cos 41^\circ}$

## TANGENTS AND RADII

CS-30. Read the following definitions, add them to your tool kit, then complete parts (a) through (d) below the box.

This figure shows a circle with three lines lying on a flat surface. Line a does not intersect the circle at all. Line b intersects the circle in two points and is called a **SECANT**. Line c intersects the circle in only one point and is called a **TANGENT** to the circle.

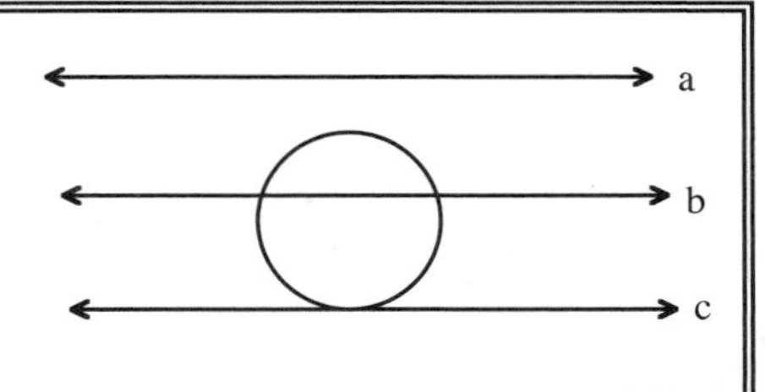


a) Draw a figure like the one at right on your paper (i.e., circle C tangent to line l at D). Draw three more circles of <u>different</u> <u>sizes</u> tangent to the line l at the same point D (that is, one line, one point D, four different size circles). Draw the radius of <u>each</u> circle to the point of tangency, D.

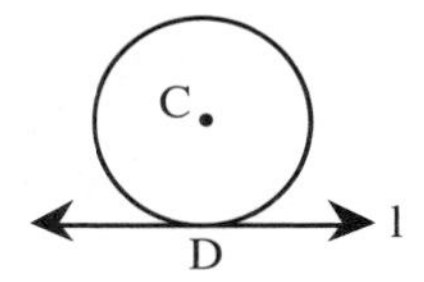


b) What appears to be true about the centers of all the circles?

c) Draw the radius of the largest circle from its center to point D. What appears to be true about the angle between the radius and the tangent?

d) After discussing your observations with your team, write a conjecture stating the relationship between a radius of a circle drawn to a point of tangency and the tangent line.

CS-31. Draw a circle and any diameter. At <u>each</u> endpoint of the diameter, draw a tangent line. After you complete the problem, add the information in the box to your tool kit.

a) What appears to be the relationship between a diameter and tangent lines that intersect at the end(s) of a diameter?

b) What appears to be the relationship between a pair of tangent lines, one at each end of a diameter?

**TANGENT/RADIUS THEOREM:** Any tangent of a circle is perpendicular to a radius of the circle at their point of intersection and any pair of tangents drawn at the endpoints of a diameter are parallel to each other.

c) Prove that the tangents drawn at the endpoints of a diameter are parallel.

CS-32. Solve the following problems.

a) If $CT = 5$, find AC and the area of ⊙C.

b) If $GI = 2\sqrt{3}$, find PG and the area of ⊙G.

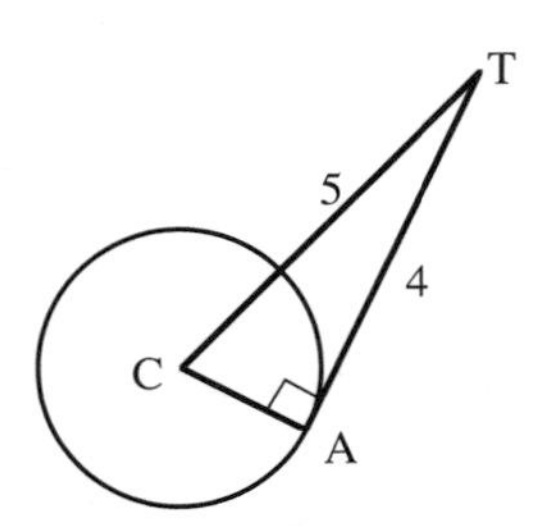


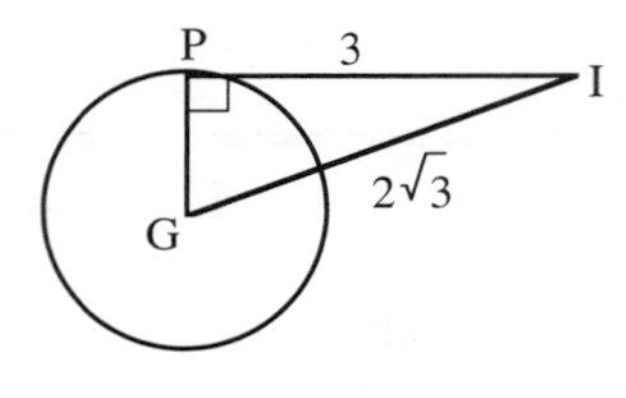


CS-33. In the figure at right, $\overleftrightarrow{PA}$ is tangent to ⊙R at E and PE = EA. Is $\Delta PER \cong \Delta AER$? Prove it or show why not.

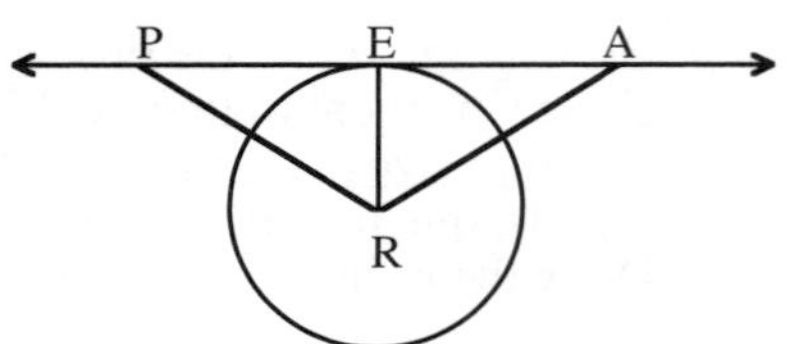


CS-34. Use the figure at right to find $m\overset{\frown}{DAB}$ and $m\overset{\frown}{DCB}$.

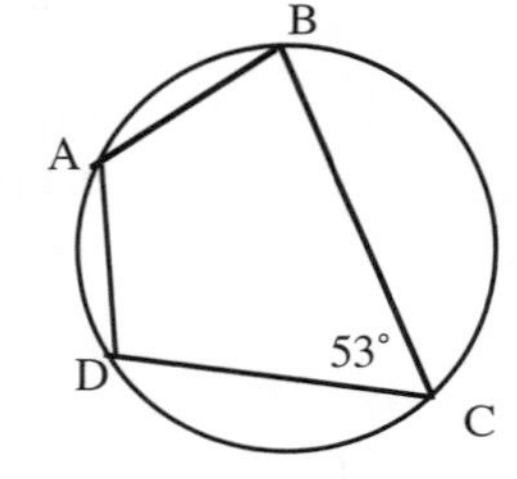


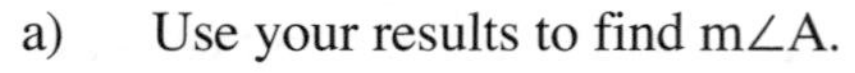

a) Use your results to find m∠A.

b) Fred has a shortcut because he notices a relationship between the measures of angles A and C. What is the relationship?

c) Do you think ∠B and ∠D will have the same relationship?

d) Prove that the opposite angles of any quadrilateral inscribed in a circle must be supplementary.

CS-35. Quadrilateral ABCD is inscribed in a circle as shown in the figure at right, that is, all four vertices are on the circle and all sides are chords of the circle.

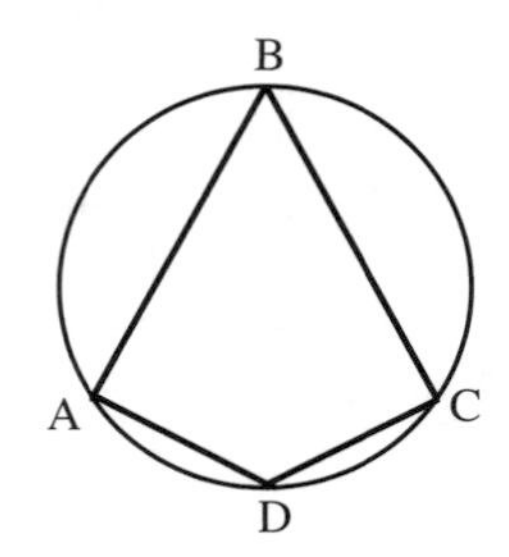


a) If m∠B is 70°, find m∠D. Explain.

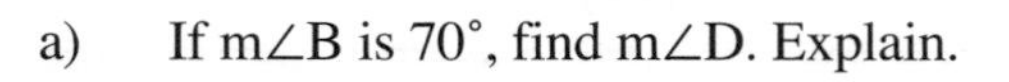

b) If $\overset{\frown}{DAB}$ is a semicircle, find m∠C and m∠A. Explain.

CS-36. In ⊙R, find the measures of $\overset{\frown}{SE}$, $\overset{\frown}{HS}$, and $\overset{\frown}{HO}$, and the measures of all the numbered angles if the measure of $\overset{\frown}{OE} = 75°$.

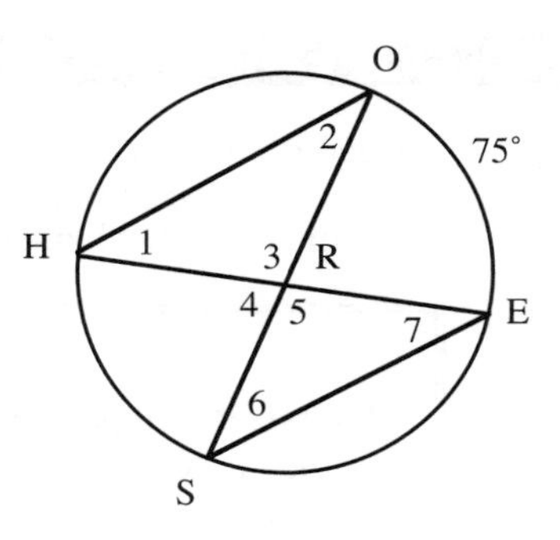


CS-37. The interior angles of a triangle are in the ratio of 2:3:5. Find each angle.

CS-38. Find the perimeter and area of the figure at right.

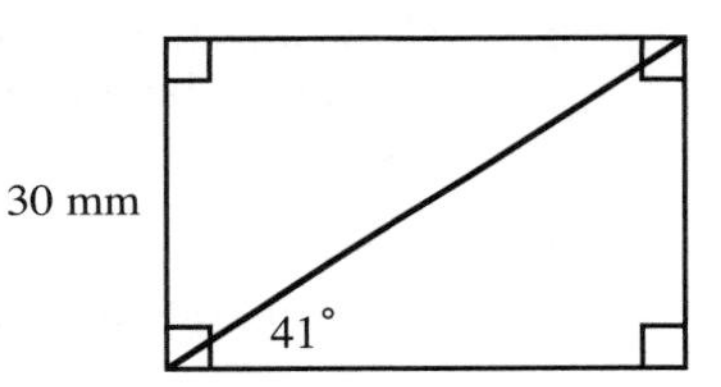


CS-39. What is the measure of the angle formed by the two hands (the minute and hour hands) of a clock?

a) at 12 o'clock?

b) at 1 o'clock?

c) at 2 o'clock?

d) at 3 o'clock?

## DIAMETER-CHORD INVESTIGATION

CS-40. Read the following definition and add it to your tool kit. Make sure that everyone in your team understands why a diameter is a chord but a radius is not. Then complete parts (a) through (f) below the box.

A **CHORD** of a circle is a line segment with its endpoints on the circle.

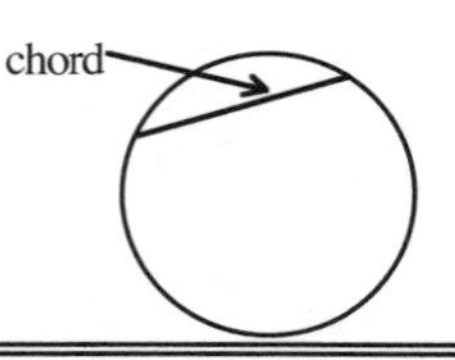


a) You will need a ruler, pencil, and a paper circular disk. Take your circular disk and draw a chord on it (not near the center). Fold the circle in half so that the ends of the chord touch each other.

b) How does the folded line appear to relate to the circle and to the chord? What appears to be true about the parts of the chord? What appears to be true about the arcs? Write a conjecture for each of these questions.

c) Use your ruler to measure the two pieces of the chord. Do the measurements confirm your conjectures?

d) Draw another chord not parallel to the first one. Fold the circular disk so the ends of the chord are together. What appears to be true about the spot where the folded lines intersect?

e) If you drew another chord and did the folding again, where do you think the folded line would intersect the other two folded lines? Do the folding a third time. Were you correct about the intersection of the folded lines?

f) Polish your conjectures from part (b) about what you believe to be true if a chord is bisected by a diameter.

CS-41. In the figure at right, ⊙K was folded in half so that the chord $\overline{AE}$ is bisected.

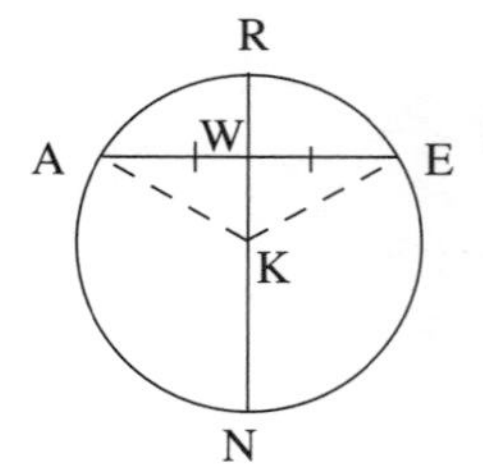

a) Is $\Delta AWK \cong \Delta EWK$? Prove your answer.

b) Tell everything you know about $\angle AWK$ and $\angle EWK$. Justify your responses.

CS-42. This time, in ⊙B, diameter $\overline{NI}$ was drawn perpendicular to chord $\overline{AR}$.

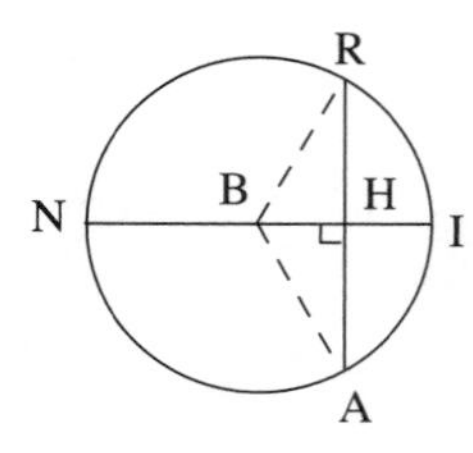

a) Is $\Delta BRH \cong \Delta BAH$? Prove your answer.

b) What do you know about $\overline{RH}$ and $\overline{AH}$? Justify your answer.

c) What can you conclude about any diameter drawn perpendicular to a chord?

CS-43. The box below states the theorems you proved in the two previous problems. Read the theorems, add them to your tool kit, then complete parts (a) and (b) below the box.

**DIAMETER/CHORD THEOREMS:**

1) If a diameter bisects a chord, then it is perpendicular to the chord.

2) If a diameter is perpendicular to a chord, then it bisects the chord.

a) Is your theorem valid if the term radius replaces the term diameter? Explain why or why not.

b) Could ΔRBA in the previous problem be equilateral? If so, state the conditions necessary to make it equilateral. If not, explain why not.

CS-44. In ⊙P, $\overline{PE} \perp \overline{RN}$. If PU = 4 and m∠ PNU = 30°, find the lengths of the sides below. Show all steps leading to your solution.

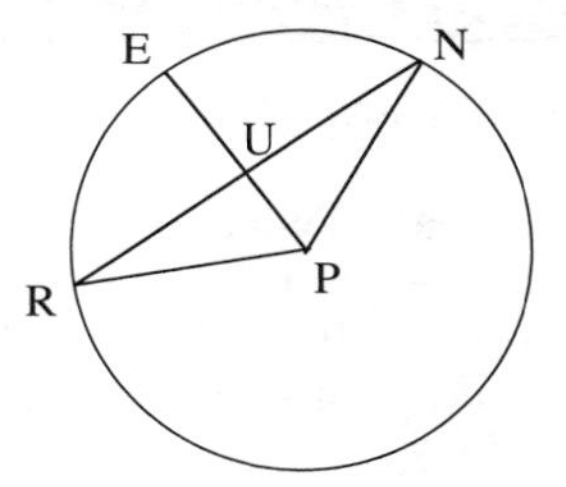


a) UN

b) UR

c) PE

d) UE

CS-45. In ⊙B, EC = 8 and AB = 5. Find BF. Show all subproblems.

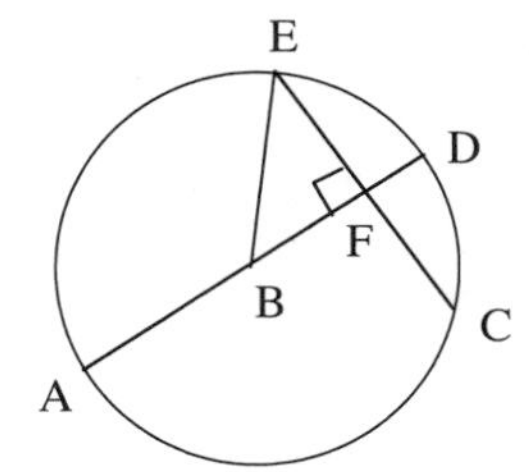


CS-46. The perimeter of a rectangle is 56 cm. If the length were cut in half and the width doubled, the perimeter would be 58 cm. What are the dimensions of the original rectangle? Draw diagrams and write equations.

CS-47. Make a sketch of a graph showing the relationship between the temperature outside today for a 24 hour period. Be sure your axes are clearly scaled.

CS-48. Graph these four equations on the same set of axes.

$y = \frac{5}{4}x + 5$ $y = -\frac{5}{3}x + 5$ $y = \frac{2}{3}x - 2$ $y = -\frac{1}{2}x - 2$

a) Calculate the area of the quadrilateral bounded by these graphs.

b) These four lines along with the x and y axes form four non-overlapping right triangles. Which of the triangles, if any, are congruent? Which, if any, are similar? Prove your answers.

CS-49. Write the following in simplest form.

a) $-2(x + 6) - x(\sqrt{2} - 1)$ b) $\frac{8 + \sqrt{2}}{\sqrt{2}}$ c) $5(x + 4) - 2(x - 3) + x$

## USING THE CIRCLE THEOREMS

CS-50. Copy each figure below on your paper, add the data, then solve.

a) Two concentric circles have their center at I. $\overline{SA}$ is a tangent chord, IN = 12", and IL = 9". Find SA. Hint: add appropriate auxiliary lines.

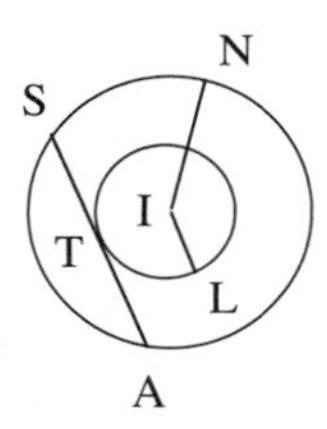


b) ⊙M has radius ME = 14', ⊙A has radius AR = 8', and NC = 17'. Find RE.

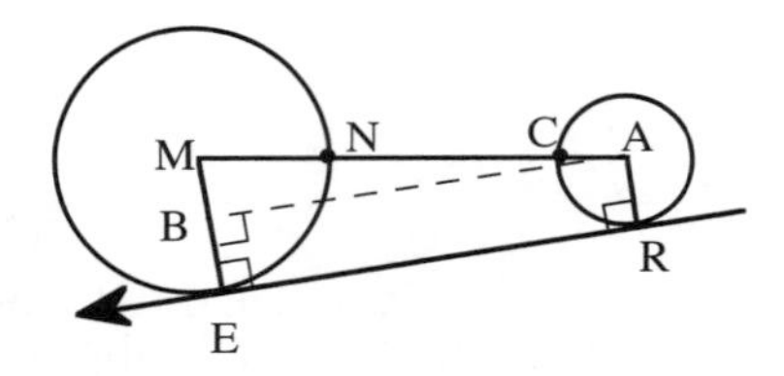


CS-51. In ⊙C at right, $\overline{MP}$ is a diameter, $\overline{MN}$ and $\overline{PO}$ are tangent to ⊙C, $\overline{NO}$ is tangent to the circle at Q, and the radius of the circle is 1.

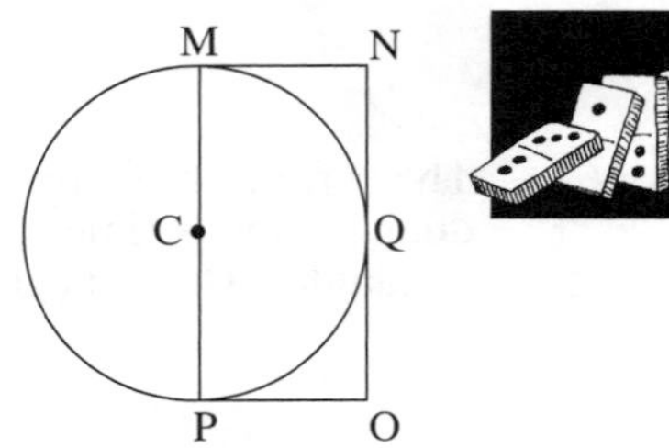


a) Which is greater, the circumference of the circle or the perimeter of the rectangle? Prove your conjecture.

b) Which is greater, the area of the circle or the area of the rectangle? Prove your conjecture.

c) Prove that the area of the circle is <u>always</u> greater than the area of the rectangle no matter what length the radius is.

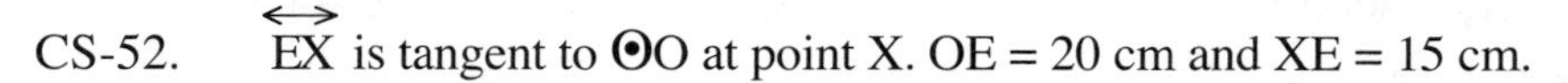

CS-52. $\overleftrightarrow{EX}$ is tangent to ⊙O at point X. OE = 20 cm and XE = 15 cm.

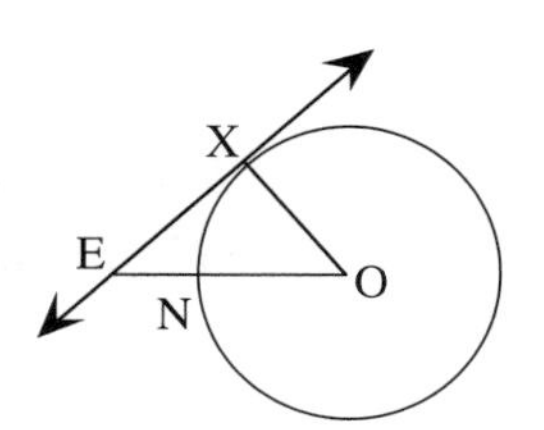


a) What is the area of the circle?

b) What is the area of the sector bounded by $\overline{OX}$ and $\overline{ON}$?

c) Find the area of the region bounded by $\overline{XE}$, $\overline{NE}$, and $\overset{\frown}{NX}$.

CS-53. In the figure of the "Munch Mouth" shown at right, the angle of the mouth has measure 60° and the radius is 1 cm. The vertex of the angle is at the center of the circle.

a) Find the area of the figure.

b) Find the perimeter of the Munch Mouth shown. Be sure to include the mouth.

CS-54. ⊙P has a diameter $\overline{AB}$ and a radius of one inch. Find the area of the shaded region.

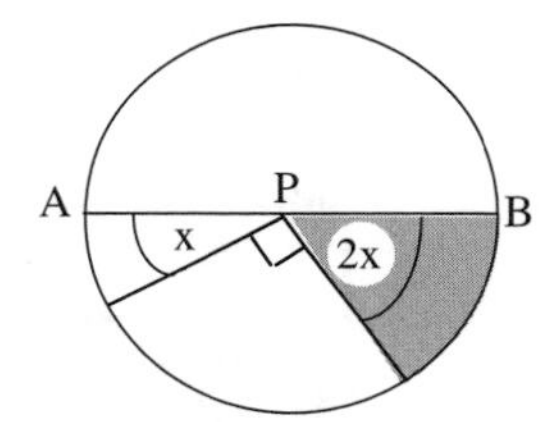


CS-55. If m∠U = 12x - 9° and m∠D = 6x + 18°, find x and m∠U.

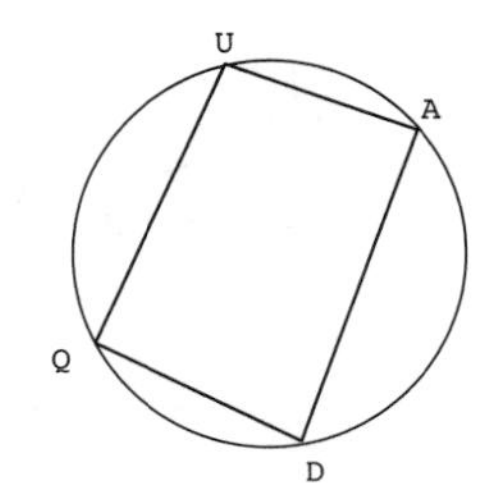


CS-56. In the figure at right, find the interior height (h) of the obtuse triangle.

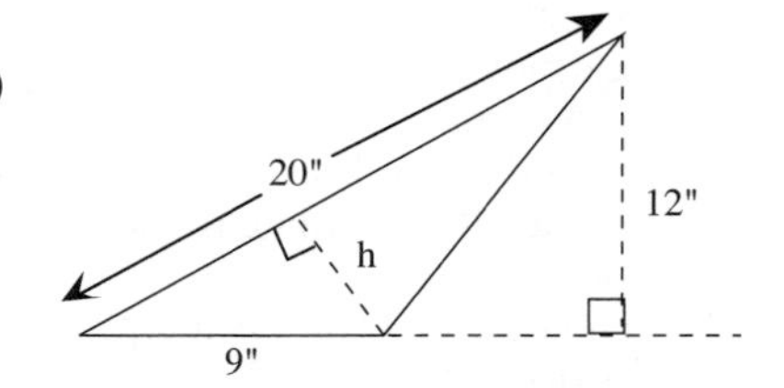


CS-57. Jill's car tires are spinning at a rate of 420 revolutions per minute. If her car tires' radii are each 14 inches, how fast is she driving?

CS-58. In New York City, vendors sell scale models of the Statue of Liberty. Each one is 23.5 cm tall, weighs 90 grams, and is covered with 110 square cm of paint. The height of each Mini-Statue of Liberty is $\frac{1}{400}$ of the height of the statue in New York Harbor.

a) How tall is the real Statue of Liberty?

b) If the Mini-statue is made from the same material as the actual statue, what is the weight of the actual statue?

c) How much paint would it take to cover the full size statue?

CS-59. The abandoned One-Eyed Jack Mine is about $3\frac{1}{2}$ miles off the main road adjacent to the Salmon River Wilderness area. There is only a rutted dirt track left where the access road used to run. It is so steep that when we hiked up it we had to pause every fifty feet or so to catch our breath.It seemed impossible but $3\frac{1}{2}$ miles farther we found remnants of the old wagons, the mineshaft, and the mill. The gold ore found in this mine was embedded in quartz and prospectors used the mill to grind up the quartz and rinse it with acid in huge shallow vats that were agitated so that the gold would sink to the bottom and the quartz could be washed away.

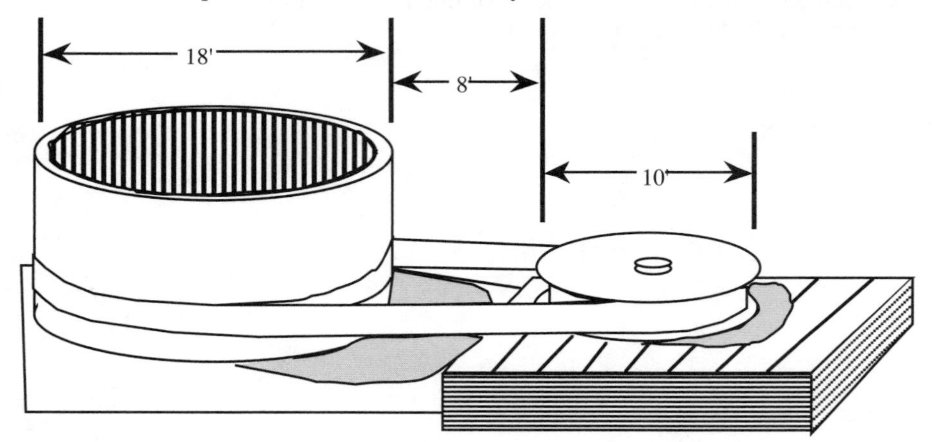


One arrangement of equipment we noticed included a circular vat about 18 feet in diameter which must have been connected by a huge belt to a smaller circular drive wheel 10 feet in diameter. The distance between the wheel and the vat was 8 feet. The equipment had been partially pre-fabricated then carried up the hill piece by piece to be re-assembled on the spot. Just the belt to connect the vat to the drive wheel would have been a major burden. We wondered how many times they had to carry new ones up to replace it. Calculate the length of belt needed to go around the drive wheel and the vat.

The figure at right is a top view of the vat (left), its drive wheel (right), and the belt. Copy the diagram on your paper and add the data from the preceding paragraph. A review of problem CS-50 (b) may be helpful to get you started. Identify all the subproblems you use.

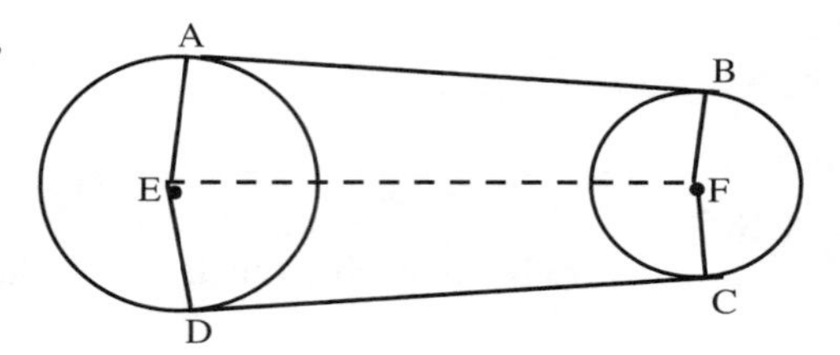

CS-60. Consider the regular hexagon at right.

a) Show how to dissect the hexagon into six congruent triangles.

b) Show how to dissect the hexagon into one rectangle and two congruent triangles.

c) If you had to find the area of the hexagon, which of the two dissections would make your task easier? Justify your answer.

d) Find the area of the hexagon.

8

CS-61. A circle of radius 6 meters is drawn inside a square so that all four sides of the square are tangent to the circle. Find the area of the region inside the square but outside the circle. Show all subproblems.

CS-62.

Eratosthenes (3rd century B.C.) was able to determine the circumference of the Earth at a time when most people thought the world was flat! He knew that Alexandria was about 500 miles north of Syrene. When the sun was directly overhead at Syrene, the sun was about 7° from being directly overhead at Alexandria.

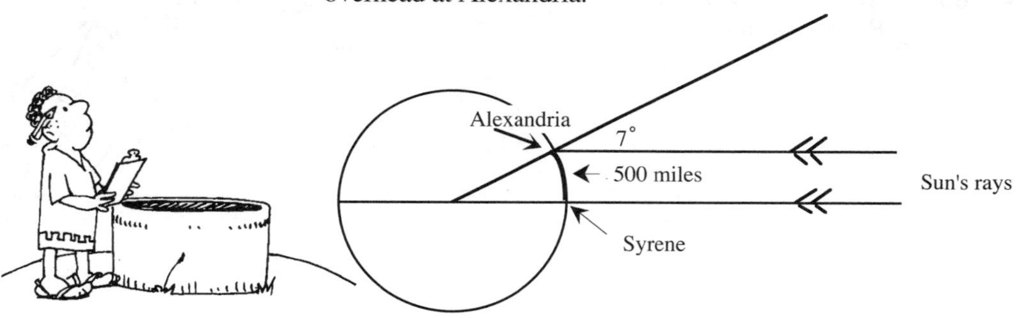


**PROBLEM:** use this information to calculate the circumference of the Earth.

CS-63. Flipper is at the west edge of a circular pool at spot A. He swims in a straight line for 12 meters. This causes him to bump his nose against the pool's edge at spot B. He turns and swims a different direction in a straight line for 5 meters and arrives at spot C on the pool's edge exactly opposite A, which is the east edge of the pool. How far would he have gone if he had swum directly from A to C? Draw a picture and show all steps leading to your solution.

CS-64. In $\odot$Y, $m\overparen{PO} = m\overparen{EK}$, is $\overline{PO} \cong \overline{EK}$? Prove your answer.
Hint: start by drawing some radii.

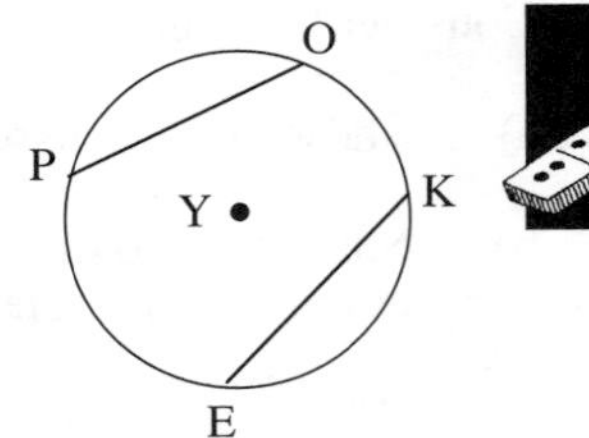

CS-65. ABCDE is a regular pentagon inscribed in $\odot$O. Calculate:

a) $m\angle EDC$.

b) $m\angle BOC$.

c) $m\overparen{BC}$.

d) $m\overparen{EBC}$.

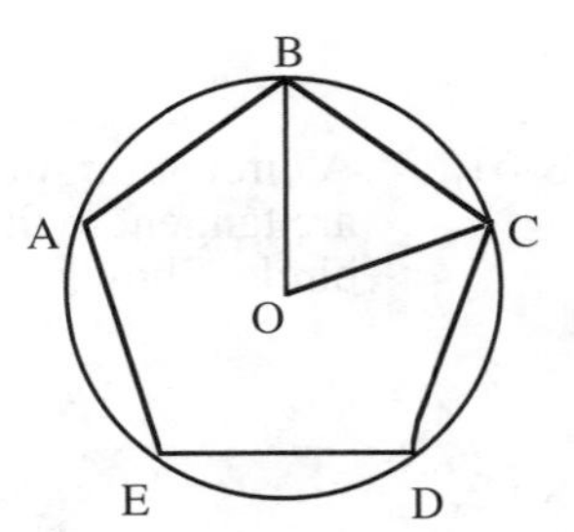


CS-66. In $\odot$T, WR = 5 and WT = 1. Find PQ. Show all subproblems. Remember that T is the center of the circle.

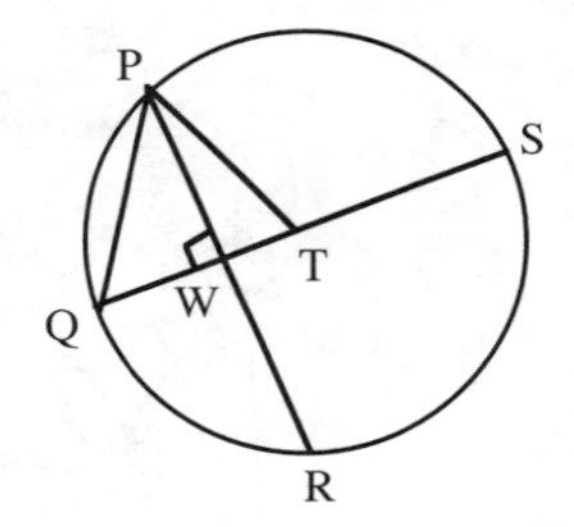


CS-67. $\odot$R has an area of $36\pi$ square units and is tangent to the x- and y-axes.

a) What are the coordinates of R?

b) What are the coordinates of the points of tangency?

c) Write the equation of the line containing the center and the origin.

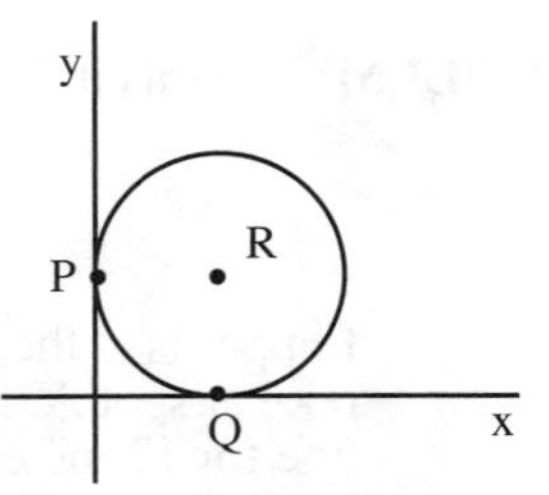

# PRISMS AND CYLINDERS

CS-68. Read the summary of the conclusion from your teacher's demonstration, add the information to your tool kit, then complete parts (a) and (b) below the box.

The volume of a prism seems to depend on only two things: the shape of the base and the height of the prism. Note that the two figures below show that either the upper or lower base may be used in the generalization. Thus, for any prism, the generalized method for finding the volume is:

**BASE AREA x HEIGHT.** **V = Bh**

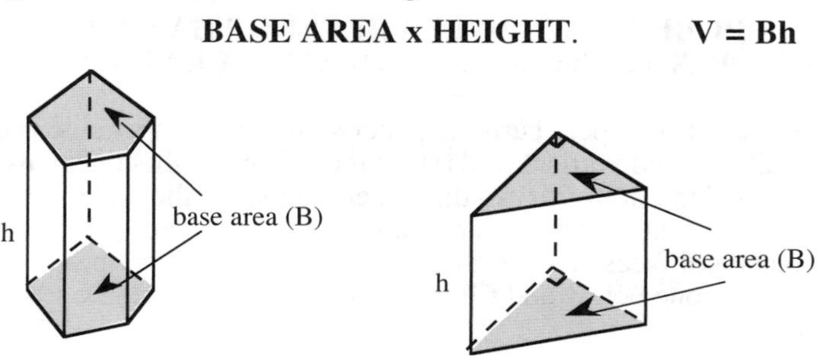


The base of a prism looks like the first picture below. All angles are right angles and all segments are 2 cm. long.

a) Find the area of the base.

b) More of these shapes are stacked to form the prism below whose height is 6 cm. Find the volume of the prism.

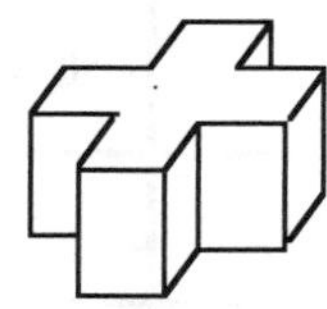

CS-69. Draw a sketch of a rectangular box with dimensions 25 mm x 18 mm x 100 mm and calculate its volume.

CS-70. **DIRECTIONS: With** your partner or in your team, take out three pieces of $8\frac{1}{2}$" x 11" paper. Complete the investigation below. All answers should be to the nearest tenth in this problem. Be sure to read the information in the box below and add it to your tool kit.

**LATERAL SURFACE AREA** is the sum of the areas of all the faces of a prism or pyramid not including the base(s).

**FOR THIS PROBLEM, WE WILL PRETEND THAT THE PAPER MEASURES 9" X 12" TO MAKE THE CALCULATIONS EASIER!**

a) Take one sheet of paper. Turn the paper so that the longer side is horizontal. Make two equally spaced vertical folds (fold the left and right edges inwards so that the paper is divided into thirds) so that when you stand the paper up it forms a triangular prism with equilateral triangular bases at each open end. (The piece of paper forms the three lateral faces. The bases of the prism will be the open ends.) Make a table similar to the one below and fill in the first row. Make sure everyone agrees on the entries!

b) Fold the second sheet of paper to make a square based prism (three vertical folds, four equal regions). Be sure that everyone agrees what the new dimensions are, then fill in the second row of the table.

c) Fold the third sheet of paper to form a hexagonal based prism (five vertical folds, six equal regions). Fill in the last row of the table. Does everyone agree on the entries?

| | Base Area | Height | Volume | Lateral Surface Area | Sum of Top/Bottom Surface Area |
|---|---|---|---|---|---|
| Triangular (Equilateral) | | | | | |
| Square | | | | | |
| Hexagonal | | | | | |

d) Compare the lateral surface area of each prism. How are they related? Explain why this makes sense.

e) Examine each of the remaining area and volume columns in the table and state what happens to the entries as the number of sides in the base increases.

f) What would you expect to happen to the entries in each category if the base is octagonal? has 20 sides? has 100 sides? If the figure did have a 100-sided base, what geometric shape would it look like? Make a sketch.

CS-71. **DIRECTIONS:** Take another sheet of paper and continue to pretend that the paper is 9" x 12". Turn the long edge so that it is horizontal, and make a cylinder with no overlapping edges and no folds by taping the 9 inch edges together. Complete the investigation below. Be sure to read the information in the box at the end of the problem and add it to your tool kit.

a) What is the circumference of the base?

b) Calculate the radius of the base.

c) Find the area of the base.

d) Find the vertical surface area of the cylinder.

e) Find the total surface area of the cylinder (including both top and bottom). Make a sketch of each part for which you are finding the area.

f) Find the volume of the cylinder.

> A prism with a circular base is called a **CIRCULAR CYLINDER.** Its volume is found by calculating the base area ($B = \pi r^2$) then multiplying the result by the cylinder's height. Thus, the formula $V = Bh$ becomes $\mathbf{V = \pi r^2 h}$.

CS-72. Calculate the base area and volume of each prism:

a) The bases are rectangles.

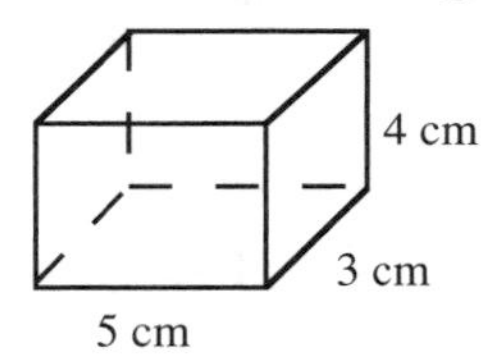


b) The bases are right triangles.

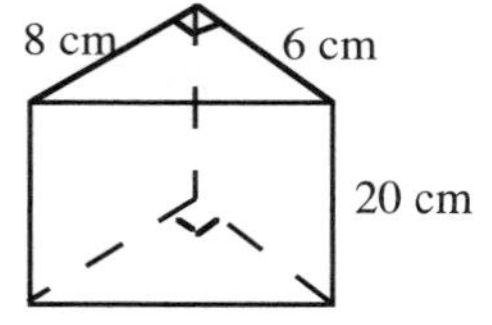


CS-73. Find the volume and total surface area of each of the following cylinders. Leave your answers in terms of $\pi$, that is, no decimal approximations.

a)

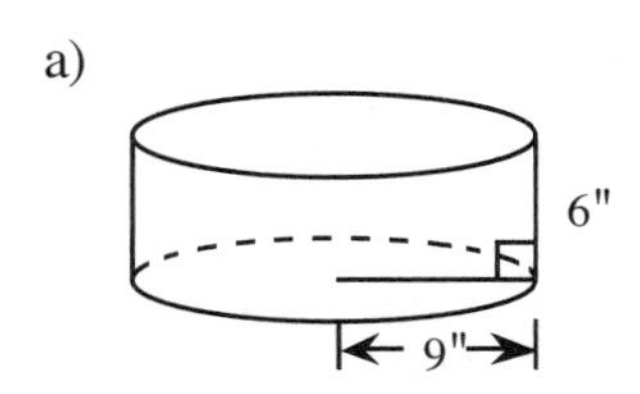


b)

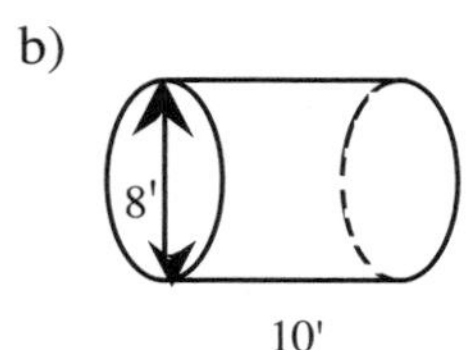


CS-74. $\Delta ABC$ is a right triangle with AB = 6", AC = 8", and BC = 10". Point D is on $\overline{AC}$ and point E is on $\overline{BC}$. Suppose AD = 2 and $\overline{DE}$ is parallel to $\overline{AB}$. Draw the figure, then find the length of $\overline{DE}$.

CS-75. In the figure at right, $\overline{AW} \parallel \overline{EN}$ and $\overline{KW} \cong \overline{ON}$. Is $\overline{AK} \parallel \overline{EO}$? Justify your answer.

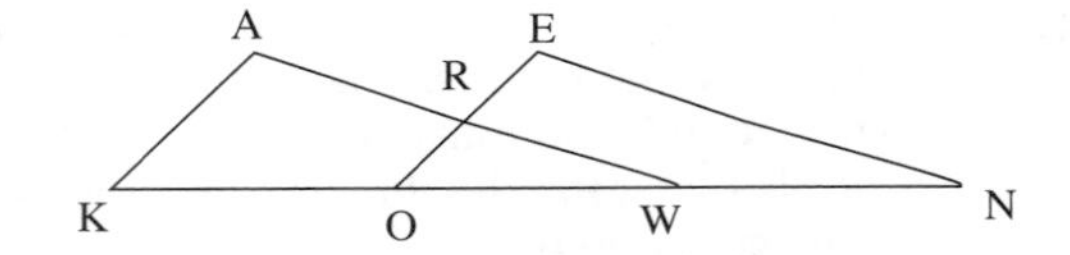


## PYRAMIDS

CS-76.

**DIRECTIONS:** Obtain three templates for a square-based pyramid for your team. Build the pyramid by folding and gluing or taping as usual. Note: if there are four people in your team, one person can cut out the three templates and the others can do the folding as the cutter finishes. Notice that the vertex of the pyramid is not directly above the center of the base. Add the definitions in the box below to your tool kit, then complete parts (a) through (e) below the box.

> Pyramids that have the vertex directly above the center of the base are called **RIGHT PYRAMIDS**. Pyramids with the vertex elsewhere are called **OBLIQUE PYRAMIDS.**

a) Examine the oblique pyramids. How do their shapes compare? How do their volumes compare?

b) In your team, assemble the three oblique pyramids so that they form a cube.

c) Measure an edge of the cube (to the nearest mm) and calculate the volume of the cube.

d) Use your observations in part (a) to find the volume of one pyramid.

e) If the volume of the cube is V = Bh, what is the volume of a pyramid?

CS-77. Recall that your teacher recently took a stack of playing cards or a stack of sheets of paper and slanted them. The new figure had the same volume as the original rectangular stack.

a) Suppose you cut the oblique pyramid which you just built parallel to its base. What would be the shape of the cross section? Draw a sketch.

b) Now suppose you took a large number of square pieces of paper of different sizes which made an oblique pyramid just like the one you constructed in the previous problem. Imagine sliding these squares so that you form any oblique pyramid, how will its volume compare to the volume of the first one?

c) Write one or two sentences that summarize the relationship between the volume of a pyramid with a square base of area B and a height h and its shape (oblique or right).

d) Recall that for a prism, we have the formula *Volume = (Base area) x (height)* or$V = Bh$. Write a formula for the volume of any square pyramid.

e) Do you think that the same conclusions would be true for a pyramid with a triangular base? a hexagonal base? Why or why not?

f) The box below formalizes the work of the last two problems. Add the formula to your tool kit.

In general, the **VOLUME OF ANY PYRAMID** may be calculated by $\mathbf{V = \frac{1}{3} Bh}$.

CS-78. A prism has a volume of 24 cubic inches. What is the volume of a pyramid with the same base area and same height as the prism?

CS-79. A pyramid has a volume of 6 cubic inches. What is the volume of a prism with the same base area and same height as the pyramid?

CS-80. Given that $UW = 6\sqrt{3}$ cm and $UX = 3$ cm, find XY, the circumference of ⊙X, and $m\widehat{YV}$.

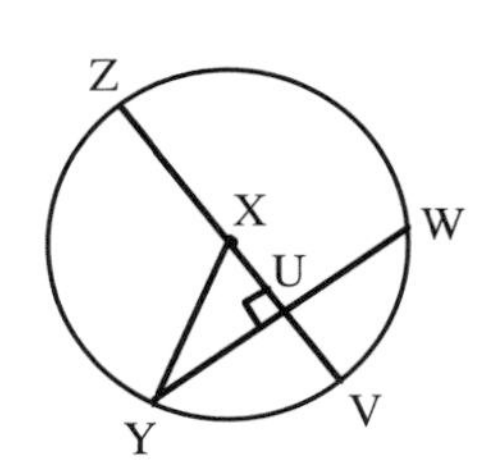

CS-81. Find the volume of each pyramid:

a) The base is rectangular.

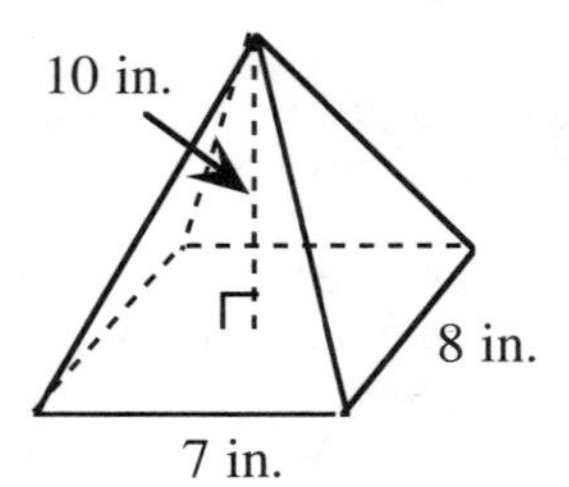


b) The base is a square.

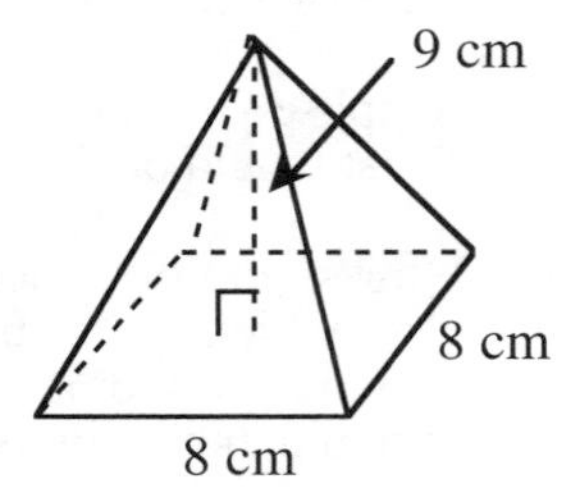


CS-82. A pyramid has a right triangular base with legs 3 inches and 4 inches long. If the height of the pyramid is 6.5 inches, find its volume.

CS-83. A pyramid has a volume of 108 cubic inches and a base area of 27 square inches. Find its height.

CS-84. Solve for x: $6x^2 + 9x - 15 = 0$

CS-85. As Antonio was walking through the park one day, he became startled when he walked into a shadow. He knew there were no clouds in the sky and he was yards away from any trees. When he looked up, he could not believe his eyes, for up in the sky, there appeared to be a UFO! He realized that his angle of sight to the UFO was approximately 60° and he calculated that he was approximately 50 feet from being directly below the UFO.

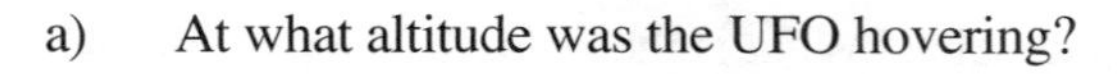

a) At what altitude was the UFO hovering?

b) The UFO's laser beam is aimed at Antonio. The laser has a range of 125 feet. Could the laser beam destroy him?

c) The laser beam has a 35% chance of hitting its target. What is the probability that Antonio will survive?

CS-86. Graph $y = -\frac{2}{3}x + 7$. Label the line "a."

a) What is the equation of the line parallel to line "a" and passing through (3, -3)?

b) What is the equation of the line perpendicular to line "a" and passing through (6, 2)?

## CONES

CS-87. Recall that the volume of a pyramid is $\frac{1}{3}$Bh. Suppose you have pyramids whose bases are all regular polygons. If you keep increasing the number of sides of the bases of these pyramids, what shape will the base become? What will the pyramids become? Draw a sketch of the last figure in the pattern on your paper.

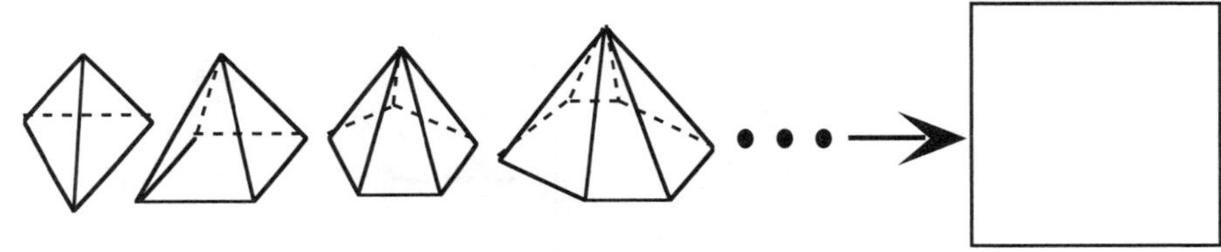


CS-88. Read the definition in the box below, add it to your tool kit, then complete parts (a) and (b) below the box.

The figure you sketched in the previous problem is called a **CONE**. It has a base (circular), a vertex and a height just like a pyramid.

The volume of a pyramid is $\frac{1}{3}$ Bh and the **VOLUME OF A CONE** is the same, but can be written $\mathbf{V = \frac{1}{3}\pi r^2 h}$.

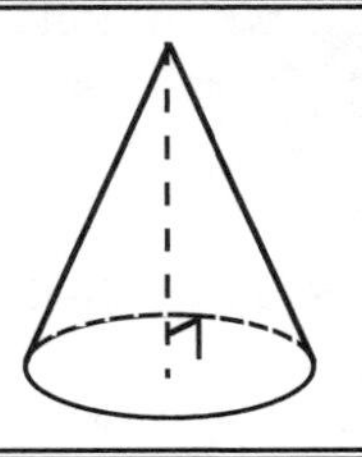

a) In the figure at right, what is the area of the base in terms of $\pi$?

b) What is the volume of the cone at right in terms of $\pi$?

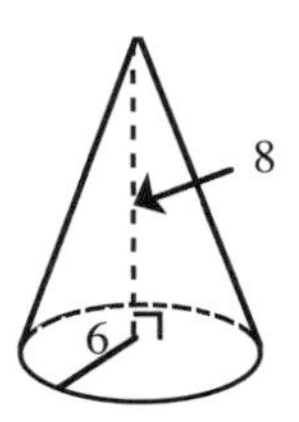

CS-89. Calculate the volumes of these cones. Units are in centimeters.

a)

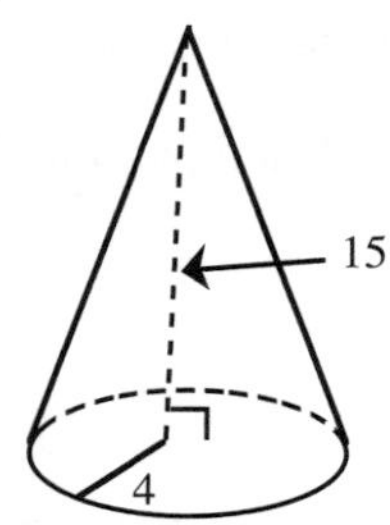


b)

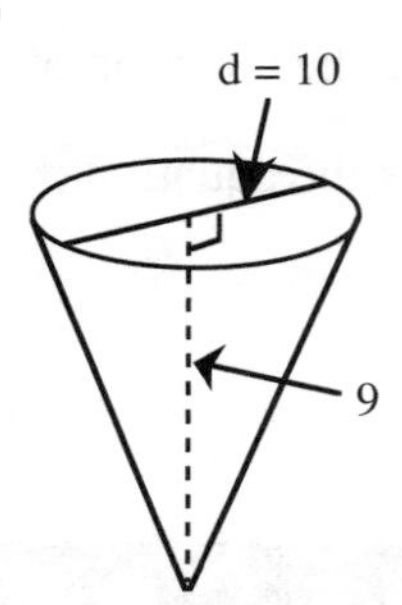


CS-90. A cone is sliced through its vertex so that the exposed surface is an isosceles triangle as shown below. Is $\Delta FED \sim \Delta BEC$? Justify your answer.

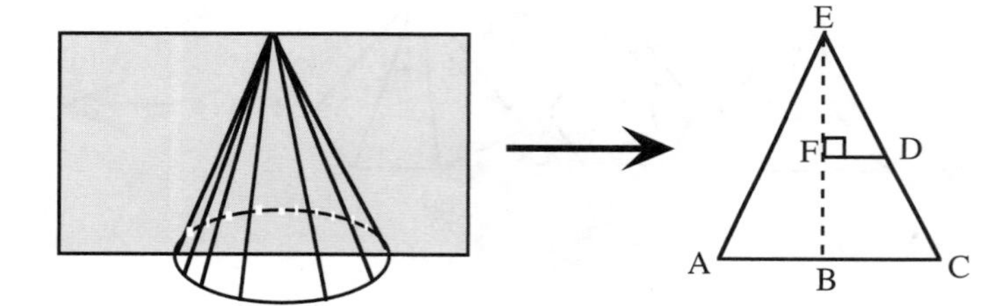


CS-91. Using the same picture above, is $\Delta AEB \cong \Delta CEB$? Justify your answer.

CS-92. Find the volume of a cone with base circumference $10\pi$ inches and height equal to the radius. Draw a sketch.

CS-93. In the figure at right, ABCD is a trapezoid.

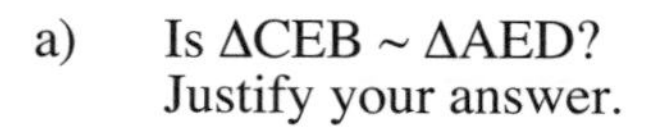

a) Is $\Delta CEB \sim \Delta AED$? Justify your answer.

b) Is $\Delta CED \sim \Delta BEA$? Justify your answer.

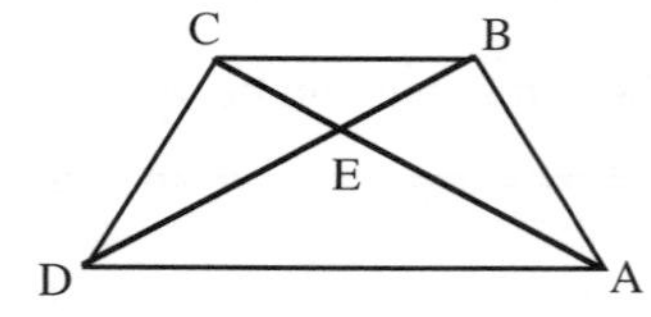

CS-94. The <u>sum</u> of the lengths of the edges of a cube is 1200 cm. Find the surface area and the volume of the cube.

CS-95. Fred claims that the volume of a cube that measures 1.2 feet on all edges is 1.2 cubic feet. Explain and use a diagram to show why you agree or disagree with Fred's claim.

CS-96. Calculate the base area and volume of each prism:

a) The bases are squares.

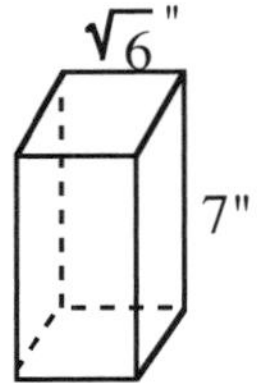


b) The bases are equilateral triangles.

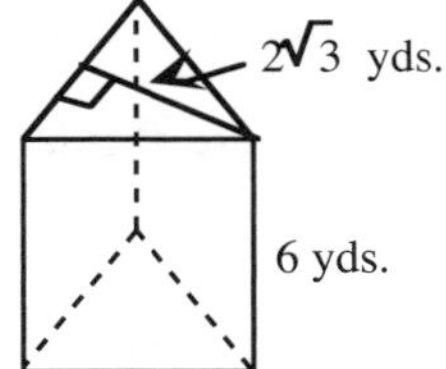


c) Six rectangular faces.

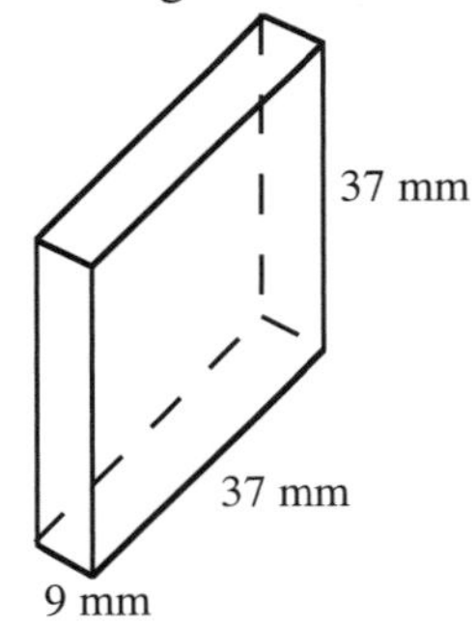


d) The base is an equilateral triangle.

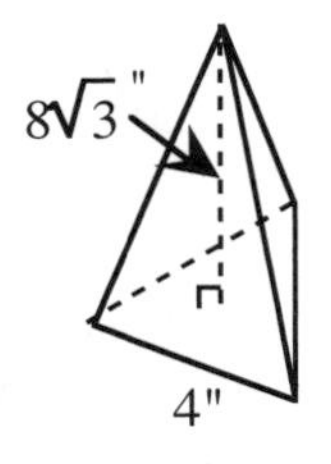


CS-97. A car goes 286 miles using 11 gallons of gasoline. How far can the car travel with 17 gallons of gasoline?

CS-98. From the given information, find the measure of all the minor arcs of ⊙C. Then find the measure of angles 1 - 9. Remember that C is the center.

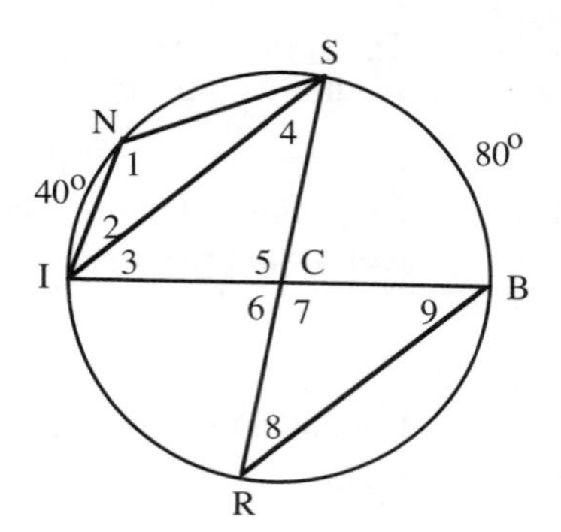


CS-99. Find the volume of each of the following solids.

a) The bases are regular pentagons with A = 54 sq.'

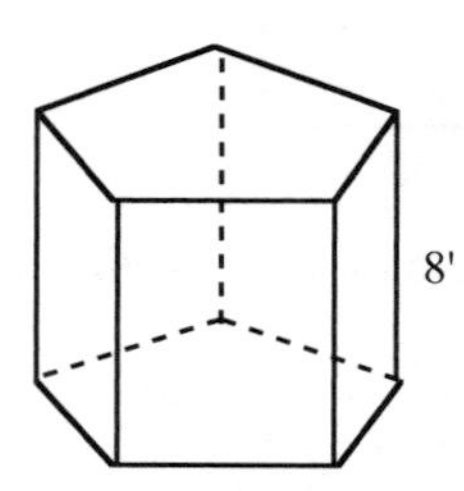


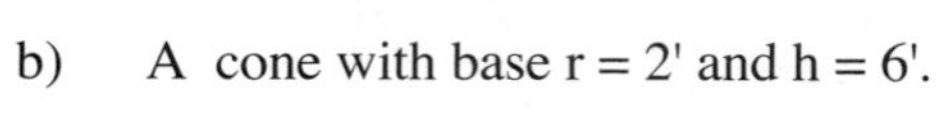

b) A cone with base r = 2' and h = 6'.

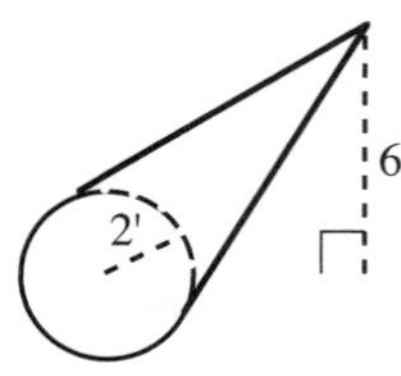


c) A circular cylinder with a circumference of $10\pi$ cm. and h = 8 cm.

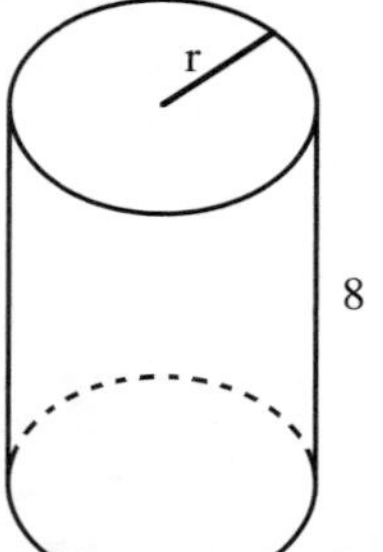


d) A pyramid with a right triangular base with legs of 4" and 7", h = 12".

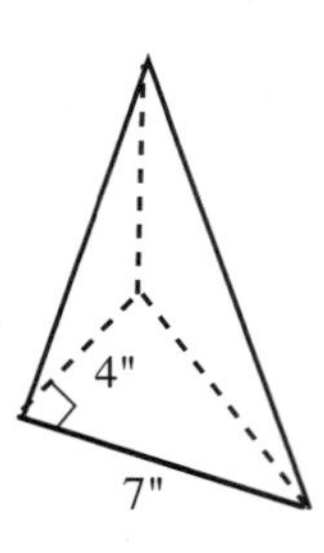


CS-100. Given ⊙O with m∠AOB = 30°, $\overline{OC}$ bisects ∠AOB, and OC = 12.

a) Find the length of $\overline{CB}$.

b) Find the length of $\overline{AB}$.

c) Find the length of $\overline{OB}$.

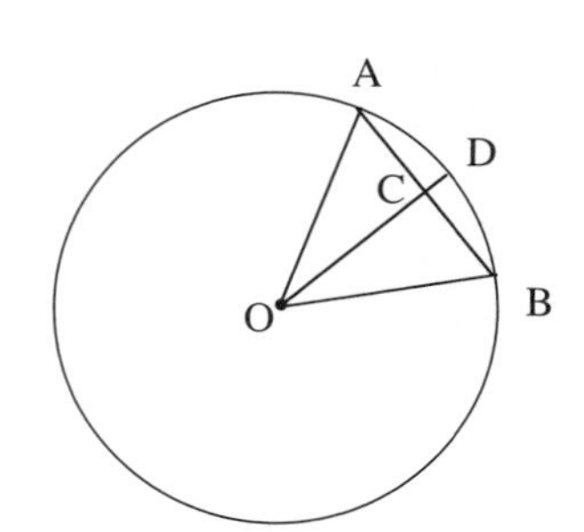

CS-101. In $\odot C$, $m\angle L = 15x - 39°$; $m\angle 1 = 5x - 14°$.
Find x, $m\angle 2$, and $m\overparen{GI}$.

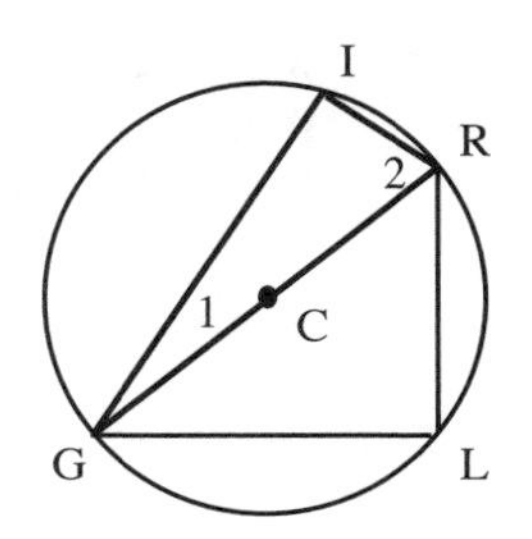


CS-102. Calculate the total surface area and volume of the prism at right. The base is a regular pentagon.

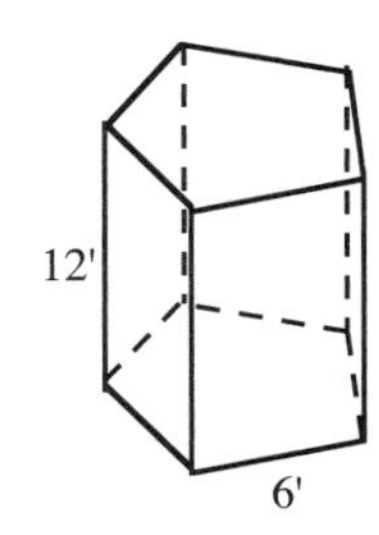


CS-103. In $\odot F$, $m\overparen{AB} = 155°$, $m\angle FAD = 42°$, and $\overline{AD} \cong \overline{BC}$.
Calculate $m\overparen{CD}$. Show your subproblems.

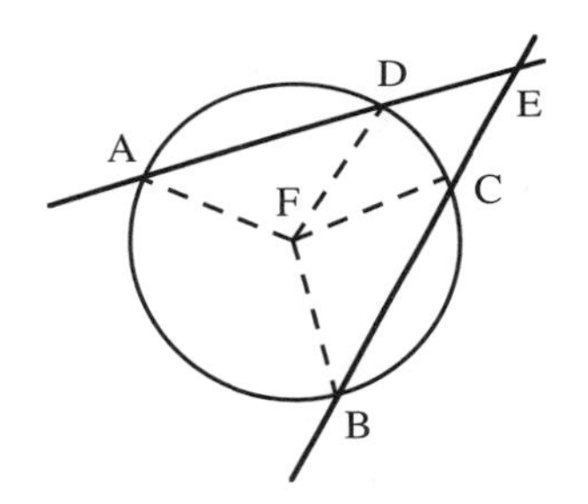


CS-104. TOOL KIT CHECK-UP

Your tool kit contains reference tools for geometry and algebra. Return to your tool kit entries. You may need to revise or add entries.

Be sure that your tool kit contains entries for all of the items listed below. Add any topics that are missing to your tool kit NOW, as well as any other items that will help you in your study of this course.

- circle
- center
- radius
- diameter
- circumference
- area
- $\pi$
- arc
- central angle
- arc measure
- arc length
- major arc
- minor arc
- concentric circles
- sector
- inscribed angle
- intercepted arc
- semi-circle
- secant
- tangent
- chord
- lateral surface area
- circular cylinder
- right pyramid
- oblique pyramid
- volume formulas
- cone

CS-105. $\Delta JEA$ is an equilateral triangle inscribed in $\odot N$ which has a radius of one foot. Is the area of the triangle larger than the area of the shaded region? Justify your answer.

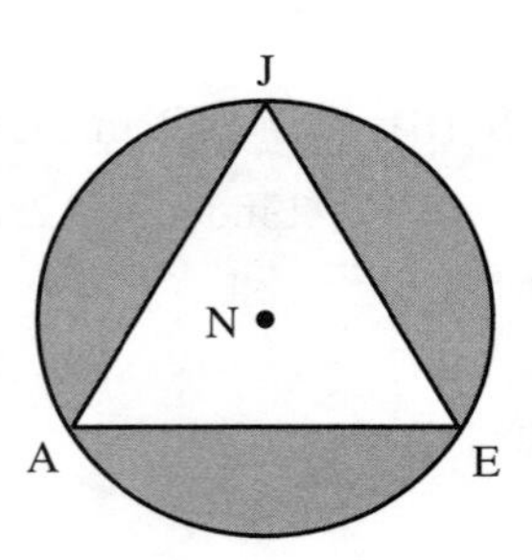


CS-106. In the figure at right,

a) find the measures of $\overparen{AB}$, $\overparen{BC}$, and $\overparen{AC}$.

b) find the diameter, circumference, and area of the circle.

c) find the lengths of $\overparen{AB}$, $\overparen{BC}$, and $\overparen{AC}$.

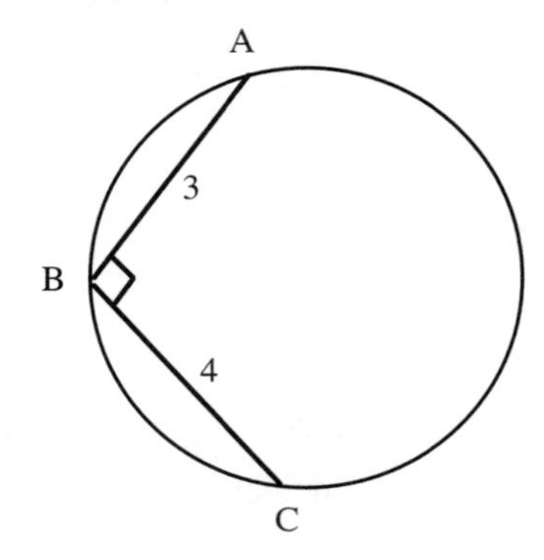


CS-107. In the figure at right, $\overline{AB}$, a chord of the large $\odot C$ with radius R, is tangent to the small concentric $\odot C$ with radius r. The length of $\overline{AB}$ is s. Find the area of the ring between the two circles in terms of s. Note: you may want to substitute numerical values for R, r, and s and look for a pattern.

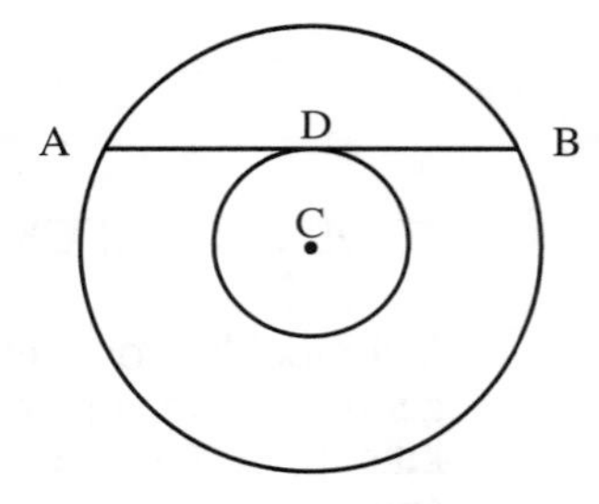

# UNIT 11

# The Poison Weed

## GEOMETRIC PROBABILITY

# Unit 11 Objectives

## *The Poison Weed:* GEOMETRIC PROBABILITY

This unit uses some of the geometry you have studied to solve probability problems. You will play various games of chance, often using area as the basis to gather data, then calculate probabilities for the games. You will learn how to determine the expected result from a game, and conclude by examining probabilities with multiple outcomes. Since area is the basis for much of our study of probability, you will have a chance to review and consolidate your work with polygons, circles, and ratios.

The algebra review topic for this unit is rational expressions. This work will give you another opportunity to work with fractions and factoring.

In this unit you will have the opportunity to:

- calculate probabilities
- apply area to gather data for probability questions.
- learn how to predict expected outcomes from games of chance.
- practice and apply the ideas about circles and solids you learned in the previous units.
- solve a theme problem that involves several major ideas from the course.
- review fractions and factoring as you work with rational expressions.

*Read the following problem and briefly discuss what you need to know to solve it. Do not attempt to solve it now. We will come back to it later.*

GP-0. Dimitri is getting his prize sheep, Zoe, ready for the county fair. He keeps Zoe in the pasture beside a barn and shed. What he does not know is that there is a single locoweed in this pasture which will make Zoe too sick to go to the fair if she eats it, and she can eat it in one bite. Zoe takes about one bite of grass or plant every three minutes for six hours a day. Each square foot of pasture provides enough food for about 40 bites. What is the probability that Zoe eats the locoweed?

*UNIT* 11

ALGEBRA
GRAPHING
RATIOS
GEOMETRIC PROPERTIES
PROBLEM SOLVING
SPATIAL VISUALIZATION
CONJECTURE, EXPLANATION (PROOF)

# *Unit 11*

## *The Poison Weed:* GEOMETRIC PROBABILITY

### SPINNERS AND PROBABILITY

GP-1. You will need a paper clip for your team, and a pencil and paper to keep score. Use the spinner at right.

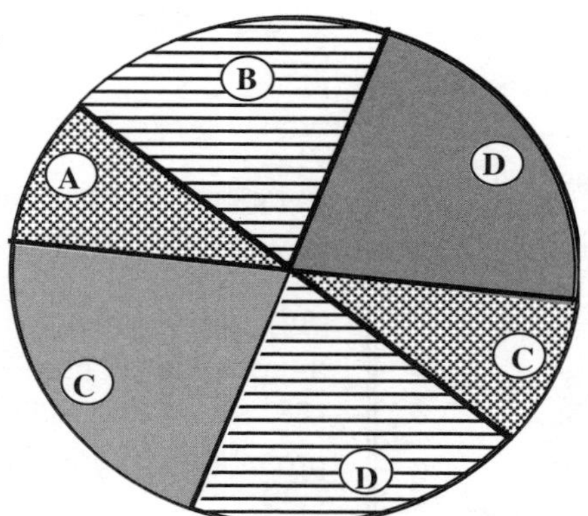


a) List the last names of the people in your team alphabetically.

b) The first person on the list is player A, the next is player B, and so on. Record this.

c) Put the spinner on a flat surface, place a paper clip at the center of the spinner, hold it in place with the point of a pen or pencil, and spin it 50 times. Each time the spinner lands on a player's letter, the player gets a point. The person with the most points wins.

GP-2. Which player, A, B, C, or D, won the game in the previous problem? Is that what you expected? Why or why not?

a) The central angles for the regions on the spinner are as follows: $A \approx 33°$, $B \approx 70°$, C(combined) $\approx 110°$, and D(combined) $\approx 147°$. Calculate the probability of landing in each region.

b) Calculate the percentage of the points scored for each region based on your results in part (c) in problem GP-1 and compare them to the probabilities you just calculated in part (a). How closely did the results from spinning match the actual probabilities? Explain any large differences.

GP-3. Use the spinner at right and a new score sheet.

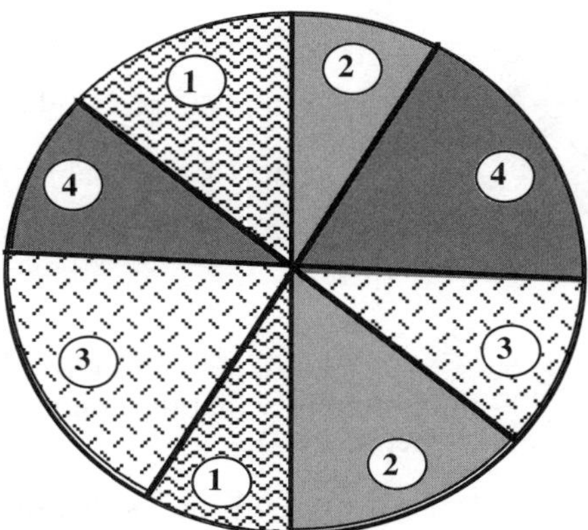


a) Choose who will be each of the players--1, 2, 3, and 4--and record the names with their numbers on your score sheet.

b) Before you start, examine the spinner. Which player do you think will win? Explain completely.

c) Spin the spinner in the same manner as in the previous problem 50 times. Score one point for each time the spinner lands on a player's number. The winner is the player with the most points. Who won in your team?

d) The combined regions marked 1 and 2 are <u>each</u> 80°, the combined regions for 3 and 4 are <u>each</u> 100°. Calculate the probabilities for each region. Compare your answers to the results of your game.

GP-4. Read the following definition and add it to your tool kit, then go on to the next problem.

The **PROBABILITY** of some event, call it A, happening is expressed as a ratio and written as:

$$P(A) = \frac{\text{number of successful outcomes}}{\text{total number of outcomes}}.$$

For example, on a standard die, P(5) means the probability of rolling a 5. The probability is equal to $\frac{1}{6}$, since a die has six faces (total possible outcomes) and only one of them is a 5 (successful, or desired, outcome). $P(\leq 3) = \frac{1}{2}$, since three of the six numbers will successfully fulfill the event description, in this case, $\leq 3$. From a shuffled deck of playing cards, P(king), read as "the probability of drawing a king," is $\frac{1}{13}$ (4 kings out of 52 cards).

The probability of drawing a heart from a shuffled deck of playing cards is $\frac{1}{4}$ (13 hearts out of 52 cards).

Note that this definition applies to sample spaces containing equally likely events.

GP-5. Suppose that you have a strip of plastic 80 cm long which is bent into a circle. The first 30 cm are painted blue, the next 30 cm painted green, and the last 20 cm painted red. If you spin an arrow in the middle of the circle,

a) what is the probability that the arrow points to blue?

b) what is the probability that the arrow points to red?

c) what is the probability that the arrow points to <u>either</u> blue or green?

GP-6. You have a spinner in the middle of a circle which is colored red, white, and blue. The probability that the spinner points to red is 40% and the probability that it points to white is 40%.

a) What is the probability that the spinner points to blue? Explain how you know.

b) Sketch a coloring on a circle that would give you these probabilities.

c) Sketch a different coloring that would give you these probabilities.

GP-7. Solve each of the following problems.

a) A circle is divided into nine congruent sectors. What is the measure of each central angle?

b) If $m\angle B = 97^\circ$, find $m\overset{\frown}{AD}$ and $m\angle C$.

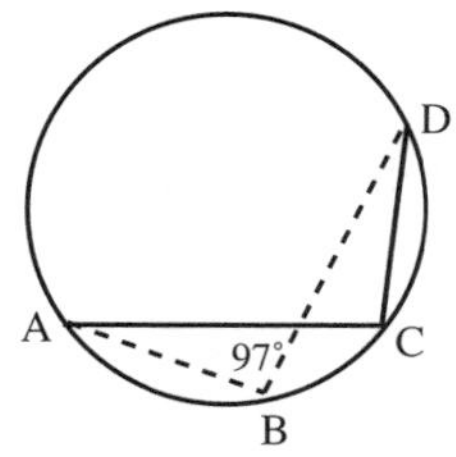


c) In $\odot C$, if $m\angle ACB = 125^\circ$ and $r = 8"$, find $m\overset{\frown}{AB}$, the <u>length</u> of $m\overset{\frown}{AB}$, and the area of the smaller sector.

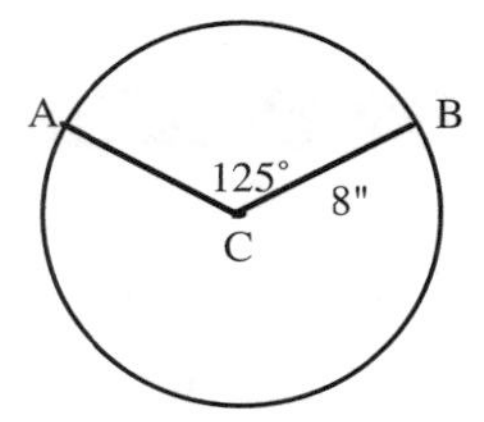


d) Find the volume of the cylinder if the diameter is 7 cm and the height is 4 cm.

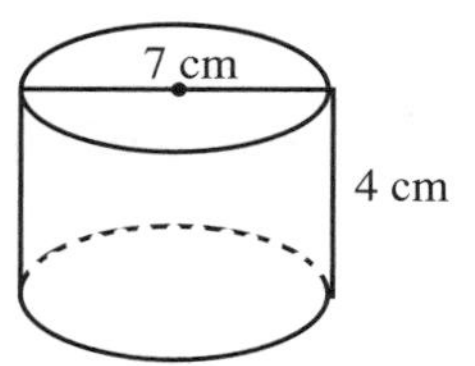


GP-8. When Erica and Ken went spelunking last weekend, they each found a gold nugget. Erica's nugget is similar to Ken's nugget. They measured the length of two matching parts of the nuggets and found that Erica's nugget is five times longer than Ken's. When they took their nuggets to the gemologist to be analyzed, they learned that it would cost $30 to have the surface area and weight of the smaller nugget calculated, and $150 to have the same analysis done on the larger nugget.

"I won't have that kind of money until I sell my nugget, and then I won't need it analyzed!" Erica says.

"Wait, Erica. Don't get upset yet. I'm pretty sure we can get all the information we need for only $30."

a) Explain how they can get all the information they need for $30.

b) If Ken's nugget has a surface area of 110 square cm, what is the surface area of Erica's nugget?

c) If Ken's nugget weighs 56 g (about 1.8 oz), what is the weight of Erica's nugget?

GP-9. Write in simplest radical form.

a) $\sqrt{27}$ b) $\sqrt{98}$ c) $\sqrt{80}$

GP-10. Find x if the angles of a quadrilateral are 2x, 3x, 4x, and 5x.

## AREA FRACTION AS PROBABILITY

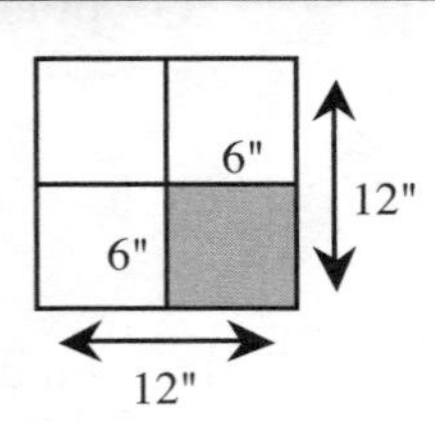


GP-11. Greg has a 12" x 12" square dart board and he likes to play darts with his friends. Greg is a good enough player that he always hits the board, but he can never be sure where on the board the dart will land.

a) What is the probability that he hits the shaded region?

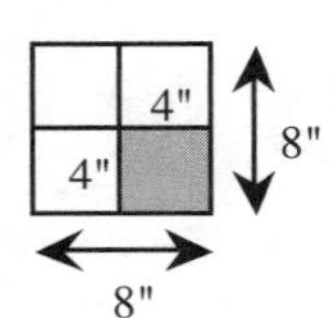


b) Greg has purchased a new square dart board; this one is 8" x 8". On this board, what is the probability of his hitting the shaded region on the board?

c) What do you notice about your answers to the last two problems? Explain completely why this makes sense.

GP-12. Greg has another new dart board. He still cannot be sure which square he hits, he only knows that he will hit the board with every toss. Under these circumstances, what is the probability of hitting a square which contains a:

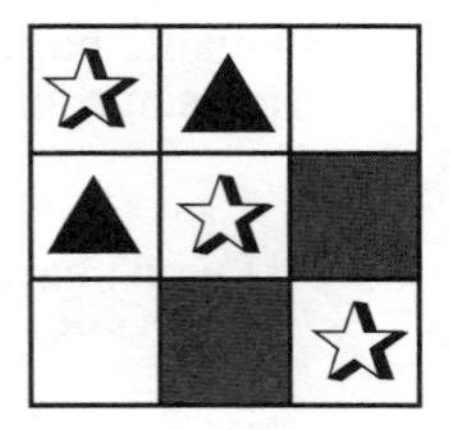

a) star?

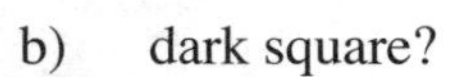

b) dark square?

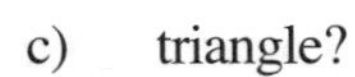

c) triangle?

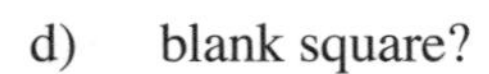

d) blank square?

e) Add your answers to parts (a) - (d). Explain why this must always be the result when you take the sum of all the probabilities of an event.

GP-13. For each of the dart boards below, calculate the probability of getting a dart in the shaded region if you hit the board. All the circles are congruent and all the squares are congruent.

a)

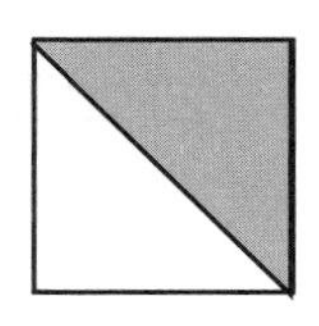

b)

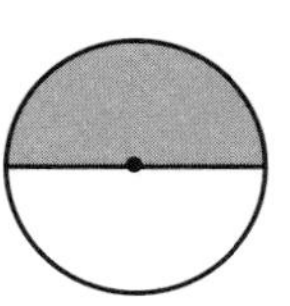

c)

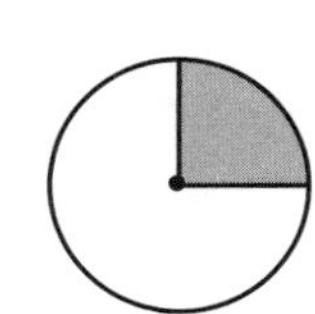

d)

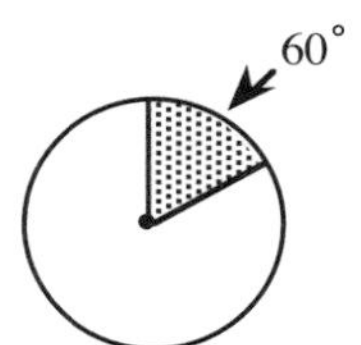


e)

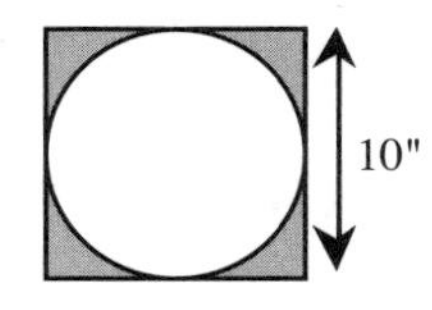


GP-14. What is the probability of getting a dart in the shaded region of the figure at right? Careful, is this figure drawn to scale?

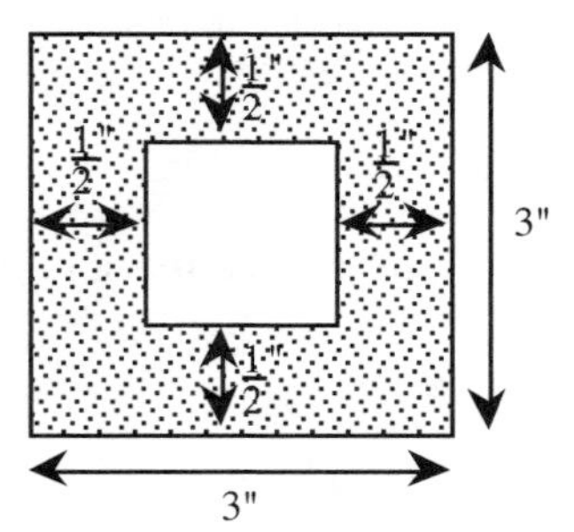


a) Would the probability change if the inner square were moved down to one corner of the outer square?

b) Would the probability be changed if the inner square were tilted but remained inside the square?

GP-15. Draw a picture of a dart board like the one above (with a small square inside a larger square) so that the probability of hitting the shaded region is:

a) exactly $\frac{1}{4}$. Justify your answer.   b) exactly $\frac{1}{2}$. Justify your answer.

GP-16. Which is greater: the probability of hitting the shaded region of figure 1 or the probability of hitting the shaded region of figure 2? Prove your answer.

Figure 1

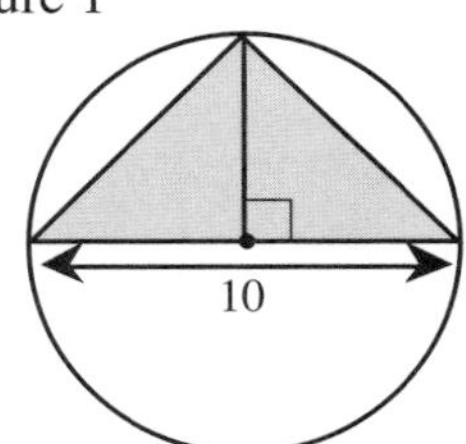


Figure 2

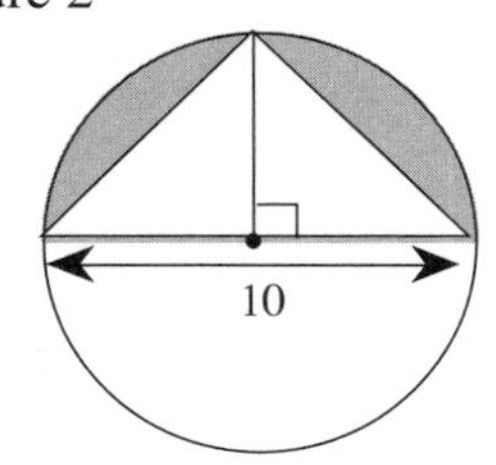

GP-17. Calculate the surface area and volume of the prism at right. Stop and think about which faces are the bases.

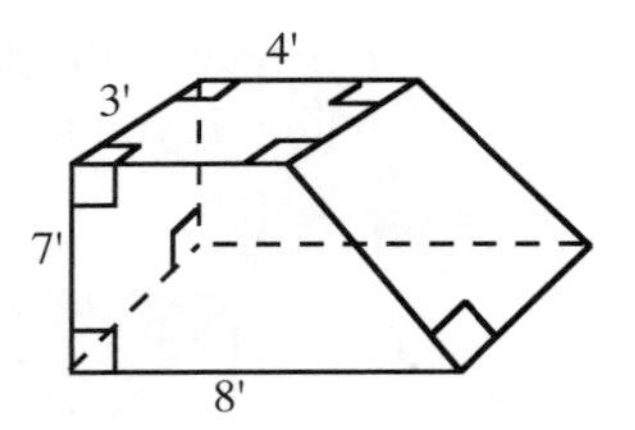


GP-18. A right triangle has legs of length 24 cm and 34 cm.

a) Find the measures of the acute angles and the length of the hypotenuse of the triangle.

b) If this triangle is inscribed in a circle (each vertex lies on the circle) what is the area of the circle?

GP-19. A 100-gon has fifty interior angles of $175^\circ$ alternating with fifty interior angles of $y$ degrees.

a) Find y.

b) Find the sum of the exterior angles of the 100-gon.

GP-20. Graph the region bounded by $y \le 3x + 2$ and $y \ge 3x - 1$.

GP-21. Two red blocks and a green block together weigh 24 ounces. Three red blocks and two green blocks together weigh 39 ounces. How much does a red block weigh? A green block?

GP-22. Simplify each expression.

a) $(2x^5y^2)^4$

b) $(3x^3)^2(2x^5 + 5y^4)$

c) $\dfrac{8x^5y^7}{(2xy^2)^3}$

GP-23. In one of your classes, your teacher gives you a project in which you need two pieces of string, each has to be at least one foot long. Your teacher gives you a five foot piece of string to use, but it is all tangled! You decide that you will be able to untangle the string if you make a cut first.

**QUESTION:** If you can only make one cut, what is the probability that when you do untangle the string, you will have the two pieces each at least one foot long?

GP-24. Chip and Dale are best friends who live next door to each other. Their bedroom windows are 75 feet apart. They have a sophisticated tin can telephone line between the two windows. During a storm the line breaks.

**QUESTION:** What is the probability that the line breaks within 20 feet of Chip's house? Make a sketch.

GP-25. You are given two pieces of wood: one is two feet long, the other is ten feet long. You are supposed to use the wood to build a triangular frame with the two foot piece as one of the sides of the frame. The ten foot piece is accidentally broken.

**QUESTION:** What is the probability it is broken into two pieces so that, along with the two foot piece, you will be able to form a triangular frame?

GP-26. During the Spring Festival, a skydiver attempted to land in a quarter circle at any one of the corners of a 1-kilometer square field, as shown in the figure at right. It is $\frac{1}{4}$ kilometer from the corner to the curved edge of each landing area. Suppose that one skydiver randomly descends onto the field. What is the probability that:

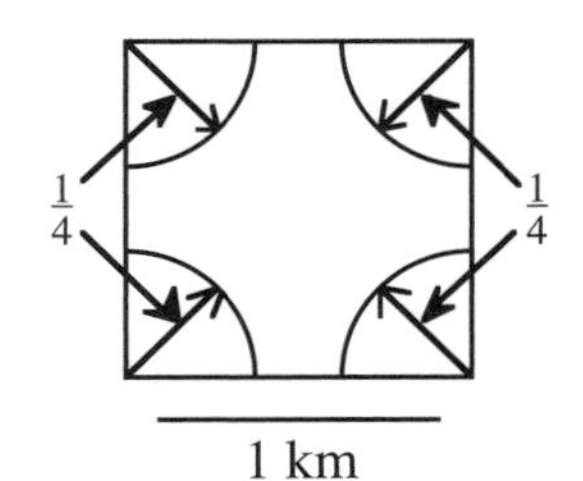


a) the skydiver lands in all four corner regions on the same jump?

b) the skydiver lands in one corner region?

c) the skydiver lands inside the square but outside the four corner regions?

GP-27. In the previous problem, suppose that the corner regions of the field were at the vertices of an equilateral triangle which is one kilometer on a side. Again, the edge of each landing area is $\frac{1}{4}$ km from the corner. Sketch the figure. Now what is the probability that:

a) the skydiver lands in one corner region?

b) the skydiver lands inside the triangle but outside the three corner regions?

GP-28. Find the volume of each of the solids below.

a) Pentagonal prism with base area 38.9 sq." and h = 13".

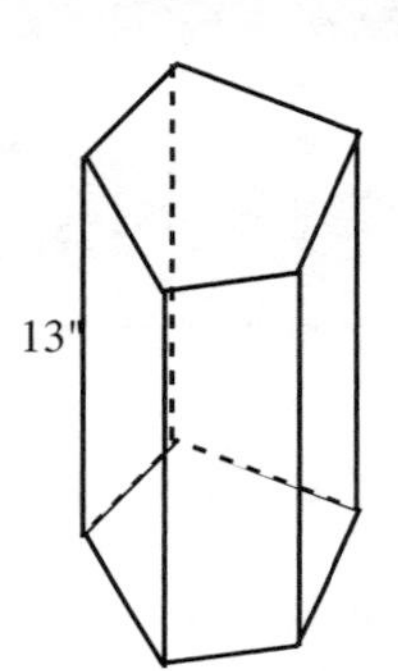


b) Pyramid with a parallelogram base, h = 12'

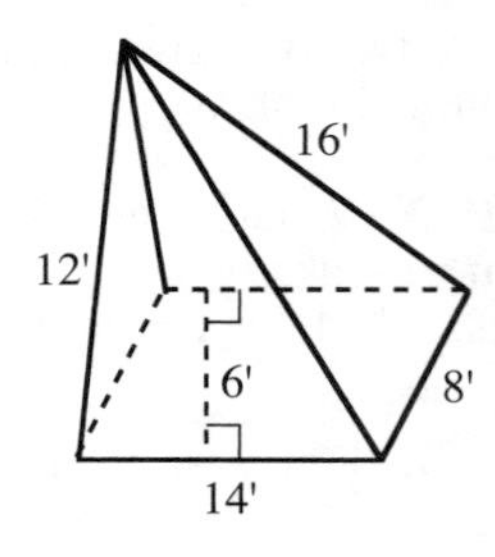


c) Cylinder with base circumference equal to $18\pi$ cm and h = 23 cm.

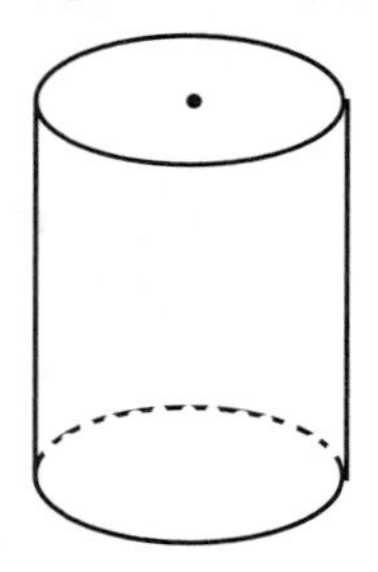

d) Cone with base area = 25 sq." and h = 11".

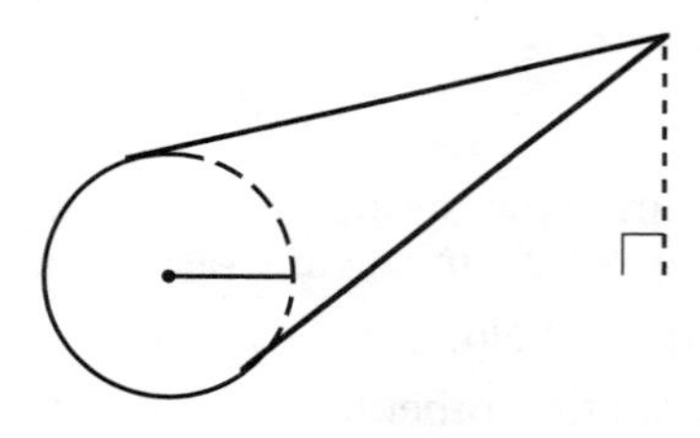

GP-29. Donnell has a bar graph which shows the probability of a colored section coming up on a spinner, but part of the graph has been ripped off.

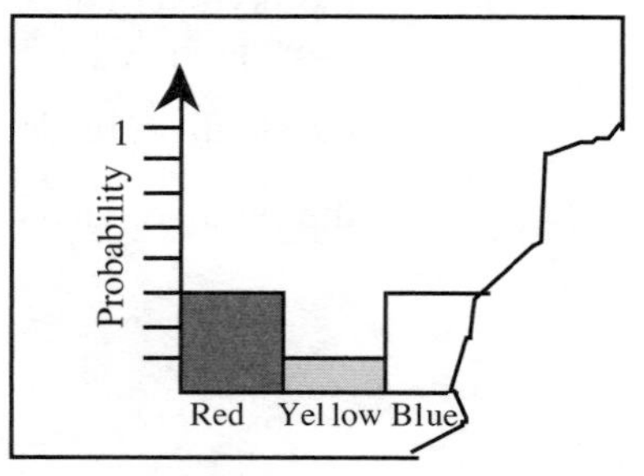


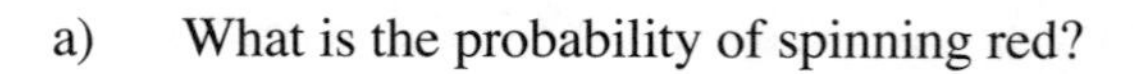

a) What is the probability of spinning red?

b) What is the probability of spinning yellow?

c) What is the probability of spinning blue?

d) If there is only one color missing from the graph, namely green, what is the probability of spinning green? Why?

GP-30. A wall running north-south is 40 feet long and 24 feet high. How far east will its shadow reach when the sun is directly west and 14 degrees above the horizon? Be sure to draw a diagram.

GP-31. Solve for w if:

a) $w(2w - 7)(3w + 8) = 0$

b) $(w - 3)(w + 5) = 9$ (careful!)

GP-32. Write in simplest radical form.

a) $\sqrt{3}(\sqrt{18} + \sqrt{32})$

b) $\sqrt{98} + \sqrt{27} + \sqrt{147} + \sqrt{18}$

c) $\dfrac{6 + \sqrt{5}}{\sqrt{10}}$

## EXPECTATION

GP-33. Use the spinner below to answer parts (a) through (d). Then read the definition in the box and add it to your tool kit.

a) What color would you expect a spinner to land on most often? Why?

b) What color would you expect a spinner to land on the least? Why?

c) Find P(red), P(blue), P(green), and P(yellow).

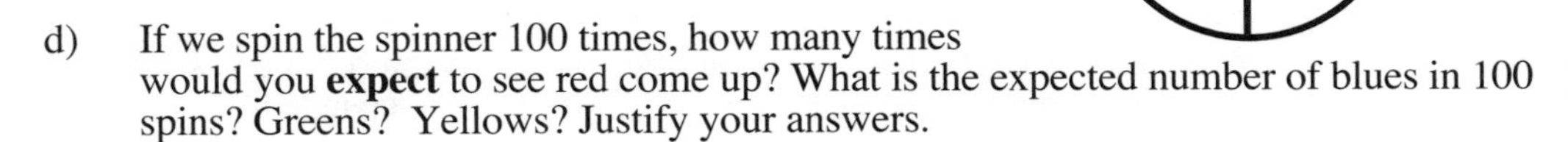


d) If we spin the spinner 100 times, how many times would you **expect** to see red come up? What is the expected number of blues in 100 spins? Greens? Yellows? Justify your answers.

By **EXPECTED VALUE** we mean the amount we would expect to win or lose each game on average if we played a game many times. Your answer to part (d) of the previous problem as well as the problems at the beginning of the unit illustrate **expected value**.

GP-34. **GAME DIRECTIONS:** You are considering playing a game using this spinner, which is divided into four equal regions. This game costs you \$4.00 to play, that is, \$4.00 per spin. The number in each region shows the number of dollars that you will be paid if the spinner lands in that part of the circle. Consider the probability of landing in each region. Do you think you should play? Explain your answer completely.

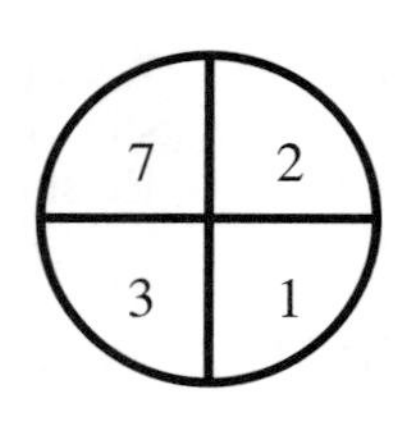


a) State each of the probabilities

1) P(7)   2) P(2)   3) P(1)   4) P(3)

b) Suppose we spin the spinner only four times. What is the <u>expected</u> number of times the spinner would land on each space?

c) After <u>four</u> spins, if we did see the expected number of 7's (that is "sevens"), 2's, 1's, and 3's, how much money would you win (or lose)?

d) Suppose we spin it eight times. What is the expected number of 7's, 2's, 1's, and 3's? What are your expected winnings (or losses)?

e) Consider 20 spins: what is the expected number of 7's, 2's, 1's, and 3's? What are your expected winnings (or losses)?

f) Finally, suppose you spin the spinner 100 times. What is the expected number of 7's, 2's, 1's, and 3's? What are your expected total winnings (or losses)?

g) Consider your answer in part (f). If you play the game 100 times, <u>will</u> this be your result, that is, <u>must</u> this be the outcome?

GP-35. Go back and read the directions to the game in the previous problem. Would it be reasonable to play the game 100 times? Explain and support your answer completely.

a) How much would you lose, <u>on average</u>, EACH time you played the game?

b) Suppose you wanted to make it a fair game, that is, if you spin 100 times, you would expect to win \$400 to balance out the money you paid to play. If all of the numbers but the 7 remain the same, what would you have to change the 7 to in order to make the game fair?

GP-36. How much is it fair to pay to play a game where the probability of winning \$50 is $\frac{1}{20}$ and otherwise you lose your money? Explain.

**GP-37.** Read the following information about rational expressions and add it to your tool kit, then do the six practice problems below the box.

**RATIONAL EXPRESSIONS** are fractions that have algebraic expressions in their numerators and/or denominators. The ability to manipulate rational expressions and write them in a simpler form is useful in solving complicated problems. The key to simplifying rational expressions is to find **factors** in the numerator and denominator that are the same so that you can write a fraction equal to 1. For example,

$$\frac{6}{6} = 1 \qquad \frac{x^2}{x^2} = 1 \qquad \frac{(x + 2)}{(x + 2)} = 1 \qquad \frac{(3x - 2)}{(3x - 2)} = 1$$

Notice that the last two examples involved binomial sums and differences. **Only** when sums or differences are **exactly** the same does the fraction equal 1. Rational expressions such as these examples **CANNOT** be simplified:

$$\frac{(6 + 5)}{(6)} \qquad \frac{x^3 + y}{x^3} \qquad \frac{x}{x + 2} \qquad \frac{3x - 2}{2}$$

Most problems that involve rational expressions will require that you **factor** the numerator and denominator. For example:

$$\frac{12}{54} = \frac{2 \cdot 2 \cdot 3}{2 \cdot 3 \cdot 3 \cdot 3} = \frac{2}{9} \qquad \text{Notice that } \frac{2}{2} \text{ and } \frac{3}{3} \text{ each equal 1.}$$

$$\frac{6x^3y^2}{15x^2y^4} = \frac{2 \cdot 3 \cdot x^2 \cdot x \cdot y^2}{5 \cdot 3 \cdot x^2 \cdot y^2 \cdot y^2} = \frac{2x}{5y^2} \qquad \text{Notice that } \frac{3}{3}, \frac{x^2}{x^2}, \text{ and } \frac{y^2}{y^2} = 1.$$

$$\frac{x^2 - x - 6}{x^2 - 5x + 6} = \frac{(x + 2)(x - 3)}{(x - 2)(x - 3)} = \frac{x + 2}{x - 2} \qquad \text{where } \frac{x - 3}{x - 3} = 1.$$

In all three examples **all parts** of the numerator and denominator--whether constants, monomials, binomials, or factorable trinomials--**MUST BE WRITTEN AS PRODUCTS BEFORE** we can look for factors that equal 1.

Note that in all cases we assume the denominator does not equal zero.

Practice simplifying rational expressions with these problems. Remember, you may need to factor before you simplify.

a) $\frac{3 \cdot 3 \cdot 7}{3 \cdot 3 \cdot 8}$

b) $\frac{x^2}{x^5}$

c) $\frac{(x - 1)(x + 3)}{(x - 1)}$

d) $\frac{(x + 2)^2}{(x + 2)(x - 5)}$

e) $\frac{12(x - 1)^3(x + 2)}{3(x - 1)^2(x + 2)^2}$

f) $\frac{x^2 - 6x + 8}{x^2 + 4x - 12}$

GP-38. Suppose you throw a dart at the square board with a square inside it as shown at right. Your score is determined by the value of the number where your dart lands. If the dart hits the board at random, how many points do you expect to get after throwing 100 darts? Explain.

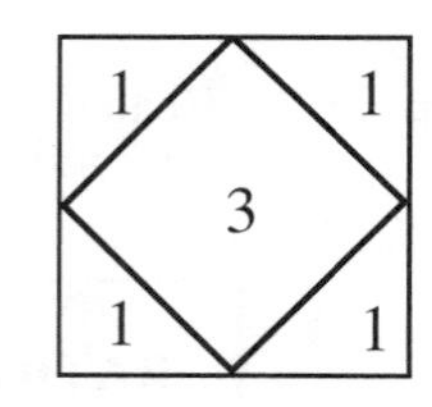


GP-39. Frank and Mabel were supposed to be going to the movies this evening, but he left her a note saying he can't go. It seems that Frank was assigned to find the expected number of times each color would come up on a five-colored spinner after 100,000 spins. She is shocked to learn that Frank intends to make a spinner and spin it 100,000 times! Mabel realizes that even spinning it once a second will take more than 24 hours. The directions for Frank's spinner include the probabilities of the different colors shown below.

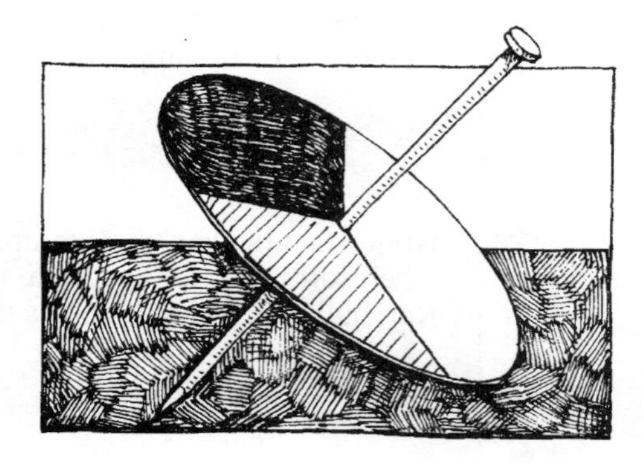

$P(\text{pink}) = \frac{1}{3}$ $P(\text{green}) = \frac{1}{20}$ $P(\text{turquoise}) = \frac{1}{6}$

$P(\text{violet}) = \frac{1}{4}$ $P(\text{yellow}) = \frac{1}{5}$

Use the probabilities to write a solution for Frank.

GP-40. The City Planning Commission has rejected a real estate developer's proposal to build luxury homes on 10,000 square foot lots. Instead, the Commission and the developer have reached a compromise to build mid-priced homes which are similarly shaped, but smaller on 6,300 square foot lots. Find the reduction factor (ratio of similarity) that the developer should use to determine the new dimensions of the smaller similarly shaped lots.

GP-41. Factor or multiply as required. Express radicals in simplest form.

a) $5\sqrt{6}(\sqrt{3} + 2x)$

b) $(\sqrt{2} + x)(\sqrt{3} - x)$

c) $(a - b)(2a^2 + 3ab - 3b^2)$

d) $48d^3 + 20d^2$

e) $x^2 - 1$

f) $3x^3 - 9x^2 - 120x$

GP-42. Circle O is inscribed in ABCD. This means the sides of ABCD are tangent to the circle. Is ABCD a square? Prove your answer.

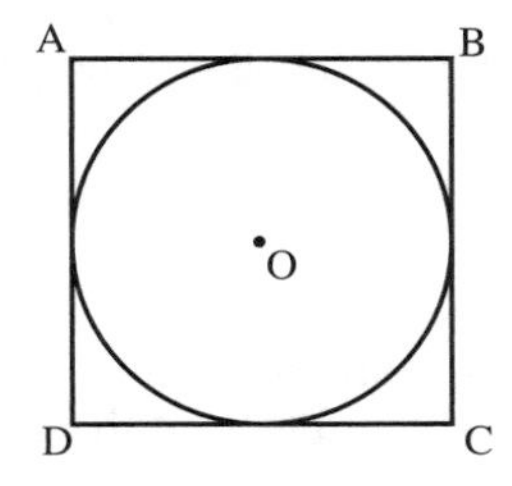

GP-43. Solve each of the equations for x.

a) Solve for x if $\frac{2}{x} + \frac{3}{x} = \frac{5}{x}$.

b) Solve for x if $\frac{2}{x^2} + \frac{3}{x} = \frac{4}{x}$.

## THE POISON WEED--AN INVESTIGATION

GP-44. Dimitri is getting his prize sheep, Zoe, ready for the county fair. He keeps Zoe in the pasture beside the barn and shed. What he does not know is that there is a single locoweed in this pasture which will make Zoe too sick to go to the fair if she eats it, and she can eat it in one bite. Zoe takes about one bite of grass or plant every three minutes for six hours a day.

About how many bites of food does Zoe take during the day?

GP-45. Suppose that the shape of the field and building are as shown at right and on the resource page provided by your teacher. Everything except where the building sits has grass growing on it. Each square foot of pasture provides enough food for about 40 bites.

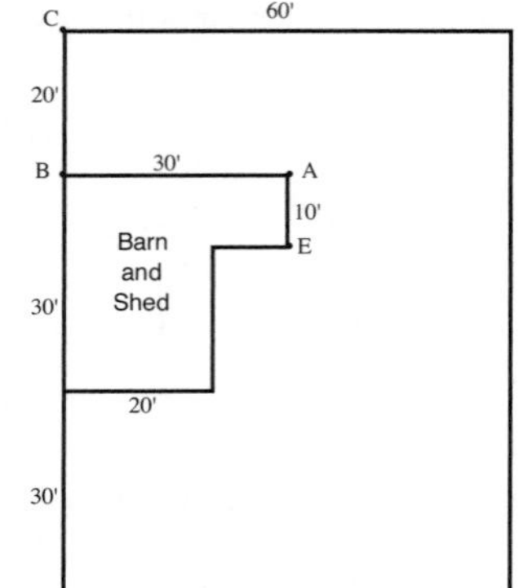


a) How many bites of food are available in the pasture?

b) What is the probability that Zoe will eat the locoweed if she grazes in the pasture for ten weeks and never eats the same patch twice?

GP-46. Dimitri has heard about the possibility of the locoweed, so he decides to tie Zoe to a rope with one end at A in the diagram.

a) Over how much area can Zoe graze if the rope is 10 feet long?

b) If we know that the locoweed is within this area, what is the probability that she eats the weed in one day?

c) Suppose the rope is 20 feet long. Over how much area can Zoe graze? What is the probability Zoe eats the weed in one day?

GP-47. Dimitri's father moves the tied end of the rope to B and lengthens the rope to 30 feet. Over what area can Zoe now graze?

a) Sketch the situation carefully, noting that it is only 20 feet from B to the north-west corner of the field, C, and the rope is 30' long.

b) Let D be the point on the north fence 30 feet from B. Find $m\angle DBC$ and CD.

c) What is the area of $\Delta BCD$?

d) What is the area of the grazing region not in $\Delta BCD$?

e) What is the total area of Zoe's grazing region for a 30' rope tied at B?

GP-48. Use the information from the previous problem to answer these questions.

a) Suppose we know that the weed lies in the region which Zoe can reach in the previous problem. What is the probability she eats the weed in a week if she never eats the same patch twice?

b) If we know that the weed does not lie in the region which Zoe can reach in the previous problem, what is the probability she eats the weed in a week if she never eats the same patch twice?

GP-49. Suppose that Zoe is tethered to a 30-foot rope attached at E. Her grazing region will extend around some barn corners. Sketch the region over which she can graze when the rope is pulled tight.

a) Identify all of the subproblems you need to solve in order to find the area of Zoe's grazing region.

b) Find the total area.

c) Find the probability that Zoe will eat the weed in four weeks if she is tied to this rope, never eats the same patch twice, and the weed is located in her grazing region.

GP-50. The information about rational expressions in problem GP-37 cautioned you that you will only find fractions with factors that equal 1 for <u>products</u> of factors, not sums (unless, of course, they are binomial sums treated as a binomial factor). Answer the following questions to verify this fact.

a) Is $\frac{5 + 7}{5} = 7$?

b) Is $\frac{5 \cdot 7}{5} = 7$?

c) Is $\frac{9 + 3}{3} = 9$?

d) Is $\frac{9 \cdot 3}{3} = 9$

e) If you tried to apply the rules for rational expressions in parts (a) and (c), the result contradicts what you know is true from your experience with arithmetic and order of operations. Compare the differences between parts (a) and (b), then parts (c) and (d). What is the <u>only</u> mathematical operation that allows you to write the same factor in the numerator and denominator so that it equals 1?

f) Use your conclusion from parts (a) through (e) to determine which of the following rational expressions is simplified correctly.

1) $\frac{(x - 4)(x + 3)}{(x + 3)} = x - 4$

2) $\frac{x^2 + x + 3}{x + 3} = x^2$

GP-51. ABCD is a rectangle with E and F as shown. $\overline{AE}$ and $\overline{CF}$ divide the rectangle into three figures of equal area.

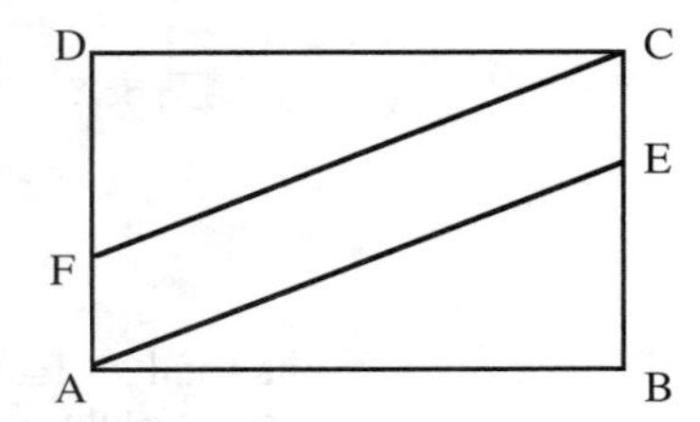


a) Explain why DF = EB.

b) If AB = 10 cm and BC = 8 cm, find EB.

c) Find $m\angle AEC$.

d) Answer the same question as part (c) except that now the rectangle is 20 cm x 16 cm.

GP-52. A mug full to the brim with coffee weighs 520 grams. When it is half full, it weighs 400 grams. How many grams of coffee will it hold?

GP-53. A new building is going up a block from the apartment building where Kim lives. On May 1, she notices the top of the new building is exactly level with her kitchen window. By May 6, the workers have added another 12 feet to the building and the top now makes an angle of 3 degrees with the horizontal from the kitchen window.

a) Draw a diagram of the situation.

b) Compute about how far the kitchen window is from the new building.

GP-54. Find the area remaining in the trapezoid at right after a circular sector of radius 5 cm is cut out as shown.

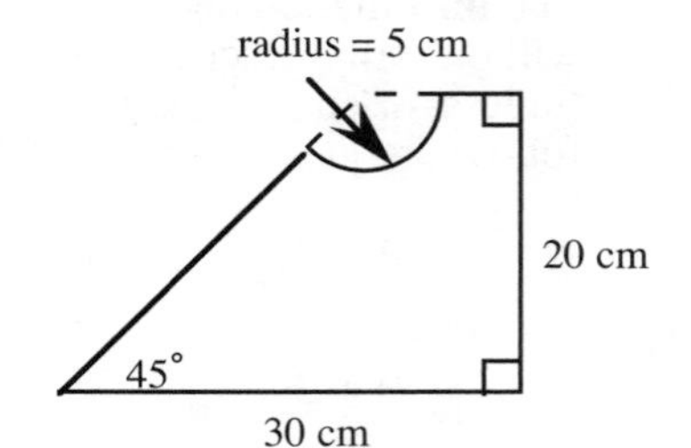


GP-55. Set up xy-coordinate axes. Graph $y = \frac{2}{3}x + 1$ and $y = -\frac{16}{11}x + 11$. Do these two lines along with the x-axis, form a right triangle? Prove your answer.

GP-56. Simplify the following rational expressions.

a) $\frac{(x + 5)(x - 1)}{(x - 1)(x + 3)}$

b) $\frac{x^2 + 4x}{3x + 12}$

c) $\frac{x^2 + 4x - 12}{x^2 + 8x + 12}$

d) $\frac{x^2 + 8x + 16}{x^2 - 16}$

GP-57. Calculate the total surface area and volume of the prism at right. The bases are equilateral triangles.

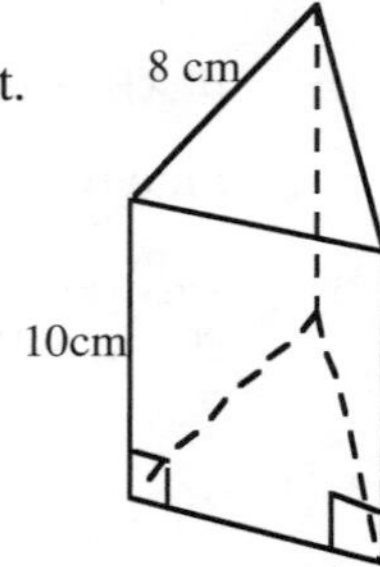

GP-58. Two similar triangular prisms are cut out of sheet metal from plates of different thicknesses. The smaller weighs 250 g and the larger weighs 400 g. If the shortest edge of the smaller prism is 26 mm, how long is the shortest side of the larger prism?

GP-59. At 4:00 PM the shadow of a 12 foot pole is 13 feet long. How tall is a building which casts a shadow 45 feet long at the same time?

GP-60. In the figure at right, ABCD is a parallelogram with heights $\overline{AF}$ and $\overline{AE}$.

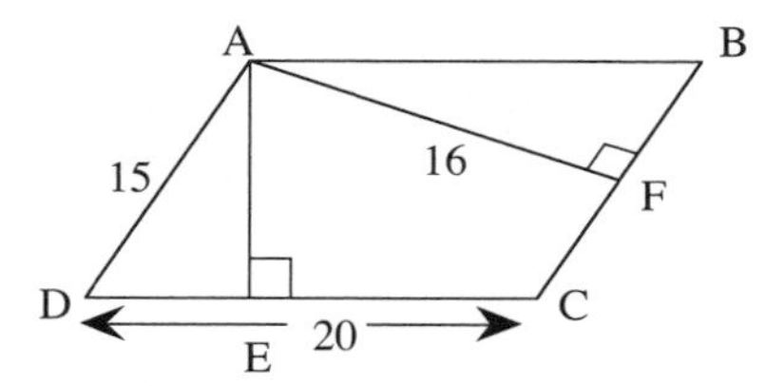


a) What is the area of parallelogram ABCD?

b) What is the length of $\overline{AE}$?

GP-61. Draw a sketch of a graph showing the relationship between the price you pay for a pizza and the diameter of the pizza.

## USING A SQUARE TO CALCULATE PROBABILITIES

GP-62. You and your best friend have a chance to win a million dollars on a game show. Your friend will have to walk through a maze and will end up in one of two rooms; she cannot backtrack. If she ends up in the room with the million dollars, then it is yours to share! Your friend has no idea what the maze looks like or where to go; she will be guessing which way to go. You, however, are given a drawing of the floor plan of the maze. Additionally, you get to decide in which room the money will be placed.

Examine the drawing at right. Where do you think you should place the money so that your friend will have the best chance of finding it, Room A or Room B? Explain your answer completely. Note: if you believe it does not matter, explain that completely.

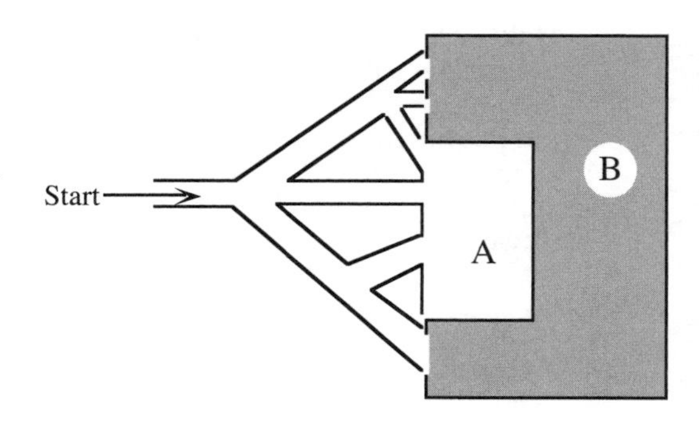

GP-63. During the last few days, we have used areas to calculate probabilities. We will use the area of a generic square (known as a unit square because it is 1 x 1, shown below, at right) to determine the probability of wandering into each of these rooms. When your friend gets to the first junction in the maze, how many choices does she have?

Assuming that each path is equally likely, we can represent this and the number of choices she has by dividing the square into three equal parts. Each part of the square now represents $\frac{1}{3}$ of the square's area.

| Upper Path |
|---|
| Middle Path |
| Lower Path |

a) Let us start with the easiest path: if she chooses the middle path, where will she end up? Is this definite? Explain. We will represent this choice in our square as shown in the figure at right.

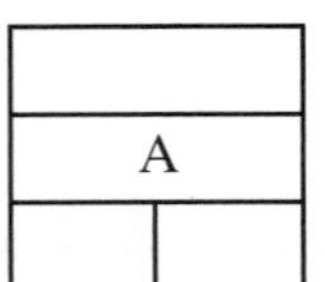


b) If she takes the lower path, where will she end up? If she is in the lower path already, are the two rooms equally likely? Represent this on the square.

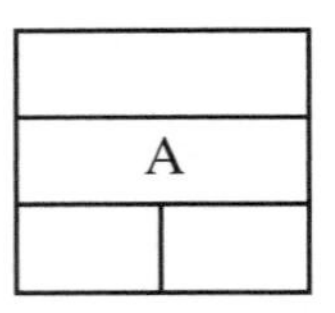


c) If she takes the upper path, she will have three choices. Represent this fact and where she could end up by using the top third of the square.

| |
|---|
| A |
| |

d) A unit square has side length 1. Use your partitioned square to calculate each of the areas of the regions with A's and B's in them. Use this to find the probability of wandering into Room A or Room B.

e) Explain why the letter representing the greatest total area of the square tells you where to put the money.

GP-64. At the last minute, the game show host changes the maze in the previous problem. Where should you put the money now? Use a unit square, show all steps in your solution process, and justify your answer.

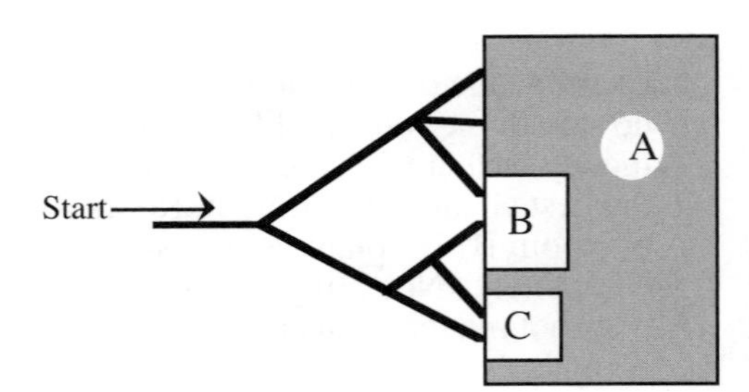

GP-65. Your friend is willing to pay you \$5.00 if you could flip a coin three times and get either all heads or all tails. "But if you don't," he says, "you have to pay <u>me</u> \$5.00!" Should you take the bet? To help answer this, you can calculate the probability of getting all heads and all tails by using a square.

a) How many possible outcomes are there for the first flip? Illustrate this fact on a square.

b) For each of those possible outcomes, how many possibilities are there for the second flip? Illustrate this on a square. Note: part of the square has been done for you as a suggestion as to how to use the square.

| HH | |
|---|---|
| HT | |

c) Illustrate the outcomes of the third flip on the square.

d) Use the square from part (c) to calculate the probability of getting all heads or all tails. Will you take your friend's bet? Explain completely.

e) Another way to solve this problem is to use a tree diagram like the one at right. Follow the "branches" left to right and see how many paths have <u>all</u> heads or <u>all</u> tails. Does this method confirm your answer to parts (c) and (d)?

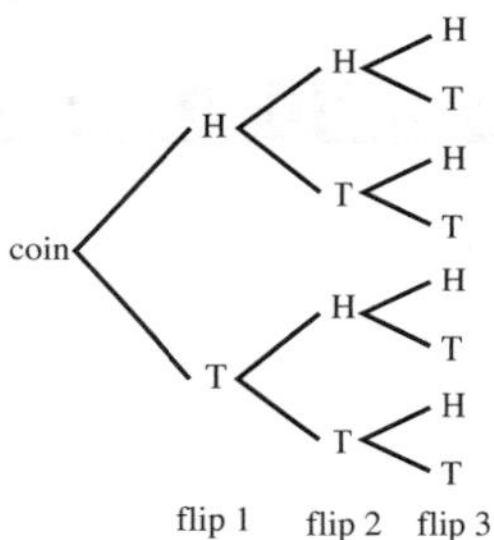


GP-66. Simplify.

a) $10(xy)^2(x - y)^2$

b) $\frac{8a^4b^2c}{(2a)^3}$

c) $\frac{4(x + 5)(x - 1)}{22(x - 2)(x + 5)}$

d) $\frac{(x + 3)^4(2x - 5)^3}{(x + 3)(2x - 5)^5}$

GP-67. Is it true that $(a + b)^2 = a^2 + b^2$? Prove your answer.

GP-68. In the figure at right, $\overline{UV}$ and $\overline{WT}$ are midsegments of $\Delta QPR$ and $\Delta SPR$ respectively.

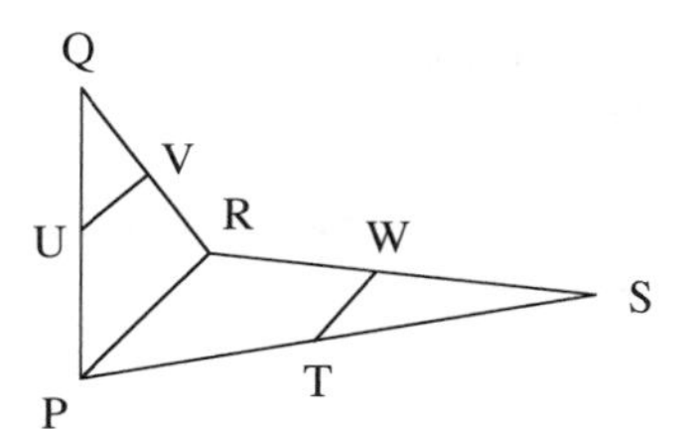

a) Is $\overline{UV} \cong \overline{WT}$? Prove your answer.

b) Is $\Delta QPR \cong \Delta SPR$? Prove your answer.

GP-69. Graph the following inequalities on the same set of axes. Darken the overlapping region.

$y \le -x + 5$ $\quad y \ge x - 5$ $\quad y \le \frac{5}{3}x + 5$ $\quad y \ge -\frac{5}{3}x - 5$

a) Is the shape a rectangle? Prove your answer.

b) Find the perimeter and area of the shape.

GP-70. Sketch a graph showing the relationship between the height of the water in the bathtub from the time a person turns on the water until his/her bath is over. Label all points that indicate an activity such as getting into the tub.

## UNIT AND TOOL KIT REVIEW

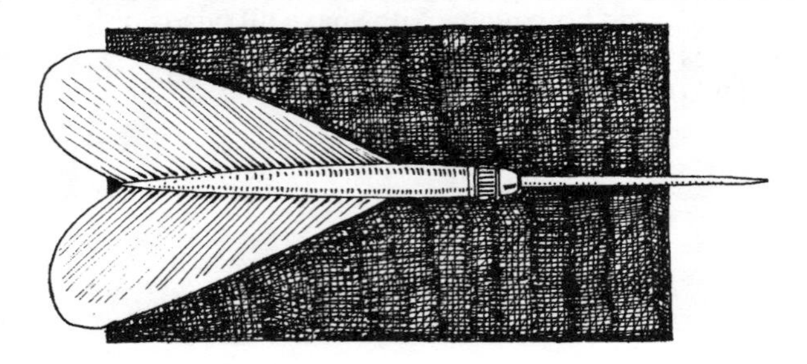

GP-71. Design a dart board which satisfies all of the following conditions simultaneously.

1) The board is a square with four colors.

2) Each colored section is connected (that is, there are **not** two or more, separate red regions, for example).

3) The smallest region takes up $\frac{1}{16}$ of the dart board. The next smallest region takes up $\frac{1}{8}$ more of the board than the first. The next takes up another $\frac{1}{8}$ more of the board than the previous region, and the last, yet another $\frac{1}{8}$ more of the board than the one before it.

a) What is the probability of hitting the color that takes up the largest portion of the board?

b) What is the probability of hitting the color that takes up the smallest portion of the board?

GP-72. Lucky you! You are going to be one of the first contestants on the new game show:

**WHEEL OF A FEW BUCKS!**

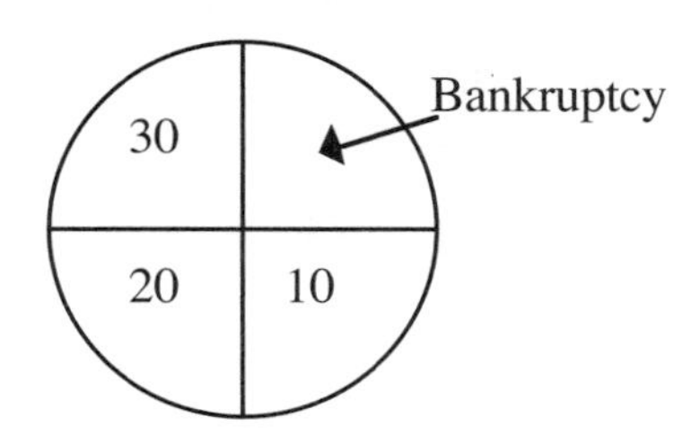


a) If you start with no money, what amount would you expect to win by spinning once? Note that bankruptcy means you lose all winnings from previous spins.

b) Suppose you have already spun the wheel eight times and have $160 and you have **not** hit a single bankruptcy. What is the probability of getting a bankruptcy on your next spin? If you already have $160 should you spin again?

c) How much money should you have before you quit spinning?

GP-73. For the maze at right, what is the probability of wandering into room A?

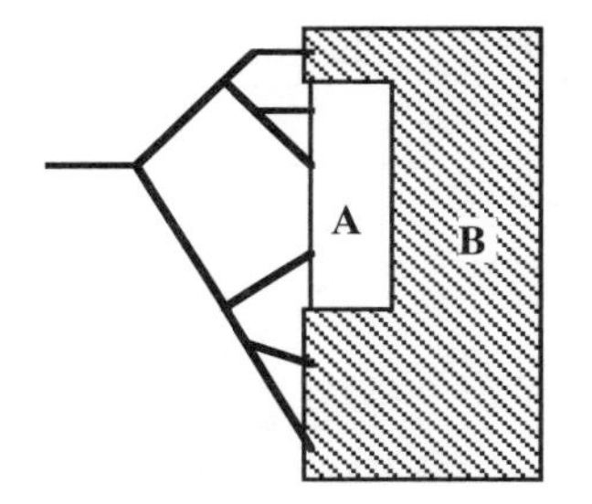


GP-74. Your friend offers you a bet: Someone will give you $5.00 if you can get at least two heads in three flips. If you do not, you give him $5.00. Should you take this bet? Explain completely.

GP-75. If the radii of the concentric circles shown at right are one inch, two inches, and three inches, what is the probability of hitting the shaded region with a dart?

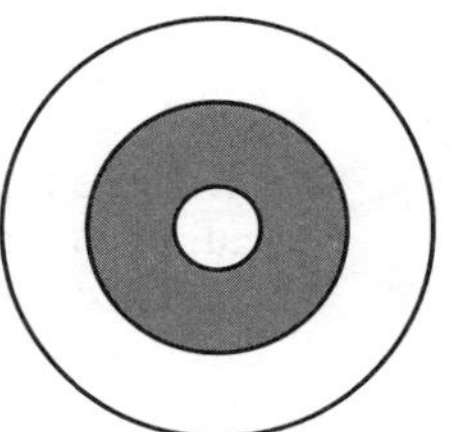

GP-76. In the figure at right, the dart board ABCD is a rectangle and points E and F are midpoints of sides $\overline{AB}$ and $\overline{BC}$ respectively. Find the probability of hitting each of the triangles: $\Delta AED$, $\Delta EDB$, $\Delta BDF$, and $\Delta FDC$.

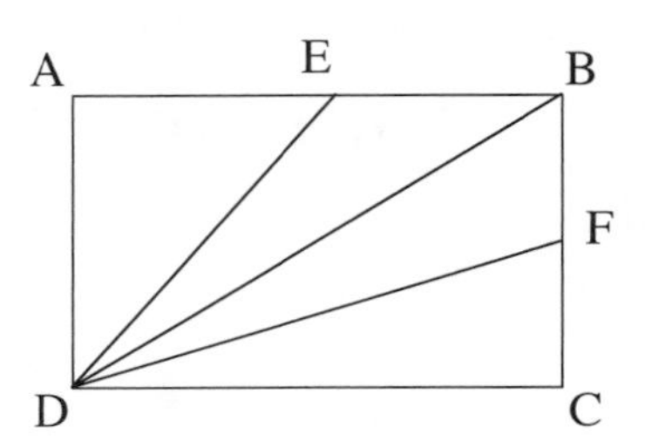


GP-77. Simplify.

a) $4x + 3[2x - 5(x - 1)]$

b) $(2x - 5)(8 - x^2)$

c) $(5xy^2)^2 \div (35x^2y)$

d) $\dfrac{(x + 4)(x - 6)}{(x - 6)(x + 3)}$

e) $\dfrac{x^2 - 3x - 10}{x^2 - x - 20}$

f) $\dfrac{2x^2 + x - 1}{x + 1}$

GP-78. The four triangles below are placed into a bag.

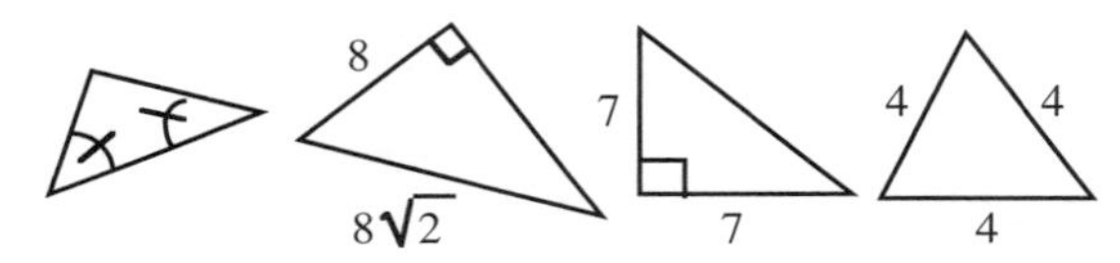


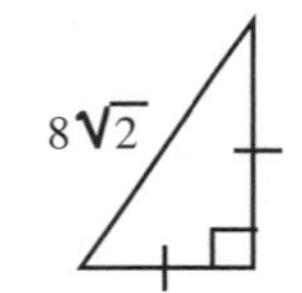


You reach into the bag and pull out a triangle. What is the probability it is similar to the triangle at right?

GP-79. TOOL KIT CHECK-UP

Your tool kit contains reference tools for geometry and algebra. Return to your tool kit entries. You may need to revise or add entries.

Be sure that your tool kit contains entries for all of the items listed below. Add any topics that are missing to your tool kit NOW, as well as any other items that will help you in your study of this course.

- probability
- expected value
- rational expressions

GP-80. About once a year, an asteroid enters the solar system in the region between the sun and the orbit of the planet Mars. What is the probability that it hits the earth? The distance from the sun to Mars is 142 million miles. The diameter of the earth is 7927 miles, and the diameter of the sun is 864,400 miles.

GP-81. On July 11, 1979, NASA's Sky Lab fell to earth after its orbit had decayed. Of all the places on earth that it could hit, it broke up scattering pieces over Australia and the Indian Ocean.

a) Calculate the probability of a single piece of Sky Lab crashing into Australia. The surface area of Australia is 2,966,155 square miles (7,682,300 sq. km) and the surface area of the earth is about 197 million square miles (510 million sq. km).

b) Lloyds of London sold insurance policies which would pay one million dollars if a piece of Sky Lab crashed into the policy owner's home. What is the probability of a piece hitting your home? Would you have bought the insurance?

c) What should the insurance policy cost to make it a worthwhile investment?

GP-82. Draw a picture of a dart board consisting of only two concentric circles, the inner one shaded, so that the probability of hitting the shaded region is exactly $\frac{1}{2}$.

GP-83. A circular dart board consists of three concentric circles. The outer circle has radius 20 cm and the innermost one has radius 2 cm.

a) Find the radius of the third circle so that the probability of hitting the middle ring is exactly one-half.

b) Now find the radius of the third circle so that the probability of hitting the middle ring is 0.4.

# Going Camping

## 3D AND CIRCLES

# Unit 12 Objectives

## *Going Camping:* 3D AND CIRCLES

This unit will give you an opportunity to use many of the ideas you have learned in the course, especially those involving polygons, area, trigonometry, volume, and circles. You will also use your visualization skills to work with three-dimensional figures. The algebra review will complete the work with rational expressions.

In this unit you will have the opportunity to:

- solve interesting problems by using subproblems and many of the ideas from the course.
- learn more about relationships between angles, chords, and secants in circles.
- practice simplifying and adding rational expressions.
- reflect on your learning and successes in the course this year.

*Read the problem below. Over the next few days you will learn what is needed to solve it. Do not attempt to solve it right now.*

3D-0. One night Kris and Karen were sitting around the campfire discussing their recent tent purchases. Kris claimed herd be divided by n.ris sleeps in a standard pup tent that has a height of 5 ft., a length of 6 ft., and a width of 5 ft. Karen, on the other hand, uses a 6-pole teepee tent with a height of 5 ft. and the greatest diagonal across the floor measuring 8 ft. Her tent looks like a regular hexagonal pyramid.

Considering floor area, surface area, and volume, which is the best backpacking tent?

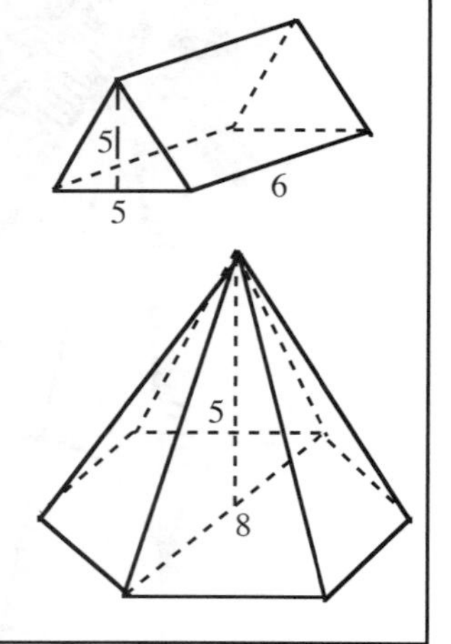


*UNIT* 12

ALGEBRA
GRAPHING
RATIOS
GEOMETRIC PROPERTIES
PROBLEM SOLVING
SPATIAL VISUALIZATION
CONJECTURE, EXPLANATION (PROOF)

# Unit 12

## *Going Camping:* 3D AND CIRCLES

## SURFACE AREA AND VOLUME

3D-1. You have found the surface area of several regular pyramids this year, including the TransAmerica Pyramid theme problem in Unit 4. In order to do so, one subproblem involved finding the height of each triangular face. This height is commonly referred to as the **SLANT HEIGHT**, abbreviated **sh**. Notice that for regular pyramids, the slant height bisects the edge of the pyramid's base and forms a right triangle with half the length of the base and a lateral edge (where two sides of two triangles meet). The slant height is also the hypotenuse of a right triangle with the height of the pyramid being one of the legs.

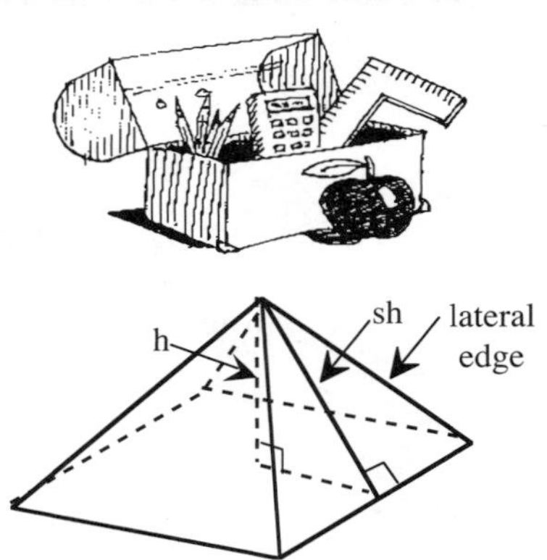


a) Find the slant height of the square-based pyramid above if the edge of the base is 10' and the height of the pyramid is 12'.

b) Find the slant height of the square-based pyramid above if the lateral edge is 15' and the edge of the base is 9'.

c) Find the total surface area of the pyramid using the data in part (a).

d) Find the volume of the pyramid using the data in part (b).

3D-2. Guess which contains more cola, the can or the bottle? Each person should write down his or her guess. Verify your answer by calculating the volume of each figure. Assume that the top part of the bottle is a complete cone and draw a sketch of the parts. Leave your answers in terms of $\pi$.

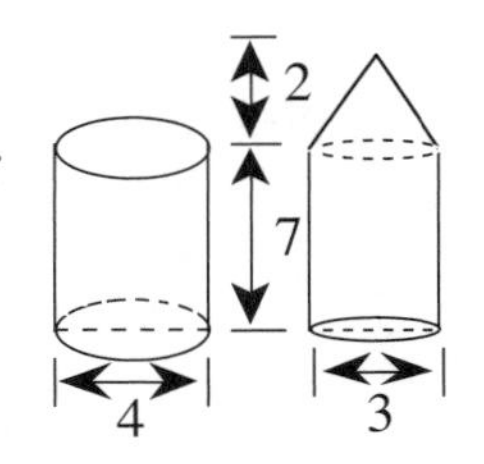


3D-3. Calculate the base area and volume of each prism:

a) The bases are regular hexagons.

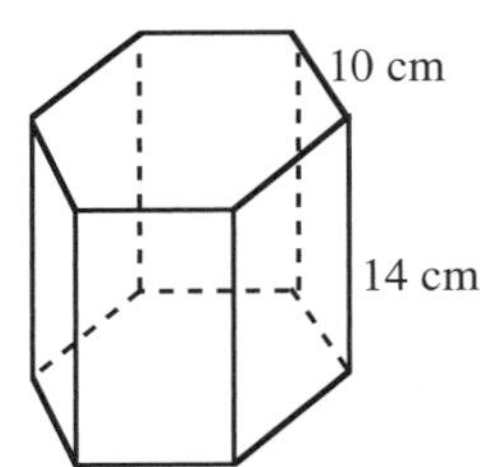


b) All corners are right angles.

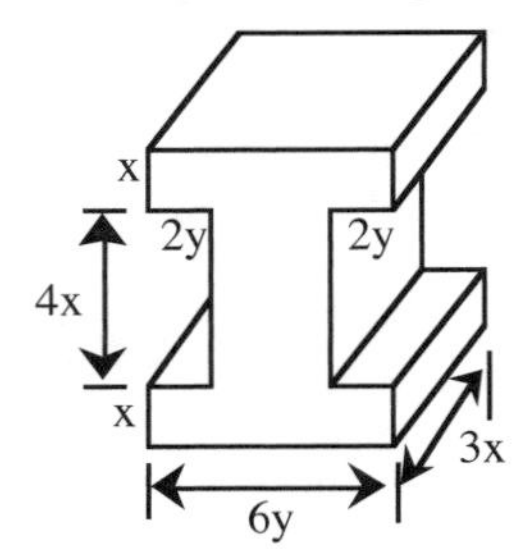

3D-4. The Sunshine Orange Juice Company wants its product in a one quart (107.75 cubic inches) can. The can manufacturer makes cans that have a base 5 inches in diameter. What will the can's height be?

3D-5. A rectangular room has dimensions 34 ft. x 19 ft. x 8 ft. If a person requires 300 cubic feet of air space to breathe properly, determine the maximum number of people who can comfortably inhabit this room.

3D-6. The four <u>triangular</u> lateral faces in this figure are all congruent to each other. Is $\overline{CD} \parallel \overline{BA}$? Justify your answer.

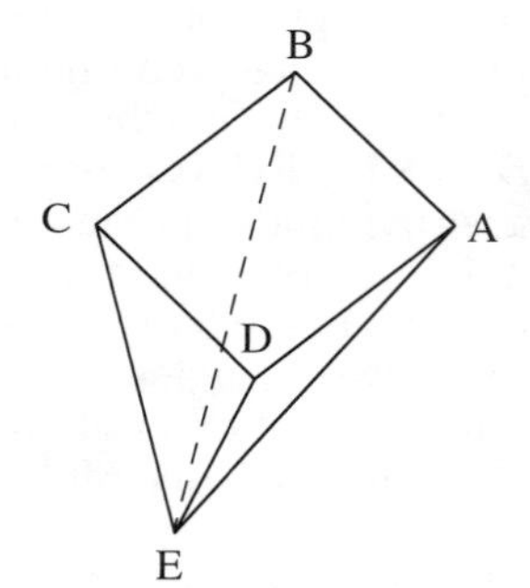

3D-7. A spinner is divided into 16 equal parts. Each part is numbered, starting at one and continuing to 16. If each of the numbers indicates the points you can earn if you land on that part of the spinner, how many points would you <u>expect</u> to get in 100 spins on average?

3D-8. Write each of the following expressions in simplest form.

a) $\frac{4\sqrt{3}}{\sqrt{6}}$

b) $(4x^2y^3)^3(2y^2)$

c) $\frac{15(x - 5)^3(x + 4)^4}{6(x - 5)^5(x + 4)}$

3D-9. Solve for x and y:

a) $2x + y = 2$
$x - y = 13$

b) $x^2 + y = 5$
$2x - y = -2$

3D-10. In part (a) in the previous problem you found that the two lines $2x + y = 2$ and $x - y = 13$ cross at (5, -8).

a) Write the equation of the line that also crosses at that point and has a slope of $\frac{1}{5}$.

b) Write the equation of the line perpendicular to the line of part (a) that also crosses the other line at (5, -8).

3D-11. Martha was playing with a hollow plastic cone that had a diameter of 12 inches and a height of 5 inches. Her brother Matt snatched the cone, cut two inches off the top, and placed the cut piece upside down in the lower portion of the cone as shown at right. How many cubic inches of water would fit inside the space of her redesigned cone?

original view

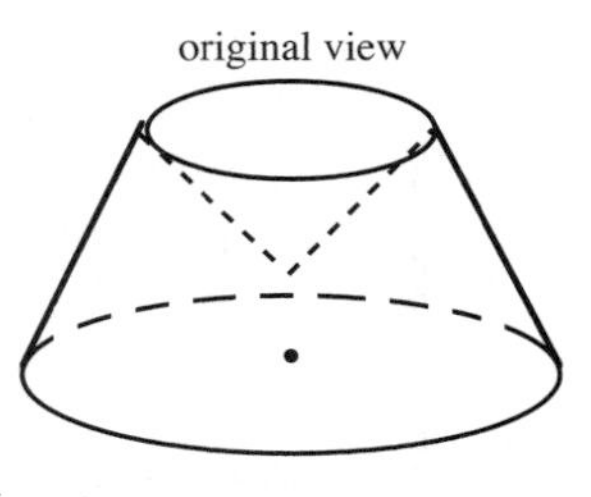

original view upside-down

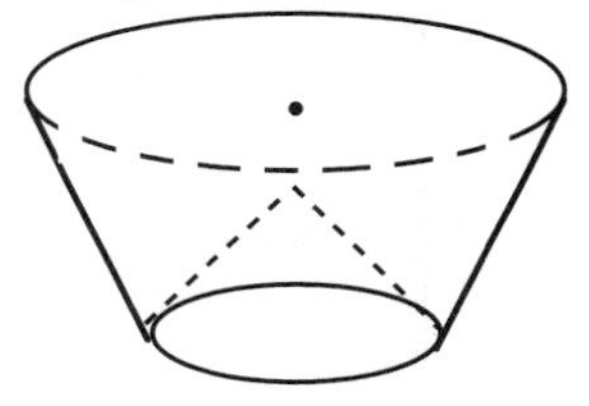

3D-12. A regular hexagonal based pyramid has a base edge of 10 cm. A lateral <u>face</u> makes an angle of 40° with the <u>base</u> at the midpoint of the base edge. Find the total surface area and volume of the pyramid. Identify the subproblems you will need to solve before starting.

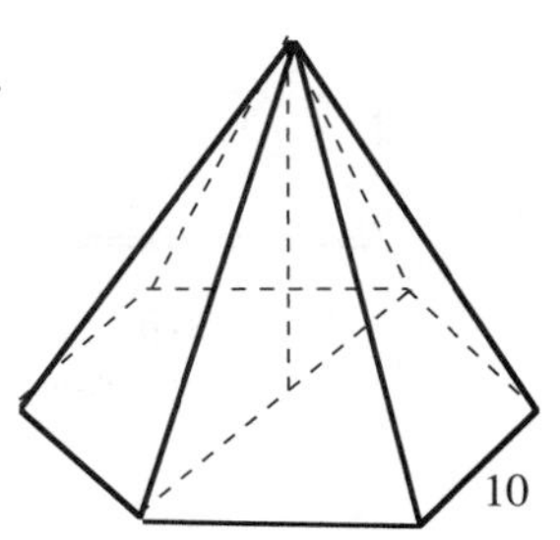


3D-13. Design a spinner that satisfies all of the following conditions simultaneously.

1) It has only four colors.

2) No two colors have the same probability of occurring.

3) The color with the smallest probability of occurring we would expect to see half as often as the color with the next smallest probability, which we would expect to see half as often as the color with the third smallest probability, which we would expect to see half as often as the color with the largest probability.

3D-14. Read the following information and summarize it in your tool kit. Then complete the problems below the box.

## MULTIPLYING RATIONAL EXPRESSIONS

Suppose you were asked to rewrite the following expression in simplest form:

$$\frac{x^2+6x}{(x+6)^2} \cdot \frac{x^2+7x+6}{x^2-1}$$

First break the problem into four factoring subproblems. Although $(x + 6)^2$ is already factored, it can be rewritten as $(x + 6)(x + 6)$. After completing the four subproblems, the factored expression is:

$$\frac{x(x+6)}{(x+6)(x+6)} \cdot \frac{(x+6)(x+1)}{(x+1)(x-1)}$$

Next, rewrite the expression so that whenever possible you have the same terms in the numerator and denominator:

$$\frac{(x+6)}{(x+6)} \cdot \frac{(x+6)}{(x+6)} \cdot \frac{x}{(x-1)} \cdot \frac{(x+1)}{(x+1)}$$

Since $\frac{(x+6)}{(x+6)} = 1$ and $\frac{(x+1)}{(x+1)} = 1$, the long product becomes

$1 \cdot 1 \cdot \frac{x}{(x-1)} \cdot 1$, or just $\frac{x}{x-1}$. Thus, $\frac{x^2+6x}{(x+6)^2} \cdot \frac{x^2+7x+6}{x^2-1} = \frac{x}{x-1}$.

By using subproblems, **each <u>part</u> was simpler** and doing several easy parts allows you to complete a complicated problem.

Note that in all cases we assume the denominator is not equal to zero.

a) Write a brief summary from this example that explains how to multiply two algebraic fractions.

b) Can $\frac{x}{x-1}$ be simplified further? Check using one or two values for $x$. Show your examples.

c) What values of $x$ cannot be used in the expression in the example above?

3D-15. Use the method above to simplify the following expressions.

a) $\frac{x+1}{x-3} \cdot \frac{x^2-4x+3}{x-1}$

b) $\frac{x^2+6x-7}{x^2-3x+2} \cdot \frac{x^2-5x+6}{x+7}$

c) $\frac{7x+42}{x^2+5x-6} \cdot \frac{x^2-x}{x^2}$

d) $\frac{x^2-3x}{x^2-4} \cdot \frac{(x-2)^2}{x^2-9}$

3D-16. In the figure at right, $\overline{BU} \cong \overline{LE}$. Is BLUE a rectangle? Prove your answer. Be careful not to make further assumptions about the figure based on its appearance.

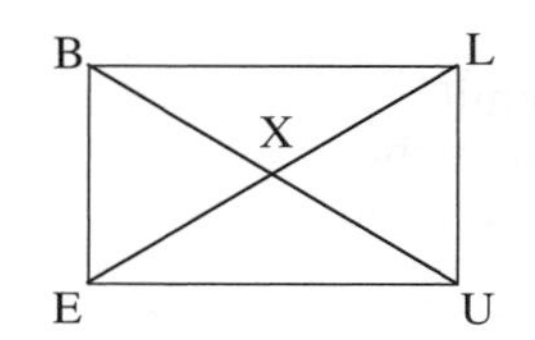

3D-17. Speed test. These problems are all straightforward. Try to have them done (correctly) in five minutes. The answers are provided below the problems for you to check your work.

a) $2x - 5 = 7$

b) $x^2 = 16$

c) $3x + 5 = 14$

d) $2(x - 1) = 6$

e) $4x - 1 = 3x + 3$

f) $\frac{x}{5} = 6$

g) $2x^2 + 5 = x^2 + 14$

h) $x + y = 5$
$2x - y = 4$

i) $x + 2x + 3x = 4x + 5$

j) $(x - 3)(x + 5) = 0$

**Answers: (a) 6; (b) 4 or -4; (c) 3; (d) 4; (e) 4; (f) 30; (g) 3 or -3; (h) (3,2); (i) 2.5; (j) 3 or -5.**

3D-18. Find the point(s) where the graph crosses the x-axis.

a) $12x^2 + 7x - 12 = 0$

b) $16x^2 + 193x + 12 = 0$

3D-19. Solve for a.

a)

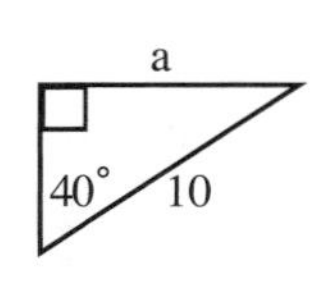


b)

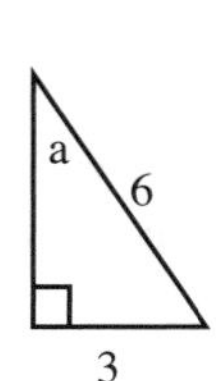


c)

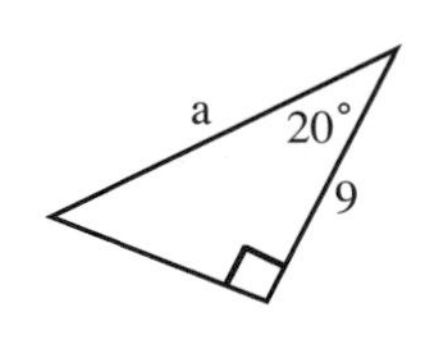


3D-20. An oblique prism has a square base whose edge length is 4 cm. The volume of the prism is 96 cubic cm. The prism makes an angle of 50° with the ground. Calculate the surface area of the prism by first drawing pictures of the prism and its parts.

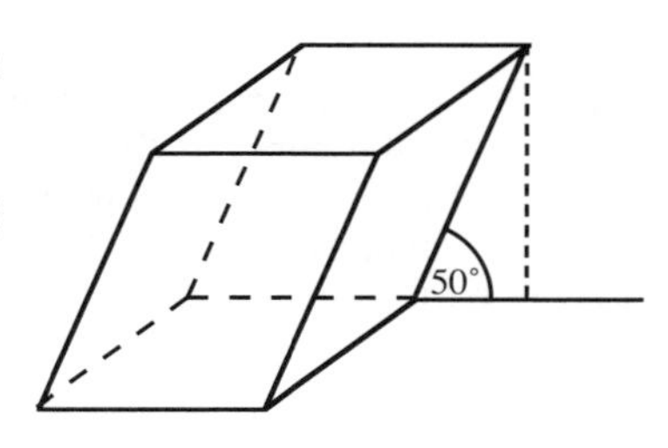

3D-21. Irma, a young inventor, has developed a revolutionary new windmill. To make the propeller blade she cut a piece of plastic in the shape of a right triangle with leg lengths three feet and five feet. Then she attached a pole to one of the lengths of the legs. The triangle is attached in such a way that it can rotate freely around the pole, and can do so at incredibly high speeds. When the triangle rotates at such speeds, it is just a blur, but it makes the propeller appear to be a three-dimensional solid.

Note: The triangle is not attached in the way shown in the illustraion at right.

a) What solid would the propeller resemble when it spins at such high speeds? Draw a sketch.

b) If the pole is attached to the shorter leg, what is the volume of the solid formed? Is it the same as the volume of the solid when the pole is attached to the other leg? Justify your answer.

3D-22. Many canned food items, such as fruits and vegetables, come in what is known as a size 300 tin. To reduce costs and increase profits, Grub Food Company has decided to reduce the dimensions of the can by 10%. The surface area of the size 300 tin label is 259 square cm and the volume of the tin is 486 cubic cm.

a) Labels currently cost 0.08 cents (that's 8/100 of a penny). What will the smaller labels cost? Careful! The answer is NOT 0.072 cents.

b) If Grub Foods currently purchases 6.8 million labels per year, how much will the company save by this change?

c) The contents of a can of peas costs the company 18 cents per can. How much will the company save if it produces 580,000 cans of peas each year?

3D-23. An aluminum cube with edge length 20 inches is melted and rolled into a long, straight wire of diameter 0.2 inches. What is the length of the wire?

3D-24. Take two pieces of paper and make two cylinders, one with the length of the paper as the circumference of its base and one with the width of the paper as the circumference of its base. Which cylinder has the greatest volume? To make calculations simpler, pretend that the paper measures 9" x 12". Prove that your answer is correct.

3D-25. Find the volume of each solid. Show all subproblems. Units are in feet.

a)

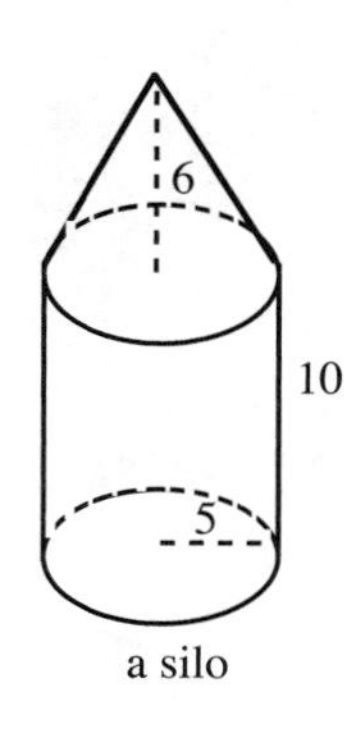

a silo

b)

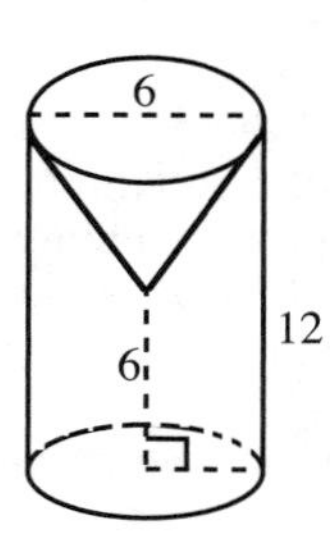


3D-26. Simplify by factoring and finding fractions equivalent to 1.

a) $\frac{6x^2 + 9x}{2x + 3}$

b) $\frac{x^2 - 3x - 10}{x^2 - 8x + 15} \bullet \frac{x^2 + 3x - 18}{x^2 + 8x + 12}$

3D-27. Solve for x and y.

a) $3x + 2y = 6$
$y = 4x + 3$

b) $y = -\frac{2}{3}x - 1$
$x = 3$

c) $3x + y = 2$
$x - 2y = 10$

3D-28. Solve for the missing sides.

a)

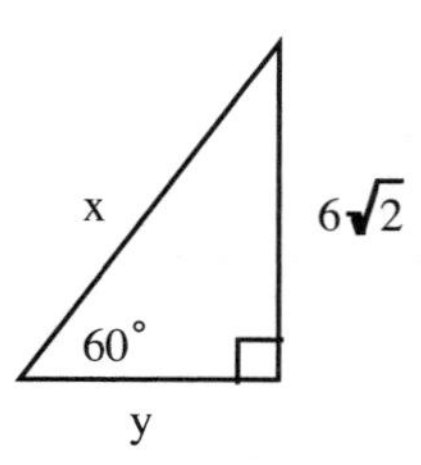


b)

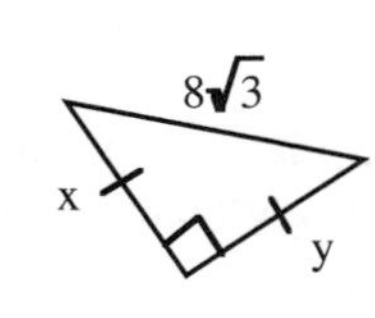

**3D-29.** One night Kris and Karen were sitting around the campfire discussing their recent tent purchases. Kris claimed her tent was larger than Karen's. Kris sleeps in a standard pup tent that has a height of 5 ft., a length of 6 ft., and a width of 5 ft. Karen, on the other hand, uses a 6-pole teepee tent with a height of 5 ft. and the greatest diagonal across the floor measuring 8 ft. Her tent looks like a regular hexagonal pyramid.

a) Which tent has the greatest floor area?

b) Which tent has the greatest volume?

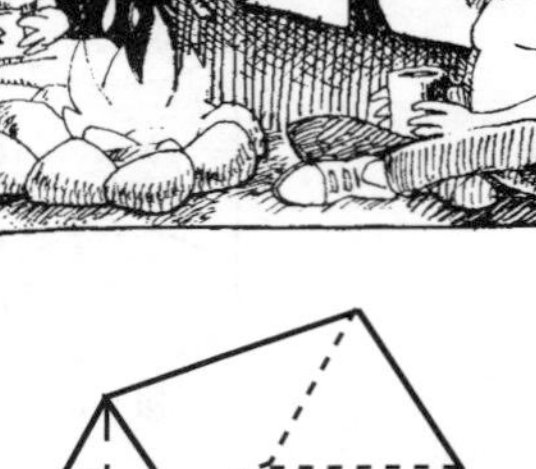


c) For camping purposes, which is more important: floor area or volume? Justify your answer.

d) For backpacking, the surface area is most important. Why?

e) Which of the tents would be the best backpacking tent? Justify your answer.

f) Calculate the measure of the angle a wall of each tent makes with the floor.

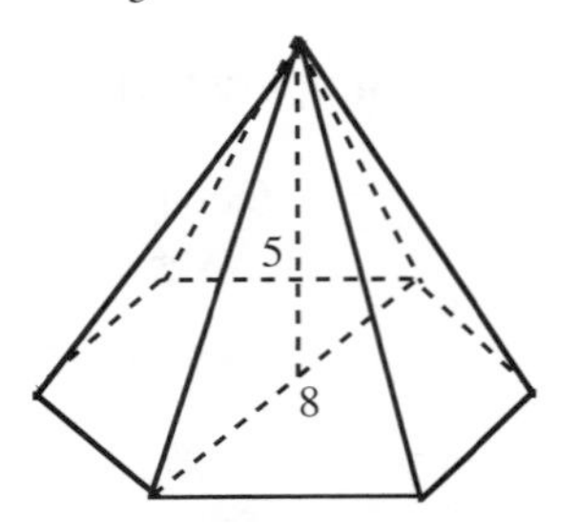


3D-30. Consider the figures at right with information about two sides and a non-included angle. Find the two possible values of x.

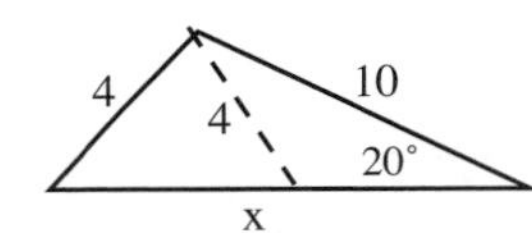


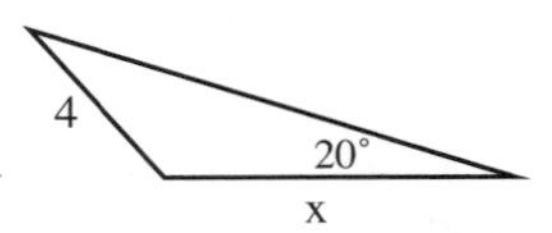


3D-31. A solid metal cylinder with radius 6 in. and height 18 in. is melted down. The entire volume of metal is then used to make a cone with a radius of 9 in. How tall will the cone be?

3D-32. Using the spinner at right, what would you expect your score to be, on average, after 300 spins?

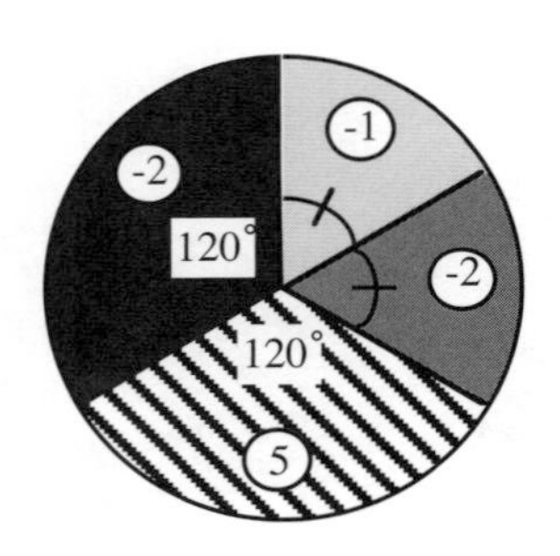

3D-33. Solve for x.

a) $\frac{x}{3} = \frac{2x + 1}{5}$

b) $\frac{x}{x + 1} = \frac{x}{4}$

c) $\frac{x}{2x + 1} = \frac{3}{5}$

d) $\frac{x}{x + 1} = \frac{x + 2}{6}$

3D-34. The Quality Cola Company has been trying all sorts of tactics to increase its sales. Recently, the Profit Division of the company presented this graph at the board meeting to explain its latest money making scheme.

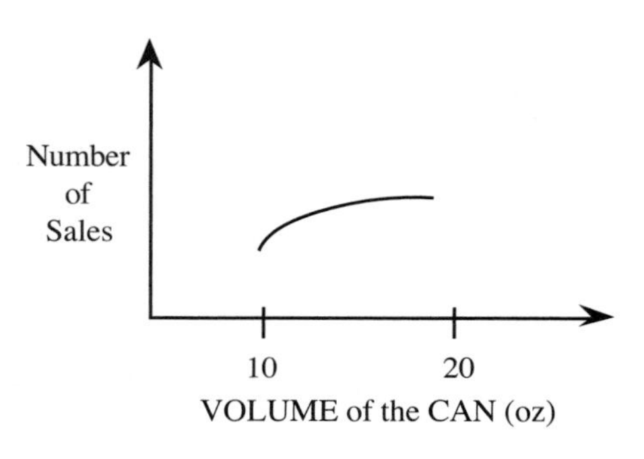


a) Explain what this graph tells you.

At the same meeting, a representative of the Costs Division presented the graph at right to show what happened to the cost of producing the cola when the Profit Division changed the volume.

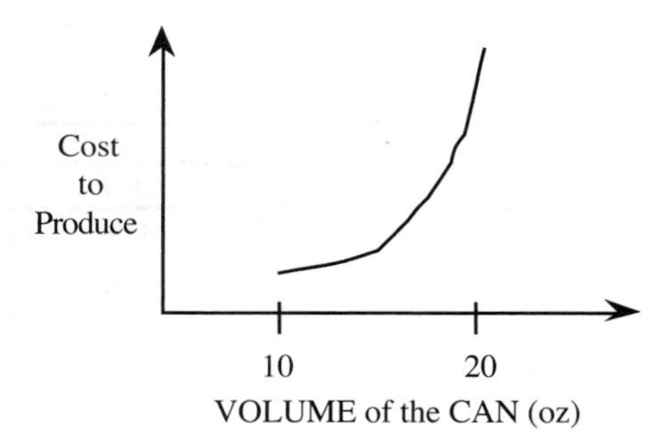


b) What does this graph tell you?

c) Based on these graphs what volume would you recommend for Quality Cola Company cans to increase profits? Support your answer.

## THE SODA CAN: GEOMETRY IN INDUSTRY

3D-35. Why is a soda can the shape (or size) it is? One reason may be that the manufacturer wants to minimize the amount of aluminum used to make the can. This requirement involves finding the smallest surface area of the can whose shape is a cylinder. Recall that the volume of a cylinder is V = Bh. The diagram below shows an expanded, flattened view of the can's surface area. The table below lists possible radii for the base of the can from one through eight centimeters. The volume of a soda can is fixed at 400 cubic cm. Use the volume with each radius to find possible heights for different sized soda cans. Once the h column is completed, calculate the surface areas. The results for r = 1 cm are already given. Round the height to tenths and the surface area to whole numbers.

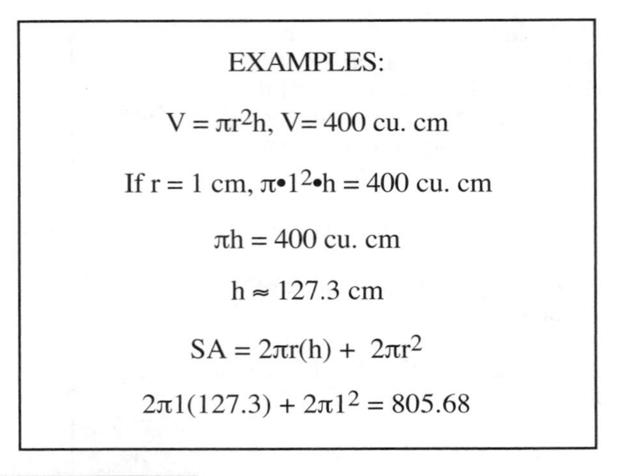

EXAMPLES:

$V = \pi r^2 h$, V= 400 cu. cm

If r = 1 cm, $\pi \bullet 1^2 \bullet h = 400$ cu. cm

$\pi h = 400$ cu. cm

$h \approx 127.3$ cm

$SA = 2\pi r(h) + 2\pi r^2$

$2\pi 1(127.3) + 2\pi 1^2 = 805.68$

| r | h | Surface Area |
|---|---|---|
| 1 | 127.3 | 806 |
| 2 | | |
| 3 | | |
| 4 | | |
| 5 | | |
| 6 | | |
| 7 | | |
| 8 | | |

a) Graph the radii and surface areas on a sheet of graph paper with the axes as shown below.

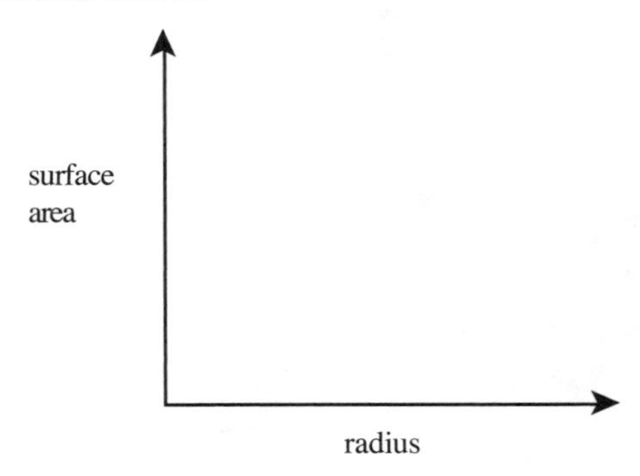


**>>>This problem continues on the next page.>>>**

b) Based on the results of your work, what radius and height should you choose to minimize the amount of aluminum used in a soda can? Explain your conclusions.

c) Measure an actual soda can and record its dimensions (r and h). How do these values compare to your results?

d) Aside from the cost of the extra aluminum, why else wouldn't a soda company want to produce cans with a 1 cm or 8 cm radius? Likewise, what reasons besides minimizing aluminum costs--in fact, using somewhat more aluminum than the absolute minimum--account for the actual size of a soda can?

3D-36. A rope 100 feet long is stretched from the top of a 70 foot flagpole to the ground. What angle does the rope make with the ground?

3D-37. Another rope is stretched from the flagpole in the last problem, so that the point where it is joined to the ground is 118 feet <u>directly</u> <u>opposite</u> from the base of the <u>other</u> <u>rope</u> (not the pole). Draw a diagram, then find the measure of the angle formed by the two ropes.

3D-38. A construction worker builds a frame in the shape of an isosceles triangle with legs 10 feet and base 12 feet. Before standing it upright, he props it up so that vertex A is four feet above the flat ground. Copy the figures below onto your paper and do parts (a) and (b).

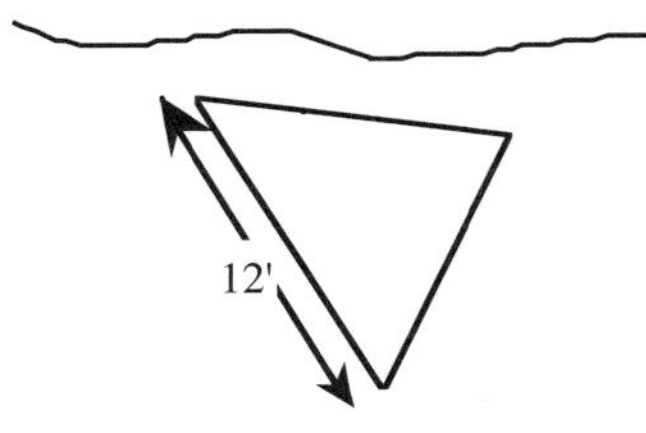


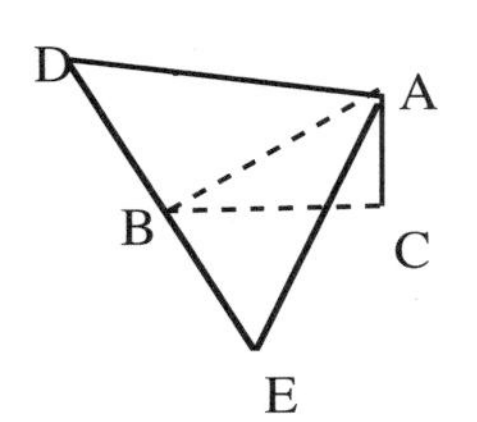


a) Identify the subproblems you would need to solve in order to find the angle the frame makes with the ground ($m\angle ABC$).

b) Redraw the two triangles needed, label what you know, and calculate $m\angle ABC$.

3D-39. Prove that the lateral edges of a regular square pyramid are congruent.

3D-40. A room is in the shape of a rectangular prism. The length of the room is 10', its width is 15', and its height is 12'. A wire is strung from a ceiling corner down to the <u>farthest</u> floor corner. Find the angle the wire makes with the floor.

3D-41. Find the volume of each of the following figures.

a) The base is an isosceles triangle with sides of lengths 13,13, and 10.

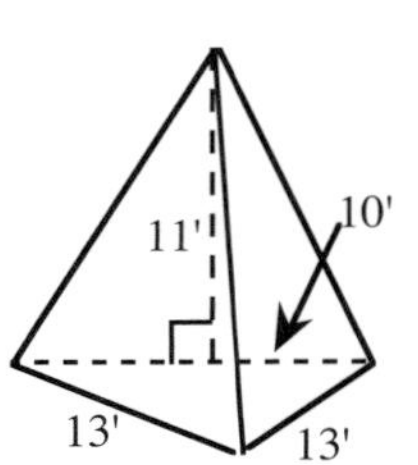


b) The base is a regular hexagon.

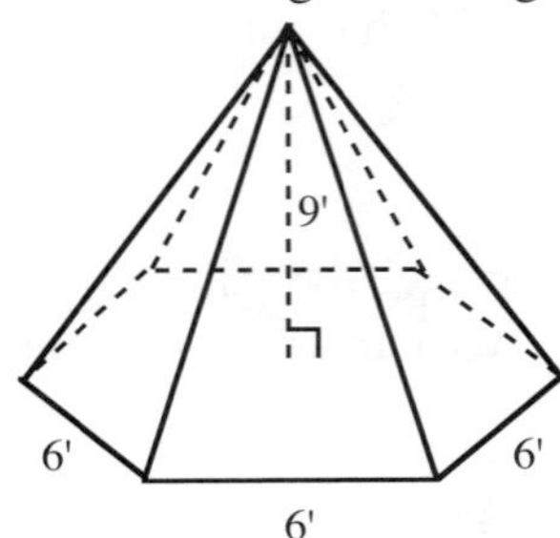


3D-42. The hexagonal pyramid in part (b) of the preceding problem is sliced through the vertex perpendicularly to the base.

a) Into what shape is the base divided?

b) The pyramid is pulled apart forming two congruent solids. Each has a volume <u>one-half</u> the original solid. Is the surface area of half the solid also one-half of the original surface area? Justify your answer.

3D-43. Solve for the variables x and/or y.

a) $5x^2 - 6x - 2 = 0$

b) $\frac{5}{3x} + \frac{2}{x} - 3 = \frac{2}{3}$

c) $3\sin x = 1$

d) $8x - y = 12$
$y = 5 - 2x$

## ANGLES, CHORDS, AND SECANTS

**INTRODUCTION:** In Unit 10 we studied the fundamentals of circles. Our work included relationships between arcs and angles, chords and diameters, and radii and tangents. Today we will extend that study by using what you know about similar triangles and exterior angles of triangles to prove four more relationships for angles and segments in relation to circles.

**3D-44.** Two chords $\overline{AC}$ and $\overline{DB}$ intersect at point E that is <u>not</u> the center of the circle. Complete parts (a) and (b) below to find the measure of $\angle AED$, then generalize your process in part (c). Auxiliary line $\overline{AB}$ has been drawn to help with the solution.

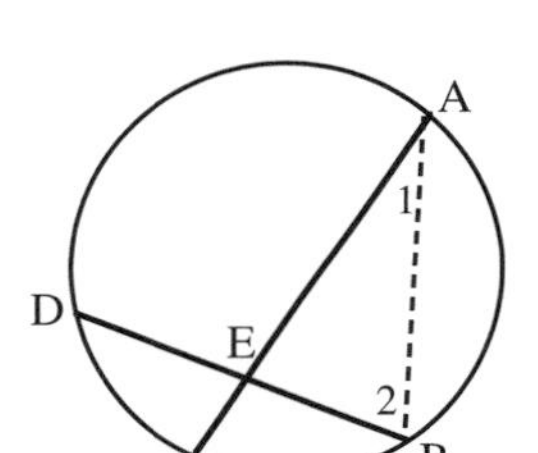


a) If $m\widehat{AD} = 160^\circ$ and $m\widehat{BC} = 48^\circ$, find $m\angle 1$ and $m\angle 2$.

b) Use $\Delta ABE$ to find $m\angle AED$.

c) Use the example you solved in parts (a) and (b) as the basis to write a general procedure for finding the measure of any interior angle in a circle formed by two intersecting chords.

**3D-45.** Suppose two secant segments, $\overline{AC}$ and $\overline{AE}$, intersect outside the circle. Follow a process similar to the previous problem to develop a general method to find the measure of $\angle A$.

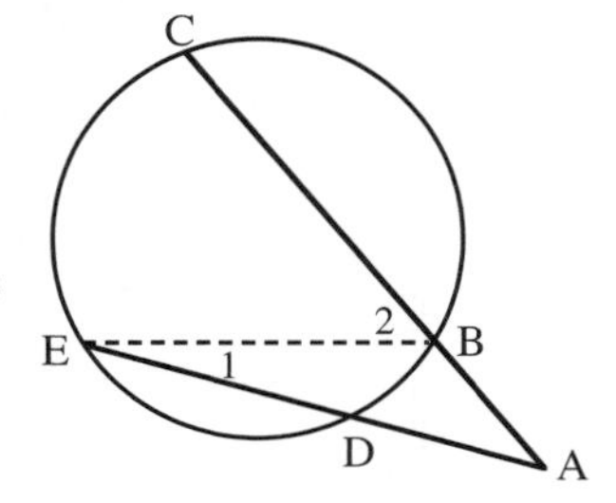


a) If $m\widehat{CE} = 100^\circ$ and $m\widehat{BD} = 30^\circ$, find $m\angle 1$ and $m\angle 2$.

b) Use $\Delta ABE$ to find $m\angle A$.

c) Use the example you solved in parts (a) and (b) as the basis to write a general procedure for finding the measure of any angle formed by two secant segments.

**3D-46.** In any circle, the diameters and radii are all equal. Sketch a circle, draw two diameters, and label each radius 5. If you multiply the two radii of one diameter (5 x 5) you get the same result for the two radii of the other diameter. Let's explore whether this relationship is true for <u>any</u> two intersecting chords.

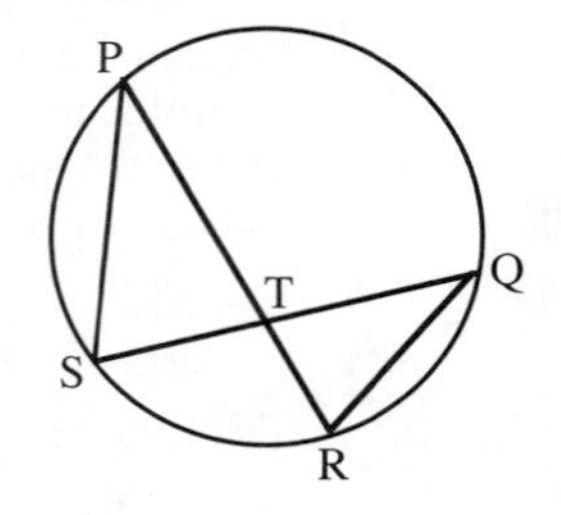


a) Chords $\overline{PR}$ and $\overline{SQ}$ intersect at T. <u>Carefully</u> measure the two pieces of chord $\overline{PR}$ in millimeters and multiply them together. Do the same for the two pieces of chord $\overline{SQ}$. What do you observe? Carefully draw one or two more circles of different sizes and repeat this experiment. Then go on to the rest of the problem to prove your observation.

b) Use what you know about inscribed angles and intercepted arcs to prove that $\Delta RQT \sim \Delta SPT$.

c) Since the triangles are similar, you know that $\frac{RT}{TS} = \frac{QT}{TP}$. Cross-multiply this proportion and the proof is complete.

**3D-47.** A similar demonstration and proof can be constructed for the lengths of secant segments (see problem 3D-50). The results of your work in the previous three problems are summarized in the box below. Add the theorems to your tool kit, then practice them in the next problem.

**ANGLE-CHORD-SECANT THEOREMS**

$$m\angle 1 = \frac{1}{2}(m\overset{\frown}{AD} + m\overset{\frown}{BC})$$

$$AE \cdot EC = DE \cdot EB$$

$$m\angle P = \frac{1}{2}(m\overset{\frown}{RT} - m\overset{\frown}{QS})$$

$$PQ \cdot PR = PS \cdot PT$$

3D-48. Use the figures in the box on the previous page to solve the following problems.

a) If $m\overset{\frown}{BC} = 54°$ and $m\overset{\frown}{AD} = 78°$, find $m\angle 1$.

b) If $m\overset{\frown}{RT} = 116°$ and $m\overset{\frown}{QS} = 42°$, find $m\angle P$.

c) If AE = 9, EC = 6, and BE = 4, find DE.

d) If PQ = 8, PR = 28, and PT = 24, find PS.

e) If $m\angle 2 = 146°$ and $m\overset{\frown}{AB} = 50°$, find $m\overset{\frown}{DC}$.

f) If DE = EB, EC = 14, and AE = 6, find DE.

3D-49. Read the following information about how to add and subtract algebraic fractions with like denominators. Add it to your tool kit, then practice it with the six problems below the box.

**ADDING AND SUBTRACTING FRACTIONS WITH LIKE DENOMINATORS**

**Add and simplify:**

Since the denominators are the same, add the numerators and simplify:

$$\frac{1}{8} + \frac{5}{8} = \frac{1+5}{8} = \frac{6}{8} \text{ or } \frac{3}{4}$$

$$\frac{4}{x+3} + \frac{5}{x+3} = \frac{4+5}{x+3} = \frac{9}{x+3}$$

**Subtract and simplify:**

$$\frac{3x^2 + 5x - 3}{x^2 + 4x + 3} - \frac{2x^2 + 2x - 5}{x^2 + 4x + 3}$$

$$= \frac{(3x^2 + 5x - 3) - (2x^2 + 2x - 5)}{x^2 + 4x + 3}$$

Change the signs: $= \dfrac{3x^2 + 5x - 3 - 2x^2 - 2x + 5}{x^2 + 4x + 3}$

Combine like terms: $= \dfrac{x^2 + 3x + 2}{x^2 + 4x + 3}$

Factor and simplify: $= \dfrac{(x+1)(x+2)}{(x+1)(x+3)} = \dfrac{x+2}{x+3}$

Note that in all cases we assume the denominator is not equal to zero.

a) $\dfrac{2x}{3x-1} + \dfrac{x-1}{3x-1}$

b) $\dfrac{x+3}{5} - \dfrac{x+4}{5}$

c) $\dfrac{6}{x+3} + \dfrac{2x}{x+3}$

d) $\dfrac{x^2}{x-5} - \dfrac{25}{x-5}$

e) $\dfrac{2x^2 + 4x + 8}{x-4} - \dfrac{x^2 + 3x + 28}{x-4}$

f) $\dfrac{a^2}{a^2 + 2a - 8} + \dfrac{10a + 24}{a^2 + 2a - 8}$

3D-50. Use the figure at right to prove that $\Delta PRS \sim \Delta PTQ$, then use this fact to demonstrate that $PQ \cdot PR = PS \cdot PT$. Refer to problem 3D-46 if you need help getting started.

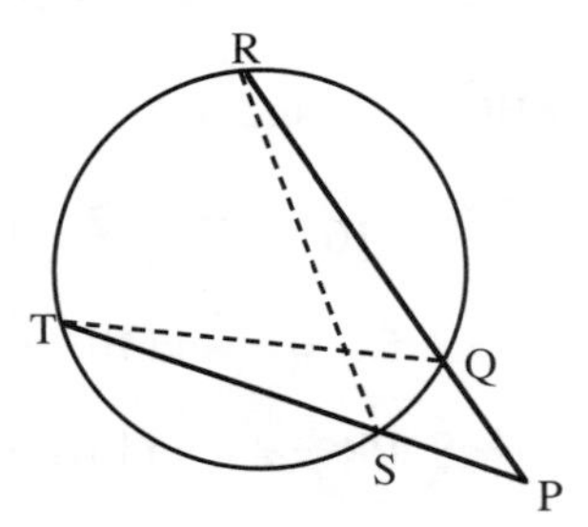

3D-51. A triangular prism has a volume of 600 cubic cm. The base is a right triangle with a hypotenuse of length 17 cm and one leg with length 15 cm.

a) Draw the figure.

b) Find the height of the prism.

c) Find the surface area of the prism.

d) Find all three angles of the triangle in the base of the prism.

3D-52. Find the area of the region at right. All dimensions are in inches.

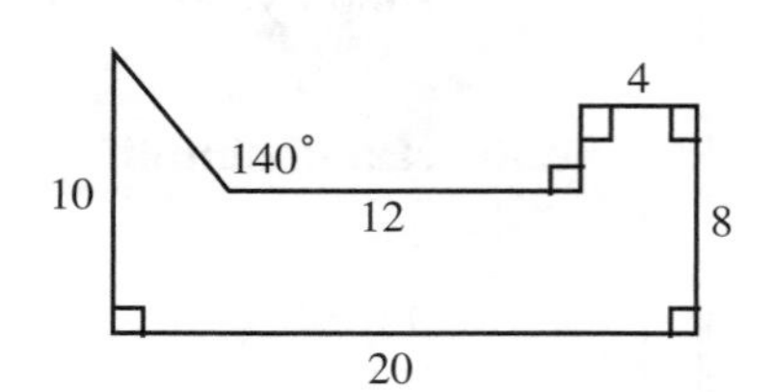


3D-53. Set up a pair of xy-coordinate axes.

a) Graph $y < -x^2 + 4$.

b) Graph $y > x + 1$.

c) Darken the intersection.

d) Approximate the area of the shaded region.

## THE EQUATION OF A CIRCLE

3D-54. For the next activity you will need to work with a partner or with your team. After you complete parts (a) through (d) on the resource page, read the definition in the box below and add it to your tool kit.

a) You are going to graph the equation $x^2 + y^2 = 1$. To do this, you must first solve the equation for y. Once you have done this, get a resource page from your teacher.

b) Complete the table on the resource page by sharing the work within your team. Be sure to look for patterns. You should not have to do very much calculating. Then plot the points on the resource page graph.

c) In a sentence, explain why the value of x cannot be larger than 1.

d) Can y be greater than 1? Try two values for y greater than 1 and see if you can find an x that will make the expression true. Explain why y can or cannot be greater than 1.

> The **EQUATION OF A CIRCLE** where r is the radius of the circle and its center is at the origin--(0, 0)--is:
>
> $$\mathbf{x^2 + y^2 = r^2}$$

3D-55. In the previous problem the radius was 1. Generally we will write the square of the radius in equations. For example, if the radius is 7, 49 would be the number on the right side of the equation, not $7^2$. Each of the equations below represents a different circle with the same center, (0, 0). Find the radius of each circle. If the radius is a radical be sure to simplify your answer.

1) $x^2 + y^2 = 4$

2) $x^2 + y^2 = 25$

3) $x^2 + y^2 = 100$

4) $x^2 + y^2 = 32$

3D-56. Get another sheet of graph paper, or use the back of the one you have.

a) Graph $y = \frac{4}{3}x$. Remember, this is a line.

b) On the same set of axes, carefully graph the circle $x^2 + y^2 = 25$.

c) Label the point of intersection in the first quadrant A and in the third quadrant B. What are the coordinates of the points A and B?

d) You have just found the intersection of $x^2 + y^2 = 25$ and $y = \frac{4}{3}x$. What is the greatest number of points in which a circle and a line can intersect?

3D-57. Sketch the graph of each circle on one set of axes. Next to each circle indicate its radius, circumference, and area.

a) $x^2 + y^2 = 9$  b) $x^2 + y^2 = 16$  c) $x^2 + y^2 = 50$

d) Now shade the area between the first and second circles. Looks like a dart board doesn't it? If you threw a dart at this board, do you think you are more likely to hit the shaded or unshaded region? Explain.

e) Find the area of the donut-shaped region outside the circle in part (b) and inside the circle in part (c).

3D-58. The following box contains examples of adding and subtracting algebraic fractions with unlike denominators. Add this information to your tool kit, then do the practice problems below the box.

**ADDING AND SUBTRACTING RATIONAL EXPRESSIONS**

| | |
|---|---|
| The Least Common Multiple of $(x + 3)(x + 2)$ and $(x + 2)$ is $(x + 3)(x + 2)$. | $\frac{4}{(x+3)(x+2)} + \frac{2x}{x+2}$ |
| The denominator of the first fraction already is the Least Common Multiple. To get a common denominator in the second fraction, multiply the fraction by $\frac{x+3}{x+3}$, a form of one (1). | $= \frac{4}{(x+3)(x+2)} + \frac{2x}{x+2} \cdot \frac{(x+3)}{(x+3)}$ |
| Multiply the numerator and denominator: | $= \frac{4}{(x+3)(x+2)} + \frac{2x(x+3)}{(x+3)(x+2)}$ |
| Distribute the numerator. | $= \frac{4}{(x+3)(x+2)} + \frac{2x^2+6x}{(x+3)(x+2)}$ |
| Add and simplify. | $= \frac{2x^2+6x+4}{(x+3)(x+2)} = \frac{(2x+2)(x+2)}{(x+3)(x+2)} = \frac{(2x+2)}{(x+3)}$ |

Note that in all cases we assume the denominator is not equal to zero.

a) $\frac{3}{(x-4)(x+1)} + \frac{6}{x+1}$  b) $\frac{5}{2(x-5)} - \frac{3x}{x-5}$

c) $\frac{x}{x^2-x-2} - \frac{2}{x^2-x-2}$  d) $\frac{3x}{x^2+2x+1} + \frac{3}{x^2+2x+1}$

e) $\frac{y^2}{y+4} - \frac{16}{y+4}$  f) $\frac{x+2}{x^2-9} - \frac{1}{x+3}$

3D-59. Graph the following inequalities on the same set of axes. Darken in the overlapping region.

a) $y < \sqrt{16 - x^2}$

b) $y > \frac{1}{4}x^2 - 4$

c) Is the shaded region a circular disk? Justify your answer.

3D-60. A rectangle is drawn inside a circle so that each vertex of the rectangle is on the circle. The rectangle has dimensions 8 cm by 6 cm.

a) How is the diagonal of the rectangle related to the circle? Explain why.

b) Find the area of the region inside the circle but outside the rectangle. Show all subproblems.

3D-61. Use the figures at right to solve the following problems.

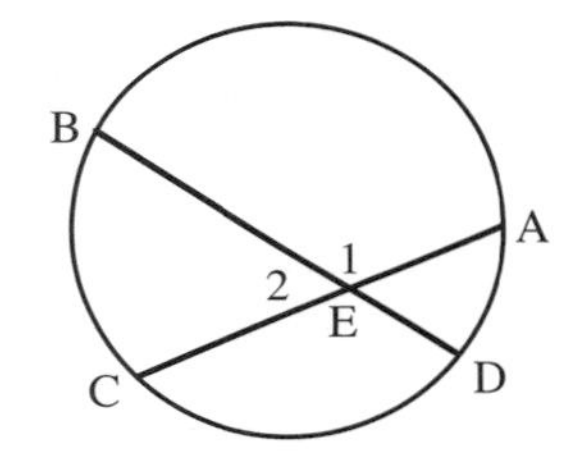


a) If $m\angle 1 = 134°$ and $m\overset{\frown}{CD} = 46°$, find $m\overset{\frown}{AB}$.

b) If $m\overset{\frown}{CB} = 86°$ and $m\overset{\frown}{AD} = 20°$, find $m\angle 1$.

c) If AE = 3.4", EC = 5.9", and DE = 1.3", find BE.

d) If $m\overset{\frown}{RT} = 126°$ and $m\overset{\frown}{SQ} = 32°$, find $m\angle P$.

e) PQ = 3", PR = 10", and PS = 4", find PT.

f) If QP = PS, PR = 12", TS = 7", find PS.

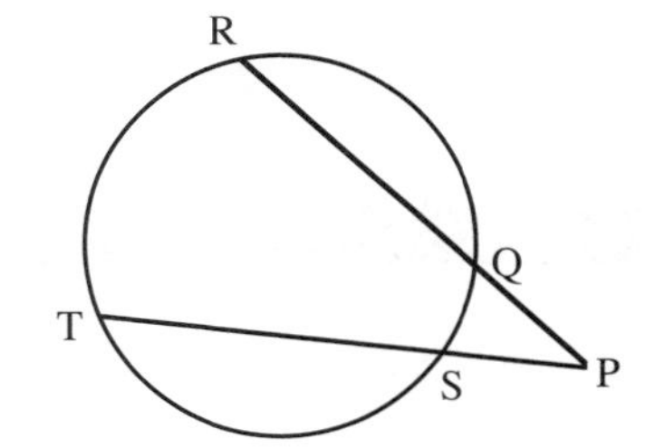

3D-62. Juan leaves from Cactus Corners driving straight east. At the same time, Denise takes the road going straight north and drives 20 miles per hour faster than Juan. After one hour they are 80 miles apart. How fast did Denise drive? Draw a diagram and write an equation.

3D-63. If $\tan \angle A = \tan \angle D$, are the triangles similar? Prove your answer.

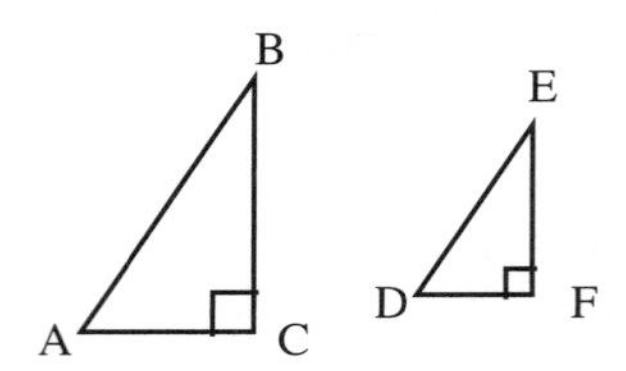

3D-64. If $\angle B$ and $\angle D$ are right angles, $\overline{AB} \cong \overline{AD}$, and $\overline{DC} \cong \overline{BC}$, is ABCD a rectangle? Prove your answer.

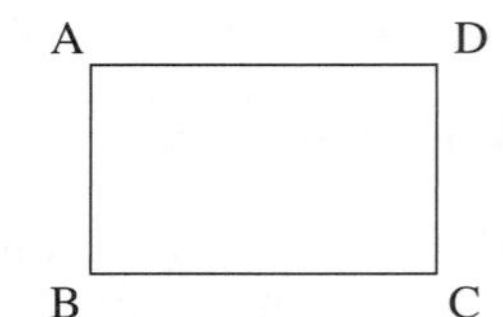

3D-65. Solve for x.

a) $x^2 - 8x - 20 = 0$

b) $\frac{5}{3x} + 6 = \frac{3}{4} - \frac{7}{2x}$

3D-66. Simplify:

a) $(\sqrt{2} + 3)(\sqrt{6} - 8\sqrt{3})$

b) $\frac{2\sqrt{6}}{5\sqrt{10}}$

## COURSE CONCLUSION AND TOOL KIT REVIEW

3D-67. COURSE REFLECTION

You have finally arrived at the end of this geometry course! Reflect back on how much you have learned throughout this year. Write a paragraph or short essay addressing your year in this course. The questions below are merely suggestions to get you started.

What topics were difficult for you to learn at first? What learning are you most proud of? Is geometry what you thought it would be? Have the study teams helped your understanding? What talent are you most proud of? How have you helped others learn? Have your study habits changed? Have your feelings about math changed? What topics are still difficult for you? What are your goals for your next math course?

**3D-68.** A LETTER OF ADVICE

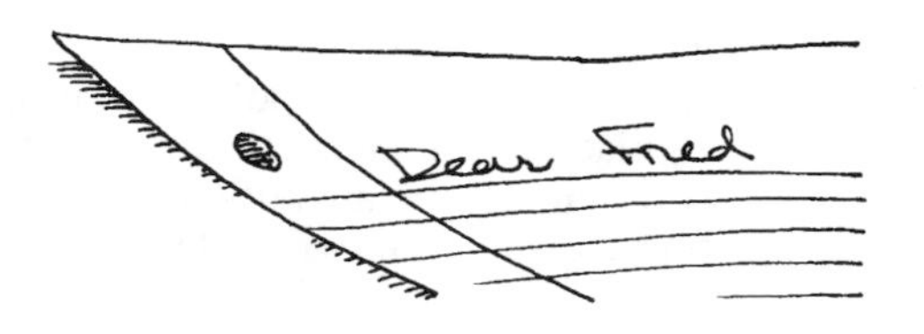


Write a letter of advice to a student entering this geometry course for the first time. Include in this letter how he or she can be successful in this course. If you could start over, what would you do differently? What types of things helped you succeed in this class? What tools were important? What general advice would you give this student?

**3D-69.** TOOL KIT CLEAN-UP

Examine the elements in your tool kit from Units 7 - 12. You should create a consolidated tool kit representing the entire course of geometry, including the algebra review topics.

Include items you value the most in your tool kit. Spend time on this. It is a good way to prepare for your final exam. You may also want to use it as a resource tool kit for next year!

a) Examine the list of tool kit entries from this unit. Check to be sure you have all of them. Add any you are missing.

b) Identify which concepts in your complete tool kit (Units 7 - 12) you understand well.

c) Identify the concepts you still need to work on to master.

d) Choose entries to create a Unit 0 - 12 tool kit that is brief, clear, and useful. You may want to consolidate or shorten some entries.

- slant height
- the equation of a circle
- angle-chord-secant theorems
- multiplying, adding, and subtracting rational expressions

3D-70. Justin, a collector of polygons, has a square and an equilateral triangle, both of which have the same perimeter. If the area of the square is 36 sq. cm., find the area of the triangle. Draw each figure, then identify the subproblems and solve them.

3D-71. A cube with edge length 1 cm is sliced through exactly three vertices, cutting off a right triangular based pyramid.

a) Make a sketch of this pyramid and label any lengths you know or can calculate.

b) Draw pictures of each face of the pyramid and calculate the area of each.

c) Using the areas you calculated above, square the area of each of the three smaller regions and add them together. Then square the area of the largest face. What do you notice?

3D-72. A circular can has a height equal to its diameter and holds one liter (1000 cu. cm).

a) What is its height?

b) If the can holds 8 liters, what is its height?

c) If the can holds V liters, what is its height?

3D-73. You are given two cubes.

a) Each edge of cube B is double that of cube A. What is the volume of cube B compared to the volume of cube A?

b) Now suppose the volume of cube B is double that of cube A. Find the ratio of the edge of cube B to the edge of cube A.

3D-74. Consider the equation $y = -2x^2 + 13x - 15$.

a) Where does the graph of this equation cross the x-axis?

b) Where does it cross the y-axis?

c) Make a sketch of this graph.

3D-75. Barbara and Marta have balloons which are similar in shape and have a red stripe around them. The only difference in their balloons is that the diameter of Marta's balloon is 1.6 times larger than the diameter of Barbara's.

a) What is the ratio of the surface areas of the two balloons?

b) What percent larger is the surface area of Marta's balloon?

c) What is the ratio of volumes of the two balloons?

d) What percent larger is the volume of Marta's balloon?

e) If Marta's balloon is blown up more than Barbara's, but they are the same otherwise, who's balloon is more likely to pop?

3D-76. You have the same situation as the previous problem except that now the volume of Marta's balloon is double the volume of Barbara's.

a) What is the ratio of the length of the stripe on Marta's balloon compared to Barbara's?

b) What is the ratio of the surface area of Marta's balloon compared to Barbara's?

3D-77. Set up a pair of xy-coordinate axes.

a) Make a table for $y = 2^x$. Use x = -3, -2, -1, 0, 1, 2, 3, 4.

b) Plot your points and connect them to make a smooth curve.

c) Graph $y \leq 2^x$.

3D-78. Begin with a regular 14-gon and connect alternate vertices to get a 7-gon. Is this 7-gon regular? Prove it.

3D-79. Your team has gone into the box business. Your company, "Boxes R Us," designs and builds the appropriate sized boxes for your clients. Your goal is always to design the correct sized box and to minimize the cost for your client. Your newest client is the Sweetness Sugar Company which processes sugar cane into sugar cubes. The sugar cubes each measure 1 cm on a side. Each box of sugar cubes contains 144 cubes. Your company needs to determine what size box to use to package the cubes. The most appropriate cardboard for these boxes costs 0.1 of a cent ($0.001) per square centimeter. Determine the exact dimensions of the box that will minimize your cost of packaging 144 sugar cubes. What will the box cost? You must show how you reach your solution as to what the best dimensions are, along with a written description in a letter to the president of the Sweetness Sugar Company.

3D-80. You will need a transparent rectangular tank (such as a fish tank) and a brick to do this problem.

a) Fill the tank about two-thirds full of water.

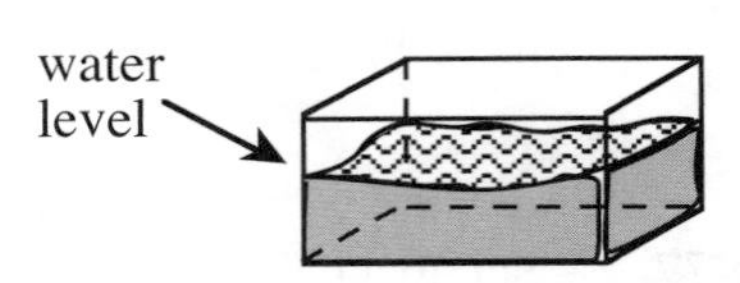


b) Find the dimensions of the brick and the rectangular prism formed by the water (measure to the nearest mm). Draw a sketch and label it.

c) Find the volume of the water and the volume of the brick.

d) Place a piece of tape on the tank at the level of the water. Gently put the brick into the tank. What happened to the level of the water? This is called **displacement** of water. Place another piece of tape at the new water level and measure the distance between the tapes. Make sure you mark the new water level with the bottom of the tape.

e) What shape does the displaced water have? Make a sketch and label any lengths you know.

f) Find the volume of the displacement. How does it compare to the volume of the brick? Write a conjecture about the relationship between the volume of the brick and the volume of water that it displaces.

g) Try this same experiment with an irregular shaped solid (rock, bottle, etc.). Make sure that the solid will sink to the bottom. Write a sentence or two about how this process could be used to find the volume of any object.

3D-81. A block of ice is placed into an ice chest containing water and it causes the water level to rise 4 cm. The base of the ice chest is 35 cm x 50 cm. When ice floats in water, one-eighth of its volume floats above the water level and seven-eighths floats beneath the water level. What is the volume of the block of ice? Draw a sketch.

3D-82. Irving wants to paint the outside of his house. He will not paint doors, windows, or the roof. Views of Irving's house are shown below with dimensions, but they are not drawn to scale.

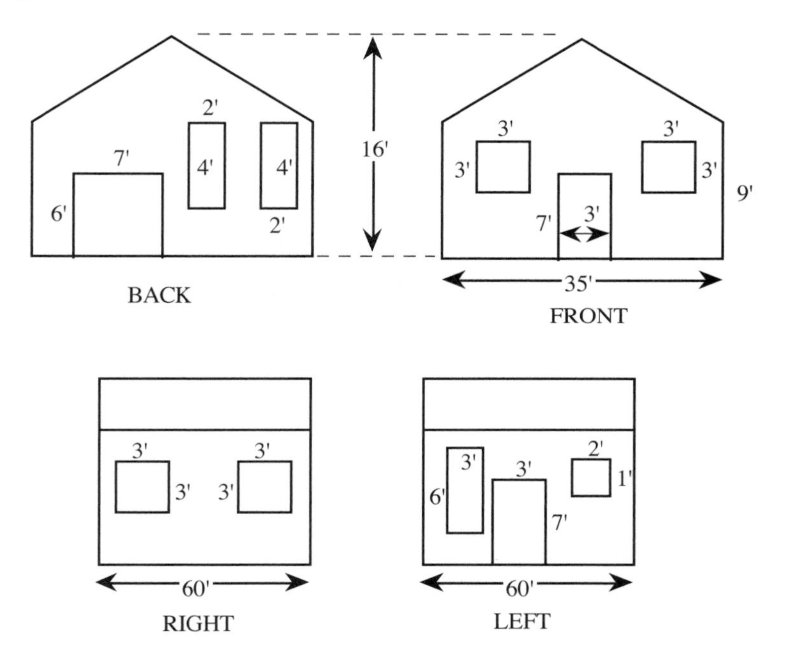


a) Find the surface area to be painted.

b) If one gallon of paint covers 250 square feet, how many gallons of paint does he need? Paint is sold by the gallon.

c) If one gallon of paint costs \$19.95, how much including $7\frac{3}{4}\%$ sales tax, will it cost Irving to paint his house?

d) Irving's dog, Sneaker, knocks over a can and spills a $\frac{1}{2}$ gallon of paint. Will Irving still have enough paint? Justify your answer.

# APPENDIX

# *Constructions*

Suppose as you came to school tomorrow, something strange happened. Suppose all your modern tools that you use in your math class were suddenly gone. No computers with powerful software, no calculators, not even rulers and protractors. Do you think your math class would continue as it had been? Could you do the problems you have been doing? Could you do something as simple as drawing a $60°$ angle?

Surprising as it may seem, there was a time when math was taught without all of our modern conveniences. As you might have guessed, that math class did look different than yours. In fact, hundreds of years ago, the ancient Greeks would try to answer certain math questions and draw geometric figures using only a straight edge and compass. A typical question might have been: can you find the midpoint of a segment using only a straightedge and compass?

When using a straightedge and compass to solve problems you produce a **CONSTRUCTION**. In these constructions, a straightedge is not a ruler (although you will most likely use a ruler as your straight edge). Finding the midpoint of a segment with a ruler would be easy. In constructions, a straightedge is just that: a straight edge, with no markings or measurements on it. A 3x5 card makes a good straightedge.

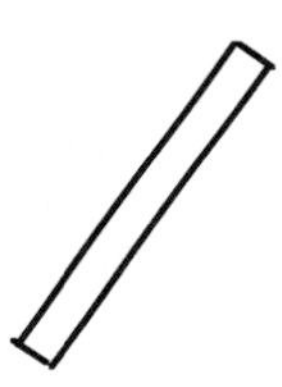

A-1. Explain why using a ruler to find the midpoint of a segment is easy while using a straightedge wouldn't be.

A-2. Experiment with your compass. What does a compass do?

A-3. Since a compass draws circles or parts of circles, many of the terms associated with constructions are terms about circles. The point of the compass is the **center** of the circle. The distance from the center to the pencil is the **radius** (pl. **radii**) of the circle. Choose and label a point O on your paper. Draw three different circles which have point O as the center but have different radii.

It may seem as if all we will be able to do is draw straight lines and circles, but there is much more possible. For instance, the compass lets us make congruent segments. Here's how:

A-4. Suppose you want to construct $\overline{CD}$ which is congruent to $\overline{AB}$. Remember! You can't just measure $\overline{AB}$ with a ruler! Discuss with your team how you might do this. If you need help, do parts (a) through (c) below.

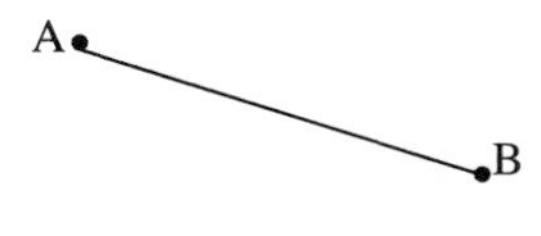


a) On your paper, use the straightedge to draw a ray with endpoint C.

b) Place your compass point (the center) on point A and open the compass to reach point B. You now have the radius of the compass set to the length of $\overline{AB}$.

c) Keeping your compass set with this radius, place the center at point C on the ray. Make an arc crossing the ray. Label this point D.

A-5. Explain completely why $\overline{AB}$ and $\overline{CD}$ of the previous problem are congruent.

A-6. Construct $\overline{EF}$ which is three times as long as $\overline{AB}$ .

In constructions, there is no guess work, approximating, or "eyeballing." Each construction should follow a set of steps that anyone else could follow that would produce the same results. As you develop your constructions, or as you follow the steps given, it might be wise to be recording your method, along with a diagram, in your tool kit. Let's look at a construction for congruent angles.

A-7. Discuss with your team how you might construct an angle Y congruent to angle X at right. If you need help, do parts (a) through (d) below.

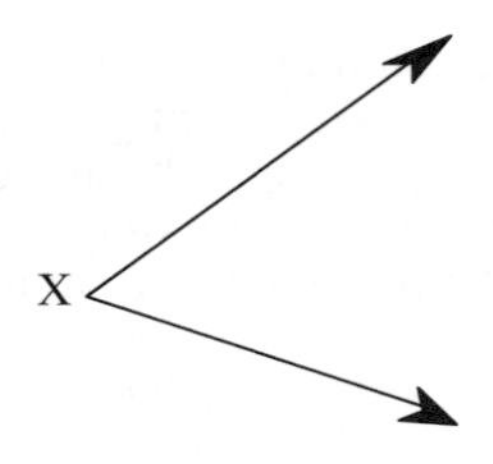


a) On your paper, draw a ray with endpoint Y.

b) With your compass point at X, draw an arc which crosses both sides of angle X. (Label these points A and B.) Use the same radius (setting) to draw a similar arc with center Y. Make sure this arc crosses the ray as shown below.

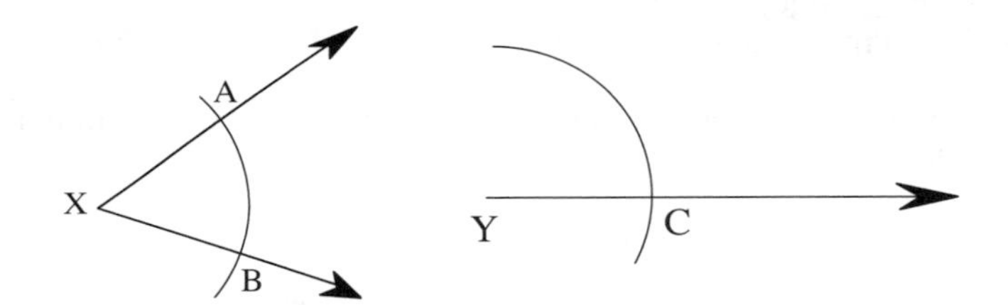


c) Place your compass point on B and adjust your compass to reach point A. Keeping this setting, place your compass point at C and make an arc crossing the first arc.

Label this point of intersection D and use your straightedge to draw $\overrightarrow{YD}$ .

d) Explain why ∠DYC is congruent to ∠AXB.

A-8. Construct two angles: first, ∠D congruent to ∠W, then ∠T so that it is 2m∠W.

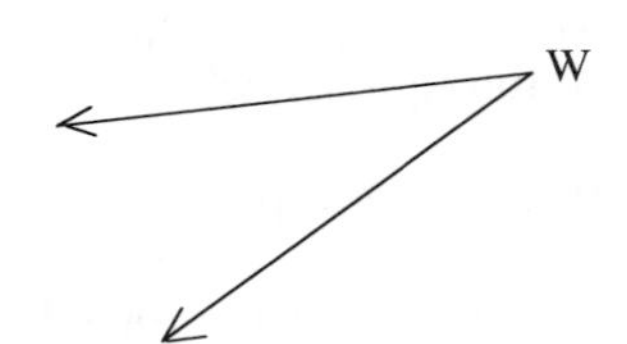

A-9. We have learned that there are several ways to prove triangles congruent.

a) What are the congruence patterns we have used?

b) These methods for proving triangles congruent also give us "guidelines" for constructing congruent triangles. Explain how we can use SSS to construct a new triangle congruent to $\Delta$IND at right.

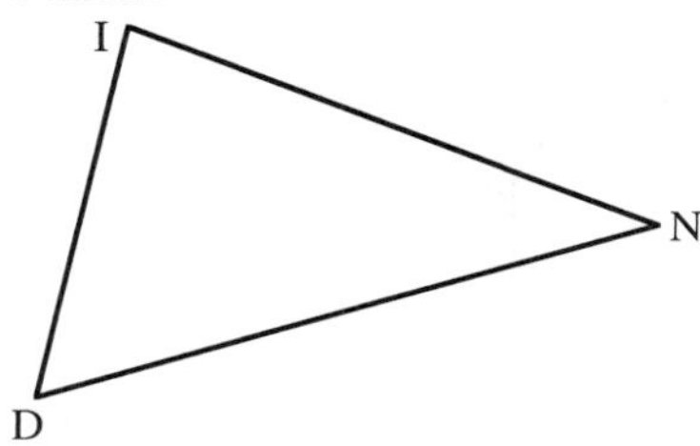


c) Construct $\Delta$CAL congruent to $\Delta$IND by SSS.

d) Construct $\Delta$ORE congruent to $\Delta$IND by SAS.

A-10. Mark a point X on your paper and use your compass to draw a circle with center X. Choose any appropriate radius. Mark any point Y on ⊙X and draw a circle with the same radius with center Y. The circles should intersect in two points. Label one of the points Z. What kind of triangle is $\Delta$XYZ? Prove your answer.

A-11. With your compass draw a circle with any appropriate radius. Using the same compass setting (radius), mark off successive points around the circle as indicated in the figure at right. You should end up back where you started. (If you didn't, try again and be more careful.) Connect each arc mark you made to the center of the circle and then each arc mark to its adjacent arc mark. You should have six triangles.

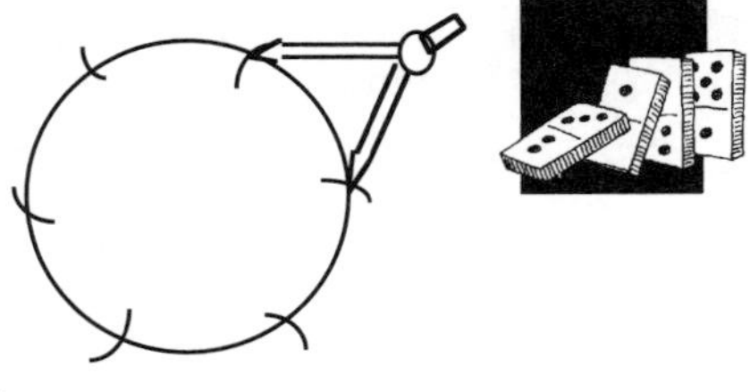

a) What kind of triangles did you construct? Prove that they are that type.

b) Construct a regular hexagon. Prove that it is regular.

A-12. Constructions can be fun and challenging to do, yet, they are not always done in math courses anymore. Why do you think that is? What problems are there with doing constructions?

A-13. Look back at the two circles you constructed in A-10. Circles X and Y intersected at two points. One was labeled Z. Label the other W and use your straightedge to connect W and Z. What appears to be true about $\overline{WZ}$ and $\overline{XY}$?

A-14. Based on what you did in the previous problem, construct the midpoint of $\overline{TS}$ and label it E.

A-15. This method of finding the midpoint actually does more than just locating the midpoint. You are, in fact, constructing the perpendicular bisector of the segment. Start with a segment congruent to $\overline{PQ}$ below.

a) Using the same radius, construct two circles, one with center P and the other with center Q. You can use any radius but the two circles must intersect in two points. Label these two points R and S and connect them with your straightedge.

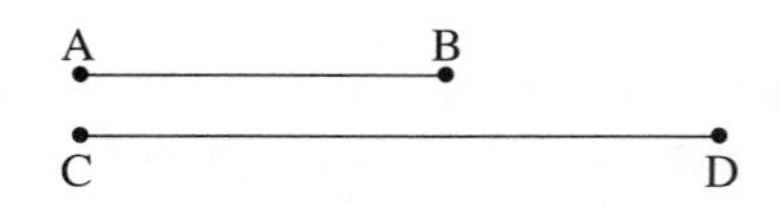


b) What kind of quadrilateral is PRQS? Prove it.

c) Prove that $\overline{RS}$ is the perpendicular bisector of $\overline{PQ}$.

A-16. Construct a right triangle with legs congruent to $\overline{AB}$ and $\overline{CD}$ at right.

A ——— B

C ————— D

A-17. Now that we can bisect a segment, we can move on to bisecting an angle. Discuss with your team a method to do this. If you need help, read parts (a) through (c) below.

a) With the compass point at D, draw an arc which crosses both sides of $\angle D$. Label the points where the arc crosses A and V.

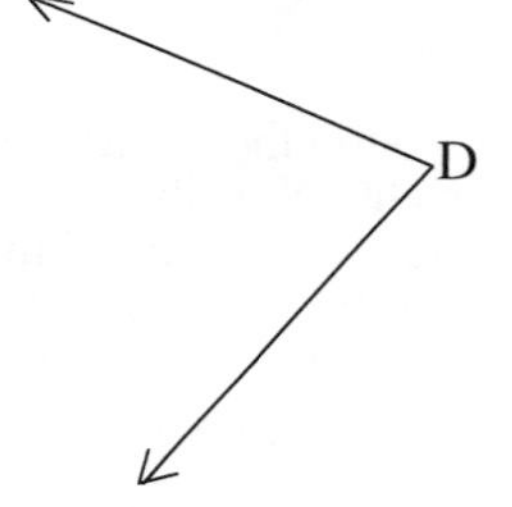


b) With A and V as the centers, draw two arcs that have the same radius and that intersect in the interior of $\angle D$. Label the point where these two arcs cross I.

c) Use your straightedge to connect D and I.

A-18. Prove that the construction in the last problem gives the angle bisector.

A-19. Construct a $45°$ angle.

A-20. Construct a $30°$-$60°$-$90°$ triangle.

A-21. In the figure at right, if l ∥ m, what is true about ∠1 and ∠2?

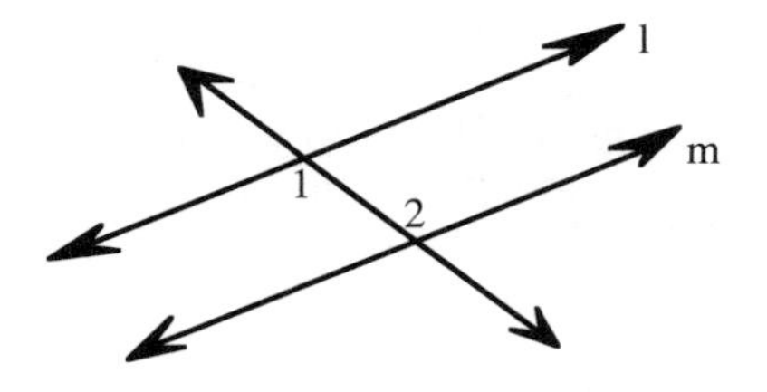


A-22. In the same figure, if ∠1 ≅ ∠2, what is true about lines l and m?

A-23. Use the ideas of the previous two problems to construct a line parallel to n and passing through point P. Hint: start by drawing a transversal through P intersecting line n.

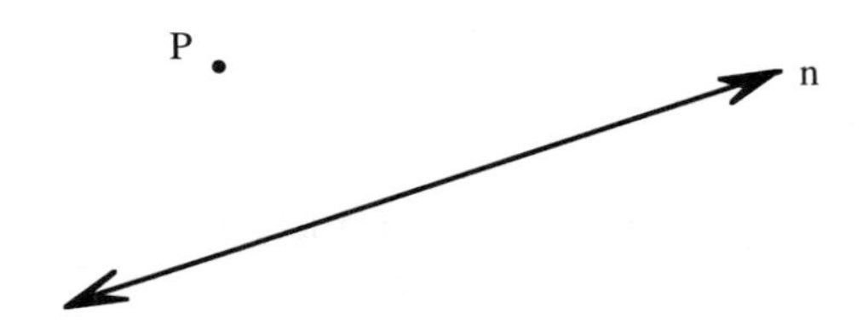


A-24. Construct a square.

A-25. Construct a regular octagon.

A-26. We have already seen how to construct the perpendicular bisector of a segment, but what if we don't have a segment? Suppose we just want to construct a perpendicular to a line at a point on the line. In the figure at right, how can we construct a line perpendicular to line l and intersecting line l at the point P? Discuss this with your team. If you need help coming up with a method, do parts (a) through (c) below.

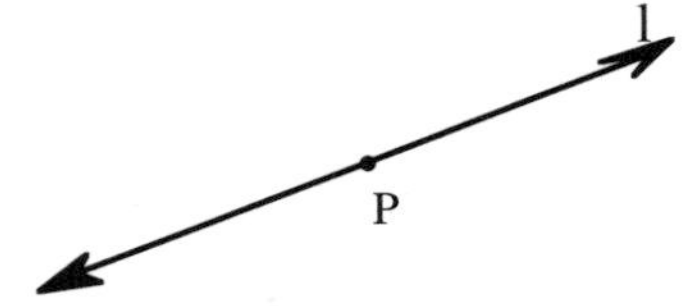


a) With P as the center and using the same radius, make an arc on either side of P. You've just created a segment which has point P as the midpoint. Label the segment $\overline{QR}$.

b) Place your compass center at Q and open the compass to reach beyond P. Make an arc above the line. Do the same thing **with the same setting** from R. Label the point where the two arcs cross S.

c) With your straightedge, connect S and P. This is your line perpendicular to l at point P.

A-27. Let's do a similar problem. This time the point P is not on line l. How can we construct a line through P perpendicular to line l? Discuss this with your team. If you need help coming up with a method, do parts (a) through (c) below.

a) Place your compass center on point P. Make a large arc which intersects line l in two places. Label these points S and T.

b) Place your compass center on S and make an arc on the other side of l from point P. Use the same setting to do the same thing from T. Label the point where these two arcs cross U.

c) Connect U and P. This is the perpendicular from point P to line *l*.

A-28. In ΔABC, construct a height (altitude) from A to the base $\overline{BC}$.

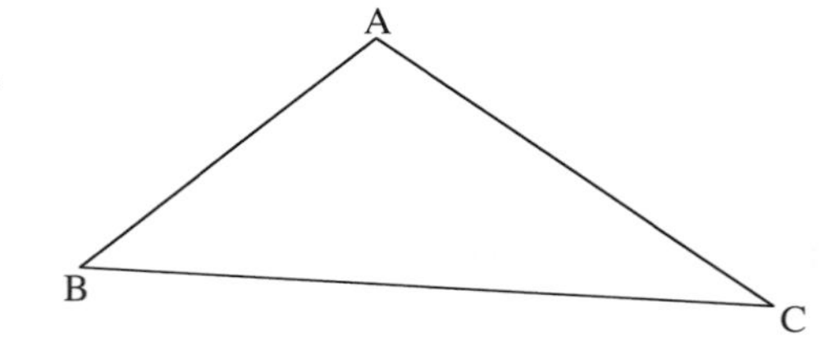


A-29. In ΔKAR, construct a height (altitude) from K to the base $\overline{AR}$. Need help? It might help to draw $\overleftrightarrow{AR}$.

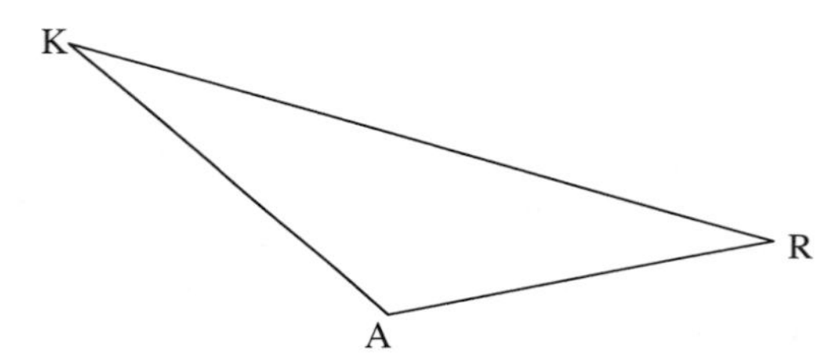


A-30. On a clean sheet of paper, draw a large triangle and label it ΔXYZ. (Each person in your team should have a different size and shape triangle.) Construct all three heights--one from each vertex. What do you notice about the three heights?

A-31. On a clean sheet of paper (preferably construction paper or cardstock), draw a large triangle and label it ΔABC. (Each person in your team should have a different size and shape triangle.) Using your straightedge and compass, find the midpoint of each side of the triangle. Connect each vertex A, B, and C to the midpoint on the opposite side. These three segments are called the **medians** of the triangle. What appears to be true about the medians? Something else is special about the medians. Research question: look up the term **centroid**. What is special about the centroid? Cut out your triangle and demonstrate this.

# GEOMETRY SKILL BUILDERS

## (Extra Practice)

## Introduction to Students and Their Teachers

Learning is an individual endeavor. Some ideas come easily; others take time--sometimes lots of time--to grasp. In addition, individual students learn the same idea in different ways and at different rates. The authors of this textbook designed the classroom lessons and homework to give students time--often weeks and months--to practice an idea and to use it in various settings. This section of the textbook offers students a brief review of 27 topics followed by additional practice with answers. Not all students will need extra practice. Some will need to do a few topics, while others will need to do many of the sections to help develop their understanding of the ideas. This section of the text may also be useful to prepare for tests, especially final examinations.

How these problems are used will be up to your teacher, your parents, and yourself. In classes where a topic needs additional work by most students, your teacher may assign work from one of the skill builders that follow. In most cases, though, the authors expect that these resources will be used by individual students who need to do more than the textbook offers to learn an idea. This will mean that you are going to need to do some extra work outside of class. In the case where additional practice is necessary for you individually or for a few students in your class, you should not expect your teacher to spend time in class going over the solutions to the skill builder problems. After reading the examples and trying the problems, if you still are not successful, talk to your teacher about getting a tutor or extra help outside of class time.

Warning! Looking is not the same as doing. You will never become good at any sport just by watching it. In the same way, reading through the worked out examples and understanding the steps are not the same as being able to do the problems yourself. An athlete only gets good with practice. The same is true of developing your algebra and geometry skills. How many of the extra practice problems do you need to try? That is really up to you. Remember that your goal is to be able to do problems of the type you are practicing on your own, confidently and accurately.

Two other sources for help with the algebra and geometry topics in this course are the *Parent's Guide with Review to Math 2 (Geometry)* and the *Parent's Guide with Review to Math 1 (Algebra 1)*. Information about ordering these supplements can be found inside the front page of the student text. These resources are also available free from the internet at *www.cpm.org*.

**Online homework help is underwritten by CPM at www.hotmath.com.**

### Skill Builder Topics

1. Writing and graphing linear equations
2. The Pythagorean Theorem
3. Area
4. Solving linear systems
5. Properties of angles, lines, and triangles
6. Linear inequalities
7. Multiplying Polynomials
8. Factoring Polynomials
9. Zero Product Property and quadratics
10. The Quadratic Formula
11. Triangle Congruence
12. Laws of Exponents
13. Proof
14. Radicals
15. Right triangle trigonometry
16. Ratio of similarity
17. Similarity of length, area, and volume
18. Interior and exterior angles of polygons
19. Areas by dissection
20. Central and inscribed angles
21. Area of sectors
22. Tangents, secants, and chords
23. Volume and surface area of polyhedra
24. Simplifying rational expressions
25. Multiplication and division of rational expressions
26. Addition and subtraction of rational expressions
27. Solving mixed equations and inequalities

## WRITING AND GRAPHING LINEAR EQUATIONS ON A FLAT SURFACE

#1

**SLOPE** is a number that indicates the steepness (or flatness) of a line, as well as its direction (up or down) left to right.

**SLOPE** is determined by the ratio: $\frac{\text{vertical change}}{\text{horizontal change}}$ between <u>any</u> two points on a line.

For lines that go **up** (from left to right), the sign of the slope is **positive.** For lines that go **down** (left to right), the sign of the slope is **negative**.

Any linear equation written as **y = mx + b**, where m and b are any real numbers, is said to be in **SLOPE-INTERCEPT FORM**. m is the **SLOPE** of the line. b is the **Y-INTERCEPT**, that is, the point (0, b) where the line intersects (crosses) the y-axis.

If two lines have the same slope, then they are parallel. Likewise, **PARALLEL LINES** have the same slope.

Two lines are **PERPENDICULAR** if the slope of one line is the negative reciprocal of the slope of the other line, that is, m and $-\frac{1}{m}$. Note that $m \cdot \left(\frac{-1}{m}\right) = -1$.

Examples: 3 and $-\frac{1}{3}$, $-\frac{2}{3}$ and $\frac{3}{2}$, $\frac{5}{4}$ and $-\frac{4}{5}$

Two distinct lines on a flat surface that are not parallel intersect in a single point. See "Solving Linear Systems" to review how to find the point of intersection.

### Example 1

Graph the linear equation $y = \frac{4}{7}x + 2$

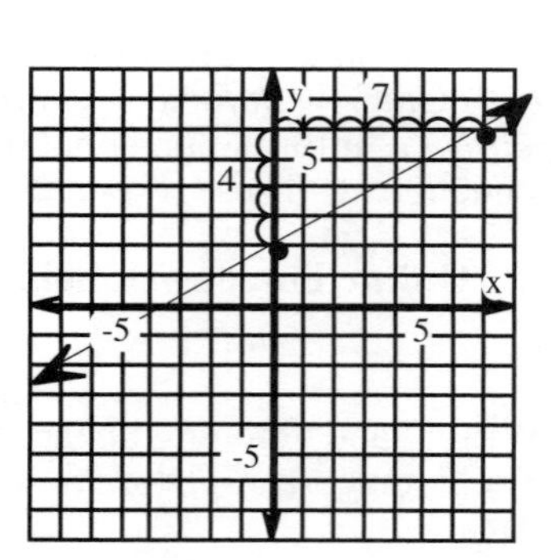


Using $y = mx + b$, the slope in $y = \frac{4}{7}x + 2$ is $\frac{4}{7}$ and the y-intercept is the point (0, 2). To graph, begin at the y-intercept (0, 2). Remember that slope is $\frac{\text{vertical change}}{\text{horizontal change}}$ so go up 4 units (since 4 is positive) from (0, 2) and then move right 7 units. This gives a second point on the graph. To create the graph, draw a straight line through the two points.

## Example 2

A line has a slope of $\frac{3}{4}$ and passes through (3, 2). What is the equation of the line?

Using $y = mx + b$, write $y = \frac{3}{4}x + b$. Since (3, 2) represents a point (x, y) on the line, substitute 3 for x and 2 for y, $2 = \frac{3}{4}(3) + b$, and solve for b.

$2 = \frac{9}{4} + b \Rightarrow 2 - \frac{9}{4} = b \Rightarrow -\frac{1}{4} = b$. The equation is $y = \frac{3}{4}x - \frac{1}{4}$.

## Example 3

Decide whether the two lines at right are parallel, perpendicular, or neither (i.e., intersecting). $5x - 4y = -6$ and $-4x + 5y = 3$.

First find the slope of each equation. Then compare the slopes.

| | | |
|---|---|---|
| $5x - 4y = -6$<br>$-4y = -5x - 6$<br>$y = \frac{-5x - 6}{-4}$<br>$y = \frac{5}{4}x + \frac{3}{2}$<br>The slope of this line is $\frac{5}{4}$. | $-4x + 5y = 3$<br>$5y = 4x + 3$<br>$y = \frac{4x + 3}{5}$<br>$y = \frac{4}{5}x + \frac{3}{5}$<br>The slope of this line is $\frac{4}{5}$. | These two slopes are not equal, so they are not parallel. The product of the two slopes is 1, not -1, so they are not perpendicular. These two lines are neither parallel nor perpendicular, but do intersect. |

## Example 4

Find two equations of the line through the given point, one parallel and one perpendicular to the given line: $y = -\frac{5}{2}x + 5$ and (-4, 5).

| | |
|---|---|
| For the parallel line, use $y = mx + b$ with the same slope to write $y = -\frac{5}{2}x + b$.<br>Substitute the point (–4, 5) for x and y and solve for b.<br>$5 = -\frac{5}{2}(-4) + b \Rightarrow 5 = \frac{20}{2} + b \Rightarrow -5 = b$<br>Therefore the parallel line through (-4, 5) is $y = -\frac{5}{2}x - 5$. | For the perpendicular line, use $y = mx + b$ where m is the negative reciprocal of the slope of the original equation to write $y = \frac{2}{5}x + b$.<br>Substitute the point (–4, 5) and solve for b.<br>$5 = \frac{2}{5}(-4) + b \Rightarrow \frac{33}{5} = b$<br>Therefore the perpendicular line through (-4, 5) is $y = \frac{2}{5}x + \frac{33}{5}$. |

Identify the y-intercept in each equation.

1. $y = \frac{1}{2}x - 2$
2. $y = -\frac{3}{5}x - \frac{5}{3}$
3. $3x + 2y = 12$
4. $x - y = -13$
5. $2x - 4y = 12$
6. $4y - 2x = 12$

Write the equation of the line with:

7. slope $= \frac{1}{2}$ and passing through (4, 3).
8. slope $= \frac{2}{3}$ and passing through (-3, -2).
9. slope $= -\frac{1}{3}$ and passing through (4, -1).
10. slope = -4 and passing through (-3, 5).

Determine the slope of each line using the highlighted points.

11. 12. 13.

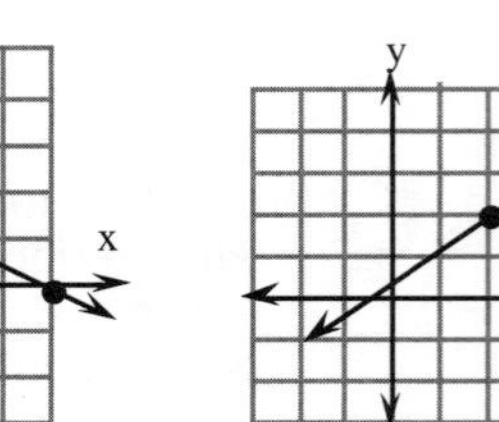

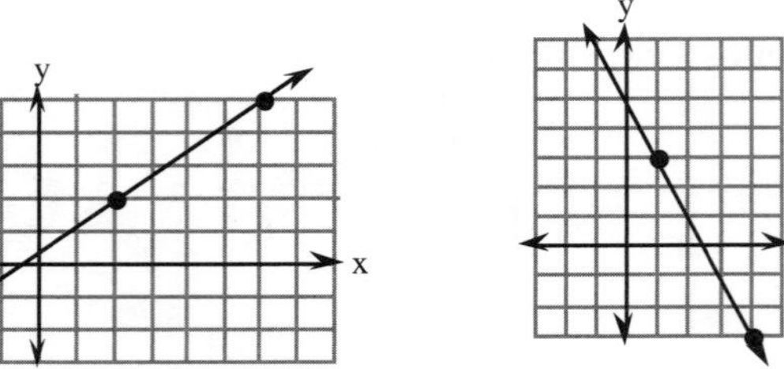

Using the slope and y-intercept, determine the equation of the line.

14. 15. 16. 17.

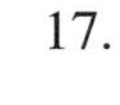

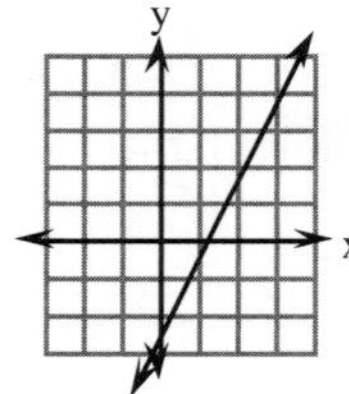

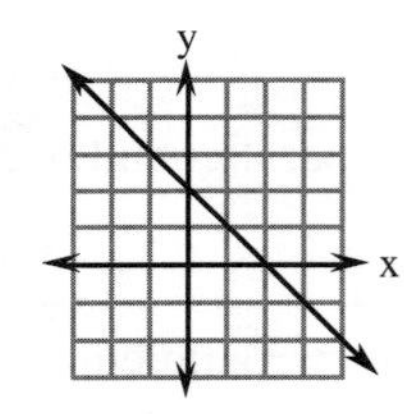

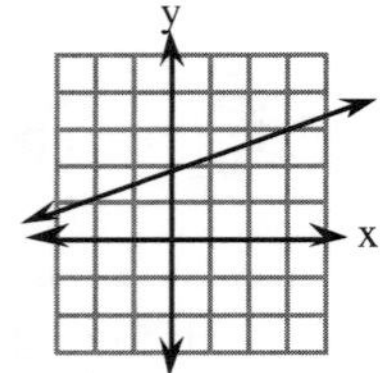

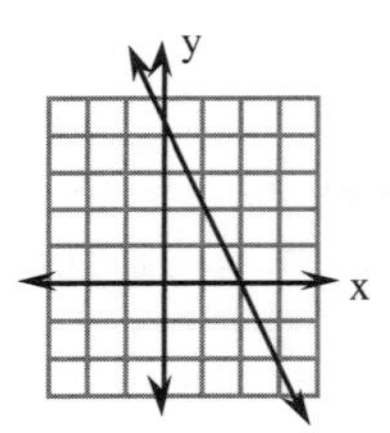

Graph the following linear equations on graph paper.

18. $y = \frac{1}{2}x + 3$
19. $y = -\frac{3}{5}x - 1$
20. $y = 4x$
21. $y = -6x + \frac{1}{2}$
22. $3x + 2y = 12$

State whether each pair of lines is parallel, perpendicular, or intersecting.

23. $y = 2x - 2$ and $y = 2x + 4$

24. $y = \frac{1}{2}x + 3$ and $y = -2x - 4$

25. $x - y = 2$ and $x + y = 3$

26. $y - x = -1$ and $y + x = 3$

27. $x + 3y = 6$ and $y = -\frac{1}{3}x - 3$

28. $3x + 2y = 6$ and $2x + 3y = 6$

29. $4x = 5y - 3$ and $4y = 5x + 3$

30. $3x - 4y = 12$ and $4y = 3x + 7$

Find an equation of the line through the given point and parallel to the given line.

31. $y = 2x - 2$ and $(-3, 5)$

32. $y = \frac{1}{2}x + 3$ and $(-4, 2)$

33. $x - y = 2$ and $(-2, 3)$

34. $y - x = -1$ and $(-2, 1)$

35. $x + 3y = 6$ and $(-1, 1)$

36. $3x + 2y = 6$ and $(2, -1)$

37. $4x = 5y - 3$ and $(1, -1)$

38. $3x - 4y = 12$ and $(4, -2)$

Find an equation of the line through the given point and perpendicular to the given line.

39. $y = 2x - 2$ and $(-3, 5)$

40. $y = \frac{1}{2}x + 3$ and $(-4, 2)$

41. $x - y = 2$ and $(-2, 3)$

42. $y - x = -1$ and $(-2, 1)$

43. $x + 3y = 6$ and $(-1, 1)$

44. $3x + 2y = 6$ and $(2, -1)$

45. $4x = 5y - 3$ and $(1, -1)$

46. $3x - 4y = 12$ and $(4, -2)$

Write an equation of the line parallel to each line below through the given point

47.

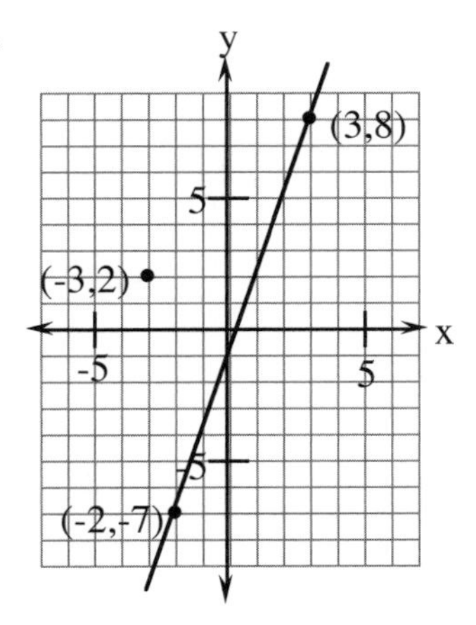


48.

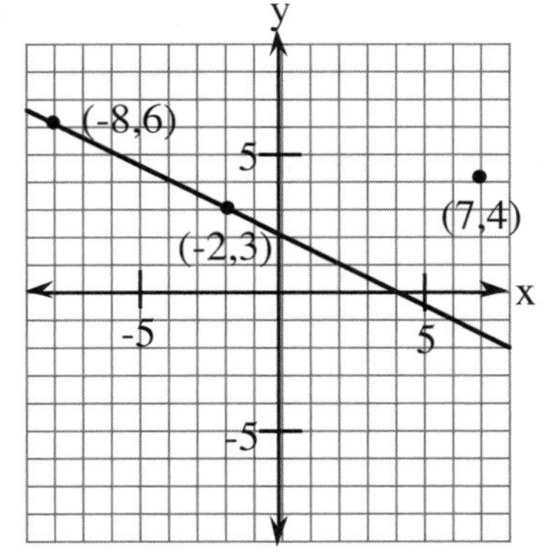


49. Write the equation of the line through $(7, -8)$ which is parallel to the line through $(2, 5)$ and $(8, -3)$

50. Write the equation of the line through $(1, -4)$ which is parallel to the line through $(-3, -7)$ and $(4, 3)$

## Answers

1. $(0, -2)$
2. $\left(0, -\frac{5}{3}\right)$
3. $(0, 6)$
4. $(0, 13)$
5. $(0, -3)$
6. $(0, 6)$
7. $y = \frac{1}{2}x + 1$
8. $y = \frac{2}{3}x$
9. $y = -\frac{1}{3}x + \frac{1}{3}$
10. $y = -4x - 7$
11. $-\frac{1}{2}$
12. $\frac{3}{4}$
13. $-2$
14. $y = 2x - 2$
15. $y = -x + 2$
16. $y = \frac{1}{3}x + 2$
17. $y = -2x + 4$
18. line with slope $\frac{1}{2}$ and y-intercept $(0, 3)$
19. line with slope $-\frac{3}{5}$ and y-intercept $(0, -1)$
20. line with slope $4$ and y-intercept $(0, 0)$
21. line with slope $-6$ and y-intercept $\left(0, \frac{1}{2}\right)$
22. line with slope $-\frac{3}{2}$ and y-intercept $(0, 6)$
23. parallel
24. perpendicular
25. perpendicular
26. perpendicular
27. parallel
28. intersecting
29. intersecting
30. parallel
31. $y = 2x + 11$
32. $y = \frac{1}{2}x + 4$
33. $y = x + 5$
34. $y = x + 3$
35. $y = -\frac{1}{3}x + \frac{2}{3}$
36. $y = -\frac{3}{2}x + 2$
37. $y = \frac{4}{5}x - \frac{9}{5}$
38. $y = \frac{3}{4}x - 5$
39. $y = -\frac{1}{2}x + \frac{7}{2}$
40. $y = -2x - 6$
41. $y = -x + 1$
42. $y = -x - 1$
43. $y = 3x + 4$
44. $y = \frac{2}{3}x - \frac{7}{3}$
45. $y = -\frac{5}{4}x + \frac{1}{4}$
46. $y = -\frac{4}{3}x + \frac{10}{3}$
47. $y = 3x + 11$
48. $y = -\frac{1}{2}x + \frac{15}{2}$
49. $y = -\frac{4}{3}x + \frac{4}{3}$
50. $y = \frac{10}{7}x - \frac{38}{7}$

## PYTHAGOREAN THEOREM

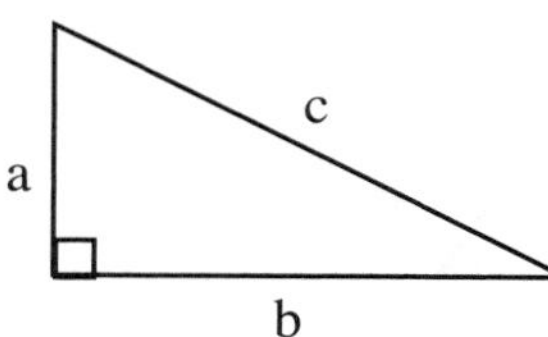


Any triangle that has a right angle is called a **RIGHT TRIANGLE**. The two sides that form the right angle, a and b, are called **LEGS**, and the side opposite (that is, across the triangle from) the right angle, c, is called the **HYPOTENUSE**.

For any right triangle, the sum of the squares of the legs of the triangle is equal to the square of the hypotenuse, that is, $a^2 + b^2 = c^2$. This relationship is known as the **PYTHAGOREAN THEOREM**. In words, the theorem states that:

$$(\text{leg})^2 + (\text{leg})^2 = (\text{hypotenuse})^2.$$

## Example

Draw a diagram, then use the Pythagorean Theorem to write an equation or use area pictures (as shown on page 22, problem RC-1) on each side of the triangle to solve each problem.

a) Solve for the missing side.

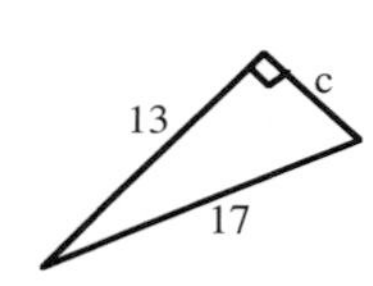


$$c^2 + 13^2 = 17^2$$
$$c^2 + 169 = 289$$
$$c^2 = 120$$
$$c = \sqrt{120}$$
$$c = 2\sqrt{30}$$
$$c \approx 10.95$$

b) Find x to the nearest tenth:

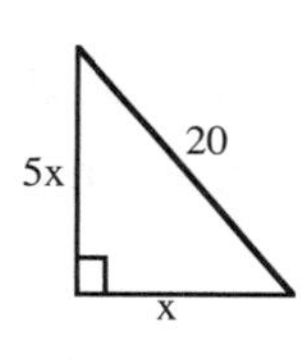


$$(5x)^2 + x^2 = 20^2$$
$$25x^2 + x^2 = 400$$
$$26x^2 = 400$$
$$x^2 \approx 15.4$$
$$x \approx \sqrt{15.4}$$
$$x \approx 3.9$$

c) One end of a ten foot ladder is four feet from the base of a wall. How high on the wall does the top of the ladder touch?

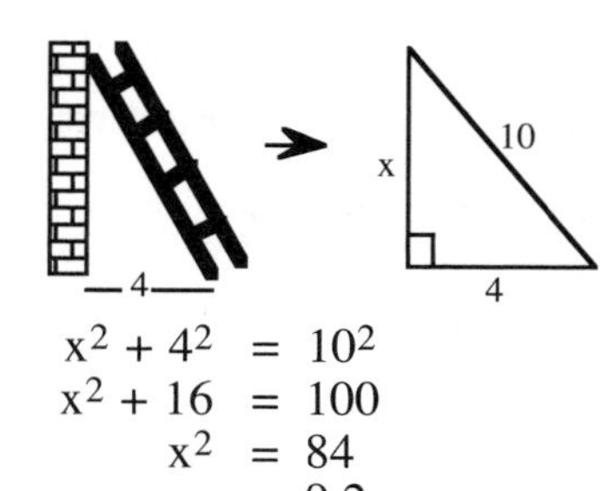


$$x^2 + 4^2 = 10^2$$
$$x^2 + 16 = 100$$
$$x^2 = 84$$
$$x \approx 9.2$$

The ladder touches the wall about 9.2 feet above the ground.

d) Could 3, 6 and 8 represent the lengths of the sides of a right triangle? Explain.

$$3^2 + 6^2 \stackrel{?}{=} 8^2$$
$$9 + 36 \stackrel{?}{=} 64$$
$$45 \neq 64$$

Since the Pythagorean Theorem relationship is not true for these lengths, they cannot be the side lengths of a right triangle.

Use the Pythagorean Theorem to find the value of x. Round answers to the nearest tenth.

1.

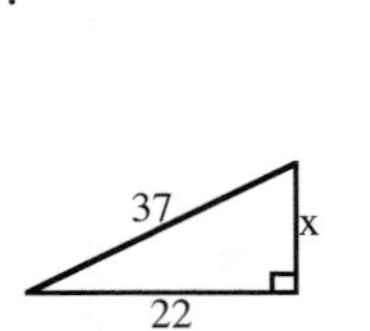


2.

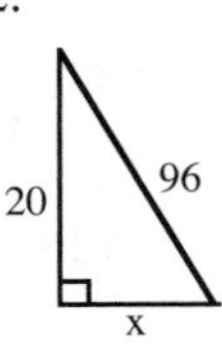


3.

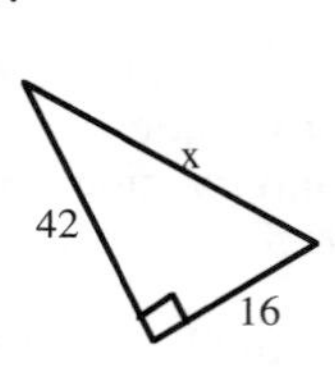


4.

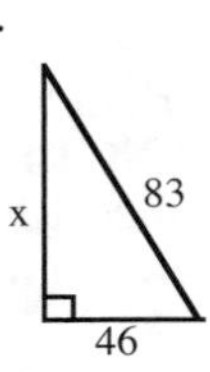


5.

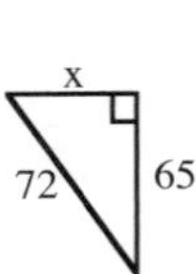


6.

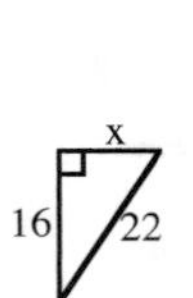


7.

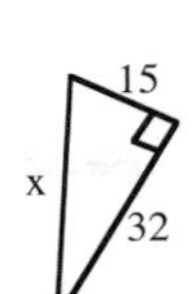


8.

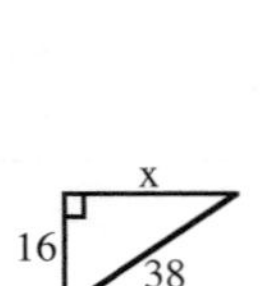


9.

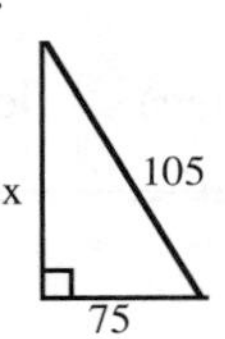


10.

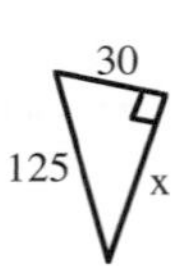


Solve the following problems.

11. A 12 foot ladder is six feet from a wall. How high on the wall does the ladder touch?

12. A 15 foot ladder is five feet from a wall. How high on the wall does the ladder touch?

13. A 9 foot ladder is three feet from a wall. How high on the wall does the ladder touch?

14. A 12 foot ladder is three and a half feet from a wall. How high on the wall does the ladder touch?

15. A 6 foot ladder is one and a half feet from a wall. How high on the wall does the ladder touch?

16. Could 2, 3, and 6 represent the lengths of sides of a right angle triangle? Justify your answer.

17. Could 8, 12, and 13 represent the lengths of sides of a right triangle? Justify your answer.

18. Could 5, 12, and 13 represent the lengths of sides of a right triangle? Justify your answer.

19. Could 9, 12, and 15 represent the lengths of sides of a right triangle? Justify your answer.

20. Could 10, 15, and 20 represent the lengths of sides of a right triangle? Justify your answer.

## Answers

| | | | | |
|---|---|---|---|---|
| 1. 29.7 | 2. 93.9 | 3. 44.9 | 4. 69.1 | 5. 31.0 |
| 6. 15.1 | 7. 35.3 | 8. 34.5 | 9. 73.5 | 10. 121.3 |
| 11. 10.4 ft | 12. 14.1 ft | 13. 8.5 ft | 14. 11.5 ft | 15. 5.8 ft |
| 16. no | 17. no | 18. yes | 19. yes | 20. no |

## AREA

**#3**

**AREA** is the number of square units in a flat region. The formulas to calculate the area of several kinds of polygons are:

RECTANGLE

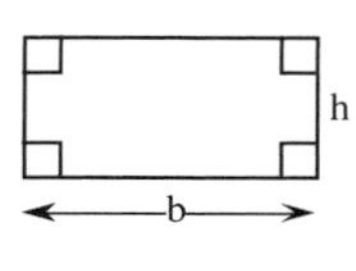


$A = bh$

PARALLELOGRAM

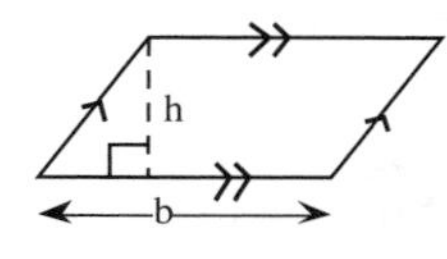


$A = bh$

TRAPEZOID

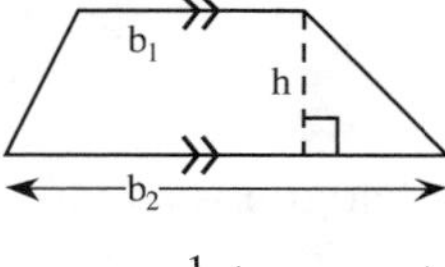


$A = \frac{1}{2}(b_1 + b_2)h$

TRIANGLE

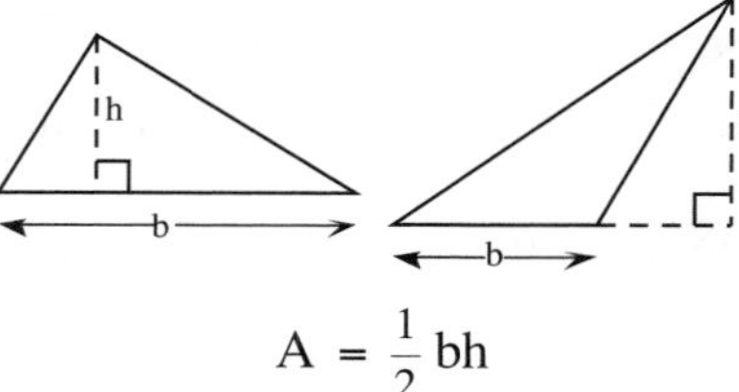


$A = \frac{1}{2}bh$

Note that the legs of any right triangle form a base and a height for the triangle.

The area of a more complicated figure may be found by breaking it into smaller regions of the types shown above, calculating each area, and finding the sum of the areas.

### Example 1

Find the area of each figure. All lengths are centimeters.

a)

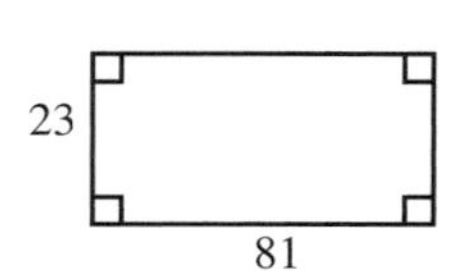


$A = bh = (81)(23) = 1863\text{ cm}^2$

b)

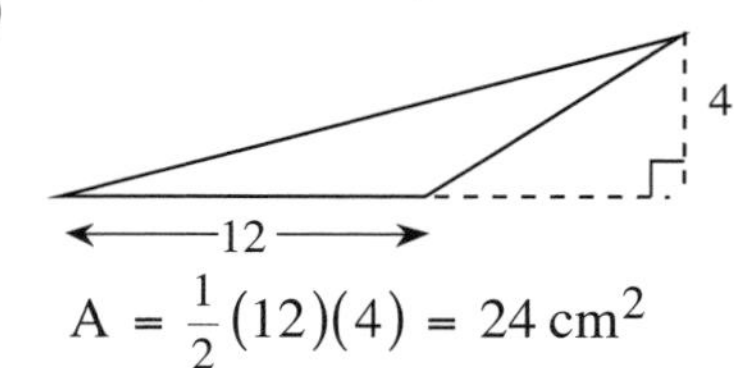


$A = \frac{1}{2}(12)(4) = 24\text{ cm}^2$

c)

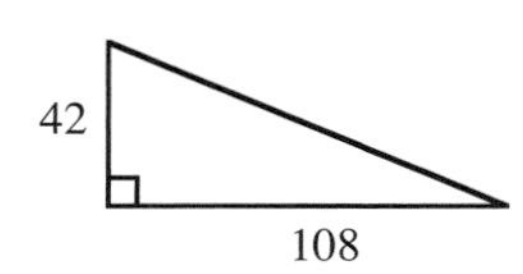


$A = \frac{1}{2}(108)(42) = 2268\text{ cm}^2$

d)

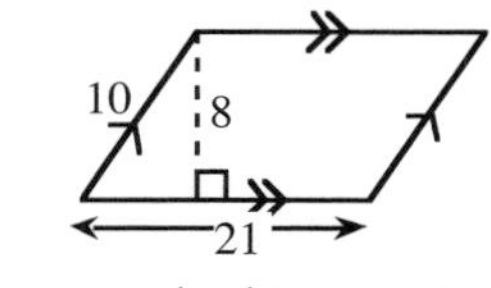


$A = (21)8 = 168\text{ cm}^2$

Note that 10 is a side of the parallelogram, not the height.

e)

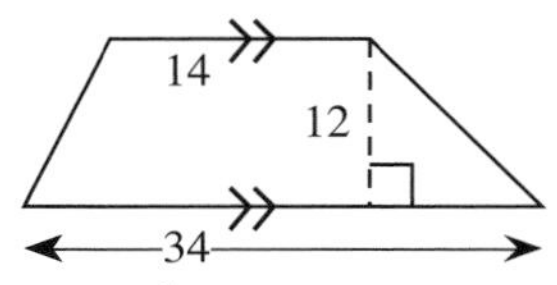


$A = \frac{1}{2}(14 + 34)12 = \frac{1}{2}(48)(12) = 288\text{ cm}^2$

## Example 2

Find the area of the shaded region.

The area of the shaded region is the area of the triangle minus the area of the rectangle.

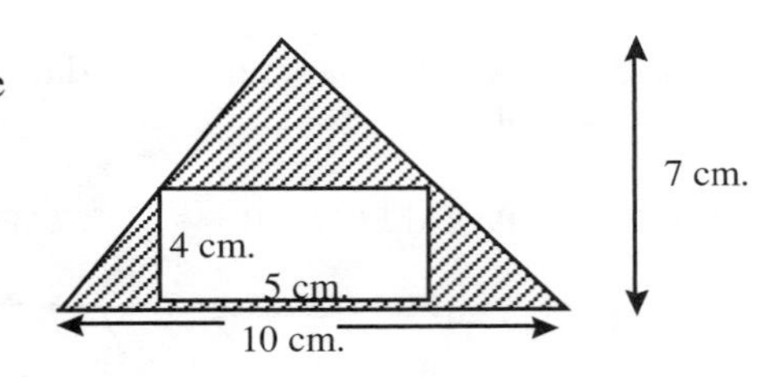


triangle: $A = \frac{1}{2}(7)(10) = 35\text{ cm}^2$

rectangle: $A = 5(4) = 20\text{ cm}^2$

shaded region: $A = 35 - 20 = 15\text{ cm}^2$

Find the area of the following triangles, parallelograms and trapezoids. Pictures are not drawn to scale. Round answers to the nearest tenth.

1.

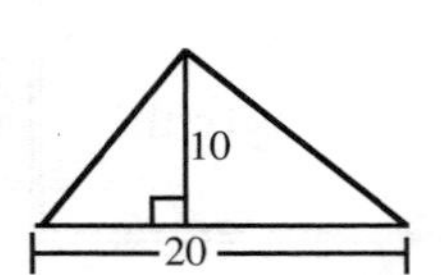


2.

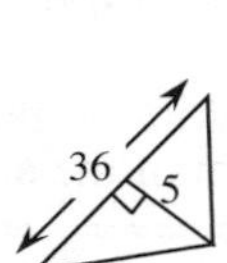


3.

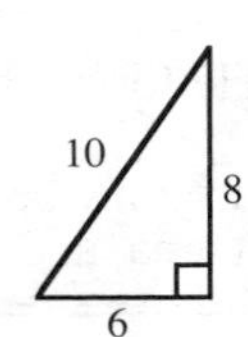


4.

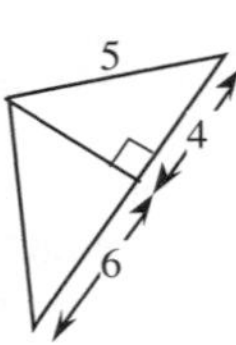


5.

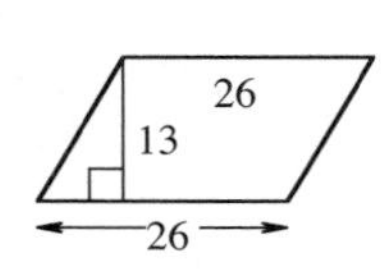


6.

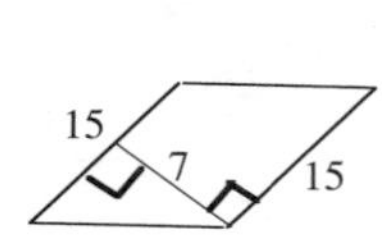


7.

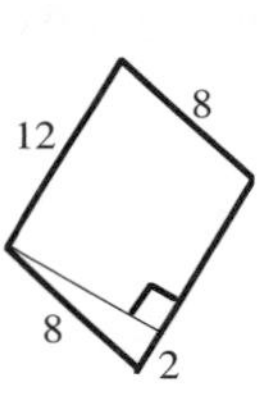


8.

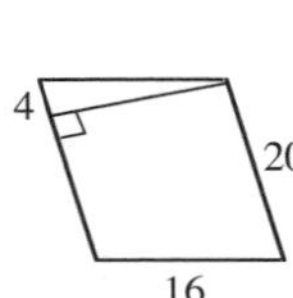


9.

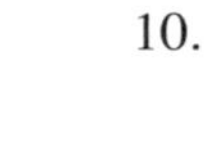

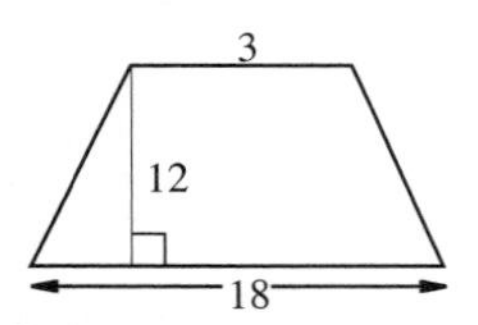


10.

9
3
4

11.

6
1.5
4
7

12.

$2\sqrt{10}$
10
8
6

13.

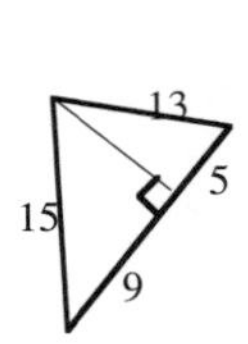


14.

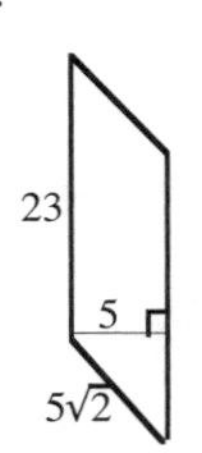


15.

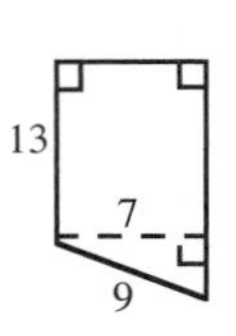


16.

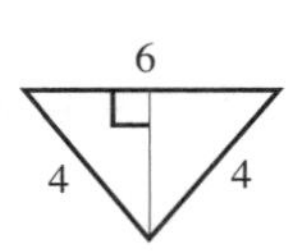

Find the area of the shaded region.

17.

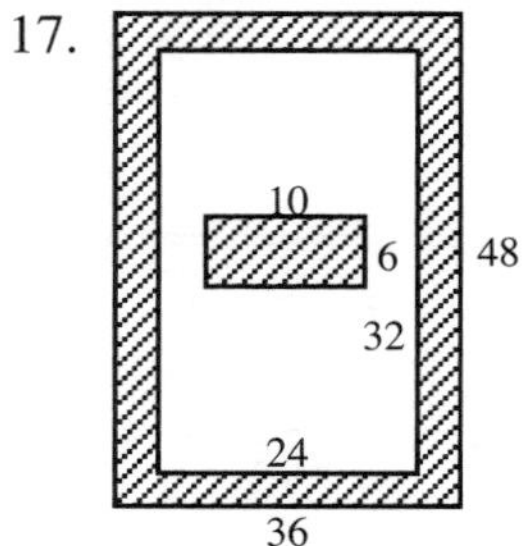


18.

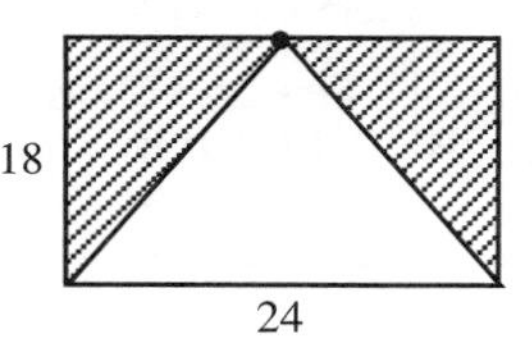


19.

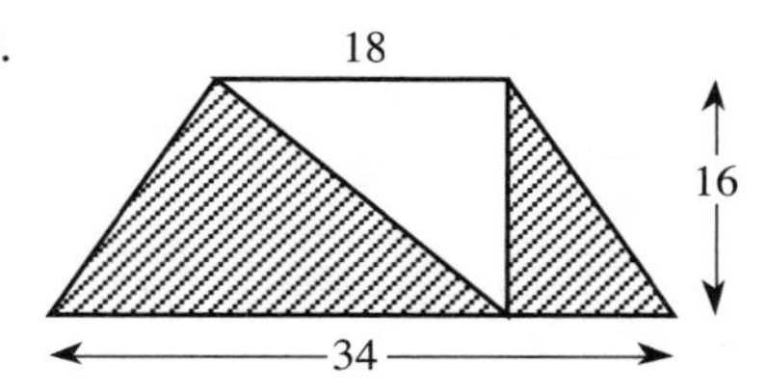


20.

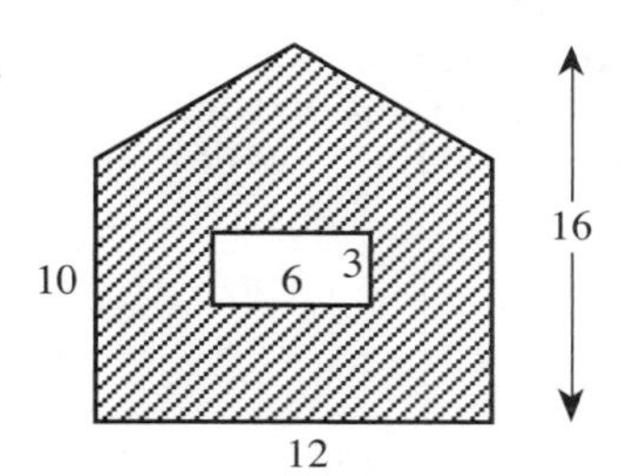


## Answers (in square units)

| | | | | | |
|---|---|---|---|---|---|
| 1. 100 | 2. 90 | 3. 24 | 4. 15 | 5. 338 | 6. 105 |
| 7. 93 | 8. 309.8 | 9. 126 | 10. 19.5 | 11. 36.3 | 12. 54 |
| 13. 84 | 14. 115 | 15. 110.8 | 16. 7.9 | 17. 1020 | 18. 216 |
| 19. 272 | 20. 138 | | | | |

## SOLVING LINEAR SYSTEMS #4

You can find where two lines intersect (cross) by using algebraic methods. The two most common methods are **SUBSTITUTION** and **ELIMINATION** (also known as the addition method).

### Example 1

Solve the following system of equations at right by **substitution**. Check your solution.

$$y = 5x + 1$$
$$y = -3x - 15$$

When solving a system of equations, you are solving to find the x- and y-values that result in true statements when you substitute them into <u>both</u> equations. You generally find each value one at a time. Since both equations are in y-form (that is, solved for y), and we know $y = y$, we can substitute the right side of each equation for y in the simple equation $y = y$ and write

$$5x + 1 = -3x - 15.$$

Now solve for x. $5x + 1 = -3x - 15 \Rightarrow 8x + 1 = -15 \Rightarrow 8x = -16 \Rightarrow x = -2$

Remember you must find x <u>and</u> y. To find y, use either one of the two original equations. Substitute the value of x into the equation and find the value of y. Using the first equation,

$$y = 5x + 1 \Rightarrow y = 5(-2) + 1 \Rightarrow y = -10 + 1 = -9.$$

The solution appears to be (-2, -9). In order for this to be a solution, it must make both equations true when you replace x with -2 and y with -9. Substitute the values in both equations to check.

$$\begin{aligned} y &= 5x + 1 \\ -9 &\stackrel{?}{=} 5(-2) + 1 \\ -9 &\stackrel{?}{=} -10 + 1 \\ -9 &= -9 \text{ Check!} \end{aligned} \qquad \begin{aligned} y &= -3x - 15 \\ -9 &\stackrel{?}{=} -3(-2) - 15 \\ -9 &\stackrel{?}{=} 6 - 15 \\ -9 &= -9 \text{ Check!} \end{aligned}$$

Therefore, (-2, -9) is the solution.

### Example 2

**Substitution** can also be used when the equations are <u>not</u> in y-form.

$$x = -3y + 1$$
$$4x - 3y = -11$$

Use substitution to rewrite the two equations as $4(-3y + 1) - 3y = -11$ by replacing x with $(-3y + 1)$, then solve for y as shown at right.

$$x = -3y + 1$$
$$4(\quad) - 3y = -11$$
$$4(-3y + 1) - 3y = -11$$
$$y = 1$$

Substitute $y = 1$ into $x = -3y + 1$. Solve for x, and write the answer for x and y as an ordered pair, (1, –2). Substitute $y = 1$ into $4x - 3y = -11$ to verify that <u>either</u> original equation may be used to find the second coordinate. Check your answer as shown in example 1.

## Example 3

When you have a pair of two-variable equations, sometimes it is easier to **ELIMINATE** one of the variables to obtain one single variable equation. You can do this by adding the two equations together as shown in the example below.

Solve the system at right:

$$\begin{aligned} 2x + y &= 11 \\ x - y &= 4 \end{aligned}$$

To eliminate the y terms, **add** the two equations together.

$$\begin{aligned} 2x + y &= 11 \\ x - y &= 4 \\ \hline 3x &= 15 \end{aligned}$$

then solve for x.

$$\begin{aligned} 3x &= 15 \\ x &= 5 \end{aligned}$$

Once we know the x-value we can substitute it into <u>either</u> of the original equations to find the corresponding value of y.

Using the first equation:

$$\begin{aligned} 2x + y &= 11 \\ 2(5) + y &= 11 \\ 10 + y &= 11 \\ y &= 1 \end{aligned}$$

Check the solution by substituting both the x-value and y-value into the other original equation, $x - y = 4$: $5 - 1 = 4$, checks.

## Example 4

You can solve the system of equations at right by elimination, but before you can eliminate one of the variables, you must adjust the coefficients of one of the variables so that they are additive opposites.

$$\begin{aligned} 3x + 2y &= 11 \\ 4x + 3y &= 14 \end{aligned}$$

To eliminate y, multiply the first equation by 3, then multiply the second equation by $-2$ to get the equations at right.

$$\begin{aligned} 9x + 6y &= 33 \\ -8x - 6y &= -28 \end{aligned}$$

Next eliminate the y terms by adding the two adjusted equations.

$$\begin{aligned} 9x + 6y &= 33 \\ -8x - 6y &= -28 \\ \hline x &= 5 \end{aligned}$$

Since $x = 5$, substitute in either original equation to find that $y = -2$. Therefore, the solution to the system of equations is $(5, -2)$.

You could also solve the system by first multiplying the first equation by 4 and the second equation by -3 to eliminate x, then proceeding as shown above to find y.

Solve the following systems of equations to find the point of intersection (x, y) for each pair of lines.

1. $y = x - 6$
   $y = 12 - x$

2. $y = 3x - 5$
   $y = x + 3$

3. $x = 7 + 3y$
   $x = 4y + 5$

4. $x = -3y + 10$
   $x = -6y - 2$

5. $y = x + 7$
   $y = 4x - 5$

6. $y = 7 - 3x$
   $y = 2x - 8$

7. $y = 3x - 1$
   $2x - 3y = 10$

8. $x = -\frac{1}{2}y + 4$
   $8x + 3y = 31$

9. $2y = 4x + 10$
   $6x + 2y = 10$

10. $y = \frac{3}{5}x - 2$
    $y = \frac{x}{10} + 1$

11. $y = -4x + 5$
    $y = x$

12. $4x - 3y = -10$
    $x = \frac{1}{4}y - 1$

13. $x + y = 12$
    $x - y = 4$

14. $2x - y = 6$
    $4x - y = 12$

15. $x + 2y = 7$
    $5x - 4y = 14$

16. $5x - 2y = 6$
    $4x + y = 10$

17. $x + y = 10$
    $x - 2y = 5$

18. $3y - 2x = 16$
    $y = 2x + 4$

19. $x + y = 11$
    $x = y - 3$

20. $x + 2y = 15$
    $y = x - 3$

21. $y + 5x = 10$
    $y - 3x = 14$

22. $y = 7x - 3$
    $4x + 2y = 8$

23. $y = 12 - x$
    $y = x - 4$

24. $y = 6 - 2x$
    $y = 4x - 12$

## Answers

1. (9, 3)
2. (4, 7)
3. (13, 2)
4. (22, -4)
5. (4, 11)
6. (3, -2)
7. (-1, -4)
8. $\left(\frac{7}{2}, 1\right)$
9. (0, 5)
10. (6, 1.6)
11. (1, 1)
12. (-0.25, 3)
13. (8, 4)
14. (3, 0)
15. (4, 1.5)
16. (2, 2)
17. $\left(\frac{25}{3}, \frac{5}{3}\right)$
18. (1, 6)
19. (4, 7)
20. (7, 4)
21. (-0.5, 12.5)
22. $\left(\frac{7}{9}, \frac{22}{9}\right)$
23. (8, 4)
24. (3, 0)

## PROPERTIES OF ANGLES, LINES, AND TRIANGLES #5

Parallel lines

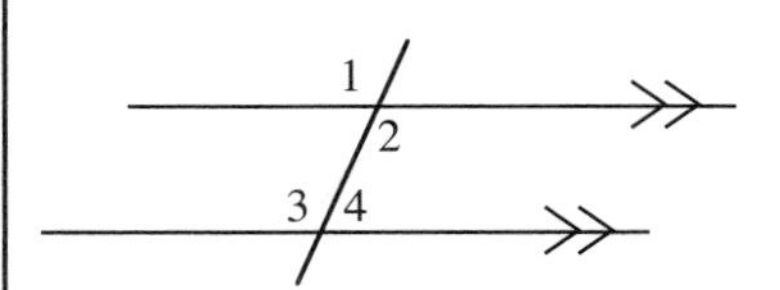


Triangles

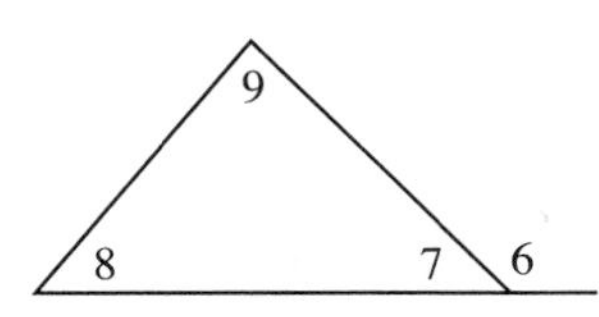


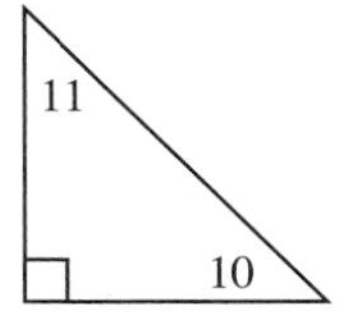


- corresponding angles are equal: $m\angle 1 = m\angle 3$
- alternate interior angles are equal: $m\angle 2 = m\angle 3$
- $m\angle 2 + m\angle 4 = 180°$

- $m\angle 7 + m\angle 8 + m\angle 9 = 180°$
- $m\angle 6 = m\angle 8 + m\angle 9$ (exterior angle = sum remote interior angles)
- $m\angle 10 + m\angle 11 = 90°$ (complementary angles)

Also shown in the above figures:
- vertical angles are equal: $m\angle 1 = m\angle 2$
- linear pairs are supplementary: $m\angle 3 + m\angle 4 = 180°$ and $m\angle 6 + m\angle 7 = 180°$

In addition, an isosceles triangle, $\Delta ABC$, has $BA = BC$ and $m\angle A = m\angle C$. An equilateral triangle, $\Delta GFH$, has $GF = FH = HG$ and $m\angle G = m\angle F = m\angle H = 60°$.

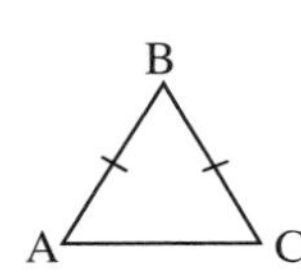


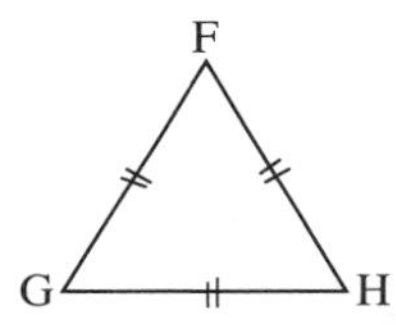


### Example 1

Solve for x.

Use the Exterior Angle Theorem: $6x + 8° = 49° + 67°$

$6x° = 108° \Rightarrow x = \frac{108°}{6} \Rightarrow x = 18°$

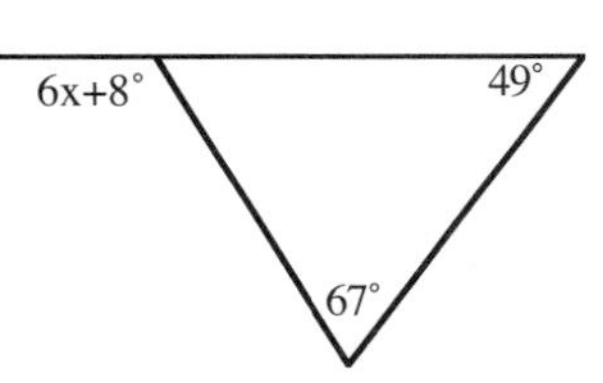


### Example 2

Solve for x.

There are a number of relationships in this diagram. First, $\angle 1$ and the 127° angle are supplementary, so we know that $m\angle 1 + 127° = 180°$ so $m\angle 1 = 53°$. Using the same idea, $m\angle 2 = 47°$. Next, $m\angle 3 + 53° + 47° = 180°$, so $m\angle 3 = 80°$. Because angle 3 forms a vertical pair with the angle marked $7x + 3°$, $80° = 7x + 3°$, so $x = 11°$.

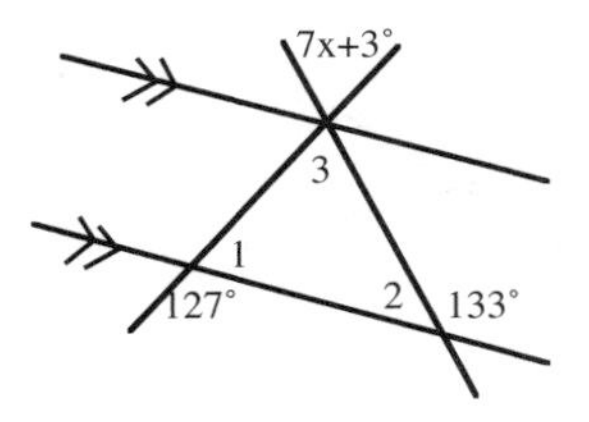


### Example 3

Find the measure of the acute alternate interior angles.

Parallel lines mean that alternate interior angles are equal, so $5x + 28° = 2x + 46° \Rightarrow 3x = 18° \Rightarrow x = 6°$. Use either algebraic angle measure: $2(6°) + 46° = 58°$ for the measure of the acute angle.

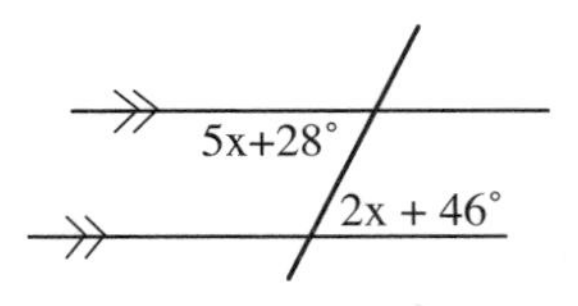

Use the geometric properties and theorems you have learned to solve for x in each diagram and write the property or theorem you use in each case.

1.

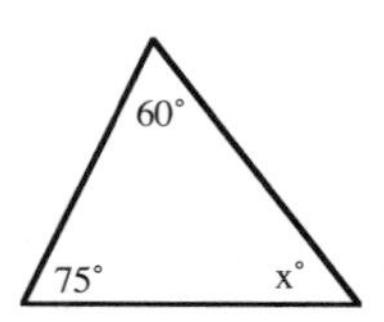


2.

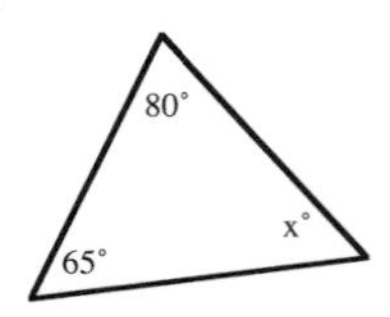


3.

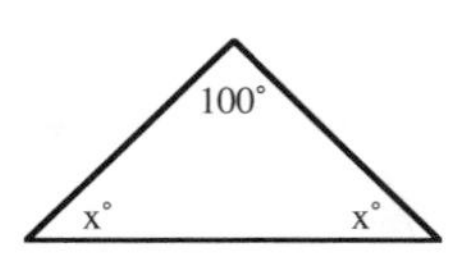


4.

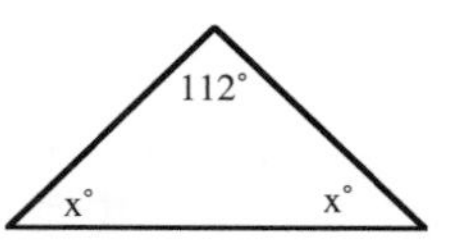


5.

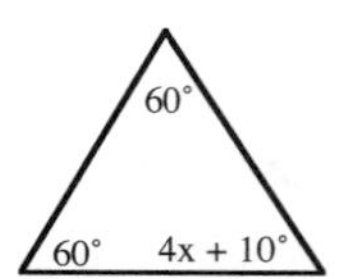


6.

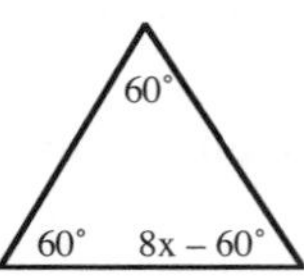


7.

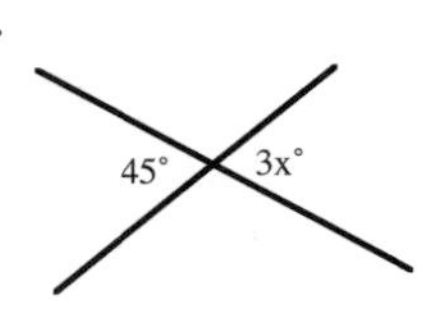


8.

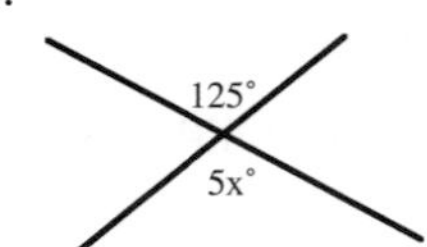


9.

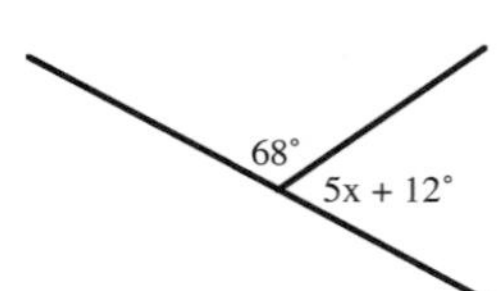


10.

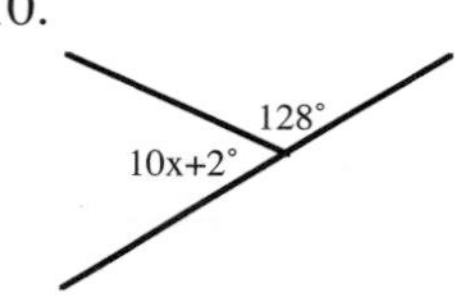


11.

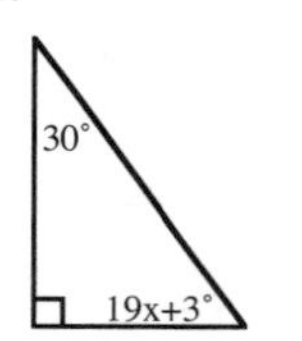


12.

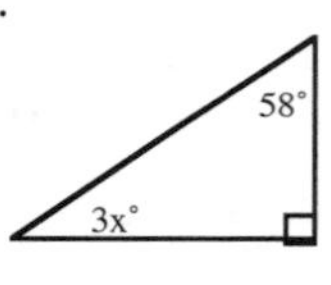


13.

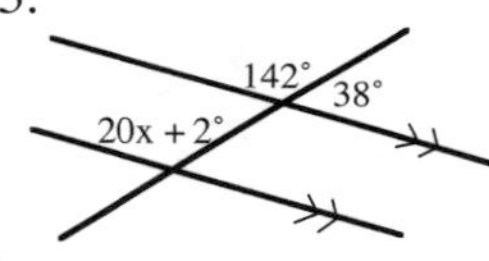


14.

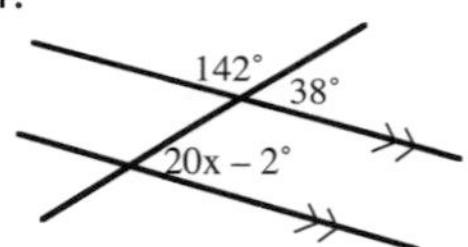


15.

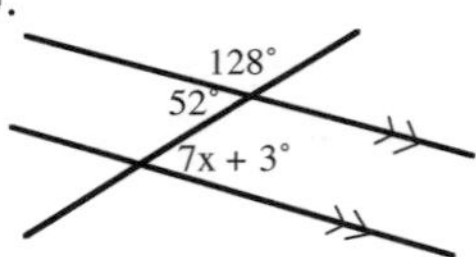


16.

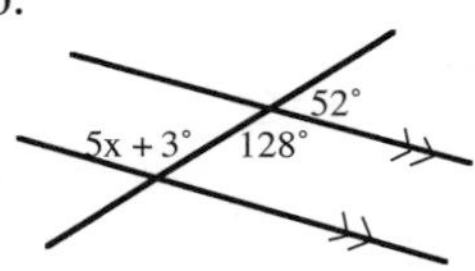


17.

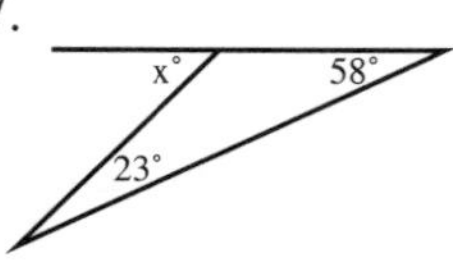


18.

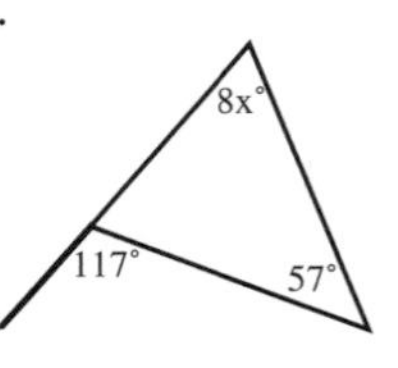


19.

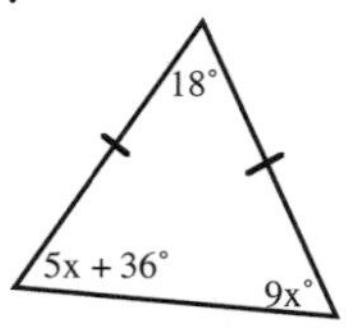


20.

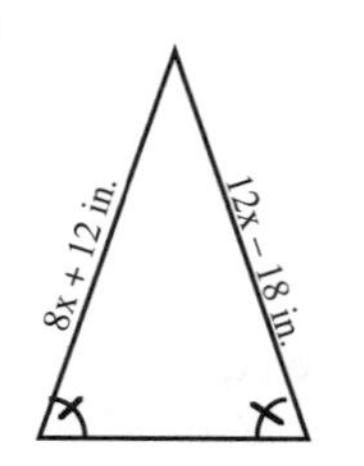


21.

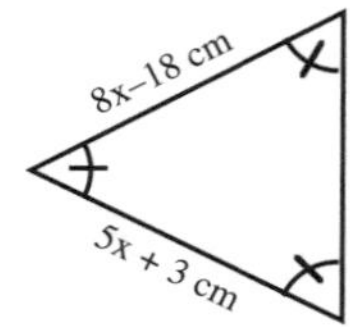

22.

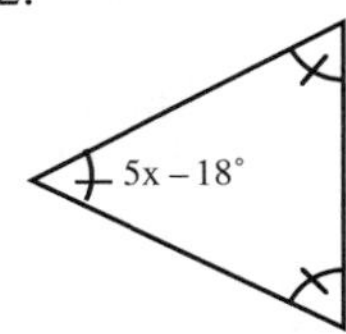


23.

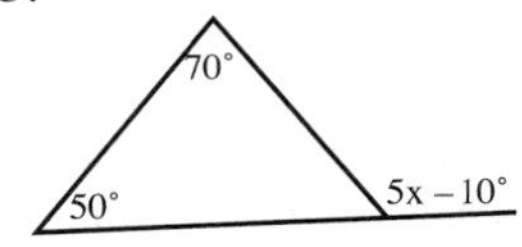


24.

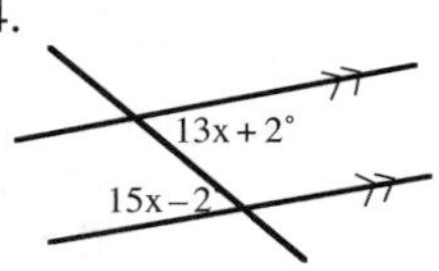


25.

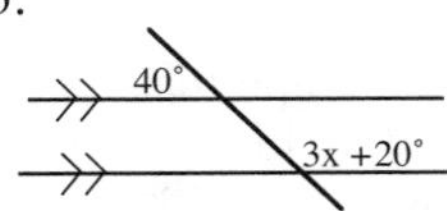


26.

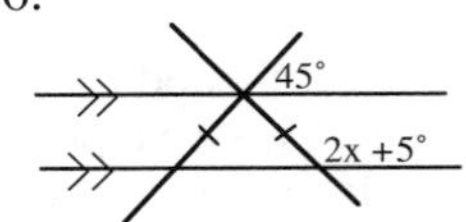


27.

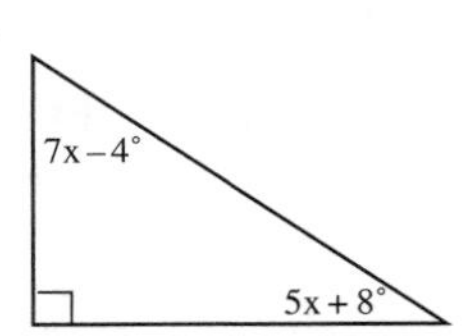


28.

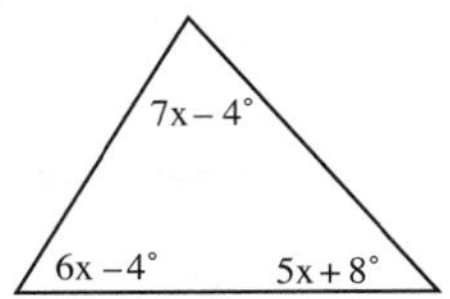


## Answers

| | | | | | |
|---|---|---|---|---|---|
| 1. 45° | 2. 35° | 3. 40° | 4. 34° | 5. 12.5° | 6. 15° |
| 7. 15° | 8. 25° | 9. 20° | 10. 5° | 11. 3° | 12. $10\frac{2}{3}°$ |
| 13. 7° | 14. 2° | 15. 7° | 16. 25° | 17. 81° | 18. 7.5° |
| 19. 9° | 20. 7.5° | 21. 7° | 22. 15.6° | 23. 26° | 24. 2° |
| 25. 40° | 26. 65° | 27. $7\frac{1}{6}°$ | 28. 10° | | |

## LINEAR INEQUALITIES #6

To graph a linear inequality, first graph the line of the corresponding equality. This line is known as the dividing line, since all the points that make the inequality true lie on one side or the other of the line. Before you draw the line, decide whether the dividing line is part of the solution or not, that is, whether the line is solid or dashed. If the inequality symbol is either ≤ or ≥, then the dividing line is part of the inequality and it must be solid. If the symbol is either < or >, then the dividing line is not part of the inequality and it must be dashed.

Next, decide which side of the dividing line must be shaded to show the part of the graph that represents all values that make the inequality true. Choose a point not on the dividing line. Substitute this point into the original inequality. If the inequality is true for this test point, then shade the graph on this side of the dividing line. If the inequality is false for the test point, then shade the opposite side of the dividing line.

CAUTION: If the inequality is not in slope-intercept form and you have to solve it for y, always use the original inequality to test a point, NOT the solved form.

### Example 1

Graph the inequality $y > 3x - 2$. First, graph the line $y = 3x - 2$, but draw it dashed since $>$ means the dividing line is not part of the solution.

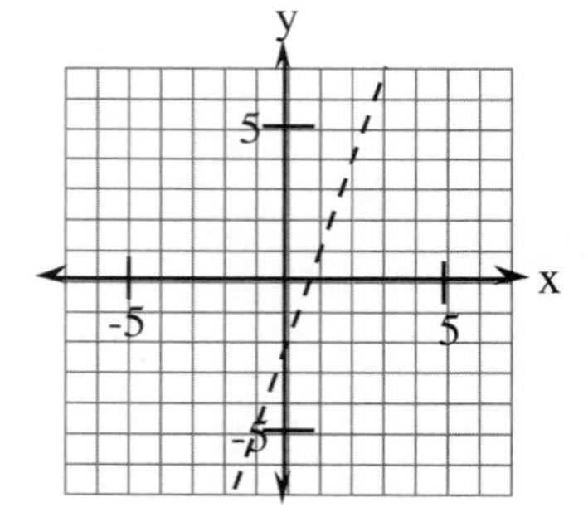


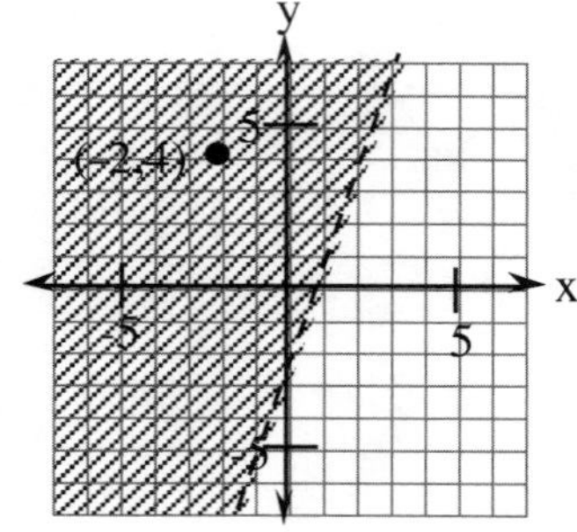


Next, test the point (-2, 4) to the left of the dividing line.

$$4 \overset{?}{>} 3(-2) - 2, \text{ so } 4 > -8$$

Since the inequality is true for this point, shade the left side of the dividing line.

## Example 2

Graph the system of inequalities $y \le \frac{1}{2}x + 2$ and $y > -\frac{2}{3}x - 1$.

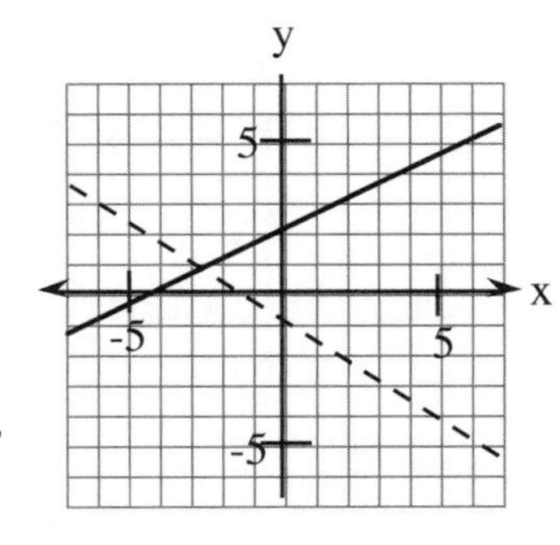


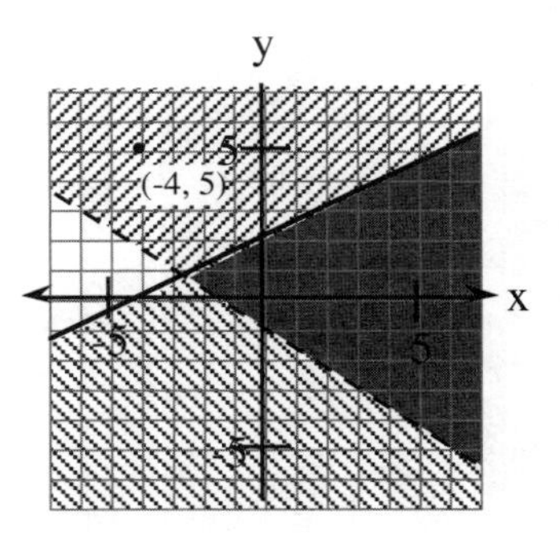


Graph the lines $y = \frac{1}{2}x + 2$ and $y = -\frac{2}{3}x - 1$. The first line is solid, the second is dashed. Test the point (-4, 5) in the first inequality.

$5 \overset{?}{\le} \frac{1}{2}(-4) + 2$, so $5 \le 0$

This inequality is false, so shade on the opposite side of the dividing line, from (-4, 5). Test the same point in the second inequality.

$5 \overset{?}{>} -\frac{2}{3}(-4) - 1$, so $5 > \frac{5}{3}$

This inequality is true, so shade on the same side of the dividing line as (-4, 5).

The solution is the overlap of the two shaded regions shown by the darkest shading in the second graph above right.

Graph each of the following inequalities on separate sets of axes.

1. $y \le 3x + 1$
2. $y \ge 2x - 1$
3. $y \ge -2x - 3$
4. $y \le -3x + 4$
5. $y > 4x + 2$
6. $y < 2x + 1$
7. $y < -3x - 5$
8. $y > -5x - 4$
9. $y \le 3$
10. $y \ge -2$
11. $x > 1$
12. $x \le 8$
13. $y > \frac{2}{3}x + 8$
14. $y \le -\frac{2}{3}x + 3$
15. $y < -\frac{3}{5}x - 7$
16. $y \ge \frac{1}{4}x - 2$
17. $3x + 2y \ge 7$
18. $2x - 3y \le 5$
19. $-4x + 2y < 3$
20. $-3x - 4y > 4$

Graph each of the following pairs of inequalities on the same set of axes.

21. $y > 3x - 4$ and $y \le -2x + 5$
22. $y \ge -3x - 6$ and $y > 4x - 4$
23. $y \le -\frac{3}{5}x + 4$ and $y \le \frac{1}{3}x + 3$
24. $y < -\frac{3}{7}x - 1$ and $y > \frac{4}{5}x + 1$
25. $y < 3$ and $y \le -\frac{1}{2}x + 2$
26. $x \le 3$ and $y < \frac{3}{4}x - 4$

Write an inequality for each of the following graphs.

27.

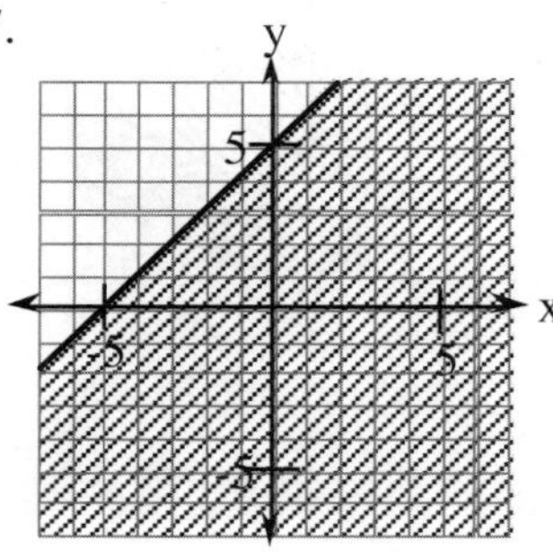

28.

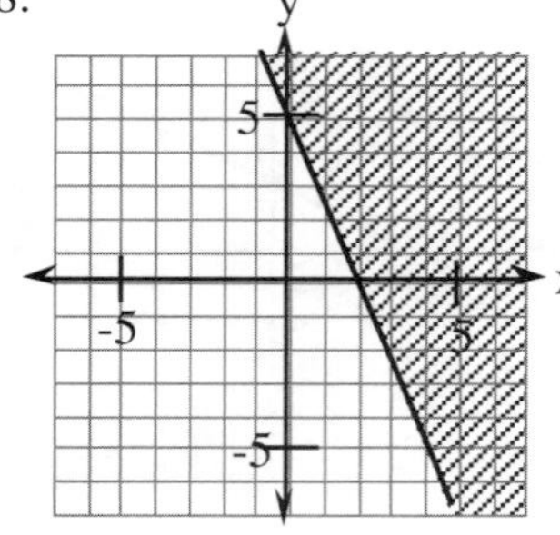

29.

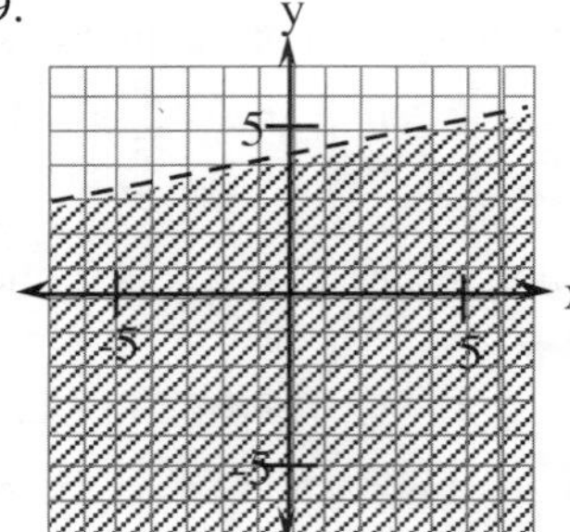

30.

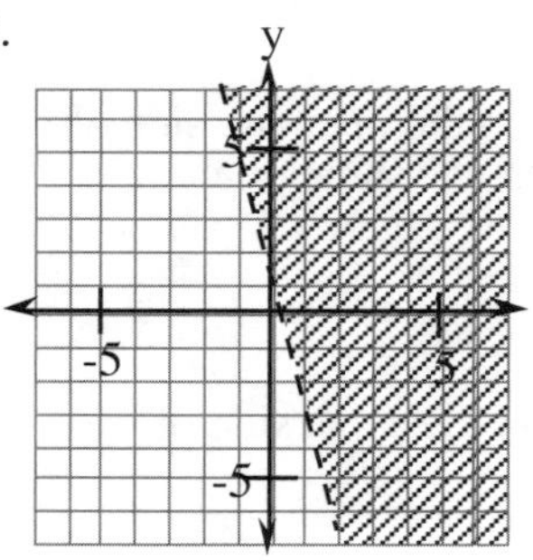

31.

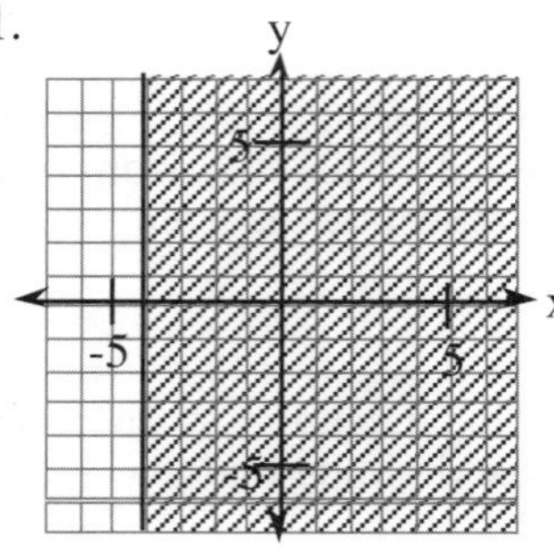

32.

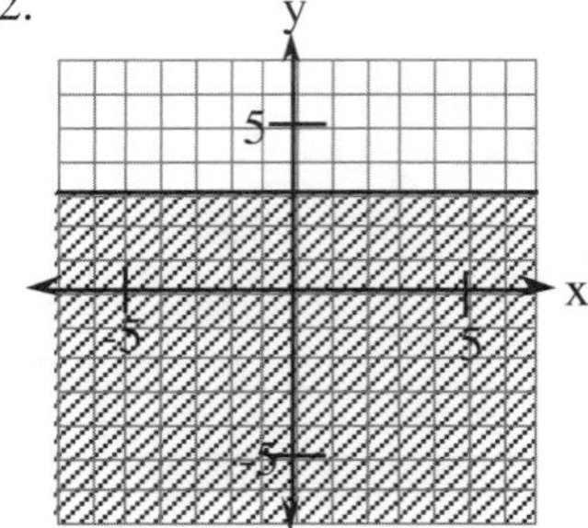

## Answers

1.

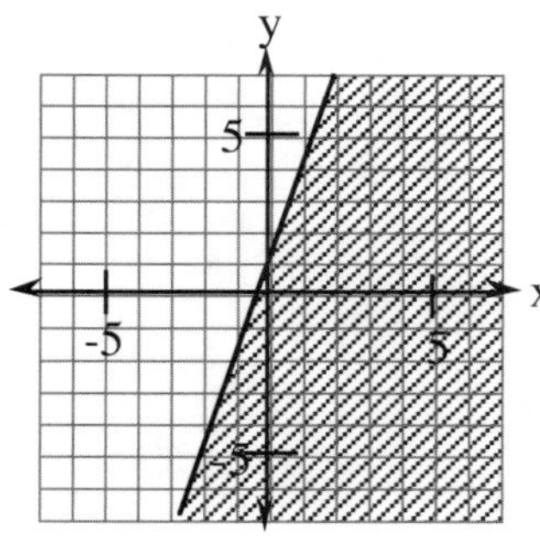

2.

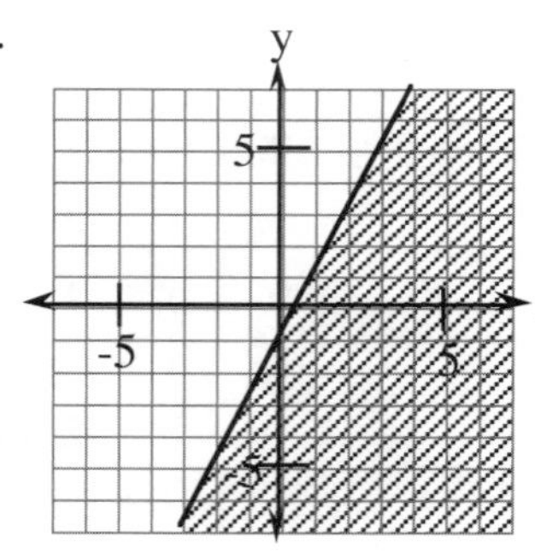

3.

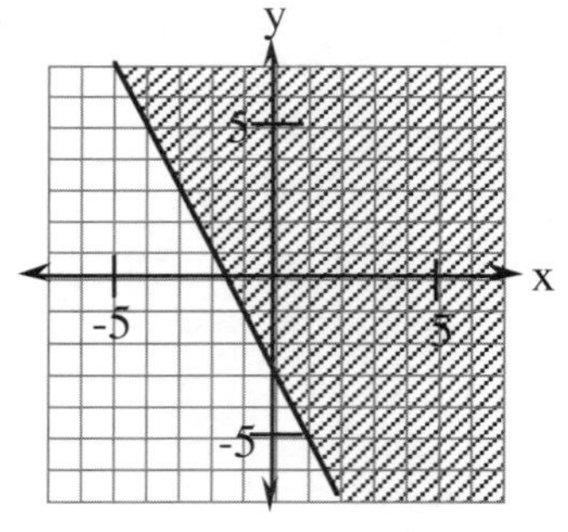

4.

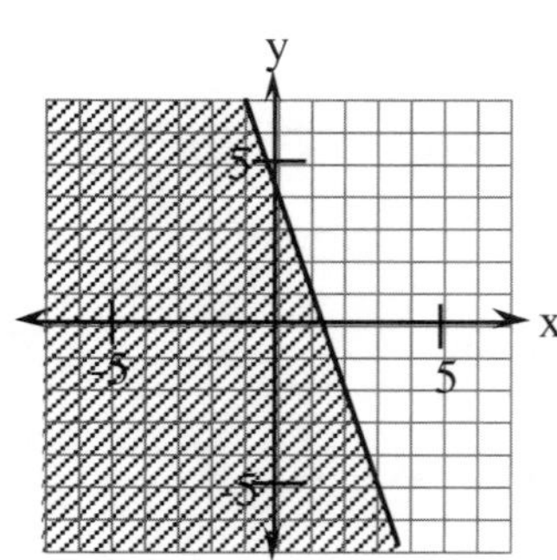

5.

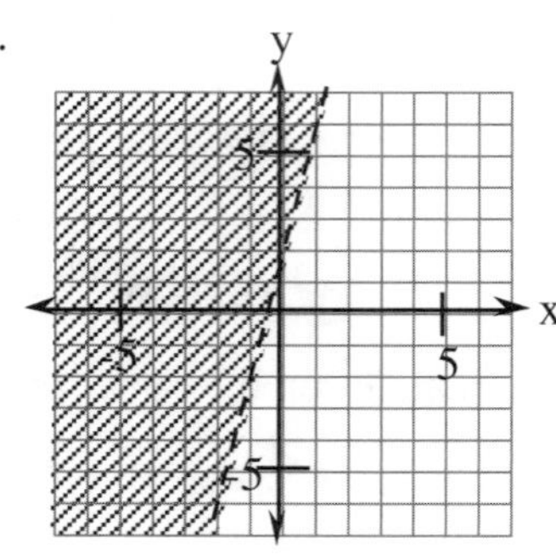

6.

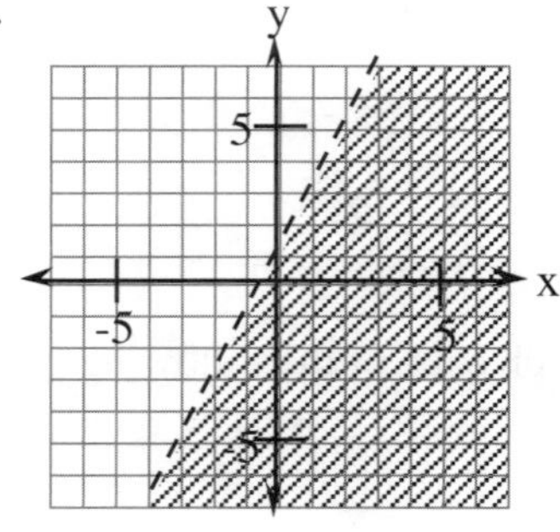

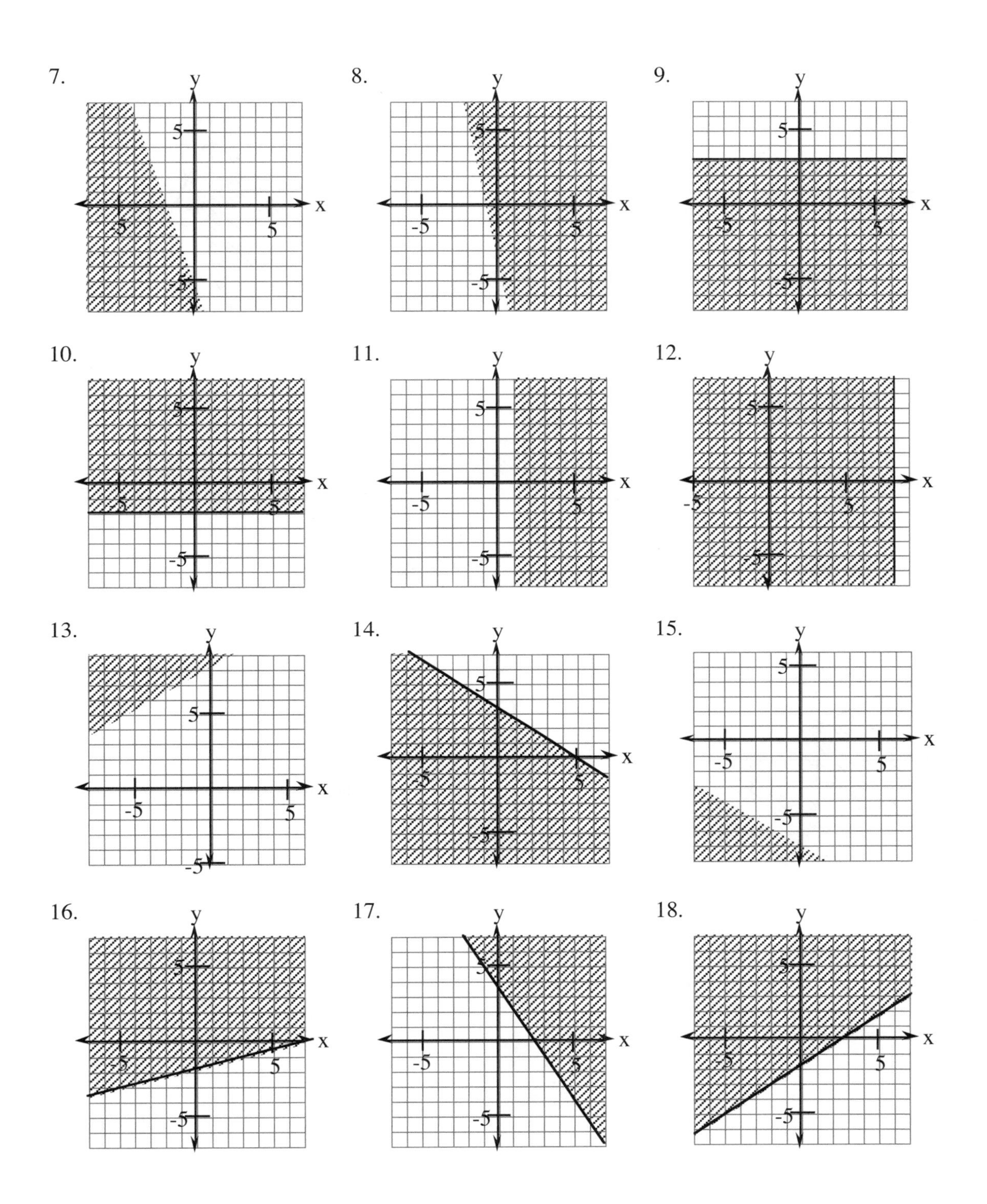
7.
8.
9.
10.
11.
12.
13.
14.
15.
16.
17.
18.
y
x
5
-5

19.

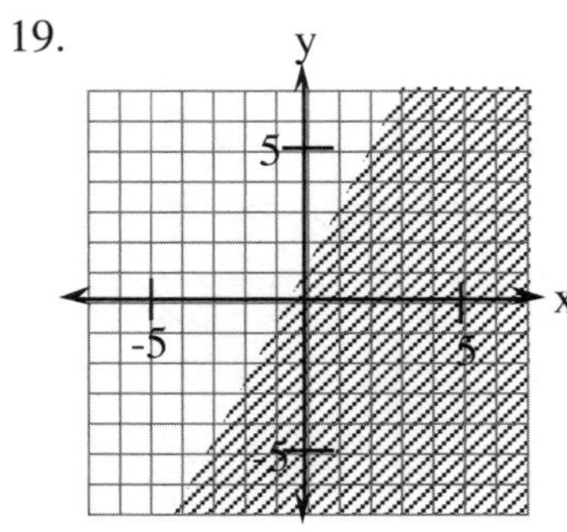

20.

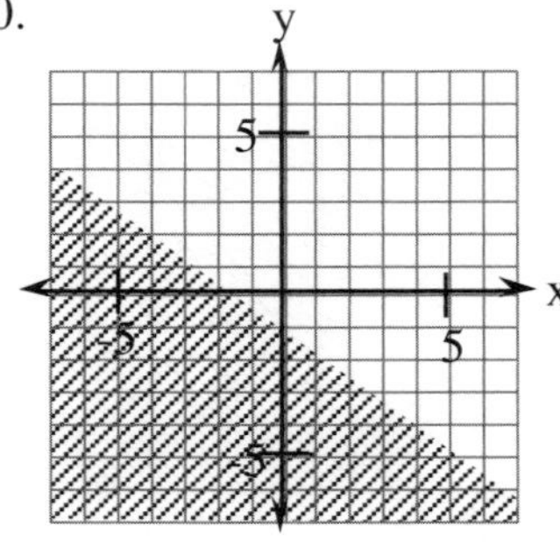

21.

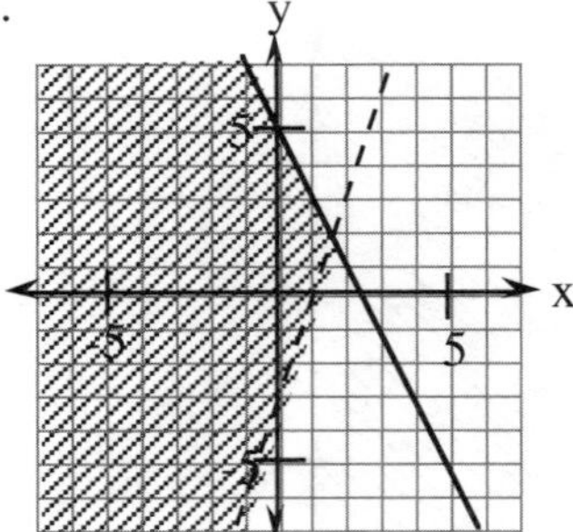

22.

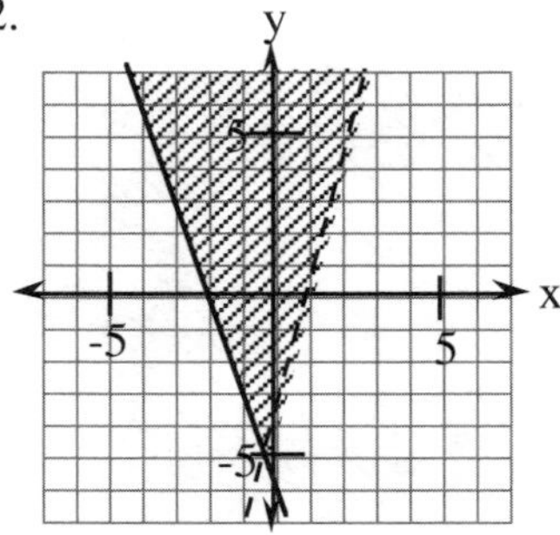

23.

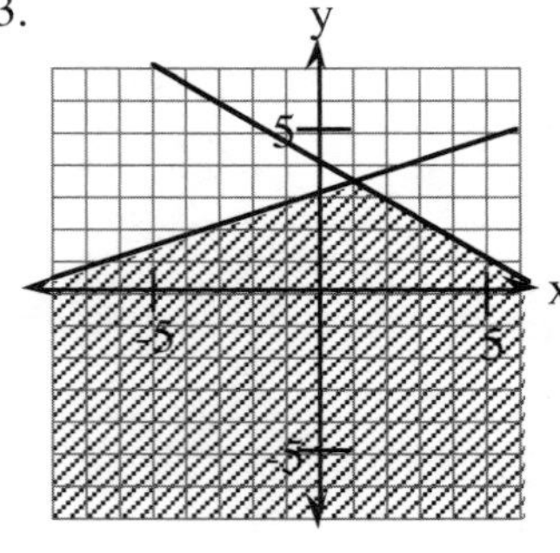

24.

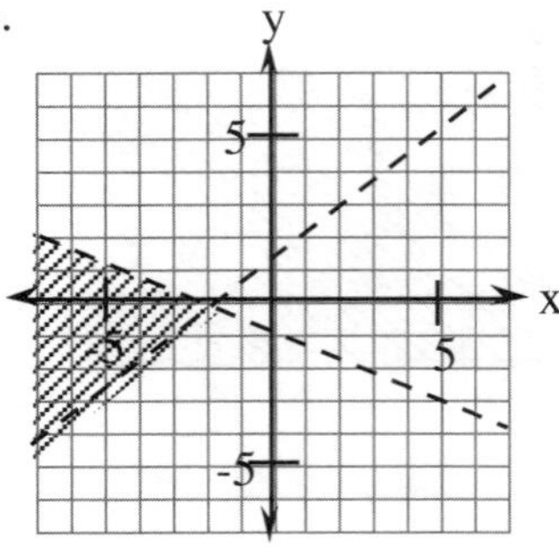

25.

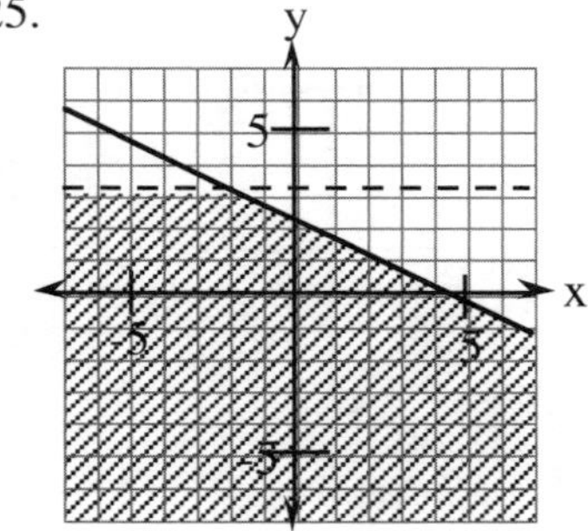

26.

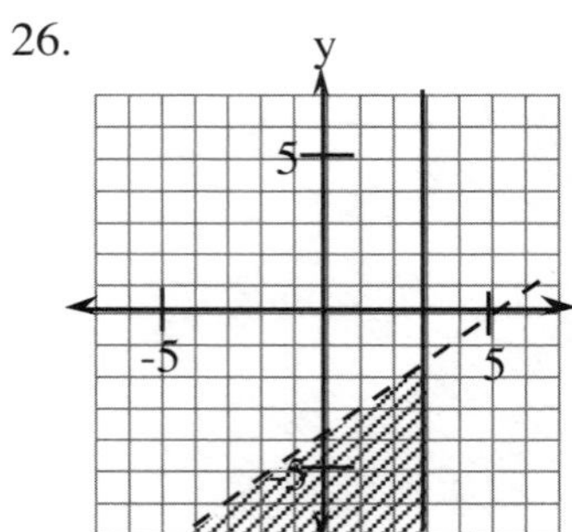

27. $y \leq x + 5$

28. $y \geq -\frac{5}{2}x + 5$

29. $y \leq \frac{1}{5}x + 4$

30. $y > -4x + 1$

31. $x \geq -4$

32. $y \leq 3$

## MULTIPLYING POLYNOMIALS #7

We can use generic rectangles as area models to find the products of polynomials. A generic rectangle helps us organize the problem. It does not have to be drawn accurately or to scale.

### Example 1

Multiply $(2x + 5)(x + 3)$

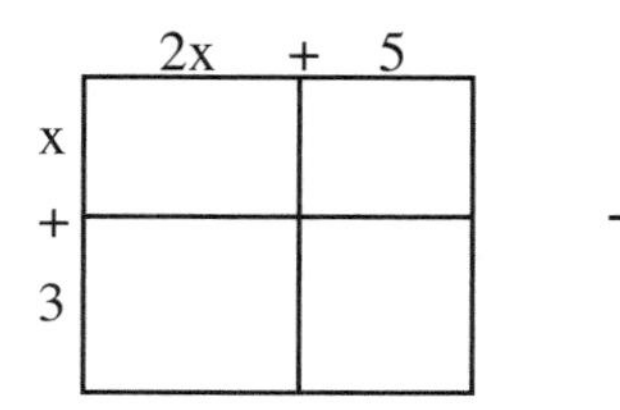

→

| | $2x$ | $+\ 5$ |
|---|---|---|
| $x$ | $2x^2$ | $5x$ |
| $+\ 3$ | $6x$ | $15$ |

$(2x + 5)(x + 3) = 2x^2 + 11x + 15$

area as a product — area as a sum

### Example 2

Multiply $(x + 9)\left(x^2 - 3x + 5\right)$

| | $x^2$ | $-\ 3x$ | $+\ 5$ |
|---|---|---|---|
| $x$ | | | |
| $+\ 9$ | | | |

→

| | $x^2$ | $-\ 3x$ | $+\ 5$ |
|---|---|---|---|
| $x$ | $x^3$ | $-3x^2$ | $5x$ |
| $+\ 9$ | $9x^2$ | $-27x$ | $45$ |

Therefore $(x + 9)\left(x^2 - 3x + 5\right) = x^3 + 9x^2 - 3x^2 - 27x + 5x + 45 = x^3 + 6x^2 - 22x + 45$

Another approach to multiplying binomials is to use the mnemonic "F.O.I.L." F.O.I.L. is an acronym for First, Outside, Inside, Last in reference to the positions of the terms in the two binomials.

### Example 3

Multiply $(3x - 2)(4x + 5)$ using the F.O.I.L. method.

| | | |
|---|---|---|
| **F.** | multiply the FIRST terms of each binomial | $(3x)(4x) = 12x^2$ |
| **O.** | multiply the OUTSIDE terms | $(3x)(5) = 15x$ |
| **I.** | multiply the INSIDE terms | $(-2)(4x) = -8x$ |
| **L.** | multiply the LAST terms of each binomial | $(-2)(5) = -10$ |

Finally, we combine like terms: $12x^2 + 15x - 8x - 10 = 12x^2 + 7x - 10$.

Find each of the following products.

1. $(3x + 2)(2x + 7)$
2. $(4x + 5)(5x + 3)$
3. $(2x - 1)(3x + 1)$
4. $(2a - 1)(4a + 7)$
5. $(m - 5)(m + 5)$
6. $(y - 4)(y + 4)$
7. $(3x - 1)(x + 2)$
8. $(3a - 2)(a - 1)$
9. $(2y - 5)(y + 4)$
10. $(3t - 1)(3t + 1)$
11. $(3y - 5)^2$
12. $(4x - 1)^2$
13. $(2x + 3)^2$
14. $(5n + 1)^2$
15. $(3x - 1)(2x^2 + 4x + 3)$
16. $(2x + 7)(4x^2 - 3x + 2)$
17. $(x + 7)(3x^2 - x + 5)$
18. $(x - 5)(x^2 - 7x + 1)$
19. $(3x + 2)(x^3 - 7x^2 + 3x)$
20. $(2x + 3)(3x^2 + 2x - 5)$

## Answers

1. $6x^2 + 25x + 14$
2. $20x^2 + 37x + 15$
3. $6x^2 - x - 1$
4. $8a^2 + 10a - 7$
5. $m^2 - 25$
6. $y^2 - 16$
7. $3x^2 + 5x - 2$
8. $3a^2 - 5a + 2$
9. $2y^2 + 3y - 20$
10. $9t^2 - 1$
11. $9y^2 - 30y + 25$
12. $16x^2 - 8x + 1$
13. $4x^2 + 12x + 9$
14. $25n^2 + 10n + 1$
15. $6x^3 + 10x^2 + 5x - 3$
16. $8x^3 + 22x^2 - 17x + 14$
17. $3x^3 + 20x^2 - 2x + 35$
18. $x^3 - 12x^2 + 36x - 5$
19. $3x^4 - 19x^3 - 5x^2 + 6x$
20. $6x^3 + 13x^2 - 4x - 15$

## FACTORING POLYNOMIALS #8

Often we want to un-multiply or **FACTOR** a polynomial P(x). This process involves finding a constant and/or another polynomial that evenly divides the given polynomial. In formal mathematical terms, this means $P(x) = q(x) \cdot r(x)$, where q and r are also polynomials. For elementary algebra there are three general types of factoring.

1) **Common term** (finding the largest common factor):

$6x + 18 = 6(x + 3)$ where 6 is a common factor of both terms.

$2x^3 - 8x^2 - 10x = 2x\left(x^2 - 4x - 5\right)$ where 2x is the common factor.

$2x^2(x - 1) + 7(x - 1) = (x - 1)\left(2x^2 + 7\right)$ where $x - 1$ is the common factor.

2) **Special products**

$a^2 - b^2 = (a + b)(a - b)$ $\qquad x^2 - 25 = (x + 5)(x - 5)$

$9x^2 - 4y^2 = (3x + 2y)(3x - 2y)$

$x^2 + 2xy + y^2 = (x + y)^2$ $\qquad x^2 + 8x + 16 = (x + 4)^2$

$x^2 - 2xy + y^2 = (x - y)^2$ $\qquad x^2 - 8x + 16 = (x - 4)^2$

3a) **Trinomials** in the form $x^2 + bx + c$ where the coefficient of $x^2$ is 1.

Consider $x^2 + (d + e)x + d \cdot e = (x + d)(x + e)$, where the coefficient of x is the sum of two numbers d and e AND the constant is the product of the same two numbers, d and e. A quick way to determine all of the possible pairs of integers d and e is to factor the constant in the original trinomial. For example, 12 is $1 \cdot 12$, $2 \cdot 6$, and $3 \cdot 4$. The signs of the two numbers are determined by the combination you need to get the sum. The "sum and product" approach to factoring trinomials is the same as solving a "Diamond Problem" in CPM's Algebra 1 course (see below).

$x^2 + 8x + 15 = (x + 3)(x + 5);\ 3 + 5 = 8,\ 3 \cdot 5 = 15$

$x^2 - 2x - 15 = (x - 5)(x + 3);\ -5 + 3 = -2,\ -5 \cdot 3 = -15$

$x^2 - 7x + 12 = (x - 3)(x - 4);\ -3 + (-4) = -7,\ (-3)(-4) = 12$

The sum and product approach can be shown visually using rectangles for an area model. The figure at far left below shows the "Diamond Problem" format for finding a sum and product. Here is how to use this method to factor $x^2 + 6x + 8$.

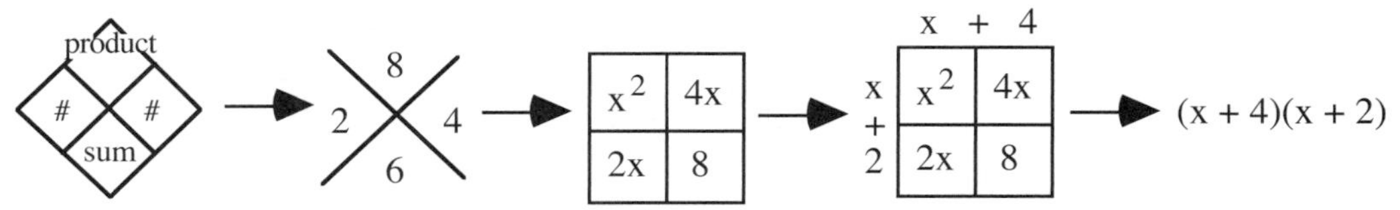


**>> Explanation and examples continue on the next page. >>**

3b) **Trinomials** in the form $ax^2 + bx + c$ where $a \neq 1$.

Note that the upper value in the diamond is no longer the constant. Rather, it is the product of a and c, that is, the coefficient of $x^2$ and the constant.

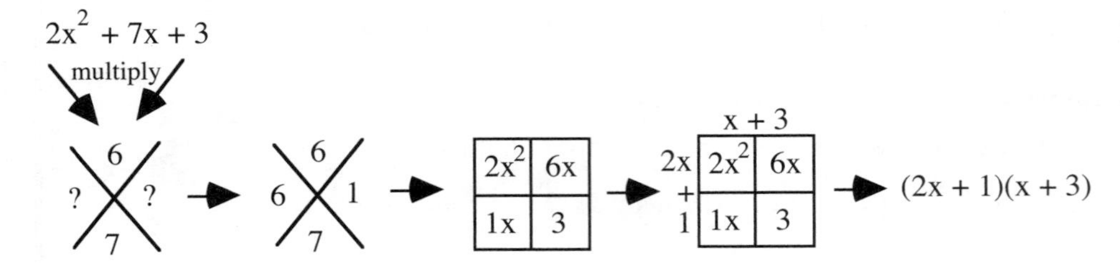


Below is the process to factor $5x^2 - 13x + 6$.

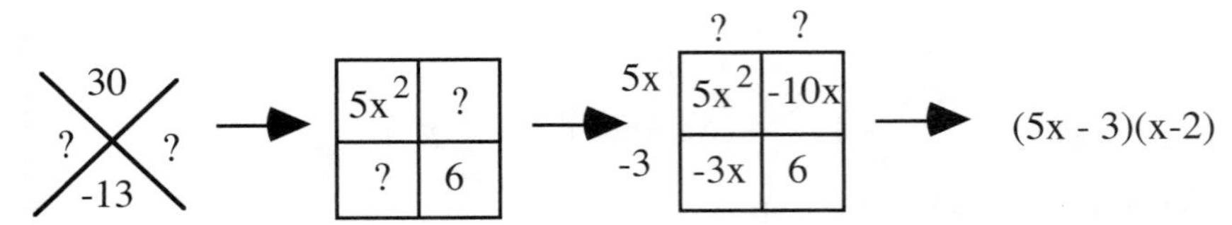


Polynomials with four or more terms are generally factored by grouping the terms and using one or more of the three procedures shown above. Note that polynomials are usually factored completely. In the second example in part (1) above, the trinomial also needs to be factored. Thus, the complete factorization of $2x^3 - 8x^2 - 10x = 2x(x^2 - 4x - 5) = 2x(x - 5)(x + 1)$.

Factor each polynomial completely.

1. $x^2 - x - 42$
2. $4x^2 - 18$
3. $2x^2 + 9x + 9$
4. $2x^2 + 3xy + y^2$
5. $6x^2 - x - 15$
6. $4x^2 - 25$
7. $x^2 - 28x + 196$
8. $7x^2 - 847$
9. $x^2 + 18x + 81$
10. $x^2 + 4x - 21$
11. $3x^2 + 21x$
12. $3x^2 - 20x - 32$
13. $9x^2 - 16$
14. $4x^2 + 20x + 25$
15. $x^2 - 5x + 6$
16. $5x^3 + 15x^2 - 20x$
17. $4x^2 + 18$
18. $x^2 - 12x + 36$
19. $x^2 - 3x - 54$
20. $6x^2 - 21$
21. $2x^2 + 15x + 18$
22. $16x^2 - 1$
23. $x^2 - 14x + 49$
24. $x^2 + 8x + 15$
25. $3x^3 - 12x^2 - 45x$
26. $3x^2 + 24$
27. $x^2 + 16x + 64$

Factor completely.

28. $75x^3 - 27x$
29. $3x^3 - 12x^2 - 36x$
30. $4x^3 - 44x^2 + 112x$
31. $5y^2 - 125$
32. $3x^2y^2 - xy^2 - 4y^2$
33. $x^3 + 10x^2 - 24x$
34. $3x^3 - 6x^2 - 45x$
35. $3x^2 - 27$
36. $x^4 - 16$

Factor each of the following completely. Use the modified diamond approach.

37. $2x^2 + 5x - 7$

38. $3x^2 - 13x + 4$

39. $2x^2 + 9x + 10$

40. $4x^2 - 13x + 3$

41. $4x^2 + 12x + 5$

42. $6x^3 + 31x^2 + 5x$

43. $64x^2 + 16x + 1$

44. $7x^2 - 33x - 10$

45. $5x^2 + 12x - 9$

## Answers

1. $(x + 6)(x - 7)$
2. $2(2x^2 - 9)$
3. $(2x + 3)(x + 3)$
4. $(2x + y)(x + y)$
5. $(2x + 3)(3x - 5)$
6. $(2x - 5)(2x + 5)$
7. $(x - 14)^2$
8. $7(x - 11)(x + 11)$
9. $(x + 9)^2$
10. $(x + 7)(x - 3)$
11. $3x(x + 7)$
12. $(x - 8)(3x + 4)$
13. $(3x - 4)(3x + 4)$
14. $(2x + 5)^2$
15. $(x - 3)(x - 2)$
16. $5x(x + 4)(x - 1)$
17. $2(2x^2 + 9)$
18. $(x - 6)^2$
19. $(x - 9)(x + 6)$
20. $3(2x^2 - 7)$
21. $(2x + 3)(x + 6)$
22. $(4x + 1)(4x - 1)$
23. $(x - 7)^2$
24. $(x + 3)(x + 5)$
25. $3x(x^2 - 4x - 15)$
26. $3(x^2 + 8)$
27. $(x + 8)^2$
28. $3x(5x - 3)(5x + 3)$
29. $3x(x - 6)(x + 2)$
30. $4x(x - 7)(x - 4)$
31. $5(y + 5)(y - 5)$
32. $y^2(3x - 4)(x + 1)$
33. $x(x + 12)(x - 2)$
34. $3x(x - 5)(x + 3)$
35. $3(x - 3)(x + 3)$
36. $(x - 2)(x + 2)(x^2 + 4)$
37. $(2x + 7)(x - 1)$
38. $(3x - 1)(x - 4)$
39. $(x + 2)(2x + 5)$
40. $(4x - 1)(x - 3)$
41. $(2x + 5)(2x + 1)$
42. $x(6x + 1)(x + 5)$
43. $(8x + 1)^2$
44. $(7x + 2)(x - 5)$
45. $(5x - 3)(x + 3)$

## ZERO PRODUCT PROPERTY AND QUADRATICS #9

If $a \cdot b = 0$, then either $a = 0$ or $b = 0$.

Note that this property states that at least one of the factors MUST be zero. It is also possible that all of the factors are zero. This simple statement gives us a powerful result which is most often used with equations involving the products of binomials. For example, solve $(x + 5)(x - 2) = 0$.

By the Zero Product Property, since $(x + 5)(x - 2) = 0$, either $x + 5 = 0$ or $x - 2 = 0$. Thus, $x = -5$ or $x = 2$.

The Zero Product Property can be used to find where a quadratic crosses the x-axis. These points are the x-intercepts. In the example above, they would be (-5, 0) and (2, 0).

### Example 1

Where does $y = (x + 3)(x - 7)$ cross the x-axis? Since $y = 0$ at the x-axis, then $(x + 3)(x - 7) = 0$ and the Zero Product Property tells you that $x = -3$ and $x = 7$ so $y = (x + 3)(x - 7)$ crosses the x-axis at (-3, 0) and (7, 0).

### Example 2

Where does $y = x^2 - x - 6$ cross the x-axis? First factor $x^2 - x - 6$ into $(x + 2)(x - 3)$ to get $y = (x + 2)(x - 3)$. By the Zero Product Property, the x-intercepts are (-2, 0) and (3, 0).

### Example 3

Graph $y = x^2 - x - 6$

Since you know the x-intercepts from example 2, you already have two points to graph. You need a table of values to get additional points.

| x | –2 | –1 | 0 | 1 | 2 | 3 | 4 |
|---|---|---|---|---|---|---|---|
| | 0 | –4 | –6 | –6 | –4 | 0 | 6 |

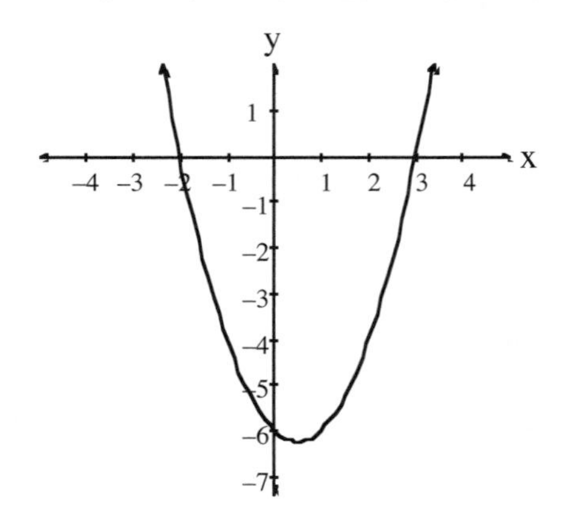


### Example 4

Graph $y > x^2 - x - 6$

First graph $y = x^2 - x - 6$ as you did at left. Use a dashed curve. Second, pick a point not on the parabola and substitute it into the inequality. For example, testing point (0, 0) in $y > x^2 - x - 6$ gives $0 > -6$ which is a true statement. This means that (0, 0) is a solution to the inequality as well as all points inside the curve. Shade the interior of the parabola.

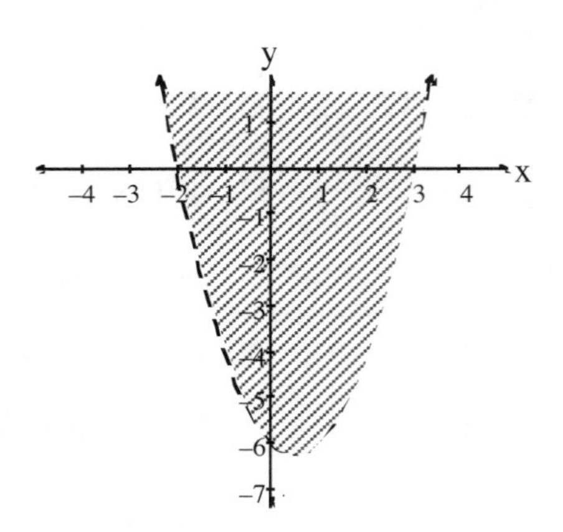

Solve the following problems using the Zero Product Property.

1. $(x - 2)(x + 3) = 0$
2. $2x(x + 5)(x + 6) = 0$
3. $(x - 18)(x - 3) = 0$
4. $4x^2 - 5x - 6 = 0$
5. $(2x - 1)(x + 2) = 0$
6. $2x(x - 3)(x + 4) = 0$
7. $3x^2 - 13x - 10 = 0$
8. $2x^2 - x = 15$

Use factoring and the Zero Product Property to find the x-intercepts of each parabola below. Express your answer as ordered pair(s).

9. $y = x^2 - 3x + 2$
10. $y = x^2 - 10x + 25$
11. $y = x^2 - x - 12$
12. $y = x^2 - 4x - 5$
13. $y = x^2 + 2x - 8$
14. $y = x^2 + 6x + 9$
15. $y = x^2 - 8x + 16$
16. $y = x^2 - 9$

Graph the following inequalities. Be sure to use a test point to determine which region to shade. Your solutions to the previous problems might be helpful.

17. $y < x^2 - 3x + 2$
18. $y > x^2 - 10x + 25$
19. $y \le x^2 - x - 12$
20. $y \ge x^2 - 4x - 5$
21. $y > x^2 + 2x - 8$
22. $y \ge x^2 + 6x + 9$
23. $y < x^2 - 8x + 16$
24. $y \le x^2 - 9$

## Answers

1. $x = 2,\ x = -3$
2. $x = 0,\ x = -5,\ x = -6$
3. $x = 18,\ x = 3$
4. $x = -0.75,\ x = 2$
5. $x = 0.5,\ x = -2$
6. $x = 0,\ x = 3,\ x = -4$
7. $x = -\frac{2}{3},\ x = 5$
8. $x = -2.5,\ x = 3$
9. (1, 0) and (2, 0)
10. (5, 0)
11. (–3, 0) and (4, 0)
12. (5, 0) and (–1, 0)
13. (–4, 0) and (2, 0)
14. (–3, 0)
15. (4, 0)
16. (3, 0) and (–3, 0)

17.

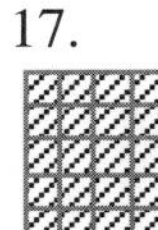

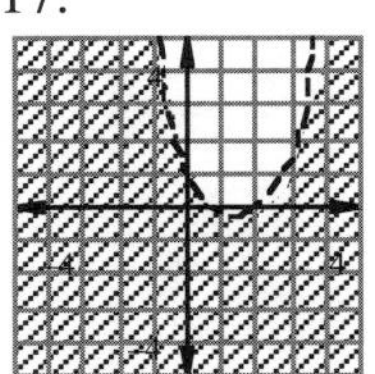

18.

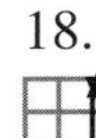

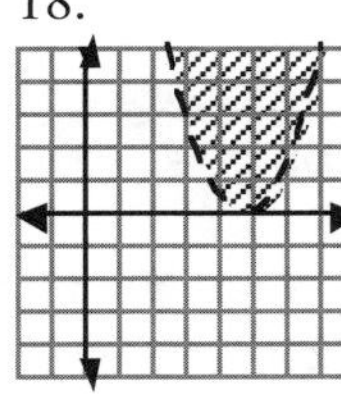

19.

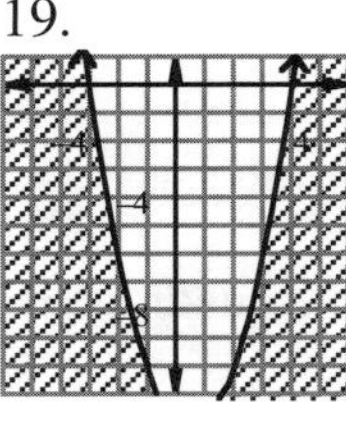

20.

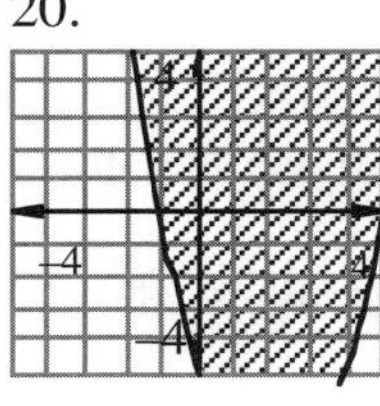

21.

22.

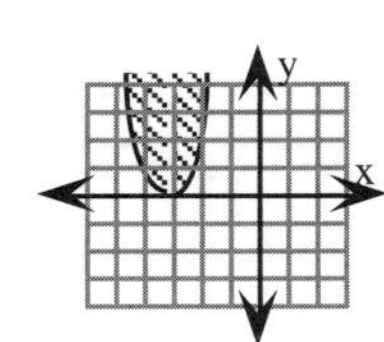

23.

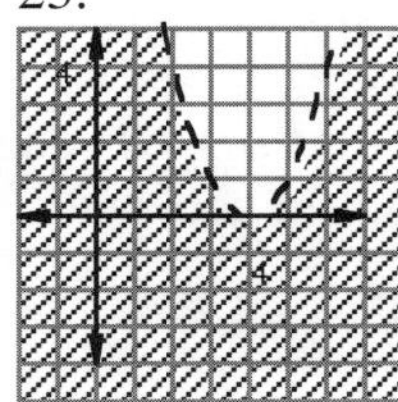

24.

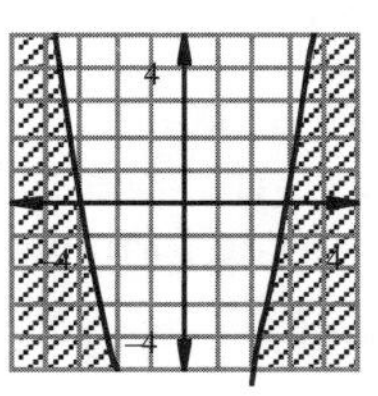

## THE QUADRATIC FORMULA #10

You have used factoring and the Zero Product Property to solve quadratic equations. You can solve any quadratic equation by using the **QUADRATIC FORMULA.**

**If $ax^2 + bx + c = 0$, then $x = \dfrac{-b \pm \sqrt{b^2 - 4ac}}{2a}$.**

For example, suppose $3x^2 + 7x - 6 = 0$. Here a = 3, b = 7, and c = -6.
Substituting these values into the formula results in:

$$x = \frac{-(7) \pm \sqrt{7^2 - 4(3)(-6)}}{2(3)} \Rightarrow x = \frac{-7 \pm \sqrt{121}}{6} \Rightarrow x = \frac{-7 \pm 11}{6}$$

Remember that non-negative numbers have both a positive and negative square root.
The sign $\pm$ represents this fact for the square root in the formula and allows us to write the equation once (representing two possible solutions) until later in the solution process.

Split the numerator into the two values: $x = \dfrac{-7 + 11}{6}$ or $x = \dfrac{-7 - 11}{6}$

Thus the solution for the quadratic equation is: $x = \frac{2}{3}$ or -3.

### Example 1

Solve $x^2 + 3x - 2 = 0$ using the quadratic formula.

First, identify the values for a, b, and c. In this case they are 1, 3, and –2, respectively. Next, substitute these values into the quadratic formula.

$$x = \frac{-(3) \pm \sqrt{3^2 - 4(1)(-2)}}{2(1)} \Rightarrow x = \frac{-3 \pm \sqrt{17}}{2}$$

Then split the numerator into the two values: $x = \dfrac{-3 + \sqrt{17}}{2}$ or $x = \dfrac{-3 - \sqrt{17}}{2}$

Using a calculator, the solution for the quadratic equation is: x = 0.56 or –3.56.

### Example 2

Solve $4x^2 + 4x = 3$ using the quadratic formula.

To solve any quadratic equation it must first be equal to zero. Rewrite the equation as $4x^2 + 4x - 3 = 0$. Identify the values for a, b, and c: 4, 4, and -3, respectively. Substitute these values into the quadratic formula.

$$x = \frac{-(4) \pm \sqrt{4^2 - 4(4)(-3)}}{2(4)} \Rightarrow x = \frac{-4 \pm \sqrt{64}}{8} \Rightarrow x = \frac{-4 \pm 8}{8}$$

Split the numerator into the two values: $x = \dfrac{-4 + 8}{8}$ or $x = \dfrac{-4 - 8}{8}$, so $x = \dfrac{1}{2}$ or $-\dfrac{3}{2}$.

Use the quadratic formula to solve each of the following equations.

1. $x^2 - x - 6 = 0$
2. $x^2 + 8x + 15 = 0$
3. $x^2 + 13x + 42 = 0$
4. $x^2 - 10x + 16 = 0$
5. $x^2 + 5x + 4 = 0$
6. $x^2 - 9x + 18 = 0$
7. $5x^2 - x - 4 = 0$
8. $4x^2 - 11x - 3 = 0$
9. $6x^2 - x - 15 = 0$
10. $6x^2 + 19x + 15 = 0$
11. $3x^2 + 5x - 28 = 0$
12. $2x^2 - x - 14 = 0$
13. $4x^2 - 9x + 4 = 0$
14. $2x^2 - 5x + 2 = 0$
15. $20x^2 + 20x = 1$
16. $13x^2 - 16x = 4$
17. $7x^2 + 28x = 0$
18. $5x^2 = -125x$
19. $8x^2 - 50 = 0$
20. $15x^2 = 3$

## Answers

1. $x = -2, 3$
2. $x = -5, -3$
3. $x = -7, -6$
4. $x = 2, 8$
5. $x = -4, -1$
6. $x = 3, 6$
7. $x = -\frac{4}{5}, 1$
8. $x = -\frac{1}{4}, 3$
9. $x = -\frac{3}{2}, \frac{5}{3}$
10. $x = -\frac{3}{2}, -\frac{5}{3}$
11. $x = -4, \frac{7}{3}$
12. $x = \frac{1 \pm \sqrt{113}}{4}$
13. $x = \frac{9 \pm \sqrt{17}}{8}$
14. $x = 2, \frac{1}{2}$
15. $x = \frac{-20 \pm \sqrt{480}}{40} = \frac{-5 \pm \sqrt{30}}{10}$
16. $x = \frac{16 \pm \sqrt{464}}{26} = \frac{8 \pm 2\sqrt{29}}{13}$
17. $x = -4, 0$
18. $x = -25, 0$
19. $x = -\frac{5}{2}, \frac{5}{2}$
20. $x = \frac{\pm \sqrt{5}}{5}$

## TRIANGLE CONGRUENCE #11

If two triangles are congruent, then all six of their corresponding pairs of parts are also congruent. In other words, if $\Delta ABC \cong \Delta XYZ$, then all the congruence statements at right are correct. Note that the matching parts follow the order of the letters as written in the triangle congruence statement. For example, if A is first in one statement and X is first in the other, then A always corresponds to X in congruence statements for that relationship or figure.

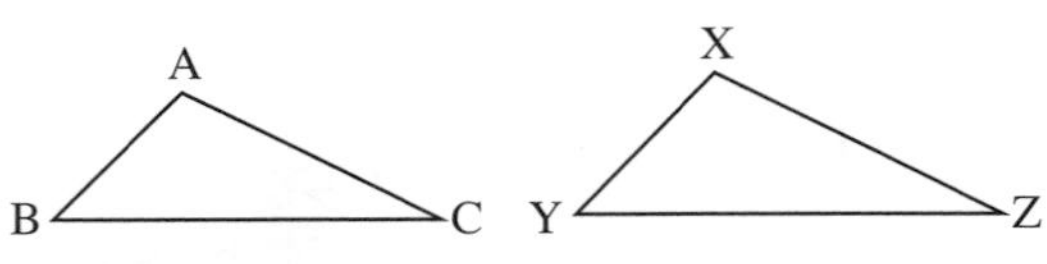


$\overline{AB} \cong \overline{XY}$  $\angle A \cong \angle X$
$\overline{BC} \cong \overline{YZ}$  $\angle B \cong \angle Y$
$\overline{AC} \cong \overline{XZ}$  $\angle C \cong \angle Z$

You only need to know that three pairs of parts are congruent (sometimes in a certain order) to prove that the two triangles are congruent. The Triangle Congruence Properties are: SSS (Side-Side-Side), SAS (Side-Angle-Side; must be in this order), ASA (Angle-Side-Angle), and HL (Hypotenuse-Leg). AAS is an acceptable variation of ASA.

The four Triangle Congruence Properties are the only correspondences that may be used to prove that two triangles are congruent. Once two triangles are know to be congruent, then the other pairs of their corresponding parts are congruent. In this course, you may justify this conclusion with the statement "congruent triangles give us congruent parts." Remember: only use this statement <u>after</u> you have shown the two triangles are congruent.

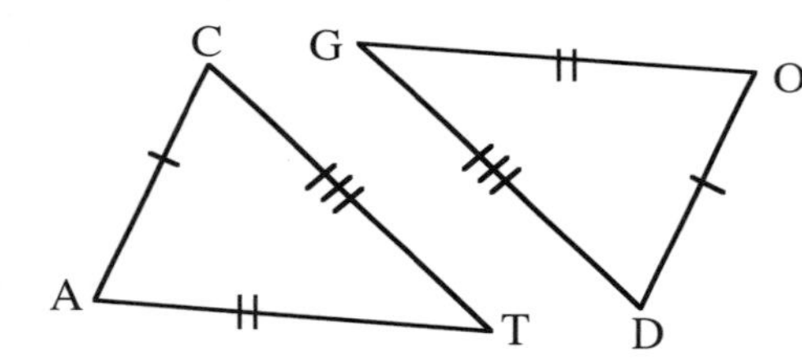


### Example 1

In the figure at left, $\overline{CA} \cong \overline{DO}$, $\overline{CT} \cong \overline{DG}$, and $\overline{AT} \cong \overline{OG}$. Thus, $\Delta CAT \cong \Delta DOG$ because of SSS. Now that the triangles are congruent, it is also true that $\angle C \cong \angle D$, $\angle A \cong \angle O$, and $\angle T \cong \angle G$.

### Example 2

In the figure at right, $\overline{RD} \cong \overline{CP}$, $\angle D \cong \angle P$, and $\overline{ED} \cong \overline{AP}$. Thus $\Delta RED \cong \Delta CAP$ because of SAS. Now $\overline{RE} \cong \overline{CA}$, $\angle R \cong \angle C$, and $\angle E \cong \angle A$.

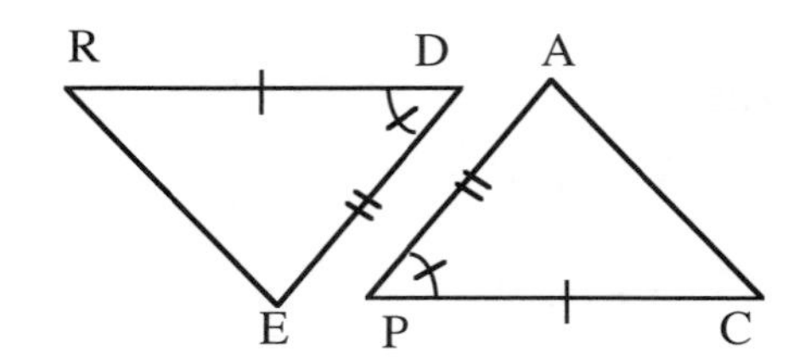


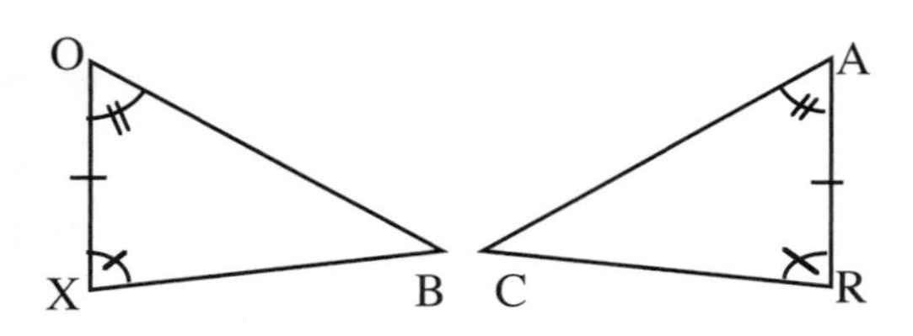


### Example 3

In the figure at left, $\angle O \cong \angle A$, $\overline{OX} \cong \overline{AR}$, and $\angle X \cong \angle R$. Thus, $\Delta BOX \cong \Delta CAR$ because of ASA. Now $\angle B \cong \angle C$, $\overline{BO} \cong \overline{CA}$, and $\overline{BX} \cong \overline{CR}$.

## Example 4

In the figure at right, there are two right angles ($m\angle N = m\angle Y = 90°$) and $\overline{MO} \cong \overline{KE}$, and $\overline{ON} \cong \overline{EY}$. Thus, $\Delta MON \cong \Delta KEY$ because of HL, so $\overline{MN} \cong \overline{KY}$, $\angle M \cong \angle K$, and $\angle O \cong \angle E$.

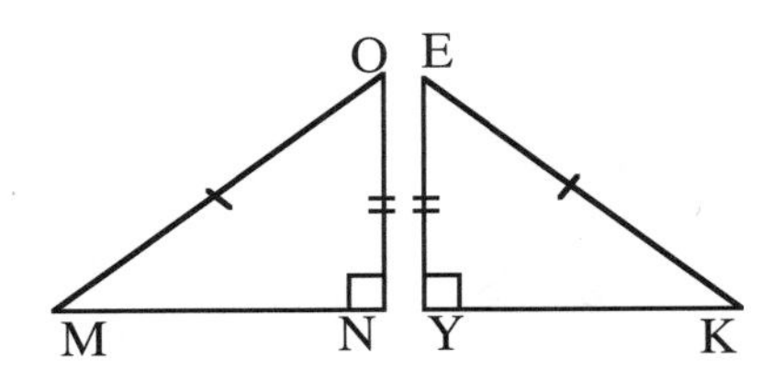


## Example 5

Refer to page 196 in the textbook (problem CG-41) and do each part, (a) through (i). Brief explanations and the answers appear below.

a) $\Delta PQR \cong \Delta ZXY$ by SSS

b) $\angle BAE \cong \angle DEC$ (vertical angles), $\Delta BAE \cong \Delta DEC$ by SAS .

c) $\overline{PR} \cong \overline{PR}$, $\Delta QPR \cong \Delta SPR$ by SSS.

d) These triangles are not necessarily congruent because we don't know the relative sizes of the sides. Remember that AAA is not a congruence property.

e) $\angle E \cong \angle H$ because the three angles in any triangle must add up to 180°. $\Delta DEF \cong \Delta GHJ$ by SAS.

f) $\Delta YXW \cong \Delta LKJ$ by SAS.

g) Not enough information; pairs of corresponding parts.

h) $\overline{QS} \cong \overline{QS}$, $\Delta RSQ \cong \Delta PQS$ by ASA.

i) Not necessarily congruent since the one side we know is not between the two angles of the second triangle.

Briefly explain if each of the following pairs of triangles are congruent or not. If so, state the Triangle Congruence Property that supports your conclusion.

1.

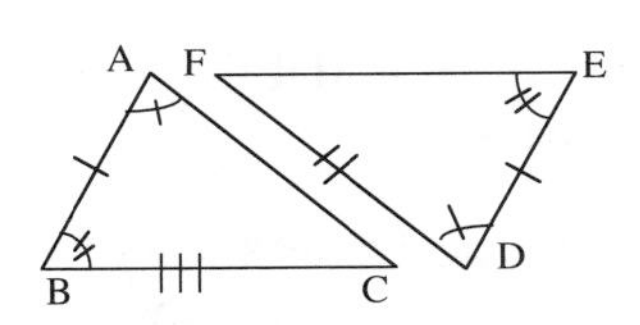


2.

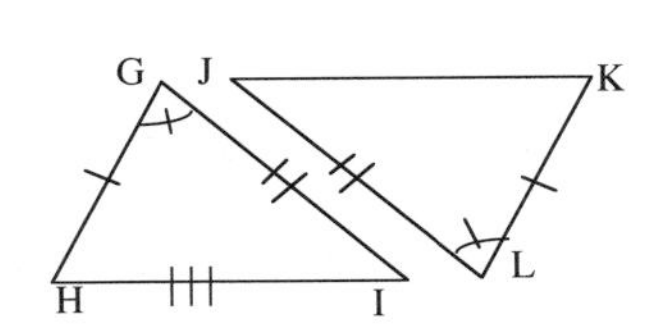


3.

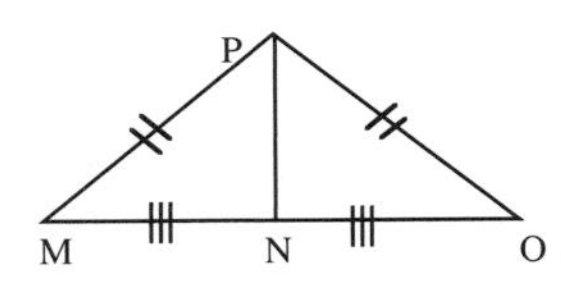


4.

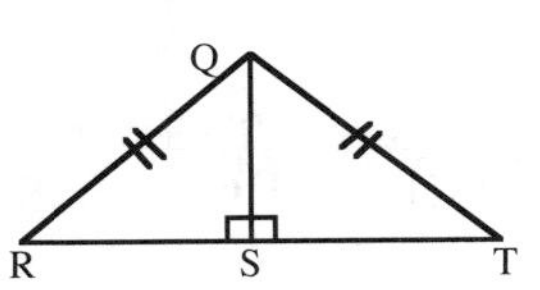


5.

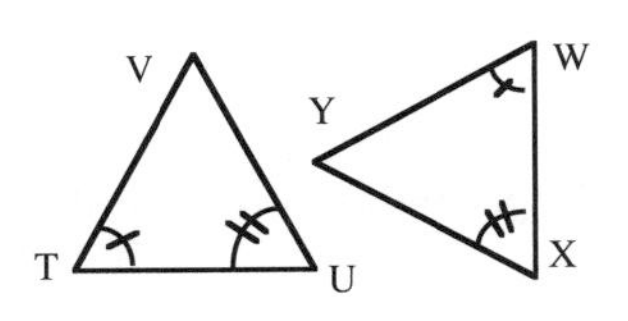


6.

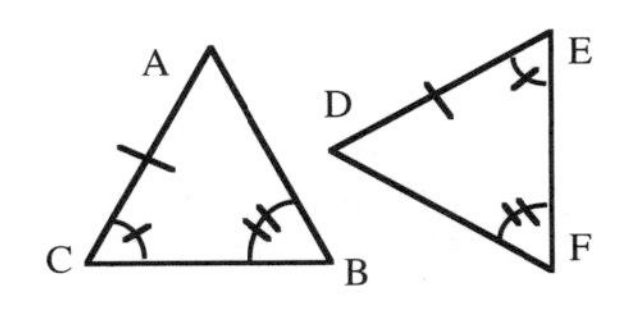


7.

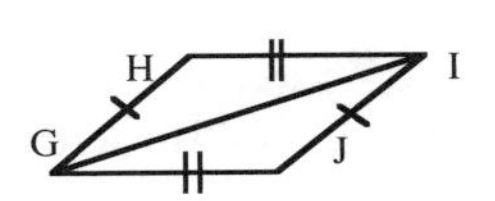


8.

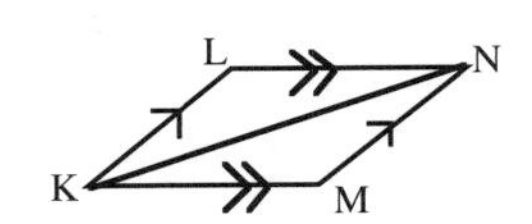


9.

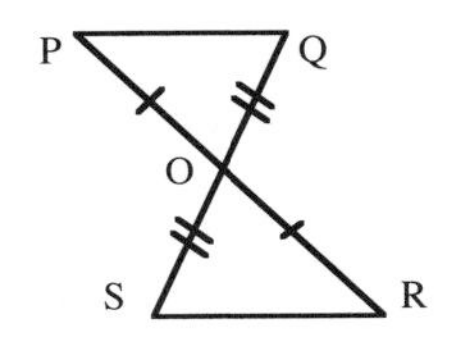

10.

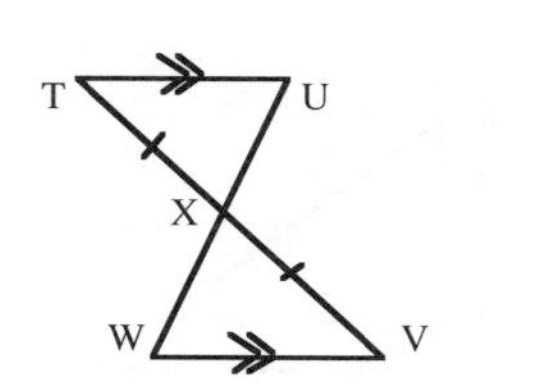


11.

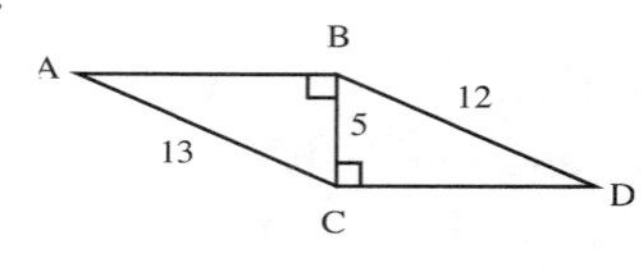


12.

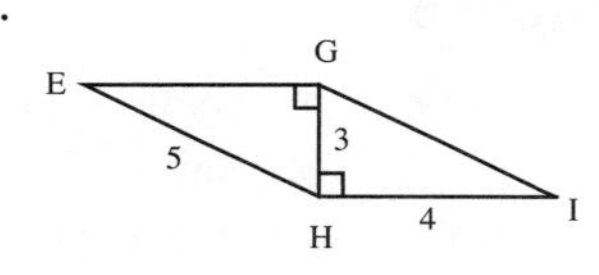


13.

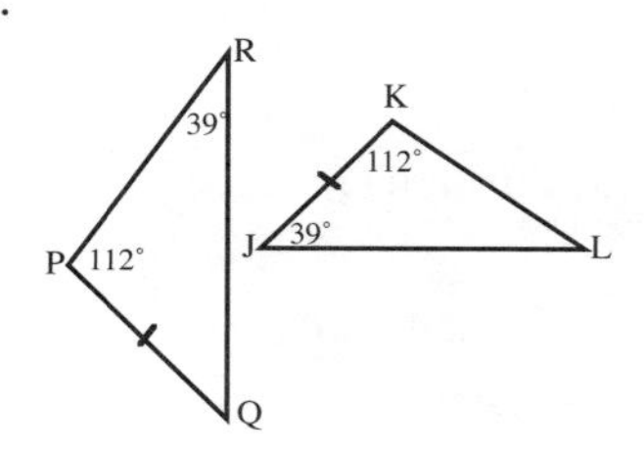


14.

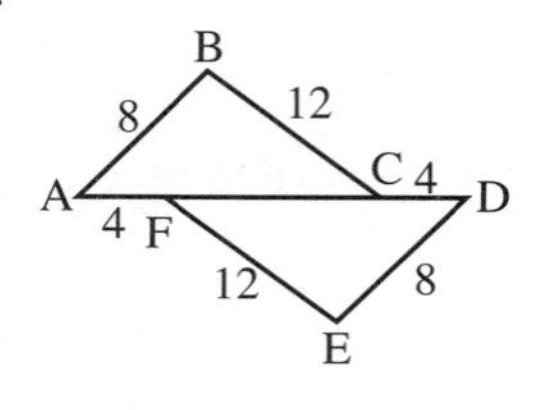


15.

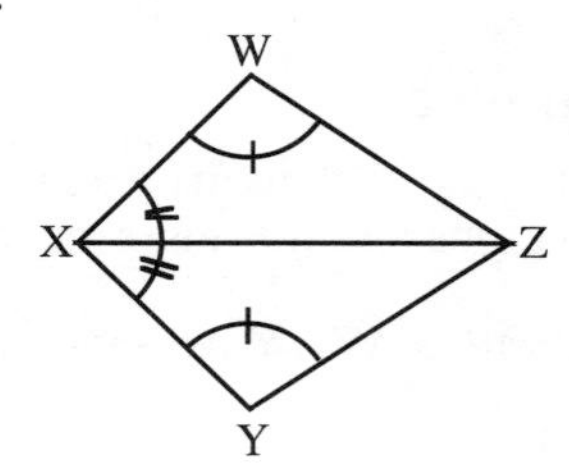


## Answers

1. $\Delta ABC \cong \Delta DEF$ by ASA
2. $\Delta GIH \cong \Delta LJK$ by SAS
3. $\Delta PNM \cong \Delta PNO$ by SSS
4. $\overline{QS} \cong \overline{QS}$, so $\Delta QRS \cong \Delta QTS$ by HL
5. The triangles are not necessarily congruent.
6. $\Delta ABC \cong \Delta DFE$ by ASA or AAS.
7. $\overline{GI} \cong \overline{GI}$, so $\Delta GHI \cong \Delta IJG$ by SSS.
8. Alternate interior angles = used twice, so $\Delta KLN \cong \Delta NMK$ by ASA.
9. Vertical angles = at 0, so $\Delta POQ \cong \Delta ROS$ by SAS.
10. Vertical angles and/or alternate interior angles =, so $\Delta TUX \cong \Delta VWX$ by ASA.
11. No, the length of each hypotenuse is different.
12. Pythagorean Theorem, so $\Delta EGH \cong \Delta IHG$ by SSS.
13. Sum of angles of triangle = 180°, but since the equal angles do not correspond, the triangles are not conguent.
14. AF + FC = FC + CD, so $\Delta ABC \cong \Delta DEF$ by SSS.
15. $\overline{XZ} \cong \overline{XZ}$, so $\Delta WXZ \cong \Delta YXZ$ by AAS.

## LAWS OF EXPONENTS #12

### BASE, EXPONENT, AND VALUE

In the expression $2^5$, 2 is the **base**, 5 is the **exponent**, and the **value** is 32.

$2^5$ means $2 \cdot 2 \cdot 2 \cdot 2 \cdot 2 = 32$ $x^3$ means $x \cdot x \cdot x$

### LAWS OF EXPONENTS

Here are the basic patterns with examples:

1) $x^a \cdot x^b = x^{a+b}$ examples: $x^3 \cdot x^4 = x^{3+4} = x^7$; $2^7 \cdot 2^4 = 2^{11}$

2) $\frac{x^a}{x^b} = x^{a-b}$ examples: $x^{10} \div x^4 = x^{10-4} = x^6$; $\frac{2^4}{2^7} = 2^{-3}$

3) $(x^a)^b = x^{ab}$ examples: $(x^4)^3 = x^{4 \cdot 3} = x^{12}$; $(2x^3)^4 = 2^4 \cdot x^{12} = 16x^{12}$

4) $x^{-a} = \frac{1}{x^a}$ and $\frac{1}{x^{-b}} = x^b$ examples: $3x^{-3}y^2 = \frac{3y^2}{x^3}$; $\frac{2x^5}{y^{-2}} = 2x^5y^2$

## Example 1

Simplify: $\left(2xy^3\right)\left(5x^2y^4\right)$

Multiply the coefficients: $2 \cdot 5 \cdot xy^3 \cdot x^2y^4 = 10xy^3 \cdot x^2y^4$

Add the exponents of x, then y: $10x^{1+2}y^{3+4} = 10x^3y^7$

## Example 2

Simplify: $\frac{14x^2y^{12}}{7x^5y^7}$

Divide the coefficients: $\frac{(14 \div 7)x^2y^{12}}{x^5y^7} = \frac{2x^2y^{12}}{x^5y^7}$

Subtract the exponents: $2x^{2-5}y^{12-7} = 2x^{-3}y^5$ OR $\frac{2y^5}{x^3}$

## Example 3

Simplify: $\left(3x^2y^4\right)^3$

Cube each factor: $3^3 \cdot \left(x^2\right)^3 \cdot \left(y^4\right)^3 = 27\left(x^2\right)^3\left(y^4\right)^3$

Multiply the exponents: $27x^6y^{12}$

Simplify each expression:

1. $y^5 \cdot y^7$
2. $b^4 \cdot b^3 \cdot b^2$
3. $8^6 \cdot 8^2$
4. $\left(y^5\right)^2$
5. $(3a)^4$
6. $\frac{m^8}{m^3}$
7. $\frac{12x^9}{4x^4}$
8. $\left(x^3y^2\right)^3$
9. $\frac{\left(y^4\right)^2}{\left(y^3\right)^2}$
10. $\frac{15x^2y^7}{3x^4y^5}$
11. $\left(4c^4\right)\left(ac^3\right)\left(3a^5c\right)$
12. $\left(7x^3y^5\right)^2$
13. $\left(4xy^2\right)(2y)^3$
14. $\left(\frac{4}{x^2}\right)^3$
15. $\frac{\left(2a^7\right)\left(3a^2\right)}{6a^3}$
16. $\left(\frac{5m^3n}{m^5}\right)^3$
17. $\left(3a^2x^3\right)^2\left(2ax^4\right)^3$
18. $\left(\frac{x^3y}{y^4}\right)^4$
19. $\left(\frac{6y^2x^8}{12x^3y^7}\right)^2$
20. $\frac{\left(2x^5y^3\right)^3\left(4xy^4\right)^2}{8x^7y^{12}}$

## Answers

1. $y^{12}$
2. $b^9$
3. $8^8$
4. $y^{10}$
5. $81a^4$
6. $m^5$
7. $3x^5$
8. $x^9y^6$
9. $y^2$
10. $\frac{5y^2}{x^2}$
11. $12a^6c^8$
12. $49x^6y^{10}$
13. $32xy^5$
14. $\frac{64}{x^6}$
15. $a^6$
16. $\frac{125n^3}{m^6}$
17. $72a^7x^{18}$
18. $\frac{x^{12}}{y^{12}}$
19. $\frac{x^{10}}{4y^{10}}$
20. $16x^{10}y^5$

## PROOF #13

A proof convinces an audience that a conjecture is true for ALL cases (situations) that fit the conditions of the conjecture. For example, "If a polygon is a triangle on a flat surface, then the sum of the measures of the angles is 180°." Because we proved this conjecture near the end of Unit 3, it is always true. There are many formats that may be used to write a proof. This course explores three of them, namely, paragraph, flow chart, and two-column.

## Example

If $\overline{BD}$ is a perpendicular bisector of $\overline{AC}$, prove that $\Delta ABC$ isosceles.

### Paragraph proof

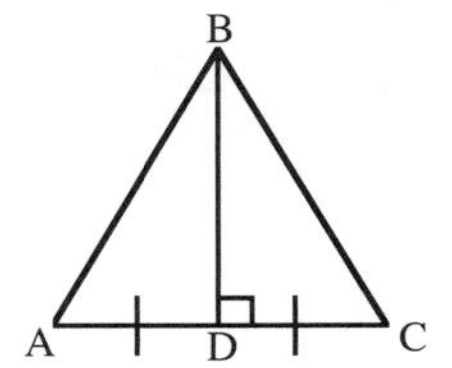


To prove that $\Delta ABC$ is isosceles, show that $\overline{BA} \cong \overline{BC}$. We can do this by showing that the two segments are corresponding parts of congruent triangles.

Since $\overline{BD}$ is perpendicular to $\overline{AC}$, $m\angle BDA = m\angle BDC = 90°$.

Since $\overline{BD}$ bisects $\overline{AC}$, $\overline{AD} \cong \overline{CD}$. With $\overline{BD} \cong \overline{BD}$ (reflexive property), $\Delta ADB \cong \Delta CDB$ by SAS.

Finally, $\overline{BA} \cong \overline{BC}$ because corresponding parts of congruent triangles are congruent. Therefore, $\Delta ABC$ must be isosceles since two of the three sides are congruent.

### Flow chart proof

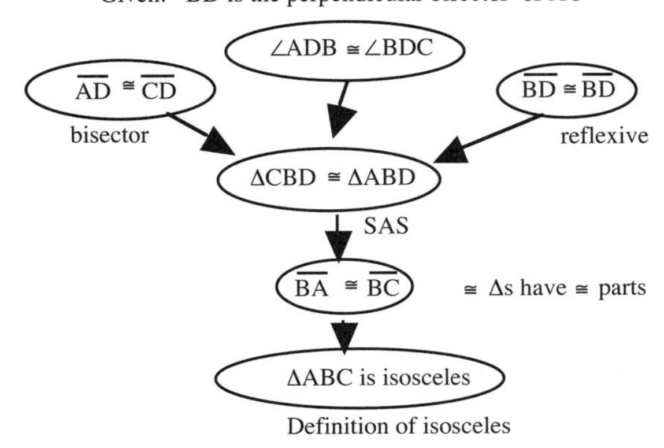


### Two-Column Proof

Given: $\overline{BD}$ is a bisector of $\overline{AC}$.
$\overline{BD}$ is perpendicular to $\overline{AC}$.

Prove: $\Delta$ ABC is isosceles

| Statement | Reason |
|---|---|
| $\overline{BD}$ bisects $\overline{AC}$. | Given |
| $\overline{BD} \perp \overline{AC}$ | Given |
| $\overline{AD} \cong \overline{CD}$ | Def. of bisector |
| $\angle ADB$ and $\angle BDC$ are right angles | Def. of perpendicular |
| $\angle ADB \cong \angle BDC$ | All right angles are $\cong$. |
| $\overline{BD} \cong \overline{BD}$ | Reflexive property |
| $\Delta ABD \cong \Delta CBD$ | S.A.S. |
| $\overline{AB} \cong \overline{CB}$ | $\cong$ Δ's have $\cong$ parts |
| $\therefore \Delta ABC$ is isosceles | Def. of isosceles |

In each diagram below, are any triangles congruent? If so, prove it. (Note: It is good practice to try different methods for writing your proofs.)

1.

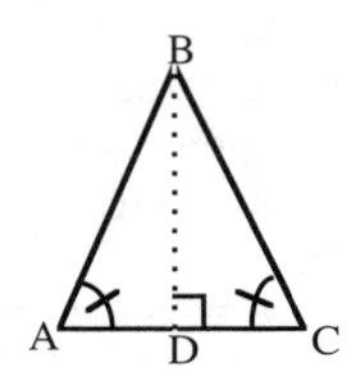


2.

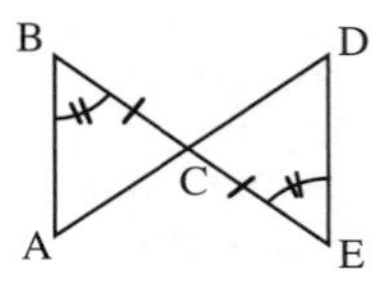


3.

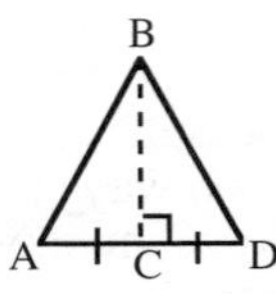


4.

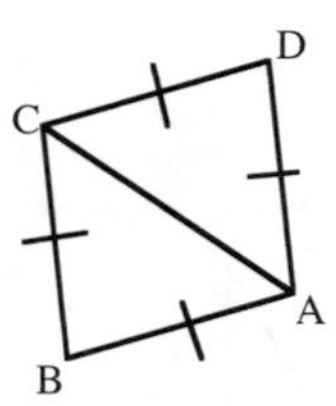


5.

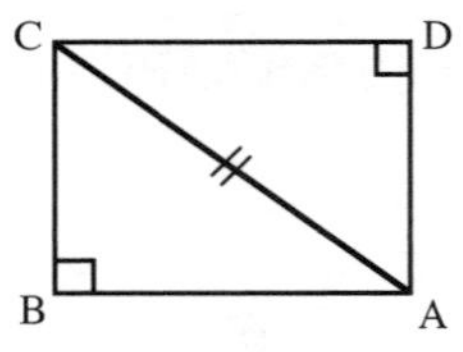


6.

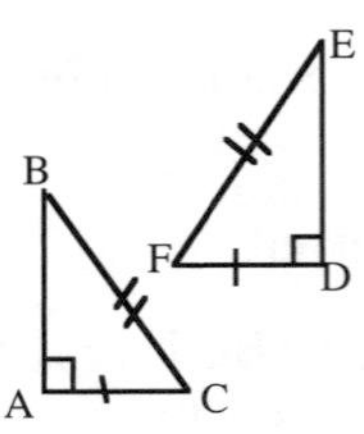


Complete a proof for each problem below in the style of your choice.

7. Given: $\overline{TR}$ and $\overline{MN}$ bisect each other.
   Prove: $\Delta NTP \cong \Delta MRP$

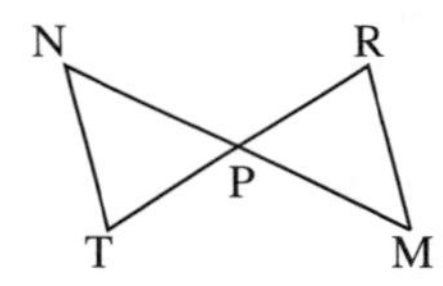


8. Given: $\overline{CD}$ bisects $\angle ACB$; $\angle 1 \cong \angle 2$.
   Prove: $\Delta CDA \cong \Delta CDB$

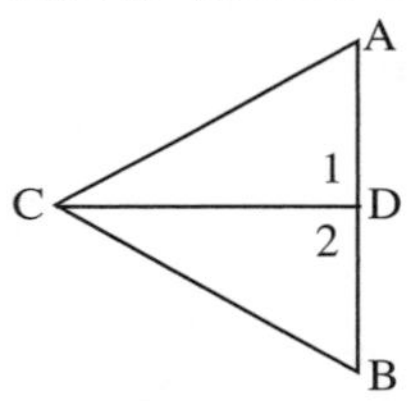


9. Given: $\overline{AB} \parallel \overline{CD}$, $\angle B \cong \angle D$, $\overline{AB} \cong \overline{CD}$
   Prove: $\Delta ABF \cong \Delta CED$

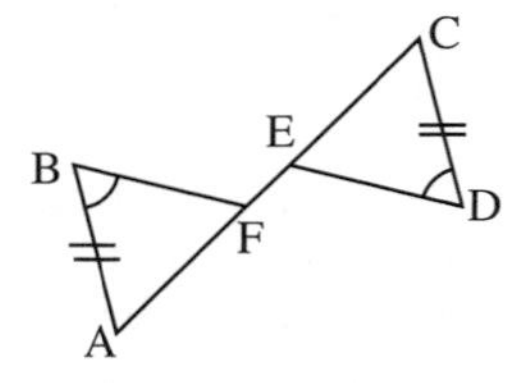


10. Given: $\overline{PG} \cong \overline{SG}$, $\overline{TP} \cong \overline{TS}$
    Prove: $\Delta TPG \cong \Delta TSG$

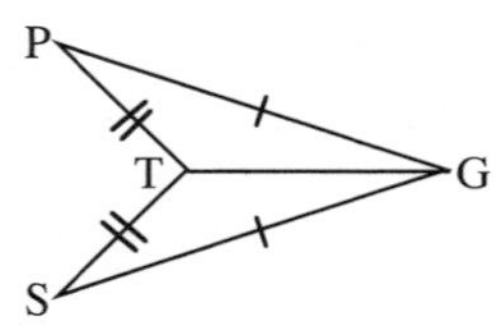


11. Given: $\overline{OE} \perp \overline{MP}$, $\overline{OE}$ bisects $\angle MOP$
    Prove: $\Delta MOE \cong \Delta POE$

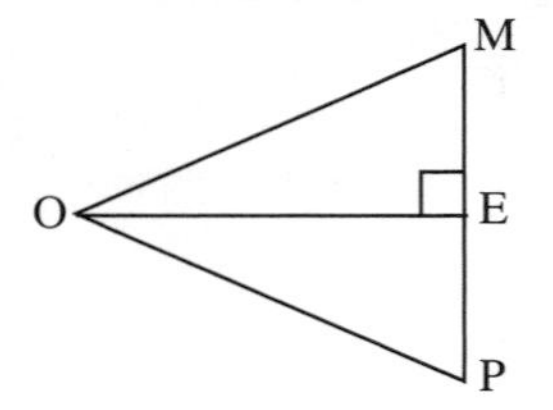


12. Given: $\overline{AD} \parallel \overline{BC}$, $\overline{DC} \parallel \overline{BA}$
    Prove: $\Delta ADB \cong \Delta CBD$

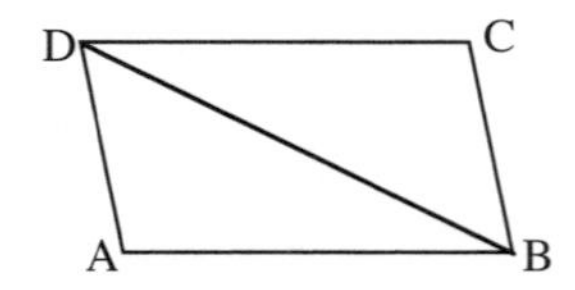

13. Given: $\overline{AC}$ bisects $\overline{DE}$, $\angle A \cong \angle C$
Prove: $\Delta ADB \cong \Delta CEB$

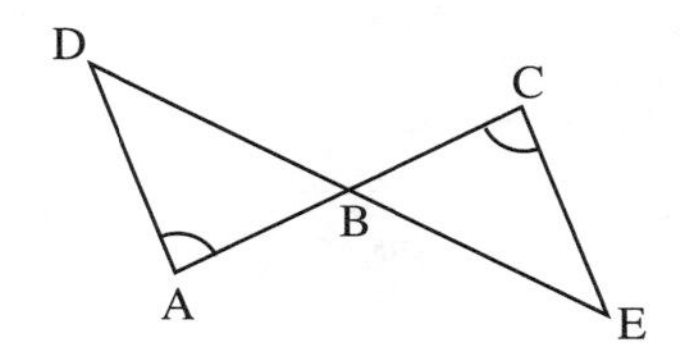


14. Given: $\overline{PQ} \perp \overline{RS}$, $\angle R \cong \angle S$
Prove: $\Delta PQR \cong \Delta PQS$

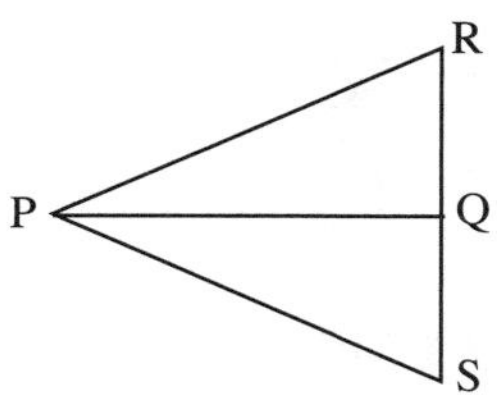


15. Given: $\angle S \cong \angle R$, $\overline{PQ}$ bisects $\angle SQR$
Prove: $\Delta SPQ \cong \Delta RPQ$

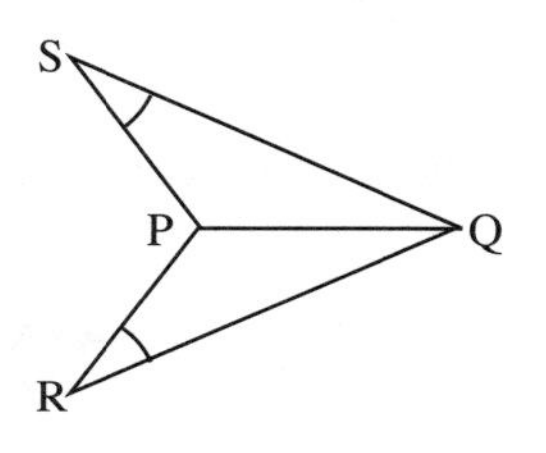


16. Given: $\overline{TU} \cong \overline{GY}$, $\overline{KY} \parallel \overline{HU}$,
$\overline{KT} \perp \overline{TG}$, $\overline{HG} \perp \overline{TG}$
Prove: $\angle K \cong \angle H$

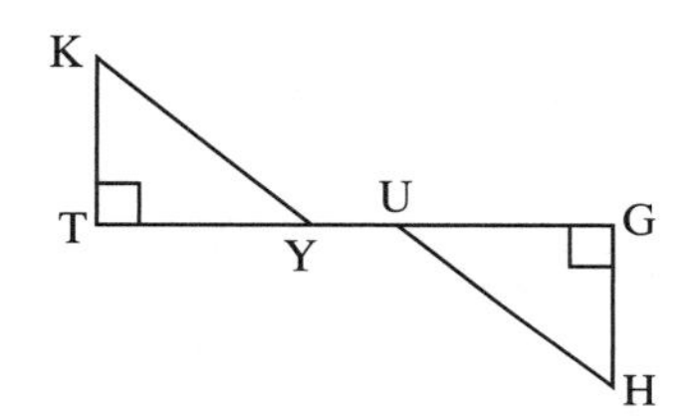


17. Given: $\overline{MQ} \parallel \overline{WL}$, $\overline{MQ} \cong \overline{WL}$
Prove: $\overline{ML} \parallel \overline{WQ}$

Q W
M L

Consider the diagram below.

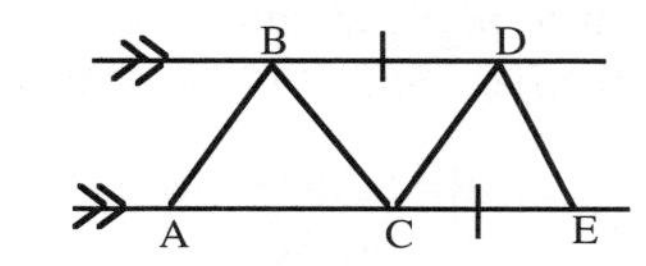


18. Is $\Delta BCD \cong \Delta EDC$? Prove it!

19. Is $\overline{AB} \cong \overline{DC}$? Prove it!

20. Is $\overline{AB} \cong \overline{ED}$? Prove it!

## Answers

1. Yes

$\angle BAD \cong \angle BCD$ — Given
$\angle BDC \cong \angle BDA$ — right $\angle$'s are $\cong$
$\overline{BD} \cong \overline{BD}$ — Reflexive
$\Delta ABD \cong \Delta CBD$ — AAS

2. Yes

$\angle B \cong \angle E$ — Given
$\angle BCA \cong \angle BCD$ — vertical $\angle$s are $\cong$
$\overline{BC} \cong \overline{CE}$ — Given
$\Delta ABC \cong \Delta DEC$ — ASA

3. Yes

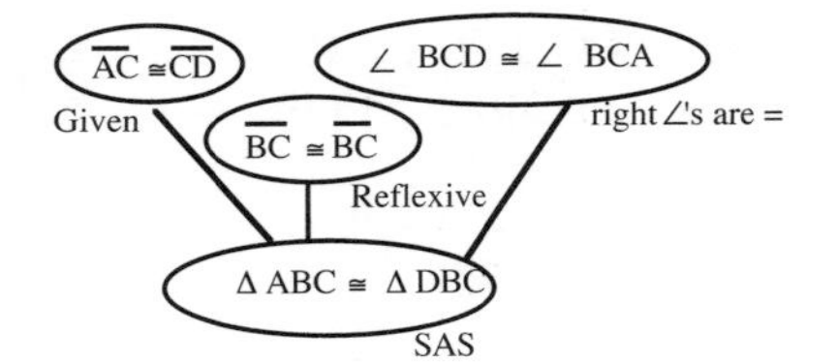


4. Yes

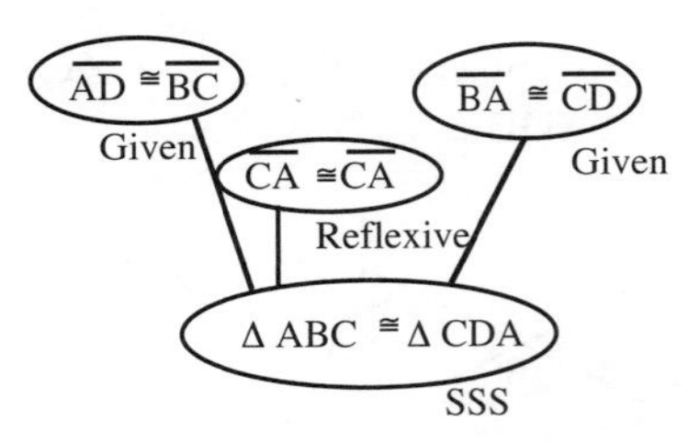


5. Not necessarily. Counterexample:

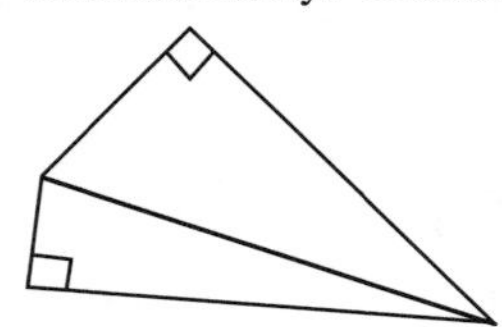

6. Yes

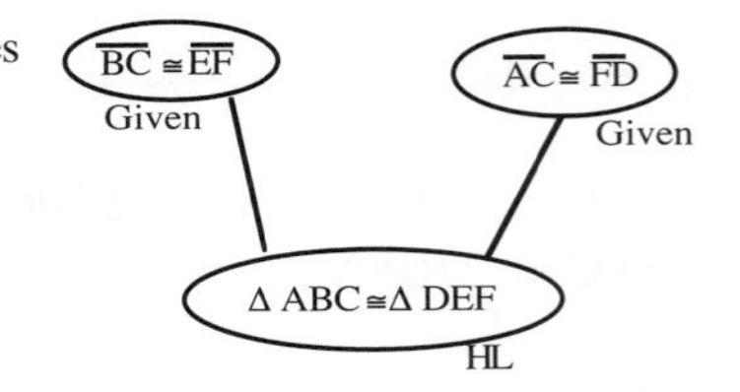


7. $\overline{NP} \cong \overline{MP}$ and $\overline{TP} \cong \overline{RP}$ by definition of bisector. $\angle NPT \cong \angle MPR$ because vertical angles are equal. So, $\Delta NTP \cong \Delta MRP$ by SAS.

8. $\angle ACD \cong \angle BCD$ by definition of angle bisector. $\overline{CD} \cong \overline{CD}$ by reflexive so $\Delta CDA \cong \Delta CDB$ by ASA.

9. $\angle A \cong \angle C$ since alternate interior angles of parallel lines congruent so $\Delta ABF \cong \Delta CED$ by ASA.

10. $TG \cong TG$ by reflexive so $\Delta TPG \cong \Delta TSG$ by SSS.

11. $\angle MEO \cong \angle PEO$ because perpendicular lines form $\cong$ right angles $\angle MOE \cong \angle POE$ by angle bisector and $\overline{OE} \cong \overline{OE}$ by reflexive. So, $\Delta MOE \cong \Delta POE$ by ASA.

12. $\angle CDB \cong \angle ABD$ and $\angle ADB \cong \angle CBD$ since parallel lines give congruent alternate interior angles. $\overline{DB} \cong \overline{DB}$ by reflexive so $\Delta ADB \cong \Delta CBD$ by ASA.

13. $\overline{DB} \cong \overline{EB}$ by definition of bisector. $\angle DBA \cong \angle EBC$ since vertical angles are congruent. So $\Delta ADB \cong \Delta CEB$ by AAS.

14. $\angle RQP \cong \angle SQP$ since perpendicular lines form congruent right angles. $PQ \cong PQ$ by reflexive so $\Delta PQR \cong \Delta PQS$ by AAS.

15. $\angle SQP \cong \angle RQP$ by angle bisector and $\overline{PQ} \cong \overline{PQ}$ by reflexive, so $\Delta SPQ \cong \Delta RPQ$ by AAS.

16. $\angle KYT \cong \angle HUG$ because parallel lines form congruent alternate exterior angles. $TY + YU = YU + GU$ so $TY \cong GU$ by subtraction. $\angle T \cong \angle G$ since perpendicular lines form congruent right angles. So $\Delta KTY \cong \Delta HGU$ by ASA. Therefore, $\angle K \cong \angle H$ since $\cong$ triangles have congruent parts.

17. $\angle MQL \cong \angle WLQ$ since parallel lines form congruent alternate interior angles. $\overline{QL} \cong \overline{QL}$ by reflexive so $\Delta MQL \cong \Delta WLQ$ by SAS so $\angle WQL \cong \angle MLQ$ since congruent triangles have congruent parts. So $\overline{ML} \parallel \overline{WQ}$ since congruent alternate interior angles are formed by parallel lines.

18. Yes

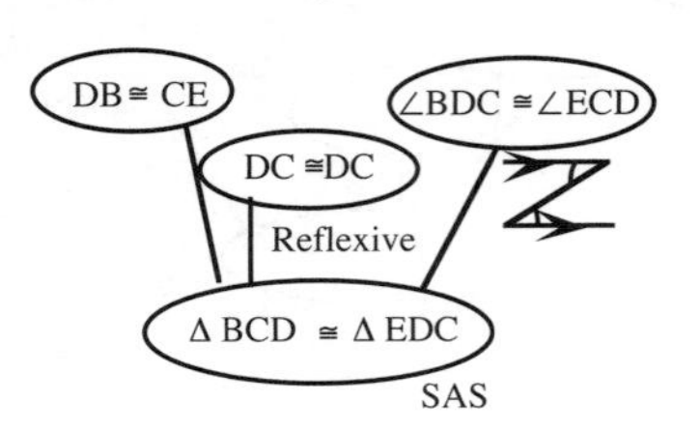


19. Not necessarily.

20. Not necessarily.

## RADICALS #14

Sometimes it is convenient to leave square roots in radical form instead of using a calculator to find approximations (decimal values). Look for perfect squares (i.e., 4, 9, 16, 25, 36, 49, ...) as **factors** of the number that is inside the radical sign (**radicand**) and take the square root of any perfect square factor. Multiply the root of the perfect square times the reduced radical. When there is an existing value that multiplies the radical, multiply any root(s) times that value.

For example:

$\sqrt{9} = 3$

$5\sqrt{9} = 5 \cdot 3 = 15$

$\sqrt{18} = \sqrt{9 \cdot 2} = \sqrt{9} \cdot \sqrt{2} = 3\sqrt{2}$

$3\sqrt{98} = 3\sqrt{49 \cdot 2} = 3 \cdot 7\sqrt{2} = 21\sqrt{2}$

$\sqrt{80} = \sqrt{16 \cdot 5} = \sqrt{16} \cdot \sqrt{5} = 4\sqrt{5}$

$\sqrt{45} + 4\sqrt{20} = \sqrt{9 \cdot 5} + 4\sqrt{4 \cdot 5} = 3\sqrt{5} + 4 \cdot 2\sqrt{5} = 11\sqrt{5}$

When there are no more perfect square factors inside the radical sign, the product of the whole number (or fraction) and the remaining radical is said to be in **SIMPLE RADICAL FORM**.

Simple radical form does not allow radicals in the denominator of a fraction. If there is a radical in the denominator, **RATIONALIZE THE DENOMINATOR** by multiplying the numerator and denominator of the fraction by the radical in the original denominator. Then simplify the remaining fraction. Examples:

$$\frac{2}{\sqrt{2}} = \frac{2}{\sqrt{2}} \cdot \frac{\sqrt{2}}{\sqrt{2}} = \frac{2\sqrt{2}}{2} = \sqrt{2}$$

$$\frac{4\sqrt{5}}{\sqrt{6}} = \frac{4\sqrt{5}}{\sqrt{6}} \cdot \frac{\sqrt{6}}{\sqrt{6}} = \frac{4\sqrt{30}}{6} = \frac{2\sqrt{30}}{3}$$

In the first example, $\sqrt{2} \cdot \sqrt{2} = \sqrt{4} = 2$ and $\frac{2}{2} = 1$. In the second example, $\sqrt{6} \cdot \sqrt{6} = \sqrt{36} = 6$ and $\frac{4}{6} = \frac{2}{3}$.

### Example 1

Add $\sqrt{27} + \sqrt{12} - \sqrt{48}$. Factor each radical and simplify.

$\sqrt{9 \cdot 3} + \sqrt{4 \cdot 3} - \sqrt{16 \cdot 3} = 3\sqrt{3} + 2\sqrt{3} - 4\sqrt{3} = 1\sqrt{3}$ or $\sqrt{3}$

### Example 2

Simplify $\frac{3}{\sqrt{6}}$. Multiply by $\frac{\sqrt{6}}{\sqrt{6}}$ and simplify: $\frac{3}{\sqrt{6}} \cdot \frac{\sqrt{6}}{\sqrt{6}} = \frac{3\sqrt{6}}{6} = \frac{\sqrt{6}}{2}$.

Write each of the following radicals in simple radical form.

1. $\sqrt{24}$
2. $\sqrt{48}$
3. $\sqrt{17}$
4. $\sqrt{31}$
5. $\sqrt{75}$
6. $\sqrt{50}$
7. $\sqrt{96}$
8. $\sqrt{243}$
9. $\sqrt{8} + \sqrt{18}$
10. $\sqrt{18} + \sqrt{32}$
11. $\sqrt{27} - \sqrt{12}$
12. $\sqrt{50} - \sqrt{32}$
13. $\sqrt{6} + \sqrt{63}$
14. $\sqrt{44} + \sqrt{99}$
15. $\sqrt{50} + \sqrt{32} - \sqrt{27}$
16. $\sqrt{75} - \sqrt{8} - \sqrt{32}$
17. $\frac{3}{\sqrt{3}}$
18. $\frac{5}{\sqrt{27}}$
19. $\frac{\sqrt{3}}{\sqrt{5}}$
20. $\frac{\sqrt{5}}{\sqrt{7}}$
21. $4\sqrt{5} - \frac{10}{\sqrt{5}}$

## Answers

1. $2\sqrt{6}$
2. $4\sqrt{3}$
3. $\sqrt{17}$
4. $\sqrt{31}$
5. $5\sqrt{3}$
6. $5\sqrt{2}$
7. $4\sqrt{6}$
8. $9\sqrt{3}$
9. $5\sqrt{2}$
10. $7\sqrt{2}$
11. $\sqrt{3}$
12. $\sqrt{2}$
13. $3\sqrt{7} + \sqrt{6}$
14. $5\sqrt{11}$
15. $9\sqrt{2} - 3\sqrt{3}$
16. $5\sqrt{3} - 6\sqrt{2}$
17. $\sqrt{3}$
18. $\frac{5\sqrt{3}}{9}$
19. $\frac{\sqrt{15}}{5}$
20. $\frac{\sqrt{35}}{7}$
21. $2\sqrt{5}$

## RIGHT TRIANGLE TRIGONOMETRY #15

The three basic trigonometric ratios for right triangles are the sine (pronounced "sign"), cosine, and tangent. Each one is used in separate situations, and the easiest way to remember which to use when is the mnemonic SOH-CAH-TOA. With reference to one of the acute angles in a right triangle, <u>S</u>ine uses the <u>O</u>pposite and the <u>H</u>ypotenuse - SOH. The <u>C</u>osine uses the <u>A</u>djacent side and the <u>H</u>ypotenuse - CAH, and the <u>T</u>angent uses the <u>O</u>pposite side and the <u>A</u>djacent side -TOA. In each case, the position of the angle determines which leg (side) is opposite or adjacent. Remember that opposite means "across from" and adjacent means "next to."

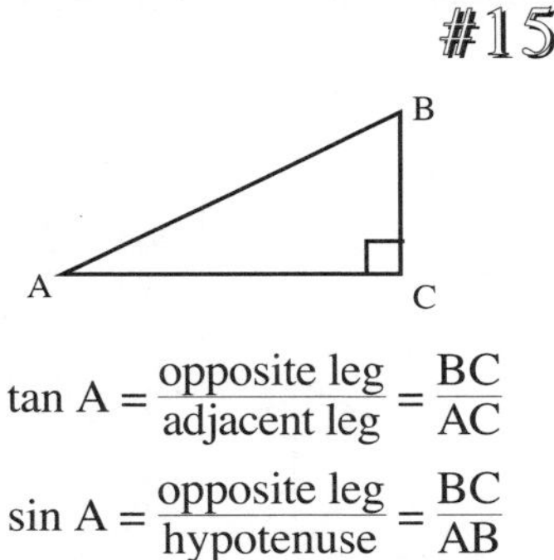


$$\tan A = \frac{\text{opposite leg}}{\text{adjacent leg}} = \frac{BC}{AC}$$

$$\sin A = \frac{\text{opposite leg}}{\text{hypotenuse}} = \frac{BC}{AB}$$

$$\cos A = \frac{\text{adjacent leg}}{\text{hypotenuse}} = \frac{AC}{AB}$$

### Example 1

Use trigonometric ratios to find the lengths of each of the missing sides of the triangle below.

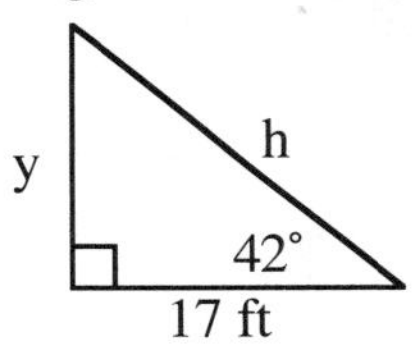


The length of the adjacent side with respect to the 42° angle is 17 ft. To find the length y, use the tangent because y is the opposite side and we know the adjacent side.

$$\tan 42° = \frac{y}{17}$$

$$17 \tan 42° = y$$

$$15.307 \text{ ft} \approx y$$

The length of y is approximately 15.31 feet.

To find the length h, use the cosine ratio (adjacent and hypotenuse).

$$\cos 42° = \frac{17}{h}$$

$$h \cos 42° = 17$$

$$h = \frac{17}{\cos 42°} \approx 22.876 \text{ ft}$$

The hypotenuse is approximately 22.9 feet long.

### Example 2

Use trigonometric ratios to find the size of each angle and the missing length in the triangle below.

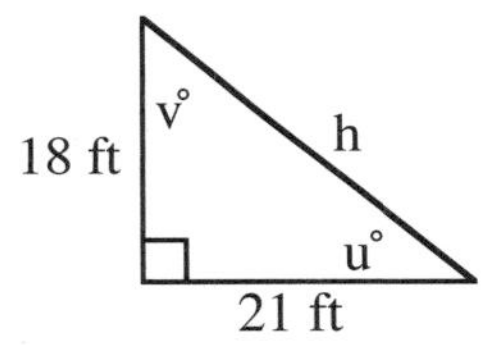


To find m∠u, use the tangent ratio because you know the opposite (18 ft) and the adjacent
(21 ft) sides.

$$\tan u° = \frac{18}{21}$$

$$m\angle u° = \tan^{-1}\frac{18}{21} \approx 40.601°$$

The measure of angle u is approximately 40.6°. By subtraction we know that m∠v ≈ 49.4°.

Use the sine ratio for m∠u and the opposite side and hypotenuse.

$$\sin 40.6° = \frac{18}{h}$$

$$h \sin 40.6 = 18$$

$$h = \frac{18}{\sin 40.6°} \approx 27.659 \text{ ft}$$

The hypotenuse is approximately 27.7 feet long.

Use trigonometric ratios to solve for the variable in each figure below.

1. 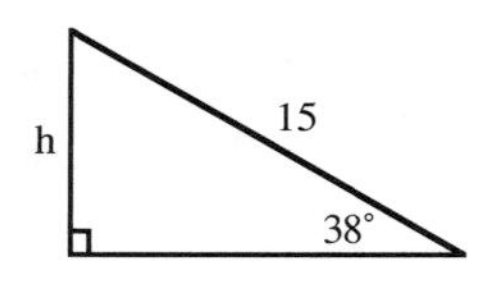


2. 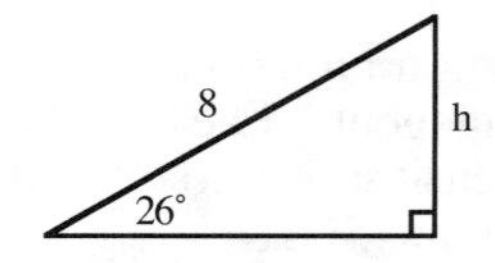


3. 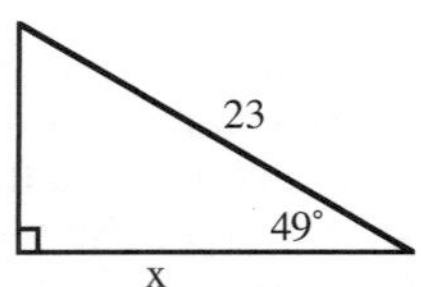


4. 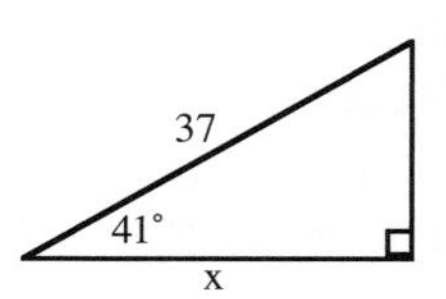


5. 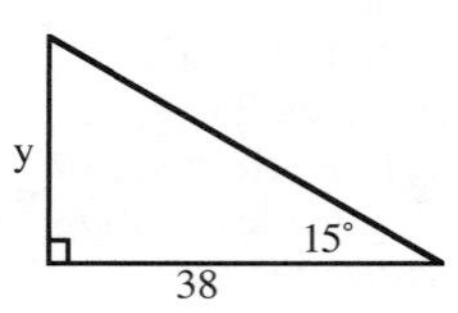


6. 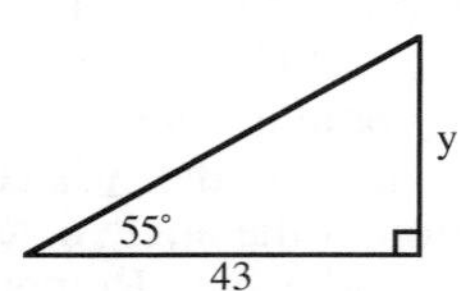


7. 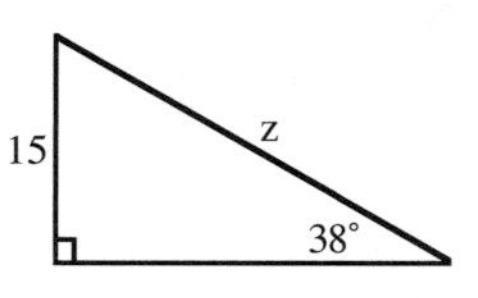


8. 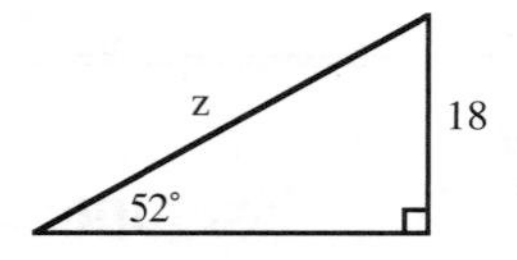


9. 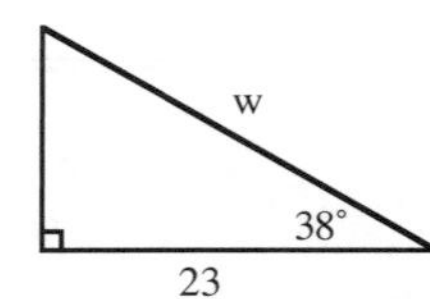


10. 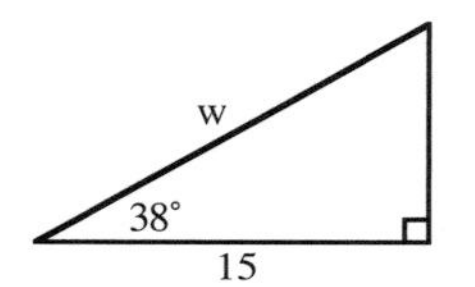


11. 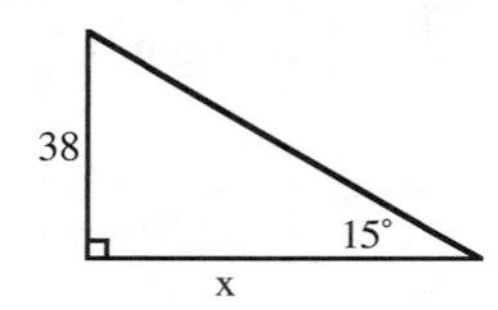


12. 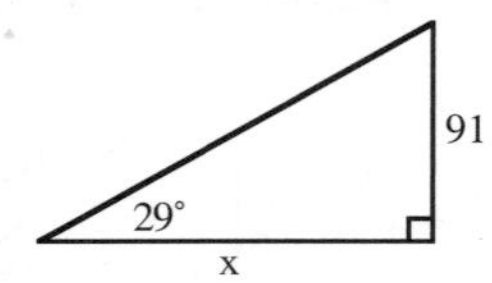


13. 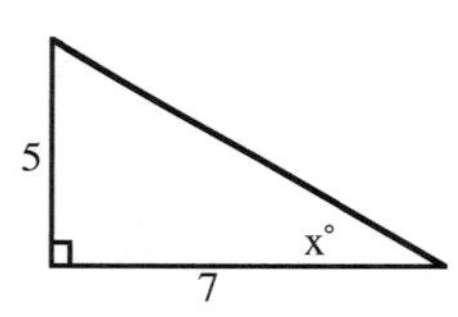


14. 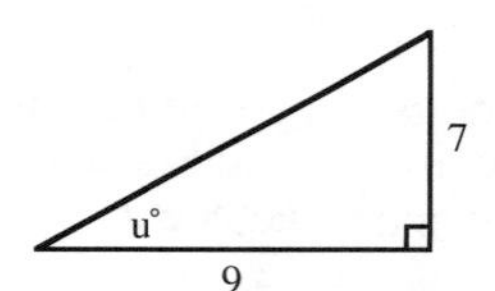


15. 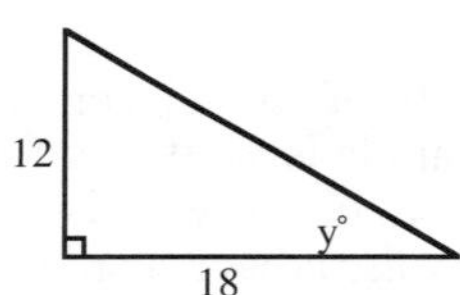


16. 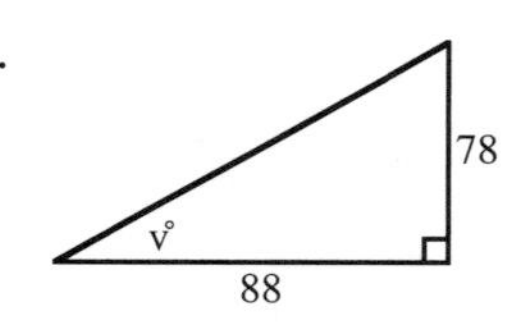


Draw a diagram and use trigonometric ratios to solve each of the following problems.

17. Juanito is flying a kite at the park and realizes that all 500 feet of string are out. Margie measures the angle of the string with the ground with her clinometer and finds it to be 42.5°. How high is Juanito's kite above the ground?

18. Nell's kite has a 350 foot string. When it is completely out, Ian measures the angle to be 47.5°. How far would Ian need to walk to be directly under the kite?

19. Mayfield High School's flagpole is 15 feet high. Using a clinometer, Tamara measured an angle of 11.3° to the top of the pole. Tamara is 62 inches tall. How far from the flagpole is Tamara standing?

20. Tamara took another sighting of the top of the flagpole from a different position. This time the angle is 58.4°. If everything else is the same, how far from the flagpole is Tamara standing?

## Answers

1. $h = 15 \sin 38° \approx 9.235$
2. $h = 8 \sin 26° \approx 3.507$
3. $x = 23 \cos 49° \approx 15.089$
4. $x = 37 \cos 41° \approx 27.924$
5. $y = 38 \tan 15° \approx 10.182$
6. $y = 43 \tan 55° \approx 61.4104$
7. $z = \dfrac{15}{\sin 38°} \approx 24.364$
8. $z = \dfrac{18}{\sin 52°} \approx 22.8423$
9. $w = \dfrac{23}{\cos 38°} \approx 29.1874$
10. $w = \dfrac{15}{\cos 38°} \approx 19.0353$
11. $x = \dfrac{38}{\tan 15°} \approx 141.818$
12. $x = \dfrac{91}{\tan 29°} \approx 164.168$
13. $x = \tan^{-1} \dfrac{5}{7} \approx 35.5377°$
14. $u = \tan^{-1} \dfrac{7}{9} \approx 37.875°$
15. $y = \tan^{-1} \dfrac{12}{18} \approx 33.690°$
16. $v = \tan^{-1} \dfrac{78}{88} \approx 41.5526°$

17.

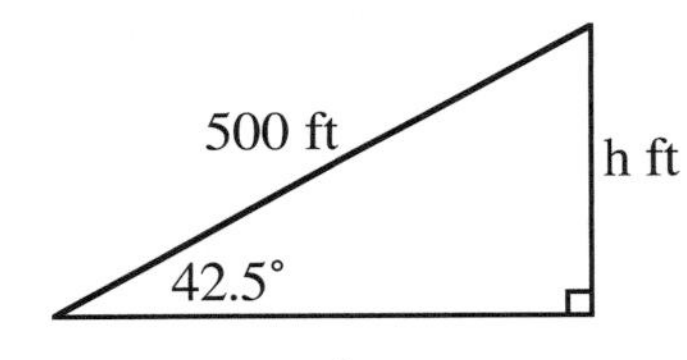


$\sin 42.5 = \dfrac{h}{500}$

$h = 500 \sin 42.5° \approx 337.795$ ft

18.

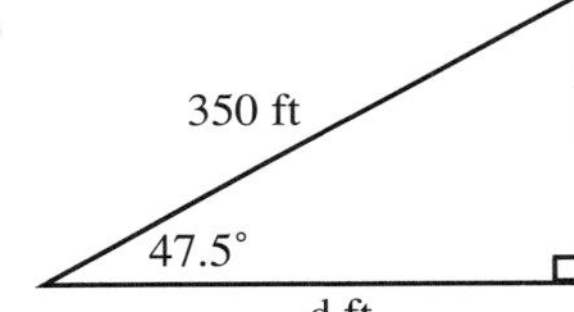


$\cos 47.5° = \dfrac{d}{350}$

$d = 350 \cos 47.5° \approx 236.46$ ft

19.

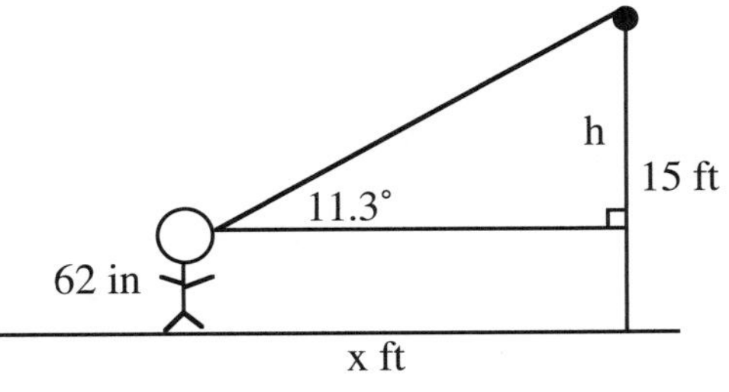


15 feet = 180 inches, 180" – 62" = 118" = h

$x \approx 590.5$ inches or 49.2 ft.

20.

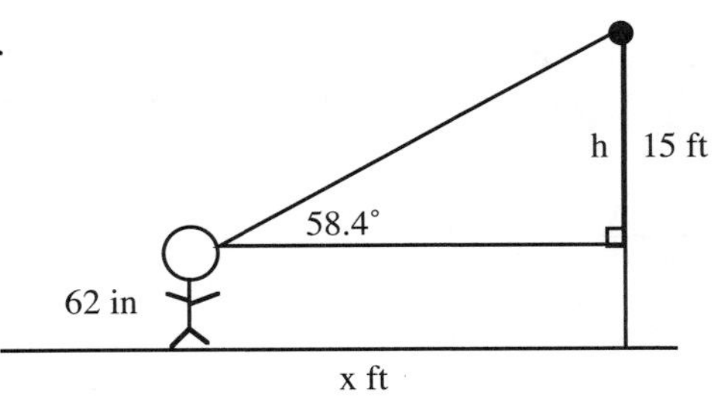


$h = 118''$, $\tan 58.4° = \dfrac{118''}{x}$,

$x \tan 58.4 = 118''$, $x = \dfrac{118''}{\tan 58.4°}$

$x \approx 72.59$ inches or 6.05 ft.

## RATIO OF SIMILARITY #16

The **RATIO OF SIMILARITY** between any two similar figures is the ratio of any pair of corresponding sides. Simply stated, once it is determined that two figures are similar, all of their pairs of corresponding sides have the same ratio.

The ratio of similarity of figure P to figure Q, written P : Q, is $\frac{3}{5}$.

$\frac{4.5}{7.5}$ and $\frac{12}{20}$ have the same ratio.

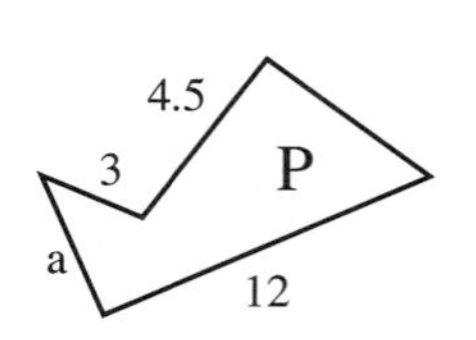


~

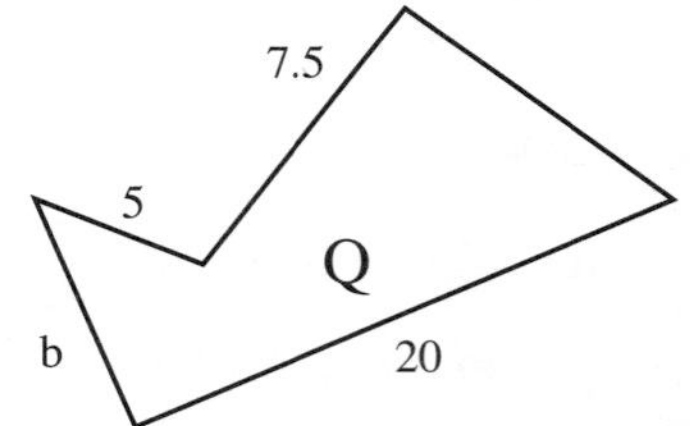


Note that the ratio of similarity is always expressed in lowest possible terms. Also, the order of the statement, P : Q or Q : P, determines which order to state and use the ratios between pairs of corresponding sides.

An Equation that sets one ratio equal to another ratio is called a **proportion**. An example is at right. $\frac{3}{5} = \frac{12}{20}$

### Example

Find x in the figure. Be consistent in matching corresponding parts of similar figures.

$\Delta ABC \sim \Delta DEC$ by AA ($\angle BAC \cong \angle EDC$ and $\angle C \cong \angle C$). Therefore the corresponding sides are proportional.

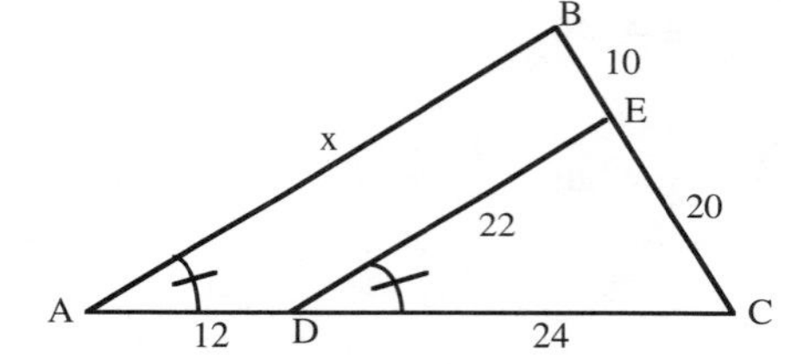


$\frac{DE}{AB} = \frac{CD}{CA}$ ==> $\frac{22}{x} = \frac{24}{36}$ ==> $24x = 22(36)$ ==> $24x = 792$ ==> $x = 33$

Note: This problem also could have been solved with the proportion: $\frac{DE}{AB} = \frac{CE}{CB}$.

For each pair of similar figures below, find the ratio of similarity, for large:small.

1.

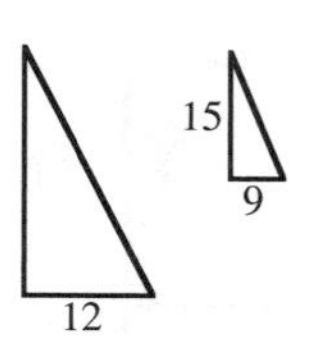


2.

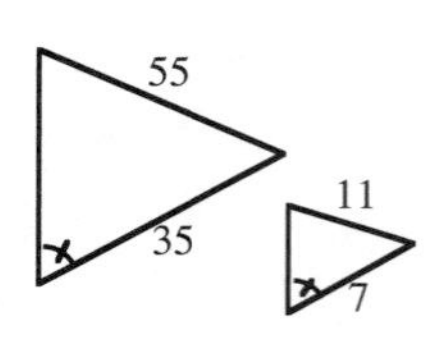


3.

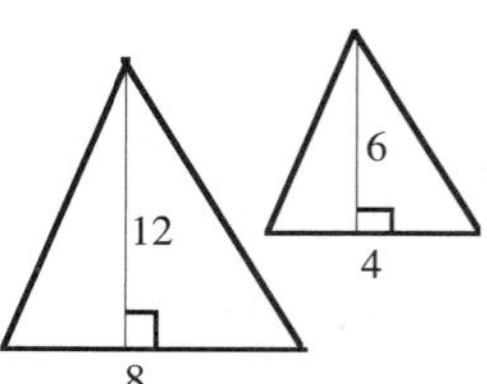


4.

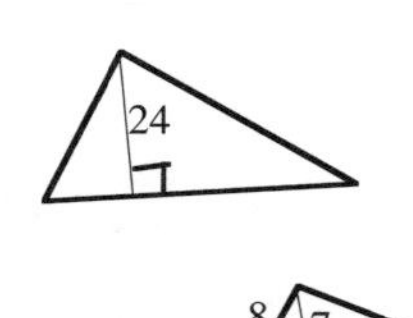


5.

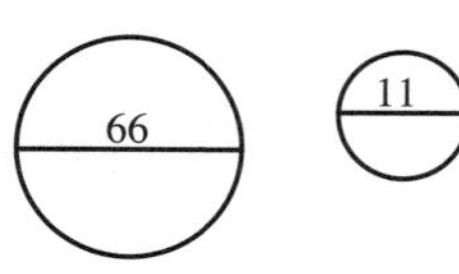


6.

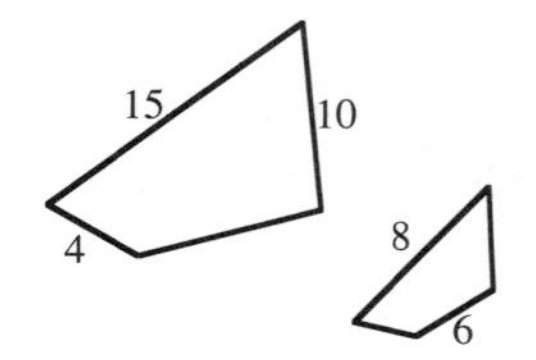

For each pair of similar figures below, state the ratio of similarity, then use it to find x.

7.

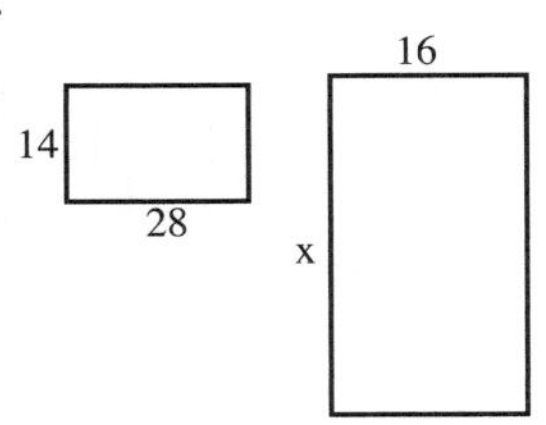


8.

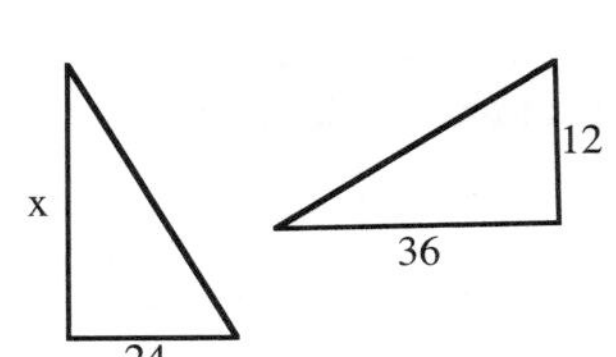


9.

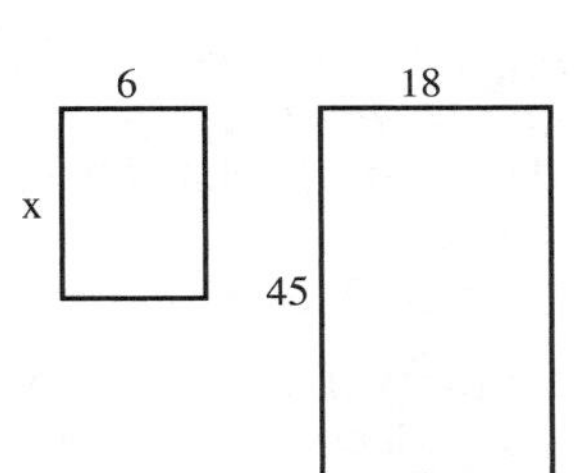


10.

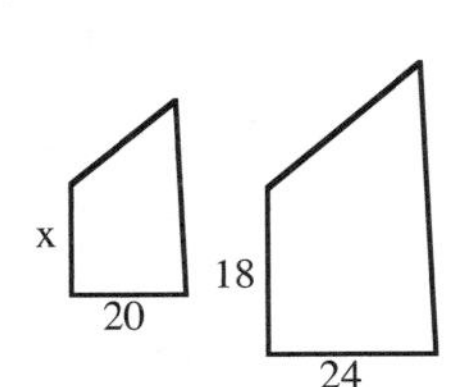


11.

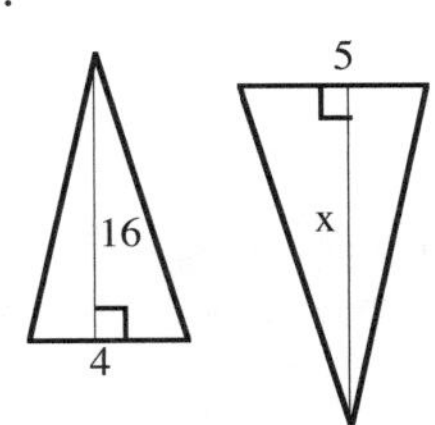


12.

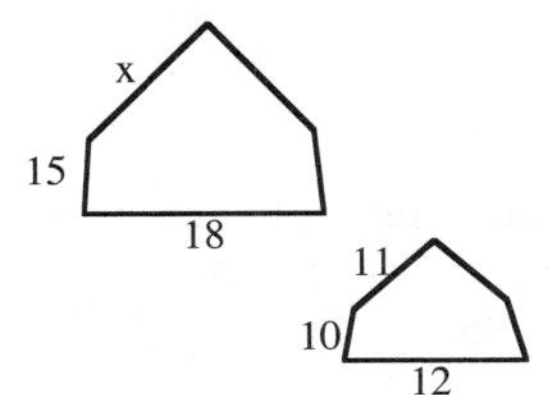


For problems 13 through 18, use the given information and the figure to find each length.

13. JM = 14, MK = 7, JN = 10 Find NL.
14. MN = 5, JN = 4, JL = 10 Find KL.
15. KL = 10, MK = 2, JM = 6 Find MN.
16. MN = 5, KL = 10, JN = 7 Find JL.
17. JN = 3, NL = 7, JM = 5 Find JK.
18. JK = 37, NL = 7, JM = 5 Find JN.

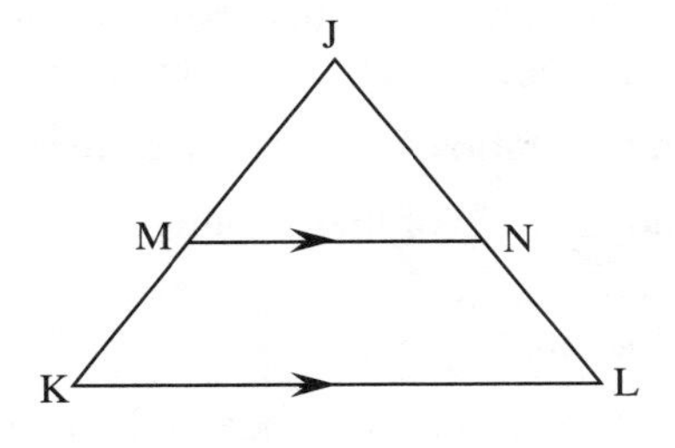


19. Standing 4 feet from a mirror laying on the flat ground, Palmer, whose eye height is 5 feet, 9 inches, can see the reflection of the top of a tree. He measures the mirror to be 24 feet from the tree. How tall is the tree?

20. The shadow of a statue is 20 feet long, while the shadow of a student is 4 ft long. If the student is 6 ft tall, how tall is the statue?

## Answers

1. $\frac{4}{3}$
2. $\frac{5}{1}$
3. $\frac{2}{1}$
4. $\frac{24}{7}$
5. $\frac{6}{1}$
6. $\frac{15}{8}$
7. $\frac{7}{8}$ ; x = 32
8. $\frac{2}{1}$ ; x = 72
9. $\frac{1}{3}$ ; x = 15
10. $\frac{5}{6}$ ; x = 15
11. $\frac{4}{5}$ ; x = 20
12. $\frac{3}{2}$ ; x = 16.5
13. 5
14. 12.5
15. 7.5
16. 14
17. $16\frac{2}{3}$
18. $11\frac{25}{32}$
19. 34.5 ft.
20. 30 ft.

## SIMILARITY OF LENGTH, AREA, AND VOLUME #17

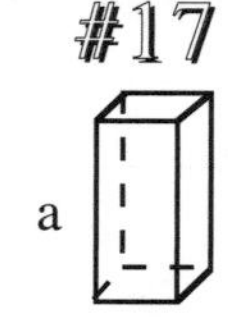


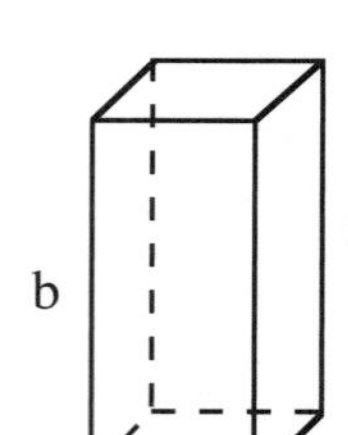


The relationships for similarity of length, area, and volume are developed in Unit 8 and summarized in problem S-68 on page 322. The basic relationships of the $r:r^2:r^3$ Theorem are stated below.

Once you know two figures are similar with a ratio of similarity $\frac{a}{b}$, the following proportions for the SMALL (sm) and LARGE (lg) figures (which are enlargements or reductions of each other) are true:

$$\frac{\text{side}_{sm}}{\text{side}_{lg}} = \frac{a}{b} \qquad \frac{P_{sm}}{P_{lg}} = \frac{a}{b} \qquad \frac{A_{sm}}{A_{lg}} = \frac{a^2}{b^2} \qquad \frac{V_{sm}}{V_{lg}} = \frac{a^3}{b^3}$$

In each proportion above, the data from the smaller figure is written on top (in the numerator) to help be consistent with correspondences. When working with area, the basic ratio of similarity is squared; for volume, it is cubed. NEVER square or cube the actual areas or volumes themselves.

### Example 1

The two rectangular prisms above are similar. Suppose the ratio of their vertical edges is 3:8. Use the $r:r^2:r^3$ Theorem to find the following without knowing the dimensions of the prisms.

a) Find the ratio of their surface areas.

b) Find the ratio of their volumes.

c) The perimeter of the front face of the large prism is 18 units. Find the perimeter of the front face of the small prism.

d) The area of the front face of the large prism is 15 square units. Find the area of the front face of the small prism.

e) The volume of the small prism is 21 cubic units. Find the volume of the large prism.

Use the ratio of similarity for parts (a) and (b): the ratio of the surface areas is $\frac{3^2}{8^2} = \frac{9}{64}$; the ratio of the volumes is $\frac{3^2}{8^2} = \frac{27}{512}$. Use proportions to solve for the rest of the parts.

(c)

$$\frac{3}{8} = \frac{P}{18}$$
$$8P = 54$$
$$P = 6.75 \text{ units}$$

d)

$$\left(\frac{3}{8}\right)^2 = \frac{A}{15}$$
$$\frac{9}{64} = \frac{A}{15}$$
$$64A = 135$$
$$A \approx 2.11 \text{ sq. units}$$

e)

$$\left(\frac{3}{8}\right)^3 = \frac{21}{V}$$
$$\frac{27}{512} = \frac{21}{V}$$
$$27V = 10752$$
$$V \approx 398.22 \text{ cu. units}$$

Solve each of the following problems.

1. Two rectangular prisms are similar. The smaller, A, has a height of four units while the larger, B, has a height of six units.

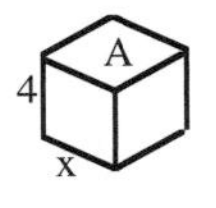


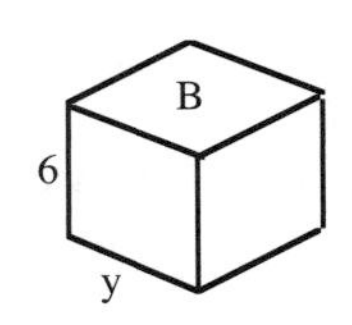


   a) What is the magnification factor from prism A to prism B?

   b) What would be the ratio of the lengths of the edges labeled x and y?

   c) What is the ratio, small to large, of their surface areas? their volumes?

   d) A third prism, C is similar to prisms A and B. Prism C's height is ten units. If the volume of prism A is 24 cubic units, what is the volume of prism C?

2. If rectangle A and rectangle B have a ratio of similarity of 5:4, what is the area of rectangle B if the area of rectangle A is 24 square units?

3. If rectangle A and rectangle B have a ratio of similarity of 2:3, what is the area of rectangle B if the area of rectangle A is 46 square units?

4. If rectangle A and rectangle B have a ratio of similarity of 3:4, what is the area of rectangle B if the area of rectangle A is 82 square units?

5. If rectangle A and rectangle B have a ratio of similarity of 1:5, what is the area of rectangle B if the area of rectangle A is 24 square units?

6. Rectangle A is similar to rectangle B. The area of rectangle A is 81 square units while the area of rectangle B is 49 square units. What is the ratio of similarity between the two rectangles?

7. Rectangle A is similar to rectangle B. The area of rectangle B is 18 square units while the area of rectangle A is 12.5 square units. What is the ratio of similarity between the two rectangles?

8. Rectangle A is similar to rectangle B. The area of rectangle A is 16 square units while the area of rectangle B is 100 square units. If the perimeter of rectangle A is 12 units, what is the perimeter of rectangle B?

9. If prism A and prism B have a ratio of similarity of 2:3, what is the volume of prism B if the volume of prism A is 36 cubic units?

10. If prism A and prism B have a ratio of similarity of 1:4, what is the volume of prism B if the volume of prism A is 83 cubic units?

11. If prism A and prism B have a ratio of similarity of 6:11, what is the volume of prism B if the volume of prism A is 96 cubic units?

12. Prism A and prism B are similar. The volume of prism A is 72 cubic units while the volume of prism B is 1125 cubic units. What is the ratio of similarity between these two prisms?

13. Prism A and prism B are similar. The volume of prism A is 27 cubic units while the volume of prism B is approximately 512 cubic units. If the surface area of prism B is 128 square units, what is the surface area of prism A?

14. The corresponding diagonals of two similar trapezoids are in the ratio of 1:7. What is the ratio of their areas?

15. The ratio of the perimeters of two similar parallelograms is 3:7. What is the ratio of their areas?

16. The ratio of the areas of two similar trapezoids is 1:9. What is the ratio of their altitudes?

17. The areas of two circles are in the ratio of 25:16. What is the ratio of their radii?

18. The ratio of the volumes of two similar circular cylinders is 27:64. What is the ratio of the diameters of their similar bases?

19. The surface areas of two cubes are in the ratio of 49:81. What is the ratio of their volumes?

20. The ratio of the weights of two spherical steel balls is 8:27. What is the ratio of the diameters of the two steel balls?

## Answers

1. a) $\frac{6}{4} = \frac{3}{2}$ b) $\frac{4}{6} = \frac{2}{3}$ c) $\frac{16}{36}$ or $\frac{4}{9}$, $\frac{64}{216}$ or $\frac{16}{27}$ d) 375 cu unit

2. 15.36 $u^2$
3. 103.5 $u^2$
4. $\approx$ 145.8 $u^2$
5. 600 $u^2$
6. $\frac{9}{7}$
7. $\frac{6}{5}$
8. 30 u
9. 121.5
10. 5312
11. $\approx$ 591.6
12. $\frac{2}{5}$
13. $\approx$ 18 $u^2$
14. $\frac{1}{49}$
15. $\frac{9}{49}$
16. $\frac{1}{3}$
17. $\frac{5}{4}$
18. $\frac{3}{4}$
19. $\frac{343}{729}$
20. $\frac{2}{3}$

## INTERIOR AND EXTERIOR ANGLES OF POLYGONS #18

The sum of the measures of the interior angles of an n-gon is sum $= (n - 2)180°$.

The measure of <u>each</u> angle in a <u>regular</u> n-gon is $m\angle = \frac{(n-2)180°}{n}$.

The sum of the exterior angles of any n-gon is 360°.

### Example 1

Find the sum of the interior angles of a 22-gon.

Since the polygon has 22 sides, we can substitute this number for n:

$$(n - 2)180° = (22 - 2)180° = 20 \cdot 180° = 3600°.$$

### Example 2

If the 22-gon is regular, what is the measure of each angle? Use the sum from the previous example and divide by 22: $\frac{3600}{22} \approx 163.64°$

### Example 3

Each angle of a regular polygon measures 157.5°. How many sides does this n-gon have?

a) Solving algebraically: $157.5° = \frac{(n-2)180°}{n} \Rightarrow 157.5°n = (n - 2)180°$

$$\Rightarrow 157.5°n = 180°n - 360 \Rightarrow -22.5°n = -360 \Rightarrow n = 16$$

b) If each interior angle is 157.5°, then each exterior angle is $180° - 157.5° = 22.5°$.
Since the sum of the exterior angles of any n-gon is 360°, $360° \div 22.5° \Rightarrow 16$ sides.

### Example 4

Find the area of a regular 7-gon with sides of length 5 ft.

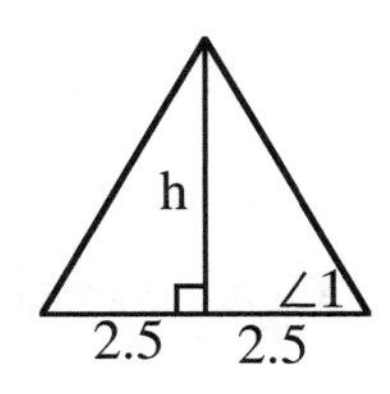


Because the regular 7-gon is made up of 7 identical, isosceles triangles we need to find the area of one, and then multiply it by 7. (See page 349 for addition figures and details that parallel this solution.) In order to find the area of each triangle we need to start with the angles of each triangle. Each interior angle of the regular 7-gon measures $\frac{(7-2)180°}{7} = \frac{(5)180°}{7} = \frac{900}{7} \approx 128.57°$. The angle in the triangle is half the size of the interior angle, so $m\angle 1 \approx \frac{128.57°}{2} \approx 64.29°$. Find the height of the triangle by using the tangent ratio: $\tan \angle 1 = \frac{h}{2.5} \Rightarrow h = 2.5 \cdot \tan \angle 1 \approx 5.19$ ft. The area of the triangle is: $\frac{5 \cdot 5.19}{2} \approx 12.98\ ft^2$. Thus the area of the 7-gon is $7 \cdot 12.98 \approx 90.86\ ft^2$.

Find the measures of the angles in each problem below.

1. Find the sum of the interior angles in a 7-gon.

2. Find the sum of the interior angles in an 8-gon.

3. Find the size of each of the interior angles in a regular 12-gon.

4. Find the size of each of the interior angles in a regular 15-gon.

5. Find the size of each of the exterior angles of a regular 17-gon.

6. Find the size of each of the exterior angles of a regular 21-gon.

Solve for x in each of the figures below.

7.

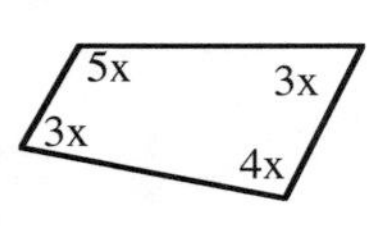


8.

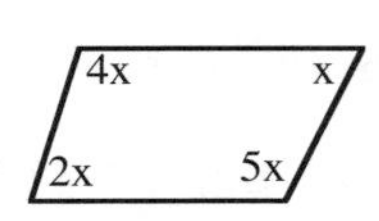


9.

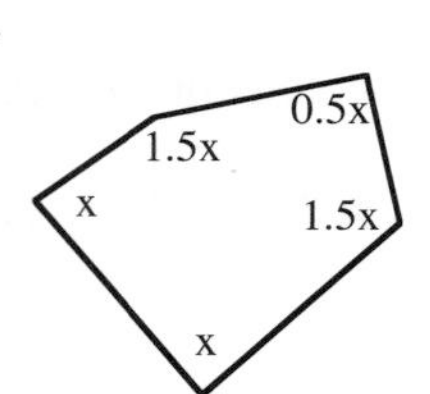


10.

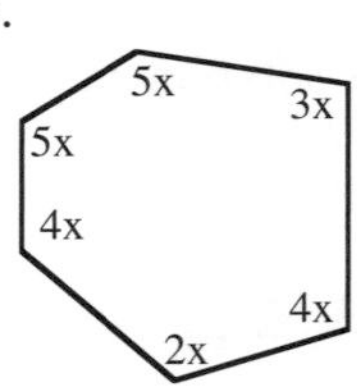


Answer each of the following questions.

11. Each exterior angle of a regular n-gon measures $16\frac{4}{11}^{\circ}$. How many sides does this n-gon have?

12. Each exterior angle of a regular n-gon measures $13\frac{1}{3}^{\circ}$. How many sides does this n-gon have?

13. Each angle of a regular n-gon measures 156°. How many sides does this n-gon have?

14. Each angle of a regular n-gon measures 165.6°. How many sides does this n-gon have?

15. Find the area of a regular pentagon with side length 8 cm.

16. Find the area of a regular hexagon with side length 10 ft.

17. Find the area of a regular octagon with side length 12 m.

18. Find the area of a regular decagon with side length 14 in.

## Answers

1. 900°
2. 1080°
3. 150°
4. 156°
5. 21.1765°
6. 17.1429°
7. x = 24°
8. x = 30°
9. x = 98.18°
10. x = 31.30°
11. 22 sides
12. 27 sides
13. 15 sides
14. 25 sides
15. 110.1106 $cm^2$
16. 259.8076 $ft^2$
17. 695.2935 $m^2$
18. 1508.0649 $in^2$

## AREAS BY DISSECTION #19

**DISSECTION PRINCIPLE**: Every polygon can be dissected (or broken up) into triangles which have no interior points in common. This principle is an example of the problem solving strategy of **subproblems**. Finding simpler problems that you know how to solve will help you solve the larger problem.

### Example

Find the area of the figure below.

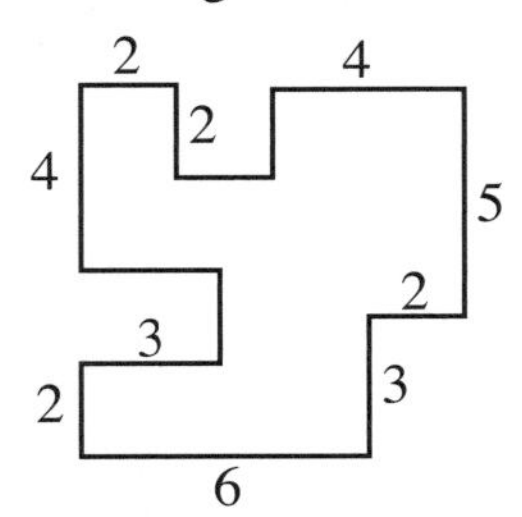


First fill in the missing lengths. The right vertical side of the figure is a total of 8 units tall. Since the left side is also 8 units tall, $8 - 4 - 2 = 2$ units. The missing vertical length on the top middle cutout is also 2.

Horizontally, the bottom length is 8 units $(6 + 2)$. Therefore, the top is 2 units $(8 - 4 - 2)$. The horizontal length at the left middle is 3 units.

Enclose the entire figure in an 8 by 8 square and then <u>remove</u> the area of the three rectangles that are not part of the figure. The area is:

| | |
|---|---|
| Enclosing square: | $8 \cdot 8 = 64$ |
| Top cut out: | $2 \cdot 2 = 4$ |
| Left cut out: | $3 \cdot 2 = 6$ |
| Lower right cut out: | $2 \cdot 3 = 6$ |
| Area of figure: | $64 - 4 - 6 - 6 = 48$ |

### Example

Find the area of the figure below.

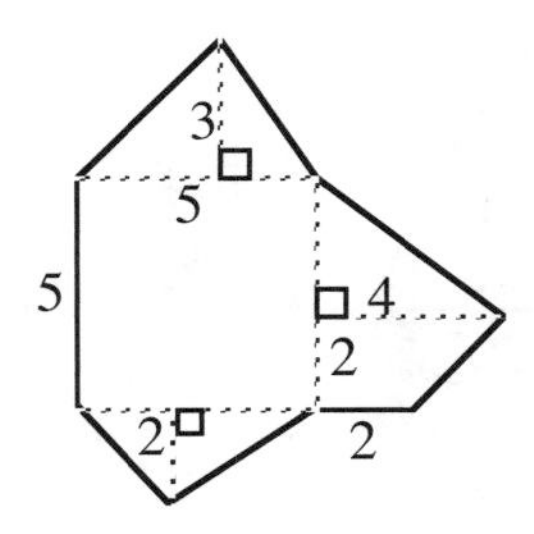


This figure consists of five familiar figures: a central square, 5 units by 5 units; three triangles, one on top with $b = 5$ and $h = 3$, one on the right with $b = 4$ and $h = 3$, and one on the bottom with $b = 5$ and $h = 2$; and a trapezoid with an upper base of 4, a lower base of 2 and a height of 2.

The area is:

| | | |
|---|---|---|
| square | $5 \cdot 5$ | $= 25$ |
| top triangle | $\frac{1}{2} \cdot 5 \cdot 3$ | $= 7.5$ |
| bottom triangle | $\frac{1}{2} \cdot 5 \cdot 2$ | $= 5$ |
| right triangle | $\frac{1}{2} \cdot 3 \cdot 4$ | $= 6$ |
| trapezoid | $\frac{(4+2) \cdot 2}{2}$ | $= 6$ |
| total area | | $= 49.5 \text{ u}^2$ |

Find the area of each of the following figures. Assume that anything that looks like a right angle is a right angle.

1\.

4, 2, 3, 2, 3, 6, 2

2\.

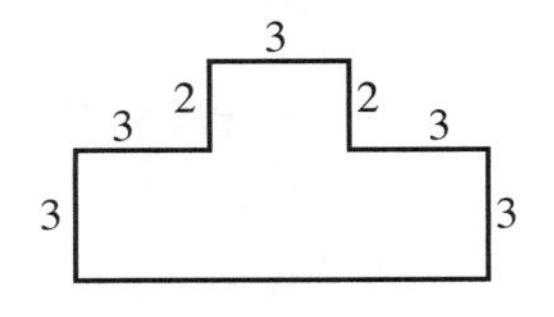


3\.

7, 5, 9, 5

4.

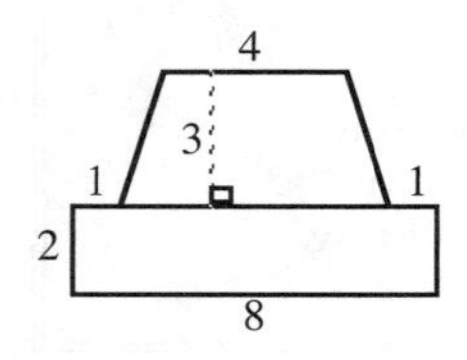


5.

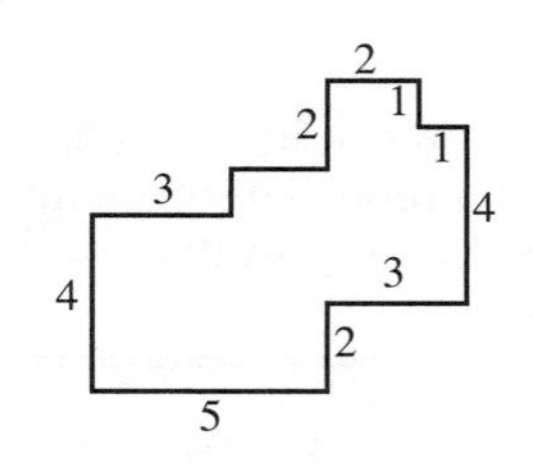


6.

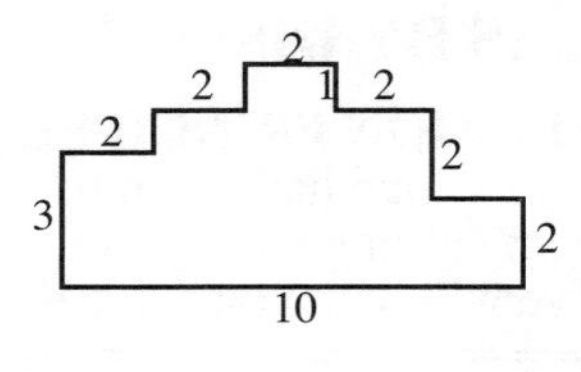


7.

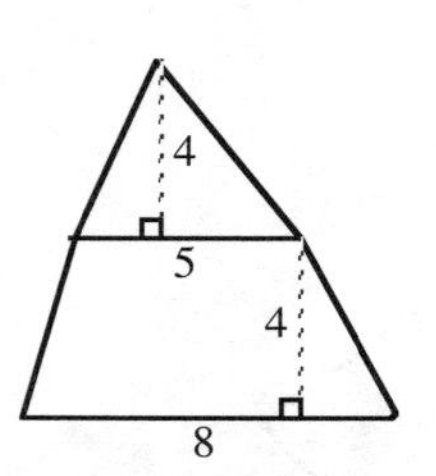


8.

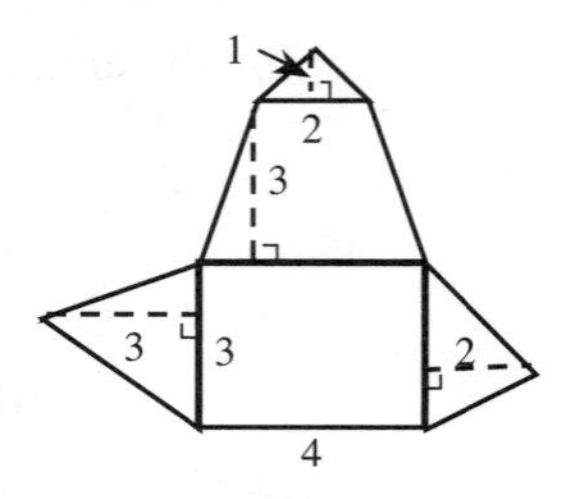


9.

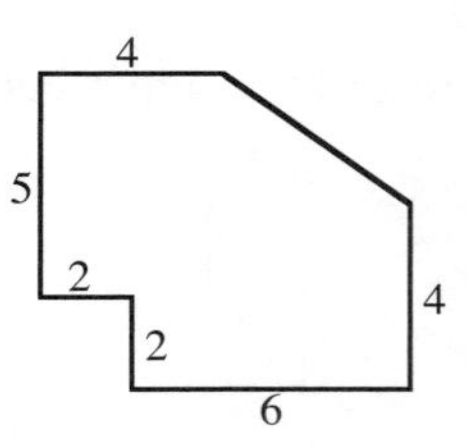


10.

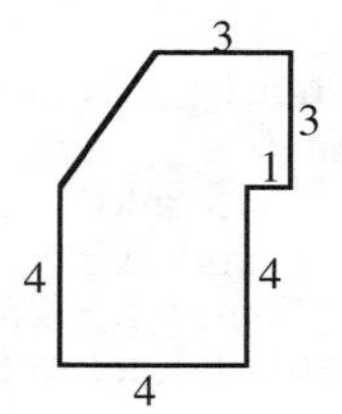


11 and 12: Find the surface area of each polyhedra.

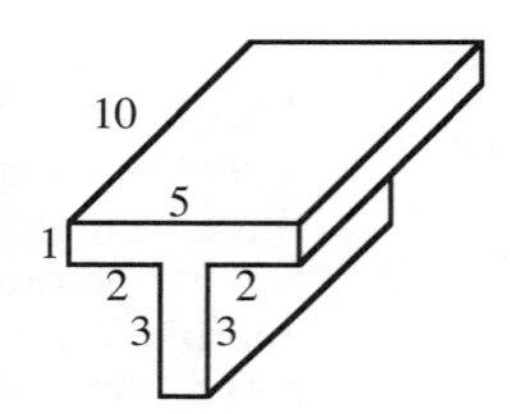


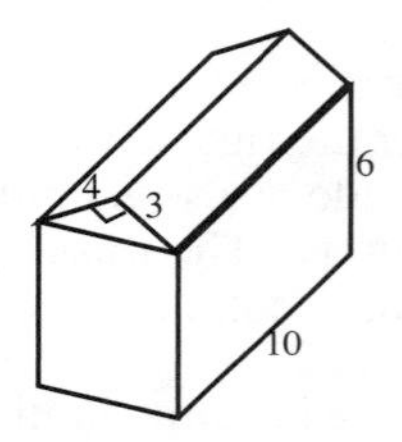


Find the area and perimeter of each of the following figures.

13.

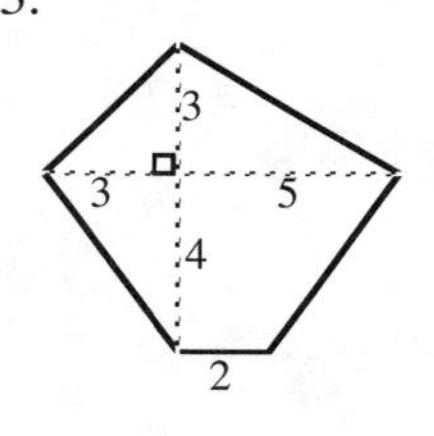


14.

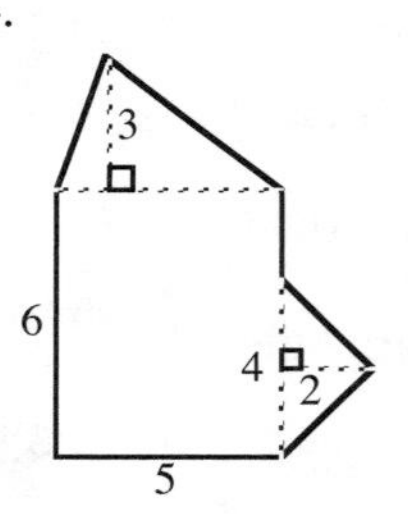


15.

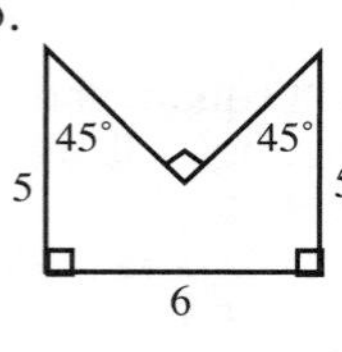


16.

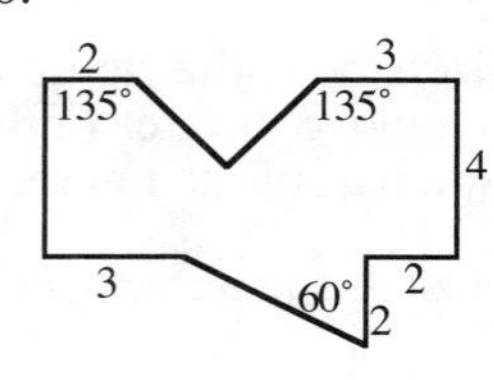


## Answers

1. 42 $u^2$
2. 33 $u^2$
3. 85 $u^2$
4. 31 $u^2$
5. 36 $u^2$
6. 36 $u^2$
7. 36 $u^2$
8. 29.5 $u^2$
9. 46 $u^2$
10. 28 $u^2$
11. SA = 196 sq.un.
12. 312 sq.un.
13. $A = 32u^2$
    $P = \sqrt{34} + 3\sqrt{2} + 12 \approx 38.3162u$
14. $A = 41.5u^2$
    $P = 18 + 4\sqrt{2} + \sqrt{10} \approx 26.8191u$
15. $A = 21u^2$
    $P = 16 + 6\sqrt{2} \approx 24.4853u$
16. $A = 14 + 6\sqrt{3} \approx 24.3923u^2$
    $P = 24 + 2\sqrt{6} \approx 28.899u$

## CENTRAL AND INSCRIBED ANGLES

**#20**

A central angle is an angle whose vertex is the center of a circle and whose sides intersect the circle. The degree measure of a central angle is equal to the degree measure of its intercepted arc. For the circle at right with center C, $\angle ACB$ is a central angle.

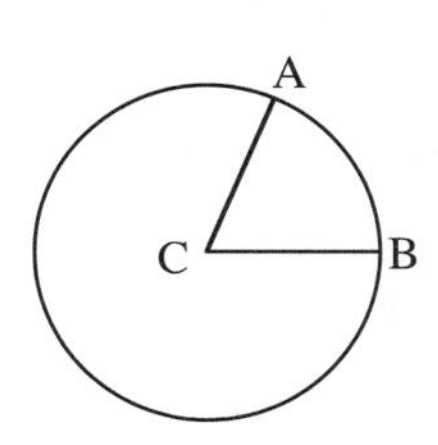


An **INSCRIBED ANGLE** is an angle with its vertex on the circle and whose sides intersect the circle. The arc formed by the intersection of the two sides of the angle and the circle is called an **INTERCEPTED ARC**. $\angle ADB$ is an inscribed angle, $\overset{\frown}{AB}$ is an intercepted arc.

The **INSCRIBED ANGLE THEOREM** says that the measure of any inscribed angle is half the measure of its intercepted arc. Likewise, any intercepted arc is twice the measure of any inscribed angle whose sides pass through the endpoints of the arc.

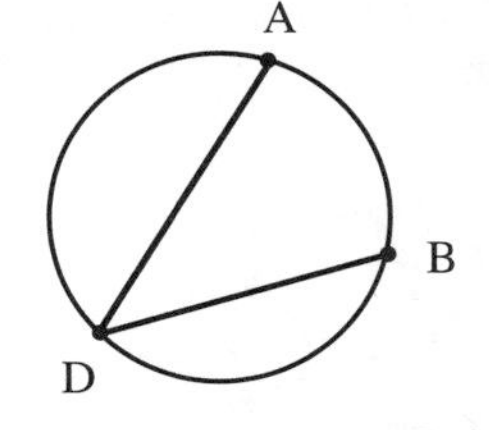


$$\mathbf{m\angle ADB = \frac{1}{2}\overset{\frown}{AB} \text{ and } \overset{\frown}{AB} = 2m\angle ADB}$$

### Example 1

In $\odot P$, $\angle CPQ = 70°$. Find $\overset{\frown}{CQ}$ and $m\overset{\frown}{CRQ}$.

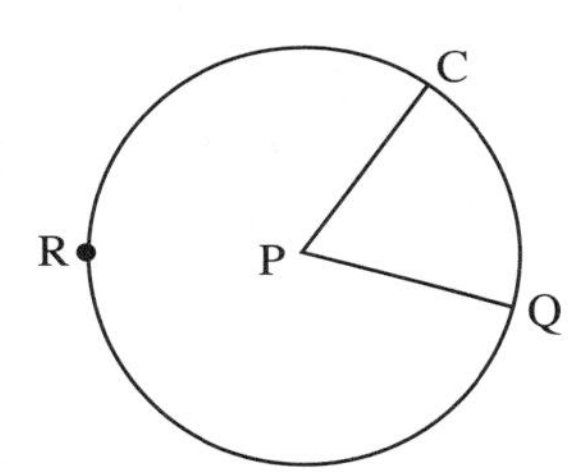


$m\overset{\frown}{CQ} = m\angle CPQ = 70°$

$m\overset{\frown}{CRQ} = 360° - \overset{\frown}{CQ} = 360° - 70° = 290°$

### Example 2

In the circle shown below, the vertex of the angles is at the center. Find x.

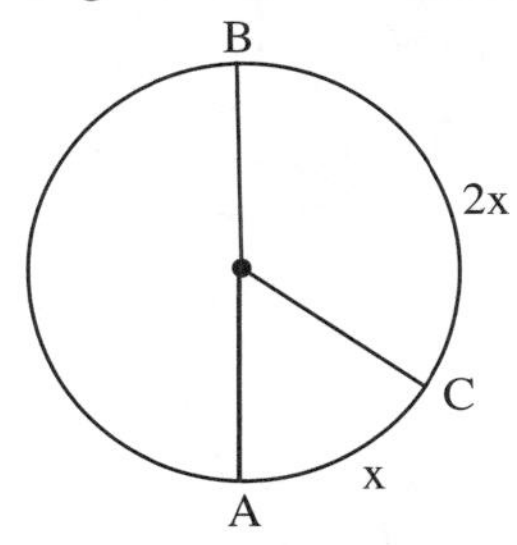


Since $\overline{AB}$ is the diameter of the circle,
$x + 2x = 180° \Rightarrow 3x = 180° \Rightarrow x = 60°$

### Example 3

Solve for x.

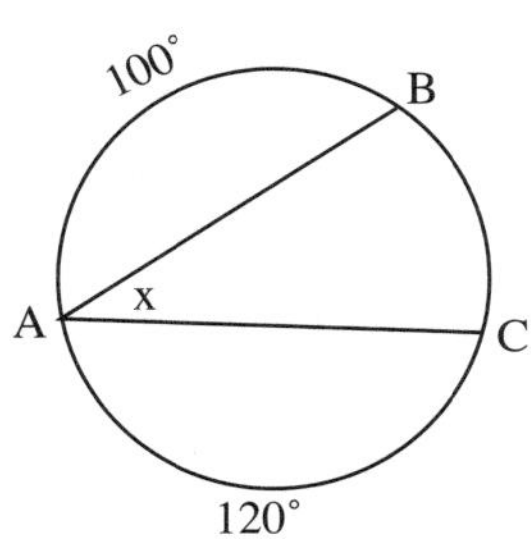


Since $m\overset{\frown}{BAC} = 220°$, $m\overset{\frown}{BC} = 360° - 220° = 140°$. The $m\angle BAC$ equals half the $m\overset{\frown}{BC}$ or $70°$.

Find each measure in $\odot P$ if $m\angle WPX = 28°$, $m\angle ZPY = 38°$, and $\overline{WZ}$ and $\overline{XV}$ are diameters.

1. $\overparen{YZ}$
2. $\overparen{WX}$
3. $\angle VPZ$
4. $\overparen{VWX}$
5. $\angle XPY$
6. $\overparen{XY}$
7. $\overparen{XWY}$
8. $\overparen{WZX}$

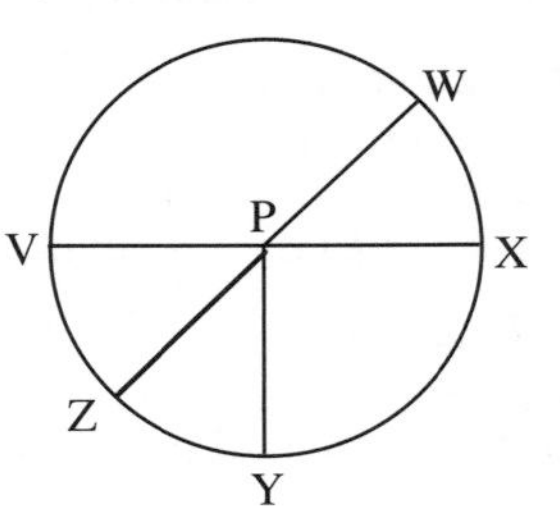


In each of the following figures, O is the center of the circle. Calculate the values of x and justify your answer.

9.
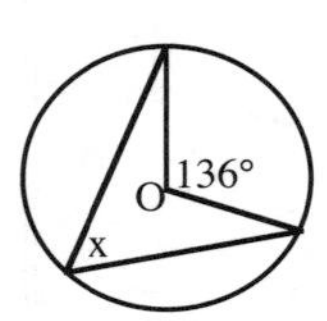


10.
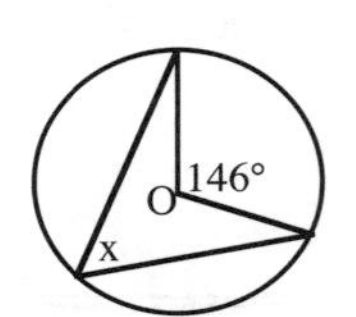


11.
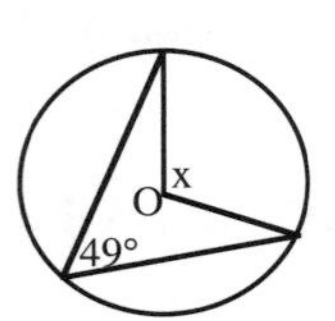


12.
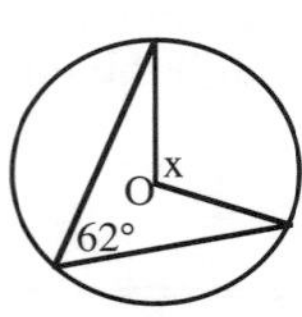


13.
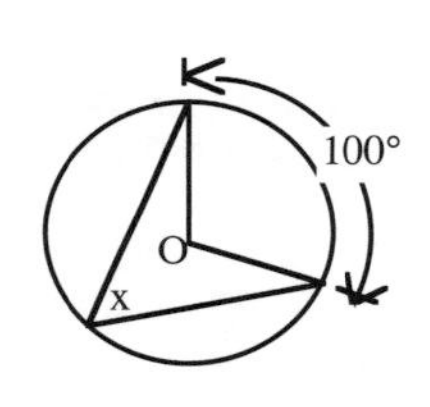


14.
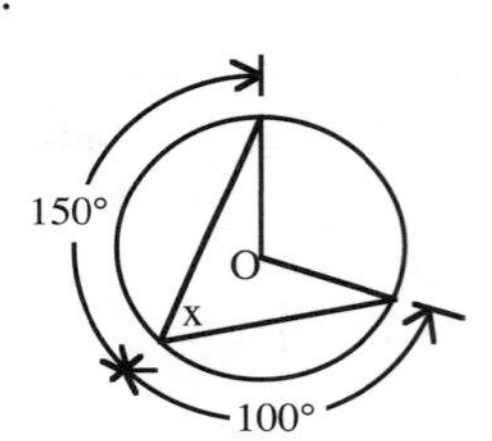


15.
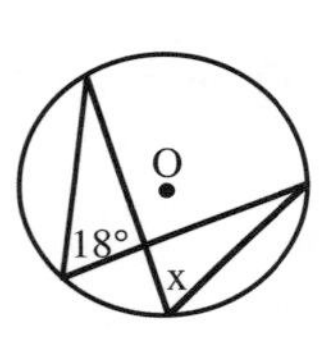


16.
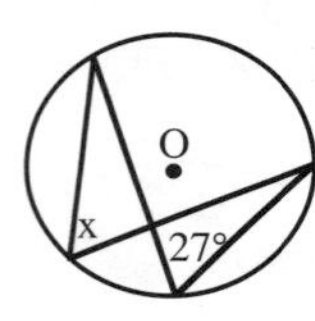


17.
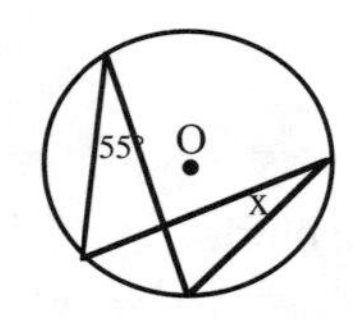


18.
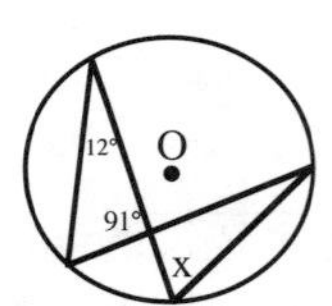


19.
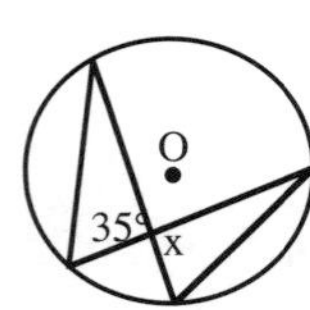


20.
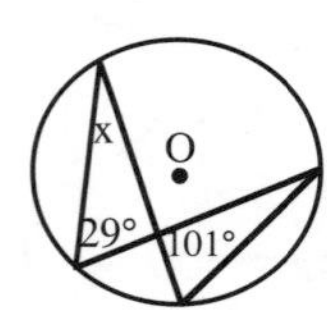


## Answers

| | | | | |
|---|---|---|---|---|
| 1. 38° | 2. 28° | 3. 28° | 4. 180° | 5. 114° |
| 6. 114° | 7. 246° | 8. 332° | 9. 68° | 10. 73° |
| 11. 98° | 12. 124° | 13. 50° | 14. 55° | 15. 18° |
| 16. 27° | 17. 55° | 18. 77° | 19. 35° | 20. 50° |

## AREA OF SECTORS

**#21**

**SECTORS** of a circle are formed by the two radii of a central angle and the arc between their endpoints on the circle. For example, a 60° sector looks like a slice of pizza. See page 376, problem CS-16 for more details and examples.

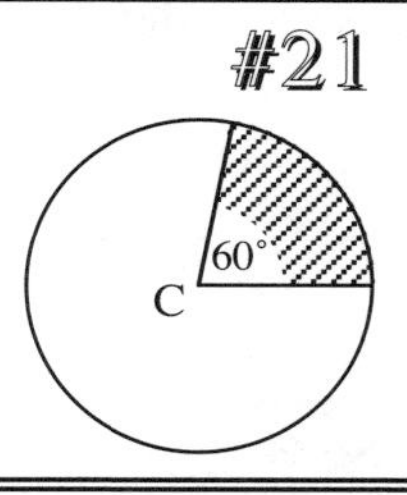


### Example 1

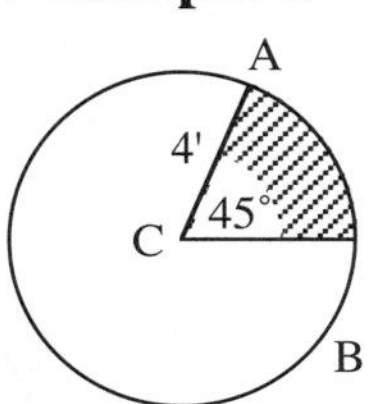


Find the area of the 45° sector. First find the fractional part of the circle involved. $\frac{m\overset{\frown}{AB}}{360°} = \frac{45°}{360°} = \frac{1}{8}$ of the circle's area. Next find the area of the circle: $A = \pi 4^2 = 16\pi$ sq. ft. Finally, multiply the two results to find the area of the sector.

$\frac{1}{8} \cdot 16\pi = 2\pi \approx 6.28$ sq. ft.

### Example 2

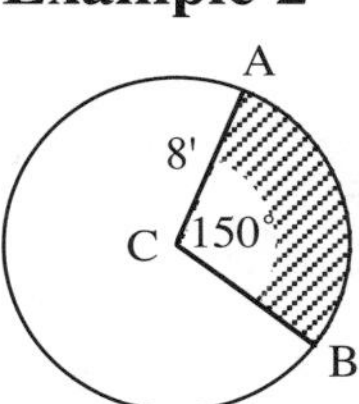


Find the area of the 150° sector.

Fractional part of circle is $\frac{m\overset{\frown}{AB}}{360°} = \frac{150°}{360°} = \frac{5}{12}$

Area of circle is $A = \pi r^2 = \pi 8^2 = 64\pi$ sq. ft.

Area of the sector: $\frac{5}{12} \cdot 64\pi = \frac{320\pi}{12} = \frac{80\pi}{3} \approx 83.76$ sq. ft.

1. Find the area of the shaded sector in each circle below. Points A, B and C are the centers.

a) 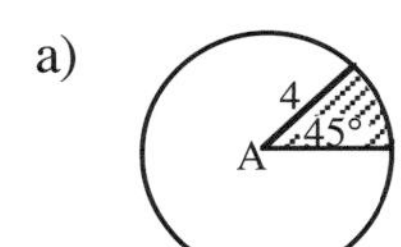


b) 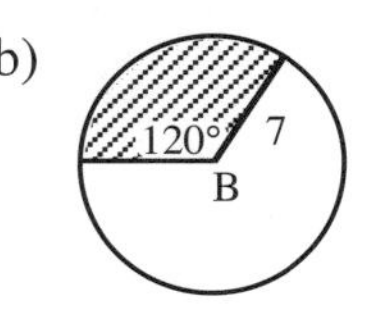


c) 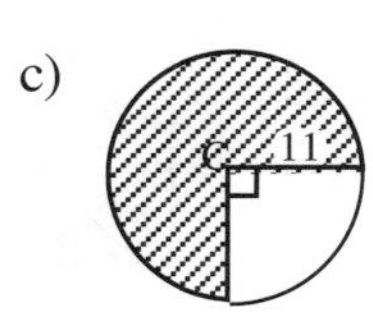


Calculate the area of the following shaded sectors. Point O is the center of each circle.

2. 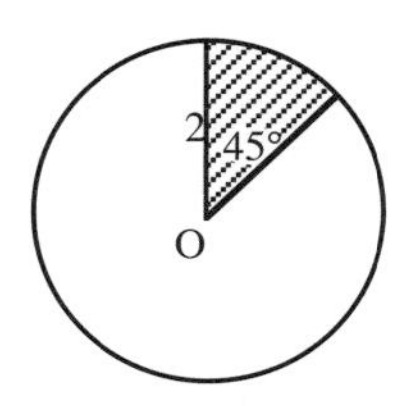


3. 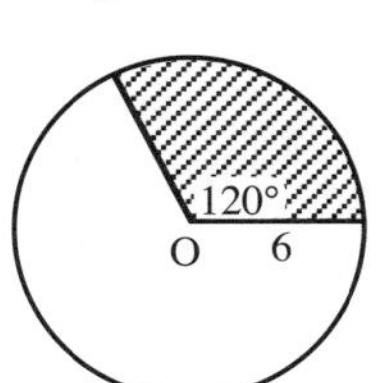


4. 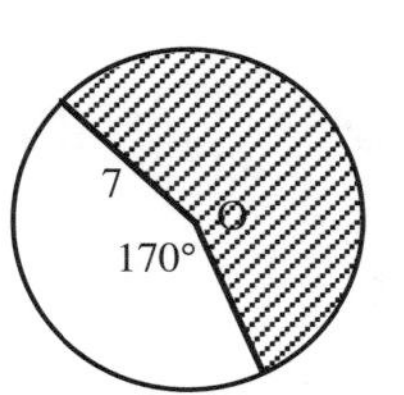


5. 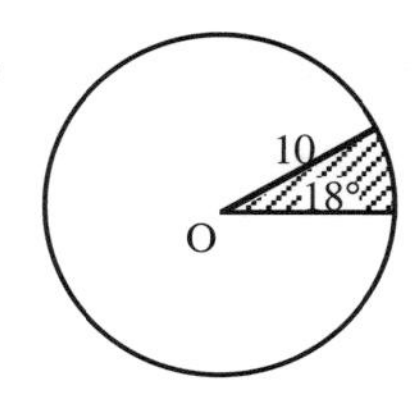


6. The shaded region in the figure is called a segment of the circle. It can be found by subtracting the area of ΔMIL from the sector MIL. Find the area of the segment of the circle.

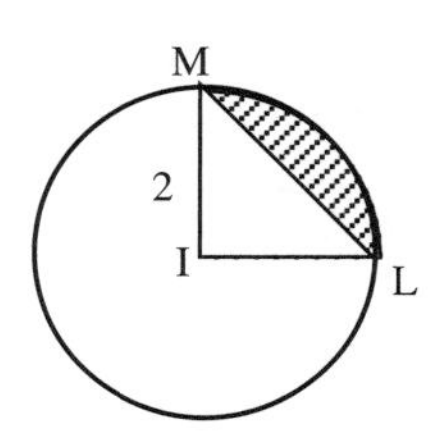

Find the area of the shaded regions.

7.

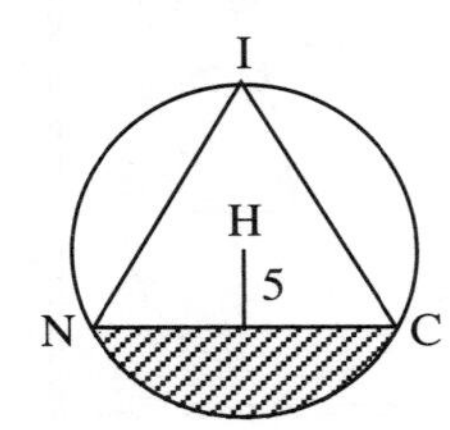


8.

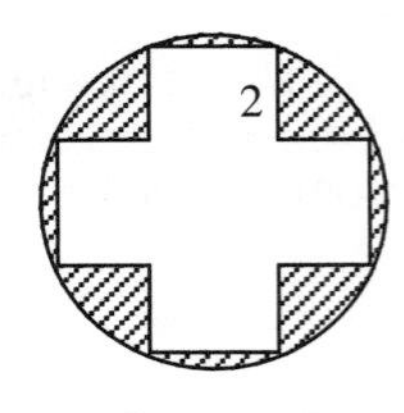


9. YARD is a square; A and D are the centers of the arcs.

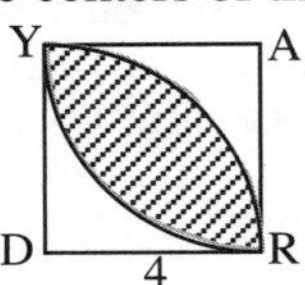


10. Find the area of a circular garden if the diameter of the garden is 30 feet.

11. Find the area of a circle inscribed in a square whose diagonal is 8 feet long.

12. The area of a 60° sector of a circle is $10\pi$ m$^2$. Find the radius of the circle.

13. The area of a sector of a circle with a radius of 5 mm is $10\pi$ mm$^2$. Find the measure of its central angle.

Find the area of each shaded region.

14.

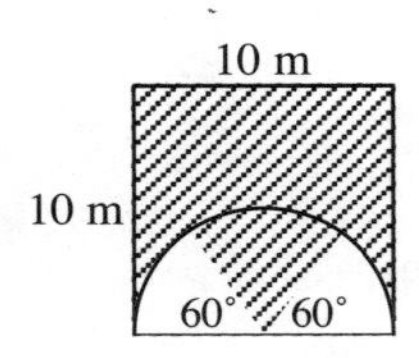


15.

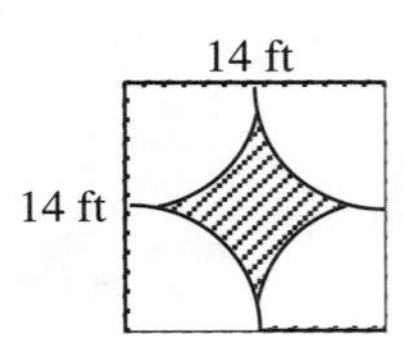


16.

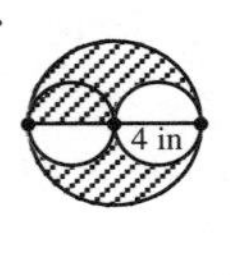


17.

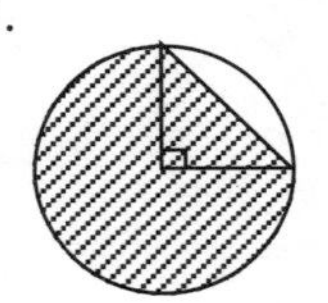

r = 8

18.

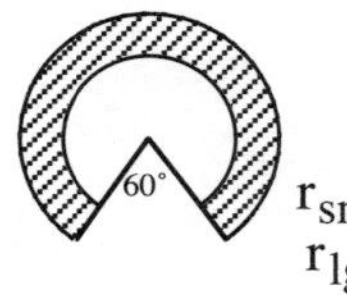


$r_{sm} = 5$
$r_{lg} = 8$

19.

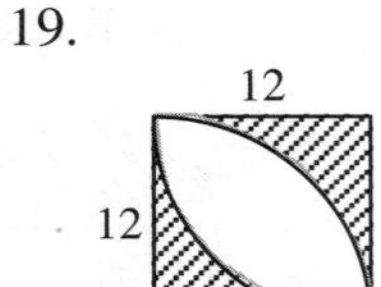


20. Find the radius. The shaded area is $12\pi$ cm$^2$.

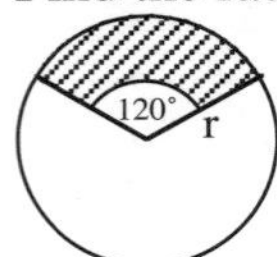


## Answers

1. a) $2\pi$ un$^2$ b) $\frac{49}{3}\pi$ un$^2$ c) $\frac{363\pi}{4}$ un$^2$

2. $\frac{\pi}{2}$ 3. $12\pi$ 4. $\frac{931\pi}{36}$ 5. $5\pi$

6. $\pi - 2$ sq. un. 7. $\frac{100}{3}\pi - 25\sqrt{3}$ un$^2$ 8. $10\pi - 20$ sq. un. 9. $8\pi - 16$ sq. un.

10. $225\pi$ ft$^2$ 11. $8\pi$ ft$^2$ 12. $2\sqrt{15}$ m 13. 144

14. $100 - \frac{25}{3}\pi$ m$^3$ 15. $196 - 49\pi$ ft$^2$ 16. $10\pi$ in$^2$ 17. $48\pi + 32$ un$^2$

18. $\frac{65}{2}\pi$ un$^2$ 19. $\approx 61.8$ un$^2$ 20. 6 un.

## TANGENTS, SECANTS, AND CHORDS #22

The figure at right shows a circle with three lines lying on a flat surface. Line a does not intersect the circle at all. Line b intersects the circle in two points and is called a **SECANT**. Line c intersects the circle in only one point and is called a **TANGENT** to the circle.

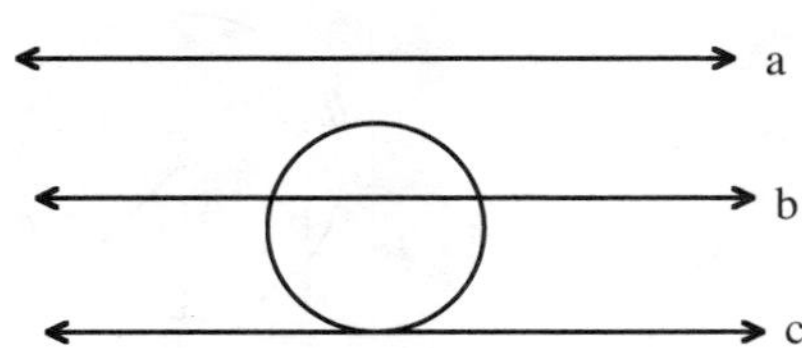


**TANGENT/RADIUS THEOREMS:**

1. Any tangent of a circle is perpendicular to a radius of the circle at their point of intersection.
2. Any pair of tangents drawn at the endpoints of a diameter are parallel to each other.

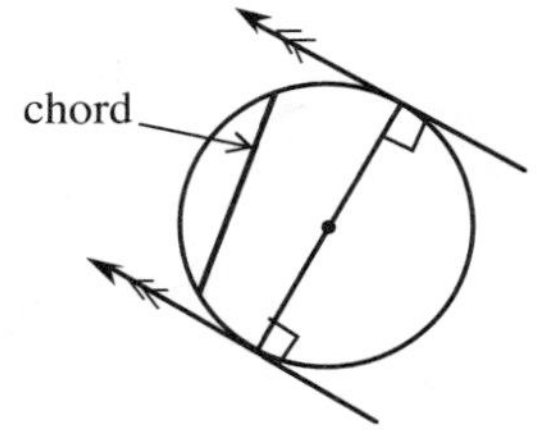


A **CHORD** of a circle is a line segment with its endpoints on the circle.

**DIAMETER/CHORD THEOREMS:**

1. If a diameter bisects a chord, then it is perpendicular to the chord.
2. If a diameter is perpendicular to a chord, then it bisects the chord.

**ANGLE-CHORD-SECANT THEOREMS:**

$m\angle 1 = \frac{1}{2}(m\overparen{AD} + m\overparen{BC})$

$AE \bullet EC = DE \bullet EB$

$m\angle P = \frac{1}{2}(m\overparen{RT} - m\overparen{QS})$

$PQ \cdot PR = PS \cdot PT$

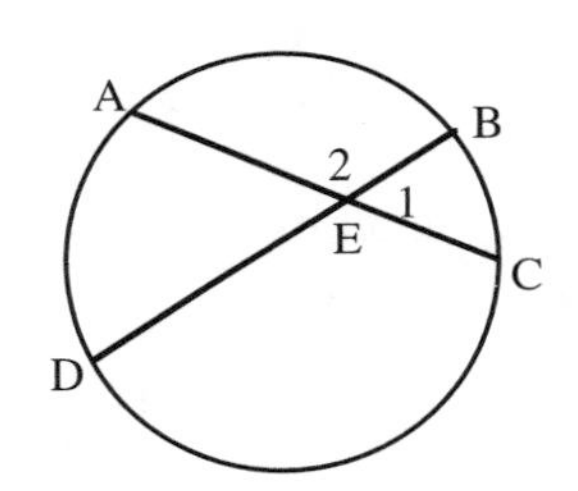


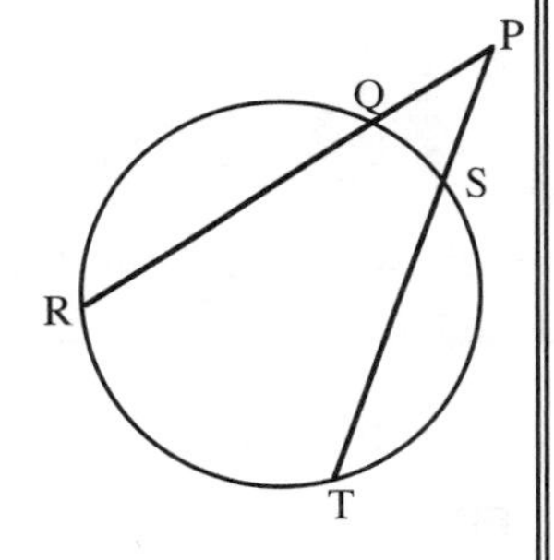


### Example 1

If the radius of the circle is 5 units and AC = 13 units, find AD and AB.

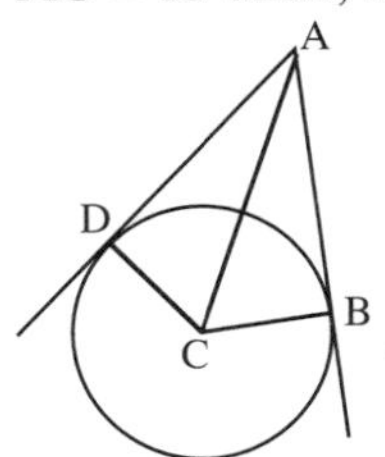


$\overline{AD} \perp \overline{CD}$ and $\overline{AB} \perp \overline{CD}$ by Tangent/Radius Theorem, so $(AD)^2 + (CD)^2 = (AC)^2$ or $(AD)^2 + (5)^2 = (13)^2$. So AD = 12 and $\overline{AB} \cong \overline{AD}$ so AB = 12.

### Example 2

In ⊙B, EC = 8 and AB = 5. Find BF. Show all subproblems.

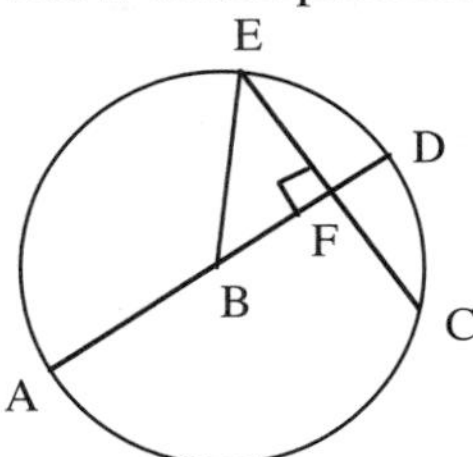


The diameter is perpendicular to the chord, therefore it bisects the chord, so EF = 4. $\overline{AB}$ is a radius and AB = 5. $\overline{EB}$ is a radius, so EB = 5. Use the Pythagorean Theorem to find BF: $BF^2 + 4^2 = 5^2$, BF = 3.

In each circle, C is the center and $\overline{AB}$ is tangent to the circle at point B. Find the area of each circle.

1. 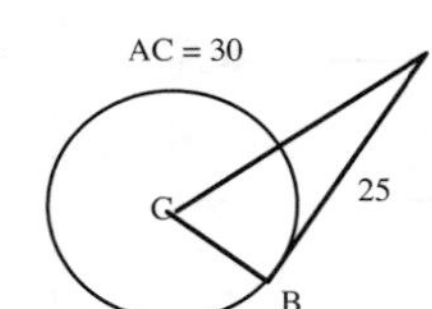


2. 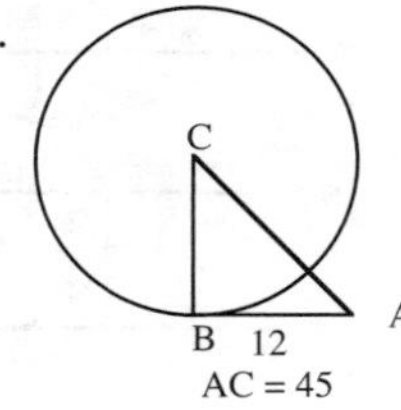


3. 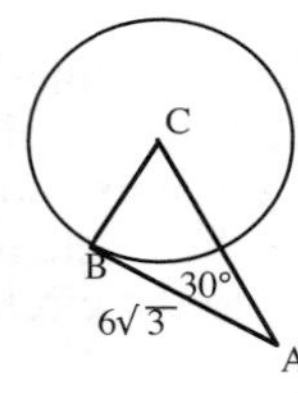


4. 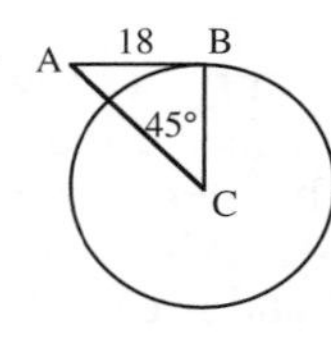


5. 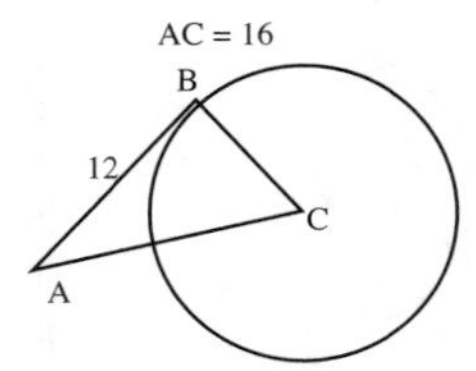


6. 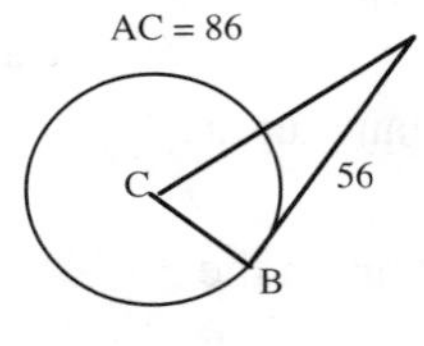


7. 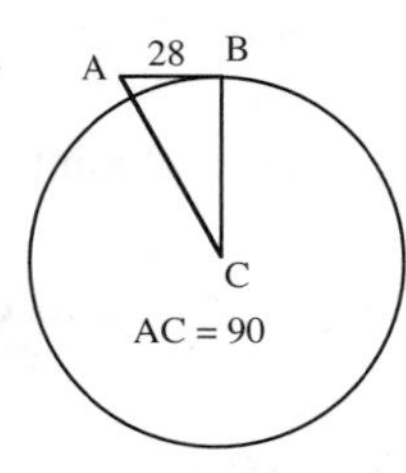


8. 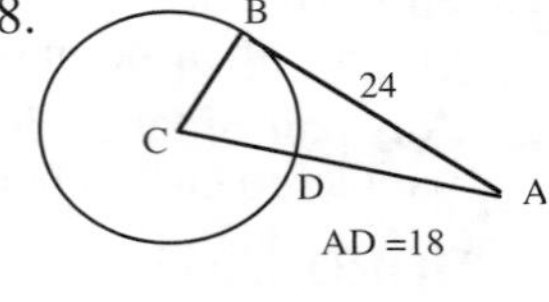


9. In the figure at right, point E is the center and $m\angle CED = 55°$. What is the area of the circle?

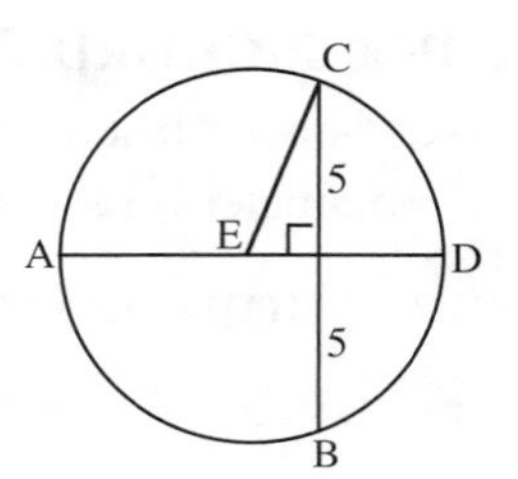


In the following problems, B is the center of the circle. Find the length of $\overline{BF}$ given the lengths below.

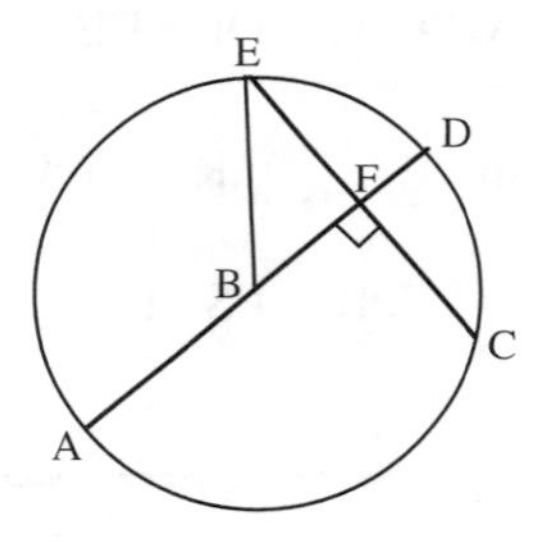


10. EC = 14, AB = 16

11. EC = 35, AB = 21

12. FD = 5, EF = 10

13. EF = 9, FD = 6

14. In ⊙R, if AB = 2x − 7 and CD = 5x − 22, find x.

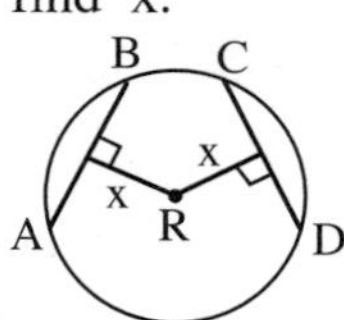


15. In ⊙O, $\overline{MN} \cong \overline{PQ}$, MN = 7x + 13, and PQ = 10x − 8. Find PS.

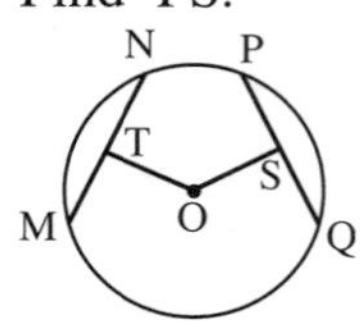


16. In ⊙D, if AD = 5 and TB = 2, find AT.

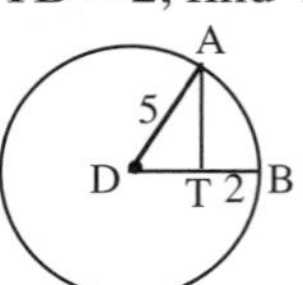

17. In $\odot J$, radius $JL$ and chord $MN$ have lengths of 10 cm. Find the distance from $J$ to $\overline{MN}$.

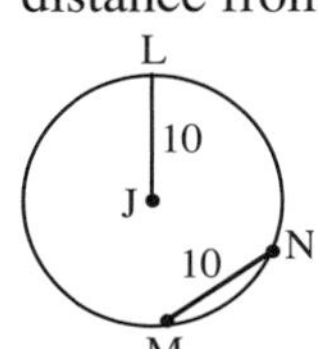


18. In $\odot Q$, $QC = 13$ and $QT = 5$. Find $AB$.

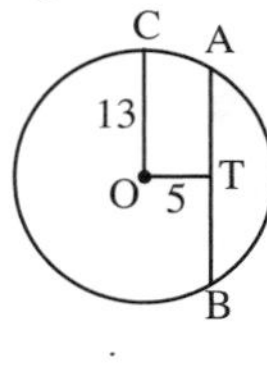


19. If $\overline{AC}$ is tangent to circle E and $\overline{EH} \perp \overline{GI}$, is $\Delta GEH \sim \Delta AEB$? Prove your answer.

20. If $\overline{EH}$ bisects $\overline{GI}$ and $\overline{AC}$ is tangent to circle E at point B, are $\overline{AC}$ and $\overline{GI}$ parallel? Prove your answer.

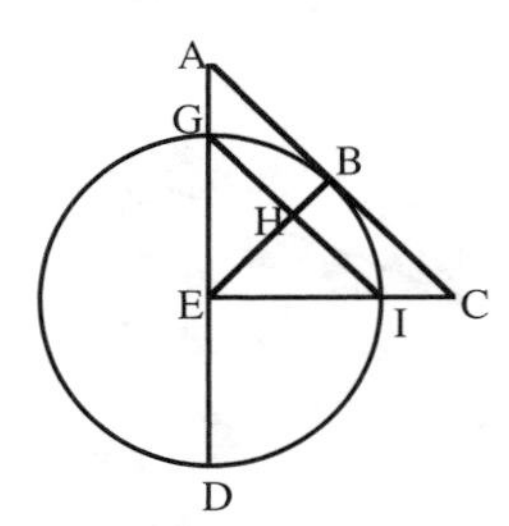


Find the value of $x$.

21.

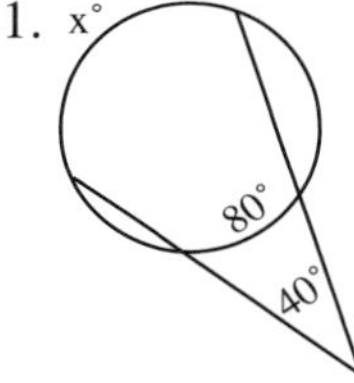


22.

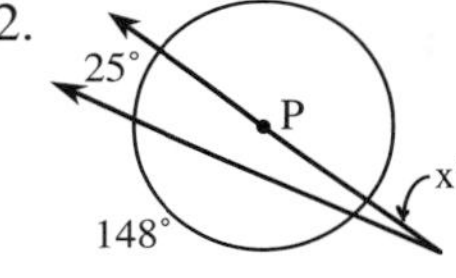


23.

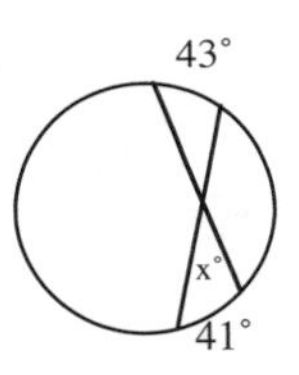


24.

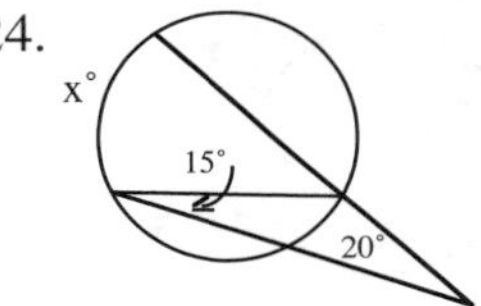


In $\odot F$, $m\widehat{AB} = 84$, $m\widehat{BC} = 38$, $m\widehat{CD} = 64$, and $m\widehat{DE} = 60$. Find the measure of each angle and arc.

25. $m\widehat{EA}$

26. $m\widehat{AEB}$

27. $m\angle 1$

28. $m\angle 2$

29. $m\angle 3$

30. $m\angle 4$

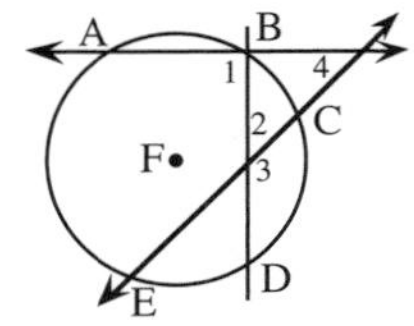


For each circle, tangent segments are shown. Use the measurements given find the value of $x$.

31.

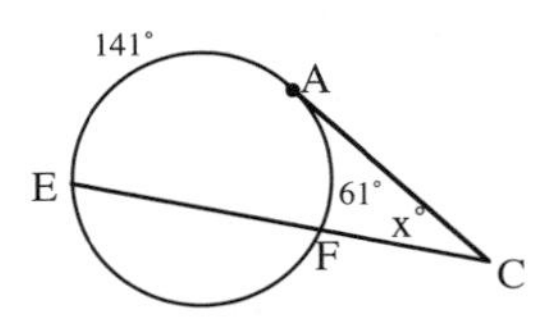


32.

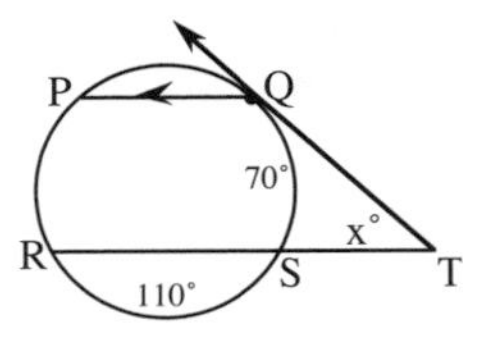


33.

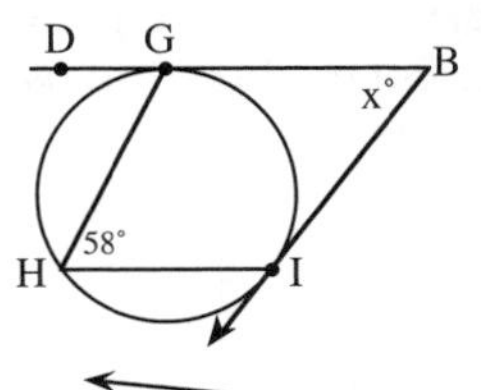


34.

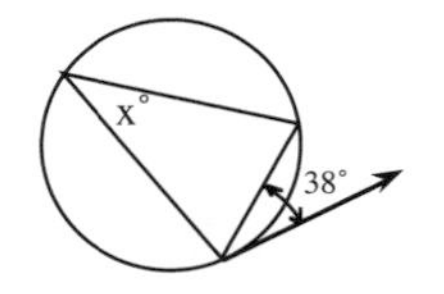


35.

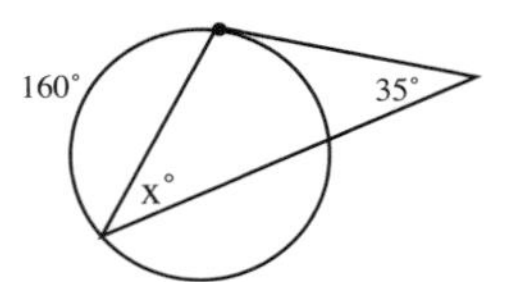


36.

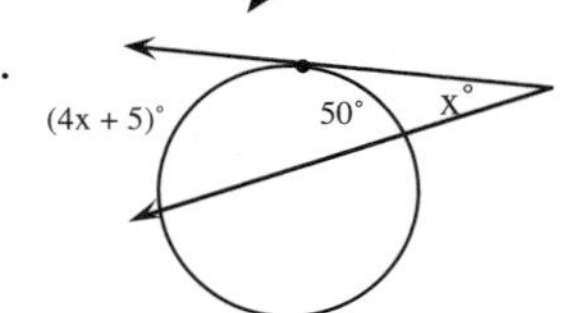

Find each value of x. Tangent segments are shown in problems 40, 43, 46, and 48.

37.

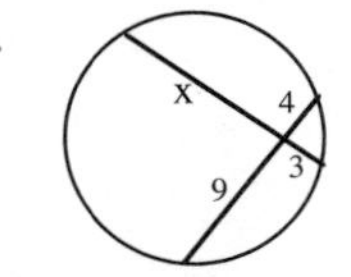


38.

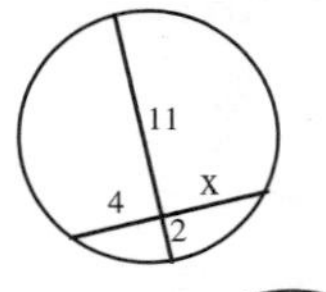


39.

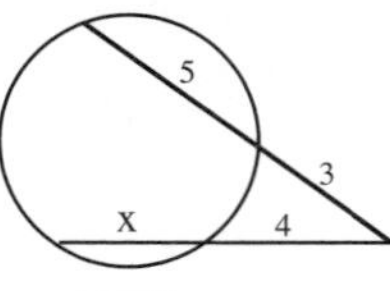


40.

41.

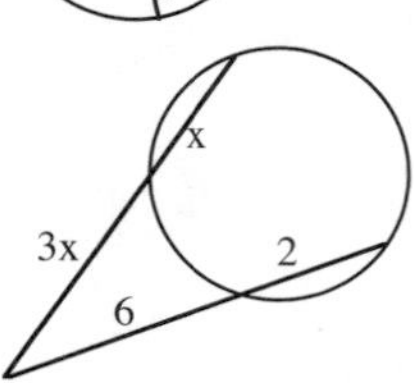


42.

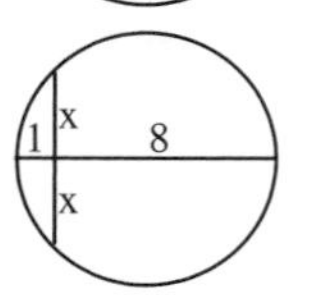


43.

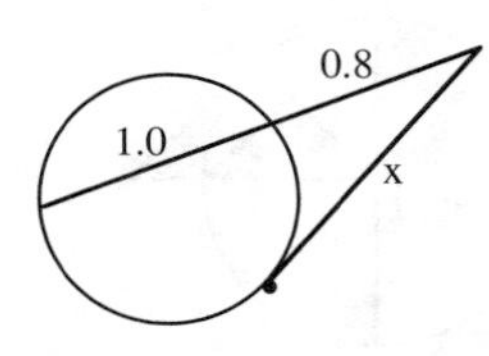


44.

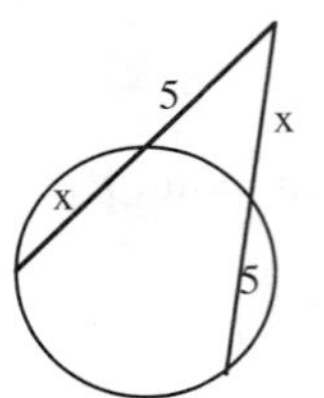


45.

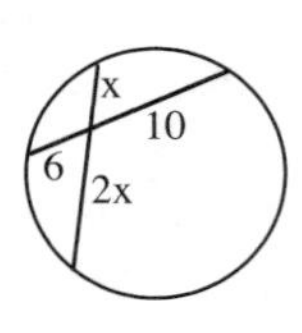


46.

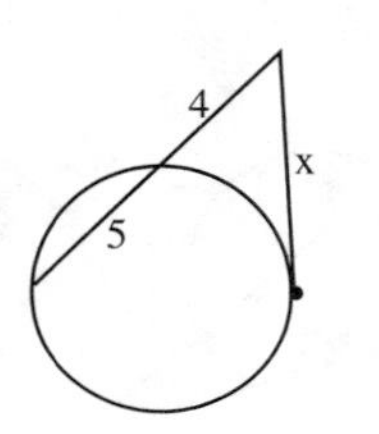


47.

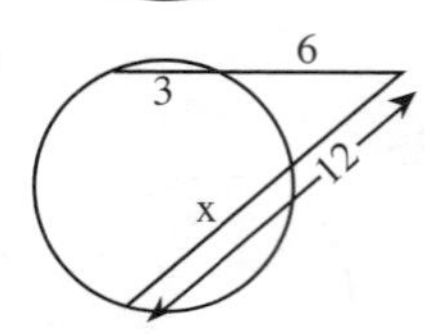


48.

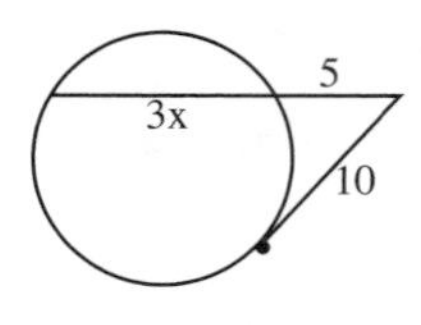


## Answers

1. $275\pi$ sq. un.
2. $1881\pi$ sq. un.
3. $36\pi$ sq. un.
4. $324\pi$ sq. un.
5. $112\pi$ sq. un.
6. $4260\pi$ sq. un.
7. $7316\pi$ sq. un.
8. $49\pi$ sq. un.
9. $\approx 117.047$ sq. un.
10. $\approx 14.4$
11. $\approx 11.6$
12. $\approx 7.5$
13. 3.75
14. 5
15. 31
16. 4
17. $5\sqrt{3}$ cm.
18. $5\sqrt{3}$
19. Yes, $\angle GEH \cong \angle AEB$ (reflexive). $\overline{EB}$ is perpendicular to $\overline{AC}$ since it is tangent so $\angle GHE \cong \angle ABE$ because all right angles are congruent. So the triangles are similar by AA~.
20. Yes. Since $\overline{EH}$ bisects $\overline{GI}$ it is also perpendicular to it (SSS). Since $\overline{AC}$ is a tangent, $\angle ABE$ is a right angle. So the lines are parallel since the corresponding angles are right angles and all right angles are equal.

| | | | | |
|---|---|---|---|---|
| 21. 160 | 22. 9 | 23. 42 | 24. 70 | 25. 114 |
| 26. 276 | 27. 87 | 28. 49 | 29. 131 | 30. 38 |
| 31. 40 | 32. 55 | 33. 64 | 34. 38 | 35. 45 |
| 36. 22.5 | 37. 12 | 38. $5\frac{1}{2}$ | 39. 2 | 40. 30 |
| 41. 2 | 42. $2\sqrt{2}$ | 43. 1.2 | 44. 5 | 45. $\sqrt{30}$ |
| 46. 6 | 47. 7.5 | 48. 5 | | |

## VOLUME AND SURFACE AREA OF POLYHEDRA

#23

The **VOLUME** of various polyhedra, that is, the number of cubic units needed to fill each one, is found by using the formulas below.

for prisms and cylinders

$V$ = base area × height, $V = Bh$

for pyramids and cones

$V = \frac{1}{3}Bh$

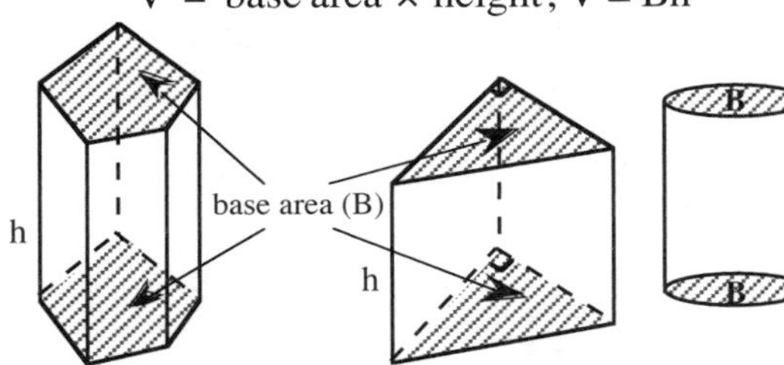


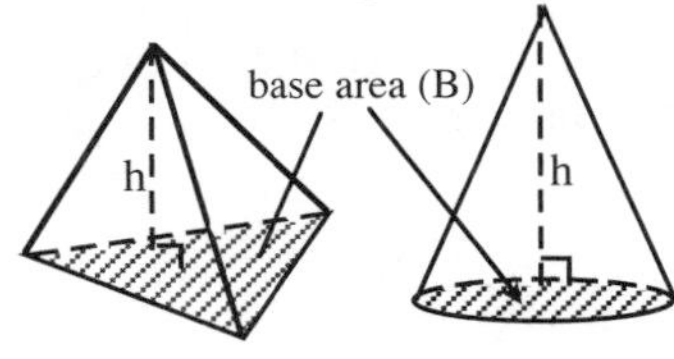


In prisms and cylinders, you may use either base, since they are congruent. Since the bases of cylinders and cones are circles, their area formulas may be expressed as:

cylinder $V = \pi r^2 h$ and cone $V = \frac{1}{3}\pi r^2 h$.

The **SURFACE AREA** of a polyhedron is the sum of the areas of its base(s) and faces.

### Example 1

Use the appropriate formula(s) to find the volume of each figure below:

a) 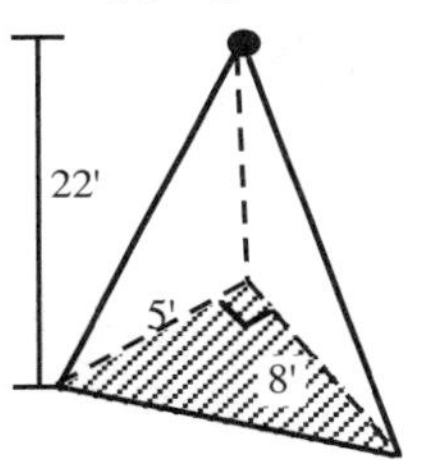


b) 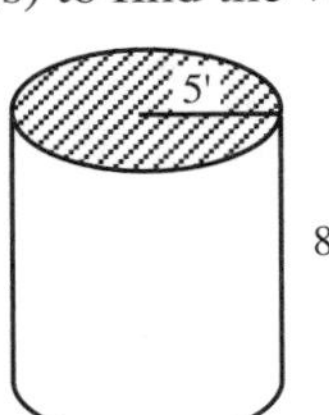


c) 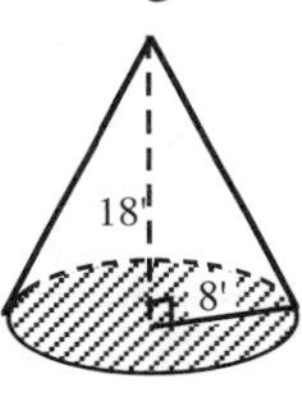


d) 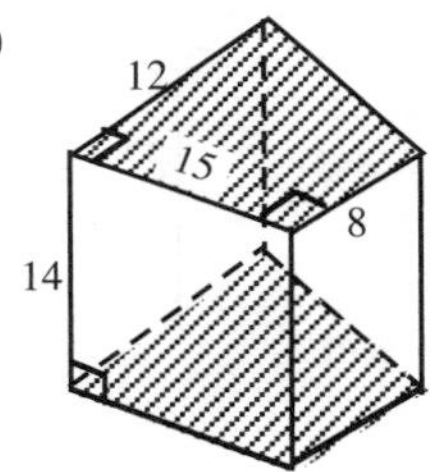


a) This is a triangular pyramid. The base is a right triangle so the area of the base is $B = \frac{1}{2} \cdot 8 \cdot 5 = 20$ square units, so $V = \frac{1}{3}(20)(22) \approx 146.7$ cubic feet.

b) This is a cylinder. The base is a circle, so $B = \pi 5^2$, $V = (25\pi)(8) = 200\pi \approx 628.32$ cubic feet.

c) This is a cone. The base is a circle, so $B = \pi 8^2$. $V = \frac{1}{3}(64\pi)(18) = \frac{1}{3}(64)(18)\pi \Rightarrow$ $= \frac{1}{3}(1152)\pi = 384\pi \approx 1206.37 \text{ feet}^3$

d) This prism has a trapezoidal base, so $B = \frac{1}{2}(12 + 8)(15) = 150$.
Thus, $V = (150)(14) = 2100$ cubic feet.

This figure is for Example 2, on next page:

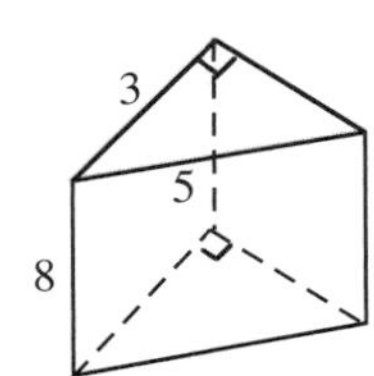

## Example 2

Find the surface area of the triangular prism shown on the bottom of the previous page. The figure is made up of two triangles (the top and bottom) and three rectangles as shown at right. Find the area of each of these shapes. To find the area of the triangle and the last rectangle, use the Pythagorean Theorem to find the length of the second leg of the right triangular base. Since $3^2 + \text{leg}^2 = 5^2$, leg = 4.

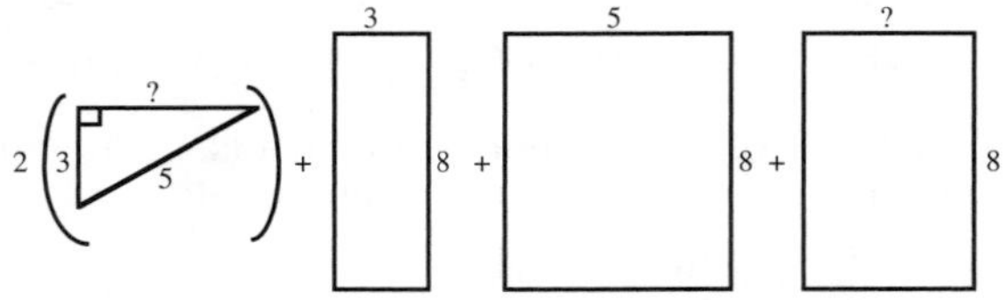


Calculate all of the areas, and find their sum.

$$SA = 2\left(\frac{1}{2}(3)(4)\right) + 3(8) + 5(8) + 4(8)$$

$$= 12 + 24 + 40 + 32 = 108 \text{ square units}$$

## Example 3

Find the total surface area of a regular square pyramid with a slant height of 10 inches and a base with sides 8 inches long.

The figure is made up of 4 identical triangles and a square base.

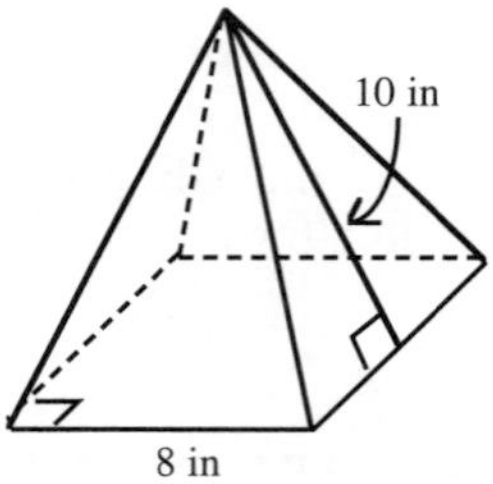


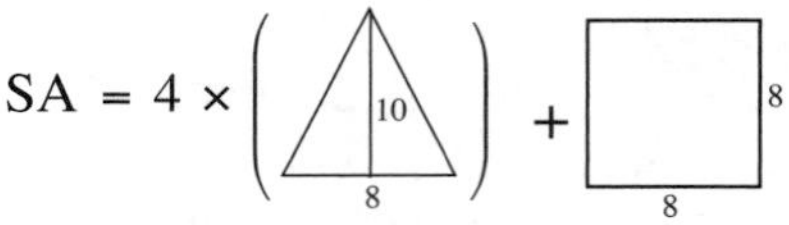


$$= 4\left(\frac{10 \times 8}{2}\right) + 8(8)$$

$= 160 + 64$

$= 224$ square inches

Find the volume of each figure.

1.

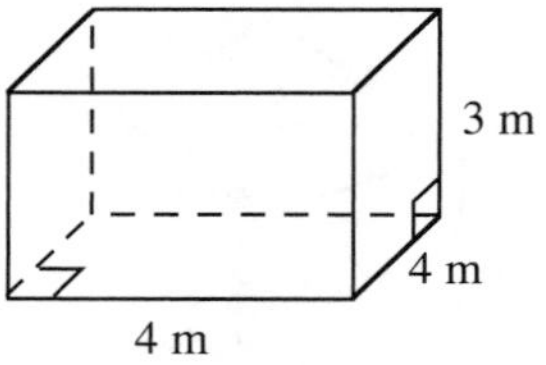


2.

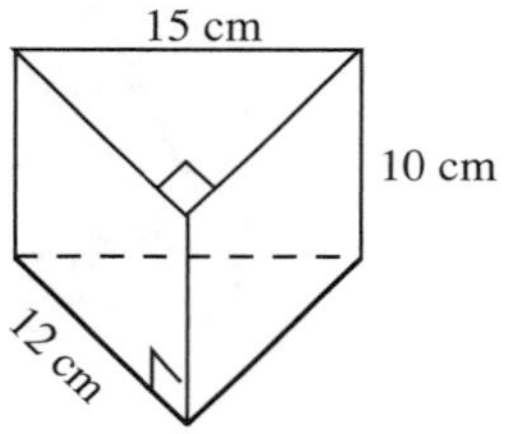


3.

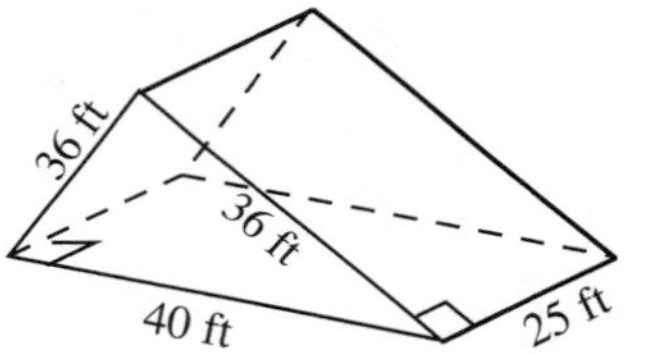


4.

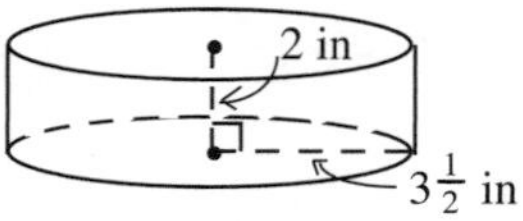


5.

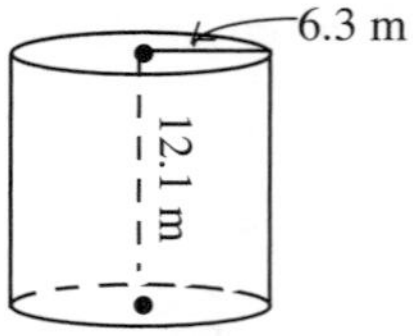


6.

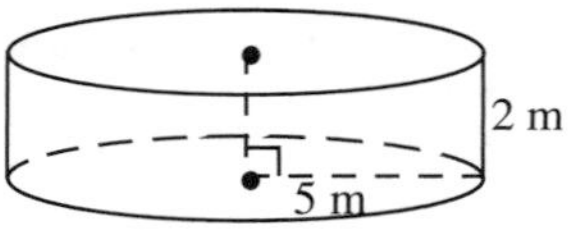


7.

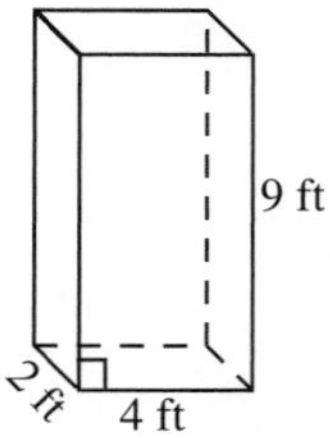


8.

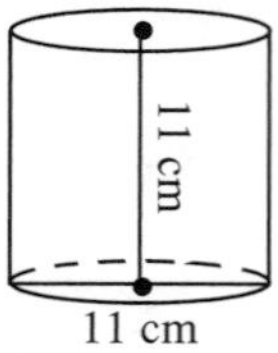


9.

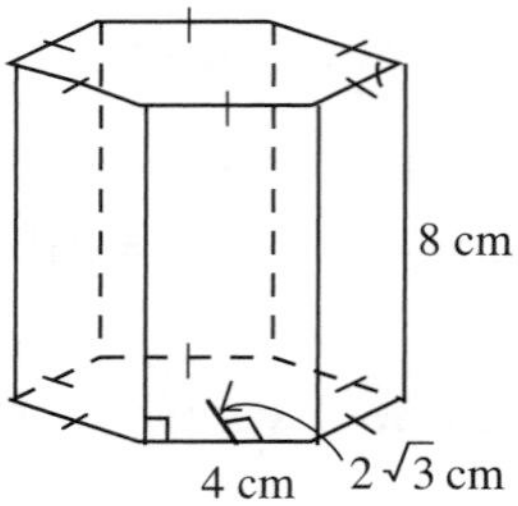

10.

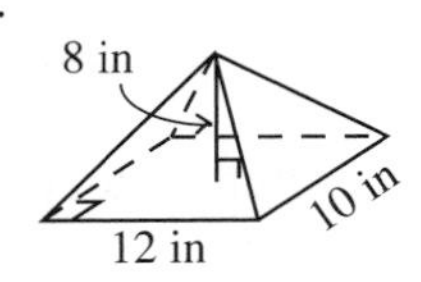


11.

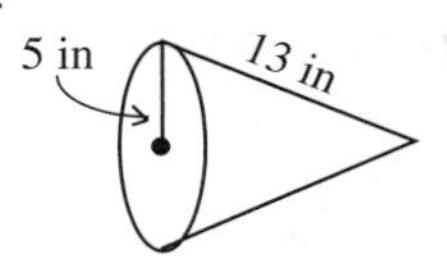


12.

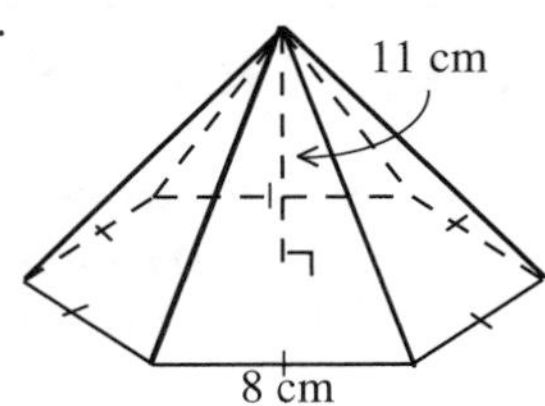


13.

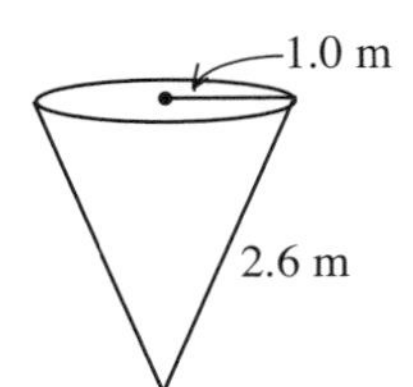


14.

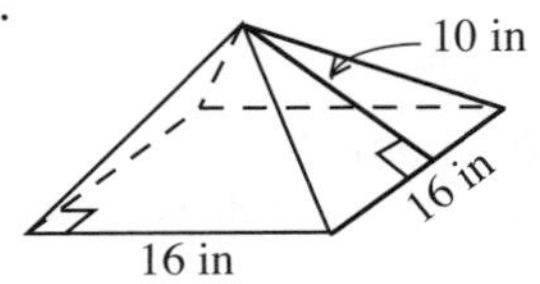


15.

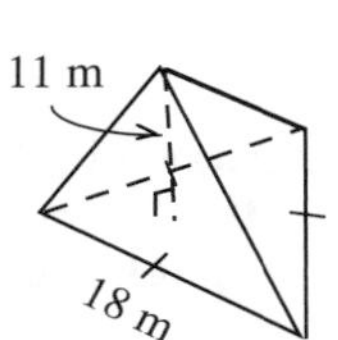


16.

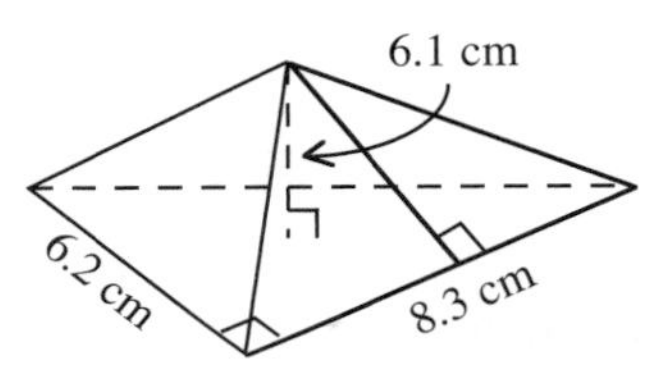


17.

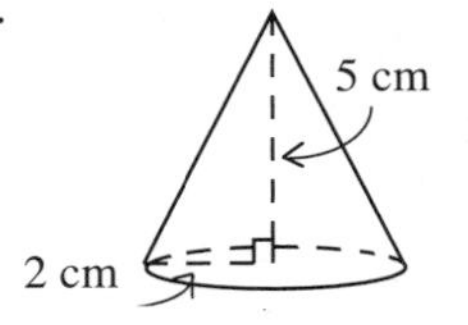


18.

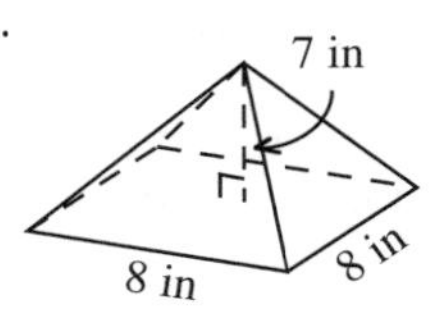


19. Find the volume of the solid shown.

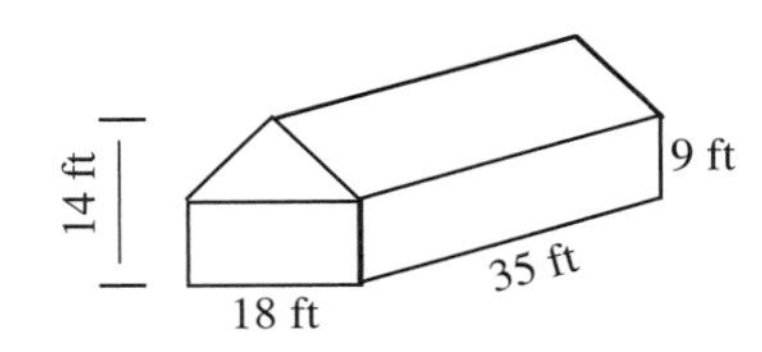


20. Find the volume of the remaining solid after a hole with a diameter of 4 mm is drilled through it.

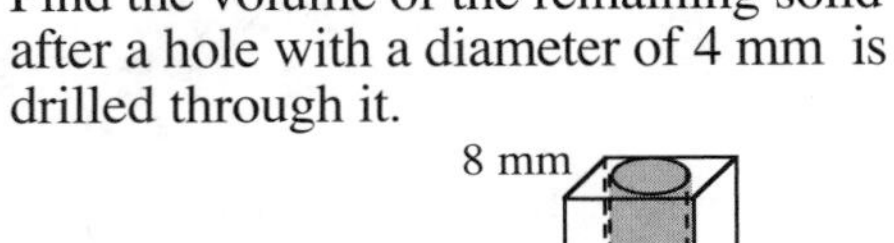

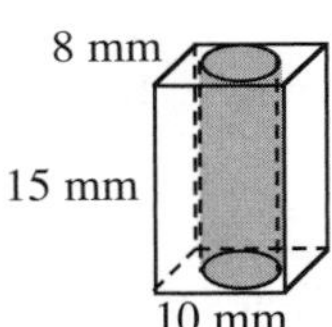


Find the total surface area of the figures in the previous volume problems.

21. Problem 1
22. Problem 2
23. Problem 3
24. Problem 5
25. Problem 6
26. Problem 7
27. Problem 8
28. Problem 9
29. Problem 10
30. Problem 14
31. Problem 18
32. Problem 19

## Answers

1. $48\ m^3$
2. $540\ cm^3$
3. $14966.6\ ft^3$
4. $76.9\ m^3$
5. $1508.75\ m^3$
6. $157\ m^3$
7. $72\ ft^3$
8. $1045.4\ cm^3$
9. $332.6\ cm^3$
10. $320\ in^3$
11. $314.2\ in^3$
12. $609.7\ cm^3$
13. $2.5\ m^3$
14. $512\ m^3$
15. $514.4\ m^3$
16. $52.3\ cm^3$
17. $20.9\ cm^3$
18. $149.3\ in^3$
19. $7245\ ft^3$
20. $1011.6\ mm^3$
21. $80\ m^2$
22. $468\ cm^2$
23. $3997.33\ ft^2$
24. $727.98\ m^2$
25. $50\pi + 20\pi \approx 219.8\ m^2$
26. $124\ ft^2$
27. $121\pi + 189.97 \approx 569.91\ cm^2$
28. $192 + 48\sqrt{3} \approx 275.14\ cm^2$
29. $213.21\ in^2$
30. $576\ in^2$
31. $193.0\ in^2$
32. $2394.69\ ft^2$

## SIMPLIFYING RATIONAL EXPRESSIONS #24

**RATIONAL EXPRESSIONS** are fractions that have algebraic expressions in their numerators and/or denominators. To simplify rational expressions find **factors** in the numerator and denominator that are the same and then write them as fractions equal to 1. For example,

$$\frac{6}{6} = 1 \qquad \frac{x^2}{x^2} = 1 \qquad \frac{(x + 2)}{(x + 2)} = 1 \qquad \frac{(3x - 2)}{(3x - 2)} = 1$$

Notice that the last two examples involved binomial sums and differences. **Only** when sums or differences are **exactly** the same does the fraction equal 1. Rational expressions such as the examples below **CANNOT** be simplified:

$$\frac{(6 + 5)}{6} \qquad \frac{x^3 + y}{x^3} \qquad \frac{x}{x + 2} \qquad \frac{3x - 2}{2}$$

Most problems that involve rational expressions will require that you **factor** the numerator and denominator. For example:

$$\frac{12}{54} = \frac{2 \cdot 2 \cdot 3}{2 \cdot 3 \cdot 3 \cdot 3} = \frac{2}{9} \qquad \text{Notice that } \frac{2}{2} \text{ and } \frac{3}{3} \text{ each equal 1.}$$

$$\frac{6x^3y^2}{15x^2y^4} = \frac{2 \cdot 3 \cdot x^2 \cdot x \cdot y^2}{5 \cdot 3 \cdot x^2 \cdot y^2 \cdot y^2} = \frac{2x}{5y^2} \qquad \text{Notice that } \frac{3}{3}, \frac{x^2}{x^2}, \text{ and } \frac{y^2}{y^2} = 1.$$

$$\frac{x^2 - x - 6}{x^2 - 5x + 6} = \frac{(x + 2)(x - 3)}{(x - 2)(x - 3)} = \frac{x + 2}{x - 2} \quad \text{where } \frac{x - 3}{x - 3} = 1.$$

All three examples demonstrate that **all parts** of the numerator and denominator--whether constants, monomials, binomials, or factorable trinomials--must be written as products **before** you can look for factors that equal 1.

One special situation is shown in the following examples:

$$\frac{-2}{2} = -1 \qquad \frac{-x}{x} = -1 \qquad \frac{-x - 2}{x + 2} = \frac{-(x + 2)}{x + 2} = -1 \qquad \frac{5 - x}{x - 5} = \frac{-(x - 5)}{x - 5} = -1$$

Note that in all cases we assume the denominator does not equal zero.

### Example 1

Simplify $\frac{12(x-1)^3(x+2)}{3(x-1)^2(x+2)^2}$ completely.

Factor: $\frac{4 \cdot 3(x-1)^2(x-1)(x+2)}{3(x-1)^2(x+2)(x+2)}$

Note: $\frac{3}{3}$, $\frac{(x-1)^2}{(x-1)^2}$, and $\frac{x+2}{x+2} = 1$, so

the simplified form is $\frac{4(x-1)}{(x+2)}$, $x \neq 1$ or $-2$.

### Example 2

Simplify $\frac{x^2 - 6x + 8}{x^2 + 4x - 12}$ completely.

Factor: $\frac{x^2 - 6x + 8}{x^2 + 4x - 12} = \frac{(x - 4)(x - 2)}{(x + 6)(x - 2)}$

Since $\frac{x-2}{x-2} = 1$, $\frac{x-4}{x+6}$ is the simplified form,

$x \neq -6$ or $2$.

Simplify each of the following rational expression completely. Assume the denominator is not equal to zero.

1. $\frac{2(x+3)}{4(x-2)}$

2. $\frac{2(x-3)}{6(x+2)}$

3. $\frac{2(x+3)(x-2)}{6(x-2)(x+2)}$

4. $\frac{4(x-3)(x-5)}{6(x-3)(x+2)}$

5. $\frac{3(x-3)(4-x)}{15(x+3)(x-4)}$

6. $\frac{15(x-1)(7-x)}{25(x+1)(x-7)}$

7. $\frac{24(y-4)(y-6)}{16(y+6)(6-y)}$

8. $\frac{36(y+4)(y-16)}{32(y+16)(16-y)}$

9. $\frac{(x+3)^2(x-2)^4}{(x+3)^4(x-2)^3}$

10. $\frac{(x+3)^4(x-2)^5}{(x+3)^7(x-2)}$

11. $\frac{(x+5)^4(x-3)^2}{(x+5)^5(x-3)^3}$

12. $\frac{(2x-5)^2(x+3)^2}{(2x-5)^5(x+3)^3}$

13. $\frac{(5-x)^2(x-2)^2}{(x+5)^4(x-2)^3}$

14. $\frac{(5-x)^4(3x-1)^2}{(x-5)^4(3x-2)^3}$

15. $\frac{12(x-7)(x+2)^4}{20(x-7)^2(x+2)^5}$

16. $\frac{24(3x-7)(x+1)^6}{20(3x-7)^3(x+1)^5}$

17. $\frac{x^2-1}{(x+1)(x-2)}$

18. $\frac{x^2-4}{(x+1)^2(x-2)}$

19. $\frac{x^2-4}{x^2+x-6}$

20. $\frac{x^2-16}{x^3+9x^2+20x}$

## Answers

1. $\frac{x+3}{2(x-2)}$

2. $\frac{x-3}{3(x+2)}$

3. $\frac{x+3}{3(x+2)}$

4. $\frac{2(x-5)}{3(x+2)}$

5. $-\frac{x-3}{5(x+3)}$

6. $-\frac{3(x-1)}{5(x+1)}$

7. $-\frac{3(y-4)}{2(y+6)}$

8. $-\frac{9(y+4)}{8(y+16)}$

9. $\frac{x-2}{(x+3)^2}$

10. $\frac{(x-2)^4}{(x+3)^3}$

11. $\frac{1}{(x+5)(x-3)}$

12. $\frac{1}{(2x-5)^3(x+3)}$

13. $\frac{(5-x)^2}{(x+5)^4(x-2)}$

14. $\frac{(3x-1)^2}{(3x-2)^3}$

15. $\frac{3}{5(x-7)(x+2)}$

16. $\frac{6(x+1)}{5(3x-7)^2}$

17. $\frac{x-1}{x-2}$

18. $\frac{x+2}{(x+1)^2}$

19. $\frac{x+2}{x+3}$

20. $\frac{x-4}{x(x+5)}$

## MULTIPLICATION AND DIVISION OF RATIONAL EXPRESSIONS #25

To multiply or divide rational expressions, follow the same procedures used with numerical fractions. However, it is often necessary to factor the polynomials in order to simplify.

### Example 1

Multiply $\frac{x^2+6x}{(x+6)^2} \cdot \frac{x^2+7x+6}{x^2-1}$ and simplify the result.

After factoring, the expression becomes: $\frac{x(x+6)}{(x+6)(x+6)} \cdot \frac{(x+6)(x+1)}{(x+1)(x-1)}$

After multiplying, reorder the factors: $\frac{(x+6)}{(x+6)} \cdot \frac{(x+6)}{(x+6)} \cdot \frac{x}{(x-1)} \cdot \frac{(x+1)}{(x+1)}$

Since $\frac{(x+6)}{(x+6)} = 1$ and $\frac{(x+1)}{(x+1)} = 1$, simplify: $1 \cdot 1 \cdot \frac{x}{x-1} \cdot 1 \Rightarrow \frac{x}{x-1}$.

Note: $x \neq -6, -1$, or 1.

### Example 2

Divide $\frac{x^2-4x-5}{x^2-4x+4} \div \frac{x^2-2x-15}{x^2+4x-12}$ and simplify the result.

First, change to a multiplication expression by inverting (flipping) the second fraction: $\frac{x^2-4x-5}{x^2-4x+4} \cdot \frac{x^2+4x-12}{x^2-2x-15}$

After factoring, the expression is: $\frac{(x-5)(x+1)}{(x-2)(x-2)} \cdot \frac{(x+6)(x-2)}{(x-5)(x+3)}$

Reorder the factors (if you need to): $\frac{(x-5)}{(x-5)} \cdot \frac{(x-2)}{(x-2)} \cdot \frac{(x+1)}{(x-2)} \cdot \frac{(x+6)}{(x+3)}$

Since $\frac{(x-5)}{(x-5)} = 1$ and $\frac{(x-2)}{(x-2)} = 1$, simplify: $\frac{(x+1)(x+6)}{(x-2)(x+3)}$

Thus, $\frac{x^2-4x-5}{x^2-4x+4} \div \frac{x^2-2x-15}{x^2+4x-12} = \frac{(x+1)(x+6)}{(x-2)(x+3)}$ or $\frac{x^2+7x+6}{x^2+x-6}$. Note: $x \neq -3, 2$, or 5.

Multiply or divide each pair of rational expressions. Simplify the result. Assume the denominator is not equal to zero.

1. $\frac{x^2+5x+6}{x^2-4x} \cdot \frac{4x}{x+2}$

2. $\frac{x^2-2x}{x^2-4x+4} \div \frac{4x^2}{x-2}$

3. $\frac{x^2-16}{(x-4)^2} \cdot \frac{x^2-3x-18}{x^2-2x-24}$

4. $\frac{x^2-x-6}{x^2+3x-10} \cdot \frac{x^2+2x-15}{x^2-6x+9}$

5. $\frac{x^2-x-6}{x^2-x-20} \cdot \frac{x^2+6x+8}{x^2-x-6}$

6. $\frac{x^2-x-30}{x^2+13x+40} \cdot \frac{x^2+11x+24}{x^2-9x+18}$

7. $\dfrac{15-5x}{x^2-x-6} \div \dfrac{5x}{x^2+6x+8}$

8. $\dfrac{17x+119}{x^2+5x-14} \div \dfrac{9x-1}{x^2-3x+2}$

9. $\dfrac{2x^2-5x-3}{3x^2-10x+3} \cdot \dfrac{9x^2-1}{4x^2+4x+1}$

10. $\dfrac{x^2-1}{x^2-6x-7} \div \dfrac{x^3+x^2-2x}{x-7}$

11. $\dfrac{3x-21}{x^2-49} \div \dfrac{3x}{x^2+7x}$

12. $\dfrac{x^2-y^2}{x+y} \cdot \dfrac{1}{x-y}$

13. $\dfrac{y^2-y}{w^2-y^2} \div \dfrac{y^2-2y+1}{1-y}$

14. $\dfrac{y^2-y-12}{y+2} \div \dfrac{y-4}{y^2-4y-12}$

15. $\dfrac{x^2+7x+10}{x+2} \div \dfrac{x^2+2x-15}{x+2}$

## Answers

1. $\dfrac{4(x+3)}{(x-4)}$
2. $\dfrac{1}{4x}$
3. $\dfrac{(x+3)}{x-4}$
4. $\dfrac{(x+2)}{(x-2)}$
5. $\dfrac{x+2}{x-5}$
6. $\dfrac{x+3}{x-3}$
7. $\dfrac{-x-4}{x}$
8. $\dfrac{17(x-1)}{9x-1}$
9. $\dfrac{3x+1}{2x+1}$
10. $\dfrac{1}{x(x+2)}$
11. 1
12. 1
13. $\dfrac{-y}{w^2-y^2}$
14. $(y+3)(y-6)$
15. $\dfrac{x+2}{x-3}$

## ADDITION AND SUBTRACTION OF RATIONAL EXPRESSIONS #26

Addition and subtraction of rational expressions is done the same way as addition and subtraction of numerical fractions. Change to a common denominator (if necessary), combine the numerators, and then simplify.

## Example

The Least Common Multiple (lowest common denominator) of (x + 3)(x + 2) and (x + 2) is (x + 3)(x + 2).

$$\frac{4}{(x+2)(x+3)}+\frac{2x}{x+2}$$

The denominator of the first fraction already is the Least Common Multiple. To get a common denominator in the <u>second</u> fraction, multiply the fraction by $\frac{x+3}{x+3}$, a form of one (1).

$$=\frac{4}{(x+2)(x+3)}+\frac{2x}{x+2}\cdot\frac{(x+3)}{(x+3)}$$

Multiply the numerator and denominator of the second term:

$$=\frac{4}{(x+2)(x+3)}+\frac{2x(x+3)}{(x+2)(x+3)}$$

Distribute in the second numerator.

$$=\frac{4}{(x+2)(x+3)}+\frac{2x^2+6x}{(x+2)(x+3)}$$

Add, factor, and simplify. Note: $x \neq -2$ or $-3$.

$$=\frac{2x^2+6x+4}{(x+2)(x+3)}=\frac{2(x+1)(x+2)}{(x+2)(x+3)}=\frac{2(x+1)}{(x+3)}$$

Simplify each of the following sums and differences. Assume the denominator does not equal zero.

1. $\frac{7x}{3y^2}+\frac{4y}{6x^2}$
2. $\frac{-18}{9xy}+\frac{7}{2x}-\frac{2}{3x^2}$
3. $\frac{7}{y-8}-\frac{6}{8-y}$
4. $y-1+\frac{1}{y-1}$
5. $\frac{x}{x+3}-\frac{6x}{x^2-9}$
6. $\frac{6}{x^2+4x+4}+\frac{5}{x+2}$
7. $\frac{2a}{3a-15}+\frac{-16a+20}{3a^2-12a-15}$
8. $\frac{w+12}{4w-16}-\frac{w+4}{2w-8}$
9. $\frac{3x+1}{x^2-16}-\frac{3x+5}{x^2+8x+16}$
10. $\frac{7x-1}{x^2-2x-3}-\frac{6x}{x^2-x-2}$
11. $\frac{3}{x-1}+\frac{4}{1-x}+\frac{1}{x}$
12. $\frac{3y}{9y^2-4x^2}-\frac{1}{3y+2x}$
13. $\frac{2}{x+4}-\frac{x-4}{x^2-16}$
14. $\frac{5x+9}{x^2-2x-3}+\frac{6}{x^2-7x+12}$
15. $\frac{x+4}{x^2-3x-28}+\frac{x-5}{x^2+2x-35}$

## Answers

1. $\frac{14x^3+4y^3}{6x^2y^2}$
2. $\frac{-12x+21xy-4y}{6x^2y}$
3. $\frac{13}{y-8}$
4. $\frac{y^2-2y+2}{y-1}$
5. $\frac{x(x-9)}{(x+3)(x-3)}$
6. $\frac{5x+16}{(x+2)^2}$
7. $\frac{2(a-2)}{3(a+1)}$
8. $-\frac{1}{4}$
9. $\frac{4(5x+6)}{(x-4)(x+4)^2}$
10. $\frac{x+2}{(x-3)(x-2)}$
11. $\frac{-1}{x(x-1)}$
12. $\frac{2x}{(3y+2x)(3y-2x)}$
13. $\frac{1}{x+4}$
14. $\frac{5(x+2)}{(x-4)(x+1)}$
15. $\frac{2x}{(x+7)(x-7)}$

# SOLVING MIXED EQUATIONS AND INEQUALITIES #27

Solve these various types of equations.

1. $2(x - 3) + 2 = -4$
2. $6 - 12x = 108$
3. $3x - 11 = 0$
4. $0 = 2x - 5$
5. $y = 2x - 3$<br>$x + y = 15$
6. $ax - b = 0$<br>(solve for x)
7. $0 = (2x - 5)(x + 3)$
8. $2(2x - 1) = -x + 5$
9. $x^2 + 5^2 = 13^2$
10. $2x + 1 = 7x - 15$
11. $\frac{5 - 2x}{3} = \frac{x}{5}$
12. $2x - 3y + 9 = 0$<br>(solve for y)
13. $x^2 + 5x + 6 = 0$
14. $x^2 = y$<br>$100 = y$
15. $x - y = 7$<br>$y = 2x - 1$
16. $x^2 - 4x = 0$
17. $x^2 - 6 = -2$
18. $\frac{x}{2} + \frac{x}{3} = 2$
19. $x^2 + 7x + 9 = 3$
20. $y = x + 3$<br>$x + 2y = 3$
21. $3x^2 + 7x + 2 = 0$
22. $\frac{x}{x + 1} = \frac{5}{7}$
23. $x^2 + 2x - 4 = 0$
24. $\frac{1}{x} + \frac{1}{3x} = 2$
25. $3x + y = 5$<br>$x - y = 11$
26. $y = -\frac{3}{4}x + 4$<br>$\frac{1}{4}x - y = 8$
27. $3x^2 = 8x$
28. $|x| = 4$
29. $\frac{2}{3}x + 1 = \frac{1}{2}x - 3$
30. $x^2 - 4x = 5$
31. $3x + 5y = 15$<br>(solve for y)
32. $(3x)^2 + x^2 = 15^2$
33. $y = 11$<br>$y = 2x^2 + 3x - 9$
34. $(x + 2)(x + 3)(x - 4) = 0$
35. $|x + 6| = 8$
36. $2(x + 3) = y + 2$<br>$y + 2 = 8x$
37. $2x + 3y = 13$<br>$x - 2y = -11$
38. $2x^2 = -x + 7$
39. $1 - \frac{5}{6x} = \frac{x}{6}$
40. $\frac{x - 1}{5} = \frac{3}{x + 1}$
41. $\sqrt{2x + 1} = 5$
42. $2|2x - 1| + 3 = 7$
43. $\sqrt{3x - 1} + 1 = 7$
44. $(x + 3)^2 = 49$
45. $\frac{4x - 1}{x - 1} = x + 1$

Solve these various types of inequalities.

46. $4x - 2 \le 6$

47. $4 - 3(x + 2) \ge 19$

48. $\frac{x}{2} > \frac{3}{7}$

49. $3(x + 2) \ge -9$

50. $-\frac{2}{3}x < 6$

51. $y < 2x - 3$

52. $|x| > 4$

53. $x^2 - 6x + 8 \le 0$

54. $|x + 3| > 5$

55. $2x^2 - 4x \ge 0$

56. $y \le -\frac{2}{3}x + 2$

57. $y \le -x + 2$
$y \le 3x - 6$

58. $|2x - 1| \le 9$

59. $5 - 3(x - 1) \ge -x + 2$

60. $y \le 4x + 16$
$y > -\frac{4}{3}x - 4$

## Answers

1. $0$
2. $-8.5$
3. $\frac{11}{3}$
4. $\frac{5}{2}$
5. $(6, 9)$
6. $x = \frac{b}{a}$
7. $\frac{5}{2}, -3$
8. $\frac{7}{5}$
9. $\pm 12$
10. $\frac{16}{5}$
11. $\frac{25}{13}$
12. $y = \frac{2}{3}x + 3$
13. $-2, -3$
14. $(\pm 10, 100)$
15. $(-6, -13)$
16. $0, 4$
17. $\pm 2$
18. $\frac{12}{5}$
19. $-1. -6$
20. $(-1, 2)$
21. $-\frac{1}{3}, -2$
22. $\frac{5}{2}$
23. $\frac{-2 \pm \sqrt{20}}{2}$
24. $\frac{2}{3}$
25. $(4, -7)$
26. $(12, -5)$
27. $0, \frac{8}{3}$
28. $\pm 4$
29. $-24$
30. $5, -1$
31. $y = -\frac{3}{5}x + 3$
32. $\approx \pm 4.74$
33. $(-4, 11)$ and $\left(\frac{5}{2}, 11\right)$
34. $-2, -3, 4$
35. $2, -14$
36. $(1, 6)$
37. $(-1, 5)$
38. $\frac{1 \pm \sqrt{57}}{4}$
39. $1, 5$
40. $\pm 4$
41. $12$
42. $\frac{3}{2}, -\frac{1}{2}$
43. $\frac{37}{3}$
44. $4, -10$
45. $0, 4$
46. $x \le 2$
47. $x \le -7$
48. $x > \frac{6}{7}$
49. $x \ge -5$
50. $x > -9$
51. below
52. $x>4, x<-4$
53. $2 \le x \le 4$
54. $x>2$ or $x< -8$
55. $x\le 0$ or $x\ge 2$
56. below
57. below
58. $-4 \le x \le 5$
59. $x \le 3$
60. below

51.

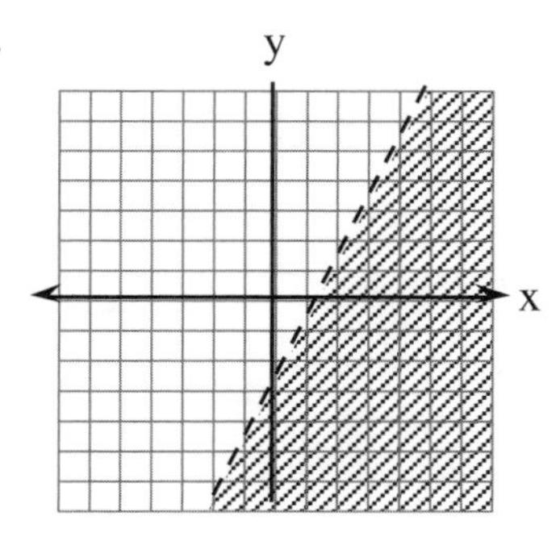


56.

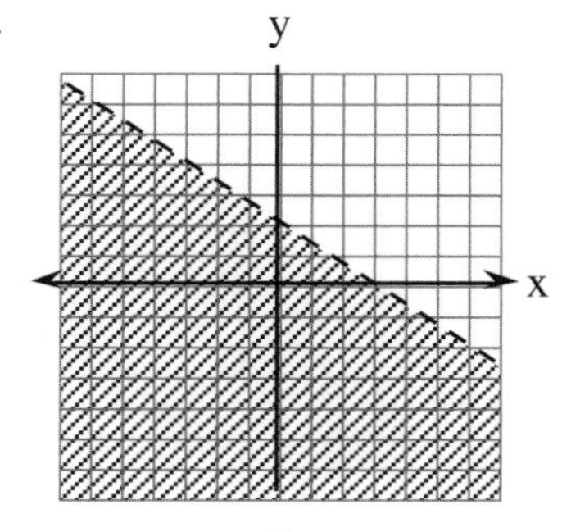


57.

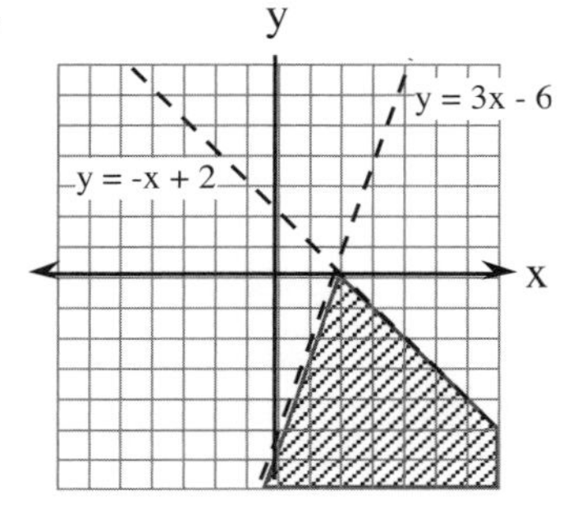


60.

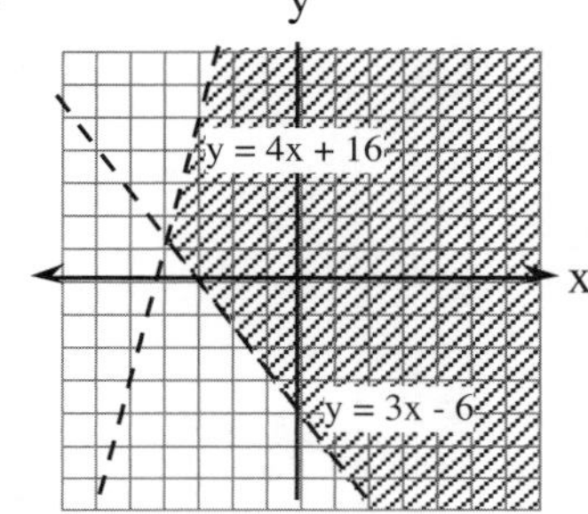

# Glossary

**AAA Triangle Similarity Theorem** Given $\Delta ABC$ and $\Delta A'B'C'$, with $\angle A \cong \angle A'$, $\angle B \cong \angle B'$, and $\angle C \cong \angle C'$, then $\Delta ABC \sim \Delta A'B'C'$. It is sufficient to show that two pairs of corresponding angles are congruent. Thus, the theorem may be referenced as the AA Similarity Theorem. (306)

**acute angle** Any angle with measure <u>between</u> (but not including) 0° and 90°. (110)

**addition method** (systems of equations) A method for solving a system of equations by adding or subtracting the equations to eliminate one of the variables. (88)

**adjacent angles** For two angles to be adjacent they must satisfy these three conditions: (1) The two angles must have a common side (they share a side); (2) They must have a common vertex; and (3) They can have NO interior points in common. This means the common side must be between the two angles; no overlap is permitted. (119)

**adjacent leg** In a right triangle, in reference to one of the acute angles, the adjacent leg of the angle is the side of the angle other than the hypotenuse. (262)

**alternate interior angles** Angles between a pair of lines that switch sides of a third intersecting line (transversal). Note that alternate interior angles are equal only when the two lines are parallel. (115)

**Alternate Interior Angle Theorem** If two parallel lines are cut by a third line then the alternate interior angles are congruent. (232)

**altitude** See height. (37)

**angle** An angle is formed by two rays joined at a common endpoint. Note that some angles are formed by segments, which can be extended as rays. (108)

**angle of incidence** The angle of approach of a moving object or a beam of light. The best example is to think of a pool ball bouncing off the side of a pool table. The smaller angle made by the path of the ball with the side of the table as it approaches the side of the table is the angle of incidence. The angle of reflection is the angle the path of the ball makes with the side of the table once it rebounds. (313)

**angle of reflection** The angle of rebound. The best example is to think of a pool ball bouncing off the side of the pool table. The angle made by the path of the ball with the side of the table as it approaches the side of the table is the angle of incidence. The angle of reflection is the angle the path of the ball makes with the side of the table once it rebounds. (313)

**arc** An arc is a connected part of a circle. Remember that a circle does not contain its interior so an arc is more like a piece of a bicycle tire than a piece of pizza. (373)

**arc length** The length of the arc (in inches, centimeters, feet, etc.) is the distance from one point on the circle to another measured around the circle. This is not the same thing as the arc measure. (373)

**arc measure** The arc measure is the number of degrees that an arc's central angle measures. Arc measure is measured in degrees and is different from the arc length. (373)

**area** For this course, the number of square units needed to fill up a region on a flat surface. The idea can be extended to cones, spheres, and more complex surfaces. (36)

**arrow diagram** Arrow diagrams are pictorial representations of definitions and conditional statements. (219)

**arrowheads** Arrowheads at the end of lines indicate that the lines extend indefinitely. Marks on pairs of lines or segments like >> and >>> indicate that the lines (or segments) are parallel. Also see the symbol page following the index. (36, 113)

**ASA (congruence)** Two triangles are congruent if two angles and their included side of one triangle are congruent to the corresponding two angles and included side of another triangle. Also equivalent to the AAS Congruence Property. (195, 202)

**auxiliary lines** Segments and lines that are added to existing figures are auxiliary lines. (246)

**axioms** Statements accepted as true. Also known as postulates. (232)

**base** The bottom side or face of a geometric figure. For a triangle, the base may be any side, although usually it is the bottom one. For a trapezoid, the two parallel sides are the bases. For a prism or cylinder, either one of the two congruent parallel faces may be the base while for a pyramid or cone, the base is the (flat) face which does not contain the vertex (where all the sides come together). (37, 168, 391)

**binomial** An polynomial of two terms; for example, $x - 5$, $3x + 2y$, $3x^2 + 34$.

**bisect** To divide into two equal parts. (64)

**center of a circle** The fixed point from which all points are equidistant on a flat surface. (370)

**central angle** A central angle is an angle with its vertex at the center of a circle. (373)

**chord** A line segment with its endpoints on a circle. A diameter is a chord which passes through the center. (384)

**circle** The set of all points that are the same distance from a fixed point. If the fixed point (center) is O, we write $\odot$ O as the symbol to represent the circle with the center at O. If r is the radius and d the diameter of a circle, the circumference of a circle C is $C = 2\pi r$ or $C = \pi d$. The area is $A = \pi r^2$. (370)

**circle (equation of a)** The equation of a circle with radius r and center at (0,0) is $x^2 + y^2 = r^2$. (447)

**circular cylinder** A prism with a circular base. The volume is $V = Bh$. Since the base is a circle and $B = \pi r^2$, $V = \pi r^2 h$. (393)

**circumference** The perimeter of (distance around) a circle. (370)

**clinometer** A device used to measure angles of elevation and depression. (283)

**complementary angles** Complementary angles are two angles whose measures sum to 90°. (127)

**concentric** Circles are concentric if they have the same center. Think of a bulls eye. (373)

**conditional statement** A statement that is written in the form "if... then..." (71)

**cone** A pyramid with a circular base. (397)

**congruent** Two shapes (for example, triangles) are congruent if they have exactly the same size and shape. The symbol for congruence is $\cong$. (188)

**conjecture** An educated guess. Many theorems begin as conjectures. Once they are proven, they become theorems. (61)

**construction** Using only a straightedge and compass to solve problems and/or create geometric figures. (457)

**converse** You get the converse of a conditional statement, hypothesis implies conclusion (H => C) by writing a new conditional statement, conclusion implies hypothesis (C => H). Knowing the original statement is true does not tell you <u>anything</u> about whether the converse is true. (130)

**convex polygon** In a convex polygon each pair of interior points can be connected by a segment without leaving the interior of the polygon. (338)

**corresponding angles** Angles on the <u>same</u> side of two lines <u>and</u> on the <u>same</u> side of a third line (transversal) that intersect the two lines. Note that corresponding angles are equal only when the two lines are parallel. (114)

**Corresponding Angle Theorem** If two parallel lines are cut by a transversal (a third line), then the corresponding angles that are formed are congruent.

**corresponding parts (of two polygons)** Points, edges, or angles in the figures which are arranged in similar ways. If the polygons are congruent, then each pair of corresponding parts will be congruent to each other. (188, 190)

**cosine ratio** In a right triangle, the cosine ratio of an acute angle A is $\cos A = \dfrac{\text{length of adjacent leg}}{\text{length of hypotenuse}}$. (267-68)

**counterexample** An example showing that a generalization has at least exception; that is, a situation in which the statement is false. (241)

**cube** A cube is a rectangular polyhedron with square faces. (163-64)

**decagon** A polygon with ten sides. (335)

**decahedron** A polyhedron with ten faces. (165)

**diagonal** Any segment which connects two vertices of a polygon and is not a side. (336)

**diameter** A line segment drawn through the center of a circle with both endpoints on the circle. Usually denoted d. Note: $d = 2r$. (370)

**Diameter/Chord Theorem** (1) If a diameter bisects a chord, then it is perpendicular to the chord. (2) If a diameter is perpendicular to a chord, then it bisects the chord. (385)

**displacement** If a non-porous object is placed into a tank of water, the water level rises. The amount of water displaced is equivalent to the volume of the object. (454)

**dissection principle** Every polygon can be dissected (or broken up) into triangles which have no interior points in common. (336)

**dividing line** A line on a two dimensional graph that divides the graph into two regions. We use a dividing line when graphing linear inequalities such as $y > 3x - 1$. (125)

**dodecahedron** A polyhedron with twelve faces. (165)

**double column proof** See T-proof. (244)

**edge** An edge of a three-dimensional polyhedron is a line segment formed by the intersection of two faces. (163-64)

**elimination method** (systems of equations) A method for solving a system of equations by adding or subtracting the equations to eliminate one of the variables. (88)

**ellipsis** The symbol "..." indicates that values have been intentionally left out to save time and/or space. The missing values follow the same pattern as those values that are listed. (44)

**endpoints** See line segment. (39)

**equilateral** A polygon with all sides of equal length. Note that "equi" means equal and "lateral" means side. (85)

**expected value** The amount we would expect to win or lose each game on average if we played a game many times. (413)

**exponents** Exponents are used to represent repeated multiplication. For $x^a$, x is the base and a is the exponent. If a is a positive integer, it tells us to multiply x by itself a times. (234)

**exterior angle** An exterior angle of a polygon is formed by extending a side of the polygon to form an angle outside of the polygon with the adjacent side. Note that an exterior angle is always adjacent and supplementary to an interior angle of the polygon. (133)

**Exterior Angle Theorem** An exterior angle of a triangle is equal to the sum of the two remote interior angles. (222)

**face** A face of a three-dimensional polyhedron is a "side" of the figure; it is always a polygon. (164)

**factoring** "Unmultiplying" a polynomial into the product of two or more factors. (181)

**factors** Factors or divisors of an integer are the integers that divide it evenly. (273)

**figure dissection** Breaking the area of a figure into smaller pieces that, when reassembled, are equivalent to the area of the original figure. (40)

**flowchart proof:** A particular style of proof where statements are written in ovals, then connected by arrows to show the logical flow of the proof. The reason(s) for each statement (conclusion) are written next to each oval. (229)

**fraction busters** A method of simplifying equations involving fractions that uses the Multiplication Property of Equality to rearrange the equation in such a way that no fractions remain. (260, 269)

**grid triangles** Similar to slope triangles, these right triangles have been cut out of graph paper, and dropped, face up, on the table. Grid triangles have two possible slopes for the hypotenuse. (72)

**height** (a) Two dimensions: a perpendicular segment from a vertex of a triangle, trapezoid, parallelogram, rectangle, rhombus, or square to a line containing the opposite base. Three dimensions: a perpendicular segment from a vertex of a pyramid to the flat surfaces containing the opposite base. (b) The length of such a segment or the distance between the flat surfaces of two parallel bases of a prism or cylinder. Also called an altitude. (37, 168)

**heptagon** A polygon with seven sides. (335)

**heptahedron** A polyhedron with seven faces. (165)

**hexagon** A polygon with six sides. (335)

**hexahedron** A polyhedron with six faces. A regular hexahedron is a cube. (165)

**HL (congruence)** Two triangles are congruent if the hypotenuse and leg of one right triangle are congruent to the hypotenuse and corresponding leg of another right triangle. Note that this only works for right triangles. (202)

**hypotenuse** The longest side of a right triangle (the side opposite the right angle). (21)

**hypothesis** A conjecture (or educated guess) in science. (61)

**icosahedron** A polyhedron with twenty faces. (165)

**if-then statement** A statement that is written in the form "if... then..." Also known as a conditional statement. (71)

**image** The resulting congruent shape under a transformation is the image of a shape. (180)

**indirect proof** This is a particular style of proof where, in general, the proof begins by assuming that something is true and then showing that such an assumption eventually leads to a contradiction of a known fact (usually something in your algebra or geometry tool kit). It is also known as a proof by contradiction. (248)

**inequality symbols** The symbols $<$ (less than), $>$ (greater than), $\leq$ (less than or equal to, and $\geq$ (greater than or equal to). (12)

**inscribed angle** An angle with its vertex on the circle and whose sides intersect the circle. (377)

**Inscribed Angle Theorem** The measure of an inscribed angle is half the measure of its intercepted arc. Likewise, the measure of an intercepted arc is twice the measure of the inscribed angle whose sides pass through the endpoints of the arc. (378)

**integers** All positive and negative whole numbers, including zero. They can be expressed as $\{ ..., -4, -3, -2, -1, 0, 1, 2, 3, 4, ... \}$.

**intercepted arc** The arc of a circle bounded by the points where the two sides of an inscribed angle meet the circle. (377)

**interior angle (of a polygon)** An interior angle of a polygon is an angle formed by two consecutive sides of the polygon. The vertex of the angle is a vertex (corner) of the polygon. The sum of the measures of the interior angels of an n-gon is $(n - 2)180°$.

**isometric drawings** Isometric drawings are three-dimensional drawings done on isometric dot paper of three dimensional structures made out of cubes. (155)

**isometric view** The isometric view of a cube solid is the three dimensional view showing the front, right and top views of the cube solid. (158)

**isosceles trapezoid** A trapezoid with a pair of equal base angles (from the same base). Note that this will cause the non-parallel sides to be congruent. (354)

**isosceles triangle** A triangle with two sides of equal length. (85)

**Isosceles Triangle Theorem** If $\Delta ABC$ is isosceles with segment BA ≅ segment BC, then the angles opposite these sides are congruent, that is, $\angle A \cong \angle C$. (225)

**justify** To give a logical reason supporting a statement or step in a proof. More generally, to use facts, definitions, rules, and/or previously proven conjectures in an organized sequence that convinces your audience that what you claim (or your answer) is valid (true). (61)

**kite** A quadrilateral with two pairs of consecutive equal sides. (354)

**lateral surface area** The sum of the areas of all the faces of a prism or pyramid not including the base(s). (392)

**Law of Sines** For any $\Delta ABC$ with sides a, b and c opposite angles A, B, and C respectively, it is always true that $\frac{\sin A}{a} = \frac{\sin B}{b} = \frac{\sin C}{c}$ . (289)

**least common multiple** The smallest multiple of two integers that is evenly divisible by both of them. (269)

**legs** The two shorter sides of a right triangle that form the right angle. (21)

**line of symmetry** Any line which divides a figure so that each side folds over the line to fit the other side exactly is called a line of symmetry. The figure is said to have line symmetry. (179)

**line segment** The portion of a line between two points. We name a line segment by its endpoints, A and B, and write AB with a bar over it.. (39)

**magnification factor** A constant multiplier used to enlarge a shape into a similar shape. The magnifying factor is also the ratio of similarity. See "ratio of similarity." (316)

**major arc** An arc with measure greater than 180°. (373)

**map** Under a transformation, we say a shape is mapped to its image. (180)

**mat plan** A mat plan is a top (or bottom) view of a multiple cube solid. The number in each square is the number of cubes in that stack. (158)

**midpoint** A point that divides a segment into two equal parts (segments). (52)

**mid-segment** A segment joining the midpoints of two sides of a triangle. (245-46)

**Mid-segment Theorem** The segment that connects the midpoints of any two sides of a triangle is half the length of and parallel to the third side of the triangle. (246)

**minor arc** An arc with a measure less than 180°. (373)

**monomial** An expression with only one term. It can be a numeral, a variable, or the product of a number and one or more variables. For example, 7, 3x, -4ab, $3x^2y$.

**n-gon** A polygon with exactly n sides. (335)

**nonagon** A polygon with nine sides. (335)

**nonahedron** A polyhedron with nine faces. (165)

**oblique pyramid** A pyramid with the vertex not directly above the center of the base. (394)

**obtuse angle** Any angle which measures between (but not including) 90° and 180°. (110)

**octagon** A polygon with eight sides. (335)

**octahedron** An octahedron is a polyhedron with eight faces. In a regular octahedron, all the faces are equilateral triangles. (163-64)

**opposite leg** The leg across the triangle from an acute angle is called the opposite leg (side). (262)

**parabola** The graph of a quadratic equation is a parabola. There are several other ways to find a parabola, including the intersection of a right circular cone with a flat surface parallel to an edge of the cone. (142)

**parallel** Two straight lines on a two-dimensional surface that do not intersect no matter how far they are extended. (36)

**parallelogram** A quadrilateral with two pairs of parallel sides. (33, 354)

**pentagon** A polygon with five sides. (335)

**pentahedron** A polyhedron with five faces. (165)

**perimeter** The distance around a figure on a flat surface. For a polygon, the perimeter is the sum of the lengths of the edges (sides). (36)

**perpendicular** Two rays, line segments, or lines that meet (intersect) to form a 90° angle are called perpendicular. A line and a flat surface can also be perpendicular if the line does not lie on the flat surface but intersects it and forms a 90° angle with every line on the flat surface passing through the point of intersection. A small box at the point of intersection of two lines or segments indicates that the lines are perpendicular. (36, 113)

**pi (π)** The ratio of the circumference (C) of the circle to its diameter (d). $\pi = \frac{\text{circumference}}{\text{diameter}} = \frac{C}{d}$ for every circle. Numbers such as 3.14, 3.14159 or $\frac{22}{7}$ are approximations of π. (370)

**polygon** A two-dimensional closed figure of straight line segments (edges or sides) connected end to end (vertices). Each edge only intersects the endpoints of its two adjacent edges. (25)

**polyhedron** A polyhedron is a three-dimensional solid with no holes which is bounded by polygonal regions. (plural: polyhedra) The polygonal regions are joined at their sides, forming edges of the polyhedron. Each polygon is a face of the polyhedron. Polyhedra are classified by the number of faces they have. (164)

**postulates** Statements accepted as true. Also known as axioms. (232)

**preimage** The original shape that produced an image under a transformation. (180)

**prism** A prism is a polyhedron with two congruent parallel bases that are polygons and the lateral faces (the faces on the sides) that are parallelograms connecting the corresponding vertices of the two bases. (160, 168, 391)

**probability** The probability of some event, A, happening is expressed as a ratio and written as $P(A) = \frac{\text{number of successful outcomes}}{\text{total number of outcomes}}$. For example, when flipping a coin, the probability of getting tails, P(T), is $\frac{1}{2}$ because their is only one tail (successful outcome) out of the two possible outcomes (a head and a tail). (406)

**problem solving strategies** Specific techniques to explore problems that do not appear to have an immediate method for solving. Primary strategies include sub-problems, diagrams, guess and check, patterns, and organized tables (lists) of data.

**proof** A logical argument. Conjectures are proven by using facts, definitions, rules and/or previously proven conjectures in an organized sequence that convinces the reader that what you claim is true. (61)

**proof by contradiction** This is a particular style of proof where, in general, the proof begins by assuming that something is true and then showing that such an assumption eventually leads to a contradiction of a known fact (usually something in your algebra or geometry tool kit). It is also known as an indirect proof. (248)

**proportion** An equation stating that two fractions (or ratios) are equal. (288, 310)

**protractor** The geometric tool for physically measuring the number of degrees in an angle. (109)

**pyramid** A polyhedron with a base that is a polygon and lateral faces formed by connecting each vertex of the base to a single given point above or below the surface that contains the base. (166, 168)

**Pythagorean Theorem** The statement relating the lengths of the legs of a right triangle to the length of the hypotenuse: $(\text{leg})^2 + (\text{leg})^2 = (\text{hypotenuse})^2$. (21)

**quadratic** A polynomial is quadratic if the highest exponent in the polynomial is two (that is, the polynomial is degree 2). (142)

**quadratic formula** For any quadratic equation written in standard form as $ax^2 + bx + c = 0$, the solutions are $x = \frac{-b \pm \sqrt{b^2 - 4ac}}{2a}$. (206)

**quadratic functions** Functions (relationships between x and y) in which the highest exponent is two (for example, $y = 3x^2 - 4x + 7$ is a quadratic function). (142)

**quadrilateral** A polygon with four sides. (335, 354)

**radical form** A number is written using a square root symbol alone or as part of a product such as $\sqrt{8}$ or $2\sqrt{2}$. (273)

**radicand** The number inside a radical sign. (273)

**radius** The distance from the center of a circle to the points on the circle. Usually denoted r. Plural is radii. (370)

**ratio of similarity** The ratio of similarity between any two similar figures is the ratio of any pair of corresponding sides. This means that once it is determined that two figures are similar, all of their pairs of corresponding sides have the same ratio. See "magnification factor." (308)

**rational expressions** Fractions that have algebraic expressions in their numerators and/or denominators. (415, 434, 445, 448)

**rationalizing the denominator** Rewriting a fractional expression which has radicals in the denominator in order to eliminate them. (284)

**ray** Ray AB (drawn with an arrow drawn over AB from left to right) is part of line AB (drawn with a bar over it that has an arrow head at each end) that starts at A and contains all of the points on line AB that are on the same side of A as B, including A. Point A is the endpoint of ray AB. (108)

**rectangle** A quadrilateral with four right angles. (354)

**reflection** A transformation across a line producing a mirror image of the original shape. Also called a flip. (180)

**regular** A polygon is regular if it is a convex polygon with all angles congruent and all sides congruent. (338)

**remote interior angles** If a triangle has an exterior angle, the remote interior angles are the two angles of the triangle NOT adjacent to the exterior angle. Sometimes called opposite interior angles. (133)

**rhombus** A quadrilateral with four congruent sides. (354)

**right angle** An angle whose measure is 90°. (21, 110)

**right pyramid** A pyramid with the vertex directly above the center of the base. (394)

**right triangle** A triangle that has one right angle. (21)

**rigid motions** Movements of figures which preserve their shape and size are called rigid motions or transformations. (179)

**rotation** A rotation (or turn) is a transformation which turns all the points in the original figure the same number of degrees around a fixed center point (like the origin). (184)

**$r : r^2 : r^3$ Theorem** Once you know two figures are similar with a ratio of similarity $\frac{a}{b}$, the following proportions for the SMALL (sm) and LARGE (lg) figures (which are enlargements or reductions of each other) are true:

$$\frac{\text{side}_{sm}}{\text{side}_{lg}} = \frac{a}{b} \qquad \frac{P_{sm}}{P_{lg}} = \frac{a}{b} \qquad \frac{A_{sm}}{A_{lg}} = \frac{a^2}{b^2} \qquad \frac{V_{sm}}{V_{lg}} = \frac{a^3}{b^3}$$

For two-dimensional figures, the theorem refers to the ratios of sides, perimeters (P), and areas (A). For three-dimensional figures, the theorem refers to the ratios of edges, areas of faces or total surface area of the solids, and volume (V). (322)

**sample space** The total number of outcomes that can happen in an event. The sample space for rolling a die is the set {1, 2, 3, 4, 5, 6} because those are the only possibilities. (406)

**SAS (congruence)** Two triangles are congruent if two sides and their included angle of one triangle are congruent to the corresponding two sides and included angle of another triangle. (195)

**SAS Similarity Theorem** If two triangles have two pairs of corresponding sides proportional and the included angles congruent, then the triangles are similar by the SAS~ Theorem. (328)

**scale** The scale in a drawing shows the units in which the scale drawing is done and that converts to the units of the original. (301)

**scale drawings** A drawing that is the reduced version of the original and similar to the original is a scale drawing. (301)

**scalene triangle** A triangle with no sides of equal length. (85)

**secant** A line that intersects a circle in two distinct points. (381)

**sector** A region formed by two radii of a central angle and the arc between their endpoints on the circle. You can think of it as a portion of a circle and its interior that resembles a piece of pie. (376)

**semicircle** An arc with measure of 180°. It is half of a circle. The endpoints of any diameter divide a circle into two congruent semicircles. (379)

**similar** Two shapes are similar if they have exactly the same shape but are not necessarily the same size. The symbol for similar is ~. (302)

**simple radical form** A number $r\sqrt{s}$ is in simple radical form if no square integer divides s and s is not a fraction. For example, $5\sqrt{12}$ is not in simple radical form since 4 (a square integer) divides 12. But $5\sqrt{12} = 10\sqrt{3}$ is in simple radical form. (273)

**sine ratio** In a right triangle, the sine ratio of an acute angle A is $\sin A = \frac{\text{length of opposite leg}}{\text{length of hypotenuse}}$. (266, 268)

**slant height** The height of a triangular face of a pyramid. (431)

**slide** See translation. (183)

**slope** A number that indicates the steepness or flatness of a line, as well as its direction (up or down) from left to right. It is determined by the ratio $\frac{\text{change in y}}{\text{change in x}}$ or $\frac{y_2 - y_1}{x_2 - x_1}$ between any two points on a line $(x_1, y_1)$ and $(x_2, y_2)$. For lines going up (from left to right), the sign of the slope is positive. For lines going down (from left to right), the sign is negative. Vertical lines have an undefined slope while horizontal lines have a slope of zero. Parallel lines have equal slopes and the slopes of perpendicular lines are negative reciprocals of each other. (47)

**slope-intercept form** Any linear equation that is written in the form y = mx + b. In this form, m is the slope and the point (0, b) is the y-intercept. (49)

**slope triangle** A right triangle drawn from the endpoints of a line segment so that the line segment forms the hypotenuse of the right triangle and the legs are parallel to the x- and y-axes. Finding the lengths of the legs of the right triangle will give you the data to find the slope of the line. (47)

**special right triangles** A 45°-45°-90° triangle (also known as an isosceles right triangle) and a 30°-60°-90° triangle. (280)

**square** A quadrilateral with four right angles and four congruent sides. (354)

**square numbers** The name given to the sequence of numbers 1, 4, 9, 16, 25, ... that arise from $1^2, 2^2, 3^2, 4^2, 5^2, ....$ (95, 96)

**SSS (congruence)** Two triangles are congruent if all three pairs of corresponding sides are congruent. (195)

**SSS Similarity Theorem** If two triangles have all three pairs of corresponding sides proportional (this means one triangle is a magnification of the other), then the triangles are similar by the SSS Similarity Theorem. (328)

**straight angle** Straight angles (lines) have a measure of 180°. (110)

**subproblem** A smaller, often simpler part of a bigger problem. (40)

**substitution (systems of equations)** A method of solving a system of equations by replacing one variable with an expression involving the remaining variable(s). (81)

**substitution property** The Substitution Property allows us to replace variables, numbers, expressions, etc. with equal items. The Substitution Property allows us to evaluate expressions such as 5x - 7 if we know that x = 3, for example. (217)

**supplementary angles** A pair of angles whose sum is 180°. (110)

**system of equations** Algebraic equations with more than one unknown. (80, 81, 88)

**tangent** A line that intersects a circle in exactly one point. (381)

**Tangent/Radius Theorem (1)** Any tangent of a circle is perpendicular to a radius of the circle at their point of intersection (point of tangency). (381)

**Tangent/Radius Theorem (2)** Any pair of tangents drawn at the endpoints of a diameter are parallel to each other. (381)

**tangent ratio** In a right triangle, the tangent ratio of an acute angle A is $\tan A = \frac{\text{length of opposite leg}}{\text{length of adjacent leg}}$ . (263, 268)

**tetrahedron** A tetrahedron is a polyhedron with four faces. The faces are triangles. In a regular tetrahedron all the faces are equilateral triangles. (163-64)

**total surface area** The sum of the areas of all the faces of a polyhedron. (171)

**T-proof** A particular style of proof where statements are written in one column as a list and the reasons for the statements are written next to them in a second column. It is also known as a double (two) column proof. (244)

**transformations** Movements of figures which preserve their shape and size are called transformations or rigid motions. (179)

**translation** A translation (or slide) is a transformation which preserves size, shape and orientation of a figure while sliding it to a new location. (183)

**transversal** A line that crosses two or more other lines. We usually use the word transversal when referring to parallel lines intersected by a third line. (114)

**trapezoid** A quadrilateral with one pair of parallel sides. (354)

**triangle** A polygon with three sides. (335)

**Triangle Congruence Properties** The Triangle Congruence Properties (SSS, SAS, ASA, and HL) are theorems which let us prove two triangles are congruent with the minimum number of corresponding parts congruent. (195)

**Triangle Inequality** The Triangle Inequality states the minimum and maximum limits for the length of the third side of any triangle. (205)

**triangular numbers** The name given to the sequence of numbers 1, 3, 6, 10, 15, .... (96, 107)

**turn** See rotation. (184)

**two column proof** See T-proof. (244)

**two-point graphing method** Using only two points to graph a linear equation. Usually one of the points is the y-intercept, but it does not necessarily need to be. (45)

**undecahedron** A polyhedron with eleven faces. (165)

**unit square** The common name for a square one unit on a side.

**vertex** The point where the sides of an angle or the edges of a polygon or a polyhedron meet. Plural: vertices. (25, 108)

**vertical angles** Vertical angles are the two opposite (that is, non-adjacent) angles formed by two intersecting lines. Vertical is a relationship between pairs of angles, so you cannot call one angle a vertical angle. (120)

**Vertical Angle Theorem** If two lines (or segments) intersect, then the measures of each pair of vertical angles are equal. (222)

**volume** The number of cubic units needed to fill up a solid. The volume of a prism or a cylinder is given by the formula $A = Bh$, where B is the area of the base and h is the height (168, 391, 393). The volume of a pyramid or a cone is $A = \frac{1}{3}\ Bh$. (395, 397)

**y-intercept** The point where a graph crosses the y-axis. (45)

**Zero Product Property** If $(a)(b) = 0$ then either $a = 0$ or $b = 0$. Use this property to solve quadratic equations in factored form. (144)

# Math 2 (Geometry): Index

## About the Index

Each entry of the index is in alphabetical order and is referenced by the problem number in which it is discussed. In some cases, a term may be discussed in several of the following problems as well. The problems are listed by a one or two-letter code that references the unit and the problem number. Whenever possible, terms are cross-referenced to make the search process quicker and easier. For example, you can find *central angle* by looking it up exactly by name and also under *angle, central.*

## Table of Unit Labels

| | | |
|---|---|---|
| Unit 0 | PR | Prelude: Study Teams and Coordinate Grids |
| Unit 1 | RC | Riding a Roller Coaster: Perimeter, Area, Graphing, and Equations |
| Unit 2 | BP | Beginning Proof: Convince Your Group |
| Unit 3 | PS | Problem Solving and Geometry Fundamentals |
| Unit 4 | SV | The TransAmerica Pyramid: Spatial Visualization |
| Unit 5 | CG | Congruence and Triangles |
| Unit 6 | TK | Tool Kit: Writing Proofs |
| Unit 7 | T | The Height of Red Hill: Trigonometry |
| Unit 8 | S | The Trekee's Clubhouse Logo: Similarity |
| Unit 9 | US | Urban Sprawl: Polygons, Area, and Proof |
| Unit 10 | CS | The One-Eyed Jack Mine: Circles and Solids |
| Unit 11 | GP | The Poison Weed: Geometric Probability |
| Unit 12 | 3D | Going Camping: 3D and Circles |

## Index

# Math 2 (Geometry): Symbol Index